TRANSACTIONS
OF THE
INTERNATIONAL
ASTRONOMICAL UNION

VOLUME XVIIB – PROCEEDINGS

PRESIDENT DE L'UNION ASTRONOMIQUE INTERNATIONALE

ADRIAAN BLAAUW

PRESIDENT OF THE INTERNATIONAL ASTRONOMICAL UNION

1976 – 1979

INTERNATIONAL ASTRONOMICAL UNION

UNION ASTRONOMIQUE INTERNATIONALE

TRANSACTIONS

OF THE

INTERNATIONAL ASTRONOMICAL UNION

VOLUME XVIIB

PROCEEDINGS OF THE SEVENTEENTH

GENERAL ASSEMBLY

MONTREAL 1979

Edited by

PATRICK A. WAYMAN

General Secretary of the Union

D. REIDEL PUBLISHING COMPANY

DORDRECHT : HOLLAND / BOSTON : U.S.A. / LONDON : ENGLAND

ISBN 90-277-1159-3
LC-Number 30-10103

CIP

Published on behalf of
the International Astronomical Union
by
D. Reidel Publishing Company, P. O. Box 17, 3300 AA Dordrecht, Holland

Sold and distributed in the U.S.A. and Canada
by Kluwer Boston Inc.,
190 Old Derby Street, Hingham, MA 02043, U.S.A.

In all other countries, sold and distributed
by Kluwer Academic Publishers Group,
P. O. Box 322, 3300 AH Dordrecht, Holland

D. Reidel Publishing Company is a member of the Kluwer Group.

Printed in The Netherlands

PREFACE

The first General Assembly of the International Astronomical Union was held in the year 1922. At three or four year intervals up to 1938 and from 1948 to 1979 the General Assemblies have been held regularly in a total of thirteen different countries.

The purpose of the General Assemblies is three-fold. Firstly they are designed to provide the opportunity for the exchange by personal contact of recently-obtained scientific results. This exchange of information can take place at the series of IAU Symposia and Colloquia held in the host country or in neighbouring countries before and after the General Assembly, at scientific Commission Meetings and at Joint Discussions of several Commissions at the Assembly itself, or during informal discussions during, for instance, the events of the social programme. Secondly, many astronomers meet during the time of a General Assembly to formulate agreed opinions on matters of world-wide interest within the specialisations of the Union's Commissions and other Committees. Thirdly, strong personal links of friendship over every kind of terrestrial division - geographical, cultural and political - are regularly formed at General Assemblies of the IAU.

These tasks are difficult if the results are to endure; they have to be approached in a spirit of service to the world astronomical community. Even if the achievements are not perfect, the activities of General Assemblies foster among the rising generation of astronomers an experience of kinship with colleagues the world over, and contributions are made towards a global appreciation of the science of astronomy.

The *IAU Transactions* that record the work of the 1979 Montreal General Assembly have been published first in the form *Reports on Astronomy 1979 (IAU Transactions vol. XVIIA, D. Reidel Publishing Co., 1979, in 3 parts)* which was already available at the General Assembly as an index and summary of work done between 1976 and 1978 within the area of each Commission, and second in the form of these *Proceedings* which form a permanent record of the business of the General Assembly and of each Commission or other organ of the Union. The Joint Discussions of Commissions, which form an important part of the scientific sessions of the IAU General Assembly, are separately reported in *Highlights of Astronomy*. Volume 5 of this series, for the Montreal General Assembly is being published by D. Reidel Publishing Company during 1980. This volume also contains texts of the Invited Discourses and some papers from Joint Commission meetings. The open meetings at Montreal on the Search for Extra-Terrestrial Intelligence are being separately published by D. Reidel Publishing Company.

The business of the General Assembly itself, the highest authority of the Union, is guided by the Executive Committee and the Report of the Executive Committee for the years 1976 to 1978 was made available to the General Assembly in the pages of the IAU Information Bulletin No. 42 for June 1979. The Report also appears as an appendix to the present volume. It includes information on the finances of the Union and a proposed Budget for the forthcoming triennium 1980-1982.

The section of this volume labelled Part 2 contains the actual proceedings of the General Assembly formal sessions at Montreal, including reports of the Finance Committee, the Resolutions Committee, and the Special Nominating Committee. The last-named Committee of the Union proposes, by procedures that are largely independent of the existing Executive Committee, names for the future Officers and Members of the Executive Committee. The Nominating Committee appointed by the national Adhering Organisations, as well as forwarding the individual IAU members' names as proposals to the General Assembly for new membership, joins with the Presidents of the forty or so Commissions in determining the composition of the Special Nominating Committee. The procedure is described in the By-laws of the Union; these By-laws, together with an up-to-date version of the IAU Statutes, the Working Rules, and the Rules for Scientific Meetings are printed in this volume.

The last half of this volume is dominated by the list of approximately 4500 names and addresses of the individual IAU members. The IAU appears to be unique among the scientific Unions in maintaining this category of individual membership. This feature adds considerably to the administrative load of the Union but it also contributes in an important way to the continuation of a spirit of individual responsibility and enterprise as well as a feeling of personal commitment to the aims of the Union.

The production of this volume would not have been possible without the careful work by the many Commission Presidents who provided camera-ready manuscripts and the equally careful contributions of Mme Brigitte Manning and Ms Patricia Smiley in the work of assembling the manuscript for publication.

Patrick A. Wayman

General Secretary

TABLE OF CONTENTS

PART 3

Comptes Rendus des Séances des Commissions
Reports of Meetings of Commissions

PART 4

Astronomer's Handbook

Appendix

PART 1

REPORT OF THE EXECUTIVE COMMITTEE

1976 - 1979

The Report of the Executive Committee to the Seventeenth General Assembly of the International Astronomical Union for the term of office 1976 - 1979 is in two parts. The first part, for the Calendar Years 1976, 1977 and 1978, was published in IAU Information Bulletin No. 42 (June 1979), pp. 14 - 49, and is reproduced as Appendix III to this volume (pp. 497 et seq.) The second part, for the first half of 1979, is contained in the oral report of the General Secretary at the First Session of the XVIIth General Assembly, given on pp. 15 et seq., below.

PART 2

INAUGURAL CEREMONY

REPORT OF THE XVIIth GENERAL ASSEMBLY

INAUGURAL CEREMONY

Tuesday, 14th August 1979, 09^h45^m

The Inaugural Ceremony was held in the Wilfrid Pelletier Hall, Place des Arts, Montreal, with distinguished representatives of the Government of Canada, the Government of the Province of Quebec, the City of Montreal, l'Université de Montréal, and the National Research Council.

The ceremony was held in the Place des Arts at the invitation of M. Jean Drapeau, Q.C., Maire de Montréal, and was followed by a luncheon given by the Mayor to all those attending the General Assembly.

Dr Harvie André, M.P., represented the Government of Canada, M. Elie Fallu represented the Government of the Province of Quebec, M. René J.A. Lévesque, Vice-recteur de l'Université de Montréal, represented the Rector of l'Université de Montréal, and Dr B.A. Gingras represented the National Research Council of Canada.

The Chair was taken by Professor Alan H. Batten, Chairman of the Canadian National Committee for the XVIIth General Assembly of the IAU.

Two musical interludes were provided by Ensemble Claude Gervaise.

Professor Batten introduced the Mayor of Montreal, M. Jean Drapeau, Q.C.

ALLOCUTION DE M. JEAN DRAPEAU, Q.C. MAIRE DE MONTREAL:

Avant que ne débutent vos travaux, accordez-moi je vous prie une joie très particulière: celle de dire que tous les citoyens de Montréal sont extrêmement heureux que vous ayez choisi notre ville pour y tenir votre 17ième réunion triennale.

Ils se sentent d'autant plus honorés de vous voir parmi eux que c'est la première fois, dans l'histoire de votre Société, que vous vous réunissez au Canada. En leur nom, en celui de notre administration municipale et au mien, je tiens donc à vous souhaiter la bienvenue la plus cordiale.

You have come to us from a great number of different countries, but I am sure that you will all feel at home in our midst, in this city which is mainly French but which at the same time is one of the most cosmopolitan on the North American continent. It is also one of the most hospitable, having opened its doors wide to people from the four corners of the world at the time of the international and universal Exposition of 1967, then during the Olympic Games of 1976. Now again, during this present month we welcome the World Cup in Track and Field and, next summer, the International Floralies.

So, with Grenoble behind you and Sofia ahead, you find yourselves today in the new world, in a metropolis which, in spite of its prodigious rate of growth, has kept its proportions to a human scale. It is here that during the next few days you will exchange information in your facinating study of the stars and planets. Therefore, to my mind, your assembling in this humanistic North American city has taken on a symbolic value.

Car l'astronomie, sans doute la plus ancienne des sciences, est également celle qui a contribué le plus largement à l'évolution de la pensée humaine. Par conséquent, l'astronomie dépasse le champ des sciences exactes. Elle appartient au domaine des humanités classiques et c'est à cela que je songe, à cet aspect humaniste de l'astronomie, quand je donne une valeur symbolique à votre présence dans notre ville. Car Montréal, de son côté, est elle-même une métropole essentiellement humaniste. Que sont, en effet, l'Exposition universelle, les Jeux Olympiques ou les Floralies sinon des manifestations de fraternité humaine? Des manifestations qui reflètent le caractère véritable de notre communauté montréalaise. Voilà comment votre visite dans notre milieu m'apparaît à la fois normale et symbolique.

Nous allons suivre vos travaux avec un intérêt certain. Je ne suis pas familier, ni la plupart des Montréalais nous l'avouons humblement, avec l'oeuvre de Ptolémée, le système de Copernic, la mécanique de Galilée, les lois de Newton, ni la théorie de la relativité généralisée d'Einstein. Mais nous sommes fort conscients des progrès troublants de la science et sensibles au résultat de vos découvertes.

Nous suivrons aussi vos travaux avec une pointe de malice, en imaginant qu'ils vous paraîtra un jour opportun, peut-être, de verser dans l'astrologie. Elle est, selon Larousse, "une utilisation pratique de la science astronomique"...Kepler, après tout, ne dédaignait pas de s'en occuper.

Finalement, plusieurs vous regarderont avec une certaine envie. Parce qu'ils sont nombreux les hommes qui voudraient, comme vous, faire profession de vivre dans les étoiles. Lorsque, généralement, on utilise cette expression, c'est pour désigner les rêveurs éternels, ceux qui ont perdu pied avec la réalité. Mais quand on y réfléchit, vivre dans les étoiles veut aussi dire chercher à mieux comprendre les hommes, et la terre, en les regardant d'un peu plus loin, à distance, pour être ensuite capable de les mieux aimer et mieux les guider. Comme quoi, tous les hommes politiques devraient commencer par aller à votre école!

Mais je m'avance ici en terrain dangereux. Il vaut mieux couper au plus court. Je vous félicite donc d'avance des succès qui viendront couronner vos travaux et je souhaite que vous gardiez de votre bref passage parmi nous le souvenir le plus heureux qui soit.

ALLOCUTION DE M. HARVIE ANDRE, M.P.:

Mr Mayor, Honorable Deputy, Messrs. Presidents, Ladies and Gentlemen:

Au nom du gouvernment du Canada, je souhaite la bienvenue à tous les membres de l'Union Astronomique Internationale réunis dans notre pays à l'occasion de la dix-septième assemblée générale de l'Union.

Le Canada a joué un rôle important dans le domaine de l'astronomie internationale et en qualité de membre originaire de l'Union Astronomique Internationale, nous sommes non seulement heureux mais également fiers de nous voir confier l'honneur d'être les hôtes de cette assemblée fort distinguée.

Astronomy is one of the oldest of the sciences, a reflection, no doubt, of mankind's innate wonderment about the heavens. From prehistoric man to modern man, the desire to understand and to explain the heavens remains a constant. Even as our explanation changes from the stars being the houses of the gods to their being the remnants of a colossal original explosion; our fascination, if anything, intensifies.

There is much to be fascinated about. The comparatively recent extension of astronomical observations to both above and below the visible electromagnetic wavelengths has contributed greatly to our knowledge of the universe. These developments plus more accurate and sophisticated instruments of analysis have led to the discovery of fascinating new celestial bodies and more information about the make-up of the universe.

Not just to the astronomers but to the laymen as well, these discoveries are exciting and whet the appetite for more. Indeed, in some ways this appetite for knowledge, for the truth, is what defines our species and by catering to it we enrich all of our lives.

In Canada our efforts or work in astronomy is divided nearly equally between the Federal Government and the Universities. If I may be permitted a small commercial, prominent amongst the Universities is the University of which I am a faculty member, (though on leave of absence); the University of Calgary. The direct government effort is part of the activities of our National Research Council.

The Canadian Government contribution to astronomy got its initial big start with the construction in 1918 of the then largest in the world, 72 inch telescope at the Dominion Astrophysical Observatory near Victoria, British Colombia. While no longer the largest, through the use of innovative auxiliary instruments, this observatory remains highly efficient. Canada also maintains two large radio telescopes in Algonquin Park, Ontario and near Penticton, British Colombia.

In 1975 these and other facilities were placed under the auspices of the Herzberg Institute of Astrophysics, named after Dr Gerhard Herzberg, distinguished National Research Council scientist and 1971 Nobel Prize winner in Chemistry. Besides honouring a renowned scientist, the name also serves to underline the interdisciplinary nature of modern astronomy. The bringing together of chemists and other branches of science with traditional optical and radio astronomers has contributed significantly to more effective research.

Just as interdisciplinary co-operation has aided astronomical research, so too has international co-operation. Perhaps more than in most fields astronomy is characterized by a high degree of international co-operation and joint effort. As just one example of this co-operation, this summer marks the bringing into service the major new optical telescope in Hawaii as a joint effort of Canada, France and the State of Hawaii. This assembly is another such example.

As one who has attended a few international congresses in my own discipline, I know well the value of such meetings. There is much to be gained from meeting with one's peers and colleagues; to learn of each others work, to renew old acquaintances, to make new ones and to pick up fresh ideas and fresh stimulation.

I know that your meetings will be interesting and may I propose the hope that they will be enjoyable and pleasant as well. The Prime Minister has asked me to express his wish that you have a successful conference and that the visitors from other countries enjoy their stay in Canada and in this beautiful city of Montréal.

ALLOCUTION DE M. ELIE FALLU, REPRESENTANT LE GOUVERNMENT DE LA PROVINCE DU QUEBEC:

Monsieur le président de l'Union Astronomique Internationale, Monsieur le représentant du Canada, Monsieur le maire de Montréal, Monsieur le vice-président du Conseil national de recherches du Canada, Monsieur le vice-recteur de l'Université de Montréal, Mesdames, Messieurs les congressistes,

C'est au nom du Premier Ministre, monsieur René Lévesque, et du gouvernement du Québec que je voudrais souhaiter la bienvenue à tous les participants à la XVIIe Assemblée générale de l'Union Astronomique Internationale.

Nous sommes honorés de ce que vous ayez choisi le Québec et sa métropole, Montréal, pour tenir vos assises générales.

Nos félicitations et nos meilleurs voeux de réussite aux initiatives prises par le président canadien de l'Union Astronomique Internationale, Monsieur Ian Halliday.

Au président du comité local d'organisation, le professeur Beaudet, et à toute son équipe nous formulons le souhait que la réussite de ce congrès soit à la mesure de leurs efforts.

Vous allez pendant 10 jours faire le bilan des travaux de recherche réalisés au cours des trois dernières années tant au chapitre de l'exploration spatiale qu'au niveau des explications des phénomènes extra-terrestres.

Vous venez de nombreux pays, mais veuillez ne pas oublier que par un heureux paradoxe l'univers que vous explorez est indivis entre les hommes, qu'il est la propriété ou le pays, dirais-je , de tous les hommes.

Comme vous avez ouvert en cette Place des Arts vos frontières, ouvrez également vos connaissances à tous ceux que vous côtoyerez.

Nous vous offrons, à l'Université de Montréal, un terrain neutre sur lequel seule la science pourra rivaliser avec elle-même.

Vous allez pendant 10 jours faire des exposés, débiter des conférences, participer à des débats, dresser des exhibits, présenter des primeurs sur la planète Jupiter. Nous vous savons déjà gré du regard critique que vous posez sur l'univers et nous désirons vous remercier de la réponse que vous donnez provisoirement à l'homme et à son inquiétude de connaissance. Nous vous remercions également de nous rappeler la place, je devrais dire plutôt le temps, que l'homme occupe dans l'univers.

Mais comme l'humanité ne vit pas seulement de connaissances, je vous souhaite également de découvrir nos cidres, nos lacs, nos maisons d'antan, nos villes et nos campagnes, nos restaurants surtout et nos chansonniers de surcroît.

Bref, bon séjour au Québec. La science d'abord, le plaisir ensuite.

ALLOCUTION DU RECTEUR DE L'UNIVERSITE, PRESENTEE PAR M. RENE J.A. LEVESQUE, VICE-RECTEUR

Monsieur le Maire, Messieurs les représentants des gouvernements du Canada et du Québec, Messieurs les présidents, Distingués invités et collègues, Mesdames, Messieurs,

C'est une fierté toute particulière qui anime l'Université ces jours-ci au moment où vous nous faites l'éminent honneur de vous rassembler ici à l'occasion de votre 17e Assemblée Générale. Vous me permettrez sans doute, en vous exprimant les regrets du recteur, Monsieur Paul Lacoste, de ne pouvoir être avec nous ce matin, de vous dire la joie personnelle que j'éprouve à vous souhaiter la plus cordiale bienvenue à l'Université.

Si l'enseignement et la recherche en sciences à l'Université remonte à l'année 1920, c'est vers la fin des années cinquante qu'un certain nombre de professeurs du département de physique ont fait une place à l'astrophysique préparant ainsi la voie au groupe d'astrophysiciens dans lequel se sont concentrés plus spécifiquement, à partir de 1968, les activités et l'intérêt de notre communauté.

L'observatoire astronomique du Mont Mégantic dont la création date de 1975, permet maintenant à nos professeurs et à nos chercheurs astronomes et astrophysiciens de travailler de plein pied en complémentarité avec leurs collègues du monde entier. Cette dernière réalisation, dont nous nous réjouissons, a rendu possible la participation de l'Université à la grande aventure astrophysique moderne dont les développements et les applications technologiques ne manquent pas d'être passionnants non seulement pour les experts scientifiques que vous êtes, mais également pour l'ensemble du public. La mise en marche prochaine, à Hawaï, du téléscope issu de la coopération bénéfique entre le Canada, la France et Hawaï, permettra d'accroître le réseau d'observation et d'étude de l'univers. Si l'on ajoute à ceci, l'honneur que vous nous faites de tenir votre important congrès dans nos murs, vous comprendrez notre heureuse satisfaction de constater le dynamisme dont vous faites preuve dans votre discipline et vos activités.

L'étude de l'univers, vous le savez bien, connaît depuis quelques années, tant auprès du public que des chercheurs de nombreuses disciplines, un regain d'intérêt. Ceci n'est pas étranger aux succès que nous a fait connaître votre audace intellectuelle. Les réelles prouesses techniques et humaines dont vous avez fait preuve ces dernières années demeurent à cet égard exemplaires. Il n'est donc pas étonnant, en cette période de restrictions des ressources financières, que les critiques soient rares à l'endroit des sommes consacrées aux investissements et à la recherche en astrophysique.

Ainsi, grâce à vos longues et patientes recherches, on peut aujourd'hui déterminer avec précision certaines caractéristiques et certaines propriétés de l'univers. On commence même à avoir une idée au moins approximative de l'évolution du monde et de ses planètes. Bien sûr, les énigmes sont encore nombreuses à résoudre. Nous sommes sans doute encore loin de connaître de façon certaine la structure de l'univers, l'origine des galaxies, de la vie elle-même ou de ses formes extra-terrestres. Il y a quelques années encore, on pouvait s'attendre à ce que certaines de ces questions soient l'objet d'un sourire devant l'ampleur de leur investigation scientifique. Aujourd'hui, beaucoup d'astronomes ont l'audace d'y faire face en s'associant avec des collègues d'autres disciplines pour y répondre.

L'aventure astrophysique moderne, internationale, est désormais largement multidisciplinaire. On fait maintenant appel non seulement à des mathématiciens et à des physiciens, mais aussi à des chimistes et à des biologistes.

Au-delà de ses succès, de ses réalisations et des développements technologiques parfois surprenants auxquels elle a donné lieu, l'astronomie, en 1979, demeure la discipline qui, plus que toutes les autres, perpétue de façon dynamique une des plus anciennes activités humaines: celle d'interroger le ciel, de contempler les étoiles, d'observer l'univers et de poser des questions essentielles avec, plus que jamais, l'espoir de glaner maintenant des réponses qui permettent chaque fois de comprendre un peu mieux le monde que nous habitons.

J'espère que durant ces dix prochains jours, vos discussions nous amèneront, elles aussi, au terme de cette 17e Assemblée Générale, à une meilleure connaissance de l'univers et ce sont les voeux que je vous adresse au nom de l'Université et en mon nom.

MOT DE BIENVENUE, PAR LE DOCTEUR B.A. GINGRAS, VICE-PRESIDENT (RELATIONS EXTERIEURES) CONSEIL NATIONAL DE RECHERCHES DU CANADA

C'était un très grand honneur que l'Union Astronomique Internationale faisait au Conseil National de Recherches du Canada en 1976 en acceptant l'invitation de ce dernier à tenir sa dix-septième Assemblée Générale au Canada. Aussi en vous accueillant à Montréal, je veux vous exprimer toute la reconnaissance du Conseil pour cette marque de confiance envers non seulement l'institution membre de l'Union, mais par delà, envers tous les scientifiques canadiens oeuvrant dans les disciplines relevant de la compétence de l'Union.

It is indeed an honor and a privilege for me to welcome all the delegates to the IAU General Assembly and all the participants in the satellite symposia held across Canada in conjunction with the Assembly. That so many countries are represented bear witness to the universality of astronomy, to the growing importance of your association and to the free collaboration among astronomers.

En cette année 1979, nous célébrons le 60e anniversaire de l'Union ainsi que, je me permets de le souligner, le 59e anniversaire de l'adhésion du Conseil National de Recherches du Canada à l'Union. Les astronomes canadiens se doivent d'être fiers d'avoir pu ainsi participer aux travaux de l'Union depuis le tout début.

Aussi au nom de la communauté scientifique canadienne, je veux vous exprimer toute notre admiration pour l'excellente contribution que les institutions membres dans plus de quarante-sept pays et que plus de 3500 membres individuels ont apporté à l'avancement de la science. C'est notre voeu le plus sincère qu'une telle assiduité et qu'un tel dévouement au développement de l'astronomie se continue à un même rythme et ce dans le cadre d'une coopération internationale accrue.

En cette année-anniversaire je m'en voudrais de ne pas particulièrement rendre hommage à nos chercheurs canadiens. Vous connaissez mieux que moi leur travail et je ne saurais rendre justice à tous si j'essayais un tant soit peu d'énumérer leurs réalisations. Je ne mentionnerai donc que deux programmes d'importance auxquels le CNRC est intimement lié; mais qui sont la mesure de la contribution canadienne à la science de l'astronomie.

Later this year, the Canadian astronomers will begin to reap the benefits of their partnership in an international project to construct a large optical telescope on the Island of Hawaii. This new faculty, conceived by astronomers from Canada, France and the University of Hawaii has been built at the summit of Mauna Kea, an extinct volcano which rises 4, 250 metres above sea level. Its 3.6 meter mirror, making it the world's seventh largest optical telescope, has been cast in France and ground and polished in the optical laboratory of the National Research Council's Astrophysical Observatory in Victoria.

The project has been marked by a spirit of international collaboration as similar projects around the globe have been, and it is appropriate that on the occasion of the IAU General Assembly it serves as a reminder to present and future astronomers of the great benefits achieved through cooperative approach in exploring the universe.

Finally, I would like to remind the IAU Assembly of a second important development in the history of Canadian astronomy. In early 1975, the Herzberg Institute of Astrophysics was created to serve as a focal point for such research in Canada and was named in honour of Dr Gerhard Herzberg, distinguished NRC scientist and winner of the 1971 Nobel Prize in chemistry. Today, the institute includes not only astronomers but also several workers representing other branches of science who, by freely exchanging their ideas, provide the driving energy behind effective research. The Herzberg Institute of Astrophysics is recognized as a centre of scientific excellence in Canada and has earned a respected place in the international scientific community.

To all Canadian scientists, whether located at NRC or in other Canadian Research Institutions, working in astronomy, planetary sciences, space physics, spectroscopy and other related disciplines, I bring encouragement from the Canadian adhering body to the IAU and I urge them to take an ever growing participation in the activities of the Union.

En guise de conclusion, permettez-moi Messieurs et Mesdames Les Congressistes, de vous souhaiter les plus intéressantes et les plus productives assises de votre histoire.

ALLOCUTION DU PROFESSEUR ADRIAAN BLAAUW, PRESIDENT DE L'UAI:

Mr Mayor, Dr André, Monsieur Fallu, Mr Rector, Mr Vice-President of the National Research Council, Dr Batten,

Now that the International Astronomical Union is again about to start the General Assembly, which plays such a prominent role in our Union's life, we first of all wish to express our gratitude for having been invited to your country, and our deep admiration for the magnificent way we are being received. Since that moment, now three years ago, when, on the other side of the Atlantic Ocean, your National Committee of Astronomy invited us to hold our seventeenth Assembly in this city, we have been eagerly looking forward to this day. For many among us, astronomers, your country and your institutes are familiar territory; their contribution to our science, now and in the past, has been so important to us, that no astronomer can afford not to be aware of the accomplishments of his Canadian colleagues. Yet, on this occasion, a happening on so much larger scale than the regular symposia or colloquia, we see our hosts in a broader perspective.

We realize that Canada has been a member country of our Union from the date of the Union's establishment, almost exactly 60 years ago, in July 1919. Only 18 adhering countries were listed at the time of the first General Assembly in 1922 - against 48 now - and 201 astronomers were members - against almost 3,800 now (and probably some 700 more by the end of this Assembly!). Canada was properly represented among these early 201 with the names of C.A. Chant, B.E. de Lury, F. Henroteau, O. Klotz, J.S. Plaskett, R.M. Stewart, and R.K. Young. Of these 18 member states of the first day, Canada is the 9th that will host us for a General Assembly. Canadian astronomy has always been well represented at these Assemblies. Today, Dr André, we have come with more than 2,000 for a return visit! But then, our feelings towards your country are not only inspired by our sympathy for a member of long standing and for Canadian astronomy, not only by the many attractive features of this large country; there is for many of us also a sentiment of deep gratitude because of the great hospitality which this nation has shown in the course of many decades with respect to the people in other parts of the world. At an occasion like this, we are reminded of those years after the disastrous world wars when so many people of other countries, like the one I come from, had to look for a better future than their own land could offer them - and so many found this in Canada. At times like the present, becoming acutely aware of the problems associated with immigration, I note with amazement and admiration that in the 1950's your country, with a population of some 14 million, was able to cope with, at times, more than 200,000 immigrants per year. No wonder, then, that many family ties link the Canadian people with people all over the world.

Canada, as three years ago your delegate at our Grenoble meeting said, is a young country where astronomy in recent years has blossomed in a vigorous proliferation of research centres across the land. We look forward to renewing our acquaintance with these developments. In doing so, we are also aware of the pioneering role Canadian astronomers played in the past. It is now 61 years ago that the Dommimion Astrophysical Observatory at Victoria dedicated its 73-inch telescope, which for a while was the largest telescope in the world, until the Mt. Wilson 100-inch came into operation. It was the product of the foresight and organizational talent of that admirable Canadian astronomer, John Stanley Plaskett, whose subsequent investigations, with that telescope, on stellar radial velocities have contributed so greatly to our understanding of galactic structure and to our knowledge of double stars. This pioneering spirit we also encountered in the early developments in radio-astronomy, when already in the 1950's Covington and collaboraters initiated solar observations at 10 cm wavelength with the compound interferometer of the National Research Council at Ottawa, the beginning of developments leading to the large radio astronomical facilities at Algonquin Park and at Penticton - and, in the sixties, to the first successful efforts in Very Long Baseline Interferometry.

Monsieur le Maire,

La présence de plusieurs milliers d'astronomes ne sera qu'une petite vague, à peine perceptible, dans cette cité qui, il y a peu de temps, accueillait Expo 67 et des foules des Jeux Olympiques. Nous vous sommes profondément reconnaissants de daigner quand même nous souhaiter une bienvenue toute spéciale. Nous avons en commun avec nos prédecesseurs olympiques, l'amour des mesures précises à la fraction de seconde près, et le besoin, à certains moments, de courir vite, bien que dans notre cas ce soit pour ne pas manquer la prochaine communication intéressante! Malgré ces efforts, j'espère que nous trouverons bien le temps de profiter des nombreuses attractions que cette grande cité peut nous offrir. Ses sites historiques, ses musées, ses parcs, ses centres artistiques sauront bien nous attirer de temps en temps. En tant que physiciens - en quelque sorte - nous savons les contributions de vos universités, entre autre la première fission nucléaire réalisée par Rutherford au département de physique de l'Université McGill. En tant qu'astronomes nous avons été enthousiasmés d'apprendre l'installation du nouvel Observatoire du Mont Mégantic, avec son téléscope de 160 cm, et nous lui souhaitons tous un brillant avenir.

Nous passerons ces quelques jours dans la province de Québec et nous sommes charmés, Monsieur le député, d'avoir l'opportunité, grâce à de nombreuses excursions, d'admirer ses beaux paysages et de visiter ses sites renommés. Nous savons que le Québec occupe une place toute spéciale parmi les autres provinces de cet immense pays. Ainsi que l'exprimait un des textes qui m'a préparé pour mon voyage ici, le Québec "a toujours été un champion de l'interprétation la plus large de l'autonomie provinciale". Bien que ce texte date déjà de dix ans, je présume qu'il est encore d'actualité. Nous savons que, tout en préservant sa culture et son caractère propre, ainsi que l'héritage de ses premiers colonisateurs, cette province a joué un rôle d'envergure dans le développement culturel et économique de la fédération canadienne. Nombreux d'entre nous sont ceux qui retrouveront ici le visage familier de leur terre natale. Et comme le dit si bien la devise de cette province: "Je me souviens"; ainsi chérirons-nous, de retour au pays après ces réunions, notre souvenir des jours que nous aurons passés ici, où nous aurons apprécié l'hospitatlité de votre gouvernement et de vos gens.

Monsieur le Recteur,

L'Union Astronomique Internationale est bien heureuse d'avoir été invitée à tenir son Assemblée Générale à l'Université de Montréal. Le nombre croissant de nos membres, et, conséquemment, les délégations accrues à ces assemblées, tout agréables qu'elles soient, imposent aux institutions qui les reçoivent des fardeaux qui sont de plus en plus difficiles à supporter. Ceci est d'autant plus vrai que nous voulons compter parmi nos hôtes les pays membres qui, bien que leur population totale soit relativement modeste, sont des membres astronomiquement actifs de notre famille internationale. Mais pas de tels problèmes à Montréal! Que votre célèbre université ait bien voulu nous ouvrir ses portes, ajoute encore plus d'éclat à cet événement. Chef-d'oeuvre d'architecture contemporaine et gardienne d'un système d'éducation dont les racines se retrouvent chez les universités classiques de nos ancêtres, l'Université de Montréal est un symbole de l'admirable qualité de la fédération canadienne toute entière où le haut respect des traditions se combine aux approches modernes et à la vitalité de la jeunesse.

If finally, I may return to the future of astronomy proper in this country, we cannot but note that this vitality is also evident in our Canadian colleagues' effort to remain up to date - and in what they have accomplished. It is, for sure, no small task, with that big brother in our astronomical family just across the southern border and his leadership in many branches of our science, for a country so much smaller in national income to stay in the front line of our - admittedly rather expensive - research and yet maintain its own character. That it is able to do so must be due to the way astronomy has organized itself here. I remember how I was puzzled, in my early years as a student, noting as a subtitle under the publications of the Dominion Observatories at Ottawa and Victoria, the words "Department of Mines and Technical Surveys". How to arrive at astronomical research starting from mining seemed to stretch too much the imagination of a young Dutch student, but not so for our Canadian colleagues. The procedure was really quite simple, as I understood from reading the history of the Victoria 73-inch. For proper mining so vital to this country - one needs proper triangulation; for proper triangulation a proper reference system; for it a determination of the meridian; for the meridian a good astronomical telescope - and once you have it, you make it a 73-inch one, turn it to the sky rather than to the horizon and observe stellar radial velocities!...Meanwhile, the Department of Mines has disappeared from the titles and we feel better at home reading National Research Council of Canada. Your presence here, Dr Gingras, representing this Council, is deeply appreciated by all of us. That our Canadian colleagues' imagination in planning new great facilities has not withered is obvious. Naturally, turning to the future we think first of all of that new adventure, the Canadian-French-Hawaii telescope, the operational stage of which has now arrived. It will allow Canadian astronomy to stay right in the foreground of the incredibly fascinating research on the most fundamental problems of our universe. It at the same time symbolizes the spirit of international co-operation that always has been so characteristic for the great projects in our science. Our Union congratulates your Research Council at having accomplished this great project, and at the prospect of seeing it used by so competent a team of astronomers as our Canadian colleagues are - and I should like to wish you all, those who allow the astronomers to do their work and our colleagues, the best of luck in their future work.

REPORT OF THE XVIIth GENERAL ASSEMBLY OF
THE INTERNATIONAL ASTRONOMICAL UNION

(RESOLUTIONS INCLUDED)

AGENDA

First Session

1. Formal opening of the General Assembly, by the President
2. Appointment of official interpreters
3. Report of the Executive Committee:
   (a) Discussion of the printed report, as published in the IAU Information Bulletin No. 42 of June 1979.
   (b) Report of decisions taken at the meeting of the Executive Committee at Montreal, in particular as regards the admission of new Members
4. Oral report of the General Secretary
5. Report by the President on the proposals for membership in the Executive Committee for the period 1979 - 1982

The Special Nominating Committee has proposed the appointment of the following Officers for the term 1979 - 1982:

President: Prof. Dr. M.K.V. Bappu
Indian Institute of Astrophysics
Bangalore- 560034
India

Vice-Presidents:
Dr. M.W. Feast, South Africa (to continue until 1985)
Dr. L. Kresak, Czechoslovakia (to continue until 1985)
Prof. R. Wilson, United Kingdom (to continue until 1985)

The Vice-Presidents continuing from the previous term are:
Dr. D.S. Heeschen, U.S.A. (until 1982)
Academician E.K. Kharadze, U.S.S.R. (until 1982)
Prof. S. van den Bergh, Canada (until 1982)

General Secretary:
Prof. P.A. Wayman, Director, Dunsink Observatory, Castleknock, Co. Dublin, Ireland

Assistant General Secretary:
Dr. R.M. West, ESO Sky Atlas Laboratory
ESO TP Division, CERN, CH-1211 Geneva, Switzerland

Advisers: Prof. Dr. A. Blaauw - former President
Prof. Dr. Edith A. Muller, - former General Secretary

6. Announcement of:
   (a) the names of representatives of Adhering Organizations empowered to vote on their behalf
   (b) the names of representatives to serve on the Nominating Committee

(c) the names of Acting Presidents of Commissions
7. Appointment of the Finance Committee
8. Appointment of the Resolutions Committee
9. Consideration of motion presented by the Executive Committee:

The Executive Committee proposes that Statute 18 (a), as in IAU Transactions, vol. XIVB, be reworded as follows:

"18 (a) For each ordinary General Assembly the Executive Committee prepares a budget proposal covering the period to the next ordinary General Assembly, together with the accounts of the Union for the preceding period. It submits these to the Finance Committee for consideration: this Finance Committee consists of one member nominated by each adhering body and approved by the General Assembly. At the first meeting during the General Assembly, the Finance Committee elects a Chairman from among its members.

"18 (b) The Finance Committee examines the accounts of the Union from the point of view of responsible expenditure within the intent of the previous General Assembly, and it considers whether the proposed budget is adequate to implement the policy of the General Assembly as interpreted by the Executive Committee. It submits reports on these matters to the General Assembly which approves the accounts and decides on the budget."

Paragraphs (b), (c), (d), and (e) of Statute 18 in force to become paragraphs (c), (d), (e) and (f) respectively.

To become valid, this change of Statutes requires the approval of at least two thirds of the votes of Adhering Countries having the right to vote by virtue of Statute 11(a).

10. Consideration of motions presented by Adhering Countries:

(a) The French National Committee of Astronomy has brought forward the following motions for adoption by the General Assembly:

(i) 1st Motion
Le Comité National Francais d'Astronomie souhaite que le Comité Exécutif de l'Union Astronomique Internationale essaie, dans toute la mesure du possible, d'augmenter la proportion des dépenses scientifiques par rapport aux dépenses administratives. Il félicite Madame le Secrétaire Général pour les efforts déjà accomplis dans ce sens.

(ii) 2nd Motion
Compte tenu du développement important de l'Astronomie en République Populaire de Chine, le Comité National Francais d'Astronomie demande au Comité Exécutif de l'Union Astronomique Internationale de poursuivre, au sein des statuts actuels de l'U.A.I., tous les efforts nécessaires à l'entrée de ce pays à l'U.A.I.

(iii) 3rd Motion
Le Comité National Francais d'Astronomie demande que soit soumise à l'approbation de l'Assemblée Générale de l'Union Astronomique Internationale de Montréal la proposition suivante:

"Les Assemblées Générales de l'U.A.I. ont lieu tous les 4 ans; la décision pourrait prendre effet en 1985 afin d'éviter la simultanéité avec les Assemblées Générales des disciplines voisines (COSPAR, UGGI)."

(b) The Swedish National Committee of Astronomy proposed that:

"The IAU should, possibly in collaboration with other organizations, establish a number of fellowships and/or grants, reserved for students in astronomy from developing countries. The financial aspects and the formulation of rules for such fellowships or grants should be considered by the relevant IAU committees and commissions. The individual IAU members and member organizations are urged to investigate their possibilities to find financial support for this project."

11. Consideration of motions presented by IAU Commissions:

(a) IAU Commission No. 6 "Astronomical Telegrams" proposes that the General Assembly increase the triennial IAU contribution to the IAU Telegram Bureau to SFrs 5,700.--.

(b) IAU Commission No. 10 "Solar Activity" proposes that the General Assembly adopt the following resolution:

"The General Assembly of the International Astronomical Union

-*noting* that there are plans to discontinue the continuous observations from which the Zurich sunspot numbers are derived,

-*strongly recommends* to the appropriate national organization the continuation of these long term observations which are vital and irreplaceable for improvement of our understanding of solar-terrestrial relationship."

(c) IAU Commission No. 20 "Positions and Motions of Minor Planets, Comets and Satellites" proposes that the General Assembly increase the triennial contribution to the Minor Planet Center over and above SFrs 6,220.--.

(d) IAU Commission No. 22 "Meteors and Interplanetary Dust" moves that the Commission consider the following resolution for adoption:

"*Recognizing* the major significance of comets for the study of the origin and evolution of the solar system in general and the interplanetary solid material in particular,

-*noting* the extreme diversity among individual comets in terms of composition, structure and physical behaviour, and

-*conceding* that ground-based observations are largely inadequate for establishing unequivocally the nature of the relationship between comets and other objects in the solar system

Commission 22 *recommends* that a vigorous program of space exploration of comets be initiated. To insure the maximum science return the program should include more than one comet and should proceed from rendez-vous missions to sample-return missions. A rendez-vous mission to P/Temple 2 in 1988 with an en route flyby of P/Halley in 1985, currently considered by NASA, is regarded as an ideal first step in the programme." the programme."

(e) IAU Commission No. 24 "Photographic Astrometry" proposes tentatively that the General Assembly adopt the following resolution:

"The IAU General Assembly strongly supports the independent and complementary astrometric programmes proposed for the ESA Astrometry Satellite and the NASA Space Telescope."

(f) IAU Commission No. 30 "Radial Velocities" provisionally proposes that the General Assembly approve to change the name of the Commission to "Stellar Radial Velocities."

(g) IAU Commission No. 44 "Astronomical Observations from Outside the Terrestrial Atmosphere", IAU Commission No. 10 "Solar Activity" and IAU Commission No. 12 "Radiation and Structure of the Solar Atmosphere" propose the following resolution:

"Noting the coming of Solar Maximum, Commission 44 wishes to encourage a sustained and coordinated effort in its study during the next 3 years."

12. Proposal and presentation by the Executive Committee of the budget of the IAU for the term 1980 - 1982.

Second Session

13. Report of the Finance Committee
    (a) Accounts of the IAU for 1976 - 1978
    (b) The residual budget for 1979
    (c) The budget for the ensuing term of office 1980 - 1982
    (d) The unit of contribution for the triennium 1980 - 1982
    The Executive Committee will propose to the Finance Committee that the unit of contribution be maintained at 1,465.- Swiss francs.

14. Appointment of the Special Nominating Committee

15. Proposals for resolutions presented by the Executive Committee

16. Proposals for resolutions presented by Adhering Organizations, subject to the recommendation by the Resolutions Committee

17. Proposals for resolutions proposed by IAU Commissions, subject to the recommendation by the Resolutions Committee

18. Announcement of new individual Members of the IAU

19. IAU Commissions:
    (a) Structural changes in Commissions, if any
    (b) The election of Presidents and Vice-Presidents
    (c) The membership of Organizing Committees

20. Place and date of the XVIIIth General Assembly

21. Election of President, three Vice-Presidents, General Secretary, and Assistant General Secretary

22. Addresses by retiring and newly elected Officers

23. Closing Ceremonies

At the first session of the General Assembly, held on Tuesday 14 August, items 1 through 12 were considered. The remaining items were considered at the second session, held on Thursday, 23 August.

FIRST SESSION

1. Formal Opening

The President welcomed the Members of the Union, Invited Participants, Registered Guests, the representatives of adhering countries and National Committees of Astronomy, and the representatives of sister Unions and other organizations, and formally opened the first session of the seventeenth General Assembly of the International Astronomical Union in the Wilfrid-Pelletier Hall, Place des Arts, Montreal, Quebec, Canada, on Tuesday, 14 August 1979 at 13.00 hours.

He wished a particularly hearty welcome to the official representatives of ICSU and sister Unions as follows:

ICSU (International Council of Scientific Unions):	Prof. J.-C. Pecker
IUGG (International Union of Geodesy and Geophysics):	Prof. P. Pâquet
IUGS (International Union of Geological Sciences):	Dr. W.W. Hutchinson
IUPAC (International Union of Pure and Applied Chemistry):	Prof. C. Sandorfy Prof. D.B. Tonks
IUWDS (International Ursigram and World Days Service):	Dr. P. Simon
SCOSTEP (Scientific Committee on Solar Terrestrial Physics):	Prof. M.R. Kundu
COSPAR (Committee on Space Research):	Prof. L.E. Peterson
CCIR (International Radio Consultative Committee)	Prof. J.P. Hagen
FAGS (Federation of Astronomical and Geophysical Services):	Dr. H. Enslin
IPMS (International Polar Motion Service)	Dr. S. Yumi Dr. K. Yokoyama
IUCAF (Inter-Union Commission on Frequency Allocation for Radio Astronomy and Space Science):	Dr. G. Westerhout
IUCS (Inter-Union Commission for Spectroscopy):	Prof. J.G. Philips
BIH (Bureau International de l'Heure):	Dr. B. Guinot

ESA (European Space Agency): Dr E. Peytremann

UNESCO (United Nations Educational, Scientific and Cultural Organization) regretted not being able to be represented and sent best wishes.

The President suggested that messages of good wishes be sent to former Presidents and General Secretaries of the Union prevented from attending the present meeting. This suggestion was approved by acclamation. The meeting also approved by acclamation the proposal of the President that a special word of good wishes should be sent to J.G. Bolton, a Vice-President, absent due to illness.

The President then asked those present to stand while the General Secretary would read the names of Members who had died since the XVIth General Assembly, or whose death had not been known at the time of that meeting. The General Secretary read the following list:

G. Adamopoulos	M. Gentili de Giuseppe	A.A. Nefd'ev
C.S. Beals	J.E. Gordon	J.J. Nissen
P.L. Bhatnagar	H. Haffner	P. Th. Oosterhoff
P.M.S. Blackett	Y. Hagihara	D.R. Palmer
M. Bloch	M. Harwood	R. Platzeck
N. Boneff	J.P. Heard	C. Popovici
M. Bonneau	S.S. Huang	A. Postoiev
Z. Brkic	J. Hunaerts	H.H. Rabben
H.E. Butler	M. Itzigsohn	K. Reinmuth
G.J. Carpenter	S.A. Kaplan	G. Righini
G. Cecchini	D. Koelbloed	V. Riives
D. Chalonge	A.K. Korol	H.U. Sandig
V.L. Chenakal	B.V. Kukarkin	H.K. Sen
A. Couder	I. Kupo	S.F. Smerd
C. Cruz-Gonzales	G.V. Lacchini	B. Sticker
I. Curea	C.G. Lacombe	E. Stiefel
G.Z. Dimitroff	A. Lallemand	N. Stoyko
H. Dingle	D.N. Limber	A.D. Thackeray
C.T. Elvey	Y.H. Lipakij	P.J. Treanor
V.I. Ezerskij	J.A. Madeira	J. Ueta
V.V. Fedynskij	S.W. McCuskey	E. Vandekerkhove
E. Fichera	D.H. Menzel	H. von Klüber
W.S. Finsen	V-V. Michkovitch	

PRESIDENTIAL ADDRESS

The President:

Before proceeding to the business of this Assembly, I should like to make a few general remarks. I know that most of you have come here in anticipation, first of all, of the many scientific meetings scheduled for the coming ten days. Naturally, this is the principal substance of the Assembly, and we are extremely pleased to see the scientific menu that lies before us. Just to mention a few of the attractive dishes: there are, outside the regular commission meetings, the eight joint discussions, most of them covering a full day, and all of them dealing with topics of great current interest, ranging from the most recent discoveries in the solar system to the rapidly developing field of extragalactic high-energy astrophysics. Adding to these the invited evening discourses to be presented to us by three very distinguished speakers, we see a programme that in itself would provide ample justification for our being here these ten days. But then, this is only part of it; meetings, sometimes joint meetings, of commissions will keep us equally busy, and let us not forget that it is one of the principal functions of the Union, to bring astronomers together from all over the world, to renew acquaintances or make new ones, and to promote the many private talks that are so essential to our research work.

With a programme so rich, and of such variety, we cannot but regret that financial and other circumstances prevent many more members of the Union being with us. The limited availability of travel grants, be it on the part of our sponsoring organizations or from the IAU itself, is one of the worries of the Executive Committee, and this is felt even more strongly now that almost everywhere in the world support for pure science, including astronomy, tends to be reduced. Naturally, this affects particularly the attendance from countries remote from the location of the General Assembly. It is, therefore, a very good tradition that successive assemblies are held in different parts of the world and we encourage by all means invitations for future assemblies to continue this tradition wherever possible.

The world wide character of the IAU's activities is not, however, restricted to just the triennial assemblies, and the symposia and colloquia; there is much more to it. As this may not be so apparent, especially not to the many younger ones among you, let me remind you of some of these other activities. Since some ten years ago, the IAU, in the spirit of what I have just said, has encouraged regional astronomy meetings to be held in different parts of the world, in order to promote international contact for IAU members and students, especially in those countries where the financial difficulties are great or even prohibitive for attending assemblies or the regular IAU symposia and colloquia. The past three years have seen three such regional events - all of them very successful. In all these cases the initiative came from within the geographical regions, but it was the IAU's relatively small subsidy and its support as an international organization, that helped turn these local initiatives into successful international gatherings. The IAU also sees it as its task, to help develop teaching in astronomy in those parts of the world where, in the struggle for economic and cultural development, virtually no expertise and money can be set aside for the promotion of a science like ours, from which the returns cannot be more than a promise for a distant future. I am referring to the young astronomers' schools for which, again, successful initiatives were honoured in recent years. As a third activity, beneficial from the points of view of both education and research, let me mention the funding of the exchange of astronomers, in this case through one of our commissions specially created for this task. I would like you to be well aware of this possibility offered by our Union, and make good use of it. Of all these examples of IAU activities, and of many more, the report of the General Secretary, whose office is an action centre for many of these activities - will give more particulars. Let me mention that, for the realization of some of these special projects, the IAU benefits much from grants received from the International Council of Scientific Unions.

The International Astronomical Union, as has been said before at our Assemblies, symbolizes the unity and common goals of astronomers and astronomy from its 48 adhering countries, virtually covering the whole world.

In this context, I wish to say a few words on an item of special importance; that item is the restoration of full participation of Chinese astronomers in the Union. The absence of astronomers from the mainland of China at our General Assemblies since 1961 has been a matter of great concern for members of the Union, and therefore also of the Executive Committee; and that is particularly true for the present Executive Committee and that of the preceding period. Many of you will remember the words spoken with regard to this problem by the previous president of the IAU, Professor Leo Goldberg, at the Grenoble Assembly. To me, and to my fellow officers, it seemed one of the most urgent problems to take up when we assumed our duties three years ago.

I am therefore extremely pleased that I am in a position to tell you that good progress has been made lately, and that perhaps it is not too optimistic to say that the restoration of the full adherence of astronomers of all parts of China is in sight. That we have been able to make this step forward, we owe above all to the contructive spirit in which this problem has been approached by all parties involved. We have been very fortunate in having had with us here in Montreal, for almost a week now, delegations of the astronomical communities of both the mainland of China and Taiwan, so as to approach the solution further, following the preparatory discussions and correspondence; in these preceding activities also, many astronomers outside the officers of the Union have participated for which we are most grateful.

The presence of the delegation of our colleagues from the mainland of China, under the chairmanship of Prof. Y.C. Chang of Purple Mountain Observatory, during the scientific meetings which we are now about to start, will, I am sure, render extra flavour to these meetings. Not only will we become much better aware of the progress of astronomy on the mainland of China and, at last, be in a better position to share with them the progress in astronomy in other parts of the world, but I believe this experience should also strengthen our awareness of the rewarding, albeit modest, role we scientists, and particularly we astronomers, can play in bridging the gaps that unfortunately tend to keep this world divided. I sincerely hope that by the time of the conclusion of this General Assembly, I may be in a position to inform you in more detail on the arrangements that we have been been able to make in this matter.

2. Appointment of Official Interpreters

The General Assembly appointed by acclamation S. Demers as official interpreter from English to French, and R. Racine as official interpreter from French to English.

3. Report of the Executive Committee

The General Secretary invited the national representatives to discuss the Report of the Executive Committee to the General Assembly, as published in IAU Information Bulletin No. 42 (see also pp. 497-523 below). There was no discussion and the Report was approved unanimously.

4. Report of the General Secretary

The President gave the word to the General Secretary who presented to the General Assembly her supplementary report for the period from 1 January to 13 August 1979 as follows:

This brief report summarizes the activities of the Union from 1st January 1979 until now.

The President, General Secretary and Assistant General Secretary held their last Officers' meeting in Geneva, Switzerland, on 19 May 1979, during which they discussed the most important problems of the IAU such as its finances, the forthcoming General Assembly and its scientific content, applications for membership to the Union, etc.

The Executive Committee continued its work by correspondence. It was twice informed of the amendments to the comprehensive budget of the IAU and it approved these amendments. It also approved the triennial report prepared by the General Secretary which has been published in Information Bulletin No. 42 as Report of the Executive Committee to the XVIIth General Assembly, and appears in Appendix I of this volume.

The agenda of the XVIIth General Assembly, as formulated by the Executive Committee, was distributed to Adhering Organizations and/ or National Committees of Astronomy before the statutory deadline. To assist attendance about 170 Young Astronomers' Grants have been allocated, as previously approved by the Executive Committee. Their amount varied with geographical location of the beneficients. The greater part of these grants are to be paid in cash here in Montreal. The Canadian National Committee offered to exempt the recipients of IAU Young Astronomers' Grants from paying the registration fee. In the name of the IAU and the recipients of the grants, I thank the Canadian National Committee most sincerely for their generosity.

Cooperation with the Canadian Organizers of this Assembly was most pleasant and fruitful, and for this I wish to thank in particular Dr. Alan H. Batten, Chairman of the Canadian Committee for the XVIIth General Assembly, Dr. Gilles Beaudet, Chairman of the Local Organizing Committee, and Mlle. Madelaine Bergevin, Secretary of the Committee.

The reports of Presidents of Commissions, prepared in camera-ready typescript, were edited and sent to the printer. They have appeared, in 3 parts, as IAU Transactions volume XVIIA, subtitled Reports on Astronomy 1979. On behalf of the Union I wish to thank our Presidents of Commissions, and all those who cooperated with them, for their fine reports. My thanks are also due to our publisher, the D. Reidel Publishing Company, for having brought out the Reports well before the General Assembly.

Apart from the many scientific meetings and sessions planned for the forthcoming 10 days of the present General Assembly - which as you can see from the programme promises to be a very interesting one - a number of Symposia and Colloquia have been planned around the General Assembly. Three Symposia and two Colloquia were held immediately before the General Assembly, and four Symposia and one Colloquium will be held shortly after its termination.

Commission No. 46 (Teaching of Astronomy) is organizing a Young Astronomers' School which is to be held at the Instituto de Astrofisica de Canarias in La Laguna, Tenerife, Spain, from 10 to 22 September 1979. This school has the financial support of a UNESCO through ICSU contract in the amount of US $ 4 000.-- which the IAU will match with the same amount. The director of the School is Prof. Francisco Sanchez, and its Secretary is Dr. Josip Kleczek who has a long experience in the organization of these schools. The lectures will be given in Spanish.

One of the tasks of the General Secretary during her term was to plan for the permanent seat of the IAU Secretariat on the premises of the Observatoire de Paris, in Paris, France, and to select the new staff members. Mesdames, Messieurs: Vous vous rappelez peut-être que lors de la dernière Assemblée Générale à Grenoble, Madame Alice Saunier-Séité, Ministre des Universités, invita l'Union Astronomique Internationale à installer son secrétariat permanent à l'Observatoire de Paris dans de locaux spécialement mis a notre disposition. J'ai le plaisir de vous annoncer que le pavillon ouest de l'Observatoire de Paris a été aménagé pour accueillir en permanence notre secrétariat; les travaux de réfection du pavillon sont achevés et le nouveau mobilier installé, l'ensemble donnant un cadre de travail agréable et accueillant. Je tiens à remercier au nom de l'UAI les responsables du Ministère des Universités de France et de l'Observatoire de Paris pour l'aménagement de ces locaux. En particulier, je remercie vivement Monsieur J. Boulon, Président de l'Observatoire de Paris, Monsieur J. Delhaye, Directeur de l'INAG, et Monsieur J.-C. Pecker, Professeur au Collège de France et ancien Secrétaire Général de l'UAI, pour toutes les interventions et l'appui qu'ils m'ont apportés dans le choix et l'aménagement des locaux pour l'emplacement permanent du secrétariat de l'UAI. De plus, je remercie très sincèrement le Collège de France qui, par l'intermédiaire de J.-C. Pecker, a offert au Secrétariat de l'UAI une machine à écrire IBM, - un somptueux cadeau. Un grand MERCI!

The permanent seat of the IAU Secretariat has also space for the Union's archives which, therefore, will be moved from Belgium and from Czechoslovakia to Paris. In the name and on behalf of the IAU I thank Prof. A. Velghe, Director of the Observatoire Royal de Belgique, and Dr. V. Bumba, Director of the Astronomical Institute of Czechoslovakia, for having kindly hosted the IAU archives, for many years, at their Observatories in Bruxelles and in Ondrejov respectively.

As regards the finances of the IAU I am happy to report that the meeting of the ICSU General Committee held in Brussels, Belgium, from 2 to 6 July, allocated to the IAU a UNESCO through ICSU subvention of US $ 8 000.-- instead of the expected US $ 6 000.-- for 1980, and that, in addition, it granted the Union for the same year some UNESCO/ICSU contracts in the amount of US $ 8 000.--. The comprehensive draft budget for 1980-82 was, therefore, slightly revised accordingly. The revised version of the budget will be presented to the Finance Committee.

I am also happy to announce that at the same recent meeting the ICSU General Committee allocated to the IAU a further grant of US $ 3 000.-- to be used as travel subsidy for the participation of Dr. S.E. Okoye of Nigeria and Dr. E. Bajaja of Argentina at the world Administrative Radio Conference (WARC) to be held this fall of 1979, over about 10 weeks, in Geneva, Switzerland.

I am pleased to inform you that at present the finances of the Union are in a very satisfactory state.

It is with great pleasure that I have received, shortly before the General Assembly the application of Indonesia to become a Member Country of the Union. The Executive Committee examined this application and arrived at the conclusion that the development of astronomy in Indonesia had reached a level that satisfies the conditions for admission to the Union.

In conclusion, I wish to express my sincere thanks to the Chairmen of National Committees of Astronomy, to the Presidents and Vice-Presidents of Commissions, and to all those who contributed to the successful work of the IAU during my term of office.

On the proposal of the General Secretary, the Assembly then approved unanimously to include in the agenda the application of Indonesia for admission to the Union as an adhering country.

5. Report of the President on the proposals for membership in the Executive Committee for the term 1979 - 1982.

The President informed the Assembly that Article 12 (a) of the By-laws provides that proposals for election to the President of the Union, six Vice-Presidents, the General Secretary and the Assistant General Secretary are submitted to the General Assembly by the Special Nominating Committee. The present Special Nominating Committee, consisting of the President in office, the past President L. Goldberg, a member nominated by the former Executive Committee, J.P. Wild, and four members elected by the Nominating Committee: A. Dollfus, B.A. Lindblad, A.G. Massevich and A. Reiz, had made the following proposals as to the composition of the next Executive Committee:

As President for the period 1979-1982:	Prof. M.K.V. Bappu, India
As Vice-Presidents for the period 1979-1985:	Dr. M.W. Feast, S. Africa Dr. L. Kresak, Czechoslovakia Prof. R. Wilson, U.K.

As General Secretary: Prof. P.A. Wayman, Ireland

As Assistant General Secretary: Dr. R.M. West, Denmark

The full Executive Committee is thus to consist of the above six together with:

Continuing Vice-Presidents: Dr. D.S. Heeschen, U.S.A.

Academician E.K. Kharadze, U.S.S.R.

Prof. S. van den Bergh, Canada

In an advisory capacity: Prof. A. Blaauw, outgoing President, The Netherlands

Prof. Edith A. Müller, outgoing General Secretary, Switzerland

According to the By-laws the formal election will take place at the final session of the General Assembly.

6. Announcements

Called on by the President, the General Secretary announced (a) the names of the representatives of adhering organizations to vote at the General Assembly
(b) the names of representatives of adhering organizations on the Nominating Committee:

COUNTRY	TO VOTE	NOMINATING COMMITTEE
A.R. of Egypt	A.S. Asaad	R. Ghobros
Argentina	J.L. Sérsic	R. Colomb
Australia	S.C.B. Gascoigne/ A.R. Hyland	B.Y. Mills
Austria	H. Haupt	H. Haupt
Belgium	P. Swings	M.P. Ledoux
Brazil	S. Ferraz Mello	P. Kaufmann
Bulgaria	M. Popova	N. Nikolov
Canada	A.I. Halliday	F.D.A. Hartwick
Chile	C. Anguita	A. Gutierrez-Moreno
Colombia		
Cuba		
Czechoslovakia	V. Bumba	V. Vanysek
Denmark	A. Reiz	H. Jørgensen
Finland	K. Lumme/ I. Tuominen	I. Tuominen
France	E. Schatzman/ J.-C. Pecker	N. Bel
GDR	G. Ruben	G. Ruben
Germany, D.R.	K. Hunger	G. Traving

COUNTRY	TO VOTE	NOMINATING COMMITTEE
Hungary	L. Dezsö	B. Szeidl
India	G. Swarup	S.K. Trehan
Iran	Y. Sobouti	Y. Sobouti
Iraq	M.A. Kaftan	M. Kaftan
Ireland	P.K. Carroll	P.K. Carroll
Israel	Yu Ne'eman/ A. Kovetz	Yu. Ne'eman
Italy	M.G. Fracastoro	M. Hack
Japan	Y. Kozai	D. Sugimoto
Korea PDR		
Korea (Republic)	Y. Key-Minn	K. Chol-Chou
Mexico	P. Pismis	S. Torres-Peimbert
Netherlands	H. van der Laan	T. de Jong
New Zealand	P.J. Edwards	P.J. Edwards
Norway	E. Jensen	R. Stabel
Poland	W. Iwanowska	A. Opolski
Portugal	J. Pereira Osorio	A.J.F. Pereira Pascoal
Roumania		
South Africa	G.D. Nicolson	E.E. Baart
Spain	J. Pensado	F. Sanchez
Sweden	A. Elvius	P.O. Lindblad
Switzerland	U.W. Steinlin	M. Mayor
Taiwan	Chun-Shan Shen	Chan-Shiann Tsai
Turkey		
Uruguay		
U.K.	M.J. Seaton	R.J. Tayler
U.S.A.	E. von P. Smith	J.E. Gaustad
U.S.S.R.	E.R. Mustel	E.V. Kononovich
Vatican City State	M. McCarthy	G. Coyne
Venezuela	D.J. MacConnell	D.J. MacConnell
Yugoslavia	G. Teleki	G. Teleki

N.B. Where two representatives are shown for one adhering organization, it is understood that the persons would divide their attendance according to their being available as alternatives.

(c) The General Secretary announced that the Executive Committee had appointed the following Vice-Presidents of IAU Commissions to act for Presidents of Commissions unable to attend the General Assembly:

Commission No. and Name	Acting President	For President
15 Physics of Comets Minor Planets and Meteorites	B.D. Donn	N.B. Richter
28 Galaxies	B.E. Westerlund	B.E. Markarian

7. Appointment of the Finance Committee

In accord with Statute 18 (a), the General Assembly appointed the following Finance Committee consisting of one representative of each adhering country:

Country	Representative on Finance Committee
Arab Republic of Egypt	S. Moustafa
Argentina	G. Carranza
Australia	W.H. Robertson
Austria	H. Haupt
Belgium	A.G. Velghe
Brazil	L. Muniz Barreto
Bulgaria	M. Kalinkov
Canada	J. Tatum
Chile	F. Noel
Colombia	not represented
Cuba	not represented
Czechoslovakia	J. Sykora
Denmark	P.E. Nissen
Finland	P.E. Markkanen
France	F. Spite
Germany, D.R.	G. Ruben
Germany, F.R.	W. Fricke
Greece	S. Svolopoulos
Hungary	L. Dezso
India•	J.C. Bhattacharyya
Iran	Y. Sobouti
Iraq	M. Kaftan
Ireland	P.A. Wayman
Israel	G. Shaviv
Italy	G. Godoli
Japan	Y. Kozai
Korea PDR	not represented
Korea (Republic)	K. Chol-Chou
Mexico	L. Carraso
Netherlands	H. van Woerden
New Zealand	P.J. Edwards
Norway	Ø. Elgaroy
Poland	S. Piotrowski
Portugal	J. Pereira Osorio
Roumania	not represented
South Africa	B. Warner
Spain	L. Quijana
Sweden	T. Elvius
Switzerland	A. Zelenka
Taiwan	Chun-Shan Shen
Turkey	not represented
Uruguay	not represented
U.K.	D. Walsh
U.S.A.	P.B. Boyce
U.S.S.R.	A.A. Boyarchuk
Vatican City State	J. Sasanovas
Venezuela	J. Stock
Yugoslavia	I. Pakvor

8. Appointment of the Resolutions Committee

At the request of the President, the General Secretary informed the Assembly that the Resolutions Committee is to advise the Executive Committee and General Assembly as regards the proposals for resolutions submitted for consideration. The General Assembly appointed F.G. Smith, B. Hauck and A.A. Boyarchuk to serve on the Resolutions Committee, with the General Secretary and Assistant General Secretary attending the meetings in an advisory capacity.

9 through 11. Consideration of motion presented by the Executive Committee, of motions presented by adhering countries, and of proposals for resolutions submitted by Commissions.

The General Secretary proceeded by reading the proposals for resolutions as under points 9, 10 and 11 of the agenda. As regards the motion of the Executive Committee to change statute 18 (a) she said that, in correspondence, the British National Committee supported the change but had also suggested that the wording of the proposed statute 18 (b) might be clarified by changing its final sentence as follows: "It submits reports on these matters to the General Assembly for approval of the accounts and decision on the budget." The Executive Committee accepts this rewording, and the motion before the General Assembly would therefore read as follows:

That Statute 18 (a), as in the IAU Transactions, vol. XIV B, be reworded as follows:

"18 (a) For each ordinary General Assembly the Executive Committee prepares a budget proposal covering the period to the next ordinary General Assembly, together with the accounts of the Union for the preceding period. It submits these to the Finance Committee for consideration: this Finance Committee consists of one member nominated by each adhering body and approved by the General Assembly. At its first meeting during the General Assembly, the Finance Committee elects a Chairman from among its members.

"18 (b) The Finance Committee examines the accounts of the Union from the point of view of responsible expenditure within the intent of the previous General Assembly, and it considers whether the proposed budget is adequate to implement the policy of the General Assembly as interpreted by the Executive Committee. It submits reports on these matters to the General Assembly for approval of the accounts and decision on the budget."

Paragraphs (b), (c), (d), and (e) of Statute 18 in force to become paragraphs (c), (d), (e) and (f) respectively.

The General Secretary announced that to become valid this change of Statutes would require the approval of at least two-thirds of the votes of adhering countries, having the right to vote by virtue of Statute 11 (a), at the Second Session of the General Assembly.

The representative of the French National Committee of Astro-

nomy, J.-C. Pecker, said that the Executive Committee motion is of paramount importance for the life of the Union and the proposed resolution should therefore be carefully studied word by word.
For the same reason it would be necessary to provide a French translation of the motion.

The General Secretary promised that the Executive Committee would provide the French translation for the next session. Similarly, French and English translations, as the case may be, would be available for the motion of the Swedish National Committee, as under point 10 (b) of the agenda, and of the three motions of the French National Committee of Astronomy, as under point 10 (a) (i, ii, iii) of the agenda.

12. Proposal and presentation by the Executive Committee of the Budget of the IAU for the term 1980-1982.

Called on by the President, the General Secretary said that the budget proposal for 1980-1982, as in the agenda, is a slightly changed version of the budget proposal circulated earlier to adhering organizations, and that attached to the agenda sent to National Committees of Astronomy. The change became necessary because ICSU had increased its contribution to the IAU for 1980. The budgetary items which benefited from this increase are (in Payments):

4.1 Exchange of Astronomers

7. Publications for developing countries

10. Symposia and Colloquia

15. Young Astronomers' Schools

As an amendment to the budgetary proposal published in the agenda distributed earlier, this revised budget proposal will have to be brought on to the agenda of the General Assembly by a vote of not less than two-thirds of the adhering countries voting.

The President asked the Assembly to vote on this proposal by a show of hands. The General Assembly was unanimous in its vote that the revised version of the budget for 1980-82 be included in the agenda.

The General Secretary announced that all other items of the agenda would be deferred until the next session of the General Assembly.

The President then formally adjourned the meeting at 15.00 hours.

SECOND SESSION

Held in the Pavillon Principal of the Université de Montréal, Montréal, on Thursday, 23 August, from 10.15 to 13.30.

Professor A. Blaauw, President, in the chair

Before passing to the agenda, the General Secretary, called upon by the President, established the quorum and found 5 adhering countries not represented. The General Assembly then appointed R. Gaizen and M. Ovenden as Tellers. The President read to the Assembly a message of good wishes received from Professor Cornelis de Jager, President of ICSU.

The General Secretary proposed that point 18 of the agenda be amended as follows:

18. Membership

(a) adherence of Indonesia

(b) membership of the Democratic Peoples' Republic of China

(c) individual membership

This proposal was carried unanimously.

13. The Finance Committee appointed at the first session was to examine the accounts of the Union for the years 1976-1978, the budget estimates of the Union for 1979 and for the period 1980-1982, and to consider the unit of contribution payable to the IAU by Adhering Countries for the next term of Office. The President called upon Dr. H. van Woerden, member of the Finance Committee, to present to the General Assembly the report of that Committee as prepared by its Chairman, Prof. W. Fricke. Dr. van Woerden read as follows:

REPORT OF THE FINANCE COMMITTEE

1. At the First Session of the XVIIth General Assembly on 14th August 1979, the Finance Committee, consisting of the nominated representatives of Member Countries, was duly appointed to perform its duties in accordance with Article 18 (a) of the Statutes. The Finance Committee met on that date when the IAU General Secretary welcomed the members and outlined the function of the Committee. The IAU General Secretary explained that for each ordinary General Assembly the Executive Committee prepares an estimate of the budget for the period to the next ordinary General Assembly which it submits to the Finance Committee, together with the accounts for the preceding period. The Finance Committee examines the accounts and the estimate of the budget on which it prepares reports and submits them to the General Assembly for approval.

2. The Finance Committee appointed W. Fricke as its Chairman, and it appointed a Sub-Committee to make a detailed examination of the accounts and of the estimate of the budget. The Sub-Committee consisted of the Chairman ex officio, A.A. Boyarchuk (U.S.S.R.), P.B. Boyce (U.S.A.), A.G. Velghe (Belgium), with Ms. E.A. Müller, P.A. Wayman and A. Jappel members *ex officio* as General Secretary, Assistant General Secretary and Executive Secretary of the Union respectively.

The budgetary estimates for 1979, summarized in the Report of the Executive Committee, were examined by the Finance Committee at its first session, and it was noted that the projected expenditure was within the financial provision of the Budget accepted by the

General Assembly, and that it was consistent with the policy of the Union embodied in the Budget.

The Finance Committee recommends to the General Assembly that the budgetary estimates for 1979 be accepted.

3. The Sub-Committee of the Finance Committee was convened by the Chairman on the 15th and 17th August, and it presented a report to the Finance Committee at its second meeting on 18th August. The Chairman reported to the Finance Committee that the accounts for the period 1976-78, had been subjected to detailed examination by the Sub-Committee with verifications supplied by the Secretariat during the examination. It was noted that the accounts had been kept at a very high standard and that all the relevant documents were available. The Sub-Committee had concluded that the accounts were in order, both on the basis of its own inspection and on the summary statement of the Auditor to the Union, M. Roger Bacle of Paris.

The Finance Committee, on the basis of the inspection of the accounts by the Sub-Committee, recommends to the General Assembly that the accounts for 1976-78 be approved.

4. The Sub-Committee reported to the Finance Committee that it had examined in detail the comprehensive budgetary proposals for the period 1980-82 submitted by the Executive Committee. It was noted that

(a) the Executive Committee had proposed to maintain the Unit of Contribution from adhering countries at the present level (S. Frs. 1 465.00);

(b) the revenue from the IAU-publications and from interest on accounts was estimated with all possible care, and that

(c) UNESCO subventions through ICSU and UNESCO will certainly be not less than S. Frs. 47 040.00.

In considering the assessments of payments for the period 1980-82 it was noted that

(a) an amendment of the following items may be expected from UNESCO funds: (4.1) Exchange of Astronomers, (7) Publications for Executive Committee and eveloping countries, and (15) Young Astronomers' Schools;

(b) the amount for Symposia and Colloquia was increased by about 20% compared to that of the previous budget period, and that

(c) all the individual budget items and the adjustment of these items appeared to be proper and adequate to implement the policy of the Union.

5. The Finance Committee, on the basis of the report of the Sub-Committee, recommends for adoption by the General Assembly the following budget presented by the Executive Committee:

BUDGET FOR 1980-1982

Unit of contribution 1 456.00 Swiss Francs
(One Swiss Franc equal to 0.2175926 g of fine gold)

RECEIPTS in Swiss Francs

1. Contributions from Adhering Countries	966 900
2. Revenue from the IAU publications	51 000
3. Interest on accounts	82 000
4. UNESCO contribution through ICSU & UNESCO contracts 1980	47 040
Total receipts	1 146 940
Total payments	1 146 050
Excess of receipts over payments	890

Payments under item 4.3
(Other projects of Commissions)

Commissions and Projects	SFrs
6. IAU Telegram Bureau	5 700
16. Planetary Documentation	5 310
17. Lunar Documentation	5 310
20. Minor Planet Center	7 200
27. Variable Stars Catalog	9 330
Total	32 850

PAYMENTS in Swiss Francs

1. Administrative Office	380 000
1.1 Secretariat in Paris	
1.2 Secretarial help for the General Secretary and Assistant General Secretary	30 000
1.3 Furniture, Typewriters	20 000
2. Subscription to ICSU	24 200
3. Commission expenses	15 000
4. Projects of Commissions	58 000
4.1 Exchange of Astronomers	
4.2 Teaching of Astronomy	3 000
4.3 Other projects	32 850
5. General Assembly	140 000
6. IAU publications	55 000
7. Publications for Executive Committee & developing countries	49 000
8. Executive Committee meeting	60 000
9. Officers' meeting	21 000
10. Symposia & Colloquia	138 000
11. Inter-Union Commissions	20 000
12. Projects of the Executive Committee	13 000
13. Representation	20 000
14. Bank Charges	1 000
15. Young Astronomers' schools	37 000
16. Regional meetings	24 000
17. Expenses of Special Nominating Committee	5 000
Total Payments	1 146 050

6. The Sub-Committee noted that the recommendation of the Finance Committee made to the XVIth General Assembly regarding the transfer of 60 000 dollars from the Current Accounts into Savings Accounts was met by a transfer of 40 000 dollars to American Savings Accounts and of 100 000 S.Frs. to two Savings Accounts at the Union de Banques Suisses. The Finance Committee expresses its satisfaction with that action.

7. The Sub-Committee noted with satisfaction an increase of the earnings from interest. The Finance Committee concurred with the view that the continuation of efforts in this direction should be encouraged.

8. The Sub-Committee stressed that the efforts for the exchange of astronomers must continue to be pursued. The funds might be augmented by UNESCO through ICSU grants. The Finance Committee considered such efforts as very important and recommends their continuation.

9. The Finance Committee supports the rewording and amendment of article 18 (a) of the Statutes as proposed by the Executive Committee in new articles 18 (a) and 18 (b) and recommends adoption by the General Assembly.

W. Fricke
Chairman, Finance Committee

The President thanked Dr. van Woerden and put the report of the Finance Committee to the vote, first as regards the unit of contribution, then as regards the accounts of the IAU 1976-1978, and eventually as regards the budget estimates for the years 1979 and the next term of Office 1980-1982. Voting on all these points was unanimously affirmative.

The President expressed the Union's appreciation of the work of the members of the Finance Committee and the Finance Sub-Committee, and especially of the work of the Chairman, Prof. W. Fricke.

14. Appointment of the Special Nominating Committee

The President referred to article 12 of the By-laws providing that proposals for election to the President of the Union, six-Vice-Presidents, the General Secretary and the Assistant General Secretary, are submitted to the General Assembly by the Special Nominating Committee. Proposals by Presidents of Commissions for membership of the SNC are reviewed by the representatives of adhering organizations on the Nominating Committee. Together with one proposal by the Executive Committee, this has resulted in a proposal that the General Assembly appoint the following members to serve on the Special Nominating Committee for the period 1979-1982:

A. Batten (Canada)
V. Blanco (Chile)
A.A. Boyarchuk (U.S.S.R.)
F.G. Smith (U.K.)
B.E. Westerlund (Sweden)

This list was unanimously approved by the General Assembly.

15. Proposal for Resolution presented by the Executive Committee

The President announced that this item refers to the change of Statute 18 as read at the First Session of the General Assembly and as supported in the report of the Finance Committee. The text, in French, is as follows:

L'article 18 (a) des Statuts, tel qu'il apparaît dans les Transactions de l'UAI vol. XIV B, soit reformulé comme suit:

18 (a) Pour chaque Assemblée Générale ordinaire, le Comité Exécutif prépare un projet de budget pour la période à courir jusqu'à l'Assemblée Générale ordinaire suivante, ainsi que les comptes de l'Union pour la période précédente. Il les soumet au Comité des Finances pour examen; ce Comité des Finances est composé de membres nommés par les organismes adhérents, à raison d'un membre par organisme, et il est approuvé par l'Assemblée Générale. Lors de sa première séance pendant l'Assemblée Générale, le Comité des Finances élit un Président parmi ses membres.

18 (b) Le Comité des Finances examine les comptes de l'Union pour voir si les dépenses engagées ont été conformes aux voeux émis lors de la précédente réunion de l'Assemblée Générale; il s'assure que le budget proposé vise à la poursuite de la politique de l'Assemblée Générale, telle qu'elle est interprétée par le Comité Exécutif. Il présente des rapports sur ces questions qu'il soumet a l'Assemblée Générale pour approbation des comptes, et pour décision sur le budget."

Les paragraphes (b), (c), (d) et (e) du présent article 18 des Statuts deviennent les paragraphes (c), (d), (e) et (f) respectivement.

(See p. 15 for text in English)

He then put this motion to the vote of the General Assembly which carried it with one abstention, and no contrary vote.

16. Proposals for resolutions presented by Adhering Organizations

The General Secretary reported that the proposals for resolutions submitted to the General Assembly by Adhering Organizations and by Commissions had been passed to the Resolutions Committee for consideration. Before voting took place, Prof. F.G. Smith, Chairman of the Resolutions Committee, would present to the Assembly the Report of this Committee.

The President called upon Prof. Smith, who noted that Item 16 of the Agenda refers to Resolutions of National Committees as they are given under Item 10 of the Agenda of the First Session of the General Assembly.

The French National Committee, according to information before the Resolutions Committee, had withdrawn the 1st and 2nd motions of Item 10 (a), leaving only Item 10 (a) (iii) to be voted on at this Session.

The President remarked that this question had been discussed by National Representatives at their meeting on 18 August at Estérel. He then put motion (iii) to the General Assembly. The motion was defeated by a substantial majority.

Professor Smith then referred to the motion proposed by the Swedish National Committee as given in Agenda Item 10 (b). This also, according to the Committee's information, had been withdrawn.

17. Proposals for resolutions proposed by IAU Commissions

At the request of the President, Professor F.G. Smith, Chairman of the Resolutions Committee, reported that proposals listed under Agenda Item 11 were considered by the Resolutions Committee together with all other resolutions proposed by Commissions. He reported further as follows:

The Resolutions Committee is required to assemble all the resolutions from Commissions and to present them in a suitable form at the Second Session of the General Assembly. This is a large task that must be carried out in a short time and it is impossible to consult Commission Presidents on the details of all resolutions. The Committee acted on two principles: first to restrict the number of resolutions to be adopted as IAU resolutions; and second, to publish, as far as possible, all the resolutions to be placed before the General Assembly, whether they are for adoption or merely to be noted. Publication was possible through the valuable co-operation of the editor of the Assembly newspaper METEORE, who arranged that the texts would be available on the morning of the Second Session, including a preliminary translation into French.

Resolutions for adoption at the General Assembly are those that *either* require specific IAU action *or* seem to be of wider interest than is implied by a Commission resolution. In the latter category, the Committee placed the resolution formerly listed under Agenda Item 11 (e). Also requiring adoption by the General Assembly are the resolutions concerning the organization of Commissions and their names.

The Chairman of the Resolutions Committee, Professor F. Graham Smith, presented the resolutions as follows:

(i) RESOLUTION No. 1

L'Assemblée Générale prend note du soutien des Commissions 4, 7, 8, 19, 20, 24, 25, 26, 33, 36, 37 et 45 pour les programmes astrométriques indépendants et complémentaires proposés pour le Satellite Astrométrique de l'ESA et le Téléscope Spatial de la NASA.

The General Assembly notes the support of Commissions 4,7, 8, 19, 20, 24, 25, 26, 33, 36, 37, and 45 for the independent and complementary astrometric programmes proposed for the ESA Astrometry Satellite and the NASA Space Telescope.

(N.B. This resolution was the result of an amendment to the motion presented by Commission 24 in consequence of the proposal under Agenda Item 11 (e). The amendment was proposed by M.J. Seaton, supported by C.A. Murray, and adopted.)

(ii) RESOLUTION No. 2

L'Union Astronomique Internationale

Considérant

a) La proposition originale de plusieurs pays en voie de développement de réaliser et de faire fonctionner, en collaboration, un Radiotélescope Equatorial Géant (GERT) en Afrique, près de l'équateur:

b) La possibilité pour GERT de fournir à des longueurs d'onde métriques des données de haute qualité comme complément essentiel aux données obtenues ailleurs en ondes centimétriques;

c) La remarquable utilisation d'un site équatorial pour obtenir une bonne résolution et une grande sensibilité;

d) Les avantages éducatifs, technologiques et sociaux attendus de la collaboration des pays engagés dans une recherche de pointe astronomique et spatiale:

Recommande que l'ICSU entreprenne des démarches pour obtenir le soutien de l'UNESCO dans la réalisation du Radiotélescope Equatorial Géant.

The International Astronomical Union

Considering

a) The imaginative proposal by several developing countries to construct and operate, in collaboration, a Giant Equatorial Radio Telescope (GERT) in Africa, near the equator;

b) the ability of GERT, operated at meter wavelengths, to provide high quality data as essential complement to data obtained elsewhere at cm wavelengths;

c) the noteworthy utilisation of an equatorial location for obtaining high resolution and sensitivity;

d) the educational, technological and social benefits expected for the collaborating nations involved in participation at the research front of astronomy and space science;

Recommends that the ICSU take steps to secure UNESCO support for the realisation of the Giant Equatorial Radio Telescope.

(iii) RESOLUTION No. 3

L'Union Astronomique Internationale

Considérant

a) l'apport que constituent pour l'humanité les resultats scientifiques obtenus par la Radioastronomie dans l'exploration de l'Univers;

b) l'utilisation croissante du domaine de fréquences radio surtout par les émetteurs embarqués;

c) que le CCIR a déterminé les niveaux d'intérference nuisibles dans son rapport CCIR 224-4;

d) que le CCIR et l'IUCAF ont résumé les besoins de la Radioastronomie dans divers rapports et documents;

e) que la "World Administrative Radio Conference" de 1979 (WARC) allouera des fréquences radio dont on peut espérer qu'elles seront maintenues pendant environ 20 ans;

Recommande que les administrations adhérentes à l'UAI présentes à la Conférence WARC, reconnaissent les besoins suivants pour la Radioastronomie:

1) L'obtention et/ou le maintien de bandes de fréquences ayant des largeurs de bande d'au moins 1-2% à des intervalles d'environ une octave dans le spectre radio.

2) L'obtention et/ou le maintien de bandes plus étroites aux fréquences des raies spectrales les plus importantes en Astrophysique, avec des largeurs de bande appropriées pour la recherche scientifique.

3) La préservation de ces bandes de fréquences de toute interférence nuisible provenant d'émissions dans la bande, en bord de bande, ou hors bande.

The International Astronomical Union

Considering

a) The value to mankind of the scientific results achieved by radio astronomy through the exploration of the universe;

b) the increasing use of the radio spectrum, especially by space and air-borne transmitters;

c) that the CCIR had specified the levels of harmful radio interference in CCIR Report 224-4;

d) that the CCIR and IUCAF have summarized the needs of the radio astronomy service in their various reports and documents;

e) that the 1979 World Administrative Radio Conference (WARC) will allocate radio frequencies which can be expected to remain in force for about 20 years;

Recommends that administrations adhering to the IAU, present at the WARC, recognize the following needs of the radio astronomy service:

1. The acquisition and/or maintenance of frequency bands with bandwidths of at least 1-2 per cent at approximately octave intervals throughout the radio spectrum.

2. The acquisition and/or maintenance of narrower bands at frequencies of the astrophysically most important spectral lines having bandwidths appropriate to the scientific investigation.

3. The preservation of these frequency bands free of harmful interference from in-band, band-edge, and out-of-band transmissions.

(iv) RESOLUTION No. 4

L'Assemblée Générale approuve les recommandations des Commissions 16 et 17 que la Commission 16 (Etude Physique des Planètes et des Satellites) et la Commission 17 (La Lune) fusionnent pour former une nouvelle Commission qui conserve le titre et le numéro de la Commission 16.

The General Assembly approves the recommendations of Commissions 16 and 17 that Commission 16 (Physical Study of Planets and Satellites) and Commission 17 (The Moon) be merged to form a new Commission, retaining the title and number of Commission 16.

(v) RESOLUTION No. 5

L'Assemblée Générale approuve les changements suivants dans le nom des Commissions:

Comm. 5	de	Documentation
	à	Documentation et Données Astronomiques
Comm. 14	de	Données Spectroscopiques Fondamentales
	à	Données Atomiques et Moléculaires
Comm. 44	de	Observations Astronomiques au-dehors de l'Atmosphère Terrestre
	à	l'Astronomie à partir de l'Espace
Comm. 45	de	Classifications Spectrales et Indices de Couleur à Plusieurs Bandes
	à	Classification Stellaire

The General Assembly approves the following changes in the names of Commissions:

Comm. 5	from	Documentation
	to	Documentation and Astronomical Data
Comm. 14	from	Fundamental Spectroscopic Data
	to	Atomic and Molecular Data
Comm. 44	from	Astronomical Observations from Outside the Terrestrial Atmosphere
	to	Astronomy from Space
Comm. 45	from	Spectral Classification and Multiband Colour Indices
	to	Stellar Classification

Each of the resolutions listed above for adoption by the General Assembly was so adopted.

Professor Smith, continuing the Report of the Resolutions Committee, invited the General Assembly to note the following Commission Resolutions:

(1)

Commission 4

recognizing (a) that timings of occultations of stars by the Moon will continue to be of value in studies of the lunar motion and figure, the rotation of the Earth, and the stellar reference frame, and

(b) that it is desirable that the observations should continue to be collected and processed by one organization,

considering that, beginning with January 1981, HM Nautical Almanac Office, Royal Greenwich Observatory, will no longer be able to act as the international centre for the receipt and processing of timings of occultations,

recommends that an organization with the appropriate experience and commitment to the occultation programme be requested to take over this important work.

(2)

Commission 4

expresses its full support in favour of the proposal of the Space Research Center of the Polish Academy of Sciences, and the Smithsonian Astrophysical Observatory to convene the Second International Colloquium on Reference Systems for Earth Dynamics.

(3)

Commissions 4,8, 19 and 31

endorse the recommendations given in the Report of the Working Group on Nutation, as set out below, and recommend that they shall be used in the national and international ephemerides for the years 1984 onwards, and in all other relevant astronomical work.

Recommendations of the Working Group on Nutation

Whereas the complete theory of the general nutational motion of the Earth about its centre of mass may be described as the sum of two components, (i) astronomical nutation, commonly referred to as nutation, which is motion with respect to a space-fixed coordinate system, and (ii) polar motion, which is motion with respect to a body-fixed coordinate system, it is recommended that:

(a) astronomical nutation be computed for the "Celestial Ephemeris Pole" using a non-rigid model of the Earth such that there are no nearly diurnal motions of this celestial pole with respect to either space-fixed or body-fixed coordinates, which can be calculated from torques external to the Earth and its atmosphere.

(b) the numerical values given in Table 1 of the complete report be used for computing astronomical nutation of the "Celestial Ephemeris Pole".

(4)

Commissions 4, 19 and 31

considering that it is planned to introduce the IAU (1976) System of Astronomical Constants, the IAU (1979) Theory of Nutation, and the equinox of the FK5 on 1984 January 1,

recommend that (a) the relationship between mean sidereal time and UT1 be modified so that there is no change in either value or rate of UT1, due to a correction to the zero point of right ascensions of the FK4 and to a correction for the motion of the zero point, to be introduced in FK5

(b) the new (provisional) expression for Greenwich mean sidereal time of 0^h UT be -

$$\text{GMST of } 0^h \text{ UT1} = 6^h\ 38^m\ 45^s\!.832 + 8640184^s\!.628\ T_u + 0^s\!.0929\ T_u^2$$

where T_u is the number of Julian centuries of 36525 days of Universal Time elapsed since 1900 January 0, 12^h UT1 (JD 2415020.0). This expression is rigorously equivalent to the following:

$$\text{GMST of } 0^h \text{ UT1} = 6^h\ 41^m\ 50^s\!.5529 + 8640184^s\!.8138\ T_u + 0^s\!.0929\ T_u^2$$

where T_u is measured from 2000 January 1, 12^h UT1 (JD 2451545.0).

(5)

Commissions 4, 19 and 31

endorse the proposal of the joint working group on the determination of the rotation of the Earth for a special period of international collaboration in the monitoring of Earth-rotation and in the intercomparison of the techniques of observation and analysis,

recognize that the responsibility for the organization of this Project MERIT should be shared with the International Union of Geodesy and Geophysics, and

request that the national and international organizations

concerned give full technical and financial support to the development of the proposal and to the implementation of the project.

(6)

Commissions 4 and 16

endorse the Report of the Joint Working Group on Cartographic Coordinates and Rotational Elements of the Planets and Satellites,

recommend that the Report be used as the basis for computing the physical ephemerides of planets and satellites in the international and national ephemerides, and

request that a small working group continues its activity in order to provide improved rotational elements.

(7)

Commission 5

considering the steadily increasing number of publications in the field of Astronomy and Astrophysics,

recommends the introduction and extensive use of proper key words by authors and publishers of primary journals and other publications in order to facilitate indexing and retrieval work,

welcomes the initiative of Astronomical and Astrophysical Abstracts (AAA) in preparing a draft vocabulary, and

invites AAA and other astronomical abstracting services to cooperate in the preparation of an agreed vocabulary.

(8)

Commission 5

considering the present unsatisfactory situation of the identification of astronomical objects in astronomical publications,

supported by representatives of Commissions 8, 24, 25, 26, 27, 28, 29, 30, 34, 37, 40, 42, and 45,

requests strongly that

(a) editors of scientific journals and other publications impose stricter standards of object indentification upon the authors through appropriate instructions to referees,

(b) catalog abbreviations be made explicit either in footnotes or in the bibliography tables

(c) two identifications be quoted for each object in order to provide a check against errors and misprints, especially for faint objects.

(9)

Commission 5

<u>considering</u> the confusion existing at the present time in the literature concerning the abbreviations of catalogues,

<u>supported by</u> representatives of Commissions 8, 24, 25, 26, 27, 28, 29, 30, 34, 37, 40, 42, and 45,

<u>desires</u> that, under IAU auspices, a list of catalogue abbreviations be published. This list should be a new version of the work of Fernandez, Lortet and Spite, revised by IAU Commission representatives and edited by C. Jaschek.

(10)

Commission 10

<u>recognising</u> that the International Ursigram and World Days Service (IUWDS) has rendered valuable service to the scientific community during the coordination of such international programs as the IGY, the IQSY and the IMS,

that the services provided by the IUWDS are not provided by any other agency and are relevant to the work of several IAU Commissions,

that these services will be particularly important for such programs as the Solar Maximum Year,

and that the IUWDS includes the prompt reporting of geophysical and solar events which will be occuring with increased frequency in the next few years,

<u>recommends</u> that the IAU continue to support the activities of the IUWDS to ensure the continuance of this valuable service.

(11)

Commission 10

<u>recognizing</u> that the long series of Relative Sunspot Numbers is a unique link with the course of solar activity in the past

<u>recommends</u> that all institutions that have demonstrated interest and competence in the work of obtaining sunspot numbers should continue the series.

(12)

Commission 10

<u>recognizing</u> that the Quarterly Bulletin of Solar Activity has for many years provided a valuable reporting service to the international scientific community,

that this service continues to be relevant to the activities of Commissions 10, 12, 40, 44, and 49 of the IAU,

that the continuation of this service will be of particular value during the coming Solar Maximum Year,

<u>recommends</u> that the IAU support the Quarterly Bulletin of Solar Activity to ensure that this service continue into the future.

(13)

Commission 10

<u>recognizing</u> that the 2800 MHz solar flux provides a standard quantitative index of solar activity for use in solar and solar-terrestrial studies,

<u>recommends</u> that those institutions making such measurements should continue to do so.

(14)

Commission 10

<u>recognizing</u> the important contribution made to solar physics by very high frequency (greater than 9 GHz) radio flux observations of the sun,

<u>recommends</u> that such observations should continue to be strongly supported during the coming solar maximum.

(15)

Commissions 10, 12, and 44

draw attention to the coming of Solar Maximum and recommend a sustained and coordinated effort in its study during the next three years.

(16)

Commissions 15, 20, 21, and 22

<u>recognizing</u> the major significance of comets for the study of the origin and evolution of the solar system in general and the interplanetary solid material in particular,

noting the extreme diversity among individual comets in terms of composition, structure, and physical behaviour, and

conceding that ground-based observations are largely inadequate for establishing unequivocally the nature of the relationship between comets and other objects in the solar system,

recommend that a vigourous programme of space exploration of comets be initiated. To ensure the maximum science return, the programme should include more than one comet and should include fly-by, rendez-vous and, ultimately, sample-return missions. A rendez-vous mission to P/Temple 2 in 1988 with an en-route fly-by of P/Halley in 1985, currently considered by NASA, is regarded as an ideal first step in the programme.

(17)

Commissions 25 and 45

noting the strong support of 12 IAU Commissions for space astrometric programs,

recommends that complementary ground-based support in astrometry, photometry, spectroscopy, etc. should be encouraged.

(18)

Commission 26 notes with deep concern that the Sproul Observatory is in danger of being closed by Swarthmore College. The termination of the astrometric program would cut off a highly productive source of double star data built up over a long period of time. The photographic plate collection with its continuing growth from the Sproul refractor is unique and continues to yield valuable results.

(19)

Commission 30

considering the need for optimum combination of independent series of data

recommends that the publication of new radial velocity measurements should be accompanied by the following essential information:

1. For each Series of measurements -

Description (or reference to published description) of the instrument characteristics, i.e., measuring principle, resolution, detector, etc.

Measuring technique (for plates)

Stellar lines employed and adopted wavelengths

Average number of lines measured

Measured velocities of IAU standard stars

Period of Observation

2. For each star or other object -

Preferably: individual velocity measures
individual times of obervation
mean radial velocity
weighting procedure
external mean error estimated

Minimum: mean radial velocity
period of observation
number of measurements
external mean error estimated

Any previous measurements used in the mean should be clearly identified.

The star should be unambiguously identified, either by catalogue number (catalogues in order of priority recommended by Commission 5), or by precise position (error 1 - 2 arcsec.), or, if necessary, by a finding chart.

(20)

Commission 31

considérant a) que le temps Atomique International, TAI, est établi par le BIH d'après les données fournies par plusieurs laboratoires où fonctionnent les horloges et étalons de fréquence indépendants

et b) que des données d'autres laboratoires, nécessaires pour améliorer le TAI, ne peuvent pas être utilisées car les liaisons horaires internationales existantes ne conviennent pas,

recommande a) que les laboratoires qui s'occupent du temps et des fréquences en Extrême Orient et Asie, en Afrique, en Australie et en Amérique du Sud, établissent des liaisons horaires internationales de grande exactitude

et b) qu'une aide leur soit apportée par les organismes nationaux et internationaux pour qu'ils puissent utiliser des techniques de grande précision, telles que celles qui impliquent l'usage du LORAN-C, des horloges transportables, des satellites artificiels de la terre, et de l'interférométrie à très longue base.

(21)

Commission 37

recommends that the designation of star clusters within the

Galaxy should be based on the equatorial coordinates (1950.0), prefixed by the letter C. The right ascension element should consist of hours and minutes of time (4 digits) and the declination element by the algebraic sign, followed by degrees and one decimal (3 digits). In both quantities appropriate leading zeros should be inserted. Once a designation has been assigned to a particular cluster it should not be changed, even if later investigations indicate that the coordinates used in the designation differ from the centre of the cluster.

(22)

Commission 44

takes note of the growing interest in 1-m. class Space Schmidt telescopes and encourages members of the IAU to make their interests known to the Chairman of the Working Group for Surveys from Space in Commission 28.

(23)

Commission 44

notes with interest, and encourages participation in, the various efforts to coordinate observations of the Sun near its activity maximum such as the Meudon Flare Buildup Study and the NASA Solar Maximum Mission.

(24)

Whereas the IAU supports the organisation of Regional Astronomy Meetings, Commission 46 recommends that the IAU urge the holding of sessions on teaching of astronomy at these meetings. Members of Commission 46 are available to assist.

Professor Smith noted, also, that the proposals implicit in items 11 (a) and 11 (c) of the Agenda were responded to in the Budget 1980-82, accepted by the General Assembly.

Professor Smith concluded by noting that a Working Group on the nomenclature of features of solar system bodies (WGPSN) had presented a Report, approved by the Executive Committee and by the Resolutions Committee. This Report now becomes an official document of the IAU.

18. Membership

a) Admission of Indonesia to the IAU

The General Secretary referred to her narrative report presented at the First Session. She reaffirmed that the Executive Committee had carefully examined the application of Indonesia for admission as an Adhering Country to the Union and that it had found that the degree of astronomical development in that country met the standards required for IAU Membership. The General Secretary therefore moved that the General Assembly admit Indonesia as a new Adhering Country. This motion was adopted unanimously.

The President invited Dr. B. Hidayat, the official representative of Indonesia, to be seated among the representatives of Adhering Countries. Dr. Hidayat then took the floor, expressed his thanks to the General Assembly for having admitted Indonesia to the IAU, and emphasized that Indonesia would spare no effort to promote astronomical cooperation throughout the world.

b) The President presented to the General Assembly the following report concerning the restoration of the Membership in the IAU of the Democratic People's Republic of China as follows:

The President reported as follows: At the first session of this General Assembly I have informed you about correspondence and discussions leading to the presence among us of Chinese colleagues from all parts of China, giving us opportunity to further work out arrangements for the formal adherence to the Union. What has been achieved up till now?

National representatives have been kept fully informed on the proceedings during the past few days and have been given copies of the following documentation: (a) an exchange of letters dated August 22 between the president of the Chinese Astronomical Society, Prof. Y.C. Chang of Purple Mountain Observatory, and myself as president of IAU; and (b) a draft for a resolution to be submitted to this General Assembly. In this letter, Prof. Chang in accordance with what we agreed upon, proposes that for the time being the full membership of China be established through dual adherence, with two adhering bodies: one being the Chinese Astronomical Society which has its seat at Nanking, and one representing the astronomical community of Taiwan, the name for which has not yet been filled in. It is furthermore proposed that for the time being at our General Assemblies each of these two adhering organizations appoint its own delegate and that the total financial contribution of China will consist of two components, one for each of these two adhering organizations. As to the voting power of the two delegates, in accordance with our statutes (article VII, 11) these will be of equal weight and the same as those of all other adhering organizations as far as administrative and scientific matters are concerned, whereas for budget matters the weight will be determined by the financial contributions according to the established system. Thus, the only point left open for final settlement of the full participation is the choice of the name for the adhering organization for Taiwan. In my letter of acknowledgement to Prof. Chang I have expressed the great satisfaction with which the Executive Committee has taken note of this proposal, and assured Prof. Chang of our readiness to take up this question of the missing name and our intention to do so in the near future, in consultation with the astronomical community of Taiwan and the Astronomical Society at Nanking.

Thus, whereas we cannot say that at this moment the full membership of China has been restored, we consider that considerable progress has been achieved and we have good hopes that the completion is forthcoming.

What I consider as one of our main accomplishments, is the fact that during these meetings at Montreal our Chinese colleagues and we have, after many years, again mutually benefited from their participation at our scientific discussions - and the prospect of continuation of this participation at our forthcoming symposia and other IAU meetings.

You may wonder why it is that the final settlement of the full membership had to be left pending, with only the question of a name to be filled in. Let me therefore explain that it is in this matter that we touch upon principles of general policy of the government authorities in these two geographical areas of China, which set limits to the range of possible names considered acceptable to all parties. We did not consider it proper to restore the full participation of China with the name of one of its adhering organizations still pending; and this the more so, now that we have full confidence that this question will be resolved soon when possibilities for solutions will be further explored after this Assembly.

This brings me to the resolution submitted to you on behalf of the Executive Committee. Its purpose is, to allow the Executive Committee to fully implement the full membership of China once the question of the name has been settled - possibly already before the end of this year. It would be a great pity if the full membership could become effective only at the next General Assembly. The resolution reads as follows:

"The General Assembly,

in view of the reported progress in restoring full adherence of China;

noting the letters on this subject exchanged between the President of the IAU and the President of the Chinese Astronomical Society, Prof. Y.C. Chang;

authorizes the Executive Committee to implement the anticipated conclusion of the current negotiations subject to ratification by the General Assembly at its next meeting."

This resolution then was submitted to the General Assembly and carried unanimously.

The President then continued as follows: Before we proceed to the next item of our agenda, I would like to express my deep gratitude to all our colleagues through whose, sometimes very considerable, effort we have been able to reach this point. It could not have been so without the fully cooperative spirit that, from the outset, has characterized the contributions to our discussions from both the astronomical communities, of the main-land of China and of Taiwan. It would have been impossible without the presence here at Montreal of the delegation of the Chinese Astronomical Society, and with regard to the delegation from Taiwan, may I express my deep admiration for the untiring effort by which its representative, Prof. C.S. Shen, has contributed to this purpose. And may I remind you also how much we owe to the many other colleagues who, not only during this Assembly but already during the preceding years, have helped so much by paving the road that has led to the present result.

Finally, would the representative from Taiwan care to say a few words at this moment?

C. S. Shen: (abstract only):

The mainland Chinese astronomers have been absent from the IAU for almost twenty years. Many attempts have been made to restore their membership, particularly during the past ten days. Many of our colleagues must be wondering why this task is still not accomplished. As the representative of an adhering organization, I would like to offer a few words on this matter:

Let me first point out that the astronomers and authorities on the island of Taiwan have never opposed the re-joining of mainland China to the IAU. In fact we have done our best, in co-operation with the officers of the IAU, especially with Professor Blaauw, to facilitate their re-joining. We have repeatedly stated (orally and in writing) that the Chinese Astronomical Society located at Taipei represents the scientific community in the geographical area of Taiwan. We do not pretend and have never claimed to represent the scientific activities in mainland China...

At present, there exists a significant gap between the two parts of China, not only in the style of living, but also in the political and economic structures that form the foundations of the two parts. Peking is trying to claim Taiwan as a province of the People's Republic of China and to imply that the authorities in Taipei are subordinate to those in Peking, and hence that the Nanking organization is a national one, while that in Taipei is regional. Taiwan is doing everything to avoid such an implication.

Of course, one has to be realistic. Mainland China has 900 million people with a potentially great astronomical community. They are our fellow Chinese; that is why we are doing what we can to solve the problem; however, we cannot accept a mode of adherence with semantic implications of a regional flavour vis-à-vis the name of the adhering body of mainland China. We represent an independent scientific community.

One need, however, not be too pessimistic about the China issue. It will usually be found that policies change if results cannot be achieved. Let us hope that well before the next General Assembly we shall have our fellow society from Nanking join us with due respect for the principle of equality of membership.

c) New Individual Members of the IAU

The General Secretary announced that the Executive Committee had, on the proposal of the Adhering Bodies and with the advice of the Nominating Committee, admitted 780 new Members to the Union and deleted 36 Members from its Membership List. The names of the new Members had been displayed in a prominent place in the course of the General Assembly. The General Secretary informed the General Assembly that these names would be incorporated in the Alphabetical List of IAU Members to appear in Transactions, Volume XVIIB.

The President then congratulated the new Members on their election and expressed his hope that their cooperation with and within the Union would be pleasant and fruitful.

19. Commission Presidents and Vice-Presidents

The General Secretary presented, on behalf of the Executive Committee, the following list of Presidents and Vice-Presidents of IAU Commissions for election by the General Assembly:

No.	Commission Name	President	Vice-President
4	Ephemerides	A.M. Sinzi	T. Lederle
5	Documentation and Astronomical Data	W.D. Heintz	G.A. Wilkins
6.	Astronomical Telegrams	J. Hers	M.P. Candy
7.	Celestial Mechanics	Y. Kozai	J. Kovalevsky
8.	Positional Astronomy	E. Høg	K.N. Tavastsherna
9.	Astronomical Instruments	E.H. Richardson	W.C. Livingston
10.	Solar Activity	V. Bumba	E. Tandberg-Hanssen
12.	Radiation and Structure of the Solar Atmosphere	Y. Uchida	R.W. Noyes
14.	Atomic and Molecular Data	J.G. Phillips	A.H. Gabriel
15.	Physical Study of Comets Minor Planets and Meteorites	B. Donn	B.J. Levin and C. Chapman
16.	Physical Study of Planets and Satellites	B.A. Smith	V.G. Teifel
19.	Rotation of the Earth	P. Pâquet	Ya. S. Yatskiv
20.	Positions and Motions of Minor Planets, Comets and Satellites	G. Sitarski	E. Roemer
21.	The Light of the Night Sky	H. Tanabe	R.H. Giese
22.	Meteors and Interplanetary Dust	W.G. Elford	O.I. Belkovich
24.	Photographic Astrometry	H.K. Eichhorn	W. Gliese
25.	Stellar Photometry and Polarimetry	J.A. Graham	J. Tinbergen
26.	Double Stars	O.G. Franz	M. Fracastoro
27.	Variable Stars	J.D. Fernie	N.H. Baker
28.	Galaxies	B.E. Westerlund	Vera Rubin
29.	Stellar Spectra	W.K. Bonsack	J. Jugaku
30.	Radial Velocities	M. Duflot	A.G.D. Philip
31.	Time	S. Iijima	G. Hemmleb

No.	Commission Name	President	Vice-President
33.	The Structure and Dynamics of the Galactic System	G.G. Kuzmin	R. Wielen
34.	Interstellar Matter and Planetary Nebulae	V. Radhakrishnan	M. Peimbert
35.	Stellar Constitution	R.J. Tayler	A.N. Cox
36.	The Theory of Stellar Atmosphere	G. Traving	B. Gustafsson
37.	Star Clusters and Associations	G. Lynga	K.C. Freeman
38.	Exchange of Astronomers	J. Delhaye	F.B. Wood
40.	Radio Astronomy	G. Swarup	K.I. Kellermann
41.	History of Astronomy	M.A. Hoskin	O. Pedersen
42.	Close Binary Stars	B. Warner	A.H. Batten
44.	Astronomy from Space	R. Van Duinen	M. Oda
45.	Stellar Classification	A. Slettebak	V.L. Straizys
46.	The Teaching of Astronomy	D.G. Wentzel	L. Houziaux and M. Rigutti
47.	Cosmology	G.O. Abell	J. Audouze
48.	High Energy Astrophysics	F. Pacini	
49.	The Interplanetary Plasma and The Heliosphere	H.J. Fahr	I.W. Roxburgh
50.	Protection of Existing and Potential Observatory Sites	F.G. Smith	A.A. Hoag

This list was approved by the General Assembly.

The slates of the Organizing Committees of IAU Commissions, as approved by the Executive Committee, were projected on a screen. The General Secretary announced that the composition of Commissions, general members included, would be published in IAU Transactions Vol. XVIIB.

20. Place and Date of the XVIIIth General Assembly

Called upon by the President, the General Secretary announced that there was no invitation by an Adhering Country before the General Assembly to host the XVIIIth General Assembly of the IAU in 1982. The place and date of the next ordinary General Assembly will therefore have to be determined by the Executive Committee and communicated to Adhering Bodies. Under consideration are the following countries: Greece or USA for 1982 and Spain for 1983.

21. Election of the New Executive Committee

The General Assembly approved by acclamation the formal proposal of the President that Prof. M.K.V. Bappu be elected the new President of the Union for the term 1979-1982.

The President then moved that Dr. M.W. Feast, Dr. L. Kresak and Prof. R. Wilson be elected Vice-Presidents for the term 1979-1985. This motion was approved by acclamation.

The President continued by proposing that Prof. P.A. Wayman be elected General Secretary for the Union and Dr. R.M. West be elected Assistant General Secretary of the Union for the term 1979-1982. This proposal was also approved by acclamation. The President then invited Prof. Bappu, Dr. Feast, Dr. Kresak, Prof. Wilson and Dr. West to join the Executive Committee on the platform.

22. Addresses by the retiring and newly elected Officers

The retiring President, Prof. Adriaan Blaauw spoke as follows:

We now approach the conclusion of the administrative part of this, 17th, General Assembly, and in a few minutes I shall call on the newly elected President; however, before doing so I wish to express my sincere thanks to my colleagues on the Executive Committee with whom I had the privilege to work during these three years - colleagues who thereby have become even closer friends than they were before.

I first want to thank the now outgoing vice presidents, John Bolton, Charles Fehrenbach and Wilhelmina Iwanowska for their fine contributions, and hope that there will be many occasions where we will meet again and further enjoy this friendship. We are very sad that John Bolton could not be with us these days. I am sure I will speak on behalf of all of us if, after this meeting, I write to him to tell him how very much his devotion to the affairs of the Union has been appreciated.

I also wish to thank the outgoing advisors, Leo Goldberg and George Contopoulos, for the way they have let us benefit from their rich experience in matters of the I.A.U. Their leaving the Executive Committee means a great loss to the built-in wisdom of the Committee, but we trust it will remain available if we ever need it again.

It has been said before that it is the General Secretariat where the heart of our Union beats. We all, members of the Union, know with how much care the Secretariat handles its many-fold affairs. Now that its crew, too will be leaving us, we have the oppurtunity to express our deep felt thanks for all you, Edith Müller, Arnost Jappel and Jarka Dankova have done for the IAU. Many of us have often wondered how you, Edith, could manage to do so many things - not only for the IAU but also in your many other obligations - and do them with such intensity, combining your scientific interests with your wonderful devotion to, and experience in, our Union. To express it in terms of an observational astronomer, you have shown a dynamic range that many of us envy! We will continue to benefit from it in your new, but more relaxed, position as advisor to the Executive Committee.

Edith will be succeeded by Patrick Wayman, an appointment that assures us that the affairs of the Union are in the best possible hands. It is not only the experience and knowledge gained in the specific task of the Assistant General Secretariat, the planning of colloquia, symposia and other meetings which Patrick Wayman has performed so admirably, but also the way he has already been involved in other important current affairs of the Union and his broad administrative experience that make him eminently suitable for this task.

May I suggest that the General Assembly show its appreciation for the IAU's Secretariat in the usual way?

At the request of Professor Blaauw, the newly elected President, Prof. M.K.V. Bappu, addressed the General Assembly:

Professor Blaauw, Members of the Union, Ladies and Gentlemen,

I would like to express my deep appreciation of the great honour you have done me by electing me President of the Union for the next three year term. At this moment, a twin emotive feeling of gratitude and responsibility overwhelms me; gratitude at the trust and confidence you have displayed by choosing me for this high office on which so many of my eminent predecessors have added lustre, and a feeling of great responsibility in a realization of the implications of this trust. That it is an union of astronomers makes the task lighter, for the astronomical fraternity is a brotherhood that unites men and takes them above the little world of nationalism into a truly international beyond. And when, to it, we add the extremely pleasant prospect of the continued availability of your wisdom, experience and counsel, Prof. Blaauw, I am sure that we face the future with serene confidence. The manner in which you have steered the Union through the occasional troubled waters of a world with political affinities, demands our admiration and gratitude. You have led a splendid team of dedicated individuals; to all the members of the Executive Committee, the Union is very deeply indebted, and especially to its charming and efficient General Secretary, Prof. Edith Müller, who has continued in the distinguished tradition of performance of General Secretaries of the past.

The last decade and more have witnessed spectacular happenings of fruitful work in new spectral domains, major strides of technological progress and great achievements by individuals and groups of individuals, both in theory and in observation. This dynamic state of affairs, of expanded horizons and activity, has brought along a manifold increase in the number of active astronomers. An increase in size of our Union is natural and inevitable; our only concern need be to ensure that we permit ourselves an adaptability of methodology and procedure whereby we can fulfill the aims of the Union. An increased reliance on symposia, more regional meetings that have the international flavour of a General Assembly, the fostering of an enlarged programme of exchange of astronomers between astronomical institutions - these are some measures we have already taken, and which, in the years to come, will hopefully justify the faith we have placed in them. Our decisions in this regard need to be tempered by a recognition of the fact that meetings of the General Assembly serve a twofold requirement. Besides being occasions when we recount the accomplishments of the past three years and project and formulate new plans for the future, perhaps the most important aspect is the opportunity provided for personal contracts and discussion between members of all ages.

The progress made by our science is a saga of accomplishments by the different individuals who make up the Union membership; it should, therefore, be the Union's utmost concern to see that opportunities for such stimulation and inspiration are plentiful.

It is a gratifying trend to witness an increase in membership of countries. We have come a long way from the initial seven who adhered to the Union in its first year of existence. This tendency heartens us, for we have a high opinion of our science and would like many others to share our enthusiasm for exploration of the Universe. And as each member, individual and country, contributes stone by stone to the vast edifice of astronomical knowledge, we rejoice without reservation and in all friendliness at the triumph of the human mind over a facet of Nature. In encouraging opportunities and providing that little extra fillip needed to cross the threshold into the domain of creative productivity, the Union can play an important role.

These are some thoughts that come to mind and which I would like to share with you. With your abundant goodwill and cooperation, my colleagues on the Executive Committee and I approach these possibilities with much optimism. May the International Astronomical Union, in the years ahead, continue to be a source of strength to astronomy and to astronomers.

23. Closing Ceremonies

Prof. Blaauw then invited Prof. Ch. Fehrenbach to propose a comprehensive vote of thanks.

Chers amis Canadiens,

Vous êtes presqu'au terme de votre peine. Mais nous savons que lorsque les invités sont partis, après un moment de soulagement et de satisfaction lorsque tout s'est bien passé, il reste à ranger et la salle à manger et le salon et il faut quelque temps pour que tout redevienne normal.

Au nom du Comité Exécutif et de tous les astronomes j'ai le grand plaisir de vous adresser nos vifs remerciements pour la façon dont vous avez préparé et mené cette grande Assemblée Générale qui a réuni plus de 2000 astronomes et 1000 de leurs invités.

Votre organisation a été parfaite à tous points de vue à la fois pour la teneur des réunions plenières à l'Université ou dans les diverses grandes salles de la Ville. Vous nous avez toujours accueilli avec le sourire et vous avez résolu tous nos problèmes.

Le seul que j'ai eu personnellement était celui de la découverte des salles dans l'Université. Mais à la fin, j'ai appris qu'on pouvait atteindre le bureau du président en prenant l'ascenseur situé près des cubicules des étudiants avancés, dont le sens m'échappe sinon sa relation avec l'ascenseur. D'ailleurs une de vos hautes autorités m'a confié la façon expéditive de faire passer les examens à vos étudiants: vous leur donez rendez-vous à une salle numérotée et ceux qui sont assez chanceux pour trouver la salle sont automatiquement reçus.

Excusez-moi si je vous "charrie" un peu, vous devez aussi utiliser ce mot pour plaisanter.

Je dois maintenant remercier d'abord le Comité d'organisation Canadien, qui sous la présidence de notre ami Alan Batten a eu la fonction essentielle d'obtenir le concours des autorités Canadiennes. Et son succès a été complet. Grâce au Conseil National de la Recherche du Canada, au Ministre de l'Education du Canada, de la Cité de Montréal et de nombreuses societés commerciales privées, grâce à l'aide des Societés astronomiques Canadiennes de nombreuses contributions matérielles ou en argent ont été obtenues. Et de nombreux jeunes astronomes ont pu recevoir des subventions de l'Union Astronomique, augmentées par vos fonds. Nous devons spécialement remercier la Municipalité de Montréal et son maire M. Drapeau qui a mis la salle des Arts à notre disposition et qui nous a offert l'excellent dejeûner qui a suivi la Cérémonie Inaugurale.

Les intermèdes musicaux de très haute qualité de l'ensemble Claude Gervaise ont été très applaudis.

Nous devons spécialement remercier Monsieur le Recteur de l'Université, representé en son absence par le Vice-Recteur, M. René Lévesque, pour toutes les facilités mises à notre disposition dans l'unité de Physique notamment, et je remercie le directeur de cette unité.

Je dois, et je le fais avec grand plaisir, remercier le Comité Local présidé par notre collègue G. Beaudet. En le remerciant personnellement je remercie aussi les secrétaires des deux Comités, Mlle Bergevin et Mlle Guibault. Le Dr Michaud, Mme Beaudet et tous ces aimables jeunes hommes et jeunes femmes qui nous ont tant facilité la vie.

Nous remercions aussi les personnes qui ont fait fonctionner la caféteria, le bistrot et qui ont servi 37 500 tasses de café. Les manifestations artistiques et culturelles ont été nombreuses. Je remercie particulièrement le "Concept Neuf" qui nous a donné un récital de musique de percussion et Mme Monique Legrac pour son récital. Tous les concerts ont été fort appréciés.

A la fin d'une assemblée de ce type on se pose toujours la question: est-elle un succès? C'est naturellement au succès scientifique qu'on pense d'abord, mais ce succès n'est pas possible si l'organisation n'est pas bonne à tous points de vue et si le sourire n'accompagne pas les petites difficultés de chaque jour.

Cette assemblée générale est un grand succès et, chers amis Canadiens, nous vous remercions pour votre excellente organisation, pour votre hospitalité, pour votre présence amicale à tous moments.

Un grand merci à tous, à ceux que j'ai cités, mais aussi à ceux, plus anonymes, qui les ont aidé. Nous garderons un excellent souvenir de cette assemblée générale à Montréal, mais nous garderons le même excellent souvenir du beau pays que vous nous avez fait visiter dans des nombreuses excursions organisées à Montréal, dans les provinces de Québec et d'Ontario. Nous avons été frappés par l'accueil si cordial et si souriant de tous les habitants de ce pays.

On behalf of the Registered Guests, Mrs Mavis Wayman expressed her thanks as follows:

Mr. President and Members of the Union, on behalf of all the guests at this General Assembly I would like to thank Mme Beaudet and her Committee for the wonderful way they have entertained us. Many thanks for the welcomes, starting at the guests' reception in Place des Arts, and continuing throughout the week in the hospitality room; also, their hard work in organizing a wide range of interesting tours for us is greatly appreciated. As hostesses, they did this beautiful city of Montreal proud. Please, Mme Beaudet, accept a small token of our appreciation for everything.

The President then called upon the retiring General Secretary, Professor Edith A. Müller, who spoke as follows:

Dear Colleagues: - You may have read in one of the issues of METEOR that this evening I shall take a deep breath and then eclipse myself for a long vacation to clear my head from all administrative, organizational and diplomatic problems in order to get ready to resume my research work and my teaching duties with full strength. First of all I shall have to catch up with the recent astronomical literature and, thus, I look forward to learning about all the work you have done and the interesting results you have found. The past six years (three as Assistant General Secretary, three as General Secretary) were for me very interesting, stimulating and, indeed full of hard work. Having attended many international meetings and visited numerous astronomical institutes in different countries, I had the opportunity of meeting many scientists and make a number of new friends. Though loaded with work the years serving the IAU were very enriching and enlightening for me.

My task as General Secretary was greatly facilitated by the kind support and advice I received from our retiring President, Professor Adriaan Blaauw, from the members of the Executive Committee and from its Advisors. I wish to express here my sincere thanks to them.

In the IAU Secretariat I appreciated the cooperation of the long-standing IAU staff members Arnost Jappel and Jarka Dankova. Furthermore, I was fortunate to have been able to secure the devoted cooperation of two part-time secretaries: Mme Roswita Läubli who is here with us, and Mme Ruth Bertschi. To all four collaborators I owe my most cordial thanks for their devoted and efficient work. The IAU Secretariat enjoyed the kind hospitality of the Astronomical Institute of the University of Lausanne, and it received substantial financial support from the Swiss Academy of Sciences through the National Committee of Astronomy, and from the Observatory of the University of Geneva. In the name and on behalf of the IAU, I thank the above-mentioned institutions for their great support. I also wish to express my gratitude to my colleagues both in Geneva and Lausanne for their moral aid and understanding and to my graduate students for their patience when pressing IAU affairs prevented me from attending immediately to their research problems.

I now wish to turn your attention to an important chapter in the history of the IAU and that of our two staff members, - a chapter which now is coming to an end. Arnost Jappel and Jarka Dankova have faithfully served several General Secretaries, - Arnost five, Jarka four General Secretaries, - moving every three years to another country with another language and other habits. Now they both will return to their home country, Czechoslovakia, Arnost Jappel starting his well deserved retirement, and Jarka Dankova resuming her work at the Czechoslovak Academy of Sciences. Arnost - with all the many interests you have, I am sure you will find yourself as busy in your retirement as before. Jarka - undoubtedly you will be happy to settle in your home country where you have your family and your friends and where you finally will feel "at home" again. To both of you the IAU and, in particular, the General Secretaries you served, express their most sincere thanks. Our best wishes accompany you in your future life and activities.

Our new General Secretary, Patrick Wayman, is to start a new chapter in the history of the IAU Secretariat. As I mentioned during the first session of this General Assembly, the IAU Secretariat will, from now on, function at its permanent seat in Paris, where a new secretary and a typist will be serving the General Secretary. Patrick - you have certainly a difficult task ahead of you by having to introduce two new staff members to the work of the IAU Secretariat, and by having to do much of this from a distance, - by some sort of a "remote control". I expect you will frequently commute between Dublin and Paris and, thus, become the "flying General Secretary". But I am sure you will handle everything most efficiently. I wish you much success.

The IAU is flourishing; - many fascinating fields of astronomical research have opened up and added exciting new results to our knowledge of the stars, the galaxies, the universe. A direct consequence of this is the rapid growth of the IAU. At the present General Assembly we are happy to welcome one new member country, Indonesia, and about 800 new Members. At this point I should remind you that proposals for new individual Members are primarily the responsibility of the National Committee of Astronomy of each member country. Evidently they are in a position to judge the competence and the scientific work and achievements of a proposed new Member. It would be a full-time job for the General Secretary if he/she were to look into each individual case in detail.

Finally, with the future of the IAU in mind, I should like to say a few words to the new Members of the IAU and also to all our Members. We are gathered here at a General Assembly, coming from different parts of the world, from countries with different governments, different cultures, different religions, - but we are all astronomers eager to learn more about the Universe and the objects which populate it. Let me remind you of our statutes which say that the IAU is a non-governmental, non-political organization whose objectives are a) to develop astronomy through international cooperation, b) to promote the study and development of astronomy in all its aspects, and c) to further and safeguard the interest of astronomy. Let the spirit of the IAU prevail in our minds and guide our work of international cooperation for the safeguard of astronomy and astronomers throughout the world.

Finally, the Chairman invited the newly elected General Secretary, Professor P.A. Wayman, to address the General Assembly. Professor Wayman spoke as follows:

Mr. President, members of the Executive Committee, fellow members of the Union,

We come from all continents and sub-continents to discuss what we love most - the science and practice of astronomy - and to enjoy, perhaps above all, the fellowship of astronomers across every kind of political and geographical divide.

Certain questions force themselves upon us at these times, as we prepare to leave our General Assembly. What does the Union mean to us? How will future ages regard our period?

We are indeed reaching a turning point, a watershed, in the affairs of the Union. Whatever problems lie ahead, we can be sure that a new era begins in 1979; new permanent staff, a new regime, with a permanent seat in the magically beautiful city of Paris. My message to IAU members on this occasion is - If you wish to make suggestions about how things should be done in the future, please write to me about it, in order to contribute to the evolution of the Union.

Arising from my work in the Union in the last three years I have some messages of thanks to convey.

I wish particularly to express admiration, in my recent experience, to the work of our retiring President, which I have witnessed at close hand, in achieving a triumph in the diplomatic fringe of our activities. If the solution is not complete, that is certainly not due to any negligence on his part. His patience in this task has been unbelievable and I do not see how any more could have been achieved.

Also I must thank organizers of Symposia and Colloquia for their understanding attitude towards difficult points that have had to be discussed from time to time in presenting their valuable proposals to the Executive Committee.

Et maintenant, j'ai le grand plaisir d'exprimer à ma collègue, Edith Müller, mes remerciements les plus chaleureux pour ces trois dernières années, pendant lesquelles nous avons consacré bien des heures aux affaires de l'Union, et dont nous nous souviendrons avec plaisir. Elle m'a mis au courant des traditions de l'Union Astronomique Internationale, de sa structure, et aussi des espoirs que nous formulons pour notre Union. En tant que Secrétaire Générale, la lourde tâche de l'installation à Paris du Secrétariat, dans ses nouveaux locaux, incombait à Edith Müller - Merci beaucoup, ma chère Edith.

Professor Blaauw, in the Chair, then declared the XVIIth General Assembly of the International Astronomical Union closed.

PART 3

REPORTS OF MEETINGS OF COMMISSIONS

COMPTES RENDUS DES SEANCES DES COMMISSIONS

COMMISSION 4: EPHEMERIDES (EPHÉMÉRIDES)

Report of Meetings 14, 15, 16, 20 and 21 August 1979

PRESIDENT: V.K. Abalakin SECRETARY: J.H. Lieske

14 August 1979

ADMINISTRATIVE MATTERS

Participants stood in silence in memoria of Dr. E.W. Woolard and Prof. Y. Hagihara. The President summarized the current status of the Commission and distributed the draft agenda for the Commission meetings during the General Assembly. The agenda was approved with some amendment. The President proposed lists of the new officers of the Commission for the next three years and of the new members of the Commission. The meeting agreed that both lists should be put to the Executive Committee (see "Membership of Commissions"). New Commission members are: A. Bandyopadhyay, P. Bretagnon, J. Chapront, B. Emerson, A. Fiala, M.A. Fursenko, N.I. Glebova, J. Henrard, R.W. King Jr., H. Kinoshita, B. Kolaczek, I.I. Mueller, R.D. Reasenberg, H. Schwan, A.S. Sochilina, E.M. Standish Jr., and A. Yamazaki. New consulting member is M.L. Smith.

The President indicated that a symposium (or a colloquium) on Reference Systems for Earth Dynamics is being planned to be held in Poland in 1980, and that the support of the IAU was desired. The meeting agreed to consider this item in a later session.

REPORT OF ALMANAC OFFICES

G.A. Wilkins explained the background to the changes in the publications of the Royal Greenwich Observatory (RGO) and of the U.S. Naval Observatory (USNO) described in the Report of Commission 4 (Trans. IAU XVIIB, Part 1, p. 2); those who requested the ephemerides data in advance did not want high precision but needed only the data on phenomena or low precision ephemerides in most cases. He showed copies of the first issues in the revised series, that is of "Astronomical Phenomena 1981" and of "Planetary and Lunar Coordinates for 1980-1984". The next issue of the latter publication would refer to the years 1984 to 2000 and should be available in 1982. The name "Astronomical Almanac" would replace the two current names for the "AE". Those requiring the data of the Astronomical Almanac in advance would receive them as computer listings or on magnetic tape.

P.K. Seidelmann then outlined the principal changes that had been introduced into the Astronomical Almanac. The publication will be reformed and organized into sections by subject. The hourly apparent lunar ephemeris will be replaced by shorter power series for direct calculation of the lunar position for any time. Physical ephemerides will be given for all the planets and observing ephemerides for all the satellites. An ephemeris for the solar system barycentre, transformation matrices for the reduction of apparent places, and standard stellar source lists will be included. The brighter star list will be expanded, accurate locations of instruments will be given periodically, and a revised explanation and glossary of terms will be introduced. The independent Day Numbers, first differences and fixed tables for unit conversions will be eliminated.

It was pointed out that a 5-term series based upon Economized Chebyshev Polynomials would be employed, since the USNO experience with users indicates a "familiarity" with power series and a "fear" of Chebyshev polynomials. French experience is the opposite -- Chebyshev polynomials are accepted by users.

B. Morando described the new presentation of the "Connaissance des Temps" (CdT) and other publications issued by the Bureau des Longitudes (BdL). Over three hundred years after its first issue the CdT for 1980 will have an entirely new presentation for economic reason, by listing the coordinates of the Sun, Moon, planets, and Galilean satellites in Chebyshev polynomial form. The phenomena and configurations of the Galilean satellites, the mean places of the FK4 and FK4 Sup stars, and Day Numbers are no longer presented. The new CdT contains an external section of explanations and examples, including those for the star place reduction and the new system of astronomical constants. The "Annuaire du Bureau des Longitudes" does not present scientific articles on astronomy, geophysics, physics and geography. These articles are now a part of a book entitled "Encyclopédie Physique et Spatiale" issued in separate volumes every year. The "Annuaire" itself is now an extended almanac for amateurs and for astronomers that wish to look up astronomical phenomena or positions of celestial objects without higher precision.

J. Chapront spoke about the efforts at the BdL to develop new theories for the planetary and lunar motions, and suggested the replacement in the CdT of old theories by new ones. He discussed also the advantage of theories against numerical integrations as well as the defects of the present CdT for the planetary motions. The present state of the new solutions could be summarized as: (i) For the major planets, Jupiter and Saturn, comparisons of the solutions with internal numerical integrations show discrepancies less than 0"3 in longitude and 0"1 for the other elements, over a range of 1000 years, and (ii) For Mercury, Venus and the Earth, the discrepancies are less than 0"007 in longitude; 0"035 for Mars. He finally emphasized the importance of the comparison between theories and observations.

Questions were asked re the source of the difference between the Leverrier and Newcomb ephemerides; it was answered that P. Bretagnon has them available on magnetic tape.

A.M. Sinzi reported that, instead of current Besselian Day Numbers, Chebyshev coefficients for the transformation matrices will be tabulated in the "Japanese Ephemeris".

15 August 1979

Joint meeting with Commission 16

CARTOGRAPHIC COORDINATES

G. de Vaucouleurs summarized his work on the determination of the rotation period of Mars, based on an extensive review of all previous ground-based observations.

M.E. Davies, the Chairman of the IAU Working Group on Cartographic Coordinates and Rotational Elements of the Planets and Satellites, presented the Report of the Working Group. See Annex 1.

After some discussion the meeting unanimously agreed that the Report should be adopted as the basis of ephemeris computations. Although the task of the Working Group finished by the approval of the Report, it was agreed that the Group would remain in a reduced form, being composed of few persons with M.E. Davies as Chairman, to inform on improved data, since such data were expected to be available in the near future. See Resolution 1.

16 August 1979

Joint meeting with Commissions 7, 8, 19, 24 and 31

NUTATION

P.K. Seidelmann, the Chairman of the IAU Working Group on Nutation, presented the Summary Report of the Working Group. See Annex 2.

T. Sasao argued that the Report should be amended by adopting the nutation theory developed very recently by J.M. Wahr (1979) for the Earth Model 1066A of F. Gilbert and A.M. Dziewonski, instead of Moldensky's model II (1961). Sasao stated that Moldensky's model is already out-of-date while the Model 1066A is one of the most heavily constrained new models. He pointed out that, between the numerical values given in Table 1 of the Report and those proposed by Wahr, there exist discrepancies of up to 0."002 in $\Delta\psi \sin \varepsilon$ and $\Delta\varepsilon$, both for the 18.6 year and six month terms. These discrepancies could not be overlooked when new techniques, such as VLBI and laser ranging, became practical in the near future. Discussion was focussed on the choice of either theory from scientific and practical view points. The President asked for a vote and the theory contained in the Report was favoured by majority mostly due to the feeling that the Wahr's theory should "age" a bit more before being proposed for adoption.

There were also discussions regarding the wording of the Recommendation: one with reference to the Resolution adopted at the General Assembly in Grenoble and the other for the usage of the terminology of "space-fixed", "body-fixed" and "inertial", and "Celestial Reference Pole" and "Celestial Ephemeris Pole". See Resolution 2.

FUNDAMENTAL CATALOGUE (FK5)

W. Fricke presented a bibliography of work done at the Astronomisches Rechen-Institut since the last IAU General Assembly. The tasks required for the construction of the FK5 are the following:

(1) Determination and introduction of the new value of the general precession in longitude;

(2) (a) Determination of the zero point correction to $(\mu_\alpha)_{FK4}$ in order to eliminate the non-precessional motion of the FK4 equinox, and

(b) of the zero point correction to α_{FK4} at the epoch 1950 such that the zero point of α_{FK5} will be identical with the dynamical equinox;

(3) Determination of the equator of the FK5 and application of that correction (certainly small, if not zero) to the δ_{FK4};

(4) Elimination of the inhomogeneities of the FK4 system (including magnitude equation) by means of recent absolute and quasi-absolute observations, and application of these systematic corrections to the FK4 positions and proper motions;

(5) Elimination of the E-terms of aberration from FK5 mean positions, because they will be included in the reduction to the apparent places;

(6) In the computation of apparent places of FK5 stars the new theory of nutation and the relativistic effects will be introduced;

(7) Conversion to the new Standard Epoch and Equinox J 2000.0, and from tropical to Julian centuries as the unit of time interval;

(8) Determination of individual corrections to the positions and proper motions of FK4 stars and their application;

(9) Addition of new fundamental stars with extention to magnitude $\sim$9.0.

Steps (1)-(7) could be completed by 1984. In the discussion Fricke pointed out that the faint stars will be taken from existing well-observed series, and that some radio sources would also be included.

20 August 1979

Joint meeting with Commissions 19 and 31

UNIVERSAL TIME AND DYNAMICAL TIME

G.A. Wilkins read the following draft resolution proposed by D.H. Sadler who was unable to attend: "Commission 4 requests that neither the name Universal Time nor its designation UT be used to denote the time-scale of Co-ordinated Universal Time, and that the General Secretary of the Union arrange for the content of the Resolution to be communicated to the appropriate international organizations". Sadler put forward this resolution since he considered that by adopting Resolution No. 1 at the last General Assembly, Commissions 4 and 31 had unnecessarily perpetuated confusion and complication. The aim of the proposed resolution was to preserve the original meaning of Universal Time and to avoid the ambiguity that arises when the same name would be used for several kinds of Universal Time. Wilkins strongly supported Sadler's proposal. G.M.R. Winkler claimed that the matter had been fully discussed at three previous General Assemblies and that he had not heard any new argument. He proposed that the meeting should not further discuss this problem, and the meeting agreed.

W. Fricke explained the change in the expression of UT1 in terms of Greenwich mean sidereal time (GMST), in accordance with Recommendation 3(c) of the Joint Report as adopted at the last General Assembly (Trans. IAU XVIB, p. 59):

1. In order that the equinox of the FK5 corresponds as closely as possible to the dynamical equinox, a correction E to the right ascensions of the FK4 and a correction $\dot{E}$ to the centennial proper motions of the FK4 will be introduced giving the change from the FK4 equinox to the FK5 equinox

$$E(T) = E_{1950} + \dot{E}\ (T - 19.50),$$

where T is measured in centuries, and where from the current work on the FK5 the following preliminary numerical values are available

$$E_{1950} = +\ 0\overset{s}{.}040,$$

$$\dot{E} = +\ 0\overset{s}{.}086 \quad \text{per century.}$$

A distinction between tropical and Julian centuries is of no importance in this case. These preliminary values have been employed in Para. (b) of the proposed Resolution. Final numerical values which will not differ from the preliminary ones by more than 10 milliseconds will be reported by the end of 1980.

2. The change of the positions and proper motions of the fundamental stars described by E and $\dot{E}$ affects the determination of sidereal time. However, the change does not affect the determination of UT1, if the relationship between sidereal time and UT1 is redefined. Furthermore, it should be noted that the 1976 change in precession does not affect the apparent right ascensions, since such change modifies the proper motions correspondingly.

3. UT1 is currently defined as 12^h + the Greenwich hour angle of a point on the equator whose right ascension, measured from the mean equinox of date, is:

$$R_u = 18^h38^m45\overset{s}{.}836 + 8\,640\,184\overset{s}{.}542\ T_u + 0\overset{s}{.}0929\ T_u^2,$$

which becomes now

$$R_u = 18^h38^m45\overset{s}{.}833 + 8\,640\,184\overset{s}{.}628\ T_u + 0\overset{s}{.}0929\ T_u^2.$$

The old expression for R_u is identical with that given by Newcomb (Astron.Pap.Amer. Eph. 6, p. 9, 1895) for the right ascension of the fictitious mean sun. It is now recognized that this expression for R_u, while intended to represent the motion of the fictitious mean sun, is not rigorously related to the position and motion of the Sun. Newcomb, unaware of the variable rotational speed of the Earth, considered that T was measured in mean solar time. Hence, the mean sun differs from R_u by 0.002738 ΔT, where ΔT = ET - UT1.

4. A small and partly unpredictable discontinuity in UT1 may occur with the introduction of the FK5 and the 1979 Theory of Nutation. This could possibly be removed by a correction to station longitudes.

After some discussion about the definitive values of E and $\dot{E}$ the Resolution was adopted. See Resolution 3.

A. Orte stated that the clarification of the UT concept and related conventions to study the Earth's rotation is a priority task. This is necessary mainly in view of the advent of new powerful techniques for measuring the duration of a day. Limitations of the present UT are in accuracy, in precision, and in the stability of the system of reference. Moreover, today's definition contains ambiguities concerning the meridian and the pole of reference. Hence explicit operational conventions are required. Orte suggested the initiation of studies for a new definition, freezing the present conventions and terminology in the meantime.

The President gave the information to the members of Commission 4 that the Project MERIT had been already discussed in Commissions 19 and 31, and asked for a vote together with the members of Commissions 19 and 31. The Resolution was adopted unanimously. See Resolution 4.

B. Guinot made some comments on his recent paper (Proc. IAU Symp. No. 82, p.7, 1979) to advocate the use of a non-rotating reference system, taking the origin at the departure point. UT1 could then be simply defined as proportional to the hour angle of the origin, and hence could be used not only to locate angularly the Earth relative to celestial objects but also to investigate the irregularity in the Earth's rotation through the relationship with sidereal time. He therefore regretted that without changing anything in the practical definition of UT1 from sidereal time, the IAU had not given, in addition, a clear statement on the principles underlying this definition. Such a statement could have put an end to the confusion arising from the historical background of solar time.

P.K. Seidelmann described the background to his proposal on the designation of dynamical time-scales defined at the last General Assembly. To differentiate between the relativistic coordinate time referred to the barycentre of the solar system and the proper time referred to the Earth, descriptive adjectives are required. Many possibilities have been considered and the simplest would be "coordinate" and "proper", but "coordinate" is subject to confusion with "coordinated" of UTC and "proper" does not specify for what it is proper time. Therefore, using the adjectives "barycentric" and "terrestrial", the designations "Barycentric Dynamical Time" (TDB) and "Terrestrial Dynamical Time" (TDT) have been proposed. Alternative designations, such as "TB" and "TT", were suggested in the discussion but the proposed designations were favoured by the majority. See Resolution 5. It was understood that, when it is unnecessary to differentiate between a coordinate and proper time-scales, the general term Dynamical Time (TD) may be used; for instance, when the accuracy is lower than 0.1 second. Several people questioned the wisdom of introducing new names for coordinate and proper times when the primary reason for doing so seems to be solely to enable almanac offices to state that Proper Time equals Dynamical Time plus some tabular corrections to be determined after the publication of the ephemerides. It was suggested that besides being confusing, it is improper to define a dynamical coordinate time and a dynamical proper time without specifying the metric.

21 August 1979

ADOPTION OF FURTHER RESOLUTIONS

The President announced that the Commission had been requested to cosponsor the Resolution on the space astrometry project initiated by Commission 24. C.A. Murray, President of Commission 24, then explained the background to the Resolution. The meeting agreed unanimously to support the Resolution. See Resolutions adopted at the General Assembly.

The President asked the Commission to support the proposal for the Colloquium on Reference Systems for Earth Dynamics to be held in 1980 in Poland. The meeting agreed unanimously. See Resolution 6.

The President asked the Commission to consider a resolution which had been put forward following an informal discussion of the future of the lunar occultation programme. L.V. Morrison explained that, although the H.M. Nautical Almanac Office (HMNAO-RGO) had been responsible since 1943 for an international service for the prediction and reduction of lunar occultation of stars, it would be impossible for the HMNAO to continue to provide this service due to its gradual reduction of the number of staff and to the necessity of allocating staff to a new programme for satellite laser ranging. He stated that the HMNAO would be still prepared to continue the prediction service and to give all possible assistance to any organization that would take on responsibility for the collection and processing of observations. In seconding the Resolution W. Fricke thanked the HMNAO for the valuable work that had been done; in particular he pointed out that Morrison's recent analysis was highly useful for his investigation on the equinox of the FK5. The Resolution was adopted unanimously. See Resolution 7.

FUNDAMENTAL EPHEMERIDES

J. Chapront, on behalf of P. Bretagnon, M. Chapront-Touzé, and himself, described the progress made at the BdL in constructing new theories for the lunar problems and planetary motions. (i) A modern solution for the Earth's motion, taking into account the second-order perturbations, has been proposed and its comparison with an internal numerical integration has shown agreement within 0".01 in longitude and 0".005 for the other osculating elements. (ii) Starting with a solution for the lunar main problem, a variational method has been retained to compute the planetary perturbations of the Moon. Main limitations are: the first order terms with respect to the masses of planets, terms of periods less than 3500 years, and internal precision of 0".0005 for the longitude. A comparison with some of Brown's results has been made.

D. Standaert presented the theoretical background to a method for computing the planetary perturbations in the Moon's motion. An algorithm has been formulated on the basis of the Lie transform method, and is being implemented using Henrard's Semi-Analytical Lunar Ephemeris as a solution of the main problem and Bretagnon's planetary theory. The accuracy of the solution is intended to be around 0".001 for terms with periods up to 2000 years. He illustrated some preliminary results obtained for the direct perturbations due to Venus on the Moon's longitude. J.D. Mulholland congratulated to the BdL and Namur people on the success achieved in solving one of the most difficult problems in the theory of the motions of the solar system objects.

P.K. Seidelmann described the current situation with respect to the IAU (1976) System of Astronomical Constants: (i) It is not certain whether the IUGG will adopt the IAU value for the equatorial radius of the Earth or an improved value, at its General Assembly to be held in December 1979. For astronomical calculations, the difference is not significant. (ii) Based on the recent discovery of a satellite of Pluto, it appears that the adopted mass and radius of Pluto are not accurate. However, definitive values are not available at this time. (iii) The adopted set of values for the gravity field of the Moon are not completely self-consistent. The adoption of some new values for the sake of consistency would not necessarily mean the adoption of the best values, particularly in cases where the true value is quite uncertain. Since improved values are likely to be available in the next few years, no changes to the IAU (1976) System were introduced at the meeting. A suggestion was made that the almanac offices may use some improved values of planetary masses in computing the new fundamental ephemerides for introduction in 1984, but they should indicate those changes clearly.

P.K. Seidelmann described the basis for the new fundamental ephemerides to be introduced in 1984. The guiding principles for the new ephemerides are: (i) the ephemerides should not be based on a single or isolated computation, (ii) all available observational data should be utilized, (iii) bases, constants and reference frame should be consistent and specified for all ephemerides, and (iv) machine readable ephemerides should be available covering extended periods of time. For this purpose, a programme of ephemerides preparation, collection of observations and comparison between ephemerides and observations has been underway. Plots of such comparisons illustrated the differences between ephemerides and their comparison with observations. In the discussion it was pointed out that the differences for the inner planets were solely the result of comparing two numerical integrations and are due to model differences and truncation errors. In response to the question of getting data on the "same system" Seidelmann stated that to the best of its ability the USNO tries to reduce data to the FK4 system.

At the conclusion of the meeting a vote of thanks to the President, V.K. Abalakin, was moved by T. Lederle and was carried with acclamation. The meeting was adjourned.

RESOLUTIONS

Resolution 1 of Commissions 4 and 16 on cartographic coordinates.

IAU Commissions 4 and 16 endorse the Report of the Joint Working Group on Cartographic Coordinates and Rotational Elements of the Planets and Satellites,

recommend

that the Report be used as the basis for computing the physical ephemerides of planets and satellites in the international and national ephemerides, and

request

that a small working group continues its activity in order to provide improved rotational elements.

Resolution 2 of Commissions 4, 7, 8, 19, 24 and 31 on nutation.

IAU Commissions 4, 7, 8, 19, 24 and 31 endorse the recommendations given in the Report of the Working Group on Nutation (1979 IAU Theory of Nutation), and

recommend

that they shall be used in the national and international ephemerides for the years 1984 onwards, and in all other relevant astronomical work.

Resolution 3 of Commissions 4, 19 and 31 on the expression of UT1 in terms of GMST.

In considering that it is planned to introduce the IAU (1976) System of Astronomical Constants, the 1979 IAU Theory of Nutation, and the equinox of the FK5 on

1984 January 1, it is recommended that:

(a) the relationship between mean sidereal time and UT1 be modified so that there is no change in either value or rate of UT1, due to a correction to the zero point of right ascensions of the FK4 and a correction for the motion of the zero point, to be introduced in the FK5;

(b) the new (provisional) expression for Greenwich mean sidereal time of 0^h UT1 be

$$\text{GMST of } 0^h \text{ UT1} = 6^h38^m45^s.833 + 8\,640\,184^s.628\, T_u + 0^s.0929\, T_u^2,$$

where T_u is the number of Julian centuries of 36525 days of Universal Time elapsed since 1900 January 0, 12^h UT1 (JD 2 415 020.0). This expression is rigorously equivalent to the following

$$\text{GMST of } 0^h \text{ UT1} = 6^h41^m50^s.5539 + 8\,640\,184^s.8138\, T_u + 0^s.0929\, T_u^2,$$

where T_u is measured from 2000 January 1, 12^h UT1 (JD 2 451 545.0).

Note: The followings are frequently used quantities which are also affected by the Resolution:

a) The interval of mean sidereal time in a mean solar day becomes

$$24^h + \frac{8\,640\,184^s.628 + 0^s.185\,8\, T_u}{36525} = 86\,636^s.555\,362\,8 + 0^s.000\,005\,087\, T_u\,,$$

while the current value is $86\,636^s.555\,360\,5 + 0^s.000\,005\,087\, T_u$.

b) The ratio of a sidereal day of 86400 mean sidereal seconds to this interval becomes

$$\frac{\text{mean sidereal day}}{\text{mean solar day}} = 0.997\,269\,566\,388 - 0.586 \times 10^{-10}\, T_u,$$

while the current ratio is $0.997\,269\,566\,414 - 0.586 \times 10^{-10}\, T_u$.

c) The ratio of the mean solar day to the mean sidereal day becomes

$$\frac{86\,636.555\,362\,8 + 0.000\,005\,087\, T_u}{86400} = 1.002\,737\,909\,292 + 0.589 \times 10^{-10}\, T_u,$$

while the current ratio is $1.002\,737\,909\,265 + 0.589 \times 10^{-10}\, T_u$.

d) Disregarding the inappreciable secular variations, the equivalent measures of the lengths of the days at 1900 are

1 mean sidereal day: $23^h56^m04^s.090\,536$ of mean solar time,
1 mean solar day : 24 03 56.555 363 of mean sidereal time,

while the current values are

1 mean sidereal day: $23^h56^m04^s.090\,54$ of mean solar time,
1 mean solar day : 24 03 56.555 36 of mean sidereal time.

Resolution 4 of Commissions 4, 19 and 31 on Project MERIT.

IAU Commissions 4, 19 and 31 endorse the proposal of the joint working group on the determination of the rotation of the Earth for a special period of international collaboration in the monitoring of Earth-rotation and in the intercomparison of the techniques of observation and analysis, recognize

that the responsibility for the organization of this project MERIT should be shared with the International Union of Geodesy and Geophysics, and

request

that the national and international organizations concerned give full technical and financial support to the development of the proposal and to the implementation of the project.

Resolution 5 of Commissions 4, 19 and 31 on the designation of dynamical times.

IAU Commissions 4, 19 and 31 recommend that the time-scales for dynamical theories and ephemerides adopted in 1976 at the 16th General Assembly be designated as follows:

(1) the time-scale for the equations of motion referred to the barycentre of the solar system be designated Barycentric Dynamical Time (TDB),

(2) the time-scale for apparent geocentric ephemerides be designated Terrestrial Dynamical Time (TDT).

Resolution 6 of Commission 4 on a Colloquium on Reference Systems.

IAU Commission 4 expresses its full support in favour of the proposal of the Space Research Centre of the Polish Academy of Sciences and the Smithsonian Astrophysical Observatory to convene the Second International Colloquium on Reference Systems for Earth Dynamics.

Resolution 7 of Commission 4 on processing of occultation data.

IAU Commission 4 recognizing

(a) that timings of occultations of stars by the Moon will continue to be of value in studies of the lunar motion and figure, the rotation of the Earth, and the stellar reference frame, and

(b) that it is desirable that the observations should continue to be collected and processed by one organization, and

considering

that beginning with January 1981, H.M. Nautical Almanac Office, Royal Greenwich Observatory, will no longer be able to act as the international centre for the receipt and processing of timings of occultations,

recommends

that an organization with the appropriate experience and commitment to the occultation programme be requested to take over this important work.

Annex 1.

REPORT OF THE IAU WORKING GROUP ON CARTOGRAPHIC COORDINATES AND ROTATIONAL ELEMENTS OF THE PLANETS AND SATELLITES

Chairman: M.E. Davies
Members: V.K. Abalakin, R.L. Duncombe, H. Masursky, B. Morando, T.C. Owen, P.K. Seidelmann, A.T. Sinclair, G.A. Wilkins
Consultants: C.A. Cross, Y.S. Tjuflin

1. Introduction

The IAU Working Group on Cartographic Coordinates and Rotational Elements of the Planets and Satellites was established as a consequence of the adoption of the following resolution at the IAU General Assembly at Grenoble in 1976 (Trans. IAU XVIB, p. 144):

"Commissions 4 and 16 noting that

(a) confusion exists regarding the present rotational elements of some of the planets
(b) extensive amounts of new data from radar observations and by direct imaging from spacecraft have made cartography of the surface of the Moon, Mercury, Venus, and Mars a reality
(c) there will be an extension of these techniques to the mapping of larger satellites of Jupiter and Saturn in the near future

assert that

(a) to avoid a proliferation of inconsistent cartographic and rotational systems, there is a need to define the rotational elements of the planets and satellites on a systematic basis and to relate the new cartographic coordinates rigorously to the rotational elements

and therefore recommend that

(1) Commission 4 (Ephemerides) and Commission 16 (Physical Study of Planets and Satellites) establish a Joint Working Group to study the cartographic coordinates and rotational elements of the planets and satellites and to report recommendations thereon at the next general assembly of the IAU."

In preparing the recommendations given in this report, the Working Group adopted the guiding principles that have been previously adopted by Commission 16 at the IAU General Assembly at Brighton in 1970, namely (Trans. IAU XIVB, p. 128):

"1. The rotational pole of a planet or satellite which lies on the north side of the invariable plane shall be called north, and northern latitudes shall be designated as positive.
2. The planetographic longitude of the central meridian, as observed from a direction fixed with respect to an inertial coordinate system, shall increase with time. The range of longitudes shall extend from 0° to 360°."

The technical arguments in support of, and in opposition to, both of these principles have been reviewed; these arguments were considered at the time of the adoption in the preparation of numerous maps of both planets and satellites, and the Group considers that the advantages that are claimed for other principles are not sufficient to justify the adoption of new principles. Because of historical usage, longitudes on the Moon and Earth are measured from 0° to 180° east and west of the prime meridian. Thus these bodies are exceptions to the general rule. The Group does, however, recommend that the rotational elements and cartographic coordinate systems be specified more simply and uniformly than in the past.

The rotational elements define the direction of the axis of rotation and the rate of rotation relative to an inertial coordinate system. The values of the elements given later in this report are based where possible on recent observational determinations. These elements, especially those for the satellites, vary with time, but it is sufficiently accurate to adopt simplified models of these motions; in particular, short-period nutations are ignored.

Each cartographic coordinate system is defined by reference to the adopted axis of rotation and an arbitrarily chosen prime meridian, whose position on the surface is specified where possible by the adoption of the longitude of a suitable observable feature. For some of the planets and most of the satellites it is sufficient, at present, to assume that the reference surface is spherical, but for others it is necessary to adopt a reference spheroid, with the principal axis of inertia along the axis of rotation.

The following sections of this report describe the ways in which the rotational elements and cartographic coordinate systems are defined. The recommended values are given in a series of tables.

2. Definition of Rotational Elements

The rotational elements of a planet or satellite specify the direction of the north pole and the orientation of its prime meridian as functions of time in the following manner:

The north pole is that pole of rotation which lies on the north side of the invariable plane of the solar system. The direction of the north pole is given with respect to the standard celestial equator and equinox of 1950.0, i.e., in effect with respect to the system of the fundamental catalog FK4. Variable quantities are expressed in units of ephemeris days (or Julian ephemeris centuries of 36525 days) from the standard epoch of 1950 January 1.0, ET, or JED 2433282.5; this epoch is denoted J1950 and is slightly different from the epoch 1950.0, which refers to the beginning of the Besselian year and corresponds to JED 2433282.423 357. The values will be given with respect to the new standard equator, equinox and epoch of J2000, i.e., of 2000 January 1.5 or JED 2451545.0, when the relationship between the systems of the new catalog FK5 and that of FK4 is precisely defined.

The direction of the north pole is specified by the values of its right ascension α_0 and declination δ_0, while the orientation of the prime meridian is specified by the angle W that is measured *along* the planet's equator in the positive sense with respect to the planet's north pole (i.e., in an easterly direction on the planet's surface) *from* the ascending node Q of the planet's equator on the standard equator *to* the point B where the prime meridian crosses the planet's equator (see Fig. 1). (The point Q is the node at which a point moving around the planet's equator in a positive sense would cross the standard equator from south to north; the right ascension of the point Q is $90^\circ + \alpha_0$ and the inclination of the planet's equator to the standard equator is $90^\circ - \delta_0$.) The prime meridian is assumed to rotate uniformly with the planet, and so W varies linearly with time due to this rotation. In addition, α_0, δ_0 and W may vary with time due to a precession of the axis of rotation of the planet (or satellite). If W increases with time, the planet has a *direct* (or prograde) rotation relative to the invariable plane; if W decreases with time the rotation is said to be *retrograde*.

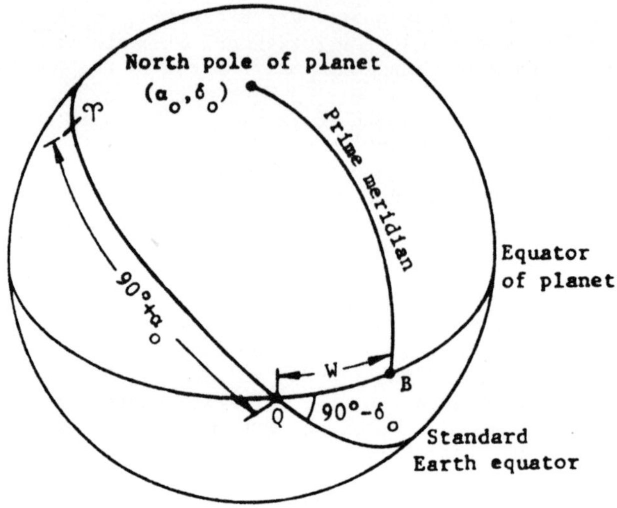

Figure 1. Reference system used to define orientation of the planet

In the absence of other information, the axis of rotation is assumed to be normal to the mean orbital plane; Mercury and most of the satellites are in this category. For many of the satellites it is assumed that the rotation rate is equal to the mean period of orbital revolution.

The angle W specifies the ephemeris position of the prime meridian, and for planets or satellites without any accurately observable fixed surface features the adopted expression for W defines the prime meridian and is not subject to correction. Where possible, however, the cartographic position of the prime meridian is defined by a suitable observable feature and so the constants in the expression $W = Wo + \dot{W}d$, where d is the interval in days from the standard epoch, are chosen so that the ephemeris position follows the motion of the cartographic position as closely as possible; in these cases the expression for W may require amendment in the future.

For the planets on which no suitable features have been observed, W has been given the value of 360° at the standard epoch (Saturn, Uranus, Neptune, Pluto); in other cases W has been chosen so that the planetographic longitude of the central point of the apparent disk as seen from the center of the Earth (i.e., the longitude of the central meridian) has an arbitrary value at some adopted epoch (Venus, Jupiter). In general, the values given here are compatible with those currently in use.

For satellites on which no suitable features have been observed, the expression for W has been chosen so that the ephemeris position of the prime meridian passes through the intersection of the satellite's equator and the plane containing the centers of the satellite, the planet, and the Sun at the time of the first superior heliocentric conjunction of the satellite and the planet after the standard epoch (Trans. IAU XVB, p. 108). An exception is the Moon, whose prime meridian passes through the mean sub-Earth direction.

Recommended values of the constants in the expressions for α_o, δ_o and W are given for the planets in Table 1 and for the satellites in Table 2. Expressions for the Sun, Earth, and Moon are given to a similar precision as those of the other bodies of the solar system for comparative purposes only.

Table 1. Recommended Values for the Direction of the North Pole of Rotation and the Prime Meridian of the Sun and Planets (1979)

α_o, δ_o are standard equatorial coordinates of 1950.0. Approximate coordinates of the north pole of the invariable plane are $\alpha_o = 272^\circ.40$, $\delta_o = +66^\circ.99$.
T = interval in Julian ephemeris centuries (of 36525 days) from the standard epoch.
d = interval in ephemeris days from the standard epoch.
The standard epoch is 1950 January 1.0 ET, i.e., JED 2433282.5.

Sun	α_o = 286°.0				
	δ_o = 63.8				
	W = 240.9	+	14°.18440 d		
Mercury	α_o = 280.9	-	0.033 T		
	δ_o = 61.4	-	0.005 T		
	W = 184.74	+	6.1385025 d		(a)
Venus	α_o = 272.8				
	δ_o = 67.2				
	W = 213.63	-	1.4814205 d		
Earth	α_o = 0.0	-	0.64032 T		
	δ_o = 90.0	-	0.55669 T		
	W = 99.87	+	360.985612 d		(b)
Mars	α_o = 317.342	-	0.108 T		
	δ_o = 52.711	-	0.061 T		
	W = 11.50	+	350.891983 d		(c)
Jupiter	α_o = 268.00	-	0.008 T		
	δ_o = 64.50	+	0.003 T		
	W_I = 17.7	+	877.900 d	System I	(d)
	W_{II} = 16.8	+	870.270 d	System II	
	W_{III} = 80.6	+	870.536 d	System III	
Saturn	α_o = 38.50	-	0.034 T		
	δ_o = 83.31	-	0.004 T		
	W_I = 360.0	+	841.558 d	System I	(d)
	W_{III} = 360.0	+	822.00 d	System III	
Uranus	α_o = 256.72				
	δ_o = -15.04				
	W = 360.0	-	554.913 d		
Neptune	α_o = 294.91				
	δ_o = 40.53				
	W = 360.0	+	468.750 d		
Pluto	α_o = 305				
	δ_o = 5				
	W = 360.0	-	56.367 d		

Note (a) The 20° meridian is defined by the crater Hun Kal.
(b) The 0° meridian is defined by the transit circle at Greenwich, England.
(c) The 0° meridian is defined by the crater Airy-0; its longitude in the system of the *American Ephemeris*, 1968 to present, (de Vaucouleurs' NA3), was 358°.4 ± 0°.3 (m.e.) on 1909 January 15.5 UT.
(d) System I refers to the atmospheric equatorial rotation.
System III refers to rotation derived from radio emissions.
System II refers to atmospheric rotation north of the south component of the north equatorial belt, and south of the north component of the south equatorial belt.

Table 2. Recommended Values for the Direction of the North Pole of Rotation and the Prime Meridian of the Satellites (1979)

α_o, δ_o, T and d have the same meanings as in Table 1.

Earth: Moon

$$\alpha_o = 270^\circ - 3^\circ\!.878 \sin E1 - 0^\circ\!.120 \sin E2 + 0.070 \sin E3 - 0.017 \sin E4$$

$$\delta_o = 66.534 + 1.543 \cos E1 + 0.024 \cos E2 - 0.028 \cos E3 + 0.007 \cos E4$$

$$W = 244.375 + 13.17635831\,d + 3.558 \sin E1 - 0^\circ\!.120 \sin E2 + 0.121 \sin E2 - 0.064 \sin E3 + 0.016 \sin E4 + 0.025 \sin E5$$

where

$$E1 = 12^\circ\!.112 - 0^\circ\!.052992\,d, \quad E2 = 24^\circ\!.224 - 0^\circ\!.105984\,d$$

$$E3 = 227.645 + 13.012000\,d, \quad E4 = 261.105 + 13.340716\,d$$

$$E5 = 358.00 + 0.985600\,d$$

Mars: Phobos

$$\alpha_o = 317^\circ\!.329 + 1^\circ\!.674 \sin M1$$

$$\delta_o = 52.717 + 1.014 \cos M1$$

$$W = 270.202 + 1128.844483\,d - 3^\circ\!.310 \sin M2 - 1^\circ\!.332 \sin M1$$

Deimos

$$\alpha_o = 316.307 + 3.051 \sin M3$$

$$\delta_o = 53.367 - 1.821 \cos M3$$

$$W = 70.832 + 285.161807\,d - 2.448 \sin M3$$

where

$$M1 = 201^\circ\!.605 - 0^\circ\!.435427\,T, \quad M2 = 93^\circ\!.440 + 1128^\circ\!.409143\,T$$

$$M3 = 23.054 - 0.018143\,T$$

Jupiter: Amalthea

$$\alpha_o = 268^\circ\!.0$$

$$\delta_o = 64.5$$

$$W = 50.2 + 722^\circ\!.6303746\,d$$

Io

$$\alpha_o = 268.002 - 0.0085\,T + 0^\circ\!.094 \sin J1 + 0^\circ\!.024 \sin J2$$

$$\delta_o = 64.504 + 0.0033\,T + 0.040 \cos J1 + 0.011 \cos J2$$

$$W = 262.7 + 203.4889538\,d - 0.085 \sin J1 - 0.022 \sin J2$$

Europa

$$\alpha_o = 268.029 - 0.0085\,T + 1.086 \sin J2 + 0.060 \sin J3 + 0.015 \sin J4 + 0.009 \sin J5$$

$$\delta_o = 64.516 + 0.0033\,T + 0.468 \cos J2 + 0.026 \cos J3 + 0.007 \cos J4 + 0.002 \cos J5$$

$$W = 156.9 + 101.3747235\,d - 0.980 \sin J2 - 0.054 \sin J3 - 0.014 \sin J4 - 0.008 \sin J5$$

Ganymede

$$\alpha_o = 268.149 - 0.0085\,T - 0.037 \sin J2 + 0.431 \sin J3 + 0.091 \sin J4$$

$$\delta_o = 64.574 + 0.0033\,T - 0.016 \cos J2 + 0.186 \cos J3 + 0.039 \cos J4$$

$$W = 195.8 + 50.3176081\,d + 0.033 \sin J2 - 0.389 \sin J3 - 0.082 \sin J4$$

Callisto

$$\alpha_o = 268.678 - 0.0085\,T - 0.068 \sin J3 + 0.590 \sin J4 + 0.010 \sin J6$$

$$\delta_o = 64.830 + 0.0033\,T - 0.029 \cos J3 + 0.254 \cos J4 - 0.004 \cos J6$$

$$W = 158.0 + 21.5710715\,d + 0.061 \sin J3 - 0.533 \sin J4 - 0.009 \sin J6$$

where

$$J1 = 19^\circ\!.2 + 4850^\circ\!.7\,T, \quad J2 = 120^\circ\!.8 + 1191^\circ\!.3\,T, \quad J3 = 349^\circ\!.5 + 262^\circ\!.1\,T$$

$$J4 = 198.3 + 64.3\,T, \quad J5 = 241.6 + 2382.6\,T, \quad J6 = 317.7 + 6070.0\,T$$

Table 2. (continued)

Saturn:	Mimas	$\alpha_0 = 38^{\circ}.5 + 13^{\circ}.1 \sin S1$
		$\delta_0 = 83.3 - 1.5 \cos S1$
		$W = 207.6 + 381.9952887\,d - 13^{\circ}.0 \sin S1$
	Enceladus	$\alpha_0 = 38.5$
		$\delta_0 = 83.3$
		$W = 301.8 + 262.7315302\,d$
	Tethys	$\alpha_0 = 38.5 + 9.4 \sin S2$
		$\delta_0 = 83.3 - 1.1 \cos S2$
		$W = 33.7 + 190.6981682\,d - 9.3 \sin S2$
	Dione	$\alpha_0 = 38.5$
		$\delta_0 = 83.3$
		$W = 121.6 + 131.5347179\,d$
	Rhea	$\alpha_0 = 38.2 + 3.0 \sin S3$
		$\delta_0 = 83.3 - 0.4 \cos S3$
		$W = 14.1 + 79.6900944\,d - 3.0 \sin S3$
	Titan	$\alpha_0 = 34.3 + 2.6 \sin S4$
		$\delta_0 = 83.7 - 0.3 \cos S4$
		$W = 79.1 + 22.5769734\,d - 2.6 \sin S4$
	Hyperion	$\alpha_0 = 33.4 + 4.9 \sin S5 + 2.7 \sin S4$
		$\delta_0 = 83.8 - 0.6 \cos S5 - 0.3 \cos S4$
		$W = 336.0 + 16.9199489\,d - 4.9 \sin S5 - 2^{\circ}.7 \sin S4$
	Iapetus	$\alpha_0 = 320^{\circ}.2 - 3^{\circ}.9\,T$) rocky or $\alpha_0 = 289^{\circ}.3$) icy
		$\delta_0 = 75.4 - 1.1\,T$) rocky or $\delta_0 = 78.7$) icy
		$W = 275.5 + 4.5379589\,d$

where $S1 = 68^{\circ}.6 - 36504^{\circ}.9\,T$, $S2 = 314^{\circ}.5 - 7226^{\circ}.0\,T$, $S3 = 134^{\circ}.9 - 1017^{\circ}.7\,T$
$S4 = 57.4 - 53.5\,T$, $S5 = 22.6 - 239.2\,T$

Uranus:	Miranda	$\alpha_0 = 256^{\circ}.7$
		$\delta_0 = -15.0$
		$W = 59.2 - 254^{\circ}.5968883\,d$
	Ariel	$\alpha_0 = 256.7$
		$\delta_0 = -15.0$
		$W = 47.3 - 142.8356047\,d$
	Umbriel	$\alpha_0 = 256.7$
		$\delta_0 = -15.0$
		$W = 146.4 - 86.8688136\,d$
	Titania	$\alpha_0 = 256.7$
		$\delta_0 = -15.0$
		$W = 202.0 - 41.3513623\,d$
	Oberon	$\alpha_0 = 256.7$
		$\delta_0 = -15.0$
		$W = 3.2 - 26.7394375\,d$
Neptune:	Triton	$\alpha_0 = 294.89 - 20.087 \sin N$
		$\delta_0 = 36.93 + 15.264 \cos N$
		$W = 132.3 - 61.2575147\,d + 10^{\circ}.521 \sin N$

where $N = 158^{\circ}.3402 + 61^{\circ}.9803\,T$

Pluto:	Charon	$\alpha_0 = 305^{\circ}$
		$\delta_0 - 5$
		$W = -$

3. Definition of Cartographic Coordinate Systems

Both planetocentric and planetographic systems of coordinates are used in the study of the planets and satellites. These systems are based on the same fundamental reference axis but differ, as explained below, in the definitions of latitude and longitude. Planetocentric coordinates are used for general purposes and are based on a right-handed system of axes, whereas planetographic coordinates are used for cartographic purposes and depend on the adoption of additional parameter to define a reference surface, usually a spheroid, that approximates an equipotential surface of the planet.

For these systems, the fundamental reference z-axis is the mean axis of rotation and the planetary equator is the plane that is normal to this axis and passes through the center of mass of the planet. The x-axis is defined by the intersection of the equatorial plane with the plane of the prime meridian, whose position is defined in an arbitrary manner. The y-axis of planetocentric rectangular coordinates is defined so as to form a right-handed system.

Latitude is measured north and south of the equator; north latitudes are designated as positive. The planeto*centric* latitude (ϕ) of a point is the angle between the equatorial plane and the line connecting the point to the center of mass. The planeto*graphic* latitude (ϕ') of a point on the reference surface is the angle between the equatorial plane and the normal to the reference surface at the point.

Longitude is measured around the equatorial plane from the prime meridian from 0° to 360°. Planeto*centric* longitudes (λ) are measured positively to the east, whereas planeto*graphic* longitudes (λ') are measured in the direction opposite to the rotation, i.e., positively to the west in the case of direct rotation. Planetocentric longitudes are measured from the ephemeris position of the prime meridian as defined by the adopted longitude of some clearly observable surface feature. These two positions may normally be assumed to coincide but it is conceivable that errors in the rotational elements may be such that the cartographic position may drift away from the ephemeris position by a small amount ΔW, where ΔW is measured positively to the east of the ephemeris position.

Planetocentric *radius* (R) is measured from the center of mass to the point concerned. In the planetographic system the position of a point (P) not on the reference surface is specified by the planetographic longitude and latitude of the point (P') on the surface at which the normal passes through P and by the *height* (h) of P above P'.

The reference surfaces for most of the planets are spheroids for which the radius of the equator (A) is larger than the polar semiaxis (C). For some planets and most satellites the reference surface is a sphere ($A = C$), and the planetocentric and planetographic latitudes are then numerically the same. The polar axis of each reference surface is assumed to be the mean axis of rotation as defined by the adopted rotational elements since the accuracy of measurements is, at present, such that a motion of the axis of rotation with respect to the axis of figure cannot be observed.

The recommended values of the parameters for the reference surfaces for planets and satellites are given in Table 3. Radii for irregular-shaped satellites are given in Table 4.

It should be noted that east longitude on the Sun, Earth, and Moon is commonly considered to be in the positive direction.

Table 3. Recommended Reference Spheroids for Mapping the Planets and Major Satellites (1979)

Planet	Satellite	Equatorial Radius (km)	Flattening
Mercury		2439	0
Venus		6052	0
Earth		6378.140	0.00335281
	Moon	1738	0
Mars		3393.4	0.0051865
Jupiter		71398	0.0648088
	Io	1819	0
	Europa	1563	0
	Ganymede	2637	0
	Callisto	2424	0
Saturn		60000	0.1076209
	Mimas	200	0
	Enceladus	275	0
	Tethys	520	0
	Dione	500	0
	Rhea	800	0
	Titan	2900	0
	Hyperion	112	0
	Iapetus	725	0
Uranus		25400	0.0165
	Ariel	400	0
	Umbriel	275	0
	Titania	500	0
	Oberon	450	0
	Miranda	150	0
Neptune		24300	0.0259
	Triton	1600	0
Pluto		1500	0
	Charon	600	0

Note: The equatorial radii for Mercury, Venus, Moon, and Mars are used in current mapping programs, and those for Jupiter and Saturn are used in sequencing and analyzing data from current flight missions. The values for Mars and Pluto differ from those recommended by the IAU in 1976 (Trans. IAU XVIB, p. 60). The reference spheroid for Mars (3393.4 km radius) has been used in all mapping programs since 1973, although the IAU 1976 radius (3397.2 km) is probably a better value. In 1976 Pluto's satellite, Charon, had not been discovered.

Table 4. Recommended Reference Shapse for Mapping Irregular Satellites (1979)

Planet	Satellite	Equatorial Radius, A (km)	Equatorial Radius, B (km)	Polar Radius, C (km)
Mars	Phobos	13.5	10.7	9.6
	Deimos	7.5	6.0	5.5
Jupiter	Amalthea	140	105	80

Annex 2.

SUMMARY REPORT

of

THE IAU WORKING GROUP ON NUTATION

(1979 IAU THEORY OF NUTATION)

1. The president of IAU Commission 4, V.K. Abalakin, established the Working Group on Nutation in response to the request made at IAU Symposium No. 78 on *Nutation and the Earth's Rotation* held in Kiev in May 1977. The final membership of the Working Group comprise the authors of this report.

2. The theory of nutation, currently adopted by the IAU, is that of Woolard and has the following characteristics:

(a) It is based on a rigid model of the Earth with dynamical symmetry ($A = B$).
(b) The "constant of nutation" is an empirical value and is not consistent with other adopted astronomical constants.
(c) Eulerian motion and forced nearly diurnal polar motion are not included in the current theory of nutation, but are assumed to be part of polar motion.
(d) The pole of reference is the instantaneous celestial rotation pole.

3. The current theoretical developments and the observational data of various types indicate the following problems with the currently adopted theory of nutation:

(a) The determinations of UT1 and polar motion using optical observations of stars, Doppler or laser range tracking of satellites, laser ranges to the Moon, and radio interferometric measurements are sufficiently accurate that their usefulness can be degraded by use of the present theory of nutation in the data reduction process.
(b) The Earth is not a rigid body and the effects of the non-rigid body can be observationally significant.
(c) As pointed out by Jeffreys and Atkinson, the currently adopted axis of rotation rotates relative to an Earth-fixed coordinate system with a quasi-diurnal period. For accurate observation reduction, this rotation cannot be ignored and a resolution was passed at the Sixteenth General Assembly of the IAU in 1976 in Grenoble to adopt a different pole of reference.
(d) Observational data indicate that, with the current theory of nutation and a redefined pole of reference, a body-fixed coordinate system would still rotate with respect to the reference pole; therefore, the theory of nutation should be revised.

4. The goal of this report is the adoption of a set of nutation coefficients so as to provide a working standard for determination of UT1 and polar motion, the reduction of optical observations of stars, Doppler or laser range tracking of satellites, laser ranges to the Moon, radio interferometric measurements and other high precision requirements. This report should not be considered as the selection or endorsement of a particular Earth model.

5. Therefore, the proposed solution incorporates the following changes:

(a) A non-rigid model of the Earth with axial symmetry ($A = B$) is used.
(b) The constants are consistent with the IAU (1976) System of Astronomical Constants and are in agreement with available observational data of various types.
(c) The reference pole is selected so that there are no diurnal or quasi-diurnal motions of this pole with respect to either a space-fixed or Earth-fixed coordinate system. The phenomenon of diurnal variation of latitude, otherwise

known as forced diurnal polar motion, is included implicitly in the new nutation theory. The new nutation theory thus includes all externally-forced motions of the Earth's rotation axis; no geophysical or free motions are included.

6. *Resolution*:

We request that the following draft resolution be submitted to Commissions 4, 8, 19 and 31, with the view of its being adopted at the Seventeenth General Assembly of the IAU. "The IAU endorses the recommendations given in the Report of the Working Group on Nutation and recommends that they shall be used in the national and international ephemerides for the years 1984 onwards, and in all other relevant astronomical work."

7. *Recommendation*:

Whereas, the complete theory of the general nutational motion of the Earth about its center of mass may be described by the sum of two components, astronomical nutation, commonly referred to as nutation, which is motion with respect to a space-fixed coordinate system, and polar motion, which is motion with respect to a body-fixed coordinate system, it is recommended that:

(a) Astronomical nutation be computed for the "Celestial Ephemeris Pole" using a non-rigid model of the Earth such that there are no nearly diurnal motions of this celestial pole with respect to either space-fixed or body-fixed coordinates which can be calculated from torques external to the Earth and its atmosphere.

(b) The numerical values given in Table 1 of the complete report be used for computing astronomical nutation of the Celestial Ephemeris Pole.

P.K. Seidelmann, Chairman
V.K. Abalakin, H. Kinoshita, J. Kovalevsky, C.A. Murray,
M.L. Smith, R.O. Vicente, J.G. Williams, Ya.S. Yatskiv.

(The complete Report will be published in *Celestial Mechanics*.)

Table 1. Nutation in Longitude and Obliquity referred to Ecliptic of Date

Epoch: J2000.0 (JED 2451545.0), T in Julian centuries, Unit: 0″0001

No.	Period (days)	Argument multiple of l	l'	F	D	Ω	Longitude coefficient of sine argument		Obliquity coefficient of cosine argument	
1	6798.4	0	0	0	0	+1	−172058	−174.2T	+92044	+8.9T
2	3399.2	0	0	0	0	+2	+2063	+0.2T	−895	+0.5T
3	1305.5	−2	0	+2	0	+1	+46	0.0T	−24	0.0T
4	1095.2	+2	0	−2	0	0	+11	0.0T	0	0.0T
5	1615.7	−2	0	+2	0	+2	−3	0.0T	+1	0.0T
6	3232.9	+1	−1	0	−1	0	−3	0.0T	0	0.0T
7	6786.3	0	−2	+2	−2	+1	−2	0.0T	+1	0.0T
8	9413.2	+2	0	−2	0	+1	+1	0.0T	0	0.0T

Table 1. (continuation)

No.	Period	Argument					Longitude		Obliquity	
	(days)	multiple of					coefficient of sine argument		coefficient of cosine argument	
		l	*l'*	*F*	*D*	Ω				
9	182.6	0	0	+2	−2	+2	−13152	−1.5*T*	+5719	−3.1*T*
10	365.3	0	+1	0	0	0	+1411	−3.4*T*	+49	−0.1*T*
11	121.7	0	+1	+2	−2	+2	−515	+1.2*T*	+224	−0.6*T*
12	365.2	0	−1	+2	−2	+2	+217	−0.5*T*	−95	+0.3*T*
13	177.8	0	0	+2	−2	+1	+129	+0.1*T*	−70	0.0*T*
14	205.9	+2	0	0	−2	0	+48	0.0*T*	0	0.0*T*
15	173.3	0	0	+2	−2	0	−22	0.0*T*	0	0.0*T*
16	182.6	0	+2	0	0	0	+17	−0.1*T*	0	0.0*T*
17	386.0	0	+1	0	0	+1	−15	0.0*T*	+8	0.0*T*
18	91.3	0	+2	+2	−2	+2	−15	+0.1*T*	+7	0.0*T*
19	346.6	0	−1	0	0	+1	−12	0.0*T*	+6	0.0*T*
20	199.8	−2	0	0	+2	+1	−5	0.0*T*	+3	0.0*T*
21	346.6	0	−1	+2	−2	+1	−5	0.0*T*	+3	0.0*T*
22	212.3	+2	0	0	−2	+1	+4	0.0*T*	−2	0.0*T*
23	119.6	0	+1	+2	−2	+1	+4	0.0*T*	−2	0.0*T*
24	411.8	+1	0	0	−1	0	−4	0.0*T*	0	0.0*T*
25	131.7	+2	+1	0	−2	0	+1	0.0*T*	0	0.0*T*
26	169.0	0	0	−2	+2	+1	+1	0.0*T*	0	0.0*T*
27	329.8	0	+1	−2	+2	0	−1	0.0*T*	0	0.0*T*
28	409.2	0	+1	0	0	+2	+1	0.0*T*	0	0.0*T*
29	388.3	−1	0	0	+1	+1	+1	0.0*T*	0	0.0*T*
30	117.5	0	+1	+2	−2	0	−1	0.0*T*	0	0.0*T*
31	13.7	0	0	+2	0	+2	−2260	−0.2*T*	+972	−0.5*T*
32	27.5	+1	0	0	0	0	+709	+0.1*T*	−7	0.0*T*
33	13.6	0	0	+2	0	+1	−384	−0.4*T*	+199	0.0*T*
34	9.1	+1	0	+2	0	+2	−299	0.0*T*	+128	−0.1*T*
35	31.8	+1	0	0	−2	0	−157	0.0*T*	−1	0.0*T*
36	27.1	−1	0	+2	0	+2	+123	0.0*T*	−53	0.0*T*
37	14.8	0	0	0	+2	0	+63	0.0*T*	−2	0.0*T*
38	27.7	+1	0	0	0	+1	+63	+0.1*T*	−33	0.0*T*
39	27.4	−1	0	0	0	+1	−58	−0.1*T*	+32	0.0*T*
40	9.6	−1	0	+2	+2	+2	−59	0.0*T*	+25	0.0*T*
41	9.1	+1	0	+2	0	+1	−51	0.0*T*	+26	0.0*T*
42	7.1	0	0	+2	+2	+2	−38	0.0*T*	+16	0.0*T*
43	13.8	+2	0	0	0	0	+29	0.0*T*	−1	0.0*T*
44	23.9	+1	0	+2	−2	+2	+29	0.0*T*	−12	0.0*T*
45	6.9	+2	0	+2	0	+2	−31	0.0*T*	+13	0.0*T*
46	13.6	0	0	+2	0	0	+26	0.0*T*	−1	0.0*T*
47	27.0	−1	0	+2	0	+1	+21	0.0*T*	−10	0.0*T*
48	32.0	−1	0	0	+2	+1	+15	0.0*T*	−8	0.0*T*
49	31.7	+1	0	0	−2	+1	−13	0.0*T*	+7	0.0*T*
50	9.5	−1	0	+2	+2	+1	−10	0.0*T*	+5	0.0*T*
51	34.8	+1	+1	0	−2	0	−7	0.0*T*	0	0.0*T*
52	13.2	0	+1	+2	0	+2	+7	0.0*T*	−3	0.0*T*
53	14.2	0	−1	+2	0	+2	−7	0.0*T*	+3	0.0*T*
54	5.6	+1	0	+2	+2	+2	−8	0.0*T*	+3	0.0*T*
55	9.6	+1	0	0	+2	0	+6	0.0*T*	0	0.0*T*
56	12.8	+2	0	+2	−2	+2	+6	0.0*T*	−3	0.0*T*
57	14.8	0	0	0	+2	+1	−6	0.0*T*	+3	0.0*T*
58	7.1	0	0	+2	+2	+1	−7	0.0*T*	+3	0.0*T*

Table 1. (continuation)

No.	Period (days)	Argument multiple of *l*	*l'*	*F*	*D*	*Ω*	Longitude coefficient of sine argument		Obliquity coefficient of cosine argument	
59	23.9	+1	0	+2	−2	+1	+6	0.0*T*	−3	0.0*T*
60	14.7	0	0	0	−2	+1	−5	0.0*T*	+3	0.0*T*
61	29.8	+1	−1	0	0	0	+5	0.0*T*	0	0.0*T*
62	6.8	+2	0	+2	0	+1	−5	0.0*T*	+3	0.0*T*
63	15.4	0	+1	0	−2	0	−4	0.0*T*	0	0.0*T*
64	26.9	+1	0	−2	0	0	+4	0.0*T*	0	0.0*T*
65	29.5	0	0	0	+1	0	−4	0.0*T*	0	0.0*T*
66	25.6	+1	+1	0	0	0	−3	0.0*T*	0	0.0*T*
67	9.1	+1	0	+2	0	0	+3	0.0*T*	0	0.0*T*
68	9.4	+1	−1	+2	0	+2	−3	0.0*T*	+1	0.0*T*
69	9.8	−1	−1	+2	+2	+2	−3	0.0*T*	+1	0.0*T*
70	13.8	−2	0	0	0	+1	−2	0.0*T*	+1	0.0*T*
71	5.5	+3	0	+2	0	+2	−3	0.0*T*	+1	0.0*T*
72	7.2	0	−1	+2	+2	+2	−3	0.0*T*	+1	0.0*T*
73	8.9	+1	+1	+2	0	+2	+2	0.0*T*	−1	0.0*T*
74	32.6	−1	0	+2	−2	+1	−2	0.0*T*	+1	0.0*T*
75	13.8	+2	0	0	0	+1	+2	0.0*T*	−1	0.0*T*
76	27.8	+1	0	0	0	+2	−2	0.0*T*	+1	0.0*T*
77	9.2	+3	0	0	0	0	+2	0.0*T*	0	0.0*T*
78	9.3	0	0	+2	+1	+2	+2	0.0*T*	−1	0.0*T*
79	27.3	−1	0	0	0	+2	+1	0.0*T*	−1	0.0*T*
80	10.1	+1	0	0	−4	0	−1	0.0*T*	0	0.0*T*
81	14.6	−2	0	+2	+2	+2	+1	0.0*T*	−1	0.0*T*
82	5.8	−1	0	+2	+4	+2	−2	0.0*T*	+1	0.0*T*
83	15.9	+2	0	0	−4	0	−1	0.0*T*	0	0.0*T*
84	22.5	+1	+1	+2	−2	+2	+1	0.0*T*	−1	0.0*T*
85	5.6	+1	0	+2	+2	+1	−1	0.0*T*	+1	0.0*T*
86	7.3	−2	0	+2	+4	+2	−1	0.0*T*	+1	0.0*T*
87	9.1	−1	0	+4	0	+2	+1	0.0*T*	0	0.0*T*
88	29.3	+1	−1	0	−2	0	+1	0.0*T*	0	0.0*T*
89	12.8	+2	0	+2	−2	+1	+1	0.0*T*	−1	0.0*T*
90	4.7	+2	0	+2	+2	+2	−1	0.0*T*	0	0.0*T*
91	9.6	+1	0	0	+2	+1	−1	0.0*T*	0	0.0*T*
92	12.7	0	0	+4	−2	+2	+1	0.0*T*	0	0.0*T*
93	8.8	+3	0	+2	−2	+2	+1	0.0*T*	0	0.0*T*
94	23.8	+1	0	+2	−2	0	−1	0.0*T*	0	0.0*T*
95	13.1	0	+1	+2	0	+1	+1	0.0*T*	0	0.0*T*
96	35.0	−1	−1	0	+2	+1	+1	0.0*T*	0	0.0*T*
97	13.6	0	0	−2	0	+1	−1	0.0*T*	0	0.0*T*
98	25.4	0	0	+2	−1	+2	−1	0.0*T*	0	0.0*T*
99	14.2	0	+1	0	+2	0	−1	0.0*T*	0	0.0*T*
100	9.5	+1	0	−2	−2	0	−1	0.0*T*	0	0.0*T*
101	14.2	0	−1	+2	0	+1	−1	0.0*T*	0	0.0*T*
102	34.7	+1	+1	0	−2	+1	−1	0.0*T*	0	0.0*T*
103	32.8	+1	0	−2	+2	0	−1	0.0*T*	0	0.0*T*
104	7.1	+2	0	0	+2	0	+1	0.0*T*	0	0.0*T*
105	4.8	0	0	+2	+4	+2	−1	0.0*T*	0	0.0*T*
106	27.3	0	+1	0	+1	0	+1	0.0*T*	0	0.0*T*

$\sin \varepsilon_{2000} = 0.397\,777\,16$

COMMISSION 5: DOCUMENTATION (DOCUMENTATION)

Report of Meetings, 14, 15 and 20 August 1979

PRESIDENT: J.-C. Pecker. SECRETARY: L. Remy-Battiau.

(The order of the reports is not the actual order in which the meetings took place, but a more or less logical order).

A. FULL COMMISSION MEETING (15 August, 09.00).

1. President's draft report.

The report is submitted for adoption. See hereafter section F.

2. Membership.

The following proposals for new members are accepted, subject to approval by the Executive Committee: H.A. Abt, N. Baker, R.D. Davies, R.H. Garstang, B. Hauck, A. Heck, V.V. Ivanov, C. Jaschek, J. Lequeux, M.C. Lortet, D.H. McNamara, J.M. Mead, P.F. Renson, H.H. Voigt.

The Consulting Members of the Commission for the period 1979-1982 are: Mme C. Bérardini, Mme A. Berthelot, M.J. Collins, P. Clague, P. Dale, Miss J. Dudley, H.J. Felber, Mme G. Grassi Conti, Mme M. Guidoni, D.A. Kemp, M. Laforge, Mme S. Laloé, Mrs S.S. Martin, Mrs P. Molholt, K. Metzner, Dr. T. Nagy, Mme A.M. de Narbonne,F.Ochsenbein, R.B. Rodman, R.A. Seal, Mr. Schiminovitch, Mme M. Vargha.

3. Officers of the Commission, 1979-1982.

W.D. Heintz will be the new President, and G.A. Wilkins the new Vice-President. The Commission adopts the following composition of the Organizing Committee: B. Hauck, J. Kleczek, P. Lantos, S.A. Mitton, J.-C. Pecker, L. Schmadel, J. Shakeshaft, I. Shcherbina-Samojlova. This composition is provisional in that J. Shakeshaft, chairman of the Working Group on Editorial Policy, and S.A. Mitton, acting chairman of the Working Group at Montreal, are both included in it. This Working Group is to be re-organized, and its president will be ipso facto member of the Organizing Committee.

4. Relations with other International Organisations.

POGSI. The work of the Policy Group on Scientific Information, set up by the International Council of Scientific Unions, and of which J. Sahade was the chairman, has terminated its work and accordingly has been discontinued.

ICSU AB. The ICSU Abstracting Board has been set up more than 20 years ago, to coordinate abstracting problems in all sciences. It has lately undergone a bad financial crisis and has recently been completely reorganized. Its next meeting will be held in Paris in October. The question is: does Commission 5 recommend to keep the adherence of IAU to ICSU AB? The commission thinks that the answer should be "yes" for the next threeyears period, since reorganization of the board has to prove its efficiency. The answer should be reexamined at the next General Assembly. The IAU representative at ICSU AB is traditionnally the president of Commission 5 even if often he designates a different representative to the meetings.

CODATA. The new IAU representative will be B. Hauck.

5. Multilingual Astronomical Dictionary.

Dr. Kleczek has almost finished its revised version of the multilingual Astronomical Dictionary. It is an extended version both in the number of languages concerned and in the scientific area covered. It should be a cheap publication, nevertheless support from the Commission is requested to obtain financial help from IAU. There is complete agreement within the Commission upon the importance of Dr. Kleczek's work. The President has sent to the General Secretary a letter of support of the request made by Dr. Kleczek.

6. Role of the Working Group on Astronomical Data.

In a letter to the President, Dr. C. Jaschek recalls the introduction of a working group on Numerical Data for Astronomers and Astrophysicists at the XIVth General Assembly of the IAU, working group dependent of Commission 5. This working group, of which the name has been often discussed (i.e. "Working Group on the Documentation, Exchange and Retrieval of Numerical Data" vs "Working Group on Numerical Data") should in C. Jaschek's opinion become now a new commission named "Astronomical Data", independent of Commission 5. But some reasons invoked (difference in membership of the commission and the working group for instance) are no more actual. On the other hand, Commission 5 seems to be the right place where "makers" of numerical data have the best chance to meet "users" of these numerical data. Moreover, as Commission 5 depends directly from the Executive Committee, it should also be so of the suggested new commission. This, the Executive Committee does not seem too keen on it. It appears more suitable to make sure that the composition of Commission 5 is extended and includes members from each Center of Astronomical Data. It should be renamed: "Commission on Documentation and Astronomical Data".

B. OPEN MEETING ON THE NOMENCLATURE OF ASTRONOMICAL OBJECTS (14 August, 15.00)
(Special session organized with representatives of other commissions)

PRESIDENT: C. Jaschek. SECRETARY: F. Ochsenbein.

1. Report of the discussion.

In his report, Jaschek proposed the three following steps to clarify and simplify the nomenclature of astronomical objects:

1. That a recommendation be sent to the editors of the astronomical journals asking them to be careful about designations: at least two identifications should be listed, their abbreviations being completely referenced; one of these identifications could be the position in a specified coordinate system at a given epoch.

2. That a list of catalog abbreviations be issued; the article by Fernandez, Lortet and Spite is a first attempt at such a publication.

3. That a guidebook of designations within the fields of the various commissions be issued.

Pecker suggested that a resolution be submitted to the Working Group on Editorial Policy for the first point, and a financial proposition to the Executive Committee for the publication mentioned in point 2.

The first step was accepted, with an extension to the editors of books and proceedings; authors should also pay attention to these recommendations, as emphasized by Mitton, especially in preparing review papers. Teleki and Pecker also suggested that the next issue of the Style Book should be examined by Commission 5 for this problem (resolution no.1 hereafter).

In order to achieve the second step, Jaschek proposed that the work by Fernandez et al. be examined by all the commissions concerned, and that remarks, additions, and corrections be forwarded to him before January 1st, 1980. The resulting list will be published by IAU (resolution no.2, hereafter).

The guidebook, proposed in the third step, will consist of different parts, each dealing with a specified type of astronomical object, and describing the present status of the nomenclature used. Dr. J.Mead agreed to sponsor this publication.

Bidelman reported on the recommendations suggested by subgroup 2 of Commission 5's Working Group on Astronomical Data:

1. For brighter stars, designations Bayer, Flamsteed, HR, HD, BD, CoD, CPD; for fainter stars, designations in specified lists or accurate equatorial coordinates for 1950.

2. For brighter non-stellar objects, Messier, NGC, IC; for other objects, equatorial position coordinate in Parkes' system for 1950.

3. For variable stars, the variable designation, together with other designation if the star is bright enough.

4. Designation schemes be proposed by concerned IAU commissions for objects in crowded regions (globular clusters, etc...).

5. For faint objects, rectangular position measures on Palomar prints are not considered satisfactory alternatives to accurate equatorial positions. (More details will be published in the Bulletin of the Centre de Données Stellaires in Strasbourg).

The question of the coordinates was discussed; equatorial coordinates seem to be better than galactic ones, and 1950 equinox should be retained (Lortet); the permanence of the designation based on the coordinates was stressed (Lynga), even if slightly different coordinates are found in more recent investigations.

The question of whether the priority should be given to the discoverer was discussed by Luyten: the name of the discoverer should be mentioned, but disappears when large surveys are published. In the system used for variable designations, the name of the discoverer has completely disappeared. For double stars (Worley) the original designation, which refers to the discoverer, is kept; but the great majority of double stars also have a BD or CoD or CPD designation. In any case, cross-identification tables are of great interest (Mead), and various designations can be found very easily and quickly with a computer. Various cross-identifications are available at the CDS (Jaschek) and at NASA (Mead); Hauck mentioned the existence of transit table for designations in open clusters. Difficulties nevertheless arise in connection with infrared catalogs (Mead): the size of the sources is a function of the wavelength.

The recommendations reported by Bidelman are accepted as resolution from subgroup 2 of the Working Group on Astronomical Data; de Vaucouleurs disagreed with the recommendations for non-stellar objects, and reported the recommendation of Commission 28 in 1973 (Trans. IAU 15B, 141, 1973). The problem of faint stars should also be carefully examined.

2. Resolution no. 1.

Considering the present unsatisfactory situation of the identification of astronomical objects in astronomical publications, the joint meeting of Commission 5 and representatives appointed by Commissions

8, 24, 25, 26, 27, 28, 29, 30, 34, 37, 40, 42, 45 held in Montreal, on August the 14th, requests strongly that:

(a) editors of scientific journals and other publications impose stricker standards of object identifications upon the authors through appropriate instructions to the referees;

(b) catalog abbreviations be explicited either in footnotes or in the bibliography tables;

(c) two identifications be quoted for each object, in order to provide a check against errors and misprints, especially for faint objects.

3. Resolution no. 2.

Considering the confusion existing at the present time in the literature concerning the abbreviations of catalogs, the joint meeting of Commission 5 and representatives appointed by Commissions 8, 24, 25, 26, 27, 28, 29, 30, 34, 37, 40, 42, 45 desires that:

under IAU auspices, a list of catalog abbreviations be published. This list should be a new version of the work of Fernandez, Lortet and Spite, revised by the IAU Commission representatives and edited by C. Jaschek.

To this resolution, the following note was added, at the intention of the Executive Committee, and in view of the financial implications of the resolution no. 2: "The planned book will be of 60 to 100 pages. A subvention (5000 $?) will be necessary even if the published document can be sold. The Commission 5 is asking this subvention for Dr. Jaschek."

C. MEETING OF THE WORKING GROUP ON ASTRONOMICAL DATA (20 August,16.00).

PRESIDENT: G.A. Wilkins. SECRETARY: L. Remy-Battiau.

The meeting was attended by 27 members and invited participants.

1. Reports.

The chairman drew attention to the printed report of the Working Group for the period 1976-78 (Trans. IAU, 17A, 7-8, 1979), and to the extensive coverage of matters of interest to the Commission in the Proceedings of IAU Colloquium no.35 on Compilation, critical evaluation and distribution of stellar data, published by Reidel in 1977, and in Bulletin d'Information du Centre de Données Stellaires, published by the Observatory of the University of Strasbourg. He stated that the CODATA Secretariat has in hand the preparation of the astronomy chapter of the CODATA Directory of Data Sources; this will be based on the survey carried out by C. Jaschek in 1976, but it requires extension and up-dating and so it will be necessary to find one or more astronomers who will assist in the editorial work. M.S. Davis reported that he had attended the meeting of the CODATA Task Group on Space and Time Dependent Data in Ottawa in 1977 and that several subgroups were now preparing reviews of relevant techniques in their disciplines.

The chairman stated that C. Jaschek had suggested that the Working Group should become a separate Commission of the Union. This had been discussed by correspondence and the suggestion had been withdrawn, but the Commission, at its first session, had recommended an extension of the name of the Commission to draw attention to its wide scope. Members of the Union with a special interest in data activities may join Commission 5 without prejudice to their membership of other Commissions.

M.S. Davis stated that his Subgroup on Computer Technology and Standards, which had been set up at the Grenoble meeting, had found it impossible to identify new developments that would be of general interest to astronomical data acquisition and processing. On the Chairman's suggestion it was agreed that the Subgroup should be disbanded, and that members of the Commission should be invited to report items of interest and suggestions for standardization to the Chairman for circulation.

2. Designation of Astronomical Objects.

W.P. Bidelman stated that his Subgroup on the Designation and Astronomical Objects had prepared a report for consideration during the meetings at Montreal. He outlined the report, which included several recommendations on such topics as the order of preference for the use of the current types of designation and the need for a list of catalogues from which numbers could be used. After discussion it was decided that it would be premature to adopt these recommendations but that the report should be referred to the Presidents of other Commissions and published for comment; attention should be drawn in the IAU Information Bulletin to the place of publication (for example, the Strasbourg Bulletin). In the meantime authors and editors should follow the more general guidance given in the resolution adopted at the special meeting on August 14.

3. Presentation of Astronomical Data.

G.A. Wilkins reported that he had circulated a first draft of an Astronomer's Guide to the Presentation of Data to the members of the Subgroup on the Presentation of Data and to some journal editors and other interested persons; a copy had also been on display on the Commission's notice-board. This draft was based on the CODATA Guide for the Presentation in the Primary Literature of Numerical Data Derived from Observations in the Geosciences, which was just about to be published as CODATA Bulletin no.32. The comments showed that the draft was generally acceptable but there were a few points on which further discussion is needed, since it is important that the Guide should be recognised by the Union. J.-C. Pecker commented that the word "paper" is ambiguous and that it should be made clear that it is not intended that the details specified in the Guide be given in short papers that are intended only to announce or summarise new results. The main discussion concerned the recommendation in the draft that "astronomers should abandon the use of metric units that are not SI units". It was agreed thatthis recommendation should be upheld but clarified. Wilkins said that he hoped to produce a separate guide on the use of SI and other recognised units in astronomical publications. It was also agreed that the Guide should follow the CODATA recommendation that the labels in tables and graphs should be given in the dimensionless form. Pecker considered that after revision the draft should be circulated to the Presidents of Commissions and to the Executive Committee of the IAU for approval, before being submitted to CODATA.

C. Worley deplored the growing practice of the publication of data at reduced sizes which are not easily read. Wilkins and others considered that use of microfiche is acceptable (since readers are readily available and full-size copies can be made if required) and preferable to non-publication of the data.

4. Astrophysical data in the Astronomical Almanac.

The Chairman introduced the next item by informing the Commission that for the editions for 1981 onwards, The Astronomical Ephemeris and The American Ephemeris and Nautical Almanac would be published with the single title The Astronomical Almanac and that many changes in content and arrangement would be introduced. In particular, the Almanac would contain tables of astrophysical data obtained from a variety of sources. D. Pascu suggested that the Commission should liaise with other Commissions in order to produce lists which would be recognised by the IAU as standard data. After discussion, it was agreed that it was neither necessary nor desirable for the Commission to act as an intermediary between the Almanac Offices and the Commissions; it was suggested that the Directors of the Offices should invite the Presidents and members of the relevant Commissions to comment on the content and basis of the tabulations in the Almanac for 1981.

5. Catalogue of Stellar Identifications.

F. Ochsenbein announced that a microfiche edition of the Catalogue of Stellar Identification would be published by the Centre de Données Stellaires, Strasbourg. It would contain data and bibliographic information.

6. Archives.

J.O. Fleckenstein drew attention to desirability of cataloguing and, perhaps, collecting in a central place the archives of the long-established observatories; he considered that there is a grave risk that much valuable information, including astronomical data, will be lost through neglect or deliberate destruction. The Chairman commented on the basis of his experience at the Royal Greenwich Observatory, that the tasks of cataloguing and conservation would be extremely time-consuming and costly, but could reveal much of interest. It was considered that the main responsibility lay with Commission 41, but that Commission 5 should consider whether it could assist in any way.

7. Officers and Membership.

The following officers and members of the Organising Committee were appointed:
Chairman: B. Hauck; Vice-Chairman: G. Westerhout.
Organising Committee: W.P. Bidelman, O.B. Dlushnevskaya, R.L. Duncombe, W.D. Heintz (ex-officio), C. Jaschek, T. Lederle, G.A. Wilkins.

It was agreed that the Subgroup on the Designation of Astronomical Objects should be continued under the chairmanship of W.P.Bidelman, who would agree any changes of membership with the Group Chairman. It was also agreed that there should be a Subgroup on the Distribution of Data; this would cover the distribution of data for use with computers and as well as in printed form; the following were nominated for membership: R.L. Duncombe, B. Hauck, A. Heck, C. Jaschek, J. Mead, J.R. Shakeshaft, Y. Terashita, G.A. Wilkins.

D. MEETING OF THE WORKING GROUP ON UDC 52 (15 August, 11.00)

PRESIDENT: D.A. Kemp
ACTING CHAIRMAN: P. Lantos. SECRETARY: L. Remy-Battiau.

The working Group on UDC 52 has been active in the past six years in first writing and compiling UDC 52 and then in implementing its use. UDC 52 has been published as "British Standard 1000: UDC 52", to be ordered from British Standard Institution, 2 Park Street, London W1, U.K.. As to the "Handbook on the use of UDC 52", it is still in its draft form. G.A. Wilkins explains that it is now been used at Greenwich Observatory. It is hoped that through this experience adequate remarks will be made to improve the draft. The use of UDC 52 will probably be mainly restricted to Europe. USA libraries use for national reasons the library of Congress classification. It would be desirable however to establish cross reference between the two classifications.

The group of Physics of the ICSU AB has put up a sub-classification for astronomy. IAU representatives (J.-C. Pecker and P. Lantos) have made efforts to establish connections between it and UDC 52. Users of ICSU AB are mainly secondary papers: they need less details than provided by UDC 52.

Commission 5 carried unanimously the following resolution:

Resolution no. 3.

Considering the steadily increasing number of publications in the field of astronomy and astrophysics, Commission 5 (Documentation and Astronomical Data) of the IAU,

- recommends the introduction and extensive use of proper key words by authors and publishers of primary journals and other publications inorder to facilitate indexing and retrieval work,
- welcomes the initiative of Astronomy and Astrophysics Abstracts (AAA) in preparing a draft vocabulary and invites AAA and other astronomical abstracting services to cooperate in the preparation of an agreed vocabulary.

During the next three years period the prime goal of the Working Group will be first to issue an agreed vocabulary and second to attract attention of people preparing thesaurus to what is going on in the Working Group and to keep close contacts. There seems to be within the commission a consensus that a good vocabulary should be bilingual, not too long but well constructed for information retrieval.

L.A. Higgs from NCR has announced the preparation of an astronomical thesaurus for data bank of NCR. P. Lantos will be kept informed about the evolution of this work.

Since work on UDC 52 is about finished, it seems that the name of the working group is no more adequate and that it should be named Working Group on "Classification Systems and Information Retrieval" (Systèmes de Classification et Recherche de l'Information). Its new President will be P. Lantos.

E. MEETING OF THE WORKING GROUP ON EDITORIAL POLICY (20 August,16.45)

PRESIDENT: J. Shakeshaft. ACTING CHAIRMAN: S. Mitton (who acted also as Secretary).

1. Organization of the Working Group.

In the absence from Montreal of J. Shakeshaft, this question will be solved by personal contact between Mitton, Shakeshaft and Heintz. Composition and chairmanship have to be reexamined.

2. Bibliographical references.

Should the Working Group to inforce more unity within the various journals? It is recommended that, first, the IAU Style Book be revised, as a result of a close cooperation between Commission 5 and the IAU General Secretary, before 1980. This first step could be followed by systematic efforts towards Journal editors in order to get an international agreement along the lines of the IAU Style Book.

3. The Working Group expresses its general regret at the proliferation of journals.

4. The Working Group notes the resolution no. 1 made by the whole Commission 5 (see hereabove § B, and hereafter § F) concerning the references to catalogues of stellar and nonstellar objects, and in general the designation of these objects. The Working Group will make these resolutions known to all editors of astronomical journals.

5. Finally, the Working Group discusses the matter of IAU publications.

It has been suggested to issue an IAU journal, which would contain the essence of IAU Symposia, and of IAU Transactions; the sale (as for all periodicals) would be increased, and the ordering by libraries more easy. One notes with regret the still high cost of IAU publications, and the newly adopted request, by Reidel, of prepayments for reprints.

F. CLOSE OF COMMISSION 5 MEETING (20 August)

The President, J.-C. Pecker, asked that his report be now adopted; this was agreed unanimously. He asked also the Commission to confirm its approval of the various resolutions (reported hereabove) discussed at the meetings of its Working Groups, as resolutions of the Commission itself. This was also agreed by consensus. He then welcomed the enlargement of the membership of Commission to well cover the activities of the Working Groups, notably the Working Group on Astronomical Data. He finally thanked the members and consultants who had assisted him during the period of office; in particular he thanked Madame Remy-Battiau, who had acted as Secretary of the Commission so efficiently.

COMMISSION 6: ASTRONOMICAL TELEGRAMS (TÉLÉGRAMMES ASTRONOMIQUES)

Report of Meeting, 15 August 1979

PRESIDENT: E. Roemer. SECRETARY: B. G. Marsden.

After welcoming those in attendance, the President asked for any comments regarding the Report of the Commission. There were none. She then asked for confirmation of the nominations for officers of the Commission during 1979-82: President, J. Hers; Vice President, M. P. Candy. It was noted that the Commission has no Organizing Committee. These nominations were accepted unanimously. No changes were proposed in the general membership of the Commission.

The President reported that, in consultation with the Director of the Telegram Bureau, she had requested a small increase in the IAU subvention for the Central Bureau for Astronomical Telegrams and had defended the importance of the subvention, even though it represents only a small fraction of the annual operating budget of the Central Bureau. [The Finance Committee subsequently approved the proposed amount of S Fr. 5700. for the 1979-82 triennium.]

The Secretary, in his capacity as Director of the Central Bureau for Astronomical Telegrams, then briefly summarized the report, given in IAU Trans. XVIIA, of the Bureau's activities during the triennium 1976-79. He pointed out in particular the fact that telegraphic communication was now handled using a pair of TWX machines owned by the Bureau; that line charges were levied on many of the items published on the IAU Circulars; and that his assumption in 1978 of the Directorship of the IAU Minor Planet Center had enabled the Bureau to operate more efficiently, much of the information--on both comets and minor planets--that had previously been published on the IAU Circulars being published instead on the Minor Planet Circulars. Since the report was written there had been a change of computers at the Smithsonian Astrophysical Observatory; this had necessitated a threefold increase in the Bureau's budget for computing. The principal operational problem was that of assuring that communications would be received and acted upon in an appropriate manner when the Director was away from Cambridge. Enthusiastic and valuable assistance had been received from Daniel W. E. Green, who was attending to the Bureau's business singlehandedly during the ten days of the IAU meetings, but such student help was available during only part of the year, and, because of inevitably limited experience, there would always be some question as to the decision-making capacity of any student assistant.

The Secretary then reported on behalf of the *ad hoc* committee appointed at the 1976 meeting of the Commission. This committee had been appointed to examine whether there might be more appropriate media for the publication of many of the short observational notes that were not obviously related to objects or events in really urgent need of observation, but where publication in one of the conventional astronomical journals would be inappropriate or would involve an unacceptable delay. The action of Astronomy and Astrophysics, which has introduced a Letters section using camera-ready material provided by the authors, was regarded as a step in the right direction, but the final distribution of the Letters section was still dependent on the printing, addressing and mailing (generally by surface mail outside of Europe) of the rest of the journal. Preparation of a common set of such pages for incorporation in a number of the leading national journals was perhaps worthy of consideration but difficult to put into practice; in any case, several journals, particularly those for which the American Astronomical Society is responsible, insist that full refereeing, rather than a quick screening for obvious errors, was necessary in order to maintain standards. Actually, much of the material appearing on the IAU Circulars is checked more completely than may be generally realized, and it was therefore decided to continue to publish as at present, but to levy line charges on all items relating to observations at other than optical wavelengths and on those relating to optical observations other than discovery announcements of comets, unusual minor planets, novae and supernovae and necessary follow-up information. Such charges help maintain subscription costs to the Circulars at a low level. This is an important consideration, for perhaps half of the subscribers are nonprofessional astronomers with no particular interest in much of the current astrophysical research.

After some discussion on the two preceding reports, the Commission recognized that, while the situation was perhaps not ideal, the Central Bureau did accomplish its essential work well, and that while the Bureau's Director was aware of potential difficulties, the financial support derived from line charges had in fact become an important factor in the continuing smooth operation of the Bureau.

The final item of business concerned proposed changes in the telegraphic code. Since it was planned to reprint the code in an updated "Astronomer's Handbook" section in IAU Trans. XVIIB, an opportunity existed to incorporate revisions in the code. One specific proposal, involving an extension of the code to permit the specification of type of magnitude over a wide range of effective wavelengths, had been circulated to Commission members before the General Assembly. Comments about this proposal, and suggestions for any other changes, were to be reviewed by the Director of the Central Bureau following the closing date of 1 October 1979.

The retiring President then extended good wishes to the incoming President, J. Hers, who had served the Commission well as Vice President during two separate terms. The meeting was adjourned following an exchange of compliments between the retiring President and the Director of the Central Bureau.

COMMISSION 7: CELESTIAL MECHANICS

Report of Meetings, 18 and 21 August, 1979

PRESIDENT: V. Szebehely SECRETARY: J. Kovalevsky

Business Meeting

OFFICERS OF THE COMMISSION

The President proposed the following names for officers during the next period of three years:

President: Y. Kozai, retiring Vice-President

Vice President: J. Kovalevsky

Organizing Committee: E.P. Aksenov, V.A. Brumberg, S. Ferraz-Mello, J.D. Hadjidemetriou, P.J. Message, F. Nahon, J. Schubart, P.K. Seidelman and V. Szebehely.

Recommendation for this list was moved and unanimously approved.

MEMBERSHIP OF THE COMMISSION

The following members of the IAU were proposed as new members of Commission 7: Drs. C.A. Altavista, C.J. Brookes, P. Cefola, S. Debarbat, B. Erdi, P. Farinella, G.M. Gaposchkin, R.J. Greenberg, P.M. Janiczek, H. Kinoshita, J.H. Lieske, C. Marchal, W. Markellos, W.G. Melbourne, V.F. Myachin, C. Oesterwinter, D.A. Pierce, I. Stellmacher, S.N. Vashkovlyak, R. Vilhena De Moraes, R.L. Wagner and K. Zare.

Their application for membership was unanimously approved (minus one abstention).

The following persons were proposed as consultants: R.R. Allan, V.R. Bond, J. Moser, J. Roels, and C. Simo.

The President sorrowfully announced that Drs. Y. Hagihara and E. Stiefel died during the last three years.

ASTROMETRY IN SPACE

After a short introduction by J. Kovalevsky, Commission 7 unanimously approved the recommendation of Commission 24 in support of ESA and NASA space astrometry programs on the grounds that they should provide an inertial reference frame for the mechanics of the Solar System.

FIRST SCIENTIFIC SESSION

T.B. Omarov presented a generalization of the Lagrange-Jacobi equation for some non-classical n-body problems that represents an effective means of analysis of the motion. This can be applied to large scale systems, for instance to the gravitating systems in a Metagalaxy. One of the generalizations concerns the case of a system with a variable number of members, permitting the analysis of the disintegration of the system. Another case refers to gravitating systems with a variable length scale. The speaker showed that the kinetic energy of peculiar motions of galaxies is equal to one half of the module of the potential energy of the inhomogeneous part of the distribution of density in the system.

A.S. Sochilina has investigated the effects of resonant tesseral and sectorial harmonics of the gravitational field on the motion of a close-by satellite. In particular, she has studied the libration mode for NAVSTAR and MOLNYA type satellites (12-hour satellites with critical inclination). The first order perturbations due to the resonance harmonics were computed for several models of the geopotential. The amplitude and the phase of the oscillations have been derived. Apart from the stable points, there are secular trends that limit the validity of the theory.

R. Meire presented a new property of systems of linear periodic ordinary differential equations for which the coefficient matrix has the so-called Hale's E-symmetry.

The monodromy matrix for such systems can be computed from the transition matrix after half a period. This decreases the computer time in the integration of such systems by 50%. As an application, the linear stability of the triangular points in the elliptic restricted three-body problem was investigated. The transition curves in the (μ,e) plane separating stable and non-stable regions were computed using the two independent second order systems due to TSCHAUNER. The new results are more accurate than the previous ones.

CELESTIAL MECHANICS IN CHINA

Dr. Yi Zhaohua reported on the work in Celestial Mechanics done in the Peoples Republic of China.

Before 1958, the main concern was in the motion of asteroids including numerical and analytical studies (Hansen-Bohlin method).

Later, the interest was extended to the following domains.

- The compilation and publication of the Chinese Astronomical Ephemeris (since 1964).
- The theory of motion of artificial satellites.
- Some basic theoretical problems including the improvement of the convergence of the development of the disturbing function and the theory of capture in the three body problem.

In the recent years, the main activities in Celestial Mechanics may be described as follows:

1) The perturbation theory of artifical satellites (Purple Mountain Observatory and Nanking University). A second order analytical theory with third order secular terms was derived including all effects due to the geopotential, air drag, solar radiation pressure, luni-solar perturbations, etc...(published in Acta Astronomica Sinica and Journal of Nanking University).

2) Qualitative investigations of the three body problem (Nanking University and the University of the Chinese Science and Technology). The topology of the M_8 manifold of the three body problem was investigated in more elementary and intuitive way than S. Smale or R.W. Easton.

- The Birkhoff conjecture that the topology of M_8 can vary only when the energy constant E passes through the admissible values of Lagrange particular solution, was strictly proved.
- The topologies of M_8 for the five corresponding intervals of E were given.
- The connectedness of M_8 has been proved (DONG JIN-ZHU, "Some problems of the regions of motion of the general three body problem") and ranges of change of the angles between the orbital planes have been investigated.

3) The stability of the Cowell's method for numerical integration was studied.

4) Investigations in the field of analytical solutions for some particular cases of the three body problem are in progress. Hill's problem and the circular restricted problem are being studied.

Second Scientific Session

This session was devoted to seven reviews concerning some of the most active domains of Celestial Mechanics.

PLANETARY THEORIES

P.J. Message, after a short reference to the important topic of comparison of theories with observations, reviewed the recent developments of analytical theories of planetary perturbations. Such theories have usually been constructed by "Poisson's method", "order by order theories" or "théories à variations séculaires".

Nowadays, computer manipulations of multiple Fourier series are used. The expressions obtained give acceptable accuracies after time intervals of the order of 1000 years. The appearance of time in the coefficients leaves open the question of the possible instability of the planetary system. However, it is known that "general theories", generalizing Laplace's and Lagrange's theory of secular variations are able to represent the actual motions for an interval of half a million years and very likely very much longer. However, reaching a similar accuracy as the classical theory would lead to series much larger than those that can be handled by present day computers. Examples were mentioned of theories of each type recently constructed in France.

RESONANCE

B. Garfinkel spoke about the ideal resonance problem that he has formulated and solved in a series of papers. This problem furnishes a reference orbit for an attack on a variety of resonance problems in the solar system. The formulation epitomizes the methodology developed by Laplace, Bohlin and Poincaré. Successful applications have been made to the following problems:

1) Critical inclination in the motion of an artifical Earth satellite including J_2 and J_4.

2) Synchronous satellite with J_{22} and $e = 0$.

3) The motion of Trojan asteroids treated as the case of 1 : 1 resonance in the restricted three-body problem.

The problem of double resonance has continued to evade a global solution, thus remaining an outstanding unsolved problem of celestial mechanics.

SATELLITE THEORY

Y. Kozai described what are the important problems in satellite motion theories by classifying the satellites into inner, outer and intermediary ones. There are many resonant pairs in the inner satellite systems. Analytical theories for outer satellites are difficult to derive when the ratio n'/n is not small. For intermediary satellites, it was noted that the motion of their orbital planes are complicated. Therefore, complete analytical theories are also difficult to derive.

POSSIBLE EXPERIMENTS ON THEORY OF GRAVITATION

C.A. Lundquist presented two possible fundamental experiments in gravitational physics that could be carried out by a space mission. This mission would begin with a near-Earth, highly eccentric orbit to test the gravitational redshift. Later, the spacecraft would be placed into a heliocentric orbit to provide a long Earth to spacecraft baseline to perform the most sensitive search thus far proposed for bursts of low-frequency gravitational waves from extragalactic sources. Both goals are natural outgrowths of previous missions to study gravitational physics. In particular, the 1976 suborbital gravity probe A (SAO and NASA) showed that using masers on spacecrafts could provide a test of equivalence principle at the 10^{-4} level. The main improvement of the proposed experiment is the use of four simultaneous sources of Doppler tracking phase data between clocks on the Earth and the probe.

PERIODIC ORBITS

J.D. Hadjidemetriou presented recent results in this field, showing that periodic orbits are useful in the study of stellar and planetary systems. There exist stable triple stellar systems even for close approaches of the three components. Also stable triple systems exist with nearly zero value of the angular momentum. For planetary systems, a global view is presented of periodic orbits of the planetary type. It is clarified which configurations are stable and what factors affect stability. Provided that the masses of the planets are small and that they are not close to a (2n - 1)/(2n + 1) resonance, nearly circular direct planetary systems are stable. If the orbits are elliptic, the system is generally unstable.

USE OF COMPUTERS IN CELESTIAL MECHANICS

M.S. Davis gave a report on this subject. Very little of a pioneering nature has been accomplished in the last three years in numerical or non-numerical uses of computers in Celestial Mechanics. The direction of modern computer science research was briefly touched upon, indicating the concentration on parallel processing on both the hardware and software levels. Then, the author gave a review of a large variety of numerical experiments in Celestial Mechanics and a survey of new or adapted analytical systems and their application to specific problems in Celestial Mechanics. Finally, attention was called to the publication by the Institute for Theoretical Astronomy in Leningrad: "Algoritmy Nebesnoj Mekhaniki", the only known publication devoted to the application of computers to Celestial Mechanics.

STABILITY IN CELESTIAL MECHANICS

A.E. Roy made a short review of this domain that is closely related to problems concerning the origin and evolution of various bodies in the solar system, multiple stars, etc... He gave a short review of the historical evolution of various concepts of stability that were originated by Laplace, Poincaré, Hill, Kolmogorov, Szebehely and others, including linear, non-linear and statistical methods. Several approaches to solve the problem were considered: local or general integrals, general perturbation methods, numerical studies, periodic orbits and topological analysis. In particular, a discussion of the relationship between resonance and stability was offered. It seems that deep resonance may be necessary for stability, but no relation seems to exist in other cases.

COMMISSION 8: POSITIONAL ASTRONOMY
(ASTRONOMIE DE POSITION)

Report of Meetings, 15, 21 and 22 August 1979

PRESIDENT: R. H. Tucker SECRETARIES: B. L. Klock, F. S. Gauss

Constitution of Commission

President: Høg. Vice-President: Tavastsherna.
Organizing Committee: Anguita, Billaud, Débarbat, Fricke, Hughes, Klock, Lopez, Nikoloff, Teleki, Tucker, Yasuda.
In order that the O.C. should adequately represent the great contribution of the U.S.S.R. to Positional Astronomy, the incoming President was asked to co-opt one or two new members to the O.C. after the G.A., under the terms of I.A.U. By-Law 21.
Deceased Members: Y. E. Gordon, A. K. Korol', H. U. Sandig, J. L. Schombert, L. S. T. Symms.
Retired: A. N. Adams, A. Bohrmann, W. Dieckvoss, B. Guinot, E. D. Hoffleit, S. Isobe, P. Rybka, R. H. Stoy, G. van Herk.
New Members: Bem, Benevides-Soares, Feissel, Hurukawa, Kovalevsky, Miyamoto, Polnitzky, Rusu, Sanchez, Yatskiv.
New I.A.U. Members: Backer, Carestia, Carrasco, Chamberlain, Chlietovsky, Corbin, Duma, Fedorova, Laclare, Loyola, Pham-Van, Pinigin, Poma, Russell, Sadzakov, Schwan, Sims, Smith.
Consultants: Blackwell, Branham, Fanselow, Helmer, Hemenway, Lindegren, Orelskaya, Pakvor, Zimmermann.
Numerical Strength (with previous values in brackets) 13 (13) Officers, 99 (84) Members, 9 (5) Consultants, Total 121 (102).

Working Groups and Study Groups (Chairman's name given first)

Refraction: G. Teleki; Altenhoff, Garfinkel, Hughes, Kolchinsky, Nefedeva, Ramsayer, Saastomoinen, Sugawa, Tengstrom, Yasuda.
Astrolabes: G. Billaud; Debarbat, Gliese, Gubanov, Noel.
Horizontal M.C.: E. Høg; Atkinson, Gietzen, Klock, Osorio, Pinigin, Timashkova, van Herk.
Co-ordination of Stellar Positional Work: (Jointly with Commission 24).
See report of Joint Meeting of 21 August for list of members.

Scientific Papers (15 August)

Débarbat presented papers by F. Chollet et al. (Astrolabe Observations of the Sun) and Requième (Photoelectric M.C., Bordeaux). A cooperative program for observing the Sun with Astrolabes, using a special filter, is being undertaken by Laclare (CERGA), Demarcq (Nice) and Chollet (Paris). The first experiments with the re-furbished M.C. at Bordeaux are expected in September, 1979. Limiting magnitude V = 12.5, expected error ±0".12 in each coordinate. Høg presented a report by himself and Høyer analysing 110,000 observations (Perth 70) and 60,000 by Nikoloff and his associates (Perth 75). Fricke remarked that this work is important for the FK4 Supp observations, rather than for proper motions. The desirability of publishing asymptotic errors was discussed. Smith gave a progress report on W_5 50, W_6 50, W_L 50, SRS, and the preliminary positions of the Sun, Moon and Planets for 1975-77 at U.S.N.O. He requested that SRS observers should send individual results. It is hoped that SRS will appear before the next General Assembly.

Teleki gave a report of the Study Group on Astronomical Refraction.

Yatskiv presented 6 papers by USSR authors. Tavastsherna has used classical and modified chain methods to reduce Melbourne R.A. observations between 1928 and 1941. Kurianova et al. report on the compilation of PFKSZ-2 from 42 catalogues reduced to the FK4 system. 22 catalogues were used for the proper motions. Duma et al. have determined corrections to FK4 equinox and equator from 18,471 planetary observations, between 1928 and 1971. Orelskaya used 10 minor planets observed from 1949 to 1968. Fricke commented on the difficulties of this type of work. Minor planets nos. 2 and 3 do not contribute at all to the equinox. The good results from 10 minor planets appeared to be fortuitous. The equinox is not easily determined from T.C. observations unless they are rigorously reduced to FK4, and this is difficult for planets. It is very hard to get anything out of Mercury and Venus. Yasuda's results from Mars agree with Duma's.

Afanasieva and Fomin used the RATA-600 radio telescope to measure astrometric positions of Mercury and Venus from their microwave radiation. The instrumental parameters are calibrated by simultaneous optical observations. The mean error quoted is $1\frac{1}{2}$ times that of the T.C. Observations are made only at elongation, and no correction for phase is made, as the illumination of the disc is not relevant to the thermal radiation. Finally, Zverev et al. proposed a new campaign of AGK3R and SRS observations, combining the lists into one as "IRS". A working group should be set up to study the problems.

Sadzakov presented a paper on Recent Observations with the Belgrade Meridian Circle, and Teleki reviewed refraction investigations between 1976 and 1978. No new studies have been made of temperature effects inside the telescope tube. Hughes reported new investigations of humidity effects at USNO. The Pulkovo Tables humidity correction works well at standard pressure, but there is considerable error at low pressure, and some attention should be given to this. Yasuda gave a progress report on the NPZT observing program. Billaud presented a report of the Working Group on Astrolabes, and Høg reported the work of the Study Group on Horizontal Meridian Circles. Teleki presented a report by Nefedeva on the new Pulkovo Refraction Tables. The functional relationships used are described in the publication. The water vapour correction methods are practically the same as those of the Fourth Edition. For zenith distances less than 70° there are no practical differences between the tables.

Hemenway reviewed the Implications of Space Astrometry on Ground Based Work. Discussion afterwards concentrated on the Space Telescope, which had a focal length of 60 m, an astrometric precision of $0\overset{''}{.}002$, and an unknown accuracy. No facility for comparison of widely separated fields was in prospect.

Tucker gave a report of IAU Colloquium No. 48 (Vienna) of which the Proceedings had been published already. Teleki reported on the Nomenclature meeting held the previous day at Montreal. Débarbat presented a Bibliography of Astrolabe Literature prepared by A. Stoyko.

Joint Meeting, Commissions 8 and 24 (21 August). Chairman: C.A. Murray

After some opening remarks by Eichhorn, Høg surveyed the prospects of Astrometry in the Eighties. Strand questioned the value of 1 micron for emulsion stability, estimating the true value to be about 0.4 micron. Murray suggested that the limiting instability was the atmosphere, but Strand said this would not be true for short-focus telescopes. Smith presented some suggestions for improvement of AGK3R and SRS. Fricke remarked that greater care must be taken in future when setting up a list. The dominance of K stars in the present lists resulted from a compromise to satisfy various groups of astronomers. Corbin reported the current state of AGK3R and SRS proper motions. In reply to Murray, he remarked that color was obviously considered more than observational history. The distribution in the weaker zones would not be improved by adding more stars. Fricke recommended that the list should not be changed. If necessary, a proper motion should be based on only 2 catalogs to avoid discarding the star.

Eichhorn presented De Vegt's paper on a proposed Fourfold Plate-Overlap Coverage of the Sky. Nicholson reported on the progress in measuring Cape Photographic Plates at Herstmonceux, and Routly on the Zodiacal Catalog Program at USNO. Corbin remarked that both the De Vegt and Routly projects deserved support. Two catalogs are better than one, as in the Yale zone $+20^{\circ}$ to $+30^{\circ}$ where the Yale and AGK2 results were combined. The two catalogs could also provide checks for each other at the fainter magnitudes. Robertson reported on the work done with the Sydney astrograph, and remarked that he would appreciate offers to measure 1000 plates.

Stock gave a paper on Magnitude and Color Errors in Refractors, which Murray remarked should be considered by the new Joint Working Group. Yatskiv reported a proposed Photographic Survey by 10 double astrographs in USSR to observe about 400,000 stars.

A Joint Working Group for coordination of work on the Stellar Reference Frame had been set up by Commissions 8 and 24, with the following composition: H. K. Eichhorn (Chairman); Høg, Hughes, Lederle, Murray, Nikoloff, Requième, Tavastsherna, De Vegt, Fresnaux, Gavrilov, Jones (B.), Potter, Robertson, Stock.

Murray raised the question of the aims of the Joint Working Group. It was agreed that these should be

(1) To consider the needs of Stellar Reference Frames for the next 20 years
(2) To formulate and implement plans for meeting these needs.

Study Group on Horizontal Meridian Circles (22 August)

Osorio reported on the status of the Oporto HMC.

Høg presented a paper on the Brorfelde design study for a horizontal instrument, formerly known as the GMC. The following points emerged during discussion:

The single taper roller bearing acted as a pivot.
The main tube will probably have additional counterweighting.
There are two auto-collimating systems.
The tilt of the main mirror will be measured with each observation.
The micrometer rotates with the main mirror.
The total weight of the moving components is 200 kg.
The mirror is small enough to have insignificant deformation imposed by the mounting springs.
The wear on the roller bearing will be negligible.
The timetable for construction has not yet been fixed.
The use of an evacuated tube would be attractive, but would raise a whole series of new problems.

Scientific Papers, Commission 8 (22 August).

Tucker formally proposed a resolution adopted at the Uppsala Symposium, calling for a special Joint Commission of IAU and IAG in the field of Refraction. This was carried by 12 votes to nil. Hughes gave a report on the status of the USNO Transit Circles, including an experimental micrometer using an image dissector, now under bench test. No details had been published as the design was still being developed. Yasuda described the projected Automatic Photoelectric MC being constructed for Tokyo Observatory. Pakvor reported on the observation of bright polar stars with the Large Transit Instrument at Belgrade. Corbin reviewed problems arising from double stars in FK4. Débarbat reported recent observations of Beta Persei, and requested information on radio source stars bright enough to be observed with an astrolabe. Tucker reported on the Anglo-Danish collaborative project for Meridian Astronomy on La Palma. It is hoped that 1377 FK4 stars will be observable, and the total number of observations in two years will be over 200,000.

In discussion, it was stated that the instrument will require two observers, working three or four days on, followed by four or three days off, for three weeks at a time. Requième (Bordeaux) has expressed interest in joining the collaboration. Stars can be observed to about 5th magnitude by day. The building is planned to permit observations down to $4\frac{1}{2}$ degrees altitude. The Spanish astronomers hope soon to take an active part in the project with the Danish and British teams.

COMMISSION 9: INSTRUMENTS AND TECHNIQUES

Report of Meetings, 15,16,17 and 22nd August 1979

PRESIDENT: J. Ring

Report of the President

Since the 16th General Assembly in Grenoble, an I.A.U. Colloquium (No.50) had been held in Maryland under the auspices of Commission 9 on the subject of "High Angular Resolution Stellar Interferometry". The proceedings of the colloquium had been published - copies were available from Dr J. Davis, Physics Department, University of Sydney, Australia.

Following a suggestion from Dr E.H. Richardson and Dr G. Odgers, discussions had begun on the possibility of a Symposium being held at Zelenchuk (the site of the U.S.S.R. 6m telescope) on the subject of "Instrumentation for Large Optical Telescopes". It had proved impossible to organise such a meeting before the 17th General Assembly but the correspondence would be passed on to the new President.

The President expressed his thanks to the Chairmen of Working Groups who had helped in the production of the "Report of Commission 9". Two mistakes had been pointed out and he took this opportunity of correcting them:

1) In the section on image intensifiers it was reported on page 7 that "On the 6m telescope the spectrum of a 16.7 magnitude star is recorded in one hour". The exposure was actually six minutes.

2) In the section on television-type systems it was stated that "At Princeton, Lowrance is continuing development of the 70 mm SEC system...." In fact, this work was discontinued in 1977.

Business Meeting, 15th August

Dr E.H. Richardson was elected President, there being no other nominations.

A ballot was held for the office of Vice-President and membership of the Organising Committee with the following result (the retiring President is customarily included):- Vice President: W.C. Livingstone. Committee members:- W.B. Burton; P. Connes; I.M. Kopylov; A. Labeyrie; D. MacMullan (Chairman of Working Group on Detectors); A.B. Meinel; N.N. Mikhel'son; J. Ring (Retiring President); N. Steshenko; G.A.H. Walker; M.F. Walker.

It was agreed after discussion that the Chairmen of Working Groups, who should be elected by the Working Groups, should be co-opted to the Organising Committee. They are as follows:- R. Albrecht (Working Group on Data Processing); J. Davis (Working Group on High Angular Resolution Interferometry); J.L. Heudier (Working Group on Photographic Problems); J. Rösch (Working Group on Large Telescope Projects).

Following the decision at the 16th General Assembly to remove inactive members from the Commission and to encourage new active members, there had been many applications for membership. The list was distributed and, after discussion, all the applicants were elected to membership.

Working Groups

The President pointed out that two of the Working Groups were still only informal ones (Data Processing and High Angular Resolution Interferometry). It was agreed that they should henceforth be formally constituted. A discussion ensued on the role and organisation of the Commission's working groups. It was agreed that their functions included:-

(a) The exchange of information between members
(b) The preparation of I.A.U. reports
(c) The organisation of meetings at I.A.U. General Assemblies and elsewhere.

The first function suggested that 'membership' of the Working Group (which might in practice mean inclusion in a mailing list) should preferably be open to all members of the I.A.U., whilst the other functions required there to be a nucleus (or Committee). It was agreed that Working Groups should elect their Chairmen and Committee members at General Assemblies and that the composition of the Working Group (or its mailing list) should be left to the discretion of individual groups.

Working Group on Detectors (D. MacMullan)

The group held a meeting on 16 August at which the following papers were presented.

CHARGE COUPLED DEVICES (CCD's)(W. Livingstone)

Availability and device characterisation are our concern here. We note the Fairchild 201 and 202 (100x100 pixels) are discontinued. The Fairchild 211 (190x244, ½ blind, NEC ∿ 20e, QE ∿ 12%) and the Fairchild 221 (488x380, ½ blind, 30(12) x 18μ pixels, QE ∿ 12%) are now in use. Fairchild proposes a modified version for astronomy in which the aluminium electrodes are removed. This should increase the QE to about 50% at 7500Å. No doubt other commercial CCD sources of interest to astronomers will soon enter the market.

Because the Fairchild pixel sites are thick (∿ 250μ), their cross-section to cosmic rays is high. Marcus, Nelson and Lynds (SPIE vol. 172, 205 (1979)) report 1 event/cm^2/min, with each event covering several pixels. Exposure time is thus limited to a few minutes. "Thinning" the array would proportionately reduce this problem.

The above CCD's have no response $\lambda < 4000$Å. Blue sensitivity may be achieved by thinning (and back illumination) or by over-coating the array with fluorescent dyes (J. Westphal). Spectral enhancement for $2000 < \lambda < 4000$Å is obtained with several organic materials: Pyrene, Liumogen, and Corronene (W. Bell, KPNO). Ruby is suggested (Appl.Optics 18, 2085 (1979)) but may be impossible to remove in case of deposition error.

Cooling the CCD to $T < -100^{\circ}$C virtually eliminates pixel-to-pixel dark current and generally improves uniformity. "Flat fielding" is unnecessary for many applications.

Providing the temperature and control voltages (and timing) are suitably regulated reproducibility to 1% is realised.

ELECTRONIC IMAGING DEVICES IN SPACE ASTRONOMY (G.R. Carruthers)

This paper reviews the current and future applications of electronic imaging devices in space astronomy: (1) current flight missions in which electronic imaging devices are the primary astronomical data gatherers, (2) currently approved future missions in which electronic imaging devices will be used, and (3) a sampling of proposed, but not yet approved, future space astronomy investigations of this type.

The current flight missions include two astronomical satellites, the International Ultraviolet Explorer (IUE), and the second High-Energy Astronomical Observatory (HEAO-2, or Einstein). IUE uses intensified SEC vidicons to record two-dimensional cross-dispersed echelle spectra of 0.1-0.2Å resolution in the 1150-1950Å and 1900-3200Å wavelength ranges, or low-dispersion (6Å resolution) spectra in the same wavelength ranges. Einstein uses a grazing-incidence telescope to image X-rays in the 0.2-4 keV range, with either of two imaging detectors: (a) an Imaging

Proportional Counter, which has about 1 arc minute angular resolution and 0.5° field of view, with high detective efficiency, and (b) a High Resolution Imager, based on a chevron microchannel plate with crossed-grid readout, having 3 arc sec resolution but somewhat lower detective efficiency.

Also among current missions are numerous sounding rocket investigations by workers at a number of institutions. Detectors include microchannel image intensifier with phosphor output and film recording (Goddard Space Flight Center), microchannel detectors with resistive anode or discrete anode readouts (University of California/Berkeley, Johns Hopkins University, Naval Research Laboratory, University of Colorado) and electrographic cameras using alkali halide photocathodes (sensitive below 2000Å), by NRL.

Approved future investigations include experiments on the Spacelab 1 and Spacelab 2 missions, and the Space Telescope. Spacelab 1 will include the FAUST wide-field all-reflecting camera (Marseilles Observatory, France) which uses an MCP intensifier with film recording, and a set of airglow spectrographs covering the 300-10,000Å wavelength range, provided by the University of Michigan, using MCP intensifiers coupled to CCD arrays. Spacelab 2 will use a CCD detector as part of the Lockheed solar polarimeter/magnetograph, and microchannel plate detector with discrete anode readout as part of an extreme ultraviolet solar spectrograph to be provided by the Appleton Laboratory and University College, London.

The Space Telescope scientific instruments use a variety of imaging detectors. The Wide Field/Planetary Camera (California Institute of Technology) will use 800x800-element CCD arrays as the image sensors. The ESA-provided Faint Object Cameras will utilize image intensifier/electron-bombarded silicon target camera tubes in photon-counting mode. The Faint Object Spectrograph (University of California/San Diego) and the High Resolution Spectrograph (Goddard Space Flight Center) will utilize 512-element linear array (individual readout) Digicon detectors, also in photon-counting mode.

Proposed investigations for future Shuttle/Spacelab missions include both small, sounding-rocket-class investigations and larger investigations based on moderate-sized telescopes. In the first class are a high-resolution spectrograph for the 900-1315Å range proposed by Princeton University using an electron-bombarded CCD detector, as well as NRL proposals based on electrographic Schmidt cameras (a direct imaging/objective spectrography sky survey experiment, and a nebular spectrograph experiment). Larger instruments include a 0.8-meter-aperture all-reflecting Schmidt telescope, covering 5° fields of view with 1 to 2 arc sec resolution (primary wavelength range 1250-2000Å) intended for deep surveys in the far UV, and a Spacelab facility telescope, STARLAB, based on a 1-meter aperture Ritchey-Chretien telescope and capable of imaging 0.5° fields of view with 0.2 arc sec resolution. These latter two instruments would use large-format (120-180 mm image diameter), high-resolution electrographic cameras as the image detectors.

MULTI-ANODE MICROCHANNEL ARRAYS (J.G. Timothy)

Multi-anode Microchannel Arrays (MAMA's) are a family of photoelectric photon-counting array detectors being developed for use on both ground-based and spaceborne telescopes. The MAMA detector system combines the sensitivity and photometric stability of a conventional photomultiplier tube with a high-resolution imaging capability. MAMA detectors can be operated in a windowless configuration at extreme ultraviolet and soft X-ray wavelengths or in a sealed configuration at ultraviolet and visible wavelengths. Prototype MAMA detectors with up to (512x512) pixels are now being evaluated in the laboratory and the first telescope observations with a (10x10)-pixel visible-light detector have just been completed. MAMA detectors with 1x512 pixels, 1x1024 pixels and 16x1024 pixels will be operated on the Cassegrain spectrograph of the 24-inch telescope at the University of Colorado during the next twelve months.

ASTRONOMICAL APPLICATIONS OF TELEVISION CAMERA SYSTEMS EMPLOYING ELECTRON BEAM READOUT (J.Lowrance)

Much of the development work on image sensors and systems for astronomy is now on solid state devices. However, the application of more conventional television camera sensors is now a major component in the instrumentation at astronomical observatories throughout the world. Perhaps even more important, these television systems have served to train astronomers in the use of electronic imagery and the data reduction associated with its application to astronomy. This experience makes the adaptation of the solid state sensors to astronomical observations much easier and therefore rapid. In most cases the computer for data acquisition is already available, and the observatory staff is attuned to the task of applying electronic image sensors to astronomical problems. Both Cerro Tololo and the University of Hawaii have developed substantial systems employing silicon vidicons and SIT/EBS tubes controlled by microprocessors or mini-computers as well as the necessary computer equipment and software for data reduction (PASP Vol.91, p.120-139, 1979). Similar equipment is in operation at a few other observatories and it appears that almost all astronomical observatories have some television-based observing system. Most of these systems employ the silicon vidicon, SIT or SEC tubes. (A summary of the current use of television cameras in astronomical data applications was given by Prof. Lowrance which is not reproduced here.)

CAMERA ELECTRONIQUE (G. Wlérick)

G. Wlérick donne des compléments relatifs a la photométrie par électronographie au télescope de 1.93 mètre de l'Observatoire de Haute-Provence. Toutes les émulsions utilisées en Astronomie son d'abord vérifiées en laboratoire par A. Sellier, G. Lelièvre et lui-même. Trois propriétés sont examinées: sensibilité, hétérogénéitié et linéarité.

Les émulsions Ilford G5 et Kodak Industrex (autre nom Definix) son généralment non linéaires pour une densité optique D > 1, mais elles obéissent à une loi simple à deux paramètres comme l'avait vérifié autrefois S. Jeffers (1968):

$$D = D_S(1 - e^{-AEt})$$

D_S densité à saturation, E éclairement, t temps de pose; pour l'émulsion Industrex, D_S varie avec les lots, augmente avec la tension d'accélération V et vaut ≈ 2.5 à 4.5 pour V = 25 KV.

Pour l'émulsion Ilford G5, $D_S \sim 4$ pour V = 25 KV. L'émulsion L4 est en général linéaire jusqu'a D = 4 pour V = 25 KV c'est à dire que la valeur de D_S est très élevée ($D_S \sim 10$ à 20); lorsque la tension d'accélération diminue ($V \sim 15$ KV), un écart à la linéarité peut apparaître.

Il est donc essentiel de mesurer les propriétés photométriques des émulsions dans les conditions d'emploi astronomique.

D'autre part, le fait que la densité optique est répresentable par une formule à deux paramètres seulement permet d'effectuer facilement les corrections photométriques pour les mesures effectuées avec les microdensitomètres tels que le P.D.S.

Working Group on Large Telescopes (Chairman: J. Rösch)

The following papers were presented on August 15th.

THE INSTRUMENTATION OF THE 6M SOVIET TELESCOPE (I.M. Kopylov)

Here is a short description of the instrumentation now in operation:

Prime focus

1. Ritchey cassette. 16x16 cm plates (≈ 15x15 arc-minutes). UBVR colour system. m_{lim} (depending on colour) 23 - 24.5.
2. Three-stage image tube device for direct photocathode Ø ≈ 40 mm, BVR colour system m_{lim} ≈ 24 - 25.
3. Photometer-polarimeter (two channels, ordinary UBVR photometry of faint objects).

3a. Elaborated photometer-computer system aimed at studying very fast brightness variations of optical events around "black holes", neutron stars and intrinsic variable stars. Timescale $10^3 - 10^{-7}$ sec. Time resolution ≈ ±40 nanoseconds.

4. Two high speed spectrographs with several types of image tubes for spectroscopy

of faint extragalactic and stellar objects. Dispersion: 60 - 270 Å/mm, from 3000Å to near-infrared (≃ 9000Å). m_{lim} (depending upon wavelength) up to 19.5 - 20.
4a. High speed spectrograph with TV 1000-channel photon-counter scanner controlled and operated by computer and display. $\Delta\lambda$ 3000-7000Å, $m_{lim} \simeq 19.5$.
5. TV and image tube devices for the guiding of faint objects on the slit of the spectrograph.

Secondary foci
1. An echelle spectrograph with image tube. Two echelle gratings: 100 and 50 gr/mm, orders 11-20 and orders 16-25. Dispersion: 30 - 60Å/mm (depends upon grating and wavelength).
2. A spectrograph with high linear resolution (along the slit) with image tube.
3. Three-camera large spectrograph. Two of these cameras are of Schmidt-type. Dispersion (depending upon camera, grating, order, wavelength): 1.3 - 9 - 28 - 56Å/mm.
3a. Fabry-Perot magnetometer attached to large camera of P.3 spectrograph. Controlled by computer. Aimed at the determination of field strength of magnetic stars, $m_{lim} \simeq 8$, accuracy of the magnetic field determinations down to ±20 gauss.
4. Speckle interferometry device with image tube. $m_{lim} \simeq 8$ (at the moment). Accuracy of determinations of stellar angular diameters: ≃ ±0"004. New instrumentation currently being designed and built will be described at the next General Assembly.

THE 3.8 METRE UNITED KINGDOM INFRARED TELESCOPE (C.M. Humphries)

The new 3.8 metre U.K. Infrared Telescope (UKIRT) has unique design features in both its structure and its optics. By using a thin primary mirror - very thin by traditional standards - it has been possible to use a relatively lightweight supporting structure and thereby obtain substantial cost savings, a factor of at least three over conventional designs of similar aperture. The CERVIT primary mirror has a mean diameter-to-thickness ratio of only 16 : 1 (compared to 6 : 1 or 8 : 1 for a conventional large mirror) giving a mass of 6 tons instead of 15 tons. It is supported axially by 80 pneumatic pads distributed on three radii and radially by 24 counterweighted lever arms. The telescope has an English yoke mounting and the structure has a mass of 80 tons, instead of 300 or 400 tons. Roller bearings have been used for the polar and declination axes.

The telescope has now been erected at the 4200 metre (13800 ft) site at Mauna Kea, Hawaii, and commissioning is nearly complete. Three focal ratios are available - f/9 Cassegrain, f/20 coudé and f/35 Cassegrain with chopped secondary. In each case the secondary mirror acts as an aperture stop and eliminates the possibility of thermal radiation from the tube structure or mirror cell being seen by the detector.

The encircled energy diameters for 90% of the incident radiation are slightly less than 1 arc sec for the f/9 image and 1.2 arc sec at the f/20 focal plane. The f/35 system is due to be tested within the next few weeks. The optical performance of the telescope has also been obtained in terms of an MTF and compared with measurements of the atmospheric seeing MTF at the site.

At wavelengths above 8 micrometers the telescope is diffraction-limited and at all wavelengths the telescope MTF is comparable with or better than the atmospheric MTF for the visible region. The absolute pointing accuracy of the telescope is ≃ 5 arc sec RMS with computer corrections applied.

The thin mirror concept has been proved sound in its application to UKIRT and can be used with advantage in other large telescopes planned for the future. This applies equally to equatorial or altazimuth mounting.

THE 3-METER INFRARED TELESCOPE FACILITY (E.E. Becklin)

The Infrared Telescope Facility (IRTF) on top of Mauna Kea, Hawaii, became operational during the summer of 1979. In May, more than twenty-five 5 μm images of Jupiter were obtained which were used to help target the Voyager II spacecraft during its July flyby of Jupiter.

The NASA-supported facility is run by the University of Hawaii; it will go into full operation in the Fall of 1979 as a nationally-available facility for infrared research.

The 3-meter telescope meets a number of criteria which makes it particularly good for infrared observations:

1. The telescope is situated on the summit of Mauna Kea (4120 meters) where there is excellent year-round infrared transmission, lower thermal sky fluctuations and good "seeing".
2. The telescope has an image quality of 0\"8 and an image stability of 0\"2; observations with better than 1" resolution are possible.
3. The telescope has a very low thermal background and therefore high sensitivity because of the small obscuration in the primary mirror and small undersized secondary mirror. Tests at 10 μm show that the telescope can reach 0.005 Jy (1σ) in one hour of interpretation.
4. The telescope has versatile f/35 Cassegrain and f/120 coudé chopping secondaries. The Cassegrain system can efficiently chop an infrared source at 40 Hz with up to a 5' throw.
5. The telescope can point very accurately across the sky to better than 5" under computer control.

LE TELESCOPE CANADA-FRANCE-HAWAII (R. Cayrel, J. Odgers)

Le télescope Canade-France-Hawaii a été monté en usine à La Rochelle-Pallice en octobre 1976 sous abri gonflable pour permettre l'intégration de l'électronique de commande réalisée au Canada et procéder aux essais avant envoi à Hawaii. La rupture du radome fin novembre arrêtait les travaux jusqu'en mars 1977. L'intégration et les essais en usine se sont ensuite poursuivis de mars 1977 à début avril 1978 et permirent de corriger quelques insuffisances structurales avant le remontage sur le site.

Le télescope a été démonté et emballé d'avril à juillet 1978 et est arrivé à Hawaii le 12 septembre 1978 à bord du navire "Lift Off". Le rémontage sur le site a été effectué de novembre 1978 à juillet 1979.

C'est le 7 août à 12 h 15 TU que la première lumière a été obtenue. Le 11 et 12 août les premières photographies du ciel étaient prises au foyer primaire.

Le télescope sera operationnel au foyer primaire dès le debut de 1980 et environ 40% du temps sera alors attribué aux astronomes visiteurs.

Le foyer coudé et le foyer Cassegrain infra-rouge à f/35 seront mis en service vers la mi-1980. Le foyer Cassegrain f/8 sera mis en service seulement à la fin de 1980 à cause du retard dans le polissage du miroir secondaire.

Une dizaine d'instruments seront progressivement mis en fonction au cours de l'année 1980 et au début de 1981.

LE TELESCOPE DE 2 METRES DU PIC DU MIDI (J. Rösch, G. Coupinot, G. Carpentier)

La construction de ce télescope a été décidée en 1964 en vue d'exploiter au mieux la qualité des images reconnue au Pic du Midi depuis les travaux de B. Lyot et de des continuateurs. On s'est efforcé de rassembler dans ce projet les solutions permettant d'espérer la meilleure résolution angulaire possible. En particulier, on a procédé à une étude en soufflerie de la forme et de la situation du sillage en fonction de la direction du vent, et choisi l'emplacement du télescope en tenant compte des vents dominants. On a aussi adopté un type de coupole concue pour éviter tout échange d'air entre l'intérieur et l'extérieur, grâce à un joint entre l'extrémité du tube et l'ouverture circulaire mobile ménagées dans la coupole. On pourra, par la suite, fermer l'avant du tube par une lame à faces parallèles.

La réalisation de l'opération, confiée à l'INAG, a débuté en 1970 et a subi de nombreux retards, notamment en raison des intempéries (travaux extérieurs de juin à octobre). La monture, étudiée en même temps que celle du Télescope de 3m 60 CFH, est du même type (fer à cheval portant la denture d'entrainement horaire). L'ensemble de la monture et de la coupole est entièrement piloté par un ordinateur, un second pilotant l'instrumentation et permettant le traitement des données. Le miroir primaire est parabolique, ouvert à f/5. Un Cassegrain f/25 renvoie l'image classiquement par le trou central du primaire; un miroir plan de 1m 10 de diamètre peut être inséré dans le tube pour renvoyer dans le même plan que le Cassegrain une image à f/5 (Montage du type Strand). On a prévu la possibilité ultérieure d'un foyer coudé, et au besoin d'un Cassegrain f/15 et de réducteurs focaux.

L'instrumentation prévue dans une première étape comprend des Caméras Electroniques, une Caméra à comptage de photons, un photomètre très élaboré, un spectrographe pour objets faibles utilisant des récepteurs CCD, et un équipement infrarouge.

On attend la "première lumière" pour la fin de 1979, et on espère que l'instrument sera operationnel dans le milieu de 1980.

THE UK NORTHERN OBSERVATORY (G. Smith)

The UK plans for a Northern Hemisphere Observatory are now materialising at the Spanish "Roque de Los Muchachos" Observatory, on La Palma in the Canary Islands. International agreements on cooperation in astrophysical research, which govern the use of the observatory, were signed between Spain, U.K., Sweden and Denmark on May 26, 1979. Construction of telescope building for Swedish and British telescopes has already begun.

The 1-metre and 2.5-metre telescopes for UK already exist: the 1-metre is approaching completion at Grubb Parsons, and the 2.5 metre is the Isaac Newton, also at Grubb Parsons for installation of a new primary mirror and other improvements.

The designs of the 4.2-metre telescope and its building are almost complete. The alt-azimuth design has led to large economies, both in the telescope mounting and in the reduced dome size. The aim is to have a large, versatile, precision telescope on an excellent site. Versatility is achieved in the alt-azimuth design through the availability of the Cassegrain and two Nasmyth foci, with a simple and rapid interchange, as well as the prime focus.

The computer control of the telescope allows stars to be tracked within $1/4^{o}$ of the zenith. Field rotation will be compensated for in the instrument mountings. Particular attention will be paid in all three telescopes to the control of conditions within the domes, so that full advantage can be taken of the outstandingly good seeing conditions on La Palma.

THE MULTIPLE MIRROR TELESCOPE (D.W. Latham)

(A joint project of the University of Arizona and the Smithsonian Astrophysical Observatory).

On May 9, 1979 the first scientific results from the Multiple Mirror Telescope (MMT) were reported at the MMT Dedication Symposium in Tucson. For these initial applications the six images from the individual 72-inch Cassegrain systems were combined manually. At times of good seeing the individual images appeared smaller than 1 arc second, and the combined image occasionally appeared smaller than 1.5 arc second. The absolute pointing accuracy is already about 5 arc seconds, and undoubtedly this will improve with experience. Even without active correction the images hold together for 10 to 30 minutes, and the tracking is very smooth. Thus many of the aspects of the MMT design, some of them innovative, have already been proven. The six lightweight primaries have been figured accurately to the same radius. The optical support structure, which must carry more than 100 optical elements, performs better than the design goal. The alt-azimuth mount, with its huge azimuth high-angular-contact ball bearing, also works well. The co-rotating building, which has allowed laboratories to be built into the space normally reserved for telescope swing in conventional domes, has proven highly successful. At a total project cost of about 8 million dollars, the MMT falls at least a factor of 2 below the conventional cost curve for a 4.5m optical-infrared telescope. The testing and refinement of the active optics are now the main engineering efforts at the telescope, with 5 nights a week scheduled for the engineers and the weekends for the astronomers. We expect this allocation to be 50/50 by the end of 1979. Until now the scientific applications have used instruments borrowed from other Smithsonian or Arizona telescopes. Major efforts are now going into the development of facility instruments, especially infrared systems and a faint object spectrograph.

THE NASA SPACE TELESCOPE (N.G. Roman)

The Space Telescope is designed to be the largest and best telescope which is economically feasible. The size, with a 2.4m mirror, is the largest which will fit comfortably within the length and weight limit of the Space Shuttle. The guaranteed quality of 0.075λ wave-front error for the total system should be achievable with care, an accuracy of 0.05 λ at λ = 6300Å is hoped for and expected. Even the guaranteed performance will give a resolution of better than 0.1 arc second throughout the visible and ultraviolet. The f/24 Ritchey-Chretien optics can be used to feed any one of five focal plane instruments. On the first launch, these will be two cameras, two spectrographs and a photopolarimeter. Instruments may be replaced in orbit by new instruments or the entire telescope can be returned to the ground for refurbishment. The cameras cover fields of view from 4 arc seconds to 2.7 arc minutes with angular resolutions increasing from 0.1 arc sec to the highest the optical system can provide in the smallest f.o.v. The spectrographs provide resolutions of 2×10^4 and 1.2×10^5 between 1050 and 3200Å, a resolution of 10^3 between 1050 and 7500Å and a resolution of 10^2 between 1200 and 7500Å. The photometer will have a time resolution of 16 μsec with a spectral resolution up to about 20Å.

NEXT GENERATION TELESCOPE PLANS (D.N.B. Hall)

We are presently in a period where a substantial number of ground-based optical and infrared telescopes are being commissioned; a number of these have been described in this session. However there are compelling arguments that astronomers will need very much larger ground-based optical/infrared collecting area in the future. Dramatic improvements in observational capability in other wavelength regimes have invariably required follow-up O/IR observations, often involving long integrations on faint sources; this trend will undoubtedly escalate with the operation of Space Telescope, various infrared satellites, the VLA and large millimeter wave dishes. Yet this explosion in demand will come just when many potential improvements associated with improved efficiency, better detectors and new auxiliary instrumentation will have reached the point of diminishing returns; at this point one can only gain in observing time or limiting magnitude by going to correspondingly larger aperture.

Careful evaluation of the types of astronomical observations for which such a telescope is most suited leads one to the conclusion that the scientific gains come solely as a result of increased area and can be achieved equally well by a single filled aperture or an array of smaller telescopes.

On the other hand, various technical requirements are best met by an array of telescopes in the 5-10 meter size range; these may be independent or attached to a common mount (as in the MMT). With elements smaller than 5 meters, diffraction becomes a factor in image quality in the infrared and the large number of instruments may well be prohibitive. Above 10 meter aperture there arc problems matching to foreseeable detectors and spectrographs.

If one scales the cost of conventional large telescopes to a 10 meter aperture using the $D^{2.7}$ power law then costs are prohibitive. However use of thin, possibly segmented, primary mirrors and fast focal ratios hold promise of very significant cost savings and there are a number of groups in the U.S. now working on designs for telescopes of this class.

Working Group on Photographic Problems (Chairman: J.L. Heudier)

Two sessions of the Working Group were held during the General Assembly meeting on August 17, 1979. The first session was held jointly with the Working Group on Data Processing, and was attended by about 100 astronomers. The second session, which included the Business Session of the Working Group on Photographic Problems, was attended by about 60 astronomers.

In the joint session, Dr I. King described the responsibilities of the Working Group on Data Processing as including the receipt, reproduction and distribution of materials and information on data processing. Distribution of information is in the form of a newsletter. Requests to be included in the mailing list for the newsletter should be addressed to Dr R. Albrecht, Institute for Astronomy, Turkenschanzstrasse 17, Vienna, Austria. Two review papers were presented in this session. In "The Present Status of Astronomical Photography" J.L. Heudier discussed the photographic emulsion as a means of counting photons; the importance of S:N considerations, calibration standards, high quality processing, storage and handling requirements, and currently-available techniques of hypersensitization and copying. Dr R. Albrecht discussed "The Measurement and Reduction of Photographic Plates", using as an example the Vienna PDS 1000 system to highlight some of the problems of handling large quantities of information automatically. Attention was drawn to the importance of the communication between such a system and its user - i.e. "user friendliness".

Papers presented in the second session were:- Electronography with the Griboval camera - H.J. Smith and N. Griboval; Calibration - I. Furenlid; UBV Dye Transfer Colour Imagery of Galaxies - J.D. Wray; Hypersensitization of Colour Films - A.G. Smith. It is expected that most of the papers read in these sessions will be published in the AAS Photobulletin.

In the Business Session, Dr R.M. West (Chairman) reported on the activities of the Working Group since the Grenoble meeting. Publications of the Working Group have included the Proceedings of the Grenoble meeting and a Bibliography of Astronomical Photography. A Workshop on Modern Techniques in Astronomical Photography held in Geneva, May 16-18 1978 was well attended, and the Proceedings of that Meeting were published in July 1978. Copies of these Proceedings and of the Working Group publications have been sent throughout the I.A.U. countries to some 400 people on the Working Group mailing list. At a meeting of the Organising Committee held in Geneva it was agreed that the American Astronomical Society Photobulletin should be the recommended journal for publication of papers relating to astronomical photography.

In view of the formal recognition of the Working Group on Data Processing it was necessary to amend the Resolution adopted by the Photographic Working Group during the Grenoble General Assembly, by deleting the clause "and extraction of data from exposed plates".

There are two planned or proposed meetings on astronomical photography. The first, organised by AAS Working Group on Photographic Materials, will be held in San Francisco on January 18 and 19, 1980. There will also be a meeting in Europe early in 1981, probably in Nice.

The membership of the new organising Committee was agreed as follows:- Chairman: Dr J.L. Heudier, Observatoire de Nice, B.P. 252, F-06007 Nice-Cedex, France. Secretary: Miss M.E. Sim, Royal Observatory, Blackford Hill, Edinburgh EH9 3HJ, UK. Members: R.D. Cannon (UK); O. Dokuchaeva (USSR); J. Grygar (Czechoslovakia); D. Latham (USA); D.F. Malin (Australia); A. Millikan (USA); K.R. Sivaraman (India); A.G. Smith (USA); B. Takase (Japan); R.M. West (Denmark). Consultant: W.C. Miller (USA).

A questionnaire concerning the transport of photographic plates had been sent to everyone on the mailing list, resulting in 71 replies from 29 countries. The main problems arise during shipment of the plates from the manufacturer to the customer, when the plates are not under the control of either the supplier of the user. E.J. Hahn and G. Brown of Eastman Kodak reported on their attempts to identify and rectify some of the problems. A Kodak report "Technical Information for Ordering Kodak Spectroscopic Films and Plates" was circulated at the meeting, and will be sent to all Kodak agents and plate users. As a result of internal reorganisation at Eastman Kodak, Mr Gordon Brown has now inherited from Mr E.J. Hahn the responsibility for spectroscopic plates and films, filters and attenuators, and consequently problems which would formerly have been addressed to Mr Hahn should now be addressed to Mr Brown.

High Angular Resolution Techniques (Chairman: Dr J. Davis)

The following papers were presented on 17th August:-

A REVIEW OF THE VERY LONG BASELINE AMPLITUDE INTERFEROMETER AT THE UNIVERSITY OF MARYLAND (D.G. Currie)

An amplitude interferometer based on the design of an earlier interferometer, but with two independent telescopes, is being built at the University of Maryland. The interferometer is being set up at a site on the grounds of the optical research facility of the Goddard Space Flight Center in order to carry out a test program. For this purpose the baseline, parallel to the Earth's axis of rotation, has been fixed at 3m but the instrument is designed to work in a regular observing program with baselines up to 50m at a permanent observing site.

AN 11M PROTOTYPE MODERN MICHELSON STELLAR INTERFEROMETER (J. Davis)

The Chatterton Astronomy Department of the University of Sydney developed and operated the Stellar Intensity Interferometer at Narrabri for many years and, based on experience gained with that instrument, proposed in 1971 a preliminary design for a very large stellar intensity interferometer. The proposed VLSI1 would have a limiting sensitivity V(limit) $\sim$7.3 and baselines up to $\sim$ 2km. There is no doubt that this instrument could be built and that it would work. However, it would be large, expensive, and the sensitivity would be marginal for some of the most interesting of the programmes listed in the introduction to this session.

Although the potential of amplitude interferometry is still to be proven, it is inherently more sensitive than intensity interferometry, and promises to be less expensive. For these reasons the VLSI1 has been put on ice while the feasibility of a modern form of Michelson's interferometer is explored by building a prototype instrument based on ideas developed by R.Q. Twiss.

The prototype interferometer will have coelostats to direct light horizontally to a central laboratory from each end of a horizontal 11m North-South baseline. In the interferometer the wavefront distortions induced by the atmosphere will be handled by (i) restricting the aperture size to select essentially flat portions of wavefront; (ii) removing wavefront tilts by means of piezo-electrically actuated tilting mirrors; and (iii) removing the effects of randomly varying phase by rapid sampling of the interference and appropriate data handling.

A laser interferometer will be used to monitor internal paths within the interferometer and control a moving retroreflector to equalise the optical paths to the plane of interference.

The design of the interferometer is essentially complete and most of the optical components are in hand. The optical path-length compensation, angular seeing compensation and coelostat guidance servo-systems are being constructed and tested in the laboratory. A potential site is being tested and it is hoped to install and align the instrument during 1980. The prototype stage is aimed solely at establishing by observations whether fringe visibility can be measured with an accuracy of $\sim$ 2% through the atmosphere using ζ Pup and α CMa as standard reference sources. It is hoped to make these observations during 1981 and to establish optimum instrumental parameters etc. It is intended that the prototype interferometer will ultimately form the heart of a major <u>high</u> resolution instrument.

RECENT RESULTS IN TWO-TELESCOPE INTERFEROMETRY AT CERGA (A. Blazit)

The small two-telescope interferometer built by Labeyrie has been operational for three years. It consists of two 26cm telescopes rolling on tracks on a North-South baseline ranging between 5 and 35m. The recombination optics and the TV cameras are mounted on a carriage which is moved by a screw controlled by a microcomputer for fringe tracking. Observations are made in white light or in light dispersed by a direct view prism.

Until now, contrasts have been estimated visually, but Laurent Koechlin has built a contrast-measuring device which is being implemented on the interferometer. It measures the contrast of the fringes of each TV image and a mini-computer gives their histogram. The histogram is compared with those obtained for reference sequences and the real contrast can be derived with a precision of 5%.

Observational results:-

(i) We have measured 8 stars and observed fringes on 9 non-resolved stars with magnitudes to 3.9.

(ii) We are determining a new orbit for Capella and our measures of separation now have a precision of 0".4x10^{-3} arc. We expect to detect perturbations of this orbit due to a planet or interstellar matter.

A LARGE APERTURE SYNTHESIS ARRAY PROJECT (A. Blazit)

We have started the construction of a large aperture synthesis array at CERGA. Several telescopes, 1.5m diameter, rolling on baselines up to 1km will make up this array. Because we need a large number of telescopes free of vibrations and movable on tracks, we have designed a cheap, lightweight telescope which does not need a dome. It consists of a concrete sphere, 3.5m diameter, with a cylinder inside. The mount is made of three rings. The optical layout is Cassegrain-Coudé with a 1.5m primary mirror and a plane tertiary mirror to give a horizontal coudé beam. The first telescope is complete and undergoing tests and the second is under construction.

RECENT DEVELOPMENTS OF THE TWO COLOUR REFRACTOMETER TECHNIQUE (D.G. Currie)

For astrometry it is necessary to correct for the atmosphere, otherwise measurements can be little better than current PZT measurements of 0".15.

A two colour refractometer has been developed to study the possibility of correction by a dispersion technique. The precision is expected to be between 0".1 and 0".02 for a two-minute observation. The accuracy per observation is expected to be essentially the same, i.e. the systematic or external errors are expected to be less than 0".003 for the PZT.

The method uses a photon-counting quadrant detector system (the automatic guide system or AGS). The AGS has been demonstrated on the 48-inch telescope at the Goddard Optical Research Facility (GORF) to be photon noise-limited to 0".003. Observations are conducted by the use of a rotating wheel with light filters (about 3500Å and 6500Å) which are changed 480 times per second. A prism system nulls the dispersion and the setting of this null position is read as the amount of dispersion.

Test measurements have been conducted and reduced and are photon noise limited (standard deviation of 0".003 in 100 sec). Initial observations of refraction have been conducted but not analyzed. A full program will be commenced soon.

ASTROPHYSICAL APPLICATION OF SPECKLE INTERFEROMETRY (R. Foy)

In collaboration with Labeyrie's group, we have carried out two kinds of program using speckle interferometry: (i) resolution of binaries, and (ii) angular diameter measurements.

We observe spectroscopic binaries both to derive stellar masses, and to check stellar evolutionary tracks, by way of composite spectrum binaries. We found that ζ^1 Reticuli is a binary; this explains the anomalies of this star. The two components of χ Dra have low masses with respect to their temperatures; this could be due to a helium deficiency. We resolved the composite spectrum binary δ Sge, and confirmed that the orbit is highly inclined; an atmospheric eclipse should occur next spring (1980).

The discrepancy, by a factor of ten, about the gravity of Arcturus, raised by Mäckle et al (1975) has been explained by Martin (1977). Giving a large weight to the angular diameter measurement of Arcturus by Blazit et al (1977), she proposed to return to the higher value of the gravity and to a mass of 1 solar mass.

Speckle observations can be used to probe the atmospheres of Mira Variables - we found that the angular diameter of Mira and R Leo are markedly larger in the strong bands of Titanium oxide than outside. The difference reaches a factor of two between 7000 and 7100Å. Using model atmospheres by Tsuji, we showed that this difference is due to the very large opacity coefficient of TiO: the optical depth in TiO bands is unity far above the stellar surface.

We noted an increase of the diameter of Mira during maximum of November '78 with respect to maximum of December '76, which implies that Mira was cooler in November '78. This agrees with the fact that Mira was then fainter by 1.5 magnitudes than in December '76.

IMAGE RECONSTRUCTION WORK AT THE AIR FORCE GEOPHYSICS LABORATORY (S.P. Worden)

Several aspects of astronomical image reconstruction work are being pursued at the Air Force Geophysics Laboratory. In collaboration with contractors at Harvard the application of speckle interferometry to studies of the solar surface is being investigated. The use of an active optics unit developed by ITEK for similar high resolution solar observations is also being studied. In concert with groups at the University of Arizona and Lockheed, systems have been developed to observe faint objects and, in some cases, completely reconstruct images using speckle interferometry techniques. Efforts to adapt the Center for Astrophysics/Arizona Multiple Mirror Telescope for speckle imaging work have also begun.

INTERFEROMETRY FROM SPACE (A. Blazit)

Ground-based interferometers have already provided results on stars at visible wavelengths, but for extragalactic objects atmospheric fluctuations will limit the integration time and prevent the observation of objects of magnitude fainter than 15, such as quasars and nuclei of Seyfert galaxies. A space interferometer would have enough angular resolution and sensitivity to observe faint extragalactic objects in the visible and UV. It will also give dimensions and perhaps mapping of stars and systems of stars over a large range of wavelengths. The most important objects are: Stellar objectives: Extension of layers of cool supergiants/hot stars; Star formation; Structures in molecular clouds; Structures in the discs of stars; Close binaries with transfer. Extragalactic objectives: Active regions in nuclei; Shocks in dusty regions of nearby galaxies; Quasars, Seyfert nuclei. Astrometry: Absolute references; Obscure companions.

The design we have selected for the space interferometer is the "flute" tube. Two telescopes, 1m in diameter, move inside a structural tube with a certain number of viewing holes. Several sections of 15m could be assembled together in orbit. Extensions to 100m or more appear feasible.

The infrared range could be covered with this instrument if the central package of optics and detectors is made interchangeable. Observation of faint objects will require offset guiding on the fringes of a bright star. It could be realised with 3 telescope carriages, and a triple interferometer. Three fields could be observed simultaneously 120° apart on a great circle of the celestial sphere. Two guide stars could be used for absolute stabilisation, while observing a faint object. This project has been submitted to E.S.A.

A SPACE APPLICATION OF THE VERY LONG BASELINE AMPLITUDE INTERFEROMETER (D.G.Currie)

Certain of the scientific and technical apsects of the operation of a Very Long Baseline Amplitude Interferometer (VLBAI) (similar to the VLBAI currently being developed by the University of Maryland for ground-based operations) were discussed.

The Space-based VLBAI (SVLBAI) will initially provide an astrophysical capability to observe and resolve much fainter objects than may be observed from the ground. Of particular interest are the cores of QSO's, BL Lac objects, Seyfert galaxies, and the inner core of "conventional" galaxies. It should also provide an astrometric capability to study fainter stars, with higher accuracy, than may be done from the ground. This would include parallax studies, the search for extra-solar planets and a study of the relation between the FK4 catalog and the QSO/BL Lac "inertial" frame of reference.

In the first instance a ten metre system (i.e. a few metres separation between the apertures) is being considered. This will serve as a prototype for later expansion to thirty and one hundred metres. Each of the end mirrors would consist of four segments, each about 0.75m diameter, which are independently moveable. Since these need point only with "diffraction" accuracy, not "interferometric" accuracy, the existing Automatic Guiding System provides the requisite pointing accuracy. The object acquisition is done with a specially modified, thinned CCD. The interferometric delay is maintained with the AGS for the low frequency and a gyro system for the high frequency. The measurement of the differential delay between two stars is done with the fringe counting laser system. A system with a graphite inner truss and a heat-pipe/multi-layer insulation outer structure guarantees sufficient insulation from the thermal effects of the sun.

The expected precision in a planetary search program or a study of the parallax of fainter stars (15th magnitude) would be 0.005 milliseconds of arc for a 1000 second integration and a 100 metre system. For the QSO studies, the precision would be 1 millisecond of arc for a 25th magnitude object for the same conditions. A preliminary study of the accuracy questions indicates that the obvious systematic errors are not incompatible with the above examples of precision.

Working Group on Infrared and General Techniques (Chairman: J. Ring) (22 August)

The following papers were given:

EARLY OBSERVATIONS WITH THE UK INFRARED TELESCOPE AND ITS INSTRUMENTS (T.J. Lee)

Commissioning work on instruments and some observations for astronomical research programmes have taken place during the later stages of commissioning. Some instrumentation was described together with its performance. Examples of observations made were shown. Some of the astronomy to be done later in 1979 was outlined.

A CASSEGRAIN, SCANNING FABRY-PEROT INTERFEROMETER WITH AN IMAGING DETECTOR (P. Atherton, D.J. Axon & K. Taylor)

TAURUS, a new Fabry-Perot imaging device, designed to obtain complete seeing-limited radial velocity field maps of extended emission-line sources, was described. A servo-controlled Fabry-Perot is used with a focal reducer and a 2-dimensional photon-counting (area detector) system, to obtain the velocity information. The system has been fully tested and the first observations of the highly complex velocity field of the irregular galaxy M82 were used to illustrate the power of the instrument.

A CASSEGRAIN, FABRY-PEROT/ECHELLE SPECTROMETER (B. Bates)

An outline was given of a Fabry-Perot echelle spectrograph to be used for first observations in Autumn 1979. The instrument is designed to record a large number of spectral elements simultaneously at a resolving power $\approx 3 \times 10^5$. The interferometer is combined with an echelle order sorter to provide a compact spectral format suitable for image intensifier recording. The technique is based on that employed successfully during a programme of UV interference spectroscopy from a moderately stabilized balloon-borne platform.

AN INTERFEROMETRIC STELLAR-OSCILLATION SPECTROMETER (A. Forrest & J. Ring)

The usefulness of accurate measurements of the radial velocities of stars was discussed, together with the limitations to the accuracy of such measurements imposed by the properties of current instruments. Two new interferometric methods were described; the first employed a Michelson interferometer preceded by a narrow-band interference filter centred on a stellar absorption line. The path-difference in the interferometer was fixed at about 5mm and servo-controlled with the aid of a stable laser. The flux of starlight emerging from the two outputs was monitored - changes in the difference in intensity of the two beams was a measure of a change in the radial velocity of the stellar atmosphere and was used to detect stellar oscillations. The instrument had been tested on the Sun and the 300s oscillation had been detected with an accuracy of a few metres per second in the measurement of the radial velocity amplitude.

The second method involved the use of a solid Fabry-Perot interferometer, the thickness of the plate being controlled by temperature using a laser as a wavelength reference. The wavelength of a stellar absorption line should be measured to an accuracy which would allow radial velocity changes of a few metres per second to be detected. This instrument was still under construction.

POSSIBLE ADVANTAGES OF AN ALT-ALT TELESCOPE SYSTEM FOR STELLAR SPECTROSCOPY (T. Dunham Jr)

Large and heavy equipment for analyzing light from stars and planets, particularly high resolution spectrographs and systems for Fourier spectroscopy, must be carried on a rigid foundation. This requires a telescope that brings light to a fixed focus. Ordinarily, an asymmetrical equatorial coudé mounting is employed for this purpose. But this design requires heavy castings and expensive machine work to provide adequate rigidity. It also requires four mirror reflection.

An alt-alt mounting (AURA Engineering Technical Report No.38, KPNO 1971 and A. Labeyrie, Ap.J. Letters, L71, 1975) involves a simpler mechanical yoke design. Cost should be considerably less, since construction does not require special machinery. And only three reflections are required to deliver light to a fixed focus.

The two axles of an alt-alt telescope must be driven at continuously changing rates by a computer which sends pulses to stepping motors, or controls the rates of D.C. motors. Such computer controls have recently been used with success for large alt-azimuth telescope mountings. Cost could be moderate if encoder feedback is not included, and if an observer provides visual-manual guiding. Performance should be better than with most large telescopes ten years ago.

For small telescopes, the computer drive for an alt-alt telescope will make the total cost for the instrument more than for an equatorial coudé system. But, since the cost of a computer drive is almost independent of the aperture of the telescope, an alt-alt mounting should be less expensive than an equatorial coudé telescope for large apertures. It seems likely that costs for the two designs may be approximately equal for an aperture of about 80cm. If so, then the cost, in terms of photons delivered per second per dollar invested, should be approximately equal for telescopes of about 70cm aperture, because the alt-alt system requires one less reflection.

An alt-alt mounting is being seriously considered for a 50-inch telescope to feed light to a stationary stellar spectrograph in Australia. Comments and suggestions from members of Commission 9 will be welcomed.

SECOND-ORDER STATISTICS FOR IMAGE CLASSIFICATION (M.L. Malagnini)

A new procedure was described for image classification using a computer-controlled scanner, designed in particular for the identification of globular clusters in M31. The approach is based on the analysis of second-order statistics of the grey-level distributions of the images of different categories of object. Classification parameters are derived from a test-set of known objects. Results were presented for a few hundred objects which showed the potentialities of the method.

COMMISSION 10: SOLAR ACTIVITY (ACTIVITÉ SOLAIRE)

PRESIDENT: G. Newkirk, Jr. SECRETARY: A. Bruzek

Scientific Sessions

I. COLLOQUIA AND SYMPOSIA

The following colloquia and symposia of specific interest to Commission 10 were held in association with the 17th General Assembly:

- IAU Colloquium No. 51, "Convection and Turbulence in Stellar Atmospheres," London, Ontario, 27-30 August 1979.
- IAU Symposium No. 86, "Radio Physics of the Sun," College Park, Maryland, 7-10 August 1979.
- IAU Symposium No. 91, "Solar Interplanetary Dynamics," Cambridge, Massachusetts, 27-31 August 1979 (co-sponsored by SCOSTEP and COSPAR).

II. JOINT DISCUSSIONS

The Commission co-sponsored the following Joint Discussions, the texts of which will appear in Highlights of Astronomy:

- Large Scale Velocity Fields on the Sun - Co-chairmen, M. Stix and P. Gilman, 21 August 1979.
- Very Hot Plasmas in Circumstellar, Interstellar, and Intergalactic Space - Chairman, W.I. Axford, 22 August 1979.
- Physics of the Chromosphere-Corona-Wind Complex and Mass Loss in Stellar Atmospheres - Chairman, D. Mihalas and Co-Chairman, R. Bonnet, 16 August 1979.

III. JOINT MEETINGS

A. "The MHD of Sunspots," Joint Meeting of Commissions 10, 12, and 44, 20 August 1979 (C. Zwaan, Reporter)

The joint meeting, organized by J.-C. Hénoux, E.N. Parker and C. Zwaan (chairman), concentrated on sunspots as the best observed magnetic features with emphasis on the dynamical phenomena present in spots. Three review papers of 25 minutes each were prepared with input sent to the reviewers at the invitation of the organizing committee. Each of the reviews was followed by a discussion of 25 minutes with a few brief and informal contributions included in the discussions.

In the first review, "Dynamic Phenomena in the Visible Layers of Sunspots," R.L. Moore summarized the observational data on non-oscillatory, oscillatory, and running-wave phenomena in the photospheric and chromospheric structure of spots. The reviewer illustrated the relations between such dynamic features as umbral flashes and penumbral waves by means of two movies, one prepared by Big Bear Observatory and one by Lites, White, Athay, Chipman, and Packman from Sacramento Peak data. During the discussion, R.J. Bray showed recent observations of umbral dots, and V. Gaizauskas drew attention to the existence of very intense umbral flashes of very short duration.

In the second review, "The Transition Region and the Corona Above Sunspots," F.Q. Orrall discussed the properties of the non-flaring corona in active regions, quasi-static and dynamic coronal loop models, and mechanisms for driving mass flow. This review brought into perspective the wealth of new information revealed by observatories in space. Special emphasis was given to the results of the Skylab Workshop on Active Regions, which will publish its proceedings in 1980. During the discussion, C. Jordan mentioned red shifts in the Si I lines near the series limit which were observed over a small area in a spot. She interpreted the shifts as resulting from the quadratic Stark effect probably caused by the V x B field in a very dense ($N \sim 10^{14}$ to 10^{15} cm^{-3}) strong field ($B \sim 3000$ gauss) region.

The third review, by H.C. Spruit, "MHD of Sunspots, Theoretical Aspects," dealt with:

(1) Problems concerning the equilibrium structure of spots--recent numerical results for the field configuration by Schmidt and Wegmann were mentioned. The problem of maintaining pressure equilibrium in a spot where blocking of the heat flux occurs is still largely unsolved.

(2) Stability problems--the conditions for stability against fluting (division) and for occurrence of overstable oscillations were reviewed. It was pointed out that no calculations of a compressible stratified spot for overstability are available except for the work by Syrovatskii and Zhugzhda.

(3) Energy transport in the spot--the problem is to explain the brightness of the umbra (20% bolometric). Non-radiative energy transport is indicated by empirical umbral models. Overstable "turbulence" was considered the best candidate for the transport mechanism.

(4) Fine-structure--such as dots, flashes, running waves--emphasis was placed on the explanation of running umbral and penumbral waves as trapped (surface) magnetoacoustic gravity waves as has been described by Nye and Thomas.

During the general discussion, the relations between umbral oscillations, flashes, and penumbral waves were considered. Apparently, umbral resonant oscillations may excite running penumbral waves, however; it is clear that a satisfactory explanation of these phenomena requires the development of a dynamical model of the deep structure of a sunspot. The question of the presence or absence of umbral oscillations during the evolution of a typical sunspot has yet to be answered as is that of the detailed relationship between specific features in the fields of sunspots and the feet of coronal loops. In many cases, the sunspot field appears to open into interplanetary space. F. Albregtsen showed evidence for two new parameters which vary with the solar cycle--umbral brightness and the occurrence rate of X-ray bright points.

Publication of the three review papers is planned for an issue of Solar Physics.

B. The Solar Maximum Year, 1 August 1979 to 28 February 1981
Joint Meeting of Commissions 10, 12, 14, and 44, 15 August 1979
(Z. Svestka, reporter)

After opening by the chairman, Monique Pick, the Project Leaders described the aims and organization of the three components of SMY: Flare Build-up Study (FBS, presented by Zdenek Svestka), Sudden Energy Release from Flares (SERF, presented by David M. Rust), and Study of Travelling Interplanetary Phenomena (STIP, presented by S.T. Wu on behalf of the project leader, Murray Dryer).

1. Flare Build-up Study (Z. Svestka, Chairman). One of the three components of the SMY is the FBS, whose objective is to carry out a comprehensive array of observations before and during the eruption of a flare in order to delineate and understand the processes which lead to the storage of energy and its ultimate release in the flare. The array of instruments which will be involved in the FBS through the cooperation of institutions and individuals around the world include too many items of both ground and space instrumentation for us to discuss here. Many of the strategies attacking the major scientific problems and for coordinating the use of these many instruments were formulated at the Falmouth Workshop in May 1979.

A key element of this strategy is that all participating experiments, on the ground as well as in space, will focus all their efforts on the detailed observation of a particular active region. Clearly, such intense observing cannot be sustained continuously but must be restricted to specific periods, designated FBS-ALERTS. The ALERT periods currently designated are:

May-June 1980
1-15 September 1980
16-31 October 1980

The selection of the particular active region(s) to be observed during the ALERT is the responsibility of the coordinating center established at Meudon under the direction of P. Simon. The information that a particular region has been selected is transmitted to all participants in the form of a telex FBS-ACTION and intense observing will continue until the region ceases to be of interest and the ACTION is discontinued. During the May-June 1980 period 4 to 6 ACTION intervals are to be expected.

In order to be fully prepared and to test the procedures and communication systems, a trial ALERT period was scheduled for 1-15 May 1979 and an evaluation meeting was held in Montreal on 12 August 1979. Of course, the trial revealed some areas requiring improvement; but it was considered generally successful.

To assist the individuals and institutions participating in the FBS the organizing committee has prepared an FBS-Manual containing information on scientific programs, communications, participants, Cooperative Observing Sequences, etc. We plan to distribute this manual to participants in October 1979.

2. Study of the Energy Release in Flares (D.M. Rust, Chairman). From 26 February to 1 March 1979, a workshop was held in Cambridge, Massachusetts, U.S.A., to devise the scientific program and objectives for the Study of Energy Release in Flares (SERF). Three teams of scientists dealt with the theoretical problems of (1) thermalization of energy, (2) acceleration of particles, and (3) origin and effects of mass motions. As a result of this workshop, a set of Solar Maximum Year (SMY) Collaborative Observing Sequences (COS's), to be followed by both ground-based and spacecraft-borne observatories, was established. The full proceedings of the workshop are to be found in UAG-72 published by the World Data Center A (WDCA) for Solar-Terrestrial Physics, NOAA, Boulder, Colorado.

One of the major problems to be solved is to determine where is the primary energy release site and what are its characteristic dimensions. Most flare models (cf. Svestka, 1976 and Sturrock, 1979, for a review) place the energy release site in the corona, although there is growing evidence for substantial local energy release at photospheric depths (Machado et al., 1978; Emslie and Machado, 1979; Canfield et al., 1979). It is possible that outward moving material ejections trigger the energy release with the energy release site above the photosphere. However, Colgate (1978) has adduced evidence for an energy release site below the photosphere. While the characteristic dimensions of Hα bright points and magnetic reconnection theories of energy release (e.g., Sturrock, 1968; Spicer, 1977a,b) point to flare dimensions of order arc seconds or less, it is extremely difficult to release the total amount of observed flare energy in such a small volume; this empirically suggests a much larger or denser energy release site.

A second major problem to be solved is to decide whether the bulk of released energy manifests itself as macroscopic ejecta or as the acceleration of particles and thermalization of solar plasma. The results of Canfield et al. (1979), Ramaty et al. (1979), and Rust et al. (1979) seem to suggest the former. The mode of thermalization of the lower atmosphere is still uncertain; the accepted choice here is between the collisional (and ohmic) heating produced by the injection of a beam of nonthermal particles (e.g., electrons) (Brown, 1973; Emslie, 1978) and the heating produced by macroscopic phenomena such as thermal conduction (Svestka, 1973) and shock fronts (Craig and McClymont, 1976). Two related questions are whether hard X-ray bursts in large events (e.g., Hoyng et al., 1976) are thermal or non-thermal in origin and what is the mechanism responsible for white-light flares.

A third problem of major concern is that of the acceleration of energetic particles. Very few theoretical models (e.g., Sturrock, 1968) of the acceleration process exist; and, as a result, the questions still unanswered at this time are necessarily very general. General features requiring definition are:

(i) The accelerated particle population: its composition, energy spectrum, and temporal behavior;

(ii) The characteristics of the acceleration region: its size, temperatures (both ion and electron), density, magnetic configuration, etc.;

(iii) The responsible accelerating agent, e.g., shocks, plasma turbulence, direct electric fields, etc.;

(iv) The effect of the interplanetary medium on ejected particle streams.

Finally, the true volume, shape, and energy contained in mass motions needs to be determined in order to discriminate between self-driven expanding magnetic flux tubes and motions induced by gasdynamic forces such as shocks. Due to the rather poor time resolution hitherto available (a limitation which we believe to be largely overcome during SMY), it is not clear whether mass ejection begins before or after the flare impulsive phase.

In summary, there are fundamental questions to be confronted in the three principal study topics of the SERF: thermalization, particle acceleration, and mass motion. The instruments available in the 1979-1981 period will provide the data for substantial progress in these areas.

REFERENCES

Brown, J.C.: 1973, Solar Phys. 31, p. 143.
Canfield, R.C., Brown, J.C., Brueckner, G.E., Cook, J.W., Craig, I.J.D., Doschek, G.A., Emslie, A.G., Hénoux, J.-C., Lites, B.W., Machado, M.E., and Underwood, J.H.: 1979, in (P.A. Sturrock, ed.) "Solar Flares," "Proceedings of the Skylab Solar Flare Workshop," Colorado Assoc. Univ. Press, Boulder, Colorado.
Colgate, S.A.: 1978, Astrophys. J. 221, p. 1068.
Craig, I.J.D. and McClymont, A.N.: 1976, Solar Phys. 50, p. 133.
Emslie, A.G.: 1978, Astrophys. J. 224, p. 241.
Emslie, A.G. and Machado, M.E.: 1979, Solar Phys. 64, 129.

Hoyng, P., Brown, J.C., and van Beek, H.F.: 1976, Solar Phys. 48, p. 197.
Machado, M.E., Emslie, A.G., and Brown, J.C.: 1978, Solar Phys. 58, 363.
Ramaty, R., Colgate, S.A., Dulk, G.A., Hoyng, P., Knight, J.W., Lin, R.P., Melrose, D.B., Patris, C., Orrall, F.Q., Shapiro, P.R., Smith, D.F., and van Hollebecke, M.A.I.: 1979, in (P.A. Sturrock, ed.) "Solar Flares," "Proceedings of the Solar Flare Workshop," Colorado Assoc. Univ. Press, Boulder, Colorado.
Rust, D.M., Hildner, E., Dryer, M., Hansen, R.T., McClymont, A.N., McKenna-Lawlor, S.M.P., McLean, D.J., Schmahl, E.J., Steinolfson, R.S., Tandberg-Hanssen, E., Tousey, R., Webb, D.F., and Wu, S.T.: 1979, in (P.A. Sturrock, ed.) "Solar Flares," "Proceedings of the Solar Flare Workshop," Colorado Assoc. Univ. Press, Boulder, Colorado.
Spicer, D.S.: 1977a, NRL Report 8036.
Spicer, D.S.: 1979b, Solar Phys. 53, p. 305.
Sturrock, P.A.: 1968, in (K.O. Kiepenheuer, ed.) "Structures and Development of Solar Active Regions," "I.A.U. Symp. 35," p. 471.
Sturrock, P.A. (ed.): 1979, "Solar Flares," "Proceedings of the Skylab Solar Flare Workshop," Colorado Assoc. Univ. Press, Boulder, Colorado.
Svestka, Z.: 1973, Solar Phys. 31, 389.
Svestka, Z.: 1976, "Solar Flares," D. Reidel Publ. Co., Dordrecht, Holland.

3. Study of Travelling Interplanetary Phenomena (M. Dryer, Chairman). The Study of Travelling Interplanetary Phenomena project (STIP) was established in 1973 under SCOSTEP auspices. Since that time, with the cooperation of IAU and COSPAR, it has proceeded in its mission of encouraging informal cooperation on an interdisciplinary and international scale on the study of various aspects of the quiet and disturbed solar wind. The participants include spacecraft experimenters, ground-based observers and theoreticians concerned with solar physics, interplanetary scintillations, solar radio astronomy (both passive and active, as in the cases of direct solar magnetic fields and plasma waves), cosmic ray particles (solar and galactic), solar wind interactions with celestial obstacles, and solar wind plasma interaction with interstellar gases.

STIP is currently participating as one of the subprograms of the Solar Maximum Year which will extend from 1 August 1979 to 28 February 1981. STIP Interval VI (15 April-15 June 1979), declared in 1978 as a Trial Period for the SMY, has proved valuable in that a white-light coronal transient was observed by the satellite P78-1 on 8 May 1979 with its shock wave observed (based on preliminary study) at 0.78 AU by the plasma probe on Pioneer-Venus Orbiter. A search for the transient's "signature" at 0.3 A.U. is presently being made by Helios 2 experimenters.

The period of August through September 1979 was declared one year in advance as STIP Interval VII. The long advance notice was necessary in order to give national agencies sufficient time to consider advance planning for the deep space tracking of Pioneer II, Voyager I, Voyager 2, Helios 1, Helios 2, and Pioneer-Venus Orbiter.

A number of other activities have been carried out as part of STIP. First, a special issue of Space Science Reviews was published in 1976 devoted to review articles on the exceptionally well-observed series of solar flares and their interplanetary consequences during August 1972. Secondly, a STIP Symposium was cosponsored by SCOSTEP, IAU and COSPAR in Varna, Bulgaria, in June 1975. Thirdly, another STIP Symposium (again sponsored by these organizations and dedicated to the memory of the Dutch astronomer L.D. de Feiter) was held in Tel-Aviv, Israel, in June 1977, with subsequent publication of invited papers by D. Reidel (The Netherlands) and contributed papers by AFGL. Some of the impetus for papers at these symposia was provided by the declaration of a series of special two-month "STIP Intervals" during which tracking facilities of the various national agencies were requested for coordinated studies. Contributions based on one of these Intervals were published by World Data Center A for Solar-Terrestrial Physics

(Boulder) in a UAG-61 Report entitled "Collected Data Reports for STIP Interval II: 20 March-5 May 1976." Another UAG Report for STIP Interval IV (15 October-15 December 1977) is currently in press. Most recently, IAU Symposium 91 on Solar and Interplanetary Dynamics was convened at Harvard University in August 1979 with the active participation of STIP participants among the 145 registrants from more than 15 countries. Preparation of the Proceedings is currently in progress for publication by D. Reidel. SCOSTEP and COSPAR co-sponsored this Symposium. STIP participants have also planned a Symposium (on Solar Radio Astronomy, Interplanetary Scintillations and Coordination with Spacecraft) which will be held in Culgoora, Australia, in November 1979, under the sponsorship of the same organizations. Tentative plans for a STIP Symposium on Shock Waves in the Solar Corona and Interplanetary Space in Smolenice, Czechoslovakia, in June 1980, are currently under consideration by the Slovak Academy of Sciences.

Following these introductory talks, Ernest Hildner described the experiments aboard the Solar Maximum Mission Satellite (SMM) and their scientific aims. This spacecraft has been developed as the main source of data from space during the SMY, and great efforts are being made to provide good coordination between observations aboard the SMM and on the ground. A detailed description of the SMM experiments will appear in a future issue of "Solar Physics."

Peter Landecker described the P-78 solar X-ray and white-light experiments and showed examples of the data. The P-78 satellite was launched on 24 February 1979 and has been recording solar data since that time.

The second part of the meeting was devoted to a discussion of the scientific objectives of the three projects. After a review of current observational and computational knowledge of the global structure and energy storage of sheared preflare magnetic fields, which are one principal concern of FBS, Van Hoven described the local coronal conditions necessary to initiate energy release in the short time scales observed to occur in solar flares. Three basic mechanisms, which can reduce the reconnection time for magnetic fields in a flare orders of magnitude below the classical resistive diffusion time, exist: spontaneous reduction of the spatial scale by such phenomena as the magnetic tearing instability, the inclusion of cool plasma within the reconnection region, and the induction of kinetic turbulence by field aligned electric fields. Dan Spicer emphasized that, due to the rapid profileration of flare models, the need for classification of these models has arisen and presented a classification scheme distinguishing two main categories: those driven by currents flowing parallel to $\underline{B}$ and those driven by currents flowing perpendicular to $\underline{B}$, where $\underline{B}$ represents the magnetic induction. This delineation of drivers allows a further specification in terms of the type of plasma instability which is operating and the overall geometry of the magnetic field.

Next two talks by Max Kuperus and Peter Hoyng were concerned predominantly with the SERF project. Kuperus emphasized that the energy release in a solar flare is the result of a sequence of MHD and plasma instabilities triggered when the electric current densities surpass a given threshold value. If the energy storage in the active corona occurs through a "slow" build-up phase, the magnetic field can remain to a large extent force-free because of the low value of the plasma parameter $\beta = 8\pi p/B^2$ in the corona. If the build-up occurs rapidly and once the large-scale MHD instabilities set in, deviations from a force-free field can be very large with the consequence that a tremendous amount of energy is released in the form of large scale motions. The latter condition appears to occur in the largest solar flares. Hoyng limited himself to reviewing recent reinterpretations of hard X-ray emission from flares. The "classical" picture ascribes the bursty hard X-ray emission to a beam impinging on the chromosphere or the corona. The observed X-ray fluxes demand that such a beam of energetic ejections carry in excess of 10% of the total flare energy--a situation which is difficult to explain theoretically as well as to bring into agreement with the fluxes of energetic flare particles observed in space. With the emergence of these difficulties, the thermal bremsstrahlung interpretation, with its much greater efficiency of energy into X-rays, has been revived.

Two additional talks concentrated on the problems of STIP. Robert Lin reviewed current knowledge on energetic particles within the solar system covering the range from ∿1keV (solar wind plasma) to many GeV (Galactic Cosmic Rays). Such suprathermal particles arise from a variety of sources--the galaxy, the nearby interstellar medium, solar flares, the quiet sun, planetary magnetospheres, and planetary bow shocks--and care must be exercised to distinguish these many sources. Recent analyses have shown that the interplanetary medium itself is a major source of low energy ions ($\lesssim$ 100 MeV). D. Intrilligator reviewed current information concerning interplanetary disturbances particularly those which accompany solar flares. Particular emphasis was given to the evolution of shocks in the outer solar system and the role such shocks play in accelerating particles *in situ*.

A detailed report of the meeting is planned for a future issue of "Solar Physics."

The addresses of some key persons in the operation of SMY appear below:

C. de Jager, Chairman SMY Steering Committee, Space Research Laboratory, Beneluxlaan 21, Utrecht, The Netherlands

P. Simon, Secretary SMY Steering Committee, DASOP, Observatoire, 92190 Meudon, France; Telephone 534.75.30; Telex 200590.

Z. Svestka, Chairman FBS, Astronomical Inst., Space Research Lab, Beneluxlaan 21, Utrecht, The Netherlands; Telephone (030)937145; Telex 47224.

D. Rust, SMY Science Coordinator and Chairman SERF, Code 409, Goddard Space Flight Center, Greenbelt, MD 20771, U.S.A.; Telephone (301) 344-7951; Telex 89-2554 GSFC EOF GBLT.

M. Dryer, Chairman STIP, Space Environment Laboratory, NOAA/ERL, Boulder, CO 80303, U.S.A.; Telephone (303) 499-1000; Texex 45897.

IV. GENERAL SCIENTIFIC MEETING, 20 AUGUST 1979

The following brief communications were presented:

A. A. Kattenberg, High Time-Resolution Solar Observations

A description of the new Westerbork microwave facilities with examples of high time-resolution (1/10s) heliograms of active regions and bursts obtained since 19 June 1979.

B. R.A. Munroe and G. Withbroe, Lyα and White Light Rocket Coronagraph Observations

Coronal Lyα profiles were measured in a coronal hole and in quiet coronal regions at 1.5-3.5 R_o in order to determine temperature, densities, and mass flow.

C. Gelfreich, Konovich, Peterova et al., First RATAN Results

Report on the work with the 575 m Radio Astronomy Telescope of the USSR Academy of Science (RATAN) operated at λ = 1.35-4.0 cm. First results are:

- At 1.35 cm a "radio granulation" was detected which correlates with the chromospheric network.

- At 2.0-4.0 cm enhanced intensities coincide with active Hα prominences.
- Flocculi-like structures with 20-50% contrast and <1% circular polarization (corresponding to several 10's of Gauss) appear.

D. Gergely, Kundu, and Golub, Detection of Type III Bursts Associated with Flaring of X-ray Bright Points

Results obtained with Skylab data. Type III bursts were found associated with flaring X-ray bright points which occurred inside and outside coronal holes.

E. D.A. Guidice, Radio Solar Telescope Network

A report on the Air Weather Service (U.S.A.) Radio Solar Telescope Network which measures global solar flux. Observations of some special events were presented. The data are available in the NOAA SGD.

Administrative Sessions

Business meetings were held on 14, 17, and 20 August 1979.

I. ELECTION OF OFFICERS AND NEW ORGANIZING COMMITTEE FOR 1979-1982

The President noted the tradition that the Vice President automatically succeeds the President at each General Assembly. A slate of candidates for other Offices and members of the Organizing Committee, containing all candidates proposed to date, was presented by the President and nominations from the floor were accepted. It was brought to the attention of the Commission that one major task of the Organizing Committee is the preparation of the Commission Section of Reports on Astronomy and thus disciplinary as well as geographical breadth is desired. The following were selected by ballot:

President:	V. Bumba	
Vice-President:	E. Tandberg-Hanssen	
Secretary:	R. Howard	
Past President:	G. Newkirk	
Organizing Committee:		
	H. Yoshimura	(Theory of Solar Cycle)
	R. Howard	(Empirical Aspects of the Solar Cycle)
	H. Zirin	(Active Regions)
	J. Harvey	(Non-Sunspot Magnetic Fields and Sunspots)
	O. Engvold	(Prominences)
	D. Rust	(Flares and Energetic Particles)
	M. Pick	(Radio Physics of the Sun)
	V. Stepanov	(Corona)
	M. Dryer	(Solar Wind and Solar-Terrestrial Relationships)
	R. Bray	(Ground-Based Instrumentation)

II. MEMBERSHIP IN THE COMMISSION

The President noted with regret the deaths of D.H. Menzel, C. Popovici, and S.F. Smerd. The names of Sixty newly elected and current members of the IAU who have expressed the desire to join the Commission were read and their membership in Commission 10 unanimously accepted.

III. CANDIDATES FOR REPRESENTATIVES TO OTHER ORGANIZATIONS

Z. Svestka proposed that two additional members be appointed by IAU to the SCOSTEP Steering Committee for the Solar Maximum Year. J. Harvey and M. Pick were elected by the Commission and the President was instructed to inform the IAU General Secretary of this choice.

IV. WORKING GROUPS OF COMMISSION 10 (1979-1982)

A. Working Group on the Solar Maximum Year

M. Dryer
Z. Svestka
D. Rust

B. Working Group on International Programs

Chairman: P. Simon
Co-Chairman: F. Moriyama
Members: H. Coffey, L. Dezsö, M. Dryer, S. Enome, R.S. Grevishev, J. Harvey, R. Howard, S. Kane, R.P. Lin, V. Obridko, and E. Tandberg-Hanssen.

V. REPORTS OF WORKING GROUPS (1976-1979)

A. Sunspot Numbers and Indices of Solar Activity

Chairman J. Eddy reported on the analyses which the Working Group had carried out to compare the Zürich Relative Sunspot Number R_Z with such quantitative indices of solar activity as the 2800 MHz flux and other qualitative indices such as the Sunspot Number R_A, compiled by the American Association of Variable Star Observers. He noted that the study had received some urgency with the possibility that Zürich may discontinue its issuance of the parameter R_Z after Waldmeier's retirement. Based upon their comparison of monthly means of these parameters for the last 28 years, the Working Group reported:

The two qualitative indices R_Z and R_A are related linearly with a correlation coefficient of 0.99.

Systematic Departures of about 10% between the two qualitative indices R_Z and R_A appear to occur preferentially at high solar activity. The origin of this effect is unknown.

The relation between R_Z (or R_a) and S_a is linear for $R_Z \gtrsim 45$; below $R_Z \approx 45$ it is non-linear but highly reproducible.

The Commission engaged in intense discussion concerning the advisability of adopting the quantitative index S_a as the index of solar activity, the proposal that R_Z be continued to be determined at Zürich to preserve its uniqueness, and the possibility of introducing other indices of solar activity. The consensus was that the Relative Sunspot Number is a unique link with the past course of solar activity and that the 10 cm flux provides a quantitative dimension which should be preserved for future analysis.

These convictions were embodied in Resolutions I. and II. below. The full text of the Working Group Report is planned to be published in NOAA Solar Geophysical Data and the Quarterly Bulletin of Solar Activity.

B. International Programs

The President informed the Commission that the Federation of Astronomical and Geophysical Services (FAGS), which sponsors a variety of services of which those of direct interest to the Commission are the International URSIGRAM and World Day Service (IUWDS) and the Quarterly Bulletin of Solar Activity, had requested that the Commission express an opinion concerning the continued value of these services. P. Simon, Chairman of the Working Group, reported on the history, objectives, and publications of IUWDS. In the discussion it was pointed out that the contribution which might be requested of the IAU is justified in view of the extreme usefulness of the service to such Commissions as 10, 12, and 44 and such special activities as the Solar Maximum Year. At the conclusion of the discussion the Commission passed Recommendation III.

The Commission then discussed the continued value of the Quarterly Bulletin with considerable attention directed toward possible duplication of the material which is available in the NOAA-Solar-Geophysical Data. Simon noted that the Quarterly Bulletin is considered by many as the "final" report on basic phenomena of solar activity. The consensus was that both publications will continue to be of intense value to the scientific community. Moriyama reminded the Commission that his institution (Mitaka) will continue publication of the Quarterly Bulletin until at least the end of the current solar cycle if it is given assurance from the Commission that the service is of value. At the conclusion of this discussion the Commission passed Recommendation IV.

The President reported that the editors of the Quarterly Bulletin had expressed the desire to have a formal point of contact with the scientific community to give advice concerning the scope, content, format, etc. of the Bulletin and that the Working Group on International Programs should be assigned this task with the Chairman designated as the liaison official.

V. Lincoln distributed a brochure on the activities and publications of NOAA Solar Terrestrial Physics Services and Publications and informed the Commission that Gnevishev (Kislovodsk) had agreed to prepare the coronal isophotes which had been prepared so far by Boulder. A motion was adopted to thank V. Lincoln for the past work and Gnevishev for accepting the new responsibility.

C. Mass Ejections

The charge and membership of the Working Group appear in IAU Transactions, XVI B, 1976. Chairman Tandberg-Hanssen reported that:

(1) A system of reporting mass ejection events has been established through World Data Center A (Boulder, Colorado, U.S.A.). Such phenomena will appear in a "Table of Solar Transients" in Solar Geophysical Data, Comprehensive Report published by NOAA. The tabulation will include such information as location, time, height, classification, information on the wavelengths (X-ray, EUV, Hα, white light, forbidden coronal emission line, radio, etc.) at which the event was detected, as well as remarks concerning the presence of associated activity. For details, contact J.V. Lincoln, NOAA.

(2) The Working Group has contacted a number of individual scientists and institutions capable of making observations of mass ejections by a variety of techniques at a wide variety of wavelengths and has secured their cooperation in reporting such events to the World Data Centers. These activities are expected to be particularly intense during the Solar Maximum Year.

(3) The Working Group recommends the following nomenclature for the reporting of mass ejections:

- Prominences - S, flare surge; SP, flare spray; A, eruptive active region; Q, eruptive quiescent.
- Coronal - D, depletion; C, cloud; B, bubble; EL, expanding loop; R, streamer or ray; IV_m, moving type IV radio burst.

Detailed descriptions of these classifications appear in "An Illustrated Glossary for Solar and Solar Terrestrial Physics," A. Bruzek and C.J. Durant, eds., 1977, Reidel, Dordrecht.

D. Solar Maximum Year

Scientific and organizational details appear in the report of the joint meeting. Resolutions V. and VI., pertaining directly to the SMY, were passed unanimously by the Commission.

VI. MISCELLANEOUS REPORTS

Greenwich - Debrecen Photoheliographic Results - At the General Assembly in Grenoble, the Commission was informed that the Royal Greenwich Observatory would discontinue its photoheliographic program and that Debrecen Observatory might be able to continue this work. The Commission offerred strong encouragement to Debrecen in this endeavor.

L. Deszö reported that Debrecen formally assumed responsibility for this program in January 1977 and has established collaborative arrangements with Pulkovo, Kislovadsk, Royal Greenwich, and Kodaikanal to assure a continuous daily record of high quality. To assure homogeneity in the RGO and Debrecen data, an intense comparison between data from the two stations is being carried out covering several years prior to 1977. It is planned to publish the Photoheliographic results from Debrecen and Gyula, supplemented by observations from the other cooperating stations, annually with the first volume (1977) available late in 1979. The President called for a vote of thanks to L. Deszö, Debrecen Observatory, and the Hungarian Academy of Sciences for their assumption of this responsibility: the proposal carried unanimously.

Recommendations and Resolutions

I. Commission 10, recognizing that the long series of relative sunspot numbers is a unique link with the course of solar activity in the past,

Recommends that all institutions that have demonstrated interest and competence in the work of obtaining sunspot numbers should continue the series.

II. Commission 10, recognizing that the 2800 MHz solar flux provides a standard quantitative index of solar activity for use in solar and solar-terrestrial studies,

Recommends that those institutions making such measurements should continue to do so.

III. Commission 10, recognizing that the International Ursigram and World Days Service (IUWDS) has rendered valuable service to the scientific community during the coordination of such international programs as the IGY, the IQSY, and the IMS; that the services provided by the IUWDS are not provided by any other agency and are relevant to the work of several commissions of the IAU; that these services will be particularly important for such programs as the Solar Maximum Year; and that the IUWDS included the prompt reporting of geophysical and solar events which will be occurring with increased frequency in the next few years;

Recommends that the IAU continue to support the activities of the IUWDS to assure continuance of this valuable service.

IV. Commission 10, recognizing that the Quarterly Bulletin of Solar Activity has for many years provided a valuable reporting service to the international scientific community; that this service continues to be relevant to the activities of

Commissions 10, 12, 40, 44, and 49 of the IAU; that the continuation of this service will be of particular value during the coming Solar Maximum Year;

Recommends that the IAU support the Quarterly Bulletin of Solar Activity to assure that this service continues into the future.

V. Commissions 10, 12, and 44 draw attention to the coming of Solar Maximum and recommend a sustained and coordinated effort in its study during the next three years.

VI. Commission 10, recognizing the important contribution made to solar physics by very high frequency (greater than 9 GHz) radio flux observations of the sun,

Recommends that such observations should continue to be strongly supported during the coming solar maximum.

COMMISSION 12: RADIATION AND STRUCTURE OF THE SOLAR ATMOSPHERE (RADIATION ET STRUCTURE DE L'ATMOSPHERE SOLAIRE)

PRESIDENT: M.K.V. Bappu

A short business meeting was held on August 14. The slate of new officers was approved, and the Commission elected the following Organizing Committee:

President: Y. Uchida

Vice-President: R.W. Noyes

Organizing Committee: Ph. Delache, F.L. Deubner, A.H. Gabriel, S.I. Gopasyuk, J. Harvey, R. Howard, P.R. Wilson, C. Zwaan, M.K.V. Bappu (ex. officio).

The Commission approved of the format of its triennial report. This constituted a critical review of new results obtained and an outline of trends in areas within its purview that are of topical interest.

The Chairman of the Working Group on Eclipses (Newkirk) gave a brief account of the activities of this group. Principally, activity has been confined to foster contacts between scientists interested in observing eclipses and the coordinators in the countries that are favourably located for such observation. The President thanked G. Newkirk for his splendid effort over the years as Chairman. The Working Group was reconstituted with J.L. Leroy as Chairman and A. Fiala as member.

The Commission agreed that it would be useful to plan on a symposium or colloquium in 1982 with a location in the vicinity of the venue of the next General Assembly. The Commission had co-sponsored the joint discussions scheduled for the current General Assembly on (1) Velocity Fields on the Sun (2) Stellar Instabilities (3) Chromosphere-Corona Wind Complex and Mass Loss in Stellar Atmospheres. In addition, two meetings with Commission 10 were held on (1) The Coming Solar Maximum and (2) The MHD of Sunspots. The Commission also had a three hour session on August 17, 1979 for a presentation of review talks on "Waves in the Solar Atmosphere". The session was chaired by Y. Uchida. The topics were (1) The generation of Alfven waves by the overstable convection and the cooling of sunspots (2) Is the coronal heating by waves? (3) Additional acceleration of solar wind streams in coronal holes. The speakers were E.N. Parker, M. Kuperus and T. Holzer. A brief summary follows:

Parker restressed the possibility that Alfven waves, which are generated in the over stable convection and escape from the umbral subphotosphere carry the energy out of the region that results in the cooling. The cooled gas would fall due to smaller scaleheight and a sideward collapse would follow to push the field together to form a sunspot. Admitting that the escape of the waves to the outer layer (corona) was difficult to consider by virtue of observational restrictions, he considered their downward propagation. He then introduced the idea that the fluxtube inside the umbra might be a bunch of smaller flux tubes and that the umbral subphotosphere might consist of magnetic and nonmagnetic regions. He proposed that the gas between the fluxtubes would play a vital role in such a situation, and in the presence of a down-draft, the heat which would pile up in the subphotosphere would be swept away from the region.

Kuperus pointed out that the acoustic shock wave theory had in recent years seemed unsatisfactory. The Skylab observations had shown the corona to be highly structured, suggestive of a very close relationship with the magnetic field as well as the lack of detection of any pressure wave type of motion being propagated through the transition layer. Among alternate possibilities of the heating mechanisms, that relating to Alfven waves seem attractive since these propagate along the magnetic field structure. The value and temperature distribution profile are determined by the property of heat conduction and are independent of the specific nature of the heat source. Mode coupling of such a basically non-dissipative wave to a dissipative mode was indicated as a means of converting the kinetic energy into heat. He also stressed the importance of another possibility of heating by Alfvenic surface waves which propagate along the surfaces of magnetic tubes and are termed kinetic Alfven waves because of the need to use the kinetic theory, since the gyro-radius is larger than the wavelength perpendicular to the field.

The identification of the source of the high speed wind stream with the coronal hole has raised a riddle since the original thermal expansion model of the solar wind requires a higher temperature region to maintain the faster stream, but the condition in the coronal hole suggests the contrary. In his talk on topic (3), Holzer first discussed observational results which confirmed the above-mentioned identification. Then, citing the result of one of the possible hypotheses, the calculation of the flow with the funnelled crossection, he argued that although the coronal density could be lowered to explain that of the coronal hole and the velocity could be raised near the sun by this funnelling-crossection mechanism, the velocity at 1 AU in this hypothesis was systematically lower than what was needed. He turned to another hypothesis in which either the addition of heat or the addition of momentum was taken into account, and showed that both could give rise to higher velocity around 1 AU. Momentum addition in the supersonic region seemed to be favoured more by the observations. It was also considered to be most likely due to the interaction of the flow with Alfven waves.

COMMISSION 14: ATOMIC AND MOLECULAR DATA; formerly: FUNDAMENTAL SPECTROSCOPIC DATA (DONNÈES ATOMIQUES ET MOLECULAIRES)

Report of Meeting, 14 August 1976

PRESIDENT: E. TREFFTZ

The first topic of the business session was a discussion on the structure of the Commission. In her introduction the president described the difficulty in serving the astrophysics community with the physical data needed. There are two somewhat contradictory points. On the one hand the Working Groups established by now are not able to pick up new needs of data which come up with new techniques of astronomical observation. On the other hand the material to be covered by the existing Working Groups gets out of hands. The situation is different for the different Working Groups. Working Group 1, Wavelength standards, will be discussed later, it does not post problems with the material to be covered. The material covered by Working Groups 2 and 4 (Atomic Transition Probabilities and Structure of Atomic Spectra resp.) is the same as that collected by the NBS groups under the same authors, W. Wiese and W.C. Martin. Reference to those well known collections enables the authors to write a useful report within the restricted length allotted to them. For Working Group 5, Molecular Spectra, the material to be covered is blowing up critically because not only the spectral region has been widened to the UV, IR and microwaves, but also transition probabilities are wanted besides the quantities describing the spectra. The worst situation is found in Working Group 3, Collision Cross-sections and Line Broadening. The report can by no means cover all work done during the last three years. It may quote data centres and review articles and then give some examples of present activities. But even then the restriction in length is so pressing that the question arose whether a report like this still makes sense.

The following decisions were taken: Basically the structure of the Commission is useful. The Organizing Committee is asked to enquire what can be done about physical data needed in astrophysics but not supplied by the Working Groups of Commission 14. All members of the Commission who know about such needs should consider how they can be of help (organizing work shops, referring to data centres). About this or if they themselves have collected data they should inform the President of Commission 14, J.G. Phillips, who will mention these activities in the next report. To avoid dropping valuable information because of restriction in length of the report it was suggested by J.G. Phillips to collect the material annually and have it published in some cheap form. The next report can then quote these and restrict itself to a survey and a few highlights.

A suggestion of W.C. Martin to structure the report in table form may also help to make it more useful to astrophysicists. All chairmen of Working Groups agreed that the amount of work put into the report is so high that we should do everything to make it known not only to astrophysicists but also to physicists working in related fields.

Future of Working Group 1, Wavelength Standards. It was found that the primary wavelength standard is mainly dealt with by nonastronomers. This brought up the question whether we should continue the Working Group. To astronomers good secondary standards are of vital interest especially when new spectral regions become available to astronomical observation. K.M. Baird was asked to continue the Working Group, and perhaps suggest the one or the other colleague working in this field to apply for membership to the IAU, Commission 14.

K.M. Baird gave the following additions to his report of Working Group 1, Wavelength Standards (Transactions of the IAU XVII A): (1) The International Advisory Committee for the Definition of the Metre met in June 1979 to consider progress toward a formal redefinition of the metre in terms of the velocity of light. New data were presented and a number of resolutions were passed including a proposed wording of the new definition, and wavelengths of I_2 lines used to stabilize an Argon laser at 515 nm and a He-Ne line at 612 nm. It is expected that a new definition will be approved by the General Conference of Weights and Measures in 1983 (to be reported in Procès Verbaux du B.I.P.M.). (2) An independent (the third) measurement of the velocity of light has just been completed at N.R.C. (Ottawa) yielding a value in agreement with that proposed in 1973 to within a few parts in 10^{10}. As its accuracy was largely limited by the uncertainty in the present Kr^{86} standard, the result ought to lay to rest any lingering fears that there might be a discontinuity in the length of the metre by adoption of a definition stating C = 299792458 m/s exactly (submitted to Optics Communications).

W.C. Martin gave the following addition to his report of Working Group 4, Structure of Atomic Spectra: The upcoming NBS "Bibliography on Atomic Energy Levels and Spectra" (1) will cover the literature from July 1975 through June 1979. Edlén's paper giving predictions of energy levels and wavelengths for Li I-like ions (Z=3-28) has appeared (2), and a similar treatment for the Be I-like ions is in press (3). Recent compilations of atomic energy levels at NBS include Mg I-XII (4), Al I-XIII (5), K I-XIX (6), Ca I-XX (7), and Ti I-XXII (8). The tables for O V in C.E. Moore's "Selected Tables of Atomic Spectra" have been submitted for publication (9). Bashkin and Stoner's "Atomic Energy-Level and Grotrian Diagrams, Vol. II, Sulfur I-Titanium XXII" has appeared (10), and Kelly has compiled the lines of the elements H through Kr over the range 2000-3200 Å (11). Ch. Moore-Sitterly presented predictions of wavelengths of intercombination lines $2s^2 2p\ ^2P^o - 2s2p^2\ ^4P$ of boronlike ions Mg VIII to Ni XXIV which have been determined by B. Edlén.

REFERENCES

1. Zalubas, R., Albright, A.: 1980, NBS Spec. Publ. 363, Suppl. 2, in preparation.
2. Edlén, B.: 1979, Phys. Scr. 19, p. 255.
3. Edlén, B.: 1979, Phys. Scr., in press.
4. Martin, W.C., Zalubas, R.: 1980, J. Phys. Chem. Ref. Data 9, No. 1, in press.
5. Martin, W.C., Zalubas, R.: 1979, J. Phys. Chem. Ref. Data 8, p. 817.
6. Corliss, C., Sugar, J.: 1979, J. Phys. Chem. Ref. Data 8, No. 4, in press.
7. Sugar, J., Corliss, C.: 1979, J. Phys. Chem. Ref. Data 8, p. 865.
8. Corliss, C., Sugar, J.: 1979, J. Phys. Chem. Ref. Data 8, p. 1.
9. Moore, C.E.: 1980, NSRDS-NBS 3, Sec. 9, in press.
10. Bashkin, S., Stoner, Jr., J.O.: 1978, North Holland Publ. Co.
11. Kelly, R.L.: 1979, NASA Technical Memo. 80268, Goddard Space Flight Center, Greenbelt, Md. 20771. (Sec. 1 has the lines by spectrum, Sec. 2 is a finding list.)

Report of Meeting, 15 August 1979

INFORMAL GATHERING AFTER THE JOINT COMMISSION MEETING ON THE COMING SOLAR MAXIMUM (COMM. 10, 12, 14, 44)

Since in the four sessions of the Joint Commission Meeting questions of ultraviolet spectroscopy could scarcely be handled an informal gathering was organized afterwards with R.W.P. McWhirter being chairman.

A.H. Gabriel spoke about "The Observation and Interpretation of the X-Ray Spectrum". The spectrum between 1 and 20 Å is determined largely by the resonance transitions of heliumlike and hydrogenlike ions and their satellites. The satellites are produced near the resonance transition when additional electrons occupy higher orbits, e.g.

$$1s^2(nl)^q \rightarrow 1sn'p(nl)^q \quad \text{or} \quad 1s(nl)^q \rightarrow n'p(nl)^q$$

Ions of O, Ne, Mg, Si, S, Ca, Fe have been observed. The most prominent satellites are those of the lithiumlike species: $1s^2nl \rightarrow 1s2pnl$, n = 2,3,4,etc. The levels are populated by dielectronic recombination (inverse autoionization combined with stabilizing radiative transition) and inner shell excitation by electron impact. From the intensity of lines the electron temperature T_e, the ionization balance T_z and perhaps even deviation from a Maxwellian distribution of electron velocities may be determined. For that purpose wave lengths λ, radiative transition probabilities A_r, autoionization probabilities A_a, and electron impact excitation rates C_{is} must be known. Groups in Nice (F. Bely-Dubau, P. Faucher, L. Steenman-Clark), Meudon (J. Dubau, M. Loulergue), Mons (S. Volonte), and Culham (A. Gabriel, J.G. Phillips) cooperate on this program. They use the program "Superstructure" of W. Eissner et al. to calculate λ and A_r taking account of relativistic effects (Breit-Pauli approximation). For A_a and C_{is} a distorted wave approximation is used for the electron impacting on a "Superstructure" target. Similar calculations are also done in Moscow by L.A. Vainstein, U.I. Safronova, A. Urnov et al. who use a perturbation expansion in 1/Z (Z nuclear charge in the target description and charge of the ion in the description of the impacting electron) together with Coulomb radial wave functions. An example shows the satellites of Fe XXV. If the additional electron nl is highly excited the satellite cannot be resolved from the main line. This must be taken into account in the diagnostics. Work is in progress on Fe XXV, Fe XXVI, Mg XI, Mg XII, O VII, Si XIV, Ca XIX. A special problem is posed by transient ionizing plasmas. Here all the processes mentioned above must be considered, and in addition inner-shell ionization. In the discussion some details of the plasma diagnostics calculations were explained.

G.A. Doschek discussed the solar VUV spectrum 1000 Å<λ<2000 Å. He reviewed density-sensitive spectral line intensity ratios in the 1000-2000 Å region, primarily for lines formed between ~2 x 10^4 K and 2 x 10^5 K. The results are applicable to the solar spectrum and perhaps to other stellar spectra as well. Density-sensitive line ratios involve the ratio of the intensity of an intersystem or forbidden line to the intensity of an allowed line formed in the same plasma region as the intersystem or forbidden line. The ions Si III, S IV, C III, N III, N IV, O III, O IV and O V have metastable levels that give rise to intersystem transitions in the 1000-2000 Å region. For quiet sun densities (~10^{10} cm^{-3}) only the C III 1909 Å line is useful for density determinations. For solar active regions and flares, $N_e \gtrsim 10^{11}$ cm^{-3}, many of the lines of the other ions are useful. Two techniques were discussed and compared: (1) using the intensity ratio of an intersystem line of one ion to the intensity of an allowed line of a different ion, and (2) using the intensity ratios of two lines arising from the same ion. The effect of density on intersystem and allowed line profiles was outlined. The theory was illustrated with spectra recorded by the NRL slit spectrograph flown on Skylab. A density-sensitive line ratio of S X was also discussed. S X is formed at about 1.3 x 10^6 K, so this diagnostic is useful for coronal plasmas. The S X lines appear near 1200 Å and are strong lines in Skylab spectra obtained outside the solar limb. Finally, the He II 1640 Å line profile observed in quiescent prominence spectra was shown. This line is formed at about 25,000 K. It was suggested that the excitation mechanisms of the fine structure components be investigated in detail for possible application to solar or stellar atmosphere problems.

H. Nussbaumer described the methods used to meet the needs for atomic data. He described the basic quantum-mechanical approaches and the criteria for accuracy. The latter are supplied by measured energy levels. For many astrophysically important ions, especially those of carbon, nitrogen and oxygen and their isoelectronic series a simple self-consistent-field approach is insufficient. A configuration expansion improves the results considerably. Judging the accuracy of transition probabilities is difficult. Most error estimates are hardly more than guesswork. Exact error bounds exist, cf. F. Weinhold, 1971, J. Chem. Phys. 54, 1874, but unless the wave functions are very good the bounds are too wide to be of practical use. In the isoelectronic sequence of Be I different high quality calculations agree now within 3%. The discrepancy to measurement of C III to Ne VIII is less than 20%. For the intercombination transition $2s^2\ ^1S_0 - 2s2p\ ^3P_1^o$ of C III a detailed error analysis was given which leads to a transition probability of 96±11 sec^{-1} with the largest uncertainty, 6.4 sec^{-1}, coming from spin-orbit splitting. The effect of uncertainties in the transition probabilities on the plasma diagnostics was shown for the O IV intercombination system: If new A values differ by 50% in absolute value but only by 10% relative to each other the electron density N_e inferred from intensity ratios is changed by a factor of 2.

Collision cross sections of those ions show a lot of structure. Below each excitation threshold a Rydberg series of resonances occurs. In Coulomb Born approximation these are neglected. Also the Distorted-Wave approximation which replaces the Coulomb potential by a refined ion potential neglects them. Close-Coupling calculations take them into account. But it should be kept in mind that the quality of the results depends sensitively on the quality of the target description and on a sufficiently high number of partial waves (characterized by the angular momentum of the impacting electron) taken into account. The electron impact excitation of Fe VI was shown as an example.

In the discussion C. Jordan mentioned a problem with the high excitations in Si VIII and Fe XII. Answering a question of E. Trefftz she does not think it to be an effect of dielectronic recombination. H. Nussbaumer explained his error estimate of the C III intercombination transition probability in some more detail.

Report of Meeting, 17 August 1979

MOLECULAR DATA

The following additions to the reports in Transactions of the IAU, Vol. XVII A were given: Working Group 5, Molecular Spectra. R.W. Nicholls announced that the book of Huber and Herzberg (1) has appeared. He mentioned the compilation of CNRS (2) and two review papers on the systematics of molecular intensities which R.A. Hefferlin has found in the Soviet journals Uspekhi and Optics and Spectroscopy. G. Herzberg (3) reports a new spectrum in discharges in H_2 and D_2 which can be definitely ascribed to transitions between Rydberg states of H_3 and D_3. It results from one of the radiative channels of the dissociative recombination of H_3^+ and D_3^+. J.W. Brault reports on the interferometer mounted at the Kitt Peak solar telescope. It can be used in an astronomical or laboratory mode. Recent laboratory work has been on CH_4 and HCHO. The free spectral range of this instrument is 2000A-10μ. It has a resolving power of 2.5×10^6. Its wavelength precision is 1 in 10^9 and the signal to noise ratio is in the 10^4-6×10^4 range. R.W. Nicholls has done some recent theoretical work to model photodissociation phenomena (4).

P.K. Carroll described work on the complex spectrum of FeH in the blue, green and infra-red regions out to beyond 9000 Å. Work was also reported on the 3678 Å band of CrH, a molecule which has been found in cool stars by Lindgren and others through the Keenan band at 8611 Å. Work is also in progress on the CO spectrum from 800 Å to 1200 Å using plates taken at Ottawa and about 100 bands have so far been analysed.

P.L. Smith reported work on the oscillator strengths of lines likely to be seen in interstellar clouds. These include lines in the $\tilde{C}$-$\tilde{X}$ and $\tilde{F}$-$\tilde{X}$ bands of H_2O near 1240 Å and 1115 Å, respectively, and also the line R(0) of the C-X band of HCl at 1290 Å. These lines have not yet been found in interstellar sources.

B. Petropoulos reported work on Franck-Condon factors by the RKR method for the A-X system of CO, and results of the application of this work to the study of the spectrum of Mars.

S. Sahal presented additions to the printed Report on line broadening. For Stark broadening she indicated a general agreement between theory and experiment of 20 to 30 per cent. New work was available by F. Praderie and H. Hubeny for the Lyman alpha line in Vega. She discussed the formation of eximers and collisional broadening in the far wings of atomic lines. J. Cooper has studied the problem of the redistribution of radiation, but the case of the hydrogen lines still needs to be done. There is still a need for improvements in collisional broadening for molecular lines for planetary atmospheres applications.

E. Trefftz drew attention to the Chemical Kinetics Data Center at NBS Washington. She mentioned recent work on charge exchange at the Lebedev Institute, Moscow, and a talk given at the Max Planck Institute by A. Luntz, IBM San Jose Research Lab, on the excitation of oxygen atoms by impact with hydrogen molecules and hydrocarbons.

J.G. Phillips drew attention to the large number of papers on electron impact on molecules, with some 72 papers on 28 molecules during the last year or so. Most papers have been on H_2, and CO_2 and N_2 have also been extensively studied, mostly by ab initio calculations. There have also been many papers on photoionization. A supersonic nozzle has been used to give excitation temperatures of 10 to 30° K, which has made some experiments easier.

W. Huebner pointed out that at temperatures of a few thousand degrees Kelvin, and at low density the Rosseland mean opacity of hydrogen is dominated by the total extinction on the far red side of Ly α. It is important to have a theoretical model and measurements of the absorption and scattering properties of atomic hydrogen for photons with energy between about 1 and 9 eV.

J.H. Black gave a review of the highlights of I.A.U. Symposium No. 87 on Interstellar Molecules, which had been held prior to the main I.A.U. General Assembly. Molecules are important as diagnostic tools, including searches for isotopes, and searches for unusual species. 52 interstellar molecules have been observed, 13 inorganic and 39 organic, including one with 11 atoms and four with 9 atoms; no ring molecules have been identified. Species first seen in space are HCO^+, C_2H, N_2H^+, C_4H and C_3N. About 185 radio frequency interstellar lines remain unidentified. Many molecules appear to be formed in large networks of reactions; many reaction rates are known at room temperatures but they have not been measured at 20°K. Associative reactions such as $C^+ + H_2 \rightarrow CH_2^+ + h\nu$ and $CH_3^+ + H_2 \rightarrow CH_5^+ + h\nu$ are important as the starting points of chains of reactions, for example the second of these reactions may lead to CH_4 and H_2CO, among others. Statistical theory and experiment applied to the corresponding three body reactions suggests which processes are likely to be rapid. High temperature chemistry may also be important;

observations of the H_2 quadrupole radiation in the Orion cloud lead to a kinetic temperature of 2000°K. At high temperatures many more reactions are possible. Among specific processes, the photodissociation rate of CH^+ is larger than formerly thought. Oka is working on collisional excitation of molecules by H_2, and it matters whether H_2 is in para (J=0) or ortho (J=1) state. The astronomical rate of $H^+ + O \rightarrow O^+ + H$ has been confirmed theoretically. There are many challenges to experimenters. Predictions suggest that $C^+ + H_2 \rightarrow CH_2^+ + h\nu$ is fast, and that $Cl^+ + H_2 \rightarrow HCl^+ + H$ may be slow at low temperatures. Isotopes continue to be of interest, refined abundances of isotopes are important as constraints on nucleosynthesis. HDCO has now been detected: it is probably formed by gas phase reactions. On the observational side, radio identifications of lines may be nearing the confusion limit. Infrared techniques are developing rapidly. Prospects are dim for studies of H_2 below 1100 A, even the Space Telescope will have poor response at that wavelength. The most pressing needs at present are for further spectroscopy to aid in identifications, measurement of reaction rates especially at low temperatures, rates for radiative association processes, and further study of collisional excitation mechanisms.

G. Herzberg mentioned that in addition to the new Rydberg spectrum of H_3 and D_3 he had also found a related spectrum in NH_4 and is trying to find one for CH_5. It may shed light on the dissociative recombination of CH_5^+. From radio observations it looks as though the abundance of CH_4 must be enormous. The question of how much CO and C is tied up in grains is hard to answer.

W. Huebner described his calculation of model comae of comets. The gas density just above the surface of the vaporizing icy nucleus of a moderately active comet at 1 AU heliocentric distance is several times 10^{13} molecules per cm^3. The gas streams outward at first with a speed of ~0.1 km/s but rapidly reaches its terminal, supersonic speed of ~1 km/s. Its initial temperature is typically 150 to 200°K. Solar uv radiation penetrates to ~100 km of the nucleus (at 1 AU) dissociating and ionizing the gas. It has been found that photodissociative ionization plays an important role in the inner coma, producing radicals and ions that fuel a complex chemical reactions network in which new species are formed that are not simple decay products of the initial vaporizing gas. Over 600 reactions of 100 species are included in the model, and time dependent solutions are obtained. The relative importance of various reactions changes with the distance from the cometary nucleus, so that it is difficult to simplify the network of reactions. Many species do not reach chemical steady states. A large number of photo rates in sunlight at 1 AU heliocentric distance have been computed for atomic hydrogen, carbon, nitrogen, and oxygen and for some molecules formed from these elements. Most urgently needed are branching ratios for photolysis and for chemical reactions. A problem of the present model is that we still cannot produce enough C_3.

Answering a question W. Huebner pointed out that the most needed observations are those of column densities of the various species at well defined positions of the line of sight with respect to the nucleus. A special problem is posed by the H_2O^+/ CO^+ ratio.

A.J. Sauval spoke on molecular data needs for cool stars. A survey of physical parameters of the different types of stars shows that they may serve as a laboratory of molecular spectroscopy. Temperatures and column densities are higher than in ordinary laboratory conditions. For some stars the spectrum is extremely stable, equilibrium conditions are well satisfied. Disadvantages of the stellar lab are the temperature stratification, the complicated and unknown chemical composition which yields blends, perturbations (motions, magnetic fields), and absorption in the earth's atmosphere. Of the molecules identified so far in the UV - Visible - IR about 80% are diatomic compounds of H and O. Polyatomic molecules of H, C, N, O (Si, S) are mainly observed in the microwave region and the Infrared ($\lambda \gtrsim 3\mu$). Circumstellar envelopes which give rise to radio emission lines are not in thermal equilibrium (OH, SiO, H_2O). A list of observed transitions shows how many data are still lacking - oscillator strengths, Franck-Condon factors, - even for known spectra. The abundance of a molecule depends on the abundance of its components and on its dissociation energy. This explains the high interest in hydrides and monoxides. But also monocarbides, -nitrides, fluorides and sulphides deserve laboratory study. Especially poor is the knowledge of positive molecular ions. Of metal compounds most prominent are titanium and zirconium compounds. Also here spectra and dissociation energies are lacking for some compounds. The situation for other metals is even worse. In S stars the number of free carbon and oxygen atoms is a sensitive test for the abundance ratio n_C/n_O since the less abundant atom is almost completely bound in molecules. Regularities between isoelectronic sequences of molecules may help clarifying the spectra. After describing new results for TiO and ZrO a long list of wanted work on spectra, classification of bands and rotational analysis was presented (list still available to the author).

W. Huebner asked whether the assumption of chemical equilibrium is justified in a stellar atmosphere. That is an approximation, but owing to the high temperature and the large density it surely is much better justified in stars than in comets or in the interstellar medium.

K.H. Hinkle, Kitt Peak National Observatory, discussed high resolution infrared stellar spectroscopy and its data needs. The development of Fourier Transform Spectrometers (FTS) and increasingly sensitive infrared detectors have made a large number of astronomical sources available for high resolution ($\Delta\lambda/\lambda \gtrsim 2 \times 10^{-5}$) spectroscopy in the 1 to 5 μ infrared. Currently, the FTS at the coudé focus of the 4m Mayall telescope at Kitt Peak National Observatory has obtained high resolution, high signal-to-noise spectra to K magnitude +4 and M magnitude -1. The K magnitude limit is about one magnitude fainter than the limit for the 2μ Infrared Catalog, implying that nearly 10,000 objects are bright enough to be observed at high resolution. The FTS has also achieved very high resolution (up to 0.006 cm^{-1}) for the brightest sources. In order to interpret high resolution, infrared spectra of astronomical sources, high quality laboratory spectra are needed. Astronomical infrared spectra in the 1 to 5μ region are dominated by vibration-rotation transitions of light molecules including CO, OH, H_2O, SiO, C_2, HCN, C_2H_2, CS, CH, NH, H_2, HCl, and HF. The 0-2 and 0-1 sequences from the $A^2\Pi$-$X^2\Sigma$ transition (red system) of CN and atomic lines, particularly from metals of atomic weight near Fe, are commonly found in infrared spectra. Exceedingly large column densities (up to $\sim 10^{26}$ cm^{-2} for H_2) can be observed in cool stellar atmospheres. Since typical excitation temperatures are near 3000K, lines of high excitation can be found in stellar spectra. For CO laboratory frequency measurements extend to $J'' \sim 90$ while even in the relatively CO weak K0 giants the 2-0 $J'' = 90$ line has a central depth of about 10% and 2-0 lines of $J'' \gtrsim 110$ can be identified. A few molecules are in desperate need of improved measurements. NH is a potentially important nitrogen abundance indicator. Its fundamental vibration-rotation band which is present in 4μm region spectra of the M supergiant α Ori, has no frequencies or oscillator strengths measured in the laboratory. The elements Sc, Ti, V and Ni contribute a number of strong lines in the 2-5μm infrared spectra of stars. Both accurate laboratory frequency measurements of these lines and identifications of any weaker lines of these elements present in infrared spectra are urgently needed.

E.E. Whiting introduced a discussion on the rotational line intensity factors for diatomic molecules (Hönl-London factors) and the different normalization schemes in use for them. Uncritical use of published values of Hönl-London factors in ignorance of the normalization used for them can lead to grave errors. Together with J.B. Tatum and R.W. Nicholls he advocated a unique convention on a normalization which is self-consistent. It was agreed to draft a statement incorporating the recommended convention and to circulate it for consideration to the Inter-Union Commission on Spectroscopy (IUCS which has 3 representatives from IAU, 4 representatives from IUPAP and 4 representatives from IUPAC).

F. Remy reported lifetime measurement of electronic states in molecules. In the evaluation of the decay curves care must be taken to account properly for repopulation by long living states. A mere a priori subtraction of a so-called experimental background (determined by a blank experience) may lead to large errors. An attempt to use the dependence of the lifetime on the vibrational quantum number to determine the variation of the electronic transition moment with the internuclear distance is not successful owing to errors in the measured lifetimes. Large discrepancies in the published values of the $N_2C\ ^3\Pi_u$ lifetime are due to radiation trapping between this state and the long lived upper state $B\ ^3\Pi_g$ of the first positive system.

R. Clegg discussed the unidentified Keenan bands in the 8000 to 11000 Å regime of S stars. Accurate band head wavelengths have been observed. High resolution (0.4 Å) spectra do not resolve the rotational structure indicating that the molecules are not hydrides. The 8610 Å band is not due to CrH. Spectroscopic data and chemical equilibrium calculations restrict the list of candidate molecules. Laboratory spectroscopy is wanted for: ZrS, CeS, ZrCl, YCl, LaCl, CeCl, ZrOH, LaOH, CeOH, YOH, for LaS ($\lambda > 0.9\mu m$) and for ZrO, CrO, NbO ($\lambda > 1\mu$). Two new transitions of CeO have been seen in R Gem. In the same star LaO explains part of some bands.

REFERENCES

1. Herzberg, G., Huber, K.P.: 1979, "Constants of Diatomic Molecules", Van Nostrand, New York.
2. "Diatomic Molecules, a Critical Bibliography of Spectroscopic Data", 1979, CNRS, Univ. P. et M. Curie (Paris VI), Tour 13, 4, place Jussien, F 75230 Paris Cedex 05.
3. Herzberg, G.: 1979, J. Chem. Phys. 70, p. 4806.
4. Nicholls, R.W.: 1979, Chem. Phys. Lett. 64, p. 190.

Report of meeting, 20 August 1979

ATOMIC DATA

The last session of the Commission started with some business. The Organizing Committee, the Chairmen of the Working Groups, the new members of the Commission were announced. There being no objection, all names were accepted. The president reported that the Organizing Committee was proposing a change of name of the Commission. The new name will be "Atomic and Molecular Data". Considering an objection of Ch. Moore-Sitterly that the new name opens the door to unreliable data the Organizing Committee felt that the chairmen of the Working Groups could deal with this danger, that on the other hand non-spectroscopic data are needed in astrophysics and handled by Commission 14. These data are by nature less reliable. The new name describes the activities of Commission 14 more properly than the old one. The suggestion of J.G. Phillips (our next president) to collect material for the report annually was accepted for a try.

W. Wiese's addition to the report on Atomic Transition Probabilities were presented by W.C. Martin: A compilation of transition probabilities for selected lines of H through Ni has appeared in the CRC Handbook of Chemistry and Physics (1). Transition probabilities for doubly-excited states of He I have been compiled (2). Calculations for several Be-like ions have been reported (3,4). Transitions involving doubly-excited states of Be-like ions have been treated (5), and of Fe XXIV (6). Transition probabilities of forbidden lines within the ground configuration of C-like ions have been calculated (7). Numerous allowed and forbidden transitions of Fe XXI have been treated (8). Lifetimes of excited states have been measured by various investigators, using the Hanle-effect technique in Mg I and Sr I (9), and in In I (10); stepwise laser excitation has been employed in Hg I (11), as well as in Li I (12). Oscillator strengths for lines of the principal series in Rb I have been measured in absorption. (13). Branching ratios for several lines in neutral oxygen have been measured in a hollow cathode discharge (14).

In reply to a question W.C. Martin said that the compilations on wavelengths, energy levels, transition probabilities, ect. were available from him or from Dr. R. Zalubas, A 167 Physics, National Bureau of Standards, Washington D.C. 20234.

Additions to A. Burgess' report on electron impact on and photoionization of atoms and atomic ions were presented by M.J. Seaton. He reported photoelectron spectra of atomic barium by R.A. Rosenberg et al., Lawrence Berkeley Laboratory, to be published.

J.L. Culhane reported on stellar X-ray spectroscopy. The solar X-ray emission spectra has helped in understanding the hot plasma in active regions and in solar flares. The history of the theoretical calculations and of the observations of the coronal X-ray spectra are familiar. Observations of stellar spectra are done e.g. at Naval Research Laboratory (NRL), Aerospace,Lockheed in USA, Mullard Space Science Laboratory (MSSL), Leicester in the U.K. Although the early dispersive spectrometers did not have much success, lower resolution non-dispersive spectrometers have since 1972 been discovering X-ray emission and absorption features in supernova remnants (SNR's), binaries, active stars, interstellar medium, clusters of galaxies, active galaxies. In the observed objects the low resolution spectra have allowed us to identify the following physical situations. SNR's show a shock heated interstellar plasma of a temperature of $T_e \sim 10^6$ to 5×10^7 K and a density $N_e \sim 0.1$ to 10 cm^{-3}, similar to a coronal plasma, and probably not in ionization equilibrium. Some accretion powered X-ray binaries have a photoionized stellar wind. The temperature is $T_e \sim 10^5$ to 10^6 K, but ionization stages up to Fe XXVI exist due to X-ray radiation. The density is $N_e \sim 10^{11}$ to 10^{13} cm^{-3}. One sees emission lines due to fluorescence and recombination and also photoelectric absoprtion edges. Stellar soft X-ray sources are stars with coronae and winds, and accreting white dwarfs. The temperature is $T_e \sim 10^5$ to 10^7 K, the density $N_e \sim 10^8$ to 10^{13} cm^{-3}. Seen are multi-temperature coronal type emission lines and photo-electric absorption edges. Spectroscopy of extragalactic X-ray sources has indicated that clusters of galaxies have a gravitationally heated intra-cluster medium. This is primordial material enriched by ejecta from galaxies. $T_e \sim 10^8$ K, $N_e \sim 10^{-3}$ to 10^{-4} cm^{-3}. Cooler material is found around some galaxies. We see multi-temperature coronal type emission lines. Active galaxies are very luminous with an X-ray flux $L_x \sim 10^{45}$ to 10^{46} erg sec^{-1}. The radiation comes from nuclear X-ray sources surrounded by photoionized material. Emission lines due to fluorescence and recombination are seen as are the photo-electric absorption edges. In NGC 4151 and NGC 5128 neutral iron is observed. The following survey of observations refer to the cases mentioned above. Observations of supernova remnants were demonstrated by spectra on Cas A and Tycho taken by Ariel V, Einstein SSS (Holt et al.). Spectra of Pup A observed by rocket (Culhane et al.) and by the MIT satellite (Canizares et al.) show two different temperatures. A suggestion by Itoh explains the low temperature ions by a departure from ionization balance. Spectra of X-ray binaries were shown for Vela X-1 (OSO-8) and Cyg X-3 (OSO-8, Ariel V) and for Capella (HEAO A2 experiment and the Einstein observatory). The latter comprises a G 5 III - G 0 III spectroscopic binary with a period of 104 days. Spectra show Fe XVII and Fe XVIII, Mg XI, Si XIII, S XV. The elements Fe, Mg, Si, S appear to be over-abundant by a factor 1.5 to 6 compared to solar. The X-ray flux is consistent with a magnetic field $B_{CAP} \sim 10$ $B_{\odot}$. The apparent over-abundance may be explained by a coronal loop structure with different temperatures. With respect to extra-galactic objects the most exciting Uhuru result were extended cluster X-ray sources. Fe XXV, XXVI lines resolved by Ariel V indicate $T \sim 7$ keV and cosmic abundance. Also n = 3 to n = 1 transitions of these ions are seen and faintly S XV. Einstein SSS observations (Holt et al.) of NGC 1275 confirm the presence of cooler material surrounding NGC 1275 . O VIII resonance lines were observed by the MIT-FPC satellite (Canizares et al.) from M 87 in Virgo, indicating a temperature of 2×10^6 to 10^7 K while the intercluster temperature in Virgo is 30 to 40×10^6 K.

Regarding the necessary atomic physics work, it is at present the case that cross-sections and rates appropriate to coronal plasmas are being used to interpret the new data described above. This is acceptable for the present since the spectra are mainly of low resolution and are statistically weak. However the much wider range of plasma conditions (temperature, density, optical depth) that will be encountered in cosmic plasma could show a need for improved data.

R.H. Garstang gave an account of the behaviour of atoms in high magnetic fields, and referred to his review (15). High fields have become of particular interest because of the discovery that certain white dwarf stars show megagauss fields (these stars have been discussed in a review by J.R.P. Angel (16)). Recently it has been suggested that C_2 may be the origin of a very strong absorption feature in the magnetic white dwarf star LP 790-29 (17). Further calculations are clearly needed on the effects of high magnetic fields on molecular spectra, and there is also a need to carry the calculations on simple atomic spectra to higher accuracy and to higher fields.

S. Sahal gave a short account of her work (together with J.L. Leroy, G. Ratier, V. Bommier) on the use of the Hanle effect for determination of magnetic fields on the sun's surface. The strength and direction of magnetic fields influences the degree and direction of polarization of the solar radiation. The case of a prominence on the solar limb was treated in detail. To determine the polarization which would be seen without field a model must describe the excitation mechanisms of the observed spectral line. For symmetry reasons the polarization without field is parallel to the limb. The observed polarization in the line is smaller and rotated by an angle of some 10°. This reduction together with the rotation angle allows the determination of two components of the field. The observation of several lines would allow the complete determination of the magnetic field. Diagrams have been calculated for easy determination. The helium line D_3 was used as an example (18).

C. Jordan mentioned work by S. Johansson, Lund, on the spectrum of Fe II which she is using. She stressed the need for additional oscillator strengths for Fe II in the ultraviolet and for data on polarizability needed for the interpretation of motional Stark effect.

C. Cowley notified the Commission of a new Newsletter on Ap stars, distributed by W. van Rensbergen. There is still a need for oscillator strengths of transition elements such as Ti I, Cr I to which D. Locanthi added Zr I. Damping constants are also needed.

M. Malinowski reported her calculations of intensities of Fe IX, X, XI lines in coronal and laboratory plasmas describing the different processes that had to be taken into account.

P.K. Carroll described work on a plasma produced by a ruby laser. Heavy elements (Mo to U) are up to 16 times ionized. The 40 to 200 Å region was observed. Open d shells give an enormous number of lines which can't be resolved. They form a quasi-continuum.

The president thanked all the Commission members for their interest in the work of the Commission. Special thanks are due to R. Garstang who has served in the Organizing Committee for 5 terms or 15 years and who is now retiring, and to the chairmen of Working Groups and those who helped preparing the report, to all the speakers for their contributions and all those who participated in the discussion.

REFERENCES

1 Wiese, W.L., Martin, G.A.: 1979, Atomic Transition Probabilities, "CRC Handbook of Chemistry and Physics", 60th Edition, E350-E378 (Eds. R.C. Weast and M.J. Astle, CRC Press, Inc., Boca Raton, Florida).
2 Shearer-Izumi, W.: 1977, At. Data Nucl. Data Tables 20, p. 531.
3 Glass, R.: 1979, J. Phys. B 12, p. 689, p. 697, p. 1633.
4 Nussbaumer, H., Storey, P.J.: 1979, J. Phys. B 12, p. 1647.
5 Boiko, V.A., Chugunov, A.Yu., Ivanova, T.G., Faenov, A.Ya., Holin,I.V., Pikuz, S.A., Urnov, A.M., Vainshtein, L.A., Safronova, U.I., 1978, Mon. Not. R. Astron. Soc. 185, p. 305.
6 Bely-Dubau, F., Gabriel, A.H., Volonte, S.: 1979, Mon. Not. R. Astron. Soc. 186, p. 405.
7 Nussbaumer, H., Rusca, C.: 1979, Astron. Astrophys. 72, p. 129.
8 Mason, H.E., Doschek, G.A., Feldman, U., and Bhatia, A.K.: 1979, Astron. Astrophys. 73, p. 74.
9 Kelly, F.M., Mathur, M.S.: 1978, Can. J. Phys. 56, p. 1422; 1979, 57, p. 657.
10 Lubowiecka, T., Dohnalik, T., Frasinski, L.: 1978, Acta Phys. Pol. A 54, p. 369.
11 Lukaszewski, M. and Lecler, D.: 1978, Opt. Commun. 25, p. 189.
12 Heldt, J. and Leuchs, G.: 1979, Z. Phys. A 291, p. 11.
13 Caliebe, E. and Niemax, K.: 1979, J. Phys. B 12, p. L45.
14 Christensen, A.B. and Cunningham, A.J.: 1978, J. Geophys. Res. 83, p. 4393.
15 Garstang, R.H.: 1977, Rep. Progr. Phys. 40, p. 105.
16 Angel, J.R.P.: 1978, Ann. Rev. Astron. Astrophys. 16, p. 487.
17 Liebert, J., Angel, J.R.P., Stockman, H.S., Beaver, E.A.: 1978, Astrophys. J. 225, p. 181.
18 Bommier, V., Sahal-Brechot, S.: 1978, Astron. Astrophys. 69, p. 57.

COMMISSION 15: PHYSICAL STUDY OF COMETS, MINOR PLANETS AND METEORITES (L'ETUDE PHYSIQUE DES COMETES, DES PETITES PLANETES ET DES METEORITES)

Report of Meetings, 15 and 20 August 1979

PRESIDENT: B. Donn, Acting. SECRETARY: J. Rahe.

15 August 1979

I. WORKING GROUPS

1. Filter Standards

(Members: Delsemme, Giovane, A'Hearn, Matson, Neff, Vanýsek).

It was agreed to concentrate initially on filters for photoelectric photometry of neutral radicals since these are the ones likely to be used by the largest number of observers. Agreement was reached on the approximate characteristics of a suitable set of filters. Detailed characteristics remain to be worked out. There was a consensus that standardization would be effective only if a large number of filter sets were purchased simultaneously. Regular comet observers might participate in the initial purchase, but many sets would be retained for future resale and/or distribution.

The committee also agreed on the desirability of a set of standard diaphragm size and of standard calibration stars. Agreement was reached on the approximate diaphragm sizes, but the question of calibration stars was not resolved.

The working group will investigate detailed specifications, and sources of funding in the near future.

M. F. A'Hearn

2. Coordinated Observations of Bright Comets

(Members: Biermann, Brandt, Dossin, Rahe, Roemer, Wisniewski)

A. Comet Halley

The Working Group met to discuss the techniques necessary to obtain coordinated observations of a bright comet. Obviously, detailed planning must be carried out as far in advance as possible; this poses no problem at present for Halley observations. The Working Group felt that the following would be key elements of a successful program.

(1) Standardization of observations as far as possible, preferably through issuance of a detailed instruction booklet.
(2) Observations should be coordinated at a central facility which is both able and willing to do so.
(3) Publication of results in a timely fashion.
(4) A dedicated individual (DI) willing to see the project through to completion.

All of these elements are part of the planning for the Halley Wide-Angle Network currently underway at the Goddard Space Flight Center with J. C. Brandt as DI.

The Work Group also saw the need for other Halley Networks headed by other DIs. Examples are:

(1) Photometry
(2) Infrared Observations
(3) Radio Observations
(4) Spectroscopy
(5) High-resolution imaging of the nucleus

The Halley Networks could serve as the model for coordinated observations of other bright comets or benefit from the experience gained from networks set up prior to the 1985/86 apparition of Comet Halley.

J. C. Brandt

B. Other Bright Comets

Because telescopes are scheduled far in advance, except for the smaller instruments, observations other than for Comet Halley present serious problems. It is proposed that a memo be sent to all well-equiped observatories to stress the importance of cometary observations. Detailed instructions about observational requirements, such as accurate guiding, slit height, exposure time, spectral range, and calibration, should be included.

F. Dossin

3. Cometary Archives

(Members: Delsemme, Giclas, Meisel, Rahe).

The compilation of a list of existing comet photographs and spectra is thought to be highly desirable.

An "Atlas of Comet Halley Photographs and Spectra" is presently being compiled by J. C. Brandt, B. Donn, J. Rahe.

A. H. Delsemme is preparing the notebooks of N. Bobrovnikoff for publication.

4. High Resolution Spectra

An "Atlas of High Resolution Spectra of Comets," with supplementary coverage of medium resolution optical, as well as UV-, IR-, and radio spectra, is presently being prepared for publication by C. Arpigny, B. Donn, F. Dossin, J. Rahe, S. Wyckoff.

5. Research in Space (Non-mission)

(Members: Blamont, Donn, Fechtig, Greenberg, Rahe)

A workshop on "Applications of Modern Observational Techniques to Comets--With Special Emphasis on Comet Halley" will be held in late 1980 at the Goddard Space Flight Center. It will be organized by J. C. Brandt, B. Donn, M. Greenberg, J. Rahe.

Plans were made for a workshop on "Laboratory Research for Comets," to be held in Leiden, Netherland, in 1981, and organized by B. Donn and M. Greenberg.

II. MINOR PLANET REPORTS

1. B. Zellner - The Tucson Minor Planet Symposium
2. T. Gehrels, D. L. Matson - Prospects for Coordinated Asteroid Observations

III. BUSINESS MEETING

1. Commission Activities

An evaluation was made of the expansion of Commission 15 in 1973 to include minor planets and meteorites. Several problems were recognized and discussed. It was concluded that closer interaction among the members of the Commission and the Organizing Committee representing the three areas is needed. A Commission 15 resolution supporting such action was adopted.

2. Election of Officers, Members, and Consultants

President:	B. Donn.
Vice-Presidents:	C. Chapman, B. Levin.
Secretary:	J. Rahe.
Organizing Committee:	O. V. Dobrovolsky, M. Festou, L. Grossman, L. Kresak, D. L. Matson, D. Morrison, N. Richter, Z. Sekanina, G. W. Wetherill.
New Members:	D. A. Andrienko. D. B. Beard, E. L. G. Bowell, W. Brunk, L. G. Burlaga, W. A. Deutschman, A. I. Ershkovich, H. Fechtig, P. Feldman, M. Festou, R. H. Giese, I. Halliday, M. A. Hanner, B. W. Hapke, A. W. Harris, M. F. A'Hearn, K. Ibadinov, W. M. Jackson, R. F. Knacke, L. Kohoutek, C. T. Kowal, C. F. Lillie, O. T. Matsuura, D. Meisel, P. Millman, H. A. Perez-de-Tejada, C. B. Pilcher, E. M. Pittich, F. Scaltriti, G. J. Veeder, L. L. Wilkening, D. Yeomans, V. Zappala.
Consultants:	V. A. Bronshten, R. D. Chapman, E. Grun, W. K. Hartman, R. Newburn, R. Ong, M. Shimizu, R. Zerrull.

3. Resolutions

The following resolutions were adopted for Commission 15.

A. Whereas attendance at the Commission Meetings is essential to the work of the Commission, be it resolved that only those members shall be nominated as officers or as members of the Organizing Committee whose presence at the General Assembly seems reasonably assured.

B. Whereas it is now nearly impossible to correlate, and therefore to interpret, the diverse photometric data on comets, Commission 15 hereby requests the chairman of the comet filters working group to seek funds for purchasing filters for worldwide distribution and/or resale.

20 August 1979

SCIENTIFIC SESSIONS

During the scientific sessions, 15 papers were presented:
D. Morrison: Status of Comet Mission
B. Zellner: Multicolor Photometry of Distant Comets

L. Biermann: Origin of Cometary Nuclei
T. Fäy: Light Curve of the Nucleus of Comet d'Arrest
E. Gerard: The OH Radical in Comets - Observations and Interpretation of 18 cm Emission
W. M. Jackson: Lifetime of OH in Comets
C. R. O'Dell: Statistical Equilibrium in Cometary C_2
A. F. Ershkovich: Theory of Solar Wind - Comet Tail Coupling
J. C. Brandt: Morphology of Cometary Plasma Tails
V. Vanysek: Isotopes in Comets
A. H. Delsemme: Distribution and Origin of Ions in Cometary Heads
O. V. Dobrovolsky, N. N. Kiselev, G. P. Chernova, F. A. Tupieva, N. V. Narizhnaia: On the Nature of Dust Grains in the Atmosphere of Comet Ashbrook-Jackson 1977 g.
O. V. Dobrovolsky, Kh. I. Ibadinov, S. Aliev: Sublimation of Comet Nucleus Models of Solid CO_2 and Frozen Water Solutions of Some Organic Compounds in the Conditions Imitating the Cosmic Ones
O. V. Dobrovolsky, N. N. Kiselev, G. P. Chernova: The Polarization-Phase Angle Dependence for Cometary Atmospheres
A. F. Bogomolov, N. N. Kroupenio, V. A. Nazarkin, G. I. Petrov, V. F. Volobuev, L. E. Ovchinnikov: Radiophysical System to Study Internal Structure of Comet Nuclei from Spacecraft

COMMISSION 16: PHYSICAL STUDY OF PLANETS AND SATELLITES (ETUDE PHYSIQUE DES PLANETES ET DES SATELLITES)

Report of Meetings 14, 15, 16, 17, 18 August 1979

PRESIDENT: Tobias Owen

14 August 1979

COMMISSION BUSINESS

Presiding: Tobias Owen, President of Commission 16
Edward Anders, President of Commission 17

1. Merger of Commissions 16 and 17

The first item of business was a review of the action taken to merge Commissions 16 and 17 into a single Commission. The resolution forwarded to the Executive Committee and the final talley of votes are appended. The resolution was accepted by the Executive Committee and the General Assembly of the IAU during the Montreal meetings. The new Commission continues with the title and number of Commission 16 (Physical Study of Planets and Satellites).

2. Election of Officers for 1979-1982

The following slate was presented and unanimously elected by the members present:

President: Bradford A. Smith
Vice-President: Victor G. Teifel'
Executive Secretary: Garry E. Hunt
Organizing Committee: E. Anders, J. E. Blamont, M.S. Bobrov, A. Brahic, W.M. Kaula, B.J. Levin, V.I. Moroz, D. Morrison, T. Owen, G.W. Wetherill, A. Wosczyk.

The retiring president expressed his regret on behalf of the Commission members that the ill health of C.P. Florensky prevented him from serving as an officer in the new administration.

3. Meetings of Interest to Commission Members

Attention was called to the following meetings that will be held prior to the next General Assembly:

A. The Satellites of Jupiter
Org.: Dr. David Morrison
Institute for Astronomy
University of Hawaii
2680 Woodlawn Drive
Honolulu, Hawaii 96822, USA

Place: Hawaii
Time: 13-16 May 1980

B. Infrared Astronomy

Org.:	Dr. Dale Cruikshank Dr. Gareth Wynn-Williams Institute for Astronomy University of Hawaii 2680 Woodlawn Drive Honolulu, Hawaii 96822, USA	Place: Time:	Hawaii 23-28 June 1980

C. COSPAR - General Meeting

Contact for Information:

Dr. James Arnold Department of Chemistry University of California at San Diego La Jolla, California 92093, USA	Place: Time:	Budapest, Hungary 2-14 June 1980

D. Royal Astronomical Society - Uranus and the Outer Planets

Org.:	Dr. Garry Hunt Laboratory for Planetary Atmospheres Department of Physics and Astronomy University College London Gower Street, London WC 1E 6BT, UK	Place: Time:	Bath, England 7-9 April 1981

This is by no means an exhaustive list; members wishing to bring additional meetings to the attention of the general membership should contact the incoming Executive Secretary, Dr. Hunt.

4. New Infrared Telescope Facility

William Brunk of NASA headquarters made a brief announcement of the successfrl inauguration of the three-meter infrared telescope built under NASA sponsorship on Mauna Kea in Hawaii. Astronomers interested in using this facility should contact Dr. Eric Becklin, Institute for Astronomy, University of Hawaii, 2680 Woodlawn Drive, Honolulu, Hawaii 96822, USA.

5. Saturn Ring Plane Crossings

Dr. M.S. Bobrov has called attention to the rare opportunities for unusual observations of Saturn and its system of rings and satellites that will be afforded by the three passages of the Earth through the plane of the rings during the next twelve months. These observations include studies of the rings themselves, mutual phenomena (eclipses and occultations) exhibited by the satellites, and a search for new satellites near the outer edges of the ring. The dates and times of ring plane crossings furnished by Dr. Bobrov are as follows:

1 st Crossing: 27 October 1979, 3.6 hours UT	2nd Crossing: 12 March 1980, 16.5 hours UT
Sun in Ring Plane: 3 March 1980, 3.5 hours UT	3rd Crossing: 23 July 1980, 4.2 hours UT

6. Other Commission Business

There was a brief discussion of the future structure and function of the newly merged Commission. The members present expressed support for a continuation of the triennial report on progress in Solar System studies. They endorsed the proposal by Dr. Anders that this be prepared as a balanced review of the highlights of the entire field rather than a collection of reports of activities from different observatories, institutes, or countries. There was also approval for the establishment and continuation of only those Working Groups that have a specific and well-defined task.

Members with ideas about any of these matters are invited to write to the new officers with their suggestions.

7. Resolution for the Merger of Commissions 16 and 17

Whereas the Moon is no longer the only celestial body whose surface can be studied at high resolution, the only body explored by spacecraft, or even the only well-studied satellite, a separate Commission devoted to the Moon is no longer needed.

Be it therefore resolved, that Commission 17: The Moon, be merged with Commission 16: Physical Study of Planets and Satellites, to form a new Commission that shall retain the title and number of the existing Commission 16. The membership of the new Commission shall consist of those members of the existing Commissions 16 and 17 who have expressed their intent to join as well as other members of the IAU who have been invited to join on the basis of their research interests. This Merger will permit studies of the Moon to be continued for mutual benefit, in closer association with studies of other planetary bodies.

Note: This Resolution was passed by the General Assembly; Commissions 16 and 17 are now combined as Commission 16.

15 August 1979

I. JOINT SESSION WITH COMMISSIONS 40 AND 44:

STRATEGIES FOR THE SEARCH FOR LIFE IN THE UNIVERSE

Organizing Committee: F. Drake, T. Owen, M. Papagiannis (Chmn.), B. Zuckerman (USA); N. Kardashev, I. Shklovsky, V. Troitsky (USSR); J. Ribes (FRANCE); B. Lovell, M. Rees (UK); G. Marx (Hungary); J. Jugaku (Japan).

INTRODUCTION: An opportune occasion for this meeting (M. Papagiannis)

I. ALTERNATIVE VIEWS ON THE VALUE OF N

Co-Chairmen: I. Shklovsky and P. Morrison

1. N. is very small (M. Hart)
2. N is very large (T. Kuiper)
3. N is neither very small nor very large (F. Drake)
4. N is either very small or very large (M. Papagiannis)
5. Comments and Discussion
6. Summary (P. Morrison)

II. STRATEGIES FOR SETI THROUGH RADIO WAVES
Co-Chairmen: B. Oliver and J.Jugaku
1. Progress in USA and Canada (B. Zuckerman)
2. Programs in the USSR (V. Troitsky)
3. Large and Small Alternatives for the Future (S. Gulkis)
4. Comments and Discussion

III. SEARCH FOR EARLY FORMS OF LIFE IN OUR AND OTHER PLANETARY SYSTEMS
Chairman: J. Greenstein
1. The search in our solar system (G. Soffen)
2. Future spectroscopic possibilities for other systems (T. Owen)
3. Search for planets in other solar systems through:
   a. Astrometric methods (G. Gatewood)
   b. Interferometric methods (D. Currie)
   c. Spectroscopic methods (W. Heacox)
   d. The space telescope (W. Baum)
   e. Summary (D. Black)

IV. MANIFESTATIONS OF ADBANCED COSMIC CIVILIZATIONS
Chairman: S. von Hoerner
1. Radio leakage and eavesdropping (W. Sullivan)
2. Discussion
3. Summary (S. von Hoerner)

II. JOINT SESSION WITH COMMISSION 4

ROTATION OF MARS - G. de Vaucouleurs
REPORT AND RECOMMENDATIONS OF THE JOINT WORKING GROUP ON CARTOGRAPHIC COORDINATES AND ROTATIONAL ELEMENTS OF PLANETS AND SATELLITES - M.E. Davies

16 August 1979

I. SCIENTIFIC SESSIONS

A) SATELLITES, RINGS, AND OTHER TOPICS
Chairman; D. Morrison, University of Hawaii

D. Morrison/U. Hawaii	Introduction and Overview of Recent Work
H. Masursky/USGS	Geology of Io
T. Johnson/JPL	Io Volcanoes and Volatiles
C. Pilcher/U. Hawaii	Images of Jupiter's Sulfur Ring
J. Caldwell/SUNY	Atmosphere of Titan
B. Zellner/U. Arizona	Dilter Photometry of Faint Outer Satellites
D. Cruikshank/U. Hawaii	Compositions of the Satellites of Saturn, Uranus, and Neptune.
T. Owen/SUNY	Rings of Jupiter
E. Becklin/U. Hawaii	Detection of the Rings of Jupiter at 2.2 μm
W. Irvine/U. Mass.	Azimuthal Brightness Variations in Saturn's Rings
E. Epstein/Aerospace	Saturn's Rings; 3 mm Observations
W. Brunk/NASA	NASA Infrared Telescope Facility
W. Sjogren/JPL	Venus Gravity Field

B) PLANETARY ATMOSPHERES
Chairman: G. E. Hunt, UK

1. Structure of the Clouds and Upper Atmosphere of Venus from the Pioneer Venus Orbital Infrared Radiometer (VORTEX) R. Beer (USA)
2. The 120-300 cm^{-1} Spectrum of Venus in March 1979 G. Orton (USA)
3. Chemistry of the Venus Atmosphere R. Prinn (USA)
4. The Nightside Ionosphere of Venus by the Pioneer Venus Radio Occultation A.J. Kliore and J.R. Patel (USA)
5. Martian Weather Systems Observed by Viking G.E. Hunt (UK)
6. UV Observations of Planetary Atmospheres with IUE J. Caldwell (USA)

Break

7. Spatial Distribution of Methane to Ammonia Ratio in the Jovian Atmosphere B. Lutz and T. Owen (USA)
8. 21cm Observations of Jupiter I. de Pater (NK)
9. Spectrophotometry of Jupiter and Saturn J.T. Bergstralh (USA)
10. An Observed Relationship Between Planetary Albedos and Solar Activity G.W. Lockwood (USA)
11. The Uranus Atmosphere: Increases in its Microwave Brightness Temperature M.J. Klein (USA)

II. EVENING SUMMARY SESSION ON STRATEGIES FOR SETI
Chairman: L. Goldberg
1. Review of morning sessions (F. Drake)
2. Review of afternoon sessions (M. Papagiannis)

17 August 1979

I. JOINT DISCUSSION WITH COMMISSIONS 17 and 44: EXPLORATION OF THE SOLAR SYSTEM
Organizing Committee: E. Anders, M. Marov, T. Owen

A. SMALL BODIES
Chairman: E. Anders
1. Isotopic Anomalies in Meteorites as Clues to the Early History of the Solar System -R. Clayton
2. Identification of Asteroid Classes and their Possible Relationships to Meteorites -D. Morrison
3. Chronologies for Planetary Surfaces from Bombardment Histories -G. Neukum

B. VENUS - THE SURFACE AND THE ATMOSPHERE
Chairman: B.A. Smith
4. Radar Studies of Venus:
   A) Ground-Based Observations -D. Campbell
   B) Pioneer Venus Observations -G. Pettengill
5. Crustal Evolution of Venus Compared with the Other Terrestrial Planets -H. Masursky
6. The Atmosphere of Venus - Pioneer Venus Results -D. Hunten

C. JUPITER - VOYAGER RESULTS
Chairman: T. Owen
7. The Meteorology of Jupiter -G. Hunt

8. The Satellites of Jupiter -T. Johnson
9. Vulcanism and the Crustal Structure of Io -B.A. Smith
10. Electric Current Induced Vulcanism on Io -T. Gold
11. The Magnetosphere of Jupiter -N. Ness
12. A Film of Jupiter at 21 cm -I. de Pater

18 August 1979

SCIENTIFIC SESSION: THE MOON
Chairman: J.D. Mulholland

1. Dynamical evolution of the early Earth-moon system -H.Alfven and J. Hill
2. Present-day dynamics of the Earth-moon system -R.W. King
3. Free librations and the lunar dissipation parameter Q -O. Calame
4. Lunar secular acceleration and the variation of gravitation -L.V. Morrison
5. A catalogue of lunar transient phenomena -W.S. Cameron
6. Lunar crater distribution and surface unit ages -E. Whitaker
7. Lunar surface composition from thermal emission spectra -T.H. Morgan and A. Potter

COMMISSION 19: ROTATION OF THE EARTH (ROTATION DE LA TERRE)

Report of Meetings, 14, 15, 16, and 20 August 1979

PRESIDENT: R. O. Vicente SECRETARY: D. D. McCarthy

14 August 1979

After opening the meeting, the President asked for a few moments of silence in memory of the members of the Commission who died during the last three years. The Commission then heard reports on recent research. Wm. Markowitz presented a paper entitled "Independent Polar Motions, Optical and Doppler; Chandler Uncertainties". He reported that the true Chandler path, C, is indeterminate; hence it cannot be used to detect or predict earthquakes. The true free period is also indeterminate. Independent techniques, optical and Doppler, were used to separate real variations from observational. Differences in dynamical properties of forced and free oscillations were then used to separate random variations due to the annual motion, A, from those of C. Since 1892.8 the amplitude and phase of C have changed smoothly, and, what is puzzling, independently, the amplitude remained constant during three intervals which total 45 years. Models based on a strongly damped oscillator, beating of two free periods, impulsive torques, and excitation by earthquakes are not in accord with the observations. Chandler damping is weak; Q exceeds 200. Similar secular motions are shown by ILS, IPMS, and Doppler. A theoretical, 18.6-year, forced component, nearly a libration, due to ocean tides was found in ILS and IPMS. This explains in part the empirical 24-year libration. Long series of observations are needed to separate the effects.

O. Calame reported on recent determinations of Universal Time from the EROLD (Earth Rotation from Lunar Distances) campaign. Orroral Observatory has been reporting data obtained from laser ranging to the Moon since October, 1978. These have been combined with data obtained regularly from McDonald Observatory and the accuracy of the Universal Time determinations made from this combined set are of the order of ±0.2 to ±0.3 milliseconds for five to ten-day means.

J. Fanselow, in a paper entitled "Measurements of Earth Orientation Using Very Long Baseline Interferometry", discussed the development of this technique by the Jet Propulsion Laboratory using their Deep Space Network. Twenty-eight measurements of Earth orientation have been made between 1971 and 1978, and half of these show an accuracy better than ±0".025. He also summarized a catalog of 50 source positions developed in the course of this work.

P. E. G. Paquet read two papers written by D. Djurovic. The first, "Secular and Periodical Variations of UT2-TAI", presented the results from the author's study of Bureau International de l'Heure (BIH) data. Using residuals from a second-order polynomial fit to the UT2-TAI information, he found apparent periodic variations with periods of 7.4, 3.5, 2.3, 1.6, 1.0, and 0.5 years. The author stresses that these periods are not constant and that the variations are particularly large for the 2.3 and 1.6-year terms. In the second paper entitled "Variations of the Seasonal Inequalities and Irregular Fluctuations of UT2-TAI", Djurovic investigated to what extent the annual and semi-annual terms in the rotational speed of the Earth are constant, and determined the epochs and magnitudes of the largest fluctuations in UT2. He found significant deviations in the annual and semi-annual terms and that the formula for UT2-UT1 does not correspond rigorously to the seasonal variation in the Earth's rotation. Significant non-random, non-periodic fluctuations in UT2-TAI were also found to exist.

S. K. Runcorn presented a paper, "Paleontological Determination of Lunar Tidal Friction in the Geological Past", in which he discussed the rotation of the Earth from a geophysicist's point of view. He related recent studies of the length of the day determined from Devonian coral fossils and nautiloid fossils to modern astronomical results. He has concluded that there may have been changes in the lunar tidal torque over geologic time, but that uncertainties in the data still permit the lunar tidal torque to be considered as a constant.

B. Guinot presented a report on the activity of the BIH in the determination of Earth rotation parameters using optical and Doppler data. He showed that there are presently non-uniformities in the data since fewer observations are being made on weekends and that this practice affects the results of the BIH. He also mentioned that the accuracy of both the astrometric and Doppler data appears to be deteriorating and that the new techniques might be used now, even with precisions that are somewhat worse than those which are expected to be achieved eventually. Guinot also reported on the recent adoption of the 1979 BIH System as a replacement for the 1968 system. He stated that systematic errors inherent in the adoption of the 1968 system are removed in the new system, and that, although extensive use was made of the Doppler data in the determination of the new system, purely astrometric data obtained in recent years show the same corrections to the 1968 system. Research activity at the BIH is continuing in the fields of the re-evaluation of the Earth rotation parameters since 1962, determination of polar coordinates and UT1 using artificial Earth satellites, use of EROLD data, and in the theory of the rotation of the Earth.

<u>15 August 1979</u>

R. O. Vicente opened a joint meeting of Commissions 19 and 31 (Time) with some introductory remarks concerning Project MERIT (Monitoring of Earth's Rotation and Intercomparison of Techniques). The President of Commission 19 stressed the need for communication between investigators using different techniques, and that the geophysical implications resulting from this project will be quite important for developing countries.

G. A. Wilkins, the Chairman of the Joint Working Group on the Determination of the Rotation of the Earth presented the report of the Group. He stated that the Group had been set up by the Presidents of Commissions 19 and 31 on the recommendation of IAU Symposium No. 82 (Time and the Earth's Rotation) which was held at Cadiz, Spain in May, 1978. The members of the Group were chosen to represent a wide range of interests and countries and are as follows: L. Aardoom, R. J. Anderle, W. E. Carter, B. Elsmore, M. Feissel, W. J. Klepczynski, Yu. L. Kokurin, Y. Kozai, W. G. Melbourne, F. Nouel, P. E. G. Paquet, E. C. Silverberg, D. E. Smith, G. Veiss, R. O. Vicente, G. A. Wilkins, Ya. S. Yatskiv, K. Yokoyama. The advice and assistance of several other persons has also been obtained.

The Group held its first meeting at Columbus, Ohio, on 5-6 October 1978 following the "GEOP" Conference. The meeting was attended by nine members and several other interested persons. The Group decided to prepare a proposal for a special period of international collaboration in order to <u>M</u>onitor the <u>E</u>arth's <u>R</u>otation and <u>I</u>ntercompare the <u>T</u>echniques of observation and analysis. An introduction to the project was widely circulated a few months ago. Forty copies of the "Draft Proposal for Project MERIT" were recently distributed by post. Unfortunately the 120 copies sent by air freight to Montreal were not available on 15 August, but they were distributed on 16 and 17 August.

The Group's proposal specifies two principal objectives for Project MERIT. The first objective is to provide a basis for recommendation on the future international services for Earth rotation; for this purpose the Group will intercompare the techniques, identify the requirements of users, and specify the standards to be met. The second objective is to obtain and analyze new, precise data on Earth rotation in order to improve our knowledge and understanding of this important phenomenon, which has many applications. A day-to-day knowledge of Earth rotation (i.e. the direction of the axis of rotation, or polar motion, and the orientation of the prime meridian, or Universal Time) is required for surface and air navigation, for land and geodetic surveying, for the navigation of spacecraft and for fundamental astrometry. In addition, it is required for research in th dynamics of the Earth-Moon system, for determining indirectly properties of the Earth's interior (such as density, elasticity, and viscosity), and for studying interactions between the core and mantle and the effects of earthquakes and winds. Although Project MERIT will last for only a limited period, it is hoped that the results obtained will be of both immediate and lasting value and will lead to the continuing provision of data that are more precise, more accurate, and more frequent than those currently available, bearing in mind that the services must be economical and not critically dependent on a single national organization.

The draft program envisages that the remainder of 1979 will be spent in completing the proposal, seeking support and starting the detailed planning. In the second half of 1980 there will be short campaigns of observation, reduction and analysis by all techniques with the principal aim of studying the arrangements that would be required to use the new techniques for a regular service. During 1981/2 the results of the short campaign will be reviewed, planning for the main campaign will be continued, and new equipment and software will be installed and commissioned. The main campaign will be held during 1983/4 and will last at least one year so that seasonal effects can be determined. The full analysis of the data and comparison of the results will require much additional effort, but it is expected that clear recommendations about the future services will be made in 1985.

In order that Project MERIT shall go ahead it is first of all necessary that the scientific Unions, especially the IAU and IUGG, make clear their support for its objectives and program. Then the Group will need to obtain international financial support for meetings and comparatively minor common expenses so that the work can be properly planned and coordinated. Then it will be necessary to confirm that national support for the observational work and for the reduction and analysis of the data will be obtained - already the Group has clear indications that support will be provided for such an international project. Finally the Group needs the <u>active</u> support of many individuals to take part in the organizational work of planning the project and reviewing the results, and so to ensure that the full benefits are gained from the observations obtained during the period of Project MERIT.

Following this report by Wilkins, D. Mulholland described the operation of the EROLD campaign. He explained that a two-phase operation was planned, the first phase being the operation of one station for the determination of Universal Time from lunar laser ranging data on a quasi-daily basis. This has been accomplished through the regular operation of the McDonald Observatory. The second phase calls for at least two stations to be operated with quasi-daily frequency. Although data is now available from Orroral since October, 1978, the frequency of the observations is not yet sufficient to begin the second phase. Observational data is now being contributed from the Crimean Astrophysical Observatory also.

R. J. Anderle reported on the possible use of the Doppler satellite technique for the determination of UT1. In an experiment involving the use of 110 days of data in 1977 a systematic error of about 6 milliseconds over a thirty-day period was found in the UT1 estimates derived from Doppler ranging to artificial satellites.

F. Nouel presented a paper entitled "MEDOC Experiment or Polar Motion from Satellite Doppler Tracking Network" in which he described the status of this campaign to determine polar coordinates using Doppler satellite techniques. This program is designed to determine polar coordinates by means of a civilian agency. Comparison of the MEDOC results with the DPMS (Doppler Polar Motion Service) results shows rather good agreement, with the MEDOC polar coordinates showing slightly more scatter due to the smaller number of reporting ground stations. Nouel suggested that in the future a replacement satellite solely for the determination of polar coordinates should be considered if the present navigational satellites are replaced by the Global Positioning System (GPS).

16 August 1979

A joint meeting with Commissions 4, 7, 8, and 31 was held. (See report of Commission 4.)

20 August 1979

In a business meeting, lists of new members and consultants were approved. S. Yumi announced his retirement from the International Latitude Observatory at Mizusawa, Japan, and K. Yokoyama was appointed to replace him as Director of the Central Bureau of the International Polar Motion Service (IPMS). In other business, the President reported that a letter was sent to the General Secretary requesting IAU sponsorship of the Second International Colloquium on Reference Coordinate Systems for Earth Dynamics to be held in Poland in September, 1980. A resolution in support of the astrometric work of Hipparcos was passed as was a resolution thanking S. Yumi for his work as Director of the IPMS.

In a scientific session following the business meeting, S. Yumi and K. Yokoyama presented reports on the work of the IMPS and the Working Group for Polar Coordinates. The Working Group has completed the task of re-reducing the past observations of the International Latitude Service (ILS) since 1899, and these data are now available on magnetic tape from the Central Bureau of the International Polar Motion Service at the International Latitude Observatory, Mizusawa-Shi, Iwate-Ken, 023 Japan. This information has already been sent to sixty addressees and the details of the computations are scheduled to be published soon. Yumi stressed that although the computational work was done mainly at the International Latitude Observatory of Mizusawa, many institutions and individuals cooperated in this work. Yokoyama reported that in the reduction of these data, a constant of 0".087 was added to the mean latitudes of all of the ILS stations. Such a change does not change the position of the Conventional International Origin (CIO), but it will be reflected in a change of the z-term. In discussion, Wm. Markowitz pointed out that the station latitudes used for the definition of the CIO were adopted for the standardization of the calculation of the polar coordinates and should not be considered as the astronomical mean latitudes of the stations.

In the report of the IPMS, it was mentioned that the results presently being obtained with the new observational techniques are being investigated for possible inclusion in the IPMS system at a future date. Currently the IPMS publishes polar coordinates determined only from the observational results of the ILS stations as well as polar coordinates determined from latitude observations of the stations reporting to the IPMS and a series which makes use of both time and latitude observations. A precision of ±".015 for a monthly value of the polar coordinates is typical for the IPMS data. A new observing list for the ILS visual zenith telescopes to be used from 1980-92 was presented. Yumi reported that three stations of the ILS have now been equipped with photographic zenith tubes, and that the Ukiah station will receive a photogtaphic zenith tube from the U. S. Naval Observatory. The 65-cm PZT of the U. S. Naval Observatory will take the place of the Gaithersburg visual zenith tube in the future. It was also mentioned that the IPMS has computed a series of UT1-TAI data from the reported astronomical data and that this information will be published in a forthcoming Publication of the International Latitude Observatory of Mizusawa. Reporting on the work of the Mizusawa Observatory, Dr. Yumi stated that, in addition to the regular observations with the visual zenith telescope, PZT, floating zenith telescope, Danjon astrolabe, and Doppler tracking of navigational satellites, work is also progressing on the use of the Tsubokawa automatic astrolabe. Future work in the area of laser ranging and VLBI is also being investigated.

V. Milovanovic presented a paper in which he discussed the possibility of further improvements to the newly re-reduced ILS polar coordinates. He pointed out that these data have not been corrected for differential temperature effects on the spirit levels. Milovanovic also suggested a method for determining the temperature gradient along the level, from the data, since temperatures along the level were not recorded at the time of the observations.

S. Debarbat reported on the measurement of Earth tide effects at Centre d' Etudes et de Recherches Geodynamiques et Astronomiques (CERGA). Geophysical measurements were made at different locations at the observatory and these appeared to be independent of the location.

P. Brosche discussed his research on the effect of ocean tides on the rotation of the Earth during geological times. Using models for the location and depth of the oceans at different periods in the Earth's history, he finds that the lunar tidal torque in the Cretaceous was about half of the present value. Brosche concludes that it is not possible to use just a simple power law to describe the effects of the lunar tidal torque on the rotation of the Earth in the past.

L. Randic presented a paper on the asymmetry in the determinations of polar coordinates from the observations in the northern and southern hemispheres. Comparing results from the BIH with those determined from observations at the Cape of Good Hope and Santiago, he showed that an apparent asymmetry in the polar path does exist.

S. H. Ye reported on the programs and instrumentation for the determination of Earth rotation prarmeters in the People's Republic of China. An astro-geodynamic project is in progress to measure geodynamic phenomena, maintain an accurate reference frame, and to detect geophysical motion. Observations are being made with three photo-electric transit instruments, three Danjon astrolabes, four photoelectric astrolabes, and one photogtaphic zenith tube, and a number of visual zenith telescopes. A laser ranging network is being updated, and it is expected that ranging data from the LAGEOS satellite accurate to 20 cm will be available in one to three years. A VLBI network is planned to be completed in about five years.

RESOLUTIONS

Commission 19,

Noting with regret that Dr. Shigeru Yumi is resigning as Director of the Central Bureau of the International Polar Motion Service, at Mizusawa,

Thanks Dr.Yumi for his services in directing the work and the staff of the

Central Bureau for 18 years, during which time the IPMS polar motion was established with high accuracy and the ILS polar motion since 1899.9 was revised on a homogeneous basis, and for the prompt publication of results needed for astronomy, geodesy, and geophysics.

The above resolution was unanimously approved by the Commission. Other resolutions discussed and approved by the Commission are included in the reports of other Commissions and are published in this volume.

JOINT MEETING OF COMMISSIONS 19 AND 31 ON 15 AUGUST 1979 TO CONSIDER THE REPORT OF THE WORKING GROUP ON THE DETERMINATION OF THE ROTATION OF THE EARTH

Summary of presentation by the Chairman of the Group (Dr. G.A. Wilkins)

The Chairman of the Joint Working Group on the Determination of the Rotation of the Earth stated that the Group had been set up by the Presidents of Commissions 19 and 31 on the recommendation of IAU Symposium No. 82 (on Time and the Earth's Rotation), which was held at Cadiz, Spain, in 1978 May. The members of the Group were chosen to represent a wide range of interests and countries and are as follows:

L. Aardoom, R.J. Anderle, W.E. Carter, B. Elsmore, M. Feissel, W.J. Klepczynski, Yu L. Kokurin, Y. Kozai, W.G. Melbourne, F. Nouel, P.E.G. Pacquet, E.C. Silverberg, D.E. Smith, G. Veiss, R.O. Vicente, G.A. Wilkins, Ya S. Yatskiv, K. Yokoyama. The advice and assistance of several other persons has also been obtained.

The Group had held its first meeting at Columbus, Ohio, on 5/6 October 1978, following the "GEOP" Conference. The meeting was attended by 9 members and several other interested persons. The Group decided to prepare a proposal for a special period of international collaboration in order to Monitor the Earth's Rotation and Intercompare the Techniques of observation and analysis. An introduction to the project was widely circulated a few months ago. 40 copies of the "Draft Proposal for Project MERIT" were recently distributed by post. Unfortunately the 120 copies sent by air freight to Montreal were not available on August 15, but they were distributed on August 16 and 17.

The Group's proposal specifies two principal objectives for Project MERIT. The first objective is to provide a basis for recommendations on the future international services for Earth-rotation; for this purpose the Group will intercompare the techniques, identify the requirements of users, and specify the standards to be met. The second objective is to obtain and analyse new, precise data on Earth rotation in order to improve our knowledge and understanding of this important phenomenon, which has many applications. A day-to-day knowledge of Earth rotation (i.e. the direction of the axis of rotation, or polar motion, and the orientation of the prime meridian, or universal time) is required for surface and air navigation, for land and geodetic surveying, for the navigation of spacecraft and for fundamental astrometry. In addition, it is required for research in the dynamics of the Earth-Moon system, for determining indirectly properties of the Earth's interior (such as density, elasticity and viscosity), and for studying interactions between the core and mantle and the effects of earthquakes and winds. Although Project MERIT will last for only a limited period it is hoped that the results obtained will be of both immediate and lasting value and will lead to the continuing provision of data that are more precise, more accurate, and more frequent than those currently available, bearing in mind that the services must be economical and not critically dependent on a single national organisation.

The draft programme envisages that the remainder of 1979 will be spent in completing the proposal, seeking support and starting the detailed planning. In the second half of 1980 there will be short campaigns of observation, reduction and analysis by all techniques with the principal aim of studying the arrangements that would be required to use the new techniques for a regular service. During 1981/2 the results of the short campaign will be reviewed, planning for the main campaign will be continued, and new equipment and software will be installed and commissioned. The main campaign will be held during 1983/4 and will last at least one year so that seasonal effects can be determined. The full analysis of the data and comparison of the results will require much additional effort but it is expected that clear recommendations about the future services will be made in 1985.

In order that Project MERIT shall go ahead it is first of all necessary that the scientific Unions, especially IAU and IUGG, make clear their support for its objectives and programme. Then the Group will need to obtain international financial support for meetings and comparatively minor common expenses so that the work can be properly planned and coordinated. Then it will be necessary to confirm that national support for the observational work and for the reduction and analysis of the data will be obtained - already the Group has clear indications that support will be provided for such an international project. Finally the Group needs the active support of many individuals to take part in the organizational work of planning the project and reviewing the results, and so to ensure that the full benefits are gained from the observations obtained during the period of Project MERIT.

Resolution adopted at the joint meeting of Commissions 4, 19 and 31 on 20 August 1979.

Commissions 4, 19 and 31

endorse the proposal of the joint working group on the determination of the rotation of the Earth for a special period of international collaboration in the monitoring of Earth-rotation and in the intercomparison of the techniques of observation and analysis,

recognize that the responsibility for the organization of this project MERIT should be shared with the International Union of Geodesy and Geophysics, and

request that the national and international organizations concerned give full technical and financial support to the development of the proposal and to the implementation of the project.

COMMISSION 20: POSITIONS AND MOTIONS OF MINOR PLANETS, COMETS, AND SATELLITES (POSITIONS ET MOUVEMENTS DES PETITES PLANETES, DES COMETES ET DES SATELLITES)

Report of Meetings, 14, 16, and 21 August 1979

PRESIDENT: B. G. Marsden. SECRETARY: E. Roemer.

14 August 1979

SESSION 1: ADMINISTRATION

Remarking upon the unprecedented 40-minute delay in the opening of the first session of Commission 20, the President opened the meeting and welcomed those in attendance. C. Froeschlé was appointed to serve as interpreter and E. Roemer as secretary. All present stood in silent respect as the names of members and former members who had died during the triennium were read: M. Itzigsohn, V. V. Michkovitch, K. Reinmuth, and A. Schmitt.

The report of Commission 20 for the triennium 1976-78 had been circulated to members of the Commission shortly after its submission for publication in IAU Trans. XVIIA. Several corrections were given in the Commission Circular No. 9.

The list of proposed new members of the Commission, some contingent on election of the individuals to membership in the IAU, as well as the list of consultants for the triennium 1979-82, were introduced by the President. The proposals for new officers of the Commission, G. Sitarski, President, and E. Roemer, Vice-President, and for members of the Organizing Committee, Yu. V. Batrakov, M. P. Candy, F. K. Edmondson (who replaces P. Herget at Herget's request), Y. Kozai, L. Kresák, B. G. Marsden, J. Schubart, G. E. Taylor, and P. Wild, were accepted unanimously. (Kozai, Kresák, and Taylor were confirmed in their capacities as chairmen for the triennium 1979-82 of the Working Groups on satellites, comets, and occultations, respectively.)

Noting that the membership of the Commission now numbers well over a hundred, the President wondered whether mailings to so large a number might become a problem for future presidents. He suggested that some who were no longer active participants in Commission affairs might prefer to retire. It was proposed that three names be dropped because of lack of response to two mailings, and one person resigned because of membership in too many commissions. It was emphasized, in response to a comment, that membership in commissions was a separate matter from membership in the IAU.

The President next reported on the status of several resolutions and proposals that were under active review: (1) Financial support for the Minor Planet Center. A slight increase over the amount for the triennium 1976-79 had been requested. (2) Pursuant to the recommendation approved in 1976, the President was particularly pleased to note that the 1980 volume of Minor Planet Ephemerides, just received from the Institute of Theoretical Astronomy, Leningrad, includes high-precision current osculating elements for all but the lost minor planets through (2095). (3) Separate proposals have been made by T. Gehrels and by M. B. Protitch concerning magnitude data in the opposition ephemerides of minor planets. (4) A request was made by F. K. Edmondson for endorsement of proposals for continuing financial support of the Indiana minor planet program. (5) A proposed restatement of the policy governing naming of minor planets has been submitted by B. G. Marsden in his capacity as Director of the Minor Planet Center. (6) Related proposals concerning the reference system of ephemerides have been made by S. Ferraz-Mello (satellites) and D. K. Yeomans (comets). (7) A question was raised and a proposal made by J. Kovalevsky regarding the matter of requirements and priorities for astrometric observation of minor planets. (8) A resolution concerning space missions to comets, already submitted to the General Assembly by Commission 22, was brought to the attention of Commission 20 by P. R. Weissman. Following considerable discussion of the several proposals, particularly of that concerning space missions to comets, it was agreed that consultations on the matters involved would continue and that refined versions of proposed resolutions would be considered at the second administrative session of the Commission. A straw vote on whether a less desirable alternate mission should be mentioned in the resolution on space missions to comets showed members of the Commission to be equally divided on the question.

SESSION 1 (CONTINUED): MINOR PLANETS

With the President continuing in the chair, scientific reports were then presented as follows:

(1a) T. Gehrels: Magnitude Predictions for Minor Planets — Observational Data. Gehrels first reported briefly on the status of the book on asteroids, which was based on papers presented at the conference in Tucson in March 1979, and also on the current status of the Pioneer flyby of Saturn, which was in progress at the time of the General Assembly. Using as an example observations of $\underline{V}$ vs. phase angle, $\underline{B}$-$\underline{V}$ vs. phase angle, and $\underline{U}$-$\underline{B}$ vs. phase angle for (6) Hebe, he then explained that the current definition of absolute magnitude involves linear extrapolation to zero phase, without allowance for the opposition effect. Because there are considerable differences among minor planets in the dependence of magnitude on phase, it is proposed to extrapolate the actual magnitude vs. phase relation, including the opposition effect, whenever it is known for individual objects, and otherwise to use a mean phase relation. The net effect of the proposal would be a redefinition of absolute magnitudes such that in 1982 "all asteroids would suddenly become 0.3 mag brighter." In fact, the proposed new definition of absolute magnitudes is already being used in the TRIAD file of data on asteroids and in the book from the Tucson conference.

(1b) E. Bowell: Magnitude Predictions for Minor Planets — Theory. Work by K. Lumme (Helsinki) on the theory of multiple scattering in various surfaces was described. A general theory has been developed through which the magnitude vs. phase angle behavior of objects as diverse as (1) Ceres, (4) Vesta, and the ice-covered Jovian satellite Europa can be represented. A multiple-scattering factor $\overline{Q}_V$ represents the fraction of multiply-scattered light compared with the total scattered light. A good correlation of $\overline{Q}_V$ with $\overline{p}_V$ (0°) indicates that all asteroid surfaces are similar in texture. Bowell stated that he would like to see the adoption of new absolute magnitudes based on the Lumme theory. It was agreed at the Tucson meeting in March to defer a change from the present conventional mag/phase relation of 0.023 mag/degree until 1982, but in fact the conventional relation gives a poor representation of asteroid magnitudes in general.

Because of the divergent points of view being expressed, and particularly because of actions already being taken, the President indicated that he thought it desirable for the Commission to take a position to clarify what was desired with respect to magnitude ephemerides of minor planets in the near term. On a motion by E. Roemer, the statement developed by a committee consisting of T. Gehrels, K. Lumme, B. G. Marsden, E. Roemer, E. Tedesco, and J. G. Williams at the time of the Tucson conference was adopted unanimously: "We recommend that the T. Gehrels and N. Gehrels (Astron. J. 83, 1660-1674, 1978) list of asteroid B(1,0), with a phase coefficient of 0.023 mag/degree be used until 1982. Probably, at that time, a change will be made that should be durable."

(2) Y. Kozai: Asteroids with Large Secular Orbital Variations. Secular perturbations of asteroids for which $(1 - e^2)^{1/2} \cos i < 0.86$ have been computed as functions of the arguments of perihelion. For most asteroids the eccentricities are much changed, so that also the perihelion and aphelion distances are strongly affected. However, for a few of them, including the Apollo-type asteroids, the eccentricities are little affected. These objects resemble short-period comets in their dynamical characteristics.

(3) J. Kovalevsky: What Minor Planets to Observe and How? (Presented by B. G. Marsden.) New objectives, such as the use of minor-planet positions for improvement of star catalogues, observation of occultations to improve diameters and search for companions, physical observations of any kind, and spacecraft rendezvous, have been added to such classical goals in minor-planet studies as structure of the asteroid belt, celestial mechanics, and definition of the reference system. It is necessary to track asteroids so that they can be recovered when specifically needed, but is it necessary to follow them all to high precision? Some of the new programs compete for time on major instruments that are expensive to run. It would be helpful to have priorities stated by the IAU so that those observations may be made that would yield the largest scientific return. Commission 20 might provide a list of objects for which precise positions are needed, for improvement of catalogues, for prediction of occultations, or at risk of loss.

It was agreed that the challenging questions raised by Kovalevsky should be considered at the meeting of the Commission on 21 August.

(4) P. Hemenway: Fundamental Studies of Minor Planets. Minor planets make ideal "test particles" with which to test the global uniformity of existing reference frames. Crossing-point observations were defined as referring to astrometric observation against the same reference star frames but at different times of two minor planets whose apparent paths intersect. Such observations were shown to provide powerful constraints on the solution for orbital elements and system parameters. Reduction of such observations was outlined, and the applicability of crossing-point observations to ongoing observational programs was stressed. Finally, a proposed study of fainter minor planets was described. (See also Hemenway, Herget Symposium, to be published in Celestial Mechanics.)

(5) R. L. Branham: Use of Minor Planets for Determining Equator-Equinox Corrections. Theoretical observations of minor planets 1, 2, 3, 4, 6, 7, 8, 9, and 15 were used to determine mean errors for the equinox and equator of a celestial coordinate system. The observations were calculated for a ten-year period commencing 2 Jan. 1980 under the assumptions that: (1) the minor planets could be observed from quadrature through opposition to quadrature; (2) a planet could be observed only if $m_V \leq 10.0$; (3) the mean error of unit weight was 0".350; and (4) the distribution of clear nights was determined by a random number generator. The conclusions of the study are: (1) the best site from which to determine the equinox and equator from minor planet observations is one with a minimum of 60% clear nights well distributed throughout the year; (2) an observing program should last a minimum of two years, but little is gained by observing over ten years; (3) minor planets should be observed out to quadrature, but it is not essential to observe them over their entire orbits; and (4) the best minor planets to use are numbers 1, 2, 4, 7, and 15.

(6) V. I. Orel'skaya: Preliminary Values of Corrections to the Equator and Equinox of the FK4 Obtained from Observations of Ten Selected Minor Planets. (A brief summary was given by A. S. Sochilina.) About 25 000 observations made over a time span of 25 years have been analyzed to find corrections to the equinox, corresponding to $\Delta\alpha_0$ = -0ˢ052 ± 0ˢ009, and to the equator, corresponding to $\Delta\delta_0$ = 0".00 ± 0".02. These values are in good agreement with the results found by

Fricke: $\Delta\alpha_o = -0^s.050 - 0^s.083$ T, where T is in centuries counted from 1960; $\Delta\delta_o \sim 0$.

(7) L. K. Kristensen: Equator and Equinox of FK4 from Observations of (51) Nemausa. Concluding a program initiated in 1939 by Strömgren, some 2240 observations of (51) Nemausa have been used to find corrections to the equator and equinox of the FK4. The correction to the declinations of the FK4 in the equatorial zone is $\Delta\alpha_o = +0''.03 \pm 0''.03$, and the correction affecting all right ascensions of the FK4 is $E = +0^s.044 \pm 0^s.005$ (1956). The last result relies on the corrections to Ephemeris Time given in the Astronomical Ephemeris. It may be asked whether a repeat of the campaign of 1949-1961, in which long series of accurate observations were obtained by a few observatories, would still be useful in these days of space astrometry. If the answer is in the affirmative, Kristensen offered to support those who may wish to make the necessary observations.

It was emphasized in discussion of the several preceding papers that, since minor planets are not observed from the Sun, the orbital elements of the Earth also must enter as unknowns into the solutions for corrections to the coordinate system.

SESSION 2: COMETS AND MINOR PLANETS

E. Roemer, Chairman of the Working Group on Orbits and Ephemerides of Comets, then took the chair and introduced the session with some brief remarks about the functions of the Working Group. She expressed her appreciation to those who contributed information for the section on comets of the triennial report of the Commission. Special thanks were due to E. I. Kazimirchak-Polonskaya, who had provided carefully done summaries of the work of astronomers in the U.S.S.R. for several triennial reports. Reliable predictions for the returns of periodic comets are now regularly available. And, thanks in particular to efficient work at the IAU Central Bureau for Astronomical Telegrams, ephemerides for newly-discovered comets are distributed soon after sufficient observations have been reported. A number of amateurs regularly provide for the brighter comets precise positions that are nearly as good as those measured by professional astronomers from wide-field plates. The situation with observations of the fainter comets is less secure, but, thanks in particular to the group at the Center for Astrophysics led by R. E. McCrosky, it seems that very few expected comets get through a perihelion passage totally unobserved. Good progress is being made in theoretical studies of the diffusion of comets into the inner solar system from the Oort cloud and of the subsequent dynamical evolution of some of them into short-period orbits. Puzzles remain, however, as with respect to the fate of comets that seem to drop from sight after making a first passage from the Oort cloud through the inner planetary system.

Scientific papers were then presented as follows:

(1) P. R. Weissman: Evolution of the Oort Cloud. The existence of the cloud of comets surrounding the solar system is firmly established by the distribution of the original inverse semimajor axes of the observed long-period comets. Very little is known, however, of how the cloud was formed or of its subsequent evolution. Recent dynamical studies by Weissman and B. Schreur using Monte-Carlo simulations have shed light on the means by which comets are introduced into the planetary region from the Oort cloud, and on the subsequent evolution of the comets under planetary perturbations. The application of Monte-Carlo techniques to the problems of Oort-cloud formation and evolution should be similarly illuminating. This work must also be tied to general theories of solar-system formation.

(2) L. Kresák: Instrumental Effects on the Apparent Brightness Changes of Comets. There is a broad variety of magnitude scales in use for comets, ranging from the visual integrated brightness to the photographic brightness of the nuclear condensation. In determining the photometric parameters of individual

objects, these utterly inhomogenous data (with systematic differences up to 5-7 magnitudes!) are often put together, and the results are quite spurious. Improvement of instrumentation with time, and preferential use of different scales at different heliocentric distances, result in: (1) systematic underestimates of the intrinsic brightness and their correlation with perihelion distance; and (2) systematic overestimates of the photometric exponent, especially for recent and short-period comets. The impact on statistical analysis of the available data involves, e.g., apparent dying-off of absolutely bright comets, and a complete failure of the predicted death dates for short-period comets. Some photometric peculiarities claimed for the new comets are simply due to the fact that these are normally observed in a broad range of heliocentric distances — otherwise they are not recognized as new. Since it is essentially impossible to determine the integrated brightness when the comet is very faint, regular comet observers are urged to pay more attention to the magnitude determination of the nuclear condensations of bright comets. Existing astrometric plates can well be used for this purpose, provided that appropriate rough calibration is possible. Some unification of such observations would be highly desirable.

(3) D. K. Yeomans: The Intermediate-Period Comets — Orbital Problems. The nongravitational-force model that has been so successful for short-period comets has not been completely successful for the intermediate-period comets ($20 < P < 200$ years). In order of increasingly peculiar behavior, we have, first, intermediate-period comets Crommelin and Herschel-Rigollet, both of which are well behaved and show no obvious evidence of nongravitational forces. Comets Halley and Olbers have obvious nongravitational forces that are consistent with the rocket effect of an outgassing water-ice nucleus. The orbital motion of both comets has been successfully modeled. Orbits for comets Tempel-Tuttle and Pons-Brooks were not entirely successful even when the present model of the nongravitational forces was employed. Although large residual trends remained, the situation was improved for comet Pons-Brooks when the nongravitational-force model was adjusted to include in the nucleus ices more volatile than water. The orbital motions of comets Westphal and Brorsen-Metcalf are erratic; the two apparitions of each comet could not be satisfactorily linked with one orbit even when the nongravitational force model was employed. These two erratic comets are probably small objects easily perturbed by nongravitational forces and perhaps near the end of their lifetimes. Comet Westphal disappeared during its 1913 return to perihelion and was not recovered in 1975-76. The 1989 return of comet Brorsen-Metcalf is therefore uncertain.

Referring to experiences of radio astronomers trying to observe Comet Kohoutek in 1973-74, Yeomans made two suggestions for consideration by the Commission: (1) that it be stated periodically exactly what is in the published ephemerides of comets, and (2) that light-time corrections should be included, i.e., that in the future, ephemerides should be <u>astrometric</u> rather than <u>geometric</u>. With ephemerides now published to accuracy $0\overset{m}{.}01$ and $0\overset{\prime}{.}1$, the difference can be significant, particularly for fast-moving comets close to the Earth.

(4) A. C. Gilmore: The Observing Program in New Zealand. A regular program of astrometric observation of moving objects was established at the Carter Observatory, Wellington, in mid-1973. In six years the program produced some 660 positions, equally divided between comets and unnumbered minor planets in the southern sky. Until June 1978 most plates were obtained with the 41-cm cassegrain reflector at the Carter Observatory. A 10/100-cm astrograph on the same mounting provided plates for field transfers. These telescopes have now been relocated in the Black Birch Range of the South Island. Additional plates were obtained with the 61-cm reflector at Mount John Observatory, Lake Tekapo. Limiting magnitudes of the two reflectors are about 18 and 19 respectively. With the removal from Wellington of the 41-cm cassegrain, routine work will rely upon a recently completed 23/91-cm astrograph at a site south of Wellington city. It is expected

that the different field size and brighter magnitude limit of the new instrument will alter the character of the program from that established in the past six years.

(5a) G. Sitarski: The Orbit of (2101) Adonis. Two apparitions of the minor planet, discovered in 1936 and observed again in 1977, were linked. However, the iterative process of orbit improvement, assuming a linear approximation for computation of numerical values of partial derivatives, appeared to be divergent. A simple way was found to take into account the second-order terms so as to obtain the correct values of partial derivatives. A method that gives the exact solution also was developed. This second method was applied successfully also to improvement of the orbit of the minor planet Apollo from the observations in 1932 and since the 1973 recovery. In both cases the analysis was made more difficult because of close approaches of the minor planet to Venus and to the Earth between the observed apparitions.

(5b) R. Dvorak: The Orbit of (2101) Adonis. In work done in collaboration with G. Schrutka, Vienna, the definitive elements of (2101) Adonis were derived and compared with Sitarski's results. The method used was a classical differential correction in which it was not necessary to take into account the second-order terms. The method converged after six iterations (mean error $0\overset{''}{.}97$), when the orbits for 1936 and 1977 were corrected separately and then integrated.

(6) C. Froeschlé and H. Scholl: New Numerical Experiments to Deplete the Outer Part of the Asteroidal Belt. Numerical integrations over 10^5 yr of fictitious asteroidal orbits in the region $3.6 < a < 3.9$ AU show a partial depletion of an initially uniform distribution of asteroids caused by close encounters with Jupiter. Two models were used: the three-dimensional restricted Sun-Jupiter-Saturn model and the same without Saturn. Several protection mechanisms which avoid close approaches to Jupiter were found. In particular, e-ω coupling is an effective protection mechanism if there is a significant inclination to the orbit of Jupiter. A systematic search for minor planets at high ecliptic latitude is suggested.

<u>16 August 1979</u>

SESSIONS 3 and 4: SPACE ASTROMETRY

Commission 20 joined with several other commissions in co-sponsoring two joint sessions on Space Astrometry. These joint sessions were organized by G. Westerhout at the requests of Commissions 24 and 33. The proceedings will appear in <u>Highlights of Astronomy</u>.

<u>21 August 1979</u>

SESSION 5: SATELLITES

The session was opened by S. Ferraz-Mello, Chairman of the Working Group on Satellites, who called attention to forms that had been prepared to facilitate the collection of information from everyone who was working on satellites. Three different questionnaires were addressed to (1) Observers, (2) Ephemeris Users, and (3) Ephemeris Producers and Theoreticians. Chairman Ferraz-Mello then reviewed the highlights of work on satellites during the triennium, with particular emphasis on those areas in need of attention: (1) discovery of three new satellites, 1978 P 1, 1966 S 2, and 1975 J 1 was reported, but they have not yet been confirmed; (2) work has been done during the triennium on the motion of the Martian and Galilean satellites, on some of the outer satellites of Jupiter, and on some of the satellites of Uranus and Neptune, but no recent work has been reported concerning J V, J X, J XI, J XII, U III, U IV, and N I; some effort on these objects during the next triennium would be welcome; (3) the observational effort in the triennium shows an important increase with respect to the recent past, but special efforts are still necessary to increase the amount of data concerning the outer satellites of Jupiter, J V, and the satellites of Neptune. The first introductory treatise on the theory of the Galilean satellites was published as a monograph by Ferraz-Mello. Also, L. E. Rose has recently published a new determination of the mass of Saturn based on a study of the motion of S IX (Astron. J. 84, 1067-1071, 1979).

Scientific papers were then presented as follows:

(1) V. A. Shor: Report on Studies on Natural Satellites Carried Out in the U.S.S.R. (1976-1978). (Brief summary given by Ferraz-Mello.) Observational work has been summarized in the report of Commission 20 in IAU Trans. XVIIA. Theoretical researches concerned the satellites of Mars and Jupiter and non-Keplerian intermediate orbits.

(2) D. Pascu: Summary of the Observational Program at the U.S. Naval Observatory. The long-focus photographic observations made of the planetary satellites at the U.S. Naval Observatory since 1967 were inventoried. Filter techniques were used in all cases to reduce the light of the primary. While the observations are largely unpublished and incomplete, they are being used in several programs of orbital corrections, such as that at JPL in support of the navigation of the Voyager spacecrafts. Mars' satellites have been observed at every opposition since 1967, and observations of the Galilean moons of Jupiter were obtained in 1968 and at each opposition since 1973. About 1000 exposures have been made for each of these systems. Since 1973, about 500 exposures have been made of Saturn's satellites I-VIII. Observations of the Uranian moons and of Triton made in 1975 and 1977 by Walker have been published.

(3) J. Lieske: Improved Ephemerides of the Galilean Satellites. Over 4800 Earth-based observations of Jupiter's Galilean satellites have been analyzed in order to develop improved ephemerides of the satellites for the Voyager mission, using the new theory of motion of the Galilean satellites. Included are over 1700 eclipses of the satellites by Jupiter spanning the interval 1878-1974, 85 mutual events (eclipses and occultations) observed in 1973, and over 2900 exposures on photographic plates from 1967 through 1978. The resulting ephemerides (labeled E2) were employed for the Voyager I encounter and are in error by less than 200 km at the time of Jupiter close approach. A very small ($0\overset{\circ}{.}066$) amplitude of the Laplacian free libration is indicated by the data.

(4) J.-E. Arlot and W. Thuillot: The Galilean Satellites. A report was given by Thuillot concerning recent work at the Bureau des Longitudes, Paris. (1) A campaign for the observation of the mutual phenomena in 1979 has been organized by astronomers at the Bureau des Longitudes and French observers. Four phenomena have already been observed under satisfactory conditions, and new observations are expected in October and November 1979. (2) A few photographic plates showing the Galilean satellites have been analyzed by Arlot with a PDS microdensitometer controlled by a PDP 11/40 computer at Nice. A new method that uses the light gradient from the primary for the determination of the center of the planet has been devised. This has proved to be the best of all the methods tested, giving significantly better residuals than the methods currently in use. (3) An analysis of available observations of the Galilean satellites has been made by Thuillot, who used for this purpose a numerical-integration program.

In discussion, T. Kiang mentioned that mutual phenomena of the Galilean satellites had been observed at Nanking in 1974 and a lengthy paper was published in Acta Astron. Sinica. An English translation appears in Chinese Astronomy 2, 13-39, 1978.

(5) K. Aksnes: Need for Observations of Mutual Phenomena of the Jovian and Saturnian satellites in 1979-80. Around 100 photometric observations of the

Galilean satellites in 1973 provided very accurate radii and relative positions of the satellites. In 1979 and 1980 some 75 mutual phenomena involving Jupiter I-V, and some 300 involving Saturn I-VII, are expected to occur (Aksnes and Franklin, Icarus 34, 188-207, 1978). Photometric observations of these phenomena with small and medium-sized instruments are encouraged: such observations can lead to a considerable improvement in the long-term accuracy of the theory of motion for the Galilean satellites and can provide astrometric data of unprecedented accuracy on the Saturnian satellites.

(6) R. S. Harrington: The Satellite of Pluto. Plates of Pluto taken in 1979, in addition to old plates taken for special purposes, show image elongations consistent with predictions based on the hypothesis that Pluto has a satellite with a revolution synchronous with Pluto's rotation. Analysis of one electronographic plate confirms the existence of two point light sources. The very small mass of Pluto is confirmed. (The orbital inclination to the plane of the sky, currently 105°, decreases 2°/yr because of the motion of Pluto. The 90° epoch comes in 1985; photometric phenomena may be expected in the interval 1983-1988.)

(7) T. C. Van Flandern: Satellites of Minor Planets. Observations by experienced lunar-occultation visual observer J. H. McMahon of an occultation of a star by (532) Herculina on 7 June 1978 showed six secondary events in addition to the occultation by the 230-km primary. The largest of these secondaries corresponded to a 50-km body, and was apparently confirmed in a photoelectric trace taken at Lowell Observatory. The combined observations indicate that this particular secondary event was caused by a body co-moving through space with Herculina, presumably therefore a satellite. Some 28 additional secondary events observed in conjunction with occultations of stars by nine minor planets have also been interpreted as possible satellites. Calculations indicate that such satellites would be gravitationally stable out to about 100 times the diameter of the parent, and collisionally stable over the lifetime of the solar system.

It was noted by L. H. Wasserman, Lowell Observatory, in response to a question, that the altitude of Herculina at the time of the visual observations of secondary events was 8-9°, while the Lowell photoelectric observations were made at an altitude of 2-3°.

(8) J. D. Mulholland: Solar-System Astrometry at McDonald Observatory. During the interval 1974-79, the satellite systems of the Jovian planets have been observed regularly with the 2.1-m Otto Struve reflector. A total of 113 observations were made of Jupiter's satellites V-XIII, of which about 80% are measured, reduced and published. The Saturn-satellite plates (about 50 of the system I-VIII, 30 of Phoebe) are only 25% measured and 10% reduced, but reduction priority has recently been shifted to these data, due to NASA Voyager mission requirements. About 40% of the 36 Uranus system plates are measured, and 10% of the 49 Neptune exposures. This is a continuing program whose purpose is to assure that observational coverage of the faint satellites will be maintained at an adequate level of continuity for the requirements of future dynamical studies. This work is supported by NASA, under grant NGR 44-012-282.

SESSION 6: OCCULTATIONS

The Chairman of the Working Group on the Prediction of Occultations, G. E. Taylor, reviewed the current state of work on predictions of occultations of stars by minor planets, noting that nearly 100 objects are now on the list of minor planets being monitored for occultations, and that the lead time for predictions of likely events has been increased to about 18 months. Many data are available far beyond that. Searches have been made to 1989 or 1990 for occultations by the largest minor planets and by (51) Nemausa. Predictions are also being made for occultations of non-catalogue stars by (1) Ceres, and for events involving the rings of Saturn and Uranus. Plates are being taken to search for possible occultations by Pluto to around 1985. Some 18 predictions have been issued over the past 36 months. Occultations by (6) Hebe (5 Mar. 1977), (2) Pallas (29 May 1978), (532) Herculina (7 June 1978), and (18) Melpomene (11 Dec. 1978) have been successfully observed. More photoelectric chords across Pallas have been obtained than for any other solar-system object including the Moon. The observations of Herculina led to possible detection of a satellite. Prospects are excellent for successful observation in North America of an occultation by (3) Juno on 11 Dec. 1979, when Juno is very close to its stationary point. The track for an occultation by (48) Doris on 4 Jan. 1980 is very long, and again presents a good possibility of successful observations. The possible occultation by Pluto on 6 Apr. 1980 presents a very difficult problem, and is the subject of a special meeting of the Working Group.

Scientific papers were then presented as follows:

(1) B. H. Zellner: Plans for the April 1980 Occultation by Pluto. Four occultation photometers were specially constructed by the group at the University of Arizona for the 1977 Uranus-occultation event. One of the photometers was sold to the Perth Observatory, but the Arizona group still has three and is prepared to travel anywhere for observations of occultations. With respect to the Pluto event, there are various predictions, some of which indicate that no event will be observable anywhere on the Earth. Long-focus astrometry should resolve the matter about two weeks before the event. Various observing sites in Africa and South America are being considered, the cost of manning a site being between $3000 and $4000 per station. No extra money is available for observation of the Pluto event; costs will have to be covered within the present budget.

(2) J. L. Elliot: Occultations by Ring Systems. The main advantage of using stellar occultations to study planetary ring systems is their high spatial resolution of the ring structures — about 1 km for Jupiter, Saturn and Uranus. Continued observations of Uranian-ring occultations are needed to establish for the nine rings the precession rates, which can be used to determine the higher-order terms in the Uranian gravitational potential. It may be noted that the reference system of the Uranian rings has been used to obtain the radius and oblateness of the planet. These led to a rotation period of $12\overset{\mathrm{h}}{.}8 \pm 1\overset{\mathrm{h}}{.}7$ as an unexpected byproduct that is of some importance because of the disagreement of even modern values measured spectroscopically. No photoelectric observations of occultations by the ring systems of Saturn and Jupiter have yet been made; such data would tell us whether or not these systems contain narrow components of the type found in the Uranian system. To carry on further with this work, a systematic search for future occultations by ring systems is badly needed.

(3) E. Bowell: The (non)Occultation by Ceres. The track of the occultation by Ceres on 31 July 1979 that had been predicted by Bowell and Wasserman (*Astron. J.* 84, 661-667, 1979) to cross Europe had in fact lain substantially farther south than the prediction. The level of interest aroused among astronomers whose main interests are in other fields makes it clear that a large number of competent observers are willing to help observe occultations. But some of those astronomers were badly disappointed when the event did not occur as predicted. It is clear that we have not yet succeeded in establishing adequate channels of communication for the updated predictions that can be made only on the basis of last-minute astrometry.

It was emphasized in discussion that the key point in successful predictions of occultations of stars by minor planets remains the relative position of the star and minor planet, which generally must be refined within the few days immediately preceding the event. G. E. Taylor noted that more help with the astrometry would be exceedingly useful.

SESSION 6 (CONTINUED): ADMINISTRATION

Commission President B. G. Marsden then took the chair and introduced Y. C. Chang, Director of the Purple Mountain Observatory, Nanking, China. Professor Chang gave a brief overview of the astrometric program on minor planets at the Purple Mountain Observatory. He also referred to his discovery of the minor planet originally catalogued as (1125) China during his work with G. Van Biesbroeck at the Yerkes Observatory in the 1920's.

Following several announcements, the following proposals were put to votes, all passing unanimously:

(1) Y. C. Chang should become a member of the Organizing Committee of Commission 20 as soon as negotiations for entrance of the People's Republic of China to the IAU are successful.

(2) Membership of the Commission and Consultants. The following new members were admitted: J. C. Bennett, E. L. G. Bowell, N. S. Chernykh, F. L. Garcia, R. J. Greenberg, A. W. Harris, I. Hasegawa, K. Hurukawa, P. A. Ianna, C. T. Kowal, L. K. Kristensen, M. Lovas, R. Millis, T. Nakamura, D. A. Pierce, L. Quijano, M. Rapaport, H. Rickman, H.-E. Schuster, P. K. Seidelmann, V. A. Shor, C. Torres, S. Vaghi, D. T. Vu, L. H. Wasserman, V. Zappalà, and K. Ziolkowski. The following members retired or resigned: W. J. Klepczynski, A. A. Orlov, H. Q. Rasmusen, and J. M. Torroja. Consultants for 1979-82: C. M. Bardwell, N. A. Belyaev, R. L. Branham, A. Carusi, K. I. Churyumov, E. Helin, P. D. Hemenway, V. A. Izvekov, C.-I. Lagerkvist, Z. M. Pereyra, C. F. Peters, V. Protitch-Benishek, H. J. Reitsema, and P. R. Weissman.

(3) Composition of Working Groups:

Orbits and Ephemerides of Comets: J. C. Bennett, M. P. Candy, A. C. Gilmore, E. I. Kazimirchak-Polonskaya, L. Kresák (chm), B. G. Marsden, E. Roemer, G. Sitarski, Ko. Tomita, R. M. West, P. Wild and D. K. Yeomans.

Satellites: K. Aksnes, Yu. V. Batrakov, A. Borsenberger-Bec, O. Calame, C. Cristescu, S. Ferraz-Mello, R. Greenberg, P. Ianna, Y. Kozai (chm), J. H. Lieske, R. Millis, B. Morando, J. D. Mulholland, D. Pascu, M. Rapaport, E. Roemer, J. L. Sagnier, V. A. Shor, A. T. Sinclair, and D. T. Vu.

Prediction of Occultations: D. W. Dunham, J. L. Elliot, A. C. Gilmore, A. R. Klemola, Y. Kozai, B. Morando, W. H. Robertson, V. A. Shor, G. E. Taylor (chm), L. H. Wasserman, and R. M. West; H. Reitsema (consultant).

Introducing the consideration of proposed resolutions, the President noted that the requested slight increase of the subvention for the Minor Planet Center had been approved by the Finance Committee.

A slightly modified version of the resolution on comet missions discussed in Commission 20 earlier had been approved in the interim by Commissions 15 and 22; Commission 21 also was considering the same resolution. The modified version was voted upon and carried, with none opposed and one abstention.

The resolution concerning Space Astrometry, sponsored by Commission 24 and supported by a number of other commissions, was next voted upon favorably, with none opposed and one abstention.

Consideration next turned to internal resolutions of Commission 20.

The updated policy statement concerning the naming of minor planets had been refined, in part on the basis of discussion at the first administrative meeting of the Commission. The refined version was put to a vote and adopted, with none opposed and two abstentions:

"Commission 20 affirms the tradition that the discoverer of a numbered minor planet be permitted to propose a name for the object. A proposal would normally be accepted, provided that it is consistent with the broad policy of the Working Group on Planetary System Nomenclature. The name would become official following the publication in the MPCs of a brief citation explaining its significance. The Commission defines the discovery as the earliest apparition at which an orbit useful in the establishment of identifications was calculated; in the case of double designations during the same apparition, priority will be given in order of announcement of discovery, unless the double designation follows from an orbit computation using the observations made according to the second announcement. Further, the Commission proposes that, if the discoverer is deceased, or if a minor planet remains unnamed ten years after it has been numbered, a name could appropriately be suggested by identifiers of the various apparitions of the object, by discoverers at apparitions other than the official one, by those whose observations contributed extensively to the orbit determination, or by representatives of the observatory at which the official discovery was made. In such a case, the selection of a name shall be judged by a committee of three, consisting normally of the President and Vice President of Commission 20 and the Director of the Minor Planet Center, and the final decision shall be made not less than six months following the announcement of the numbering of the minor planet."

The next matter was the proposal to include correction for the light time in ephemerides for minor planets and comets, so that ephemerides should be astrometric rather than geometric. Some careful explanations were given of the meaning of the different kinds of ephemerides, and a question was raised as to what should be done about ephemerides of satellites, the relative ephemerides usually being apparent. Should the resolution apply to ephemerides of faint satellites, such as the outer satellites of Jupiter? After some discussion, it was clear that a consensus had not been reached on exactly what should be done, but there was a consensus that the matter could not be put off until the next General Assembly. It was therefore proposed that a vote be taken on the principle that ephemerides of minor planets and comets should be astrometric rather than geometric, such matters as the date of introduction of the change and the question of what should be done about ephemerides of satellites being left to an ad hoc committee to work out. This proposal passed, with none opposed, and one abstention. The President then appointed a committee consisting of himself, K. Aksnes, Yu. V. Batrakov, B. Morando, J. D. Mulholland, P. K. Seidelmann (chm), and G. E. Taylor to compose the specific recommendation. This is now available as follows:

"Commission 20 proposes that, as soon as practicable, the published ephemerides of minor planets, comets and satellites be astrometric or apparent, rather than geometric. An astrometric position means that the light-time correction has been applied and that the ephemeris is directly comparable with star positions given in a catalogue referred to a mean equinox such as that of 1950.0, except for the application of observer-dependent effects, such as parallax, refraction and diurnal aberration; an astrometric position differs from an apparent position by the effects of precession, nutation and annual aberration. The Commission further proposes that explanations of the meaning of the ephemerides be published from time to time in the standard sources."

The resolution regarding the absolute magnitude to be used in ephemerides of minor planets until 1982 was next formally voted upon; a recommendation based on that developed at the asteroid conference in Tucson was adopted, with none opposed, and three abstentions:

> "Commission 20 recommends that the T. Gehrels and N. Gehrels (Astron. J. 83, 1660-1674, 1978) list of asteroid B(1,0), with a phase coefficient of 0.023 mag/deg, be used until 1982. Probably, at that time, a change will be made that should be durable."

The President appointed an ad hoc committee, with membership consisting of Yu. V. Batrakov, E. Bowell (chm), T. Gehrels, L. Kresák, B. G. Marsden, E. Roemer, and L. D. Schmadel, to look into the matter of magnitude ephemerides for minor planets and to make a recommendation for consideration at the next General Assembly.

With respect to the questions raised by Kovalevsky regarding priorities for observational programs on minor planets, it was noted that there is need for a substantially larger number than now available of minor planets with well determined orbits for use in such studies as that of J. G. Williams on asteroid families. It was proposed that an ad hoc committee be appointed to sharpen the goals and observational requirements of astrometric programs on minor planets. The President asked the following to be members of the committee: E. Bowell, J. Kovalevsky (chm), L. K. Kristensen, B. G. Marsden, J. Schubart, V. A. Shor, G. E. Taylor, C. J. van Houten, and J. G. Williams.

Announcements were made regarding future meetings of interest to members of Commission 20: (1) Satellites of Jupiter, 13-16 May 1980, Kailua-Kona, Hawaii; (2) Comets: Gas, Grains, and Plasma, 11-14 March 1981, Tucson, Arizona. The latter meeting is being organized by L. Wilkening in the pattern of earlier Tucson conferences directed by T. Gehrels and will be followed by publication of a book by the University of Arizona Press.

J. D. Mulholland inquired about the value of such programs as the ongoing long-focus observational program on faint satellites at the McDonald Observatory, noting the tight situations with respect to both telescope time and funding. Y. Kozai spoke in support of the program. With time running out, F. K. Edmondson proposed a vote of thanks to President Marsden for his distinguished service as the ninth president of the Commission. This was adopted by acclamation, and the meeting adjourned.

COMMISSION 21 : LIGHT OF THE NIGHT SKY (LUMIERE DU CIEL NOCTURNE)

Report of Meetings, 16, 18 and 21 August 1979

PRESIDENT : R. Dumont. SECRETARY : D.W. Schuerman.

16 August 1979

ADMINISTRATIVE SESSION

1. Welcome address

Dans son allocution de bienvenue, en français puis en anglais, le président évoqua la mémoire de ses prédécesseurs disparus Jean Dufay et Daniel Barbier, et se félicita de la présence à l'Assemblée Générale de tous les anciens présidents vivants de la Commission : F.E. Roach, M. Huruhata, H. Elsässer et J.L. Weinberg. Une minute de silence fut observée en souvenir des membres ou consultants décédés depuis trois ans : Takao Tohmatsu et John R. Roach.

In a comment on language difficulties, the president urged those participants whose mother tongue is English, to fully realize how privileged their situation is in international meetings. In return, they ought to consider a clear and not too fast diction as being a matter of courtesy towards other participants.

2. Commission membership

The names of the new officers for the triennium 1979-1982, nominated by the Executive Committee on proposal by the current officers, were given with a few comments on the activities of these scientists :

President : H. Tanabe
Vice-President : R.H. Giese.

The president-elect was recognized as one of the very few colleagues whose experience extends over practically all of the various components of the night skylight.

The new Organizing Committee recommended by the current officers was accepted by the members attending the session, and therefore transmitted for approval by the Executive Committee : R. Dumont, M. Harwit, C. Leinert, A.C. Levasseur-Regourd, K. Mattila, Yu. L. Trutse, J.L. Weinberg.

The following existing IAU members were co-opted for Commission 21 membership : A.L. Broadfoot, H. Fechtig, L.M. Fishkova, I. Halliday, T. Matsumoto, D.H. Morgan, S. Mukai. The following applicants were found admissible to Commission 21 membership under the condition (fulfilled a few days later) that they are admitted into the Union : W.J. Baggaley, P.D. Feldman, M.S. Hanner, Ph. Lamy, M. Maucherat-Joubert, N.Y. Misconi, S. Nawar, M. Rapaport, D.W. Schuerman, H.J. Staude, T.I. Toroshelidze, J.A. Tyson. Two members resigned from Commission 21 membership : D.M. Hunten and W.R. Steiger.

A new list of 26 consultants was approved for the next triennium.

3. Reports on Astronomy 1979, and the scientific interests of the Commission

The president expressed his thanks to all who provided him with information on their activities, results and aims, allowing him to improve the quality of the

Report. Whether or not some components have to be included in Commission 21 field was then discussed ; due to the 8-page limitation to the Report, priorities on the topics could not be avoided. Most of the participants agreed to exclude general airglow studies from the field of the IAU, unless they are operationally related to astronomical problems, e.g. when the knowledge of the brightness of an astronomical component directly depends on the knowledge of a telluric one.

4. The Newsletter of the Commission

Approval and satisfaction was expressed to the president with the three issues of the newsletter that he edited. The president-elect, with the help of the vice-president and the Organizing Committee, plans to continue this service. At least one issue per year is contemplated, possibly more. Members are instructed that the only useful size for the information that they are requested to send to Dr. Tanabe is of the order of half a page ; and that preprints or reprints have interest only in addition to such precise and concise summaries of their results and projects.

5. Relationship with other Commissions

Commission 50 is charged with the protection of observing sites. One of its major concerns is the ratio of artificial light to the natural light of the night sky. In this regard, our Commission has been requested to provide standard levels. H. Tanabe and R. Dumont handled most of this correspondence, from which the average figures 250 S_{10} or 11 R/nm, and 30 R, for visible continuum and NaI radiation respectively, can be pointed out. F.E. Roach and the president emphasized that the Report of Commission 50 is a document of the highest interest.

Relationship with Commission 22 mainly dealt with preparation of the IAU Symposium n° 90 "Solid particles in the solar system", held in Ottawa immediately after the General Assembly, and dedicated to our distinguished colleague Peter M. Millman. This meeting was co-sponsored by our Commission, having deputed 3 persons in the Organizing Committee and approximately 25 in the attendance. 5 invited papers and 22 contributions were given by members of our Commission.

6. Other topics

The report of the Background Starlight Working Group of the Commission was mentioned, to be given during the last scientific session (see below).

The administrative session came to an end with a short address of the incoming president, H. Tanabe, who thanked the retiring one, R. Dumont, for his efforts during the triennium, especially to improve the circulation of information - in both of the official languages of the Union - between night skylight people.

SCIENTIFIC SESSION : THE OUT-OF-ECLIPTIC MISSION AND OTHER GENERAL REPORTS

1. Photopolarimetry from International Solar Polar Mission (Out-of-Ecliptic)

R.H. Giese : *Experiment objectives and description ; Observations of the Zodiacal light.* - The Principal Investigator explained the fascinating geometry of the spacecraft trajectory and the viewing directions, and gave figures for the expected brightnesses of the zodiacal light. Mapping of the sky will be detailed and frequent, enough for a satisfactory knowledge of the gradients of brightness versus viewing direction and versus spacecraft position, allowing to apply new inversion techniques to the zodiacal brightness integral.

J.L. Weinberg : *Observations of the Background Starlight.* - A brief summary was presented on the objectives of the astronomical phases of the Zodiacal Light/ Background Starlight Experiment, by the Co-Principal Investigator. The Pionner 10-11 spin-scan technique will be used to make multi-color maps of background starlight (brightness and polarization), from spacecraft positions and in viewing directions where the zodiacal light is negligible.

2. Other scientific reports of general interest

A.S. Asaad : *Zodiacal Light observations in Egypt* (no abstract available).

F. Sánchez : *Research on the night skylight in Spain.* - Two groups recently established in Spain are active in our field : one in the Instituto de Astrofisica de Andalucia (Granada) is concerned by various airglow problems, some of which are connected to astronomical components (airglow continuum ; emission of the F level at 557.7 nm) ; the second group, in Instituto de Astrofisica de Canarias (Tenerife), works on separation of atmospheric and extraterrestrial continuums, zodiacal light (visible and IR observations ; models), origin and evolution of interplanetary dust. Both existing and planned instrumentation were described.

18 August 1979

SCIENTIFIC SESSION : NIGHT SKYLIGHT GENERATED IN THE INTERPLANETARY SPACE

1. Zodiacal Light : Optical Study of the Zodiacal Cloud

D.W. Schuerman : *The general inversion and information content of the zodiacal light brightness integral.* - Optical space probes afford the only means to determine the brightness per unit-volume of interplanetary space as a function of position and scattering angle. Observations along a viewing circle, which is a unique function of the space probe's trajectory, provide the most information possible about the brightness per unit-volume of the zodiacal light (see Space Res. 19, p. 451, 1979, and Planet. Space Sci. 27, p. 551, 1979). Such observations directly relate to dust dynamics and "in situ" measurements.

A.C. Levasseur-Regourd : *Inversion of the brightness integral in zodiacal light photometry.* - While observations made tangentially to the orbit provide a direct inversion (of interest also for future cometary missions), observations made in the symmetry plane require the knowledge of the derivatives $\partial Z/\partial \varepsilon$ (satisfactorily known) and $\partial Z/\partial R$ (never measured up to now inside the earth's orbit). It is shown that for $\varepsilon = 90°$ the latter equals $-\partial^2 Z/\partial \varepsilon^2$ in the vicinity of R = 1 AU, leading to a new determination of the decrease of density with heliocentric distance, R. The general trend resulting from Helios measurements is confirmed, but a slight departure from the $R^{-1.3}$ law could exist close to 0.7 and to 1.0 AU.

H. Tanabe, A. Takechi and A. Miyashita : *Recent results about the False Zodiacal Light.* - Results of photographic, spectroscopic and photoelectric observations of the false zodiacal light made at Iriomote Island, Japan, in 1978-79 were presented. To interprete these data, the proposed hypothesis is that the false zodiacal light seen at low latitudes is part of the subtropical arc. However, more observations are required for a complete explanation of the phenomenon.

N.Y. Misconi : *The symmetry plane of the zodiacal cloud.* - Recent results on the location in space of the plane of maximum dust density and its possible association with the orbital planes of the planets were reviewed. It is suggested that the location of this maximum depends on heliocentric distance, varying from a position near the orbital plane of Venus, to a position approaching the invariable plane near the earth's orbit.

R. Robley : *Changes in Zodiacal Light brightness with solar activity.* - Zodiacal light brightness, although poorly correlated with solar activity as defined by sunspot number, is found strongly related to the solar corona, measured at mean heliocentric latitudes. This behaviour is similar to that of the brightness of the eclipsed moon, and both are assumed to be governed by the solar wind ejected at various heliocentric latitudes. The gegenschein observed from 1964 to 1978 at Pic-du-Midi Observatory, the zodiacal light according to Weill (C.R. Acad. Sci. Paris 263 B, p. 943, 1966) and the eclipsed moon according to Danjon (ibid. 171, p. 1127, 1920) have their brightnesses very well related to the extension of the polar coronal holes.

M.S. Hanner : *On the albedo of the interplanetary dust.* - The albedo of the interplanetary dust can in principle be obtained directly from a comparison of the zodiacal light brightness with in-situ measurements of particle size distribution and spatial concentration. This comparison yields a geometric albedo of ~ 0.2, based on the lunar microcrater flux curve (Fechtig, Lect. Notes Phys. 48, p. 143, 1976), with $\rho = 3\,g\,cm^{-3}$. The composition of collected micrometeorites and IR observations suggest an albedo ~ 0.1, which would indicate that the microcrater flux curve is too low by a factor of ~ 2. The very low albedo derived by Cook (Icarus 33, p. 349, 1978) results mainly from his choice of $\rho = 0.18\,g\,cm^{-3}$ in converting in situ data to particle size.

2. Interstellar Wind : Influx of Gas and Dust from the Interstellar Medium

A. Lyle Broadfoot and S. Kumar : *Remote sensing of Solar wind/Interstellar wind interaction.* - On the basis of the weak emissions observed by Voyager from jovian magnetosphere/solar wind interaction, were described the optical signatures one might expect to detect from the solar wind/interstellar wind interaction. Results provided by the UV spectrometer on Voyager 2 in interplanetary cruise, especially the first detection of interplanetary Ly β resonance scattering, were discussed.

J.M. Greenberg : *Interstellar dust component in the solar system.* - The Zodiacal Light-Background Starlight experiment aboard ISPM was reported to afford a possibility of distinguishing between interplanetary and local interstellar dust (see Nature, 275, p. 39, 1978). These prospects were presented in connection with the author's current conjecture that those zodiacal light particles from comets which survive close to the sun may be similar to silicate particles forming the cores of interstellar core-mantle particles.

21 August 1979

SCIENTIFIC SESSION : NIGHT SKYLIGHT OF STELLAR, INTERSTELLAR OR EXTRAGALACTIC ORIGIN

J.L. Weinberg : *Report of the Commission 21 Working Group on Background Starlight.* - This group, which was reactivated at the preceding General Assembly in Grenoble, includes K. Mattila, M. Maucherat-Joubert, F.E. Roach, H. Tanabe, H.C. Van de Hulst, J.L. Weinberg and R.D. Wolstencroft. Theoretical studies and existing or anticipated data bases (star counts and photometries, including sky surveys and polarimetric studies) were discussed briefly in terms of timing, standardization and availability, and their use to evaluate the individual components of the background starlight. The recommendation was made that Commission 21 organize an IAU Colloquium on background starlight during the next triennium.

R.D. Wolstencroft : *Wide field polarimetry of the Diffuse Galactic Light.* - The state of polarization of the total light from the Milky Way at 5300 Å, corrected for foreground zodiacal light, has been derived at $|b| \leqslant 33°$ and $l = 0°$ to $240°$. The polarization is oriented predominantly normal to the galactic plane, consistent with an origin in the Diffuse Galactic Light. The degree of polarization is lowest ($p \sim 2\,\%$) close to the galactic plane ($b = 0°$ to $-10°$) and increases to $p \sim 5\,\%$ at $|b| = 30°$. Modelling of these data is in progress.

R.C. Henry : *A search for Far-UV starlight scattered from high galactic lati-*

tude interstellar dust. - The Apollo 17 wide field (12° × 12°) far-UV (1180-1680 Å) spectrometer was used to scan six large bands of the sky during transearth coast. After accounting for internal scattered light and the light expected from stars in the field of view, no significant residual flux remains. This imposes important constraints on the scattering properties of the interstellar dust grains : either they are extremely strongly forward scattering ($g > 0.9$), or their albedo is low ($a < 0.2$). The work of Jenkins and Shaya (1979) suggests that the latter is the case. The absence of a scattered light signal from dusty moderate-galactic-latitude regions indicates that the high-galactic-latitude flux reported by Henry *et al.* (1979) and by Anderson *et al.* (1979) is extragalactic in origin.

M. Maucherat-Joubert and J.M. Deharveng : *Far-UV integrated sky background.* - The discrepancy between recent measurements of the far-UV diffuse radiation leads to different interpretations as to its dominant contributor, either extragalactic light or galactic plane light scattered off by dust grains. From the ELZ measurements (D2B satellite), (i) a correlation is found between 1690 Å brightness and hydrogen column density, (ii) a spectrum of the diffuse radiation in the range 1690-3100 Å is obtained and is strikingly similar to that of the Coalsack dark nebula. These results suggest a significant contribution of Diffuse Galactic Light in the UV observed, but do not rule out an extralagactic component.

D.H. Morgan : *Diffuse Galactic Light observed by the S2-68 Survey Telescope.* - Results at 2740, 2350, 1950 and 1550 Å were given. Detailed maps of the whole sky were presented. The analysis of part of their data was described and the resulting albedo values quoted. More detailed maps of the Orion region were presented, showing the large scale UV nebulosity and Barnard's loop superimposed. Similar maps around λ Ori were described and the data interpreted in terms of a model of the dust distribution around λ Ori . Asymmetry parameter values $g > 0.75$ are ruled out ; the albedo is found to be higher at 2740 Å than at the other wavelengths ; these results are similar to those obtained for the DGL. Also described are the geocoronal Ly α radiation, and the charged particle background at 550 km as a function of geographic coordinates.

A.N. Witt : *Diffuse Galactic Light and the scattering properties of interstellar grains.* - A re-assesment of the DGL observations of Lillie and Witt (Ap. J. 208, p. 64, 1976) reveals good agreement with more recent UV surveys by Paresce *et al.* (Ap. J. 230, p. 304, 1979) and by Maucherat-Joubert *et al.* (A. & A. 70, p. 467, 1978) who used similarly small fields of view. There is evidence in all these studies for a moderately high diffuse intensity of 1000 to 1500 photons $cm^{-2}s^{-1}Å^{-1}sr^{-1}$ in the 1400-1600 Å range at high galactic latitudes, consistent with the presence of high-albedo interstellar dust with a more isotropic phase function than found in the visible. There is growing evidence that the product $a \cdot g$ (albedo × phase function asymmetry factor) is $\leqslant 0.2$ in the far-UV.

C.T. Hua : *Far-UV observation of the sky background by "Galactika" aboard Prognoz-6.* - Results from more than 4000 spectra obtained with the French-Soviet experiment aboard the Prognoz-6 satellite (1977) were presented. The highly elongated orbit (495 to 2.10^5 km) offers the ability to measure the far-UV sky background intensity for three dark regions : Pisces, Aries and Gemini-Cancer. Contamination by the geocoronal Ly α radiation is taken into account. A detailed analysis is planned for publication.

R. Dumont : *A suggested new clue to disentangle Zodiacal Light from Diffuse Galactic Light.* - In DGL studies, ZL is often subtracted under the assumption that its brightness at high ecliptic latitude is nearly independent of the longitude. Recent photometric surveys of the ZL do not support this assumption ; along the ecliptic parallels ± 60°, the zodiacal brightness oscillates during the year from ~ 50 to ~ 100 S_{10} . A safer process would consist in pointing less than 15° away from the ecliptic directions $\lambda = \lambda_{\odot} + 180°$, $\beta = \pm 75°$. Within these two circles, the zodiacal brightness is the dimmest over the sky, and practically uniform. The annual relative motion of the helioecliptic vs. galactic spheres allows observations of the DGL with the lowest uncertainty due to the ZL, in a range of galactic latitudes from - 60° to + 60°.

F.E. Roach and J.R. Roach † : *Cosmic Light and galaxy statistics with the Space Telescope.* - The plans for the Space Telescope include the observation of discrete astronomical objects as faint as apparent magnitude 28. The several components of the Light of the Night Sky will be variously affected. The atmospheric nightglow will be below the telescope and observable only if grazing traverses of the earth's limb are made. The starlight will not be integrated as is required for many ground-based observations made with telescopes of small aperture and large field. In the direction of the galactic poles individual stars will be outnumbered by discrete galaxies. The integrated cosmic light due to galaxies will actually be obtained by galaxy counts. The background light on exposures with the camera arrangements will be a combination of the zodiacal light and the diffuse galactic light, and will compete with light from discrete sources near apparent photographic magnitude 26.

En conclusion à l'ensemble des réunions de la Commission 21 pendant cette Assemblée Générale, Franklin E. Roach, doyen des anciens présidents de la Commission, souligna l'intérêt des sessions scientifiques, et notamment du présent échange sur l'un des problèmes les plus difficiles de notre domaine d'investigation : les contributions stellaire, interstellaire et extragalactique à la lumière du ciel nocturne. Il remercia enfin le président sortant, R. Dumont, et souhaita la bienvenue aux nouveaux officiers, H. Tanabe et R.H. Giese.

COMMISSION 22: METEORS AND INTERPLANETARY DUST (METEORES ET LA POUSSIERE INTERPLANETAIRE)

Report of Meeting 20 August 1979

PRESIDENT: I. HALLIDAY SECRETARY: B.A. MCINTOSH

Administrative Session, 09:00, 20 August 1979

I. OFFICERS AND MEMBERSHIP

The 25 members and guests present stood in silence for a few moments in memory of two recently deceased colleagues: V.V. Fedynsky, U.S.S.R., member, and J.W. Rhee, U.S.A., consultant.

The President noted that officers of the Commission were elected by the IAU Executive Committee. Nominations by the Commission should take into account equitable geographic distribution and ability to fulfill the requirements of the positions. Confirmed were: W.G. Elford, President, and O.I. Belkovich, Vice-President.

The President announced the names of new members; and that the total membership was now eighty. (The complete list, including members of the Organizing Committee, is reproduced elsewhere in this volume.) Consultants for the next term are: V. Benyukh, E. Grün, K.B. Hindley, F. Horz, G. Schwehm, A.K. Terentjeva, J. Trulsen, D.K. Yeomans.

II. REPORT OF THE COMMISSION

As in the past, selected members were asked to prepare sections of the report which were then put together and edited by the President. Dr. Halliday noted that restrictions on the number of pages made it impossible to include all references. The report was approved without discussion.

III. IAU/IAGA JOINT COMMITTEE

Current progress of the joint committee of IAU Commission 22 and IAGA Commission 5 — "On Radar Observations of Meteor Rates and Radiants, and Anomalies at the Base of the Thermosphere" was reported by the Convener, C.S.L. Keay. It was proving very difficult to effect standardization of radar equipments because of local constraints. New radars in various stages of planning, construction or operational testing were: Australia (two - Elford, Keay), Canada (two - Jones, McIntosh), New Zealand (Baggaley), U.K. (Kaiser). It was noted from the floor that a system was being operated in New Hampshire, U.S.A.

Belkovich described radars being operated in the U.S.S.R. at Dushanbe, Kazan, Kharkov, Kiev, and Oblinsk.

IV. RESOLUTIONS

After considerable discussion the Commission approved a resolution on comet missions. (Submitted jointly by Commission 15, 20, 21, and 22. The text may be found elsewhere in this volume.)

The Commission was polled for its opinion on the desirability of an administrative resolution regarding conflicts of meetings among IAU, COSPAR, and IUGG. Noting that COSPAR already has a non-conflict policy and also that it may change its meeting interval to 2 years, and that geographic location of meetings frequently created problems as severe as time conflicts, concensus was that such a resolution was unnecessary.

The Commission agreed to co-sponsor a Joint Discussion on the formation and early history of the Solar System.

Incoming President expressed the members' thanks to the retiring president for his work on behalf of the Commission during the past three years.

Scientific Session, 10:30, 20 August 1979

The President noted that only a short scientific session had been scheduled because of the existence of IAU Symposium No. 90, "Solid Particles in the Solar System" to be held in Ottawa, August 27-30, jointly sponsored by Commissions 21 and 22 and by COSPAR.

Under the general theme, "Observational Techniques : Present and Future", the following papers were presented:

1. Interplanetary dust - J.A.M. McDonnell.
2. Television-type photo-optical devices - R.L. Hawkes.
3. Camera networks - I. Halliday.

(Radar observing techniques were covered earlier by Keay and Belkovich.)

Summaries of three papers originally submitted to Symposium 90 were presented here because the authors were unable to travel to Ottawa. P.B. Babadzhanov, V.S. Getman. Orbits, chemical composition and fragmentation features of meteoroids in the atmosphere according to instantaneous photographs. P.B. Babadzhanov, Vu.U. Abrudov. Evolution of orbits and intersection conditions of the earth of Geminid and Quadrantid meteor streams. O.I. Belkovich, Radar meteor studies at Kazan.

The problem of anomalous sounds from bright meteors - sounds heard simultaneously with the visual sighting - was reopened by C.S.L. Keay in describing his investigations of a fireball in southeast Australia, 1978, April 7.

COMMISSION 24: PHOTOGRAPHIC ASTROMETRY (ASTROMÉTRIE PHOTOGRAPHIQUE)

Report of Meetings, 15, 16 and 18 August 1979.

PRESIDENT: C. A. Murray

SECRETARIES: A. R. Klemola
F. V. Prochazka

Business Meeting (15 August 1979)

The President welcomed members of the Commission and announced the appointment of A. R. Klemola and F. V. Prochazka as secretaries. Brief mention of the preparation and distribution of the Reports on Astronomy was made. He thanked all who contributed. The contribution from the Yerkes Observatory by K. Cudworth and E. Vilkki, which failed to arrive in time for inclusion in the Report, is given here.

> The modern Yerkes parallax program with the 40-inch refractor is continuing with 75 stars currently under observation and 67 parallaxes having been determined. Most of these have average standard errors of $\pm 0\overset{''}{.}006$ from measurements on the Yerkes photoelectric image bisector but the last 14 stars have been measured by van Altena (now at Yale) on a PDS microdensitometer and have standard errors of $\pm 0\overset{''}{.}004$. We expect that future measurements will be made with a new PDS at the planned mid-west facility in Madison, Wisconsin. Moreoever, Cudworth is continuing his proper motion work in globular clusters to determine membership and, usually, internal motions. Studies of M3, M5, M13, M92 and M15 are completed and some observations have been made on M2, M22 and M71.

As one of the main items of business, the President recalled the two resolutions passed at the IAU Colloquium No. 48 in Vienna in September 1979. These resolutions, concerning the improvement of the stellar reference frame and the call for support for the ESA Astrometry Satellite and the NASA Space Telescope astrometry programs, form the basis of joint discussions with other Commissions and are described later in this report. The texts of the resolutions passed at Vienna appear in Information Bulletin No. 41. The President also recalled the establishment of a Working Group on the identification of Radio/Optical Astrometric Sources. The recommendations of the Working Group appear later in this report.

The President reported the composition of the Officers and Organizing Committee for the triennium 1979-1982:

President: H. K. Eichhorn
Vice President: W. Gliese
Organizing Committee: Chr. de Vegt, L. W. Fredrick, G. Gatewood, R. S. Harrington, C. A. Murray, H. I. Potter and A. R. Upgren.

The President presented lists of individuals seeking membership in the Commission. Following a brief discussion, unanimous approval was given to the proposed list of new members of the Commission:

G. F. Benedict, J. W. Christy, P. Connes, T. E. Corbin, K. Cudworth, C. C. Dahn, F. W. Fallon, I. V. Gavrilov, R. B. Hanson, J-L Heudier, H. Jahreiss, I. I. Kanaev, I. G. Kolchinskij, R. S. Le Poole, P. K. Lu, T. E. Lutz, V. S. Matiagin, H. A. McAlister, D. D. Polozhentsev, N. G. Rizvanov,

J. L. Russell, K. Sims, C. A. Smith, J. D. Stock, L. H. Wasserman and, as consultant, E de Graeve.

The recent death of H. U. Sandig who had been proposed for membership, was noted with regret.

Gliese reported briefly the work of Commission 5 on the previous day concerning the nomenclature of stellar and non-stellar objects. It was noted that the decision was made to leave the matter of designation to the relevant Commissions to the extent that particular classes of objects fall within their domain of interest. It was announced that Gliese would continue to represent the interests of this Commission in such deliberations.

The President then brought up the question of approval of the Rules of the Commission. Vasilevskis pointed out that during his term of presidency a mail ballot was held with the following results: 25 approvals, 6 rejections and one abstention. Several members revived the matter of possible inconsistency of the name of the Commission with regard to the expanding range of astrometric work now being carried out.

In a brief note to the Commission Upgren called for the establishment of a Working Group on Parallax Calibration Fields. This was discussed at greater length in a later session.

Scientific Meeting (15 August 1979)

In introductory remarks Gliese, as organizer of the session on proper motion programs, mentioned three groups of programs.

a) Large proper motions measured relative to slowly-moving distant stars. Comparison of Luyten and Lowell motions reveal systematic differences amounting to several hundredths arcsec per year. But the comparison of Luyten and USNO parallax program (98 stars in common) shows agreement: $\langle\Delta\mu\rangle = +0''.027$ per year.

b) Determination of proper motions relative to extragalactic objects.

c) Proper motions in photographic catalogues referred to the system of stars observed with meridian instruments. If given in the FK4 system these require correction due to change in precession and motion of the FK4 equinox according to values given by Fricke.

Scientific reports were then presented as follows:

1) W. J. Luyten: New Catalogue of Large Proper Motions. After a brief description of earlier work in identifying stars of large proper motion, it was announced that a new version of the earlier LTT catalogue was in progress. Since then over 800 Palomar Sky Survey plates have been hand-blinked or machine processed. The new catalogue, denoted NLTT, will contain between 55,000 and 60,000 stars. All relative motions of $0''.180$ or more per year are included in order to be sure that no absolute motions of $0''.2$ are omitted. Volume I ($+90^\circ$ to $+30^\circ$) has been published, Volume II ($+30^\circ$ to 0°) should be issued before December, 1979, Volume III (0° to -30°) should be ready early 1980, and Volume IV (-30° to -45°, -45° to -90°) should be ready by July, 1980. Data for approximately 250,000 stars found in the Minnesota surveys are being put on to magnetic tape for eventual transfer to NASA where they will be available to interested users. Charts for about 3000 stars are being prepared for distribution. The speaker emphasized the consequences of the significant systematic zero-point difference, reaching as much as $0''.07$ per year, found in the comparison of the Lowell and USNO motions.

2) B. F. Jones: Status of Lick Northern Proper Motion Program. After a brief summary of the early history and scope of the proper motion program with respect to galaxies, the current status of the work was presented. Second-epoch plates have been secured for 17 per cent of the 1246 fields. Measurements are complete for the declination band 0° to $+45^\circ$ for the sky outside the zone of avoidance. Comparison of the Lick motions with the AGK3, the Russian program and the Yale zones is in progress.

3) W. F. van Altena: Status of the Yale Southern Proper Motion Program. The first-epoch phase is essentially complete for the sky from -20° southward. In addition plates for about one-half of the fields at -15° have been obtained. A small program consisting of about 100 fields is in progress. One part consists of a small pilot program for the measurement and study of stellar motions in a region near the south galactic pole. Supplementary first-epoch plates are also being taken for isolated areas as far north as $+25^\circ$ for eventual comparison with the Lick northern program. It is anticipated that the second-epoch phase for the southern sky will begin in a few years.

Joint Meeting on Space Astrometry (16 August 1979). Chairman: G. Westerhout

A discussion of space astrometry and its impact on astronomy and astrophysics was organized by G. Westerhout at the request of the Presidents of Commissions 24 and 33 and with the cooperation of several other Commissions. The purpose of the discussion was to examine work which would be needed in other fields of astronomy and astrophysics to enable full potential of the astrometric data to be realized. The first pair of papers, by Høg and by Jefferys, provided brief technical descriptions, accuracy, and proposed plans of the European Astrometry Satellite and the U.S. Space Telescope, respectively. The remaining papers provided a description of the probable impact of the expected space astrometry data and the associated possible requirements for further ground- and space-based astrophysical observations.

The following papers were presented:

1) E. Høg: Hipparcos, The European Space Astrometry Satellite.
2) W. H. Jefferys: The U.S. Space Telescope: Astrometric Capabilities.
3) B. Pagel: Astrophysical Applications of a Large Increase in Number and Accuracy of Stellar Parallaxes.
4) C. Jaschek: Spectroscopy: What are the needs once Space Astrometry has given us a new data base?
5) A. D. Code: The Impact of new astrometric data on ground-based and space-based photometric studies.
6) P. O. Lindblad: The Impact of faint-star parallaxes and very accurate proper motions on galactic structure and dynamics studies.
7) J. Andersen: The needs in the radial velocity area in view of the impending space astrometry projects.
8) J. Dommanget: Double stars: interactions of ground-based and space observations.

A brief discussion was held concerning a Joint Resolution indicating support by the IAU for the European Astrometry Satellite and the U.S. Space Telescope programs.

Joint Meeting on Nutation and FK5 (16 August 1979)

The report for this session, organized by V. K. Abalakin and held in cooperation with Commissions 4, 7, 8, 19 and 31, appears elsewhere in the Proceedings.

Scientific Meeting (18 August 1979) Chairman C. A. Murray

The President reviewed progress on the Joint Resolution concerning IAU support of the astrometric space programs. Support was expressed either in writing or verbally by twelve Commissions for the following text:
"The IAU strongly supports the independent and complementary astrometric programmes proposed for the ESA Astrometry Satellite and the NASA Space Telescope".

The President then introduced the scientific session by inviting K. J. Johnston as Chairman of the Working Group on the Identification of Radio/Optical Astrometric Sources to present his report. Other speakers included R. S. Harrington and A. N. Argue.

The primary objective of this working group was to select a preliminary list of suitable celestial sources that may be employed as a benchmark in establishing an inertial reference frame. Radio sources with suitable optical counterparts were selected forming a primary list of eighty-two and a secondary list of thirty-two extragalactic sources. These two lists were deemed necessary as the complete radio/optical structure of a large number of sources is not yet known. Extensive additional observations are required to definitely establish these sources, and the list of sources presented will serve as a starting point in this direction. In addition a core list of nineteen sources was established in order to have a small number of sources which will be put on all astrometric programs in order to precisely evaluate systematic errors in radio/optical positions. The working group made the following recommendations:

1) The objects on the primary list should receive the maximum possible optical and radio attention, to ensure the highest-quality positions and information on structure, which will therefore make these objects suitable as an inertial reference frame at the milli-arc-second level for space astrometry.

2) That the publication of radio catalogs contain clear references to the radio frequency, epoch of observation, the equator and equinox and the values of the principal astronomical constants used in the reduction process.

3) The objects on the secondary list should continue to be surveyed, both to improve knowledge of structure and to improve positional, and therefore identification, information. These objects may become of quality to be on the primary list.

4) Additional objects should be observed for radio structure and optical counterparts as facilities permit, for possible later inclusion on either of the lists.

5) The core list should be observed as intensively as possible by all groups carrying out either optical or radio astrometry of these sources, to make possible a complete analysis of the systematic errors present before the space astrometry programs go into operation.

In addition it was proposed that the life of the Working Group be extended in order to revise and extend the source list. The report was accepted by the Commission; copies may be obtained from K. J. Johnston, Code 7134, Naval Research Laboratory, Washington, D.C. 20375.

The following membership of the Working Group was confirmed: A. N. Argue, (Chairman), B. Elsmore, J. L. Fanselow, R. S. Harrington, P. Hemenway, K. J. Johnston, I. Kumkova, Ch de Vegt, H. Walter, A. Witzel.

Hemenway reported that the University of Texas Radio Astronomy Observatory is surveying the sky from declination -36° to +70° at about 360 MHz. Based on preliminary data it is expected that more than 2000 stellar-like objects brighter than magnitude 18 will be found from among the more than 50,000 sources expected from the program. Positional accuracies of ±1".0 and ±1".5 are expected for the strong and weakest sources, respectively. Finding charts will be provided for all radio sources, with radio position indicated.

The session continued with the discussion of topics associated with the proposed revision of the parallax catalogue under W. F. van Altena. The emphasis of the speakers was upon the study of errors and problems of combining parallaxes measured at different observatories. The reports are summarized briefly.

1) R. S. Harrington: Because of the demonstrated very high precision of the USNO parallaxes, now approaching one milli-arcsec, all parallaxes have been re-reduced using tighter algorithms. These will be published to 0".0001, and should replace the previously published values.

2) T. E. Lutz (with A. Marcus and W. Nicholson): Methods of modern statistical analysis have been applied to the comparison of parallaxes obtained at different observatories. Tests for significance show that most differences are not significant. Bias existing in the data available in the literature is an important factor in these comparisons. Inter-observatory corrections should not be applied.

3) R. B. Hanson: The frequency distributions of trigonometric parallaxes determined by the four major General Catalogue observatories (Allegheny, McCormick, Cape, Yale) have been used to study their absolute zero points, systematic differences, and external errors. The method is to fit the observed distributions by convolving power-law space distributions with Gaussian observational errors. The zero point of the General Catalogue parallaxes is confirmed to within ±0".001, without systematic observatory differences generally found from comparisons using multiple observations. The external error estimates agree with the General Catalogue precepts and other recent determinations. The systematic apparent magnitude effect in the Allegheny parallaxes in the blue magnitude range 2 to 8 has been confirmed by comparing Allegheny with cluster, spectroscopic, and Yale parallaxes. The effect is related to the use of very narrow sector openings to produce large magnitude reductions for the bright stars in the early Allegheny program.

4) A. R. Upgren: Spectroscopic parallaxes obtained from MK classes for more than 3000 stars in the GCTSP and Supplement were compared with the trigonometric parallaxes for each observatory separately and also for the mean parallax. A multivariate analysis of the data (with L. A. Breakiron) was made to measure the dependence of the differences π (tr) minus π (sp) on ten independent variables: sin RA, cos RA, Dec, zenith distance, b, M_V, V (or B) magnitude, B-V, π (sp), and internal error in π (tr). Few significant dependences of either systematic or external errors on any of these were found. Partial confirmation of the magnitude effect in the Allegheny parallaxes, already noted by Hanson's work, was found. From the variation between internal and external errors in the parallaxes of the four observatories with the most parallaxes, those of Allegheny and Yale appear to have some correlation whereas Cape and McCormick have none. Problems of calibrating parallax errors could be greatly reduced in the future by having each telescope observe some standard stars and regions. The relatively high weight of the parallaxes observed since the GCTSP and of the programs now in progress should be of great concern in the creation of a new parallax catalogue and plans to reduce or control systematic parallax errors should now be made with care. At least four new parallax programs are beginning or planned at latitudes not now covered by one of the current programs: Pulkovo (P) at 60°N, La Palma (R) in the Canary Islands at 28°N, CIDA (Cv) in Venezuela at 9°N, and Bosscha at Lembang (L) in Indonesia at 7°.

5) J. W. Stein: Allegheny trigonometric minus spectroscopic parallax differences ($\Delta\pi$) were compiled for 35 stars with spectroscopic parallaxes less than 10 milli-arcsec in RA zones 4^h, 14^h, and 21^h. This rather small data set shows no obvious correlation between $\Delta\pi$ and sector opening, exposure times, start and end year of the plate series, time interval covered by the series, or Allegheny publication number. Statistical testing indicates that for this data set the relationship between $\Delta\pi$ and B magnitudes is best described by a weak step function not very different from the Allegheny correction given in the Jenkins Catalogue.

A Working Group on Parallax Calibration Fields was established with the following membership: R. B. Hanson, R. S. Harrington, P. A. Ianna, S. L. Lippincott, T. E. Lutz, H. I. Potter, A. R. Upgren (Chairman), and W. F. van Altena.

Joint Meeting on Coordination of Astrometric Position Work (21 August 1979)

The report for this session, organized by H. K. Eichhorn and held in cooperation with Commission 8, appears elsewhere in the Proceedings.

COMMISSION 25: PHOTOMETRY AND POLARIMETRY (PHOTOMETRIE ET POLARIMETRIE STELLAIRES)

Report of Meetings: 14, 16, and 20 August 1979

PRESIDENT: M.F. McCarthy
VICEPRESIDENT: J.A. Graham

SECRETARY: B. Bookmyer

Activities of Commission 25 at the Montreal General Assembly consisted of 5 sessions apart from informal exchanges among members on tours, at intersessions and coffees.

Two of these five sessions were Joint Meetings with other Commissions, full details of which are communicated elsewhere in these Proceedings. The other three sessions were concerned with matters specifically connected with Commission 25 itself.

JOINT MEETINGS

The Joint Meeting on Astronomical Nomenclature was organized by Commission 5 and was presided over by C. Jaschek on 14th Aug. This session addressed the problem faced by all of us today of how adequately and accurately to specify astronomical objects observed and how to give ample and unique cross references to our colleagues in different specialized fields. The delegates of the several Commissions approved with commendation the offer of Dr. Jaschek to draw up a Reference Catalogue. This group also approved the offer of Dr. Jaylee Mead to prepare a first draft of a Users' Guide. Colleagues are invited to communicate to Dr. Mead their needs, suggestions, results and questions.

The second Joint Meeting on Aug 16 was devoted to Space Astrometry and was organized by Commission 24 and 33 with the cooperation of Commissions 4, 7, 8, 19, 20, 26, 33, 36, 37, 45 and our own Commission 25. This session was chaired by Gart Westerhout. Details are given elsewhere in these Proceedings. Of specific importance for Commission 25 members was the request to our Commission to support the proposal of Commission 24 for adoption by the Final Meeting of the General Assembly at Montreal. This proposal read as follows in the form suggested by Pres. C.A. Murray of Commission 24:

The General Assembly noting the support of Commission 4,7,8,19,20,25,26,33,36, 37 and 45 is asked to adopt the following resolution by Commission 24:

> The IAU strongly supports the independent and complementary astrometric programmes proposed for the ESA Astrometry Satellite and the NASA Space Telescope.

At the recommendation of Commission 25 and 45 a proposal listed below was attached to the original proposal given above. This was done at the business sessions of the respective Commissions. After discussion, the addendum to the original proposed resolutions was formulated and approved unanimously by the members of Commission 25. The following is the text of the addendum:

Noting the strong support of the IAU for space astrometric programmes, Commission 25 (and 45) recommends that complementary ground based support in astrometry, photometry and spectroscopy etc. should be encouraged.

At the recommendation of the General Secretariate this resolution was subsequently formulated as a separate resolution and was passed in the form stated above by the General Assembly on 25 Aug.

Dr. Westerhout and Pres. C.A. Murray of Commission 24 had earlier requested the President of Commission 25 to name a speaker to examine what work would be required in the fields of photometry and polarimetry to enable the full potential of the new astrometric data to be developed for the full benefit of astronomy and astrophysics. After consultation with the Organizing Committee the President invited A.D. Code who described the impact of new astronomical data on ground and space based photometric and polarimetric studies. Dr. Code presented clearly the needs of ground based and space based photometry and indicated the progress which could be expected from the proposed programs in space astronomy.

SCIENTIFIC SESSIONS

Three were the sessions of Commission 25 at Montreal: one a Business Session was devoted to election of new members of the Commission and new Officers and the routine business details of the Commission during the past triennium and accepted proposals for its work during the years ahead; the second and third sessions were specifically scientific and presented reports and papers of interest to the Commission. New Techniques in Stellar Photometry was the topic of the first Scientific Session and this was chaired by M.F. McCarthy. New Techniques in Stellar Polarimetry was the title of the Second Scientific session, which was chaired by Dr. J. Landstreet. Both sessions were held on 20 Aug, the first in the morning and the second in the afternoon. Since it is planned to publish elsewhere the papers presented at these two Scientific Sessions, we list here the program with the names of the speakers and the topics as presented. There was an active discussion period following each paper and at the end of each session. Many members of other Commissions were welcomed at Commission 25 at these Scientific Sessions and contributed important questions and comments. The Commission is grateful to J. Tinbergen who organized the scientific session on polarimetric techniques.

NEW OBSERVING TECHNIQUES IN STELLAR PHOTOMETRY

New Detectors in Stellar Photometry G.E. Kron (USA)

TV Reduction Techniques for Stellar Photometry R. Albrecht (Austria)

in situ Techniques for Data Reduction with New High Speed Detectors

W.E. Kunkel (Brazil)

Electronographic Stellar Photometry A.P. Penny (U.K.)

NEW OBSERVING TECHNIQUES IN STELLAR POLARIMETRY

Polarimetric Techniques for Photometrists J. Tinbergen (The Netherlands)

Multiplex Detectors and High Resolution Polarimetry I. MacLean (U.K.)

Magnetic Field Measurements E.F. Borra (Canada)

BUSINESS MEETING

President McCarthy opened the Business Meeting of the Commission 25 on Aug 16 with twenty-five members present. He began with a commemoration of Prof.H.Haffner who died since the last General Assembly; the members stood for a minute to commemorate our departed colleague.

The President invited Dr. B. Bookmyer to serve as Secretary for this session of the Commission.

The Triennial Report for Commission 25 appears in Vol 2, p. 83-96 of Reports on Astronomy. It was discussed and approved by vote of the Commission Members present. Two items which arrived too late for them to be included in the published report are given here.

REPORT FROM MEMBERS

From F. Rufener, Geneva Observatory : Geneva Photometric System

New catalogues assembling all the measurements made in the Geneva photometric system (UBV $B_1 B_2 V_1$ G system) are being published. 55,000 measurements of 13,000 stars have been obtained through June, 1978. A compilation of these data allows the establishment of

a) a list of all the measurements reduced outside the atmosphere with a weight for the colours and the V magnitude as well as the Julian date and place of observation.

b) a catalogue arranged according to increasing right ascension containing the mean colour indices for each star with the number (P) of good measurements and their dispersion (σ_c) as well as the V magnitude, its weight (Q) and dispersion (σ_V). A variant of this catalogue contains the derived indices and parameters (d, Δ, g, m_2, B_2-V_1, V_1-G, X, Y, Z, β_c).

c) a catalogue grouping in photometric boxes all the stars having the same colours within one hundredth of a magnitude.

The observations are being made from the Northern and Southern hemispheres. The selection of stars to be measured depends upon the field of interest of the various research workers at Geneva Observatory (galactic clusters, zones of the galactic poles, Am and Ap stars, stars with particular kinematics, stars in the solar neighborhood, supergiants and OB stars).

From M. Golay, Geneva Observatory:

I wish to inform you that we have started the photometry at 1900 Å (pass-band= 200 Å) of the galactic plane. The plates cover a field of 6°. I join an example containing the Andromeda Nebula. The President circulated among the members of the Commission the print photo sent by Prof. Golay; this photo is remarkable for the preeminence of one of the spiral arms of this galaxy and illustrates the new regions now available for photometric studies.) We reach magnitude 13 for B stars with 3 or 4 minutes exposure time.

In discussion of the Triennial Report the question was raised (by Vice President J. Graham) as to whether a more extensive report, one longer than the 12-14 pages which the IAU can publish in its volumes: Reports on Astronomy, should in the future be prepared and circulated to members; an example of this is the very detailed and full report of Commission 33 on Galactic Structure and Dynamics as prepared by F. Kerr and colleagues. This full report is issued independently of the shorter version submitted for Reports on Astronomy. It has the advantage that more recent data concerning research in the fields of interest to the Commission can be reported. For Reports on Astronomy, the effective closing date for research which can be reported is approximately one year before the opening of the General Assembly; the deadline for the current Reports on Astronomy was 31 Dec 1978. The fuller version is usually mimeographed or xeroxed and can have a much later deadline for information submitted; it also allows a more fulsome description. The members of Commission 25 commenting on this suggestion agreed that such a fuller report would be most desirable, would be of great value to individual members and to others who look to such a report for specific references and general coverage of material for classes and lecture purposes. It was pointed out that this fuller report should be made available to those requesting it within or outside of Commission 25 but would not be circulated generally.

Among the topics discussed at the Business Session of Commission 25 were the following:

1. There is a need to give good references to the passbands used in photometry. This is very necessary for any intercomparisons of photometric results.

2. There is a divergence in the epochs employed in different listings of observations: e.g. HD used 1900, the SAO uses 1950 some others use 2000. With modern calculators the precession to these other systems is not difficult but sometimes involved. Most tabulations at present seem to favor the 1950 epoch, though we are now closer to 2000 AD. Important it is to be clear in specifying which epoch one is using and to give the coordinates with the best possible accuracy.

3. There is an increasing need for improving and extending photometric systems for faint stars. This becomes more evident with each new break-though involving radio sources, X-ray sources, gamma-ray sources etc. Attention of telescope allocations committees should be called to the long and short range needs of astronomical photometry and polarimetry and more consideration should be given to the use of time with large telescopes for work with photometric sequences. This topic could well be discussed whenever we are speaking of the needs of ground based and space photometry. Perhaps a large size ground-based photoelectric telescope in each hemisphere to aid in photometry for space observatories would be a prudent and profitable step at this time. There were several valuable comments and objections concerning the programming and support for such a project. It was pointed out that neither the Union or Commission 25 were fund raising or fund granting organizations. The importance of the IAU is founded on the concerted motivation of its constituent members who can return to their homelands with specific recommendations for their communities.

4. Of importance for the budget and for uniformity of standards in photometry is the fabrication at the same time of several sets of photometric filters. D. Crawford (USA) may be contacted to coordinate orders in bulk. J. Graham noted that KPNO can order in bulk, then charge individual purchasers of filter sets. He stressed the value of having similar transmission characteristics for photometric researches whose results will be intercompared.

5. One of our Commission members, A. Cousins has devoted a tremendous amount of work to define the infrared photometric system in the southern hemisphere. The problem of the differences, specifically in the red-infrared wavelength bands were discussed and the importance of specifying bandpasses and of giving transmission curves to define one's system was emphasized. It should be noted in this regard, added Vice President J. Graham, that UBVRI photometry in Chile, S. Africa, Australia is different from the H.L. Johnson system. W. Wisnieski noted that a comparison between the H.L. Johnson RI and the Cousins RI is difficult. It is now

impossible to obtain the original Johnson type filters from the manufacturer and thus comparisons become impossible. J. Graham noted that there is some difficulty in reproducing magnitudes and colors for very red stars on the Cousins system but no difficulties are experienced up to spectral type about K5.

6. J. Tinbergen presented the members with his special Bibliography for Optical Polarimetry Standards. These are the result of collaboration between Tinbergen and the following colleagues: P.A. Bastiaansen, G. Knoechel, J.D. Landstreet and J.J. Michalsky. These standards are applicable to both the visual (300 to 1050nm) and for the UV and IR regions.

7. An excellent summary of recently available photometric catalogues was given at Montreal by A.D. Code in his address to the Joint Meeting on Space Astrometry, mentioned above. Attention was also called to a listing of photometric catalogues in Dudley Obs. Report no. 14 by A.G.D. Philip.

8. C.J. Van Houten asked about the definition of magnitudes and colors and specifically about possible advantages of a system such as that introduced by the Walravens where log I rather than magnitude was employed. He noted that perhaps by using a variety of logarithms to different bases etc., a differentiation could be made among systems whose similar letterings i.e. UBVRI, ubvy, can become confusing. Van Houten noted also that the old term of "Color Index" was being dropped by many in favor of the word "Color". It was pointed out that with present day calculators logarithms to different bases offer no stumbling block but the intercomparison of results might become more involved. The specification of bandpass and the publication of transmission curves for filters were suggested as further aids for avoiding confusion when writing about magnitudes and color indices.

9. A report on the comparison of stellar spectrophotometry made at the Sternberg Astronomical Institute and the Astrophysical Institute in Alma Ata with photometry in various systems was given as a present to the Commission by I.N. Glushneva and A.V. Kharitonov. Because of the heavy schedule and the consequent lack of available time it was not possible for Dr. Glushneva to read her paper at Commission meetings. The President noted for the members that the paper contains: 1) tables of the mean differences between the spectrophotometric data and the photometric data of Johnson and Mitchell, 2) the results of intercomparing spectrophotometric data with the photometric data on the Vilnius system, and at Alma Ata and 3) a comparison of data between the Moscow and Alma-Ata systems. Dr. Glushneva was invited to rise and accept the congratulations of Commission members of this interesting work in photometry carried out in the Soviet Union. The President announced that copies of this paper would be available during the time of the Montreal assembly.

The President then read the list of the names of new members of Commission 25 to be voted on by members of the Commission. These are listed in this Volume, page . Commission 25 has seventy seven members currently and thirty four new names were proposed. Since several of the newly proposed members had not as yet been admitted to the Union itself, it was decided to approve them conditionally subject to their candidacy as members of the Union being approved. All 34 newly proposed candidates were approved by the vote of the members present of Commission 25. Subsequently the General Assembly on Aug 23, voted to make the conditionally elected candidates member of the Union! So now our Commission is 111 members strong. We have 2 Consultants to the Commission: H. Butcher and E. Miller.

Dr. U. Steinlin was named and approved as Commission Representative to consult with Officials of the Working Group on Photographic Problems. This Group offers much excellent service and information to all colleagues working in photographic research.

Next the President noted the list suggested for the Organizing Committee of Commission 25 for the next term. This list was drawn up after consultation with the present members of the Organizing Committee and with the Officials and the Executive Committee of the IAU. This list was proposed to the members of the Commission present and it was approved unanimously. Our new President will be J. Graham (USA) with J. Tinbergen (The Netherlands) as Vice President. The Organizing Committee for the epoch from the end of the present General Assembly until the end of the next one will be composed of the following:

A. Behr, M.S. Bessell, P.F. Chugajnov, G.V. Coyne, A. Feinstein, B. Hauck, P.W. Hill, A.R. Hyland, A.U. Landolt, T. Markkanen, M.F. McCarthy, F.G. Rufener, W. Schoneich. The President thanked in the name of the Commission the retiring members of the Organizing Committee who have served our sciences so well for several years. They are Prof. M. Golay (Switzerland), Prof. K. Osawa (Japan) and Prof. V.L. Straizys (USSR).

Concluding the Business Session the President thanked all of the members, present and absent, for their kind cooperation and response to letters during the past triennium. He mentioned the special help and advice received from the Vice President and the Organizing Committee. Three secretaries have assisted in the past three years in sending letters and circulars to members; unlike the Secretary for the present session of the Commission, Dr. B. Bookmyer, these persons are neither photometrists nor polarimetrists but have helped both fields and this commission by their generous work. B. Epp of New York, R. Amati of Marino, Italy, and E. Bauer of La Serena, Chile, all merit our gratitude and thanks.

Now we have an increase of 33 per cent in the membership of Commission 25; we have a new President and Vice President and Organizing Committee. Sic iter ad astra.

M.F. McCarthy S.J.

COMMISSION 26: ETOILES DOUBLES (DOUBLE STARS)

Compte-rendu des séances des 16 et 21 août 1979

Président: P. Muller Secrétaire: J. Dommanget

16 août 1979

Après avoir ouvert la séance, le Président invite l'assistance à observer quelques instants de recueillement à la mémoire de notre collègue W.S.Finsen, l'un des piliers de l'astronomie des étoiles doubles visuelles dans l'hémisphère austral, décédé le 16 mai dernier.

I - QUESTIONS ADMINISTRATIVES

Sur la proposition du Président, J. Dommanget est chargé du secrétariat des séances.

Le Président entame l'ordre du jour en signalant tout d'abord qu'au niveau des relations inter-commissions:

1) à la séance "Space Astrometry", la commission avait délégué officiellement J. Dommanget, lequel y a présenté un exposé qui résume le point de vue qui est le nôtre, et dont le texte bilingue était disponible en séance;

2) en vue de coopérer à l'étude du problème de la "nomenclature" soulevé par la commission 5, et à la suite d'une suggestion de W.D.Heintz, un sous-comité de la commission 26 a été installé; il comprend au départ P. Couteau, O.G.Franz, W.D.Heintz et C.E.Worley, et il étudiera également la présentation et la publication d'un nouvel Index Catalogue.

Le Président évoque le rapport d'activité pour signaler les très riches contributions reçues de la quasi-totalité des membres, lesquelles témoignent de la diversité et de l'importance des travaux effectués depuis 1976, et qui malheureusement ne peuvent être que résumés dans le peu de place offert par les Transactions. Il remercie aussi les quatre auteurs qui se sont partagé l'essentiel de la tâche.

La parole est alors donnée à S.L.Lippincott pour exposer la situation du Sproul Observatory. Une proposition de résolution, mise aux voix, est adoptée à l'unanimité avec la rédaction suivante:

"La commission 26 note avec grande inquiétude que le Sproul Observatory est menacé de fermeture par le Swarthmore College. La clôture du programme astrométrique marquerait la fin d'une source très féconde et de très longue date de données sur les étoiles doubles. Les archives photographiques du réfracteur de Sproul et leur croissance continue sont uniques et donnent encore des résultats de grande valeur."

Le point central de l'ordre du jour concerne la formation de la commission pour le prochain triennat. Le Bureau a arrêté les procédures à appliquer, en fonction des statuts et directives de l'Union comme de notre règlement intérieur, adopté en temps voulu à la suite d'un vote général par correspondance.

1) Pour la Présidence, il propose de suivre la règle normale et de soumettre

au Comité Exécutif le nom du vice-Président O.G.Franz; ce choix est ratifié par acclamations.

2) Quatre noms ont été avancés pour la vice-Présidence; deux des intéressés s'étant récusés, il reste en présence M.Fracastoro et C.E.Worley, ce qui permettra selon la suggestion du Bureau et en application des directives de l'Union de présenter deux noms assortis d'un ordre de préférence. Le vote désigne dans l'ordre décroissant du nombre de voix M. Fracastoro et C.E.Worley qui seront donc présentés respectivement en le et en 2e ligne.

3) Suivant les règles adoptées, outre le mandat réservé au Président sortant et où P. Muller remplace S.L.Lippincott, trois membres sont à élire pour le Comité d'Organisation. Huit propositions ont été reçues; deux des intéressés n'ayant pas désiré se porter candidats, restent en présence: E.H.Geyer, W.D.Heintz, P.J. Kulikovsky, W.J.Luyten, A. Poveda et C.D.Scarfe. A.Poveda et C.D.Scarfe sont élus au premier tour, P.J.Kulikovsky au second. Il faut noter que 16 membres étaient présents et que 4 autres avaient donné une procuration régulière, vérifiée par le Président en début de séance.

Les admissions et radiations à la commission sont de la compétence du Président, en fait du Bureau. Il est donné connaissance des décisions prises; les admissions résultent de propositions des membres, ou de celles des Comités Nationaux transmises par le Secrétaire Général.

Radiations: W.S.Finsen (décédé), P.S.Thé (à sa demande), J.D.Hadjidemetriou qui a offert sa démission pour insuffisance d'activité dans notre domaine.

Admissions: C.Allen, G. Gatewood, C.L.Morbey, membres de l'Union; H.A.McAlister, E. van Dessel, G.M.Popovic, J.N.Russell, E.W.Weis, tous en instance d'entrée à l'Union donc sous réserve qu'ils soient admis.

II - PROCHAIN COLLOQUE

O.G.Franz propose que la commission organise en 1981 un colloque à l'Observatoire Lowell (Flagstaff) avec pour sujet provisoire "Current observational techniques". La commission accueille cette invitation avec un vif intérêt, et un accord général se fait en faveur d'une date à fixer en mai.

III - SPACE ASTROMETRY

J.Dommanget signale qu'à la séance inter-commissions sur le sujet le Président G.Westerhout a demandé aux commissions représentées (8, 20, 25, 26, 30, 33, 37, 44, 45) de soutenir la résolution proposée par la commission 24 souhaitant la réalisation prochaine des projets Hipparcos et L.S.T. Après lecture de cette résolution, celle-ci est adoptée à l'unanimité décision dont sera informé G.Westerhout.

21 août 1979

Cette seconde séance est consacrée à diverses communications de caractère scientifique.

Avant de passer aux exposés, le Président annonce:

1) qu'aux admissions prononcées il a ajouté le nom de R.H.Wilson, proposé comme membre de l'Union par le comité de Turquie et maintenant accepté (en même temps

que les cinq candidats déjà nommés); le Secrétariat Général l'avait omis dans les dossiers transmis. Nous sommes heureux d'accueillir un observateur déjà très connu de tous, et d'ailleurs présent à la session.

2) qu'en addition au rapport de la commission il désire faire part d'une communication de E. Vilkki reçue tardivement, annonçant le lancement au 40 pouces de Yerkes d'un programme photographique consacré surtout à des étoiles de Luyten et de Giclas (mouvements propres).

3) que le 4e catalogue d'éphémérides d'étoiles doubles est en cours de composition à l'Observatoire de Paris et devrait être distribué vers la fin de l'année.

La parole est donnée ensuite aux orateurs inscrits.

M.W.Feast - Appeal for double star observing programs at the Johannesburg refractor.

Le réfracteur de 26" de l'ancien Republic Observatory reste disponible et se trouve en état de marche; R.H.Wilson, qui y a travaillé récemment, donne quelques précisions à ce sujet. Pour y accéder, il y a lieu de s'adresser à:

National Institute for Telecommunications
P.O.Box 3718
JOHANNESBURG 2000
Republic of South Africa

pour prendre toutes dispositions locales. Contact doit être pris d'autre part avec le Dr. Feast, Directeur du Anglo-South African Observatory au Cap en ce qui concerne les programmes d'observation.

A.H.Batten - Recent results on ADS 11060 and 70 Ophiuchi.

Des observations faites au dernier périastre, en 1978, montrent que le système de 11060 est triple, avec un couple visuel d'orbite très excentrique et dont la principale est un couple spectroscopique de période 0,88j. Dans un travail à paraître (PASP), un système complet d'éléments a été déterminé.

En ce qui concerne 70 Ophiuchi, les observations de vitesses radiales n'ont pas confirmé la perturbation soupçonnée.

K.Aa Strand considère que ces résultats prouvent une fois de plus l'intérêt des observations de Vr des binaires visuelles et rappelle le catalogue des Vr relatives de J. Dommanget et O. Nys. Le premier fait savoir que le manuscrit d'une seconde édition est prêt et qu'il comporte cette fois en plus les éphémérides relatives aux couples astrométriques.

D.W.Dunham - Possibilities and value of detection and observation of satellites of asteroids by double star specialists.

L'observation d'occultations d'étoiles par certains astéroïdes a fait envisager la possibilité de duplicité de ces derniers. Des calculs récents ont permis d'étayer une telle hypothèse du point de vue dynamique. Les observateurs d'étoiles doubles serrées sont les plus aptes à pouvoir en apporter la preuve éventuelle. Le cas de 18 Melpomène est le plus intéressant parmi d'autres tels que 6 Hébé, 171 Ophélia, 532 Herculina etc..

C.E.Worley fait remarquer qu'il ne peut faire état que d'une observation personnelle négative.

C.E.Worley - Status of the card catalogue and prospects of a new Index catalogue.

Après avoir rappelé l'historique des catalogues et de leurs copies, C.E. Worley présente l'évolution des fichiers depuis la dernière Assemblée Générale, et précise la situation actuelle par les données suivantes:

Catalogue Index:	73 111 entrées
Catalogue des mesures:	396 769 entrées

Le travail accompli par un personnel trop réduit porte sur l'introduction des mesures nouvelles une fois publiées et des couples nouveaux, ainsi que sur celle des mesures antérieures à 1927 et sur la correction des nombreuses erreurs dans les fichiers originaux.

La question de l'édition d'un nouvel Index est posée. Plusieurs problèmes surgissent, ceux notamment du choix de l'équinoxe et des Notes. Il est probable, avec la très large diffusion des tables de précession et des calculateurs de poche, qu'un équinoxe unique suffira, ce qui permettra d'enrichir le sigle attaché à l'étoile. D'autre part, on se trouve devant l'annonce croissante de couples décelés soit par "speckle interferometry" (méthode des tachetures), soit au cours d'occultations par la Lune, objets presque toujours inaccessibles au contrôle visuel. C.E.Worley pense qu'il sera prudent de leur réserver, ainsi qu'à un certain nombre d'autres, une catégorie spéciale de couples douteux.

R. Harrington - Two interesting perturbation systems.

G. 107-69/70 is a double perturbation quadruple system with one pair being a red dwarf and a black dwarf, and the other two dwarfs. Details will be published in the Astronomical Journal.

The USNO astrometric data covering 8 years of Barnard's star show no perturbation at the $0\rlap{.}''0004$ level, which is 8 standard deviations below the amplitude currently predicted.

COMMISSION 27: VARIABLE STARS (ETOILES VARIABLES)

REPORT OF MEETINGS: 16, 17, 21 August 1979

PRESIDENT: J. Smak SECRETARY: Nancy R. Evans

ADMINISTRATIVE MEETING, 16 August 1979

1. In Memoriam.

The President spoke briefly in commemoration of four deceased members: P.L. Bhatnagar, B.V. Kukarkin, P. Th. Oosterhoff, and E. Vandekerkhove. Members stood in silence for a minute in memoriam.

2. Nominations.

The Commission agreed to nominate the following Organizing Committee for 1979-82:

President:	J.D. Fernie
Vice-President:	N.H. Baker
Members:	M. Breger, M.A. Cherepashchuk, A.N. Cox, M.W. Faest, H. Gursky, W.E. Kunkel, J. Smak, W. Wenzel.

3. New Members.

The President proposed that the Organizing Committee be authorized to review the list of candidates and decide on their admission. The motion was carried unanimously.

4. Archives of Unpublished Photoelectric Observations.

Dr. M. Breger reported on the working rules and the present status of the Archives. He proposed that a formal letter be sent to the editors of major astronomical journals suggesting that they should recommend to their authors to deposit lengthy tables of pe observations with the Archives rather than trying to publish them in extenso. After a long discussion the members voted against such a drastic measure. The members expressed their appreciation to Drs. W. Fitch, M. Breger, V.P. Tsesevich, and E. Makarenko for their work in maintaining the Archives.

5. Nomenclature.

Dr. M. Breger reported on a meeting of Commission 5 with representatives of several other Commissions on problems of the nomenclature. The proceedings of this meeting are published elsewhere.

6. I.B.V.S.

The Editor of I.B.V.S., Dr. B. Szeidl answered questions concerning editorial policy and other problems. The members expressed their appreciation to Dr. Szeidl and his collaborators for their extremely useful work.

7. G.C.V.S.

The President read a letter from the Editors of G.C.V.S. containing information on their work, including preparations for the next edition. A motion for the continuation of the I.A.U. subvension to the Moscow Variable Star Bureau was proposed by the President, seconded, and carried unanimously.

SCIENTIFIC MEETING, 17 August 1979
FLARE STAR WORKING GROUP

The following summary is by Dr. P.B. Byrne:

Dr. C. Jordan delivered an informal presentation on interpretation of UV spectra and Dr. B. Bopp gave some opinions on possible future directions for optical observations. Lively discussions took place and it came to light that a number of groups (at Armagh, Catania, and Sacramento Peak) are independently involved in UV spectroscopy of flare stars in both their flaring and "quiescent" states. Indeed subsequent discussions with the Boulder group showed that they and the Armagh group have contiguous programmes between September 6 and 11. People interested in cooperative ground-based observations are invited to contact P.B. Byrne (Ireland) or J. Linsky (USA).

A broader definition of the working group's activities is necessary, these being no longer confined to the study of flaring. Rather there is now interest in applying all available diagnostic tools to the study of chromospheric activity in the latest type dwarfs. With this in mind it was decided to discontinue the international cooperative programme of optical monitoring of a selected group of flare stars on a regular year-to-year basis. In its place we would urge observers to search for activity in stars not hitherto shown to be active, to expand our knowledge of its occurence in as wide a range of spectral types as possible.

With the resignation of the previous Chairman, Professor Mavridis it was decided to form an organizing committee for the future. For details of the composition of this committee members are referred to coming issues of I.B.V.S. Future plans include a possible workshop meeting on flare stars to be held during September 1980. Those interested in joining the group should contact Dr. B.W. Bopp(Univ. of Toledo), Dr. P.B. Byrne (Armagh Observatory), Dr. M. Rodono (Catania Observatory), or Dr. S.P. Worden (Sacramento Peak Observatory).

SCIENTIFIC MEETING, 21 August 1979

The following papers were presented:

W. Eichendorf - HR 4511
J. Arsenijevic - o And
M. Friedjung - Infrared Observations of Kuwano's Nova-Like Object
M. Popova, E. Vitrichenko, N. Afanasiev - KR Aur
G. Bakos - A New Short-Period Eclipsing Binary
G. Bakos - Variability of some Solar Type Stars
H. Maracco - The Period Changes of RY Sgr
H. Pedersen - Optical Bursts from the X-ray Burster MXB1636-53
R. Foy - Variation of the Diameter of Mira.

SCIENTIFIC MEETING, 21 August 1979
Joint Meeting of Commissions 27 and 35
on "STELLAR EVOLUTION AND VARIABLE STARS"

The following papers were presented:

- L.A. Willson - Pulsational Constants of Mira Variables
- A.N. Cox - The Structure of Yellow Giants and Cepheids
- D.S. King - The Structure of Am Stars, Delta Del Stars, and Delta Sct Variables
- N.R. Simon - Resonances in the Cepheid Instability Strip
- M. Breger - Multiple Radial/Non-Radial Periods in 38 Cnc
- M. Savedoff - Comment on Two-Dimensional, Rotating, Quasistatic Evolution Calculations
- D. Sugimoto, M.Y. Fumimoto, K. Nariai, K. Nomoto - Generalized Theory of Shell Flashes
- M.Y. Fujimoto, D. Sugimoto - Type I Supernovae and Helium Shell Flashes
- K. Nomoto, S. Miyaji, D. Sugimoto, K. Yokoi - Supernova Explosion Triggered by Electron Capture
- C.-C. Wu - IUE Observations of Cen X-4 and Nova Cyg 1978
- H. Bond - The Difference Between Classical and Dwarf Novae.
- P. Noerdlinger - Diffusion of Helium in Population II Stars.

COMMISSION 28: GALAXIES (GALAXIES)

Report of Meetings, 15, 16, 17, 18, 21, 22, and 23 August 1979

ACTING PRESIDENT: B. E. Westerlund SECRETARY: V. C. Rubin

Dr. Westerlund was Acting President in the absence of the Commission President, Dr. B. E. Markarian, who was ill.

Commission 28 was involved in ten sessions, four of them all-day joint discussions:

Nuclei of Normal Galaxies (Commissions 28, 33, 34, 40)

Ultraviolet Astronomy/Results from Recent Space Experiments (34, 28, 29, 44, 45)

Very Hot Plasmas in Circumstellar, Interstellar, and Intergalactic Space (34, 10, 28, 44, 48, 49)

Extragalactic High Energy Astrophysics (40, 28, 44, 47, 48)

Details of these joint meetings will appear in Highlights of Astronomy. Of the remaining sessions, two were joint with Commission 33 (Structure and Radio Astronomy). The remaining sessions were Commission 28 alone. Programs for these meetings and abstracts of some of the papers presented follow the discussion of the business meeting below.

16 August 1979

BUSINESS MEETING

I. NEW OFFICERS AND MEMBERS

The Commission unanimously approved the election of the new President, B. E. Westerlund, and Vice-President, V. C. Rubin. The President proposed, and the Commission unanimously accepted, the list of names proposed for the new Organizing Committee.

The list of IAU members who wish to join Commission 28 was read and adopted, with one addition from the floor. A second list of 79 new members to the Union who indicated an interest in joining Commission 28 was discussed. In 1976, Commission 28 had decided to accept for membership only those IAU members of three years standing. After some discussion, it was voted to set aside this previous decision, and to accept as members all new members who wish to join Commission 28, subject only to the IAU rule that they belong to only one Commission during their first three years.

II. PROPOSED RESOLUTION

Dr. J. Heidmann proposed the following resolution, "In view of the various values of the Hubble constant H it is recommended to keep track of their influence in calculations and results by writing $H = 100\ h\ km\ s^{-1}\ Mpc^{-1}$ where h is a dimensionless number." A discussion and a straw vote indicated little support for the resolution, so the resolution was not voted upon.

III. REPORT OF COMMISSIONS

Following the discussion by the Commissions' Presidents concerning the reports of Commissions, it was decided to change the format for references and to use abbreviated journal references, rather than the Astron. Astrophys. Abstracts numbering system. This change should make the reports more readable and useful.

IV. ATLAS OF GALAXIES IN COLOR PREPARED BY JAMES WRAY

Dr. Wray presented color slides of over 20 galaxies. The slides were made from color prints produced by a dye transfer technique, using image-tube photographs taken in three colors. The three colors have a linear response throughout the entire dynamic range, and the presentation was spectacular and impressive for the wealth of science contained in the prints. Dr. Wray asked for an expression of support from Commission 28 in his efforts to obtain financial support for the publication of an atlas of about 170 galaxies.

The following statement, signed by B. E. Westerlund, was given to Dr. Wray.

"Commission 28 (Galaxies) recognizes the important scientific value of the color photographs of galaxies which you have prepared, and considers that they will be a continuing source of value in many ways for astronomers working in extra-galactic research. In this regard the members of Commission 28 present at the administrative session introduced and unanimously passed by general aclaim the following resolution in support of your work:

'In view of the high scientific quality and potential usefulness of your photographs for galaxy research we hereby express our strong support for your efforts to publish an atlas of color photographs of galaxies.'"

V. ASTRONOMICAL NOMENCLATURE

Dr. G. de Vaucouleurs reported on the meeting called by Dr. C. Jaschek (Commission 5: Documentation) which he attended as a representative of Commission 28. The attempt to set up a universal IAU system of nomenclature for all types of objects was described, with special reference to the report of Dr. Bidelman's working group for non-stellar objects. After some discussion, there was general agreement among Commission 28 members that authors should give two designations for every object; one a familiar or catalogue name and the second a position conforming to the rule for designating objects adopted earlier by the IAU. Although some members expressed strong preference for coordinates for the epoch 2000, it was recognized that stellar astronomers have a preference for 1950 coordinates. Commission 28 is invited to submit a statement to Dr. Jaschek by the end of 1979 containing its suggestions on this matter.

PROGRAMS OF SCIENTIFIC SESSIONS

GALACTIC DYNAMICS August 15 Joint with Commission 33

A. Kalnajs: Bar Modes of Disks.
A. Toomre: What Amplifies the Spirals?
R. H. Miller: Numerical Simulation of Galaxies.
G. Contopoulos: Integrals of Stellar Motion and their Disappearance.
P. O. Vandervoort: Resonance Phenomenon in the Dynamics of Spiral and Barred Galaxies.
C. C. Lin: Progress in Density-Wave Theory.
W. W. Roberts: Gas Flow in Galactic Bars.
P. Schwartz: Gas Flow in Inner Disk Forced by Oval Distortion.

RADIO STUDIES OF GALAXIES August 16 Joint with Commission 40

P. C. van der Kruit: Nearby Galaxies in Continuum and the Hydrogen Line.
E. M. Berkhuysen: Radio Spectral Index and Polarized Emission of M 31 and M 33.
H. van der Laan: Progress Report on the Westerbork M 31 Study.
R. Wielebinski: Radio Haloes of Galaxies.
E. Raimond: H I in NGC 4278.
L. J. Rickard: CO Emission from Nearby Galaxies.
R. D. Davies: Integral Properties of 100 Sbc Galaxies.
D. E. Harris: Overview on Active Galaxies.
R. Wielebinski: High Frequency Observations of Very Large Radio Galaxies.
E. B. Fomalont: Radio-Tail Structure of Jets in 3C31 and NGC 315.

J. F. C. Wardle: Radio Jets in Quasars.
K. Weiler: Statistics of BL Lac Objects.
D. Stannard: BL Lac Objects.
A. Bosma: Fleurs Map of Fornax A.
B. F. Burke: VLA Observations of the Double Quasar 0957 +561 A, B.

NEW RESULTS IN EXTRAGALACTIC RESEARCH August 16
A. Yahil: The Determination of the Deceleration Parameter in the Local Supercluster of Galaxies.
G. de Vaucouleurs: H_0: 50 or 100?
F. H. Chaffee, Jr., R. J. Weymann, M. Davies, N. P. Carleton, D. Walsh, and R. F. Carswell: The Twin Quasars 0957+561A, B: A Summary of the Current Observational Knowledge.

NEW IDEAS CONCERNING LARGE SCALE STRUCTURE OF GALAXIES August 17
Morphology
J. S. Gallagher: A Relationship between Morphology and H I Content in SO Galaxies.
J. Heidmann: Giant Clumpy Irregular Galaxies.
W. Eichendorf: The Physical Parameters of Double Galaxies.
Rotation
F. Bertola: Structure and Dynamics of Early Type Galaxies.
H. van Woerden: Distribution and Kinematics of Gas in Early Type Galaxies.
V. Rubin: Rotation Curves of Sa and Sc Galaxies.
J. W.-K. Mark: Global Dynamics in Disk Galaxies: Do They Provide Indicators for Haloes?
Formation
R. B. Larson: The Early Evolution of Galaxies.
Chemical Properties
S. M. Faber: The Chemical Composition of Galaxies: Recent Results.
Clustering
J. Einasto (read by M. Longair): Cluster Chains as Basic Elements in the Cell Structure of the Universe.

GALACTIC ENVIRONMENT August 17 Joint with Commission 33
W. E. Kunkel: Galactic Environment: Optical Data.
A. N. M. Hulbosch: High-velocity Clouds.
B. M. Tinsley and J. Jaaniste: Galactic Environment: Theoretical Considerations.
R. B. Larson: Nature and Cosmogony of Population III Objects.

NEW DATA ON GALAXIES August 18
G. Wynn-Williams and E. Becklin: Infrared Studies of the Central Regions of the Spiral Galaxies NGC 253 and IC 342.
B. Tully: A Hierarchial Description of Galaxy Clustering in the Region of the NGC 1023 Group.
W. A. Sherwood and G. F. O. Schnur: Spectra of Early Type Galaxies.
M. Whittle: Emission Line Profiles in Pictor A.
J. W. Sulentic: Old and New Properties of the M87 Jet.
G. de Vaucouleurs: The Inclination of the LMC.
T. Murai: The Dynamics of the Triple System: The Galaxy, LMC, and SMC.
M. W. Feast: Cepheids, the P-L-C Relations, and the Distance to the LMC.
M. Azzopardi: Determination of the SMC Distance Modulus by Hγ Photometry.
P. Dubois: Distance Modulus and Depth of LMC.
M. Penston: Motions in the Interstellar Medium in Some External Galaxiec.

ABSTRACTS OF SCIENTIFIC PAPERS:

P. Vandervoort: The resonance regions of a galaxy are defined in the first instance by conditions under which resonances occur in the stellar orbits. Theories of the role of resonant stars in galactic dynamics must take proper account of the intrinsic nonlinearity of the resonance phenomena. Current theories of equilibrium structures of galaxies are nonlinear in this sense. Even when resonant stellar motions are unimportant, the resonance regions of a galaxy are well defined regions which play a special role in the dynamics.

W. W. Roberts: Recent steady-state and time-evolutionary gas dynamical studies help to provide a deeper understanding of barred spirals. The dark narrow dust lanes observed against the bar can be understood as a tracer of the shocks formed in the gas. Large non-circular motions occur in the calculated gas flows with magnitudes up to 100 - 150 km s^{-1}, and these can account for the strong velocity gradients observed in some bars.

R. D. Davies: A study has been made at Jodrell Bank of the integrated neutral hydrogen properties of Sbc (T = 4) galaxies. Correlations between optical and H I properties show that the radius is a better discriminant of galactic properties than is luminosity class.

G. de Vaucouleurs: The two main approaches to the Hubble constant by Sandage and Tammann (1974-75) and by de Vaucouleurs (1978-79) were compared. The need for proper treatment of interstellar extinction, for independent checks of linearity, and for allowance for distortion of the velocity field by the Local Supercluster was emphasized.

F. H. Chaffee et al.: The combined optical observations from the Kitt Peak 2.1-m, the Steward 2.3-m, and the MMT reveal that the twin QSO's, whose angular separation is 6 arc sec, are nearly indistinguishable spectroscopically. The most stringent parameter is that provided by the Mg II and Fe II absorption lines in the two objects which give a velocity difference of 7 ± 15 km s^{-1}. Such a small difference is difficult to reconcile with explanations of the objects as two distance QSO's. We discuss the possibility of the objects being the double image of a single QSO produced by a gravitational lens lying between us and the QSO.

J. Heidmann: Clumpy irregulars contain ~6 clumps, each 100 times more important than HII regions with respect to optical luminosity, total mass, UV luminosity, and early star content. These bursts of star formation of unprecedented power deserve further investigations (IR, X, CO, VLA, abundances) of their nature, age, origin, and evolution (see "First Year of IUE Symposium," London, 1979).

Bertola: It is well established that the velocity of rotation in these galaxies is remarkably less than in S 0 and spirals. The study of the shape of their isophotes shows that all the possible trends of ellipticity can be found and that often a twisting of the axes is present. These observational facts are not consistent with the conventional view of elliptical galaxies as oblate spheroids and we have to consider other solutions such as triaxial or prolate structures.

van Woerden: A survey of 54 southern lenticular (SO and SO/a) galaxies at Parkes has yielded 15 unconfused HI line detections. The gas content is not correlated with colour, luminosity or bulge-disk ratio. The "Tully-Fisher relation" (blue or yellow luminosity vs. inclination-corrected profile width) for (northern + southern) lenticulars is closely similar to that for southern late-type (Sc - Im) galaxies.

Westerbork observations of NGC 4203 (SO-) reveal HI extending over 3 optical diameters, with considerable structure and a flat rotation curve. NGC 4694 (SBOp) has an unresolved HI concentration at the centre and two clouds of $\sim 10^8 M_\odot$ 4 - 5 arc min west, at relative velocities ~60 km/s.

V. C. Rubin: For a set of 21 Sc and 10 Sa galaxies of both high and low luminosities, optical rotation curves have been determined. All rotation curves are flat. This implies a significant mass beyond the optical image. Correlations between V_{max}, radius, mass, luminosity, and morphology are discussed.

J. W.-K. Mark: Three classes of internal dynamical indicators were considered. (1) The absence of a strong global stellar bar is evidence for a halo only if the galaxy does not also have a nuclear bulge amounting to about 30% of its mass. Such a bulge is sufficient to suppress strong bar-focusing disk instabilities, even with no halo (Berman and Mark, A & A July, 1979). (2) Haloes comparable or heavier than the disk drive self-excited waves (Bertin and Mark, A & A submitted). Isolated galaxies with warps are thus possible evidence for haloes. (3) Indirect indicators may be developed from spiral structure. Absolute amplitudes of density waves are affected by halo sizes. For example, large halo masses within the disk radius would produce more chaotic spiral structure of small pitch angle.

S. M. Faber: Nearby E and SO galaxies usually show color and line-strength changes within the innermost 10 kpc, which are commonly attributed to composition gradients. Beyond 10 kpc, the gradient may flatten. The global mean composition of ellipticals is strongly correlated with luminosity and flattening. These correlations are consistent with gaseous collapse models, but harder to understand in the context of stellar collapse or merger models. The abundance differences seen among spiral galaxies are larger than expected from a single-zone model, which predicts a close correlation between mean gaseous abundance and remaining gas content. Evidently at least one additional parameter other than gas consumption must be important in controlling the chemical evolution of disk galaxies. This factor might be radial gas flows or variations in the nucleosynthetic yield, but its identity is presently unknown.

F. G. O. Schnur and W. A. Sherwood: We have 9 A resolution spectra from 3800A to 7000A of the inner 1" x 4" region of 50 galaxies classified as SO by Sandage or having T = -3 to 0 in RC2. We find 70% show emission lines, i.e., contain gas, and 18% contain [OIII]. One galaxy is a Type I Seyfert.

J. W. Sulentic: The 1956 polarization photographs of the knots in the jet of M 87 have been repeated in 1978. A comparison reveals that at least the brightest knot has varied both in total ($0.1 - 0^m.2$) and polarized ($0.3 - 0^m.4$) intensity. Relative to the nucleus, no proper motion of this knot was detected at 3σ limit ($\sim$7pc). Extensive attempts to detect features in the optical spectrum of the brightest knot using the 5-m SIT digital spectrograph were unsuccessful. A striking similarity was noted between the radio and optical properties of the knots and the BL Lac class of objects.

de Vaucouleurs: 1. The Large Cloud is a disk system: spiral structure; rotation; photometric major axis ($170^\circ \pm 5^\circ$) and kinematic line of nodes coincide; Z velocity dispersion of young stars ($\sim$10 km s^{-1}) agrees with values in galaxy and other spirals; edges of Magellanic systems are flat (NGC 55, 2188 ...).

2. If main arms are trailing, east side is near side.

3. Inclination derived from ellipticity of outer isophotes in red light, $b/a = 0.89 \pm 0.01$, is $i = 27^\circ \pm 3^\circ$; it is confirmed by other methods.

4. Stars of identical absolute magnitude should be brighter on near side by $\Delta m_1/\Delta Y = 2.17 \tan i/57.3 = 0.019$ mag/degree. However, if cosec law applies across field, differential extinction should amount to $\Delta m_2/\Delta Y = -0.010$ mag/degree if $A_B = 0.20$ (minor axis of LMC coincides with galactic meridian). Four possible cases:

Near side	Predicted slope $(\Delta m_1 + \Delta m_2)/\Delta Y$	
	With A_B = const.	with $A_B = 0.2$ cosec B
W	-0.019	-0.029
E	+0.019	-0.009

5. Use of cepheids apparent magnitudes reduced to constant period P_o via P-L relation was first applied to 3 groups of LMC cepheids from Harvard photographic data. The three period groups (I) $0.4 \leq \text{Log } P < 0.8$ (50 stars), (II) $0.8 \leq \text{Log } P < 1.3$ (25 stars), (III) $1.3 \leq \text{Log } P < 1.7$ (21 stars) were reduced to $\text{Log } P_o = 0.5$ (I), 1.0 (II) and 1.5 (III) via $\Delta m/\Delta \text{Log } P = 1.65 + 0.50 \text{ Log } P$ based on Shapley's empirical P-L relation. Least squares solutions gave:

$$(1) \quad m = 16.095 + 0.0325Y$$
$$\pm.078 \quad \pm.0234$$

$$\text{(II)} \quad m = \underset{\pm.072}{15.013} + \underset{\pm0382}{0.0059Y}$$

$$\text{(III)} \quad m = \underset{\pm.101}{13.880} + \underset{\pm.0637}{0.0483Y}$$

The weighted mean of slopes $\Delta m/\Delta Y = +0.027 \pm 0.019$ (m.e.) confirms that the east side is nearer and agrees best with A_B = const. case.

6. These results have recently been confirmed by Gascoigne and Shobbrook who made a similar analysis of 14 selected cepheids ($0.635 \lesssim \text{Log P} \lesssim 1.099$). The cepheids were chosen to form two compact groups about 2°8 W and 3°8 E of the center of the Bar. They express their results as mean distance moduli calculated via the P-L and P-L-C relations

$$M_V = -1.18 - 2.900 \text{ Log P}$$

and

$$M_V = -2.24 - 3.656 \text{ Log P} + 2.52 (\langle B \rangle - \langle V \rangle)$$

	< Log P >	< V >	< B-V >	< Y >	< μ >	N
E	0.842	14.98	0.68	-3°775	18.59 ± .034	8
W	0.910	14.93	0.71	+2°779	18.71 ± .038	6

The differential modulus $\Delta \langle \mu \rangle = +0.12 \pm 0.05$ implies $\Delta m/\Delta Y = +0.018 \pm 0.008$, is in near perfect agreement with the predicted value (+0.019), if the near side is the E side and A_B = const.

7. This analysis was again repeated by Martin, Warren and Feast for 26 cepheids with photoelectric BVI light curves and Log P < 1.7, divided into two groups E and W of the line of nodes:

	< Δμ >	< Y >	N
E	-0.036 ± 0.041	-2°7	14
W	+0.043 ± 0.026	+1°3	12

The differential modulus $\delta \langle \Delta\mu \rangle = 0.079 \pm 0.048$ corresponds to a slope $\Delta m/\Delta Y = +0.023_5 \pm 0.014_5$ which once again confirms that the east side is nearer and that differential extinction across the field is negligible.

In summary we have 3 independent determinations of $\Delta m/\Delta Y$

1955	de Vaucouleurs	+0.027 ± 0.019	(96 stars)
1978	Gascoigne and Shobbrook	+0.018 ± 0.008	(14 stars)
1979	Martin, Warren and Feast	+0.023 ± 0.014	(17 stars)

unweighted mean	$+0.022_6$
weighted mean ($W = 1/\sigma^2$)	$+0.020_2$
weighted mean (W = N)	$+0.025_5$
adopted	+0.023 ± 0.007

The slope implies an inclination $i = 31° \pm 8°$ which agrees within errors with $i = 27° \pm 3°$ from optical isophotes, $27° \pm 5°$ from HI isophotes and $25° + 9°$ from cluster distribution. The weighted mean i = 27°2 + 2°4 is essentially unchanged from the previously adopted value.

Azzopardi: The computation of the Hγ equivalent width on objective-prism plates of 210 star members issued from the lists of Azzopardi and Vigneau (1975, 1979) using the method described by Azzopardi et al. (1978) and mean visual absolute magnitudes and intrinsic colors for MK luminosity classes for galactic supergiant stars allowed us to determine 19.2 as SMC distance modulus.

Dubois: has shown that the stellar absolute magnitude calibration based on hydrogen line intensities in the SMC is different from the galactic one. The temperature has greater influence in the SMC, consequently:

1) The distance modulus of the SMC depends on the temperature of the stars used.

2) It is necessary to use a SMC calibration to determine the SMC depth.

Report of Working Groups

I. THE MAGELLANIC CLOUDS

The WG convened on August 20 with B. E. Westerlund as Chairman.

Dr. M. W. Feast was nominated as the new Chairman and elected unanimously. It was agreed that the WG should from now on consist of all astronomers working in its field of research and interested in joining it. About 50 astronomers present at the meeting joined the WG.

A monograph has recently been written on the Large Magellanic Cloud and related objects by J. V. Feitzinger in German "Galaxien vom Magellanschen Typ". Several members expressed their hopes that it should be made available to the members in English.

The following papers were presented:

M. Azzopardi and J. Breysacher: Systematic Survey for Wolf-Rayet Stars in the Magellanic Clouds.

M. Duflot and Ch. Fehrenbach: Radial Velocities of Stars in the LMC.

H. B. Richer: Carbon Stars in the Large and Small Magellanic Clouds: A Comparison.

E. Hardy: Some Colour-Magnitude Diagrams and Abundances in the Magellanic Clouds.

M. T. Brück: The Distribution of Faint Stars in the SMC.

E. Maurice: The Structure and Kinematics of the SMC.

N. Sanduleak and A. G. D. Philip: An Objective Prism Survey for Faint LMC Member Stars.

J. W. Feitzinger: Kinematics and Dynamics of the LMC.

S. C. B. Gascoigne: Ages and Abundances for Some Magellanic Cloud Clusters.

Th. Page and G. Carruthers: Distribution of Hydrogen and Hot Stars in the LMC.

N. W. Broten: Ionized Hydrogen Mass of the Magellanic Clouds.

K. Nandy: Ultraviolet Observations of LMC Supergiants.

J. Koorneef: Ultraviolet Observations of Stars and Stellar Associations in the LMC.

J. Hutchings: Groundbased and IUE Results on OB Star Stellar Winds in the Magellanic Clouds.

P. Benvenuti, M. A. Dopita and S. D'Odorico: UV Spectra of two Supernova Remnants in the LMC.

E. H. Geyer and U. Hopp: Star Counts in Spheroidal Emission Regions of the LMC.

J. V. Feitzinger, W. Schlosser, Th. Schmidt-Kaler and Ch. Winkler: The Central Object of the Supergiant Gas Nebula 30 Doradus.

ABSTRACT OF PAPERS

Azzopardi: An objective-prism survey using an interference filter led us to detect 4 new WN stars in the SMC and 17 in the LMC, increasing to 8 and 101 respectively, the numbers of known WR stars in these systems. The census of the WR population in the Magellanic Clouds can now probably be considered as complete within the survey limits.

Richer: Spectra and photoelectric VRI photometry of a sample of carbon stars found in the LMC and SMC survey of Blanco and McCarthy have been obtained. The main results are: (1) SMC carbon stars are bluer than those in the LMC by about 0.15 magnitudes in (R-I) and by 0.26 in (V-R). (2) SMC carbon stars are more luminous than those in the LMC by about 0.3 magnitudes if the difference in distance moduli of the two galaxies is 0.5 magnitudes. (3) At a given color, there are no major differences in the spectra of the LMC and SMC carbon stars. The first two results above are probably related to the relative metal deficiency of the SMC compared to the LMC, and imply that caution should be employed in using luminous late-type stars as extragalactic distance indicators to galaxies of varying metallicity.

Brück: The distribution of stars down to B = 21^m along the major axis of the SMC between 2° and 6° from the centre has been studied, using counts made by the COSMOS measuring machine on photographs taken with the UK 1·2m Schmidt Telescope. The distribution of stars brighter than 19^m (M = o), taken to be disk stars, are closely correlated with the HI brightness, whereas very little correlation exists for the fainter stars, which are identified as halo objects. There is a difference in the correlation indices for the Bar and arm regions of the Cloud, a result interpreted as due to differing stellar populations in these two parts.

Maurice: From a study of the 24 brightest stars of the SMC, Ardeberg and Maurice find a differential distance modulus of the two Magellanic Clouds of 0.50 ± 0.07 and a depth (∿40 %) larger than that of the LMC; this is supported by surface distribution of distance moduli, radial velocities and polarization; no rotation curve can be traced. The major apparent feature, a high-distance and high radial-velocity line in the direction of the SMC Wing and of the LMC, is in good agreement with a disruption model of the SMC (Toomre) in a close encounter with our Galaxy; separation time (6×10^8 years) calculated from the depth (∿15 Kpc) and radial velocity difference (∿25 kms^{-1}) in the SMC is in good agreement with the model.

Feitzinger: The presently known radial velocities of planetary nebulae, star clusters, stars, HII and HI regions have been catalogued. With this data set the basic kinematical and dynamical parameters of the LMC are determined, new rotation curves deduced and the velocity residuals (observation minus model) discussed.

Carruthers: Comparison of our far-UV (1050-1600 A) flux measurements over the whole of the Large Magellanic Cloud with Henize's H-alpha measurements and the McGee-Milton 21-cm survey of the LMC reveal that interstellar hydrogen in the LMC is often concentrated in 100-pc clouds within the 500-pc clouds detected by McGee and Milton. Furthermore, at least 25 associations of O-B stars in the LMC are outside the interstellar hydrogen clouds, four of them appear to be on the far side. Measurements of atomic hydrogen toward selected stars by means of its 1216 A (Lyman-α) absorption are in progress using the IUE satellite.

Nandy: The ultraviolet observations obtained with the IUE and the complementary visible data of the reddened and comparison stars in the Large Magellanic Cloud, obtained with the ESO 3.9m telescope at La Silla, have been used to determine the interstellar extinction law for that galaxy. The present result indicates significant differences in the far ultraviolet extinction laws of the LMC and the Galaxy. The slope of the LMC extinction curve shortward of 1500Å, normalised to $E_{B-V} = 1$ and $A_V = 0$, is found to be steeper than that of the galactic extinction curve. If this result is confirmed by future observations of a larger sample of reddened and comparison stars in the LMC, this would imply a significant difference in the content of small particles in the interstellar media of the two galaxies.

Geyer: Star counts to the limits $16^m.5$ and $20^m.5$ respectively within the boundaries of 31 spherical emission regions of the LMC have been carried out and their average diameters ∅ determined on U-Schmidt plates.

The slopes of the linear correlations between log N - the total counted stars of the embedded cluster - and log ∅ depend on the magnitude limits of the counts. This shows that less luminous stars are also contributing towards the illumination of the nebulae. A method for the determination of relative ages of emission regions was found.

Schmidt-Kaler: Short exposure photographs in various colours and spectra of the central part of the 30 Dor nebula have been obtained with the ESO 3.6 m and Bochum 0.6 m telescopes to investigate the structure by digital image analysis. R 136 consists of 3 components, R 136a being by far the bluest, covering $(0.7\ pc)^2$ and providing 90% of the excitation of the nebula. The bolometric magnitude is brighter than -14^m, $T_{eff} \simeq 52000$ K, the mass estimated at 500 $M_\odot$. The inner structure of 30 Dor is explained as a result of stellar wind from R 136.

II. GALAXY PHOTOMETRY AND SPECTROPHOTOMETRY

The Group met at two successive sessions on the afternoon of August 20. Dr. M. Capaccioli was reelected as Chairman for the next three years and Dr. S. Odamura was requested to serve as Secretary.

The following papers were presented:

J. Kormendy and D. Koo: Triaxial bulge components in barred spiral galaxies.

R. J. Talbot: Multicolor photometric study of the structure of M83.

G. Carranza: Line intensity variations across NGC 1313, 7552, 7793 and other peculiar galaxies.

K. Freeman: Galaxy photometry programs at Mt. Stromlo.

S. Okamura: Galaxy photometry programs in Japan.

P. van der Kruit: The Westerbork-Palomar survey.

G. de Vaucouleurs: Programs on photometry and spectrophotometry of galaxies at the University of Texas in Austin.

M. Capaccioli: Galaxy photometry programs at the Asiago Observatory.

During the discussion the importance of the Austin Conference on Galaxy Photometry, Kinematics and Dynamics (August 6-8, 1979) was fully acknowledged. Request was made to have it organized periodically. The problems of accessibility and archives of digital data resulting from two-dimensional mapping of galaxies were raised. It was proposed that, for the time being, the Chairman will take care to collect, file and distribute the basic information such as objects-list and data-format. It was noted that the proposal by the W. G. to provide standard luminosity profiles for a few well-observed galaxies (NGC 3115, 3379, 4486) has not yet been fully accomplished. In this context, suggestion was made to include some distant galaxies in the list of photometric standards.

III. INTERNAL MOTIONS IN GALAXIES

On the morning of August 21 the W. G. held a 90-minute meeting. It was chaired by Paris Pişmiş and was attended by 30-40 people.

During the brief administrative session the question was put forward - as in the previous XVI Assembly - whether the Working Group should continue as such. It was agreed that it was useful to have such a W. G. The election of a new chairman was postponed. A ballot was circulated later by mail and Stephen T. Gottesman was elected chairman for the next three-year period.

The Scientific part of the Meeting included:

a) A brief introduction by the Chairman where she recalled some of the problems of interest in the topic of the internal motions in galaxies.

b) Seven papers on current research. The authors and the titles of the papers are listed below.

T. G. Hawarden: "Kinematics of N 5291".

V. I. Pronik: "Gas Motions in the Nucleus of N 4151".

Y. Terzian, B. F. Jones and R. A. Sramek: "VLBI of the Nuclei of Normal Galaxies".

G. Carranza: "Interferometric Study of N 5236".

D. Helou, E. Salpeter, Y. Terzian: "HI Survey of Binary Galaxies".

V. L. Afanasyev, L. A. Matik, I. I. Pronik: "N 1275, a Pair of Interacting Galaxies".

D. Burstein and N. Krumm: "The Gaseous Disk of the SO Galaxy N 4203".

A highlight in the activities of the W. G. during the past triennium was the "Conference on Photometry, Kinematics and Dynamics of Galaxies" held in Austin (August 6-8, 1979) in conjunction with the W. G. on the Photometry and Spectrophotometry of Galaxies. 13 reviews and 80 contributed papers were presented.

On behalf of the members of the W. G. on the Internal Motions in Galaxies, the W. G. extended its gratitude to the University of Texas, at Austin, the host institution of the Conference, to Dr. Harlan Smith, Director of McDonald Observatory, to Dr. G. de Vaucouleurs, Chairman of the Scientific Organizing Committee, and to Dr. D. S. Evans, Chairman of the Local Organizing Committee.

ABSTRACTS OF PAPERS

P. Pişmiş: Spectral regions that yield velocity data of Galaxies are optical, radio, molecular, UV, etc.

Largest contribution to overall vel. field is from HI, though detailed field in nucleus is obtained optically.

Some overall properties of interest are: Rotation curves have extended maxima, waves in rotation curves are physically significant; interpretation: density waves, two component model.

Nuclear regions: activity or geometry?

Bars: gas flow along bars? Rotation can be either as a solid body or differential.

Overall velocity field: Isovelocity contours show bi-symmetry; interpretation: oval distortions, warps.

Desiderata: More data necessary to establish reliable correlations between global parameters of galaxies such as morphological type, luminosity, V_{max}, R_{max}/R, mass, etc.;

VLBI and radio recombination line studies;

Velocity fields in double galaxies; and

Velocity fields in Magellanic irregulars.

Terzian: We have performed a 3-station (Arecibo, Green Bank, Owens Valley), and a 7-station (Bonn, Arecibo, Green Bank, Illinois, Fort Davis, Owens Valley, Hat Creek) VLBI experiment at λ18 cm of the nuclei of normal galaxies. More than 30% of the observed galaxies showed central radio compact sources. Preliminary results show that the sizes of the radio components are $\lesssim 6 \times 10^{-3}$ arc second. Radio maps will soon be produced from these observations. We also plan to observe these sources with the VLA (resolution ∿ 1 arc second), and a further VLBI experiment will be performed at λ 6 cm to obtain spectral information.

Carranza: The distribution and motions of the ionized medium in M83 were studied with the 150-cm reflector at Bosque Alegre equipped with high luminosity optics and interferometric devices. The Hα plates (scale 110"/mm) allowed the identification of more than two hundred H II regions distributed on a nearly circular area of 7' diameter which roughly coincides with the galaxy main body; with the higher contrast of the interferograms a weak Hα background was also detected on the same region.

The beginning of a quite symmetrical spiral pattern is clearly drawn by the emission regions but such structure becomes rather complex in the outer parts of the galaxy; the individual H II regions are, however, larger and brighter in the NW arm. The field of radial motions over the main body was established by means of more than six hundred individual radial velocities (the majority from interferograms and a few from spectra) fall between 400 Km s^{-1} (NE) and 600 km s^{-1} (SW). The systemic radial velocity is 498±5 km s^{-1}, in fair agreement with other determinations. The main features of the resulting velocity field are:

a) an approximate symmetry with respect to the center of the galaxy,
b) the kinematic major axis changes from the neighborhood of the NS direction in the bar and nucleus to PA 45° - 50° in the outer regions,
c) local extrema of relative radial velocity (with respect to that of the nucleus) are found symmetrically disposed in the NS direction at about 1.5-2 arc sec. from the center of the galaxy.

A comparison of morphological and kinematical features of M 83 reveals: 1) the radial velocities with respect to the nucleus are larger along the arms than in between; 2) the central isovelocities from 470 km s^{-1} to 530 km s^{-1} are deformed (even twisted) when crossing the bar; 3) some intermediate isovelocity lines are deformed towards the ends of the bar. A confrontation of morphological and kinematical features over the main body of M 83 might suggest a process related to the inner structure of the galaxy, perhaps some kind of bar induced dynamical disturbance to account for the observations.

Helou: The survey used the Arecibo telescope to observe 25 galaxies in 9 groups of at most 5, as a first phase. The sample was drawn from a list of binaries obtained from an automated search of the UGCG, with further selection for HI detectability. The survey focuses on the internal dynamics of the galaxies, but a complete area map will be produced for some of the groups in the second phase of the project.

Burstein: We have observed the HI disk of N 4203 at Arecibo with 4 arc min resolution. These data yield: HI/optical diameter $\approx$ 3; M/L at 20 kpc $\approx$ 20. This S0 galaxy evidences no spiral structure or conspicuous star formation down to 27th mag. arc sec^{-2} in B.

IV. EXTRAGALACTIC SURVEYS FROM SPACE

The WG met on two successive sessions on August 22 with an attendance of about 70 astronomers. R. Barbon, as chairman, briefly recalled the circumstances that led to the formation of the WG. During the discussion on business matters the WG agreed that formal contacts shall be taken with all the IAU Commissions showing interest in the type of data provided by a wide field space telescope. Owing to its widespread nature, the new name "Working Group for Space Schmidt Surveys" has been approved for this WG. K. Henize has been elected new chairman for the next three year term, with F. Bertola as vice-chairman. The WG agreed also to nominate a subgroup of astronomers to keep in touch with the various teams in different nations involved in space survey projects. This consists of: H. Arp, R. Barbon, R. D. Cannon, G. Carruthers, G. Courtes, G. Lemaitre, H. J. Smith, D. J. Schroeder, K. van der Hucht, R. M. West, B. Westerlund, J. Wray.

In the scientific session the following papers were presented:

H. J. Smith: The Spacelab Wide Angle Telescope (SWAT).

R. M. West: The ESA Space Schmidt.

R. D. Cannon and T. G. Hawarden: The UK Schmidt: Rehearsal for the Operation of a Space Schmidt Facility?

M. F. McCarthy: Research Projects with a Space Schmidt: Low Dispersion Spectroscopy.

At the end of the meeting the WG approved the following Recommendation:

This Working Group recommends that the space agencies of the IAU member nations incorporate in their space astronomy programs support for fast wide-angle Schmidt-type telescopes capable of recording images comparable in quality to that provided by the Palomar Sky Survey. Besides exploring in the UV region inaccessible to ground based astronomy, these instruments would also reach substantially fainter background light levels at all wavelengths because of the darker night sky. For example, such telescopes will:

1. discover and study new types of objects,
2. explore for faint extensions in galaxies and clusters of galaxies,
3. detect UV rich sources, and
4. provide, in a short time, UV and IR data of large numbers of objects of use to the entire astronomical community.

Now that the Space Telescope is an approved project, the urgency of such instruments as necessary complements to it, is increased. Therefore, this Working Group strongly recommends that steps be taken as soon as possible to ensure that such a large-field telescope will be available in the same time period as the launch of the S.T.

ABSTRACTS OF THE PAPERS

H. J. Smith: Wide field telescopes will play as important a role in space astronomy as they have on the ground, for reconnaisance and survey work, for studies of large scale and/or faint surface brightness features, for statistical studies and for reference including e.g. identifications of X-ray and radio sources. Wide-field surveys of significant parts of the sky contain information of interest to a wide range of astronomers for many years. A space Schmidt telescope is needed to accomplish the above ends, and to make more efficient the work of the Space Telescope. A relatively economical all-reflecting space-Schmidt is being proposed, for cooperative international construction, initially for single pallet operation on space shuttle spacelab missions. The Schmidt will have an aperture of about 0.75m, operating at about f/3 with images better than 2 arc sec. over a 5 degree field, recorded by an electrographic camera. Early specific applications include UV studies of galaxy morphology and interacting regions, the detection of early type stars at least to m = 26 including important cases of close degenerate companions of normal stars, and the study of up to four magnitude fainter halos and extensions of galaxies - some of the missing mass? - in IR as well as UV thanks to the freedom from airglow.

R. M. West: A group of European astronomers has recently proposed to ESA an all-reflecting Schmidt telescope of medium size (70cm), fast focal ratio (f/3) and wide field (5°) for incorporation into the Spacelab programme. This telescope will be particularly suited for work in the far UV, but can also be used for very important work at other wavelengths and for (objective) spectroscopy. An assessment study is carried out at ESA and a report will be available in late 1979.

R. D. Cannon and T. G. Hawarden: Experience in operating the UK 1.2m Schmidt Telescope as a facility for outside users suggests that several important points must be considered also in the organisation of the Space Schmidt. 1) The enormous amount of information (10^9 pixel per frame) provided by a wide angle medium resolution telescope means that each frame will be of utility to many workers in different ways. 2) The dissemination of this information is a major project in itself and must be provided for by the time the telescope is operative. 3) It is suggested that this can best be achieved by provision of data in the form of a high quality photographic atlas and in digital form from a central institute or institutes whence users could request data for specific objects or areas.

M. F. McCarthy: The Space Schmidt will offer many possibilities and advantages for research done with objective prisms, gratings or grisms. Spectral studies on quasars, Seyfert galaxies and clusters will be done in the UV region although distinct advantages can also be expected in the IR (6800 - 9000 A) where a gain of 3-4 magnitude is foreseen. The problem of the determination of the ratio of Carbon to late type M giants in different galaxies has been illustrated.

COMMISSION 29: STELLAR SPECTRA (SPECTRES STELLAIRES)

Report of the Business meeting, August 17, 1979

PRESIDENT: M. HACK SECRETARY: A.B. UNDERHILL

The first item of business was the confirmation by the Commission of the present vice-president W. Bonsack as president for the next three years and the election of eight members, Y. Andrillat, G. Cayrel, C.R. Cowley, C. Jaschek, V. Khokhlova, J.P. Swings, R. Viotti and K.O. Wright to form the Organizing Committee. It was confirmed that J. Jugaku should be vice-president for the next three years.

The president drew the Commission's attention to three atlases that have been published recently: (1) "A Photographic atlas of northern Be stars" which has been prepared by A.M. Hubert-Deplace and H.Hubert and published with the assistance of the International Astronomical Union, (2) "Revised MK spectral atlas for stars earlier than the sun" by W.W. Morgan, H.A. Abt and J.W. Tapscott, and (3) "A photometric atlas of the spectrum of Procyon" prepared by R.E.M. and R.F. Griffin. These may be ordered by contacting the authors at the Meudon Observatory, Kitt Peak Observatory and The Observatories, Cambridge, England respectively. A 4th catalogue of stars having MK spectral types will be published by the Dept. of Astronomy, Northwestern University, Evanston, Illinois in November 1979. Information may be obtained from W. Buscombe.

The question of working groups was discussed and a reorganization is to take place. Pro-tempore Chairmen have been appointed to coordinate membership and organize the election of a Chairman and steering Committee.

Pro-tempore Chairman	Title of Working Group
M. Jaschek	Be stars
C.R. Cowley	Ap stars
L. Pasinetti	Standard stars for Equivalent Widths and Line Profiles

The working group on ultraviolet spectra has been dissolved.

It was decided that the President of Commission 29 should write a letter to R.E.M. and R.F. Griffin expressing the Commission's great interest in their Procyon Atlas and its appreciation of the Griffin's work in preparing such an excellent and valuable Atlas.

Future Symposia and Colloquia of interest to Commission 29 were discussed. Five are under consideration.

(1) W.W. Weiss (Vienna) has prepared a workshop on "Short time-scale variation in Ap stars" to be held in Vienna in late June or early July, 1980.

(2) H.G. Groth (Munich) has proposed that a Colloquium on "Be stars" be held at the Munich University Observatory in the Summer of 1980.

(3) C. Chiosi (Padua) and R. Stalio (Trieste) have proposed that a Symposium on "The evolutionary effects of Mass Loss" be held in Trieste in June or September 1980.

(4) G. Cayrel (Meudon) has proposed that a Colloquium on "G and K stars as tools for understanding Galactic Structure" be held in Paris in 1981.

Aller emphasised the value of those who have large stores of spectra needing study making this material available to astronomers from countries which had no access to large telescopes. The transcription of the needed material can be easily carried out for spectra that are already in digital format. Bonsack said that he was willing to compile and circulate a register of all astronomical spectroscopists giving a brief summary of their fields of expertise. He will be sending out a form to collect the required information and the members of Commission 29 are asked to cooperate in this venture.

H.J. Lamers gave a brief report on the meeting on emission-line stars which was held in Trieste in June 1979. The discussion emphasized the value of comparing normal stars with emission-line stars of about the same type for obtaining information about the physics of stars.

A vote of thanks to M. Hack for her capable and enjoyable presidency of the Commission was proposed and passed with enthusiasm.

Anne B. Underhill
Secretary, Commission 29

Other Activities of the Commission

Two Joint Discussions were held with the partecipation of Commission 29: Ultraviolet Astronomy. Results from recent Space Experiments (Commissions 28, 29, 34, 44, 45) on August 15, 1979, and Physics of the chromosphere-corona-wind complex and mass-loss in stellar atmospheres (Commissions 10, 12, 29, 35, 36, 44) on August 16, 1979. Commission 29 has participated in the organization of a Joint Meeting on Am stars with Commission 45 and has organized two Joint Meetings on Stellar Abundances and Stellar Rotation with Commissions 36 and 45 with the following programs:

Stellar Abundances (Aug. 18, 1979): Chairman W. Bidelman
Invited papers: D.L. Lambert, The accuracy with which stellar abundances can be determined.
B.E.J. Pagel, Red giants in globular clusters and large scale variations in the galaxy.
L.H. Aller, Abundances in outer galaxies.
Short contributed papers:
R. Foy, G. Cayrel: Why the pressure induced splitting of the curve of growth should not be ignored in abundance determinations.
A.B. Underhill: The importance of the choice of a representative model atmosphere for interpreting anomalous line intensities in B-type spectra.
M. Spite: The problem of Aluminium in 6 old halo dwarfs and giants.
R.E.S. Clegg, D.L. Lambert, R.A. Bell: Isotopes of Titanium in cool stars.
R.E.S. Clegg: Aluminium in CS molecules in the Carbon star IRC +10216.
F. Thevenin, G. Cayrel, R. Foy: Some comments on the latest papers on the so-called super-metallicity in G and K stars.
G. Cayrel, M. Bentolila: Toward a fourth edition of $\left[\frac{Fe}{H}\right]$ determinations.
L. Rossi: A tentative suggestion for He abundance in Feige 86.
I.N. Glushneva, V.T. Doroshenko: Stellar energy distribution in the spectra of main sequence B-F stars. Comparison with theoretical models.

Stellar Rotation (Aug. 21, 1979) Chairman A. Slettebak
Invited papers: L.V. Kuhi, Rotation of very young stars.
Myron Smith, Late-type star rotation.
M. Plavec, Rotation in binary stars.
Short contributed papers:

J.B. Hutchings: Determination of v and i for fast O and B-type rotators.
L. Carrasco: Specific angular momentum in late-type stars (flare stars).
G. Belvedere, L. Paternò, M. Stix: Differential rotation of main sequence stars.
R. Rajamohan: Stellar rotation of main sequence single and binary stars.

Moreover a panel discussion on Ap stars was organized by C.R. Cowley and S.J. Adelman, and a workshop on the extraordinary object SS 433 was held together with Comm. 40, 42, 48 (see report of Comm. 42).

The working group of Comm. 29 and 45 on Ap stars has elected the following organizing Committee: C.R. Cowley (chairman), V.L. Khokhlova, M. Hack, M.M. Dworetsky, S.C. Wolff, D.S. Leckrone, M. Gerbaldi, M. Jaschek.

The working group on Be stars has elected the following Organizing Committee: M. Jaschek (chairman), A.M. Delplace, A. Feinstein, P. Harmanec, L. Houziaux, A. Slettebak, A. Underhill.

COMMISSION 30: RADIAL VELOCITIES (VITESSES RADIALES)

Report of Meeting, 1979 August 15

PRESIDENT: A.H. Batten SECRETARY: J. Andersen

1. The President opened the meeting and recalled the loss by death of the following members of the Commission: J.F. Heard, A.D. Thackeray, and R.K. Young. The Commission stood in silence as a tribute to the memory of these distinguished colleagues.

2. The report by the President was approved, with a typographical correction.

3. The following changes in the composition of the Commission for the next three years were unanimously approved:

President:	M. Duflot
Vice-President:	A.G.D. Philip
Organizing Committee:	A.H. Batten, J. Andersen, E. Maurice, J. Sahade
Resigned:	H. Gollnow
New Members:	J. Andersen, A. Florsch, I.D. Karachentsev, M. Mayor, C.L. Morbey, B. Nordström, N.B. Sanwal, C.D. Scarfe

Prof. Fehrenbach found it desirable that the Organizing Committee should have a member from the U.S.S.R. The President regretted that his extensive efforts to obtain this had not been successful.

4. Brief reports were given on the Joint Meetings on Nomenclature (M. Barbier) and on Space Astronomy (J. Andersen).

5. In 1976, a working Group was established to advise on:

- suitable procedures to compute the mean velocity of a constant velocity star from published measurements, and

- the information to be given when publishing new radial-velocity data.

As Chairman of the Working Group, Dr. Barbier reported on the first subject that questionnaires had been distributed to all members of the Commission asking their opinion on the following points:

- application of systematic corrections

- weighting procedure (dispersion, number of measurements, external m.e.)

- use of the mean velocities of the Wilson catalogue

A summary of the answers, which were received from most of the institutions currently active in this field, was distributed, but unfortunately it had been impossible for the Working Group to meet before the Commission meeting. The conclusions of an extended discussion were that systematic corrections, however desirable, were often impossible to apply in practice because of inadequate standardization. Weighting should be done according to external mean error and increase more slowly than the square root of the number of measurements, if that number is large. The Wilson mean velocities should be treated as single measures with suitable weight.

There was also agreement that computation of the best mean velocity still requires a specialist judging all the available data. A general recipe for generating an overall mean value from a computerized bibliographical catalogue cannot be optimal, because even the mean velocity in a given paper may well contain measures from previous publications.

The Commission was informed that Prof. Evans' new edition of the General Catalogue is now complete from 0^h through 20^h and that part released for distribution through the Stellar Data Centre in Strasbourg. Arrangements have been made to complete it using the methods employed by Prof. Evans.

On the second subject, Dr. Duflot presented the recommendations of the Working Group for the rules to be followed. It was unanimously agreed to forward these as a resolution to the General Assembly.

The Working Group's task was thus completed and the Group dissolved. The Commission expressed its appreciation of the active work performed.

6. During the scientific session, the following papers were given:

B. Campbell:	High-Precision Measurements of Radial Velocities
A.H. Batten:	Standard-Velocity Stars: Introduction
C. Fehrenbach:	Work at Observatoire de Haute-Provence
E. Maurice:	Work at Marseille: Standard Stars and Minor Planets; Wave Lengths of Forbidden Lines
J. Andersen and B. Nordström:	The Copenhagen Programme: Need for Extended and Improved Lists of Standards
M. Mayor:	Standardization of Radial Velocities Measured with CORAVEL

Dr. F. Ochsenbein presented the microfiche version of the updated Bibliographical Catalogue now available from Strasbourg.

COMMISSION 31: TIME (L'HEURE)

Report of Meetings, 15, 16, 17, 18, and 20 August 1979

PRESIDENT: A. Orte

SECRETARIES: J. Benavente
D. D. McCarthy

General

Prior to Commission meetings, a short session of the Organizing Committee was held on August 14 afternoon, in which items and proposals to be submitted to the consideration by the Commission were revised.

Commission 31 meetings were arranged to allow members to attend most of the scientific sessions of Commissions 4 and 19.

30 members of Commission 31, 7 future members, 5 consultants, 3 future consultant members and 2 invited participants attended Commission 31 meetings.

15 August 1979

Following a joint meeting with Commission 19 (Rotation of the Earth) (See report of Commission 19.) an administrative meeting of Commission 31 was held. Members stood in silence to honor the memory of Z. M. Brkic, H. Dingle, E. Fichera, C. G. Lacombe, and N. Stoyko, deceased since the last meeting at Grenoble.

The members approved the nomination of S. Iijima as the new president and G. Hemmleb as vice-president. Organizing Committee members for 1979-1982 are D. J. Belotserkovskij, H. Enslin, H. F. Fliegel, B. Guinot (ex officio), D. D. McCarthy, A. Orte (past president), P. Pâquet, E. Proverbio, A. M. Sinzi, G. A. Wilkins. New members of the Commission are P.M. Afanaseva, C.C. Costain, V.A. Naumov, J.D.H. Pilkington, E. Smylie, G. Stanila, G. V. Starytsin, H. Yasuda, J. M. Chamberlain, I. Dominski, J. Gaignebet, H. Hellwig, W. G. Melbourne, P. Morgan, S. B. Pushkin. Consultants are C. O. Alley, J. A. Barnes, G. Becker, R. Lake, S. Leschiutta, P. Mourilhe-Silva, J. McA. Steele, J. Terrien, P. Giacomo, R. Kaarls, Y. Koga, J. Rutman, S. Starker. Representative to CCDS: Wm. Markowitz, to Directing Board of BIH: H. M. Smith and A. Orte, to CCIR: H. M. Smith, and to FAGS: H. Enslin.

B. Guinot presented the report of the Director of the Bureau International de l'Heure (BIH) in which he discussed BIH activities concerning atomic time and the problems arising from the need for an accurate scale unit for TAI. Since 1977, TAI has been established in two steps. A large number of clocks give a stable intermediate time scale through a stability algorithm. TAI is then derived from this time scale by frequency steering based on the data of primary frequency standards. Concerning the stability algorithm, he reported on the increasing number and quality of the contributing clocks. In addition to the data of about eighty industrially made cesium clocks and one Hydrogen maser, the BIH now receives the data of three laboratory cesium clocks (from NBS, NRC, and PTB). These laboratory standards are expected to have a much better long-term stability but raise problems in weighting which have not yet been solved. International time comparisons continue to limit both precision (to a few tenths of microsecond) and coverage (North America, Europe, and North Africa).

The BIH continues to operate the UTC system without any particular difficulty.

In the discussion Guinot emphasized the practical importance of the synchronization of national master clocks as well as systems such as Loran-C, OMEGA, and navigational satellites, with UTC. In addition to these current activities, the BIH participates in research concerning satellite comparisons including use of NTS-1, Symphonie, a proposal for Project LASSO, and study of the use of pulses of a high satellite in a differential mode for distance of a few hundred kilometers.

Wm. Markowitz presented a report on the 1977 Consultative Committee for the Definition of the Second (CCDS). An abstract of this report had been previously published in the IAU "Information Bulletin No. 38", p. 11. H. M. Smith reported on recent activities of International Radio Consultative Committee (CCIR) Study Group 7. IWP 7/1 which had formulated the present system of coordinated radio time signals, having satisfactorily completed its tasks, was formally dissolved. Mr. Smith had chaired this working party since its inception in 1968. The modifications proposed for the dissemination of coded information on DUT1 has been incorporated in Recommendation 460. CCIR concurred with CGPM in the use of the language-independent designations, TAI and UTC, and proposed in Recommendation 535 that the term UTC, should be used to designate time in all international communication activities. New working parties had been set up to advise on world-wide time dissemination by means of satellites (IWP 7/4) and on the accuracy and reliability of frequency standards and reference clocks (IWP 7/5). Further studies had been proposed on time transfer and dissemination. A glossary of terms had been prepared, and the list of standard frequency and time signal emissions (Report 267) had been revised.

The President brought to the consideration of the Commission the fact that some members have questioned the validity of Commission 31 inasmuch as the actual definition of time has little to do with astronomy, and that it is not possible for members to belong to more than three commissions. He proposed the following alternatives:

a) that Commission 31 remain as is;
b) that Commission 31 be modified
   b_1) by merging with another commission such as Commission 19 or 4,
   b_2) by any other decision.

J. D. Mulholland pointed out that it was necessary for him to resign from Commission 31 in order to retain his membership in three other commissions in which he is interested. In his opinion the present situation is that:

1) Commission 31, and no other, is now responsible for atomic time scales and dissemination of time.
2) Commission 19, and no other, is now responsible for polar motion.
3) Commission 4, and no other, is now responsible for dynamical time scales.
4) The only subject of interest to Commission 31 that remains is Universal Time. The controversy arises because all three commissions take jurisdiction, sometimes without consulting or even informing the others.

Since he feels that the responsibility for Universal Time logically resides with Commission 19, Mulholland proposed that Commission 31 renounce all responsibility for Universal Time and retain a purely consultative role as it does with dynamical time. Opinions were offered by G. Winkler and B. Guinot and it was agreed to defer any further discussion until the meeting of 18 August.

SCIENTIFIC PRESENTATIONS

G. Becker discussed the present and future of time and frequency metrology in which he reviewed the development of highly stable oscillators (excluding quartz oscillators). He spoke about experiments with super-conducting cavities in which very high Q factors (10^{11}) and small frequency instabilities (3×10^{-16}) have been achieved. Research on stored ion frequency standards has revealed the possibility of extremely high Q factors (10^{12} to 10^{14}) and small Doppler shifts. At different laboratories a reduction of the size of the Hydrogen maser, e.g. for space craft applications, has been achieved. The passive Hydrogen maser has turned out to be a promising device being less sensitive to frequency shifts due to detuning of the resonator. Not very much is known about the long-term instability of Hidrogen masers. Commercial Cesium beam standards of two producers are still very important for timekeeping. Due to the temperature dependence of the frequency (about 10^{-13} $^{\circ}K^{-1}$) a seasonal fluctuation of the frequency of TAI has been detected using the primary standards of NRC and PTB as references. NRC and PTB have operated their primary standards continuously (NRC since 1975, PTB since 1978) as "primary clocks". The uncertainty of the NRC standard is estimated to be 53 x 10^{-15}, and that of the PTB standard is estimated to be 7 x 10^{-15}. Each of these clocks is considerably more stable than TAI. A new method of considering the weighting of standards in the computation of TAI appears to be necessary. In discussion following this paper, G. Winkler suggested that temperature gradients within the laboratory can account for seasonal fluctuations observed in the standards.

H. Hellwig presented a paper on the present and the future of infra-red and visible radiation t/f metrology. He discussed the use of a laser stabilized by saturated absorption. At present time stabilities of 10^{-11} may be achieved.

C. C. Costain read a paper by A. G. Mungall entitled "Canadian Primary Cesium Standards, CsV, CsVIA, B, and C". The Canadian primary 2.1 meter cesium clock, NRC CsV was the first primary frequency standard designed to run continuously as a clock. It has operated in this manner since May, 1975, and both independent re-evaluations and international time comparisons made since then have demonstrated long-term stability and accuracy limits of the order of 5 x 10^{-14}. Three new, smaller, one-meter primary cesium clocks, CsVIA, B, and C have been constructed to similar specifications of accuracy and stability. In operation since late 1978, they currently serve as secondary standards. Their performance as primary clocks will be evaluated later in 1979.

B. Guinot discussed the steering of the frequency of TAI in steps of 0.2 x 10^{-13} at intervals not shorter than sixty days in such a way that the frequency of TAI does not depart by more than 1 x 10^{-13} from that of the primary frequency standards of NBS, NRC, and PTB. In the implementation of this steering, seasonal effects have been found with a peak-to-peak amplitude less than 1 x 10^{-13}. These do not seem to be entirely due to the time comparison methods. The general trend of a decreasing TAI frequency persists and was compensated by steering. The stability of TAI has been estimated to be 1 x 10^{-13} (for $\tau \approx$ 2 months); and is limited by the links between laboratories. Guinot stressed the importance of links with higher resolution.

P. Giacomo presented the main conclusions that can be drawn from the last meeting of the Consultative Committee for the Definition of the Meter, CCDM (June 1979).

The variety and quality of stabilized lasers (optical frequency standards) is continuously increasing. Results have been presented concerning to Ar^+ lasers locked to iodine (λ = 515 nm), He-Ne locked to iodine (λ = 633 nm and λ = 612 nm) or to methane (λ = 3.39 μm), He-Xe locked to HCOOH (λ = 3.51 μm), and CO_2 lasers locked to SF_6 or to OsO_4 ($\lambda \approx$ 9 to 10 μm).

Stabilities usually attain 1 x 10^{-11} to 1 x 10^{-13} for sampling times of 1 s (Allan variance). Although short term stability and frequency repeatability of these lasers are generally well known, we have but very little information on their reproducibility, which seems to be better than the 1 x 10^{-11} level.

"Direct" frequency measurements attain the λ = 1.15 um level (f = 260 THz) and, by frequency doubling, the visible (λ = 576 nm, f = 520 THz) is reached.

Two new wavelengths of absorption lines of iodine, used as references; λ = 514 673 467 fm (Ar^+ laser) and λ = 611 970 771 fm (He-Ne) laser are recommended by the CCDM as secondary standards, their relative uncertainty being 2×10^{-9}.

All recent measurements of frequency or wavelength confirm the value already recommended for the velocity of the light ($c = 299\ 792\ 458\ ms^{-1}$). It is under consideration for adoption in 1983 a new definition of the meter based on the distance covered by a plane electromagnetic wave in 1/299 792 458 of a second. Such a definition should implicitly establish the velocity of the light. Before it can be adopted it will be necessary to consider the problems which could arise when putting it into use. Access to the new definition would be possible only in a few well equipped laboratories. Therefore, to extend the benefits of this new definition of the meter to more modest laboratories it will be necessary to recommend values for wavelengths of some secondary standards and to give a reasonable estimate of their uncertainty. It will also be necessary to advise on the means that are available, to these laboratories, for checking their own implementation of such secondary standards; for example, the way to compare frequencies from different standards in the visible.

16 August 1979

A joint meeting with Commissions 4, 7, 8, 19, and 24 was held for the presentation of the report of the Working Group on Nutation and the Progress Report on the Preparation of FK 5. (See the report of Commission 4.)

17 August 1979

C. O. Alley reviewed the relativistic effects in timekeeping and reported on experiments whose results agree with the general theory of relativity to within $\pm$ 2 %.

S. Starker reported on a proposed experiment using the Space Shuttle. After presenting a short review of a package of experiments in the fields of communications and navigation, one of the experiments, "Clock Synchronization and One-Way Ranging", was described in detail. The goals of this experiment are: investigation of the behavior of atomic clocks in a low orbit, time dissemination, orbit determination, and ground station position determination with one-way time signals, and comparison of on-board and ground-based clocks by a two-way method. It was emphasized that, in low orbits, the relativistic effect of velocity prevails over the effect of gravity, and that this experiment therefore provides a good opportunity to measure the velocity effect with high accuracy. In the ensuing discussion, it was stressed that although the relativistic velocity effect is well known from particle radiation measurements, the proposed experiment would deal with macroscopic, man-made clocks as objects of observation, and that the accuracy would be much higher than in the case of particle radiation provided that no unexpected disturbing influence appears.

J. Rutman discussed two-way time transfer via Symphonie satellite between NRC and LPTF. B. Guinot mentioned and Rutman stressed that this campaign was part of a more general program begun with the active participation of PTB and including the Symphonie connection PTB-LPTF. This work is intended to test the possibility of long-term operational methods of time transfer by satellite. Instrumental delays remain as the primary problem, and Rutman asserted that a precision of $\pm$ 5 ns can be achieved for individual transfers.

P. Nuspl reported on progress with the CENSAR (Centralized Synchronization and Ranging) experiment. Using the HERMES satellite the tracking precision is about $\pm$ 1 ns, prediction error is $\pm$ 10 ns, and the bit timing can be synchronized to $\pm$ 2 ns.

S. Iijima concisely reviewed the situation in the Far East concerning time comparisons. He noted that 24 commercial Cesium clocks, two Hydrogen masers, and three laboratory standards belonging to seven institutions in the Far East are not able to provide regular contributions to the formation of TAI because of the lack of adequate links. Since January, 1976 intercomparisons between laboratories and Loran-C have been performed, including a linkage to USNO via portable clock and NTS 1 satellite, but the precision has been limited by defects in ionospheric modelling. He stressed the need for closer cooperation with western laboratories and offered Resolution No. 1 for consideration. Discussion was postponed for the final administrative session. S. Ye compared the situation described by Iijima to the similar situation in China, and further stressed the importance of having better and more permanent links in this area.

S. Leschiutta presented a paper on the future of world-wide synchronization. In his talk he stressed the need for 3-5 cm accuracy for geodesy which, in turn, requires new methods of synchronization such as VLBI, laser, and advanced navigational satellites.

W. J. Klepczynski discussed high precision clock comparisons by means of VLBI measurements. In his talk he stated that VLBI provides one of the most accurate time transfer systems currently operational. After an initial effort to install a self-calibration system at the two terminals taking part in the experiment is made, accuracies of $\pm$ 1 ns are feasible on a regular basis with only modest effort. It seems that a precision of $\pm$ 0.1 ns should be obtainable on a regular basis within the next few years. Two factors which affect this latter goal are:

a) ability to calibrate antenna systems to this precision, and
b) ability to model adequately the tropospheric water vapor.

G. Winkler discussed time dissemination capability with the Global Positioning System (GPS). A timing precision of $\pm$ 0.1 μs is expected, and the 24 satellites in high orbits will provide continuous coverage. He expressed the thought that GPS would, in the future, become the standard method for intercomparing clocks such as Loran-C is today. The cost of the first receiver is $150 000 but subsequent models may be as cheap as $50 000. He asked those who are interested in purchasing a receiver to contact him possibly to lower the price through a large order. He noted that even with a limited number of satellites, a good coverage could be obtained and he cited as a reference, Navigation, Vol. 25, No. 2, 1978.

18 August 1979

SCIENTIFIC PRESENTATIONS

J. Gaignebet presented a paper on the LASSO experiment, a project for the synchronization of clocks by using laser ranging to a geostationary satellite. The package will be mounted on the meteorological satellite, SIRIO-2, to be launched during 1981. He briefly described the technical concept of LASSO and emphasized the operational requirement for an accuracy of some hundreds of picoseconds in contrast to the normal requirement for a few nanoseconds accuracy. This leads to the need for a very stringent collaboration among all elements of the link. Event timers are therefore required at the ground stations. Although a two-way system is conceived in order to measure the propagation delay, provisions are possible to allow smaller stations to access the system at the price of a reduced accuracy. Gaignebet announced that ESA would send, in the near future, an Announcement of Opportunity to all potential participants.

J. Rutman presented a general review of the world situation in national timekeeping which was followed by brief comments on their national organizations by G. Becker for the Federal Republic of Germany, J. McA. Steele for the United Kingdom, C. C. Costain for Canada, S. Leschiutta for Italy, P. Pâquet for Belgium, and G. Winkler for the United States.

A. M. Sinzi presented his results for the determination of ET-TAI from occultation observations made at four stations in Japan over a period of fifteen years. Over nine thousand individual observations of stars to the ninth magnitude were used in this study. He showed that ET-TAI may be represented by the expression, $\mathrm{ET-TAI} = (30^{s}.74 \pm 0^{s}.04\ \mathrm{m.e.}) - (0^{s}.016 \pm 0^{s}.007\ \mathrm{m.e.})\ (\mathrm{year} - 1970)$.

ADMINISTRATIVE SESSION

The President, A. Orte, opened the session by presenting for discussion a draft of a recommendation by S. Iijima. After some debate, Resolution No. 1 was passed unanimously. A second discussion of the future of Commission 31 followed. Comments by J. Hers were presented expressing the view that he could not visualize an IAU without a commission on time since time is a basic building block of astronomy, no matter what the particular interests of Commission 31 might be at any given time. It was debated whether Commission 31 should merge with another commission or modify its terms of reference that were adopted at the Grenoble IAU meeting. By vote it was decided that Commission 31 would remain in its present form as defined by the Grenoble terms of reference.

20 August 1979

A joint meeting was held with Commissions 4 and 19 to discuss the redefinition of Universal Time and the nomenclature of Dynamical Time. (See the report of Commission 4.)

Resolution No. 1

1. The International Astronomical Union

Considering

a) that International Atomic Time, TAI, is established by the BIH on the basis of data provided by several laboratories operating independent clocks and frequency standards, and

b) that data from other laboratories needed to improve TAI cannot be used by reason of the inadequacy of existing international links,

Recommends

a) that laboratories concerned with time and frequency in the Far-East and Asia, Africa, Australia, and South America establish international links with high accuracy, and

b) that support be given by national and international bodies to these laboratories for the use of high precision techniques such as Loran-C, flying clocks, artificial Earth satellites, and VLBI.

1. L'Union Astronomique Internationale

Considérant

a) que le temps Atomique International, TAI, est établi par le BIH d'après les données fournies par plusieurs laboratoires où fonctionnent des horloges et étalons de fréquence indépendants et

b) que des données d'autres laboratoires, nécessaires pour améliorer le TAI, ne peuvent pas être utilisées car les liaisons horaires internationales existantes ne conviennent pas,

Recommande

a) que les laboratoires qui s'occupent du temps et des fréquences en Extrême Orient et Asie, en Afrique, en Australie et en Amérique du Sud établissent des liaisons horaires internationales de grande exactitude et

b) qu'une aide leur soit apportée par les organismes nationaux et internationaux pour qu'ils puissent utiliser des techniques de grande précision, telles que celles qui impliquent l'usage du LORAN-C, des horloges transportables, des satellites artificiels de la terre et de l'interférométrie à très longue base.

Other resolutions discussed and approved by the members of the Commission are included in the reports of Commissions 4 and 19 in this volume.

COMMISSION 33: STRUCTURE AND DYNAMICS OF THE GALACTIC SYSTEM (STRUCTURE ET DYNAMIQUE DU SYSTEME GALACTIQUE)

Report of Meetings on 15, 16, 17, 21 and 22 August 1979

PRESIDENT: F. J. Kerr
VICE-PRESIDENT: G. G. Kuzmin

15 August 1979

Galactic Dynamics
Joint Meeting with Commission 28

Invited papers were presented by:

A. Kalnajs: Dynamical Models of Galaxies
A. Toomre: Stability of Galactic Disks
R.H. Miller: Numerical Simulations of Galaxies
G. Contopoulos: Integrals of Stellar Motion and their Disappearance
P.O. Vandervoort: Resonance Phenomena in the Dynamics of Spiral and Barred Galaxies
C.C. Lin: Progress in Density-Wave Theory
W.W. Roberts: Gas Flow in Galactic Bars

In addition, V.C. Rubin presented the results of recent observations on a substantial sample of spiral galaxies, B.J. Robinson stated that synthesis observations of four large southern barred spirals were planned, and J. Schwarz discussed gas flows in galactic bars.

Business Meeting of Commission 33

The new officers and Organizing Committee members were elected. G.G. Kuzmin will be the new President, and R. Wielen Vice-President. Nominations for new members of the Commission were discussed, and most of those proposed were admitted. During this discussion, it was found that some new members had expressed a wish to join Commission 33, without demonstrating their interest in the work of the Commission. The following resolution was adopted:

Commission 33 resolves that the Executive Committee should request an IAU member who wishes to become a member of a Commission to give briefly a good reason for his/her wish to join that Commission, adding references to one or two recently published papers in the areas covered by the Commission.

At the suggestion of B.J. Bok, it was agreed to send a telegram from the Commission to J.H. Oort, expressing regret that for the first time since the inception of the IAU he was unable to attend a General Assembly.

The present status of the system of galactic constants was discussed. Papers on the circular rotation velocity and the distance to the center were presented by G.R. Knapp and M.W. Feast, respectively. After a lengthy discussion, it was decided that it was premature to set up a working group at this stage to consider whether any changes in the system of constants should be recommended. Instead, the retiring President was asked to prepare a document, including the main results from the two papers presented at the meeting, and drawing the attention of Commission members to the need for further research in this field.

It was agreed that the tradition of having a detailed Commission report should be continued, in addition to the shorter version in the IAU Transactions.

16 August 1979

Impact of Space Astrometry on Astronomy and Astrophysics
Joint Meeting with Commission 24 and Others

This half-day session addressed the new possibilities that will be presented in several branches of astronomy and astrophysics if astrometric data are obtained from two proposed satellites, Hipparcos and Space Telescope. A detailed account of the meeting is given in the report of Commission 24. The main presentation from Commission 33 was an invited paper by P.O. Lindblad on "The Impact of Space Astrometry on Galactic Structure and Dynamics"

17 August 1979

Galactic Environment
Joint Meeting with Commission 28

Invited papers were presented by:

W.E. Kunkel: The Galactic Environment, Optical Data
A.N.M. Hulsbosch: High Velocity Clouds
B.M. Tinsley: The Hypergalaxy
R.B. Larson: Nature and Cosmogony of Population III Objects

The papers and the associated discussion covered the available evidence for a galactic halo, the possible nature and origin of invisible matter in the outer parts of galaxies, possible interpretations for the high velocity clouds of hydrogen, and the concept that a community of galaxies (including our own) might be embedded in a single corona as a Hypergalaxy.

17 and 22 August 1979

Some Topics in Galactic Structure

The following papers were presented:

J.R. Lesh: The Gould Belt
K.A. Janes: Evidence for an Abundance Gradient in the Galactic Disk
C. Chiosi: Theoretical Results on Chemical Gradients across the Disk
R. Giovanelli: High Velocity Clouds
D. Downes: The Large-Scale Distribution of Radio Continuum Sources in the Galaxy
A.F.J. Moffat: The Outer Region of the Galactic Disk
D. Weistrop: The Distribution and Density of Dwarf M Stars in the Solar Neighborhood
L.N. Kolesnik: On the Three-Dimensional Structure of the Spiral Arms of the Galaxy (Presented by E.K. Kharadze)
R.D. Davies: Recent Jodrell Bank Radio Surveys at the Galactic Disk
J. Tinbergen: On the Local Dust and Magnetic Field
T.M. Bania: OB Stars and the Structure of the Interstellar Medium

A full account of the Joint Discussion on "Nuclei of Normal Galaxies", sponsored by Commissions 28, 33, 34, and 40, is given in the "Highlights of Astronomy".

COMMISSION 34: INTERSTELLAR MATTER AND PLANETARY NEBULAE (MATIERE INTERSTELLAIRE ET NEBULEUSES PLANETAIRES)

Report of Meetings, 17, 18, 20 and 21 August 1979

PRESIDENT: G.B. Field VICE-PRESIDENT: V. Radhakrishnan

Business Meeting

The policy of admitting to membership all members of the IAU who have published research results in the subject area of the Commission was reviewed and approved.

Approximately 120 new members were admitted to the Commission, and 30 members were retired, bringing the total membership to 419. While this is a large number of members, it was agreed that it is manageable, and that scientific sessions should continue to be held at General Assemblies in the same style as in the past. Perhaps more sessions than at the present General Assembly (six) will be necessary.

The slate of Officers and Organizing Committee for the next Triennium was approved: President, V. Radhakrishnan; Vice-President, M. Peimbert; Organizing Committee, T. de Jong, B. Donn, G. Field, L. Higgs, E. Kostyakova, J. Lequeux, U. Mebold, M. Morimoto, Y. Terzian, and B. Zuckerman.

Commission 34 will sanction a meeting on Infrared Astronomy in Hawaii, June 1980.

Commission member H. Dickel led a discussion of how astronomical objects of interest to Commission 34 should be designated in publications, particularly catalogues. She agreed to chair a group of members to review current practice and to recommend policy to Commission 5 in those areas of interest to Commission 34.

Five scientific sessions were planned for the General Assembly, and are discussed below. In addition, in response to the interest expressed by many members, a special session on Abundance Gradients in Galaxies was arranged and chaired by P. Mezger.

Supernova Remnants (Chaired by L. Higgs)

Presentations were made by L. Cowie, K. de Boer, J. Dickel, L. Higgs, S. Murray, N. Panagia, V. Radhakrishnan, and S. van den Bergh. Attention was drawn to clouds of neutral hydrogen moving at high velocity in the galactic disk, both within supernova remnants and outside them as well. Apparently acceleration of clouds by supernova shock waves can account directly for much of the kinetic energy of observed HI clouds. The first detailed x-ray images of remnants were discussed. Although they appear to be in general agreement with theoretical expectations, more detailed study is warranted.

Planetary Nebulae (Chaired by Y. Terzian)

Y. Terzian presented a review talk, and a written review by G. Khromov was made available. Other papers were presented by L. Aller, Y. Andrillat, H. Ford, S. Kwok, M. Peimbert, C. Purton, and V. Weidemann. The nuclei of planetary nebulae are complex systems. About eight (8) nuclei are known to be binaries. The occurrence of higher ionization in many planetaries than can be explained purely on the basis of the radiation emitted by the observed nucleus suggests that other phenomena, such as high speed hot winds, may be present. The apparent discrepancy between ultraviolet and optical determinations of the carbon abundance in several planetaries has been resolved by better observations; the average carbon abundance is 2 or 3 times solar. An evolutionary sequence is envisioned in which the gas expelled previously as a red giant wind is ionized progressively by the exposed hot stellar nucleus; this accounts for a number of objects having rather characteristic radio spectra. Later, a fast wind (1000 km sec^{-1}) emitted by the nucleus pushes this gas from inside, forming and accelerating outward the shell of gas observed as a planetary nebula.

Ionized Gas (Chaired by M. Peimbert)

Papers were given by E. Becklin, E. Gerard, T. Gull, M.C. Lortet, J. Mathis (for R. Reynolds), C. Montes, M. Peimbert, and M. Perinotto. It was found that in the Orion nebula the carbon abundance in the gaseous phase is solar; models of this object based on dust scattered starlight were presented. From a survey of the galactic plane in Hα + [N II], [O III], and [S II], new supernova remnants and distorted bubbles have been found. Two very faint filaments, emission measures of 3 and 11 $cm^{-6}pc$, that extend at least 12° have been detected at b= -21° with radial velocities ∿ -70 and -10 km s^{-1} (LSR); a radial velocity gradient of 20 km s^{-1} across the length of the low-velocity feature has been found. A theory based on nonlinear compton scattering has been proposed to interpret the strong variability of H_2O interstellar masers.

Atomic Hydrogen (Chaired by H. van Woerden)

Papers were given by W.B. Burton, A. Dupree, G. Knapp, U. Mebold, E. Salpeter, and U. Schwarz. A definitive model for the HI at the galactic center requires a disk tilted at ∿ 24° to the galactic plane. Measurements using lyman-α backscatter and absorption show that within 100 pc of the sun, the average value of n (HI) along various lines of sight is between 0.01 and 0.1 cm^{-3}. High-velocity clouds have been resolved at Westerbork into individual condensations with dimensions 1-10 pc, densities 5 - 50 cm^{-3}, and masses 0.1 to $2M_{\odot}$ (if the distance is 1 kpc). The spin temperatures of some high-velocity clouds are high, ranging between 60 and 900°K. Absorption measurements in the disk indicate that not all the apparent emission unaccompanied by absorption can be explained by a population of very small clouds. Instead, warm gas is definitely present, with spin temperatures as high as 5000°K. Some warm HI is located in shells around molecular clouds; the highest column density observed, ∿ $3x10^{20}cm^{-2}$, is about what is expected from theoretical calculations of the dissociation of H_2 molecules by ultraviolet starlight.

Interstellar Dust (Chaired by B. Donn)

Papers were given by P. Aannestad, B. Donn, W. Duley, J.M. Greenberg, W. Kratchmer, J. Mathis, and A. Webster. A good fit to the 3.1-micron obsorption feature observed in dark clouds can be obtained with laboratory mixtures of H_2O and CH_4 ice, which have been irradiated by ultraviolet at low temperature and then warmed up so that the radicals formed by photolysis combine to form new molecules. On the other hand, a newly discovered allotropic form of carbon, called carbyne, may form at high temperatures in stellar atmospheres rather than graphite, and theory suggests that it also may have a band at 3.1 microns. Experiments and calculations indicate that various other infrared bands are due to compounds of Si, Mg, and Fe with oxygen. While the compounds in question have usually been identified as silicates formed in near thermodynamic equilibrium in stellar atmospheres, better fits to the data are obtained if the compounds are mixtures of metallic oxides formed by accretion of atoms on grains occurring in the interstellar medium itself. Such an origin of interstellar dust would be consistent with recent ultraviolet absorption-line data, which indicate that depletion of metals is reduced in high-velocity clouds where one infers that shock waves capable of sputtering grains are present, while depletion is enhanced along lines of sight intersecting particularly dense clouds.

Abundance Gradients in Galaxies (Chaired by P. Mezger)

Papers were given by L.H. Aller, C. Chiosi, E. Churchwell, S. D'Odorico, B.E.J. Pagel, V. Pankonin, M. Peimbert and J. Schmid-Burgk. Extragalactic determinations of abundance gradients based on H II regions and supernova remnants were analyzed with special emphasis on the production of nitrogen. Galactic determinations based on optical observations of H II regions and planetary nebulae were reviewed. Radio determinations of abundance gradients across the disk of the Galaxy based on: a) He^+/H^+ abundance ratios and b) a T_e versus O/H relationship were presented. Theoretical models of the chemical evolution of the Galaxy are being undertaken to explain these observations.

COMMISSION 35: STELLAR CONSTITUTION (CONSTITUTION DES ÉTOILES)

Report of Meetings, 15, 17 and 21 August 1979

PRESIDENT: B. Paczyński VICE-PRESIDENT: R. J. Tayler

BUSINESS MEETING 15 August 1979

The following nominations were approved:

PRESIDENT: R. J. Tayler (Astronomy Centre, University of Sussex)

VICE-PRESIDENT: A. N. Cox (Los Alamos Scientific Laboratory, New Mexico)

ORGANISING COMMITTEE: D. J. Faulkner, P. Giannone, I. Iben, R. Kippenhahn, B. Paczyński, G. Ruben, D. Sugimoto, A. Tutukov, J-P. Zahn.

There was a discussion about the manner in which the new organising committee had been chosen, which followed consultations amongst the existing organising committee and some past officers. Although no objection was raised to the proposed membership, it was suggested that in future members of the commission should be invited to nominate members of the organising committee. It was also suggested that normally members of the organising committee should not serve for more than two terms. These two proposals were agreed.

Forty six new members of the commission were elected after a discussion in which several members suggested that the credentials of the candidates should be examined carefully. Finally, it was agreed that any member of the IAU who expressed a wish to take part in the work of the commission should be entitled to join it.

This led to a general discussion concerning communication within the commission. It was explained by R. J. Tayler, who had been responsible for coordinating the production of the draft reports, that these had not been circulated to members of the commission because his institution could not afford to pay the costs of the offprints and of their postage. As the commission had no budget it was difficult to see how communication with an increasingly large number of commission members could be maintained. Several members of the commission urged that the commission should try to facilitate exchange of information amongst its members, in particular concerning availability of tabular material and numerical codes. The incoming President was asked to raise the problem of commission expenses with the incoming General Secretary.

SCIENTIFIC MEETINGS

Theory of Rotating Bodies 15 August 1979

This three hour meeting was organised and chaired by B. F. Schutz and it was concerned with the effect of rotation in both stars and disks. M. Clement (Toronto) presented results of numerical calculations of the structure of differentially rotating stars and of some of their non-axisymmetric normal modes of adiabatic pulsation. He commented that the numerical problems need delicate handling particularly for modes with corotation points but that the numerical method appeared to work. B. Schutz (Cardiff) presented similar normal mode calculations for self-gravitating Bardeen disks, commenting also on the marked difference between modes with and without corotation points. He attributed this to the differences between the discrete and continuous spectrum. Y. Sobouti (Teheran) discussed problems in identifying g-modes of rotating fluids and showed that one must exercise caution when obtaining g-mode frequencies for slow rotation by a perturbation method. B. Durney (N.C.A.R., Boulder) presented some new calculations on the effect of rotation on the solar convection, taking into account both turbulent viscosity and thermal conductivity. He also discussed the loss of angular momentum by stars, involving dynamo build-up of a magnetic field followed by magnetic braking. E. Schatzman (Nice) summarized the highlights of papers presented at the Erice Workshop on Stellar Rotation (2-10 June 1979), in particular considering two questions; are laboratory experiments done so far relevant to astrophysics and what new experiments should be done? V. Icke (Minnesota) presented numerical calculations and computer generated films of accretion on to rotating disks, suggesting that well-collimated jets of plasma might be produced from the accreting cloud. R. C. Smith (Sussex) suggested that the interaction of circulation and tidal torques in close binary stars might prevent them from possessing a time independent state of rotation. D. Sugimoto (Tokyo) concluded the meeting by pointing out that the incompressible limit of a sequence of polytropes with a given (rapid) rotation is not necessarily the Maclaurin spheroid, but may be a toroid for rapid enough rotation. The Maclaurins may not be reliable analogues to compressible configurations in such cases.

Pulsars (Joint Meeting with Commissions 40, 48) 17 August 1979

This three hour meeting was organised by R.J. Tayler. The first session was devoted to pulsar observations and the second to pulsar theories.

PULSAR OBSERVATIONS

Review Talk J. H. Taylor

There are at present 328 pulsars known but it is unlikely that there will be a very large increase in number in the near future because of limitations of existing telescopes. Almost all pulsars have periods between 0.2s and 2s and there is a genuine fall off at each end. The dispersion measure ranges from 3 to 500 cm^{-3} pc. There are many different timescales associated with pulsars; pulse, pulse width, subpulse, micropulse, pulse nulling, pulsar lifetime. The average pulse shape does not change very much with period; the pulse width is typically a few per cent of the period. There are double peaked profiles with a continuum of separations, the Vela pulsar being double in optical and X-ray but not in radio. Subpulses and micropulses also seem to scale with period. Pulsars are nearly all very good clocks when account is taken of $\dot{P}$, being accurate to a few milliseconds in 10 years. A few are not such good clocks. In particular glitches have been observed for Vela, Crab and PSR 1641-45 and a few others have residuals which wander around. If $\dot{P}$ is plotted against P, there is a cut-off in the diagram which cannot be due to observational selection but which must be due to a real physical effect. This could be produced by the decay of the magnetic field at the light-cylinder to a critical value or it could arise from lack of further pair creation at the polar cap. The distances derived from dispersion measures are statistically rather reliable, with 0.03 electrons cm^{-3} being a good average. Pulsars are concentrated towards the galactic plane but not as strongly as population I objects; $\bar{z} \approx 400$ pc. The distribution of pulsars appears to peak near 5 kpc. from the galactic centre in good agreement with various indicators of current and recent star formation. It appears that one pulsar is formed about every 10 years.

Short Contributions

A. G. Lyne	Pulsar Proper Motions
G. S. Downs	Preliminary Results of Timing Measurements of Four Pulsars
M. Bartel	Simultaneous Two Station Single Pulse Observations of the Radio Pulsar PSR 0809+74
D.H.P.Jones	Changes in the Optical Light Curve of the Crab Pulsar between 1970 and 1977

Lyne reported that he and his colleagues had now measured 26 pulsar proper motions which ranged from a few milliarcseconds per year to 400. Suggestions that pulsars received a high velocity at the time of formation were confirmed. Parallax measurements were also being made. Downs reported preliminary evidence for discrete changes in P and $\dot{P}$ for four pulsars. Bartel discussed simultaneous observations of the same pulsar in the German Federal Republic and the USSR. Jones described seven years' study of the stability of and small changes in the optical light curve of the Crab pulsar. In addition to these advertised short contributions, B. Robinson reported that a second pulsar had been discovered to be a partner in a binary system with a period of about 1265 days.

PULSAR THEORIES

Review Talk on Pulsar Magnetospheres L. Mestel

Mestel started by reviewing the standard picture of a pulsar as an obliquely rotating magnetised neutron star, whose main loss of energy is through the emission of low frequency electromagnetic waves. The radio pulse energy is always less than the total energy by several orders of magnitude. As a result, explaining the radio pulses is perhaps not the primary task of the theory. It, however, appears that the pulsed γ-ray emission can sometimes be a major source of energy loss. Theory should therefore demonstrate that in some circumstances a significant fraction of the wave energy can be converted into energy of highly relativistic particles, which can then radiate γ-rays close to the light-cylinder. He then outlined progress that had recently been made in trying to produce a self consistent model of a pulsar magnetosphere. Models have been obtained for magnetospheres of pulsars in which the magnetic and rotation axes are aligned. Electrons pulled out of the polar regions of the neutron star are driven across the light-cylinder by centrifugal forces, moderated by the electric field component along the magnetic field. Just beyond the light-cylinder the electrons become highly relativistic and emit high frequency radiation, which carries away energy and angular momentum, so braking the star. The aligned pulsar cannot possess all the properties of an obliquely rotating pulsar but preliminary investigations in the latter case suggest that the essential features of a generally aligned pulsar are the emission of a low frequency wave and of very high frequency γ radiation.

Short Contributions

M. Salvati	γ-rays and Pulsars
V. Radhakrishnan	The Effect of the Diffuse γ Radiation on Pulsars
F. Pacini	γ-rays from Young Pulsars
P. A. Sturrock	Electric Fields in Pulsar Magnetospheres

Salvati reiterated that γ-ray pulsed emission is always the main form of pulsed emission and that in slow pulsars it may be the major emission. The shape of the γ-ray pulses suggests that γ-ray emission occurs near to the light-cylinder. Radhakrishnan proposed that the breakdown of pulsar gaps could be produced by the diffuse γ-ray background. This seems possible if the gap magnetic field is $\sim 10^{12}$ gauss. Pacini discussed the production of γ-rays by nuclear collisions in an expanding supernova shell and suggested there might be a strong detectable flux of γ-rays some time after a supernova explosion. Sturrock discussed some of the problems relating to the structure of electric fields in pulsar magnetospheres.

Structure and Evolution of Variable Stars
(Joint Meeting with Commission 27) 21 August 1979

This three hour meeting, which was organised by B. Paczyński and J. Smak, is reported by Commission 27.

COMMISSION 36: THE THEORY OF STELLAR ATMOSPHERES (LA THEORIE DES ATMOSPHERES STELLAIRES)

Report of Meetings, 14, 16, 18, 20, and 21 August 1979

PRESIDENT: D. Mihalas SECRETARY: J.M. Marlborough

Business Meeting, 14 August 1979, 3:30 pm

I. COMMISSION ACTIVITIES

The President outlined the activities of the Commission during the General Assembly.

II. ELECTION OF PRESIDENT AND VICE PRESIDENT

Dr. G. Traving was elected President unanimously. Two candidates, Dr. B. Gustafsson and Dr. A. Sapar, were nominated for Vice President. In a vote by secret ballot, Dr. Gustafsson was elected Vice President, Receiving 13 votes to 1 for Dr. Sapar.

III. ELECTION OF ORGANIZING COMMITTEE

The nominees proposed by the outgoing organizing committee were presented. Additional nominations from the floor were requested. After discussion the following names for the new organizing committee were approved: H. Frisch, A.G. Hearn, W. Kalkofen, K. Kodaira, J.M. Malborough, D. Mihalas, A. Sapar, M.J. Seaton, A.B. Underhill.

IV. MEMBERSHIP

a) DEATHS: Huang, S.S.; Menzel, D.H.

b) VOLUNTARY RESIGNATIONS: Busbridge, I.W.; Houziaux, L.; Kopylov, I.M.; Rucinski, S.; Schatzman, E.L.; Strömgren, B.

c) NEW MEMBERS

The president presented under several headings names of those proposed for or requesting membership in the Commission.

(i) Present IAU Members

The following names were approved as members of the Commission: Bues, I.D.; Cassinelli, J.P.; Gough, D.; Grinin, V.P.; Michaud, G.; Oxenius, J.; Peraiah, A.; Schimid-Burgk, J.; Scholz, M.; Souffrin, P.; Spite, F.; Stalio, R.; Stein, R.F.; Stepien, K.; Toomre, J.; Zahn, J.P.

(ii) New IAU Members

The following names were approved as members of the Commission: Bernat, A.P.; Bertaud, C.; Carbon, D.F.; Clegg, R.F.S.; Cram, L.E.; Domke, H.; Dufton, P.L.; Holzer, T.E.; Kamp, L.W.; Karp, A.H.; Klein, R.I.; Kontizas, E.; Kudritzki, R.P., Kurucz, R.; Leibacher, J.; O'Mara, B.J.; Panek, R.J.; Peters, G.J.; Ramsey, L.W.; Ross, J.E.R.; Sakhibullin, N.A.; Schönberner, D.; Shine, R.A.; Simonneau, E.; Shipman, H.L.; Tarafdar, S.P.; Viik, T.; Yorke, H.W.

(iii) Seven other new IAU members were not approved for membership in the commission at the present time.

V. WORKING RULE FOR MEMBERSHIP IN THE COMMISSION

The President distributed a proposed working rule concerning membership in Commission 36. After disucssion the rule was approved by the members present. The text of the approved rule is:

"In order to maintain the size of the Commission at a level where effective interaction among members can occur, and with regard to the IAU rule concerning membership in Commissions, the following procedures will be implemented starting with the XVIIth General Assembly in 1979. (a) Each incoming new President shall write, as soon as is convenient after the General Assembly at which he is elected to all current Commission members asking them to affirm by letter their desire to remain active members of the Commission. (b) The names of all members failing to affirm their membership will be reported back to the Commission at the next following General Assembly (when they will have the opportunity to affirm their membership). (c) The next incoming President shall, in addition to writing all other members as described under (a) above, write once again to those members who failed to respond earlier, informing them that they did not affirm their desire to remain members of the Commission, and asking them again to do so. (d) If a member fails to respond the second time, then his name will be dropped from the active membership list of the Commission."

The meeting adjourned at 4:10 pm.

SCIENTIFIC SESSIONS

1) JOINT MEETING: 16 August, 1979

Physics of the Chromosphere-Corona-Wind Complex and Mass Loss in Stellar Atmospheres

A. Observational Evidence for Chromospheres, Coronae and Mass Loss.
Chairman: H. Lamers.
Introduction to the Joint Discussion: D. Mihalas
Papers Presented:
- (i) Stars of Spectral Types Earlier than the Sun: T. Snow
- (ii) Solar and Late-Type Stars: C. Jordan
- (iii) Mass Loss and Winds in Massive X-ray Binaries: H. Henrichs.

B. Observations of the Solar Chromosphere-Corona-Wind Complex
Chairman: R. Bonnet
Papers Presented:
- (i) Large Scale Structure of the Corona: J. Zirker
- (ii) A High Resolution View of the Solar Chromosphere and Corona: G. Brueckner

C. Physics of Dissipative Processes in Stellar Atmospheres.
Chairman: J.P. Zahn
Papers Presented:
- (i) Convective Instabilities and their Implications for Chromospheres and Coronal Heating: J. Toomre
- (ii) Small Scale Dissipative Processes in the Solar Atmosphere: J. Leibacher.

D. Propulsion and Dynamics of Stellar Winds and Their Influence on Stellar Evolution.
Chairman: A. Underhill
Papers Presented:
- (i) The Solar Wind: T. Holzer
- (ii) Stellar Wind Theories: A.G. Hearn
- (iii) Effects of Mass Loss and Angular Momentum Loss in Stellar Winds on Stellar Evolution: S. Sreenivasan

2) JOINT MEETING: 18 August, 1979

"Stellar Abundances"
Chairman: W. Bidelman
Papers Presented:
- (i) Accuracy with which Stellar Abundances can be Determined: D. Lambert
- (ii) Red Giants in Globular Clusters and Large Scale Abundance Variations in the Galaxy: B. Pagel
- (iii) Abundances in Other Galaxies: L. Aller

3) SCIENTIFIC MEETING: 20 August, 1979
"Diffusion Phenomena in Stellar Atmospheres"
Chairmen: F. Praderie, C. Cowley
Papers Presented:
- (i) Astrophysical Context of Diffusion in Stars; Open Problems: G. Michaud
- (ii) Uncertainties in the Usual Approach to Diffusion: A Microscopic Point of View: J. Oxenius
- (iii) Confrontation Between Theory and Observations; General Spectroscopic Survey of Stars where Diffusion Occurs: W. Bonsack
- (iv) Competition Between Diffusion and Macroscopic Motions: S. Vauclair
- (v) The Role of Magnetic Fields; A Critical Evolution of Their Measurement: J. Landstreet
- (vi) One Neglected Hydrodynamical Aspect in the Presence of Radiative Forces; Horizontal Aggregation by Photoluriation: E. Spiegel

4) JOINT MEETING: 21 August, 1979
"Stellar Rotation"
Chairman: A. Slettebak
Papers Presented:
- (i) Rotation of Very Young Stars: L. Kuhi
- (ii) Rotational Studies of Lower Main Sequence Stars: M. Smith
- (iii) Rotation in Binary Stars: M. Plavec

COMMISSION 37: STAR CLUSTERS AND ASSOCIATIONS (AMAS STELLAIRES ET ASSOCIATONS)

PRESIDENT: S. van den Bergh SECRETARY: C.A. Murray

In view of the fact that IAU Symposium No. 85 on Star Clusters was scheduled to take place immediately following the General Assembly no scientific sessions were scheduled by Commission 37. G. Lyngå represented our Commission at a joint meeting with Commission 5 (Nomenclature). Members of our Commission also participated in a joint meeting on Space Astrometry.

17 August 1979

BUSINESS SESSION

1. The following new organizing committee was elected and will serve for the three year period 1979-1982.

T.A. Agekjan
B. Balázs
K.C. Freeman (Vice-president)
G.L.H. Harris
D.C. Heggie
G. Lyngå (President)
S. van den Bergh (Past President)

2. The Commission received a report from its Nomenclature working group headed by Dr. A.F.J. Moffat. Following some discussion the following resolution was adopted:

> Commission 37 recommends that the designation of star clusters within the Galaxy shall be based on the equatorial coordinates (1950.0) prefixed by the letter C. The right ascension element shall consist of hours and minutes of time (4 digits) and the declination element of algebraic sign followed by degrees and one decimal (3 digits); in both cases leading zeroes must be inserted where appropriate.
>
> Once a designation has been asigned to a particular cluster it shall not be changed, even if later investigations indicate that the coordinates used in the designation differ from the centre of the cluster.

3. Following a recommendation by our Commission the Strasbourg Data Centre has accepted responsibility for the assignment of cluster names.

4. A working group headed by Dr. Moffat was appointed to make recommendation regarding the numbering of stars in clusters.

5. The Commission received reports on the status of the Prague cluster catalogue and of the Lund computer-based catalogue.

COMMISSION 38: EXCHANGE OF ASTRONOMERS
(ECHANGE DES ASTRONOMES)

Report of Meetings, 17 and 20 August 1979

PRESIDENT: D. A. MacRae VICE PRESIDENT: J. Delhaye

Commission 38 differs from others in that it is a Committee of the Executive Committee. However it has the same status and independence as other commissions of the Union. The work of Commission 38 is carried on continuously throughout the period between General Assemblies by the President and Vice-President. They report at intervals to the Organizing Committee and depend on that body for advice and assistance when necessary.

The principal program of Commission 38 is to foster the exchange of astronomers by awarding grants to qualified individuals in order to enable them to make visits to institutions in countries other than their own. It is intended that scientific, intellectual and cultural benefits will accrue and that astronomy in the home country will be enriched after the applicant returns.

Requirements, guidelines and procedures for the operation of the program have been set forth at the Grenoble General Assembly in 1976 (Transactions, XVIB, pp. 259-261).

At the first business meeting of the Commission the President tabled copies of three reports (dated 1977 July, 1978 December and 1979 August) which had been presented to the Executive Committee of the Union and to the Organizing Committee. Each report contained details of the awards which had been made to date, supplement by a listing of applications which, for various reasons, could not be approved. Each report also contained a discussion of the experience gained by the President and Vice-President in their administration of the program.

Approximately half of the 56 applications handled during the three-year period fell within the guidelines; however 13 of these were either still pending, postponed, or withdrawn for other reasons. A total of 15 awards had been made, averaging US $760. The President emphasized other positive aspects of the exchange of astronomers program, such as influencing host institutions to improve the level of local support, referring applicants to other more appropriate sources, and providing applicants with encouragement and support in seeking help within their own countries. One advantage the IAU program offers is that there is no deadline and no official application form.

As a result of the discussion of the President's report at the second business meeting, it was decided that no basic changes were required in the operating rules of the Commission as formulated at Grenoble. However the new President and Vice-President should feel free to interpret the rules with discretion and to apply them in a flexible manner when they can do so to the benefit of the program. Some minor modifications were proposed to be included in a revised wording. For example it was proposed that study leaves should be among the purposes for which grants could be made, and that visits of two-months duration might be permitted if the circumstances were such that an applicant, perhaps because of teaching duties, could not be away for longer.

An extensive discussion took place of the resolution proposed by the Swedish National Committee (see the Report of Commission 46) and of related topics. It was generally agreed that a program of visits by experienced astronomers to countries with little or no formal astronomy would be beneficial. IAU support for such a program might be channelled through Commission 38, giving it responsibility for organizing the visits, the selection of astronomers, etc., while the host country would be expected to undertake local arrangements. The Commission decided it should co-operate with Commission 46 in its working group on Education of Astronomers from Developing Countries and appointed a small subcommittee from among those present to meet with that working group the following day.

Subject to Executive Committee approval, the officers and the Organizing Committee of Commission 38 for 1979-82 will be as follows:

President: J. Delhaye (France)

Vice-President: F. B. Wood (USA)

Organizing Committee: G. S. Kromov (USSR), D. A. MacRae (Canada), P. M. Routly (USA), J. Sahade (Argentina), F. G. Smith (U.K.), J. Tinbergen (Netherlands), H. H. Voigt (G.F.R.), J. P. Wild (Australia).

The resignation of A. Reiz from the Organizing Committee (he was the second President of the Commission) was accepted with reluctance. The resignation of M. K. V. Bappu from the Organizing Committee was accepted with understanding, but regret. Both remain as members of the Commission. New members of the Organizing Committee, the last three in the above list, were added in order to broaden its geographical representation. The following were appointed as new members of Commission 38: G. Godoli (Italy), K. C. Leung (USA), E. A. Müller (Switzerland), and E. V. P. Smith (USA).

The assistance of Dr. Edith Müller, immediate past General Secretary of the Union, in the work of Commission 38 and the operation of its program is gratefully acknowledged.

COMMISSION 40: RADIO ASTRONOMY
(RADIO ASTRONOMIE)

Report of Meetings, 14, 15, 16, 17, 18, 20, 21 and 22 August 1979

PRESIDENT: H. van der Laan SECRETARY: R.G. Strom

JOINT DISCUSSIONS

On 20 August Commission 40 cosponsored a Joint Discussion with Commissions 28, 44, 47 and 48 on Extragalactic High-Energy Astrophysics.

On 21 August Commissions 28, 33, 34 and 40 held a Joint Discussion on the Nuclei of Normal Galaxies.

The transactions of these meetings will be published in "Highlights of Astronomy".

JOINT MEETINGS

A Joint Meeting of Commissions 44, 16 + 17 and 40 on the Search for Life in Space was held on 15 August.

Commissions 35, 40 and 48 were the sponsors of a Joint Meeting on Pulsars which took place on 17 August.

A Joint Meeting on Close Binaries and Stellar Activity held on 18 August was cosponsored by Commissions 42, 10, 27, 40, 44 and 48.

These activities are outlined in one of the reports of the cosponsoring Commissions.

14 August 1979

With the President, H. van der Laan, in the chair the following business was transacted.

I. REPORT OF THE PRESIDENT

In a brief summary the President reported on Commission activities since the Grenoble General Assembly. Much of this involved preparations for the World Administrative Radio Conference (WARC) to be held this year, which was also an important discussion topic later in this business meeting. During the last three years Commission 40 (co)sponsored IAU Symposia 86 and 87, and Colloquium 49.

II. REPORTS ON ASTRONOMY

The composition and content of the latest triennial Commission Report were sketched. As had been agreed in Grenoble, purely astronomical developments are now reported on by the Commission responsible for that area of astronomy, and the membership voted to continue this format. This year's report, published in volume XVII A of the Transactions, was also issued as a Leiden astronomy preprint and distributed to all Commission members. A vote of thanks to those responsible for the report was adopted by acclamation.

III. COMMISSION ACTIVITIES IN MONTREAL

The President reviewed the joint sessions which have been organized for this General Assembly. Those which still had room for short contributions were indicated, and it was announced that an extra scientific session would be organized if there was sufficient demand. This, together with a session devoted to new developments at radio observatories, was held on 22 August.

IV. IUCAF NEWS

Westerhout summarized the work carried out by IUCAF in preparation for the WARC. This will concern the radio spectrum from 2.5 MHz to 275 GHz, and IUCAF's job has been to coordinate and communicate the radio astronomy and space science position to participating national delegations. This has been achieved by preparing a position paper, after several iterations with IUCAF correspondents, containing the reasoned wishes of the radio science community. It has generally been possible, for example, to include all spectral lines discussed in the preliminary deliberations. IUCAF has tried to identify those delegation members sufficiently knowledgeable to be able to explain the positions adopted, and will man an office at WARC. Additional comments were made by Wielebinski, who stressed the intensity of competition for radio frequencies, and Burke, who pointed out that the positions delegations adopt are most likely to be influenced by their own nationals.

Robinson reported on the activities of the working group set up by Commission 40 to decide which are the astrophysically most important spectral lines. In 1976 a questionnaire was distributed to astronomers active in radio spectroscopy asking their opinions and, because of concern that the situation might have changed, this exercise was repeated at the beginning of 1979. As Table 1, summarizing these two surveys, shows, the results are very similar. With one or two exceptions, the changes were described as "well within the noise".

After some discussion, the President suggested that Westerhout and Robinson organize an informal discussion before the second business meeting to consider some of the points raised regarding the IUCAF position paper.

V. NOMINATIONS OF OFFICERS AND ORGANIZING COMMITTEE

The slate suggested by the outgoing Organizing Committee consisted of:
President: G. Swarup
Vice President: K.I. Kellermann
Organizing Committee: Ables, Booth, Broten, Dulk, Fanti, Höglund, Kardashev, Moffet, Morimoto, Okoye, van der Laan, Weliachew.
Other nominations could be made before the second business meeting, when elections would be held.

VI. NEW MEMBERS

It was decided that the Organizing Committee would review the qualifications of the eighty-odd applicants. A list would be made available to Commission members before the holding of elections at the second business meeting.

VII. RESOLUTIONS

Commission 40 would propose one resolution for adoption by the IAU, relating to frequency protection and WARC-'79. This was to be considered at the second business meeting after the IAU resolutions committee had approved the text.

VIII. MISCELLANEOUS

Morimoto announced that Tanaka has asked for suggestions and comments on Commission J activities during the next URSI General Assembly in 1981.

16 August 1979

Joint meetings were held by Commissions 40 and 28 in the morning, and 40 and 48 in the afternoon of 16 August.

TABLE 1. ASTROPHYSICAL IMPORTANCE OF LINES

Line (1976 Survey)	Frequency (GHz)	Line (1979 Survey)	Frequency (GHz)
Hydrogen	1.420	Hydrogen	1.420
CO (J = 1→0)	115.271	CO (J = 1→0)	115.271
Hydroxyl	1.665 1.667	Hydroxyl	1.665 1.667
$^{13}C^{16}O$ (J = 1→0)	110.201	Formaldehyde	4.829
Hydroxyl	1.612	$^{13}C^{16}O$ (J = 1→0)	110.201
Formaldehyde	4.829	Water vapour	22.235
Water vapour	22.235	Hydroxyl	1.612 1.720
Hydroxyl	1.720	Hydrogen cyanide	88.631
Hydrogen cyanide	88.631	Formyl ion	89.189
CO (J = 2→1)	230.538	$^{12}C^{18}O$ (J = 1→0)	109.782
$^{12}C^{18}O$ (J = 1→0)	109.782	Silicon monoxide	86.243
Formyl ion	89.189	Ammonia	23.694 23.723 23.870
Carbon monosulphide	97.981	Silicon monoxide	42.820 43.122
Ammonia	23.694 23.723 23.870	CH	3.263 3.335 3.349
Silicon monoxide	42.820 43.122	CO (J = 2→1)	230.538
Formaldehyde	14.468	$^{13}C^{16}O$ (J = 2→1)	220.399
$^{13}C^{16}O$ (J = 2→1)	220.399	Formaldehyde	140.839
Formaldehyde	140.839	Carbon monosulphide	97.981
CH	3.263 3.335 3.349	Deuterium	0.327
Carbon monosulphide	48.991 146.9	HC_3N	36.392*
Deuterium	0.327	Formaldehyde	150.498
		$^{12}C^{18}O$ (J = 2→1)	219.560
		Nitric oxide	150.74
		Carbon monosulphide	48.991
		N_2H^+	93.17

*Not included in IUCAF 277

I. RADIO STUDIES OF NEARBY GALAXIES

This meeting, which lasted one session, was chaired by Elly Berkhuijsen. The following papers were presented:

Nearby galaxies in continuum and the hydrogen line (P.C. van der Kruit).
Radio spectral index and polarised emission of M31 and M33 (E.M. Berkhuijsen).
Progress report on the Westerbork M31 study (H. van der Laan).
Radio halos of galaxies (R. Wielebinski).
HI in the elliptical galaxy NGC 4278 (E. Raimond).
CO emission from nearby galaxies (L.J. Rickard).
Integrated properties of 100 Sbc galaxies (R.D. Davies).

II. RADIO STUDIES OF RADIO GALAXIES AND QUASARS

The talks given at this session whose chairman was A.H. Bridle were:

Radio galaxies - a review (D.E. Harris)
High frequency observations of large radio galaxies (R. Wielebinski).
Magnetic fields in radio jets (E.B. Fomalont).
A radio jet in a quasar (J.F.C. Wardle).
The radio variability of BL Lac objects (K.W. Weiler).
Fleurs maps of radio sources (A. Bosma).
BL Lac objects (D. Stannard).
Rotation axes of radio galaxies (P.A.G. Scheuer).
VLA map of a double quasar (B. Burke).

III. VARIABILITY OF FLUX DENSITY, POLARIZATION AND STRUCTURE IN RADIO SOURCES

Centimeter wavelength variability in compact radio sources (W.A. Dent).
Millimeter wavelength variability in compact radio sources (E.E. Epstein).
Low frequency variability in compact radio sources (R. Fanti).
Variability of polarization in compact radio sources (H. Aller).
Search for variable radio sources in the galactic plane (P.C. Gregory).
Superluminal motions in compact radio sources (M.H. Cohen).
Slowly varying radio sources (D. Shaffer).
Morphology and alignment in compact radio sources (T.J. Pearson).
Search for relativistic Doppler shifts in the continuum spectra of double radio sources (Yu. N. Parijskij).
X-ray variability in compact radio sources (D. Warrell).
Energetics and dynamics in relativistically expanding radio sources (M. Salvati).
Intrinsic and extrinsic models for fast variations and superluminal motions (P.A.G. Scheuer).

22 August 1979

The second business meeting was chaired by the president. After approval of the agenda, the business transacted was as follows.

I. PREPARATIONS FOR WARC 1979

Westerhout, in introducing this discussion, said that since the first Commission business meeting there had been a number of consultations and a three hour meeting of those interested. He pointed out that IUCAF represents the interests of both radio astronomy and space science and in its position paper, IUCAF 277, it proposes a table of bands to be protected for both services. IUCAF's advisory role at WARC was emphasized. A major suggestion would streamline the use of footnotes in the Radio Regulations (Article 5). The present series of footnotes scattered throughout the table would be replaced by four to cover: (a) Primary exclusive continuum bands; (b) Continuum bands shared with other services; (c) Spectral lines; (d) Regional protection (special cases).

Commenting on this proposal, Swarup suggested that only spectral lines should have a blanket footnote. Wielebinski said that one reason for the reform was a threat from European countries to abolish footnotes altogether, although he agreed with Swarup insofar as spectral lines are the main problem. As Robinson was absent, Westerhout led a discussion of spectral line priorities (see the table in the summary of the first business meeting). After due consideration, the Commission unanimously adopted a motion which would make the latest list the official viewpoint of the radio astronomy community.

Howard then presented a resolution relating to frequency protection to be adopted by the IAU as a whole. (For the text, see Resolution No. 2.) After some discussion it was unanimously adopted.

This section of the meeting then closed after the chairman had listed those members of the radio astronomy and space science community who will be present in Geneva during WARC. Problems with coordination and communication were mentioned, and after a final summary of the IUCAF role by Westerhout, Howard noted with regret that while the viewpoint of radio science as a whole has been represented, no one has put forward the case for radio astronomy uncluttered by the needs of other services.

II. REPRESENTATION IN IUCAF AND CCIR

The Commission chose Westerhout and Schilizzi as IUCAF representatives. Doherty and Robinson will be the new CCIR representatives. The terms of office for both IUCAF and CCIR begin after WARC, in 1980.

III. GERT PROPOSAL

The General Assembly had before it a proposal asking ICSU to urge appropriate bodies to support the Giant Equatorial Radio Telescope project (GERT). The IAU executive was now requesting Commission 40 endorsement of the proposal. Swarup outlined the astronomical justification for the project, which would build an Ooty-type cylindrical reflector 2 km long and 50 m wide. It would be used at 327 and 38 MHz for both lunar occultation and synthesis mapping. Okoye said that the site being considered is in Kenya, 300 km north of Nairobi, and emphasized the value of such a project to developing countries. After discussion, the meeting unanimously backed the proposal.

IV. ELECTIONS OF OFFICERS AND ORGANIZING COMMITTEE

There had been no additional nominations after the first business meeting. The slate of candidates originally proposed (see section V of the report of that meeting) was then carried without dissent.

V. FUTURE SYMPOSIA AND COLLOQUIA

Mention was made of international conferences planned for the coming three years of interest to Commission members. These include IAU Symposium 95 on Pulsars (26-31 August 1980), a symposium on millimeter astronomy (20-25 August 1980) and IAU Symposium 97 on Extragalactic Radio Sources being held in conjunction with the opening of the VLA (August 1981).

VI. MEMBERSHIP

Changes to the membership list proposed by the Organizing Committee, with several additions, were approved by the membership.

The meeting closed with a vote of thanks to all those responsible for organizing Commission activities in Montréal.

Scientific Sessions.

Commission 40 held both morning and afternoon scientific sessions on 22 August.

Status and prospects of radio observatories. At this session, chaired by G. Swarup, the following papers were presented:

Multi-telescope radio-linked interferometer at Jodrell Bank (B. Anderson).
New Arecibo interferometer (K.C. Turner).
Cosmic 10 m radio telescope in space (N.S. Kardashev).
The Very Large Array (D.E. Hogg).
European VLBI System (K.I. Kellermann).
The Westerbork Synthesis Radio Telescope (R.G. Strom).
The Cambridge 150 MHz radio telescope (A. Hewish).
From Molongolo Cross to M.O.S.T. (B.Y. Mills).
Ooty Synthesis radio telescope (G. Swarup).
Effelsberg 100 m telescope (R. Wielebinski).
The present and future of Rattan 600 (Yu.N. Parijskij).
French-German mm-wave facilities - IRAM (D. Downes).
Japanese Radio Telescope (M. Morimoto).
The Onsala 20 m telescope (A. Hjalmarson).
U.S. Naval Observatory: The 35 km interferometer at Green Bank as an astrometry instrument (G. Westerhout).
Hartebeesthoek Radio Observatory (G.D. Nicolson).
Culgoora Synthesis telescope (D.J. McLean).
Siberian solar radio telescope (G. Smolkov).
Penticton Observatory (J. Galt).
Algonquin Radio Observatory (N.W. Broten).
Clark Lake array (M.R. Kundu).
Gauri Bidanaur 34 MHz T-antenna (C.V. Sastry).
University of Massachusetts 14 m millimeter-wave telescope (W.M. Irvine).

Short contributed papers. K.I. Kellermann was the chairman of this session and the following talks were given:

Solar observations with the WSRT and the VLA (M.R. Kundu).
Jovian polarimetry with the WSRT at λ50, 21 and 6 cm (I. de Pater).
Observations with the Bonn 100 m Telescope at 7 mm (D. Downes).
Sensitive large area surveys at 2.3 GHz (E.E. Baart).
Recombination line survey made with the Bonn 100 m Telescope (T.L. Wilson).
The new 1720 MHz OH outburst in V 1057 Cyg (T.A. Pauls).
Radio flares in Cir X-1 (G.D. Nicolson).
Broad HI shells on the peripheries of extended HII regions (R.S. Roger).
Synthesis observations of HI clouds (C.H. Costain).
Possible HI detection of nearly protogalaxies (R.D. Davies).
Radio observations at 408 MHz of a sample of spiral galaxies (I.M. Gioia).
The double quasar - a gravitational lens ? (D. Walsh).
Multiwavelength polarimetry of 3C 310 and 3C 465 (W. van Breugel).
MTRLI and VLBI of 3C 380 (P.N. Wilkinson).
Quasars with resolved radio components (T. Kiang).
Radio jet in Cen A and the precessing beam model of radio sources (R.D. Ekers).
Summary of searches for short duration radio bursts from M87 (J.A. Roberts).
Plasma turbulence models of QSOs (J. Colvin).
Magnetic dipole model for cosmic radio sources (H.D. Greyber).

COMMISSION 41. HISTORY OF ASTRONOMY/HISTOIRE DE L'ASTRONOMIE

Report of Meetings, 15 and 16 August 1979

PRESIDENT: J. Dobrzycki

Business session

Silent tribute was paid to members lost by death since 1976, V.L. Chenakal, T. Przypkowski, V.V. Michkovitch and B. Sticker.

The Commission dealt with the election of its officers for the next term and with the admission of new members. In accord with the policy adopted in previous terms, the Commission sponsored the admission of several outstanding and professionally active consulting members to a full IAU membership.

A survey of the progress of various long term editorial projects was presented by J. Dobrzycki and J. Fleckenstein /editions of works of Copernicus, Kepler, Euler, Bernoulli/ and M. Hoskin /General History of Astronomy/. From M. Hoskin's report it appears that editorial work is most advanced on volume IV, dealing with the recent history of astronomy since inception of astrophysics, and on volume III /astronomers, instruments, stellar astronomy: XVII-XIX centuries/. These two volumes should be ready for processing by the publisher during 1980.

Scientific session

This session opened with the discourse by G. de Vaucouleurs on "The Early History of Astronomical Photography". Illustrated with examples demonstrating pioneer work in the consecutive photochemical techniques applied to astronomy, the lecture outlined the main phases in the evolution of astrophotography.

John A. Eddy spoke on "Recent Progress in Archeoastronomy". He presented the possibilities and restrictions of astronomical evaluation of etnographic and archeological sources using three cases studies from American Precolumbian and Indian culture. Professional organization of the field was expanded by new periodical publications.

Susan McKenna-Lawlor presented a report on her studies on "Astronomical observations in Medieval Irish Chronicles". The survey of over thirty Gaelic and Latin chronicles is now in progress. Astronomical and climatic events, noted from 6th century onwards, are analysed and put in context with local agricultural and military concerns.

It may be noted that similar surveys of cosmographical notes are advanced or in progress also for Italian and Polish sources.

"The Plenum Universe from Aristotle to Einstein" was the next subject, presented by Owen Gingerich. In a short synthetic review the speaker outlined the role of concepts like void and ether in the cosmology from ancient Greece to the times of Einstein.

Three short communications on history of Chinese astronomy followed. Dr. Y. Chang reported on the record of Halley's comet from 11th century BC. T. Kiang presented a report /by H. Siyi/ on a survey, now ready for publication, of astronomical data from local sources in ancient China /since 8th century AD/. From this survey T. Kiang selected the subject of his communiqué on the supposed identification of Cygnus X-1 with a nova, recorded in Chinese /and Japanese/ records from 1408.

Addenda to the report of the Commission

1. An exhibition of early astronomical photographs was set up during the IAU General Assembly. Authentic exhibits ranged from the early Harvard daguerotypes to dry-plate short focus survey photograms from Heidelberg. Exhibits were made available thanks to the courtesy of Harvard, Lick, Lund and Heidelberg observatories. The Merseyside County Museums, Liverpool, provided a series of panels with many reproductions of pioneer astronomical works. The co-operation of the contributing institutions is gratefully acknowledged. Special thanks are due to M. B.Drouin /Université de Montréal/ for his efficient and instrumental help in setting up the exhibition.

2. Preservation and accessibility of various types of sources for history of astronomy is a highly topical problem, discussed at the 1976 Commission meeting. For written material one can name cases of satisfactory attention given to this matter by the Royal Greenwich Observatory, Herstmonceux, and the R.A.S. /the Herschel archives/. A growing problem is that of obsolete observatory equipment. Quite a number of big-sized instruments of historical importance become redundant and may, or have already to, be dismounted. It is hoped that in each case satisfactory documentation of the instrument will be preserved for possible future needs.

3. Information services of the Commission are continued. Shortly before the Montréal meeting Information circular No. 29 was distributed by the president, and No. 17 of the "Bibliography on books and papers on history of astronomy" /for 1976/ was issued at the Sternberg Astronomical Institute in Moscow.

4. The following, highly selective list of books published in 1976 - 1979 illustrates to some extent the main and topical fields of interest within history of astronomy. In general, perhaps the most penetrating research may be noted in history of medieval and Renaissance cosmology, the Islamic mathematical astronomy and, in a subject markedly distant in time, twentieth century astrophysics and cosmology. Established journals on history of science had the main share in the publication of research results.

Āryabhata, Āryabhatiya /critical edition of the texts and of its commentaries/, Delhi 1976
R. Berendzen, R. Hart, D. Seeley, Man Discovers the Galaxies, New York 1976
D.M. Clark, F.R. Stephenson, The historical supernovae, Oxford 1977
N. Copernicus, On the Revolutions /Complete works II/, Warsaw-Cracow 1978
F. Barone /ed./, Opere di Nicola Copernico, Torino 1979
A.M. Duncan /ed./, N.Copernicus, On the revolutions of heavenly spheres, London 1976
D.O. Edge, M.J. Mulkay, Astronomy Transformed. The Emergence of Radio Astronomy in Britain, N.York - London 1976
J. Meeus, H. Mucke, Canon of Lunar Eclipses -2002 to +2526, Wien 1979
R.R. Newton, The Crime of Claudius Ptolemy, Baltimore - London 1977
J. North, Richard of Wallingford, Oxford 1976
A.G. Davis Philip, D.H. DeVorkin, In Memory of Henry Russel /Historical part of IAU Symposium No. 80, The HR Diagram, The 100th Anniversary of H.N. Russell/, Albany 1977
K. Shukla, Aryabhata, Indian Mathematician and astronomer, Delhi 1976
D.J. Warner, The Sky Explored. Celestial Cartography 1500-1800, N.York - Amsterdam 1979
P. Zambelli /ed./, Alberto Magno, Speculum Astronomiae, Pisa 1977.

COMMISSION 42: CLOSE BINARY STARS (ÉTOILES DOUBLES SERRÉES)

Report of Meetings, 15, 17, 18 and 21 August 1979

PRESIDENT: G. Larsson-Leander SECRETARY: C.D. Scarfe

As scheduled, Commission 42 convened for two business meetings, on August 15 and 21, and two scientific meetings, on August 17 and 18. The theme of the first scientific meeting was "Recent Progress in Close Binary Studies". The programme for this meeting grew to such extent that the allotted time was unsufficient, and the meeting was therefore continued on August 21, after the end of the business session. The second scientific meeting, August 18, was a Joint Meeting together with Commissions 10, 27, 40, 44 and 48 on the topic "Close Binaries and Stellar Activity". In addition, there were three informal meetings, *viz.* August 17 a meeting on VV Cep arranged together with Commission 29, and August 18 and 22 joint meetings with Commissions 29, 40 and 48 for the purpose of discussing SS 433. Reports from all these meetings are given below.

Business Meeting I - August 15, 1979

Condensed minutes of this meeting, which started at 14.15 and ended at 15.45, are as follows:

1. After having adopted the agenda, as listed on the black board by the President, the Commission elected C.D. Scarfe as Secretary for the meetings. He also agreed to act as correspondent to the General Assembly journal *Meteor*.

2. The President asked for a moment of silence in memory of the three Commission members, J.F. Heard, S.-S. Huang and C. Popovici, who had died since the last IAU General Assembly.

3. The President announced that C.M. Huffer and H. Kristenson wish to withdraw from the Commission, as they are no longer active in the field. During the past triennium K.G. Kron, B.F. Marino, E.F. Milone and R.M. Williamon were co-opted as Commission members. With these changes, the Commission counts 130 members. The President read a list of 63 new applications for membership, divided as follows: 24 individual requests from present IAU members, 3 individual requests and 36 recommendations from National Committees concerning persons now proposed as new members of the IAU. The Organizing Committee will consider all these names, and the Committee was authorized by the Commission to decide upon the question of admittance and to report on the outcome at the second business meeting. The President pointed out that most of the applications came from persons highly qualified in the field of close binaries and that the large majority were likely to be accepted. One further member, T.H. Morgan, was proposed by Kondo.

4. The President listed the members of the present Organizing Committee with their respective years of service and said that as expected, B. Warner, the Vice-President, would be the next President. The Commission is proposing A. Batten as next Vice-President, with J. Smak as an alternative. M. Plavec and T. Herczeg, who had served for a long time, asked to be allowed to stand down from the Organizing Committee. The names of L.B. Lucy and R.H. Koch were proposed as replacements. Koch agreed and Leung was asked to contact Lucy, who was absent. As expected, a positive answer was received from Lucy.

5. The President read a letter from K. Gyldenkerne, Chairman of the Coordinated Programs Committee, who wishes to resign from the Committee and recommends its abolition. This view was supported by Andersen and Merrill. The Committee provided encouragement and advice but was able to offer little else, according to Kondo and Bolton. Wood pointed out that nowadays communication between those participating in such programmes is better than formerly, and Herczeg suggested that the Organizing Committee could take over the responsibility of organizing co-ordinated programmes. Merrill stressed the importance of collecting and publishing the results of campaigns. The Commission agreed to abolish the Coordinated Programs Committee.

6. The Committee for Extraterrestrial Observations was discussed. The Chairman of this Committee, Y. Kondo, feels that it has fulfilled its purpose, but he is willing to continue as a clearing house of information and watchdog. It is quite obvious that the X-ray binary campaigns have been successful, *e.g.* Cyg X-1 (Bolton) and X Per (de Loore). The newsletter on X-ray binaries has been continued till 1978. It was agreed to keep the Committee but no longer to publish the newsletter.

7. Kondo spoke on the use of HEAO-2 (Einstein) and IUE satellites. HEAO-2 is expected to last one more year, so those who wish to propose observations should at once contact Dr. Fred Seward, Center for Astrophysics, 60 Garden Str., Cambridge, MA 02138, U.S.A. Mattei pointed out that many optically variable objects observed by HEAO-2 are being followed by AAVSO members simultaneously. IUE is geosynchronous; 2/3 of its time is controlled by the U.S. and 1/3 by Europeans. It has high and low resolution spectrometers. Prospective observers should contact Dr. Don West, Code 685, Goddard Space Flight Center, Greenbelt, MD 20771, U.S.A. Kondo supplies an IUE newsletter. The IUE lifetime should be 3 to 5 years.

8. Koch, reporting on the work on the new *Finding List*, said that this would soon be available with information on 3555 stars in our Galaxy. For over 100 stars the most recent photometric analyses are based on visual or photographic estimates. Herczeg suggested that publishing a list of these would inspire observers to obtain more accurate data. Merrill surmised that many may be of long period and hard to observe.

9. The President read a letter from I.-S. Nha, Yonsei University, Seoul, reporting that he is preparing an *Atlas of O-C Diagrams of Eclipsing Binaries*. The work will take 2 or 3 more years, and as a by-product times of minima are collected and stored on computer cards. A data centre for the times of minima will thus exist in Seoul. Hall mentioned that such data are also collected in Cracow, by Kreiner, who will soon publish lists of times of minima for all β Lyraes, to be followed by similar lists for RS CVn's, W UMa's and eventually Algols. Wood pointed out that the duplication of effort ensures that data will not be lost in any disaster.

10. Following a proposal by Merrill, it was decided to send a card with greetings to Dr. D.J.K. O'Connell, past-President of the Commission; members were invited to sign this card.

The meeting then ajourned.

Business Meeting II - August 21, 1979

The meeting started at 9.15 and ended at 9.40. Condensed minutes are as follows:

1. The President gave a brief report from the Commission 5 Joint Meeting on Nomenclature, on August 14. The main decisions of interest to Commission 42 are:
1. Authors are strongly recommended to identify stars, whenever possible, in at least two ways (*e.g.* by name and by catalogue number).
2. When abbreviations for not widely known catalogues and lists of stars are used, authors should clearly indicate the meaning by giving proper references.

2. The President described the work on the Commission Report, printed in Vol. A of the *IAU Transactions*. Preprints of the Report had been circulated to Commission members, but as the Report was prepared in camera-ready form there was no opportunity to correct mistakes. The President extended his thanks to the colleagues who wrote parts of the Report and to all Commission members who furnished information.

3. With reference to decisions mentioned under item 3 of the minutes of Business Meeting I, the President stated that 51 new members had been admitted into the Commission. The new members, the names of which were included in the membership list posted on the advertisement board, are as follows: C.T. Bolton, B.B. Bookmyer, B.W. Bopp, A.N. Bunner, J.V. Clausen, G.W. Collins, J.-P. De Grève, M.J.H. de Groot, H.W. Duerbeck, J.A. Eaton, B.P. Flannery, M. Friedjung, C.D. Garmany, D.M. Gibson, E.F. Guinan, G. Hammerschlag-Hensberge, D.P. Hube, P.C. Joss, J.M. Kreiner, M.I. Kumsiashvily, Z. Kviz, D.Q. Lamb, J.A. Mattei, E. Meyer-Hofmeister, T.H. Morgan, K. Nariai, R.E. Nather, J.P. Oliver, G.J. Peters, V. Piirola, H. Ritter, E.L. Robinson, M. Rodonò, I.W. Roxburgh, N.B. Sanwal, A. Sanyal, H.J. Schober, N.I. Shakura, G. Shaviv, F. Shu, R.C. Smith, J.-E. Solheim, W.M. Sparks, D. Sugimoto, P. Szkody, J. van Paradijs, O. Vilhu, W.S.G. Walker, R.F. Webbink, E.J. Weiler, A. Yamasaki.

4. The President agreed to edit the *Bibliography and Program Notes on Close Binaries* for another 3 years, and indicated on a query by Merrill that his university could continue to absorb the cost for doing so. Hall suggested that the *Bibliography* be sent to *Astronomy and Astrophysics Abstracts*, and asked how widely it was disseminated to non-members. The President replied that its existence is mentioned in all Commission Reports and that it is sent to some 70 astronomers, who are not members but have asked for it.

5. Recent and future meetings were discussed. Among recent meetings the President mentioned the Rochester Colloquium on "White Dwarfs and Variable Degenerate Stars", which was co-sponsored by the Commission, and our own Symposium, "Close Binary Stars", in Toronto. He also mentioned that the Commission is co-sponsoring IAU Symposium No. 93, "Fundamental Problems in the Theory of Stellar Evolution", in Kyoto, July 1980, and a proposed Symposium on "Pulsars" in Bonn, August or September 1980. As to the future, Leung suggested a Symposium in 1981 in Hong Kong on "Contact and Related Objects", and Friedjung proposed a meeting on "Symbiotic Stars" before the next General Assembly. Both suggestions were referred to the Organizing Committee for consideration. Also, it was mentioned that Commission 26 is planning a Colloquium in Flagstaff in 1981, which may be of interest to Commission 42, and there will probably be another Bamberg Colloquium.

6. Hall reported briefly on the Second RS CVn Workshop, held in Montreal on August 13, 1979. He asked for new names, addresses and telephone numbers for the RS CVn hot line. Reports will continue to be handled by Hall and Catalano. Catalano will be writing a review paper on BY Dra stars, and Hall one on photometric aspects of RS CVn stars. Preprints will be welcomed by both. Kreiner will publish eclipse ephemerides for RS CVn stars in a forthcoming *IBVS*. The *IAU Circ* will list new RS CVn stars. The Working Group on RS CVn Binaries will continue its work.

7. The President said that the General Secretary had stressed the importance of having some written Working Rules for the various Commissions. The preparation of the rules would be a task for the new President and the Organizing Committee.

8. The President asked for proposals for new co-ordinated programmes. There were none. Those who wish to propose such programmes in the next 3 years are adviced to contact the new President, B. Warner.

The meeting then ajourned.

Scientific Meeting I - August 17 and 21

The first scientific meeting, on the theme "Recent Progress in Close Binary Studies", was started with two sessions on August 17 and continued on August 21, after the end of Business Meeting II. The following papers were presented:

J. Merrill - Some aids in the solution of light curves of eclipsing binaries with elliptical orbits.
D.M. Popper - A review of stellar masses.
J. Andersen, J.V. Clausen and B. Nordström - Recent progress in the Copenhagen programme of absolute stellar dimensions.
J. Andersen, J.V. Clausen, B. Nordström and B. Reipurth - *uvby* light curves of the unique RS CVn binary TY Pyx.
K.-C. Leung - Early-type semidetached and marginal semidetached systems.
D.Ya. Martynov - RX Cas: Preliminary report on the results of the co-ordinated campaign of observations (presented by A.M. Cherepashchuk).
D. Chochol - The close binary SZ Cam - a semidetached system.
H. Al-Naimiy - Photometric observations and light-curve analysis of the peculiar system ER Vul.
A.R. Sadik - Light and colour curves of UV Psc.
S.A. Naftilan - The complicated close binary CG Cyg.
G.J. Peters - Hα observations of Algols.
C.T. Bolton - The orbit and mass flow in the BN star HDE 235679.
M.G. Fracastoro - An analysis of the Barr effect.
K. Walter - Oblique axes of rotation in close binaries and the photometric proof of precessional motions.
T. Mazeh - On the tidal evolution of close triple systems.
I.S. McLean - Optical polarimetry of close binary systems.
V. Piirola - Polarimetry of CH Cyg.
G.T. Bath - Accretion powers the brightest stars.
G. Shaviv, M. Shara and D. Prialnic - Generalized scheme for novae explosions.
I. Nevo - The eruptions of the cataclysmic variables and the mass ratios of their components (presented by T. Mazeh).
K. Nomoto and K. Nariai - Recurrence of nova explosions.
M. Fujimoto - Composition of grains and theory of nova explosions.
J. Ziółkowski and B. Paczyński - A binary model and disc parameters for Cyg X-2.
C.C. Wu - IUE observations of the X-ray nova Cen X-4.

Scientific Meeting II - August 18

The second scientific meeting was a Joint Meeting together with Commissions 10, 27, 40, 44 and 48 on the topic "Close Binaries and Stellar Activity". The proposal for this meeting originated from D.S. Hall, and it was planned by an organizing committee, as follows: G. Larsson-Leander (Com. 42, Chairman), J.L. Linsky (Com. 10), D.S. Hall (Com. 27 and 42), R.M. Hjellming (Com. 40), Y. Kondo (Com. 44).

The meeting occupied two sessions, the first chaired by Larsson-Leander and the second by Hall. After a short introduction by Hall, the following review talks were given:

B.W. Bopp - BY Dra and RS CVn stars: The discoveries of classical photometry and spectroscopy.
Y. Kondo - A review of recent ultraviolet observations of close binaries from space.
J.H. Swank - X-ray observations of stellar coronae and winds.
R.M. Hjellming - Radio aspects of stellar activity in close binaries.
I.S. McLean - Polarimetry of emission lines as a probe of stellar winds.
J.L. Linsky - On the differences at chromospheric levels between RS CVn-type binaries, active and quiet single stars, and active and quiet regions in the Sun.
J.L. Modisette - Mass transfer between binary stars.

The Introduction and shortened versions or abstracts of the review papers are published in Vol. 5 of the *Highlights of Astronomy*.

Informal Meeting on VV Cephei - August 17

This informal meeting, together with Commission 29, had been proposed by R. Faraggiana and was organized by her. K.O. Wright was Chairman, and some 25 persons especially interested in VV Cep participated.

The photometric data from the recent eclipse were first discussed. S. Kawabata presented a review of work done in Japan, including light curve analysis. Wright described his spectroscopic observations, now covering a whole period in high dispersion. Ultra-violet spectroscopy was reported by W. Hagen and by Faraggiana. I.S. McLean reviewed polarization observations. Kawabata described Hα observations and outlined an envelope model, worked out by M. Saito *et al.* A.P. Bernat made a comparison of VV Cep with 31 and 32 Cyg.

Informal Meetings on SS 433 - August 18 and 22

These two informal meetings for discussing the extraordinary object SS 433 were held jointly with Commissions 29, 40 and 48. They were organized by A. Mammano, who also served as Chairman. About 200 persons attended, and the following 20 short contributions were presented:

B. Sherwood - Distance to SS 433.
K.W. Weiler and N. Panagia - Plerionic (Crab-like) structure in W50.
R. Schilizzi - VLBI at 6 cm with Westerbork, Effelsberg and Onsala telescopes.
B. Geldzahler - VLBI detailing structure of a few milliarcsec (jet?).
W.S. Gilmore and E.R. Seaquist - High resolution band 20 cm synthesis map.
Yu.N. Parijskij - Multifrequency simultaneous monitoring and resolution of the "jet" with the 600 m telescope.
W.M. Goss - An H I absorption profile line with Westerbork telescope.
B. Zealey - Discovery of optical counterpart of SNR W50, centred on SS 433.
E. Becklin and G. Wynn-Williams - Dramatic IR changes observed at Mauna Kea.
J. Bally - Brackett alfa to Brackett beta line ratios from central lines.
B. Margon - Implementation to the kinematic model for the moving features.
C. Zirmani - Optical spectra showing structures.
A. Mammano - Multiple branches and velocity changes in optical spectra up to 11000 Å.
B. Cosmovici and F. Strafella - Profile changes of stationary O I 8446 (presented by A. Mammano).
D. Crampton and J.B. Hutchings - Periodic variations of positions of central emission and absorption lines (13 days period).
J. Kemp - Photoelectric observations concerning the 13-d period (presented by G.W. Collins).
G.W. Collins and G.H. Newsom - On the nature of SS 433; a binary model.
I.S. Shklovskij - Jets produced by a very fast spinning NS (0.001 s) in a post-supernova binary system.
V. Icke - Clouds ejected by a disk-black-hole system.
M. Milgrom - Thomson scattering model for the secondary branches (presented by A. Mammano).

COMMISSION 44 : ASTRONOMICAL OBSERVATIONS FROM OUTSIDE THE TERRESTRIAL ATMOSPHERE (OBSERVATIONS ASTRONOMIQUES AU-DEHORS DE L'ATMOSPHERE TERRESTRE)

PRESIDENT: R. M. BONNET

During the course of the XVIIth General Assembly in Montreal, Commission 44 was involved in 5 Joint Discussions (over 8), 5 Joint Meetings :

1 - "Coming Solar Maximum", with Commissions 10, 12 and 14 ;
2 - "The M.H.D. of Sunspots", with Commissions 10 and 12 ;
3 - "The Search for Life in the Universe", with Commissions 16 and 40 ;
4 - "Close Binaries and Stellar Activity", with Commissions 10, 27, 40, 42 and 48 ;
5 - "Space Astrometry", with Commissions 8, 20, 24, 25, 26, 30, 33, 37 and 45.

The reports of Joint Meetings 1 and 2 are to be found in the general report of Commission 10. That of Joint Meeting 4 is to be found in the general report of Commission 42 and that of Joint Meeting 5 in the general report of Commission 24.

14 August 1979

BUSINESS MEETING

The President opened the meeting by briefly summarizing the major events which occurred in Space Astronomy since the last General Assembly, i.e. the launches of HEAO-1, IUE, HEAO-2 (Einstein Observatory) and the successful continued operations on Copernicus, COS-B and OSO-8. He then described the activity of the Commission in the past 3 years : the edition of 4 issues of the Commission 44 Newletter, its involvement in the Middle Atmosphere Program (MAP), in the work of the International Radio Consultative Committee (CCIR), of the Scientific Ballooning and Radiation Monitoring Organization (SBARMO), in the COSPAR Panel on Potentially Environmentally Detrimental Activities in Space (PEDAS), etc...

He pointed out that Space Astronomy was more and more integrated in the general field of Astronomy as evidenced by the large number of Joint Discussions and Joint Meetings in which the Commission was involved either as principal or as co-organizer.

He then reviewed the activity report of the Commission. He initiated a discussion on the membership of the Commission. A letter had been sent to every member in the Commission asking whether he was willing to remain a member, to resign, or to become a consultant member. In parallel more than 40 new applications had been received. The meeting was then asked to express an opinion as to the necessity of keeping the membership large by,in particular not screening out the new applications and keeping those members who even did not reply to the President's circular letter.

The participants then proceeded to the election of the new Vice-President after nominations were proposed from the floor. M. Oda was elected by written ballot and his name was consequently submitted to the Executive Committee.

17 August 1979

SCIENTIFIC MEETING (Jointly with Commissions 16 and 40)

This meeting on the Strategies for the Search for Life in the Universe was opened with a brief welcoming address by M. Papagiannis who chaired the Organizing Committee.

Session I : Alternative views on the value of N.

I. Shkovsky and P. Morrison were the co-Chairmen of this section. N is the number of technologically advanced civilization currently present in our galaxy.

M. Hart, supported the view that N is very small based on the lack of any evidence for extraterrestrial presence on earth. T. Kuiper suggested that N is probably very large, because life seems to be able to operate an inverse 2nd law of thermodynamics, i.e., to use available energy to locally reduce entropy and th thus progressively build organisms and societies of high complexity and sophistication. F. Drake defended the view that N is neither very small nor very large based on the analysis that interstellar travelling would be prohibitively expensive. However the energy problem might not seem as severe after a few centuries in the light of future technological advances. M. Papagiannis supported the thesis that N is either very small or very large and that intermediate values are quite unlikely.

In addition to the above speakers, V. Troitsky dealt with the technological requirements of a radio antenna that could broadcast a beamed signal to a distance of 10,000 light years, and found that these requirements are formidable, if not prohibitive. Finally, P. Morrison gave an articulate summary of the different views on the value of N and the directions we ought to proceed in the future in view of the alternatives discussed. These directions, ought to include both the continuation of radio searches, as well as the further exploration of our solar system.

Session II : Strategies for SETI through radiowaves.

B. Oliver and J. Jugaku were the co-Chairmen of this section.

B. Oliver opened this section with a brief summary of the advantages of radio communications and the rapid progress that is being made in electronic support equipment, especially in the construction of multichannel analyzers. Analyzers with 10^6 channels are within the capabilities of current technology, he said. B. Zuckerman gave an extensive review of the six radio searches that have been performed up to now in the United States and Canada. V. Troitsky gave a similar, but less detailed summary of the several searches that have been going on in the Soviet Union and finally S. Gulkis discussed different alternate possibilities, both small and large, for the future.

Session III : Search for early forms of life both in our own solar system and in others.

J. Greenstein and N. Kardashev were the co-chairmen of this section.

G. Soffen gave an overview of the search for life in our solar system. He discussed the Viking mission to Mars and he noted that finding another independent origin of life would be of immense importance in our efforts to understand the mystery of life.

T. Owen discussed how we hope to use spectroscopic observations, in particular those revealing the presence of water and oxygen, to search for life in other solar systems.

Searching for signs of life in other solar systems, however, is still a task for the future because we have not yet been able to detect any planets even in the nearest stars. Great progress, however, is being made in this area, and many new techniques were discussed by experts in the field. G. Gatewood described the astrometric methods and said that the recent replacement of the photographic process by photoelectric techniques has led to a dramatic improvement in the sensitivity of astrometric methods which are looking for minute changes (1.0-0.1 milliarc sec) in the position of a star produced by the rotation around it of one or more planets.

Dr. Currie described the interferometric methods where efforts are being made to develop a two aperture set-up, like a radio interferometer. Single aperture techniques, using speckle interferometry, have already made it possible to measure the diameters of some large red giants stars. I. McLeon of the laboratory of K. Serkovski at the University of Arizona, described the progress being made in developing the sentitive spectroscopic techniques needed to measure the orbital velocity (10-20 m/sec) of a star about the common center of mass of the star and a large planet orbiting around it. W. Baum described the advantages the space telescope will have in this area since it will be able to detect objects up to a magnitude of 28 by avoiding the airglow and the smearing of a star's image by atmospheric seeing. It seems that the combination of the space telescope and the new photoelectric astrometric technique might be able to provide some impressive results in the next 5-10 years. Finally D. Black gave an excellent summary of the advantages and disadvantages of the different presently available methods for the detection of planets and their prospects for the future. He also discussed some direct methods that are currently being proposed, including occulting disks, apodized images and the infrared space interferometer proposed by R.N. Bracewell, which might hold a great promise for the future.

Session IV : The different manifestations of advanced cosmic civilizations.

S. von Hoerner and M. Rees were the co-chairmen of this section.

W. Sullivan discussed the possibility of eavesdropping on other neighboring civilizations by considering what could they have learned by doing the same to the earth. Since strong narrow-band signals, such as the carrier frequencies of TV stations, come from discrete locations on the earth, it is possible from the Doppler effects to determine the rotation of the earth and its orbital period around the sun. From the mass of the sun, which one could figure out from its luminosity, one could then determine the distance of the earth from the sun and hence also its average temperature. S. von Hoerner summarized astroengineering projects that advanced civilizations might undertake.

He discussed also the possibilities for interstellar travelling. Theoretical computations have shown that it is possible to propell a spaceship over interstellar distances even with present day technology.

Conclusion

An Open Evening Session, chaired by L. Goldberg, was held on August 16, in which the discussions and the conclusions of the Joint Meeting were summarized for the general membership of the IAU.

The meeting focused strongly on the controversy of the colonization of the galaxy and on the progress being made in the search for planets in other stars. The possibility of galactic colonization seems to call for a careful search inside our own solar system and for the parallel study with radio antennas of all nearby stars. These tasks, as well as the search for planets, can be comfortably weaved into the current goals of space exploration and astronomical research for the 80's. Thus it seems that the proper strategy at this time is to amalgamate the Search for Life in the Universe into normal scientific research and space exploration, keeping always a keen eye and an attentive ear for opportunities that might advance the cause for this quest.

17 August 1979

SCIENTIFIC MEETING : International Space Programmes and Future Needs of Space Astronomy.

SESSION I on International Space Programmes was organized and chaired by D. Walsh.

1. ESA Program

The program of the European Space Agency was presented by E. Peytreman. He first described the satellites already in orbit :

a. COS-B which studies the sky in Gamma rays and has been very successful in obtaining the first map of gamma-ray emissions from the galactic disk and detecting 29 point sources including 2 new gamma-ray pulsars.

b. IUE (see also § 3 below) in which ESA participates for one sixth of the total observing time.

He then described the projects in preparation and their associated experiexperiments.

c. EXOSAT (scheduled for a launch in November 1981) whose aim is twofold :

- determine the position of galactic and extragalactic X-ray sources ;
- study the spectral characteristics and time variations of individual X-ray sources.

d. SPACELAB-1 on which astronomy experiments make use of the pallet is a mission in cooperation with NASA. It is scheduled to fly in April 1982.

e. ISPM or Out-of-Ecliptic Mission (scheduled for a launch in 1983) is also a cooperative program with NASA. ESA participates in building one of the two spacecrafts, and several experiments.

2. NASA Program

N. Roman gave a complete description of the NASA program with a summary of its status. She started by the satellites already in orbit : COPERNICUS, HEAO-A and B (EINSTEIN Observatory) which has an excellent imaging quality in the X-ray regime. She then described the other projects :

a. HEAO-C, to be launched in September 1979 and which has a strong international component with the C_2 Koch and Peters experiment.

b. SMM, due to be launched in early 1980, contains 7 instruments devoted to the coordinated study of the next solar maximum :

- a Gamma Ray Spectrometer ;
- a Hard X-Ray Spectrometer ;
- a Hard X-Ray Imaging Spectrometer ;
- a Soft X-Ray Polychromator ;
- an U.V. Spectrometer and Polarimeter ;
- a Coronograph Polarimeter ;
- a Solar Constant measuring and monitoring instrument.

c. SPACELAB-2, a pallet only mission, contains :

- an Infra-Red Telescope to study individual sources ;
- four Solar instruments :
  - . the High Resolution Telescope/Spectrometer of NRL ;
  - . a Solar U.V. irradiance monitor in the near U.V. ;
  - . an experiment to measure the He/H abundance ratio in the solar corona built in the U.K. ;
  - . a magnetic and velocity field experiment.

d. The SPACE TELESCOPE is a truly international program with a 15% participation of the European Agency. The ST is a shuttle launched-and-operated satellite, scheduled for a launch in 1983, with an anticipated 10 year lifetime. It contains five focal plane instruments operating from the far ultra-violet (λ > 1200 A) to the near infra-red (1 μm) :

- the Faint Object Camera (under the responsibility of the ESA) ;
- the High Resolution Spectrometer ;
- the Faint Object Spectrograph which will reach the 24th-25th magnitude ;
- the High Speed Photometer to the construction of which Canada participates.

All these projects, she said, have a strong international implication through Guest Observers Programs.

She mentioned the non yet aproved programs envisaged for the mid 1980's or the second half of the decade :

- the Solar Optical Telescope (SOT) a 1.25 m instrument with associated focal plane instruments ;
- the Gamma-Ray Observatory (GRO) ;
- the Shuttle Infra-Red Telescope Facility (SIRTF) a 80 cm cryogenically cooled telescope.

3. The International Ultra-Violet Explorer (IUE)

R. Wilson gave a complete description of the spacecraft and the instrument, together with the characteristics of the orbit and the two ground stations. He also described in detail the allocation of observing time in the various scientific areas. A list of 12 countries besides ESA countries, the U.K. and the U.S.A. was given including 2 Eastern countries clearly showing the genuine international character of the program. This presentation was followed by a discussion during which the need for a better distribution of the information on already accomplished observations, was stressed.

4. The Infra-Red Astronomical Satellite (IRAS)

The IRAS was presented by R. van Duinen. It is a cooperative venture between the Netherlands, the U.K. and the U.S.A. The scheduled launch year is 1981. IRAS will make a survey of more than 95% of the sky in four spectral bands between 5.5 μm and 120 μm. It will have pointing and rastering capatilities allowing the study of individual sources. Spectrophotometry with moderate spectral resolution will be possible. The sensitivity of the instrument will be 100 to 1000 times that of presently available systems. The lifetime will be limited by the on-board cryogenics to 1.5 years. The satellite will be placed on a circular polar orbit.

Van Duinen also described the operations of the instrument and the division of work between the three countries involved. It is expected that a survey catalogue will be available one year after completion of the survey itself. 40% of the observing time outside the survey will be allocated to Guest Observers.

5. Discussion

N. Roman then introduced and led a discussion on the problems and procedures for selecting experiments. She first described the in-house NASA procedure (Announcement of Opportunity, Peer Group Review, NASA in-house Review and Recommendation, Selection by the Associate Administrator for Space Science). She also reviewed the procedures for selection of Guest Observers Programs and indicated that for the case of SPACELAB Instruments, the policy is to compose payloads of 80% reflown instruments.

The discussion then focused mainly on how to disseminate the information on Space Programmes to the largest community. The release of information through Scientific Journals and the system of Newsletters was mentioned as one possible efficient way.

SESSION II : Scientific Needs for Space Astronomy

This Session was chaired by R. van Duinen. The topic was introduced by L. Goldberg. The text of his presentation being reproduced in its integrity on page of the present volume, it is not summarized here.

A discussion ensued on the rationale for Space Astronomy. The question : "Can we do the same things from the ground ?" should be the first to be raised before initiating a complex and expensive program of space astronomy. R. C. Catura stressed the need for an all-Sky X-Ray Survey, using an imaging Telescope, as a resource in itself for astronomy and in guiding future large X-Ray observatories already in the planning stage in various agencies. H. Smith mentioned that up to now Radio Astronomy has not really taken advantage of observations from space and that, as a return of the of large space structures it might become cheaper in the future to build large antennas offering interesting possilities for astronomy.

SESSION III : Potentialities for Space Astronomy of the New Generation of Space Transportation Systems and Platforms.

The Session was chaired by R.M. Bonnet and was introduced by K. Pinkau. He presented the subject in continuity with the discussion of Section II. E. Peytreman kindly served as secretary. In essence, Professor Pinkau said, all presently planned or already available systems are sufficient to carry into orbit most of the envisaged scientific payloads. He identified however 3 main areas where the needs are not generally yet fulfilled and where the technology or the policy followed by the various space agencies should be improved and changed :

- there are requests for carrying many small size experiments in contrast to the availability of few large observatories ;
- the technology is not yet ripe to generally allow the external users of an instrument (in contrast to the Principal Investigator) to make use of the data which may be made available to them ;
- large observatories become more and more costly and within the fixed ceeling of the national or international budgets this means that flight opportunities tend to become more and more scarce. The question arises then of how to keep the groups alive and of how to maintain their know-how in space technology.

Concerning the first point, SPACELAB should allow for carrying small size type experiments at rather frequent intervals provided that the cost of integration, flight and operation be kept in reasonable limits. There seems to be no technical limitations at present to carry out any kind of astronomy instruments into space but there are strong money limitations instead, and it is in the direction of making space vehicles and experiments cheaper that all efforts should be made.

Following this presentation the discussion focused on two main questions :

1. Is there a need for and means to carry into space small payloads, as opposed to large observatories ?

A. Underhill said that the answer varies according to the degree of sophistification already achieved in certain disciplines.

The discussion then centered on the use of SPACELAB and its cost. Everybody agreed that SPACELAB was a good way of carrying small payloads in particular in that it offers a longer observing time than rockets. S. Jordan said that SPACELAB flights and seven day missions are sufficient for most solar physics payloads and Dr. Carruthers said that SPACELAB offered capabilities that do not exist elsewhere. According to N. Roman, the cost of an experiment is two to three times more expensive on SPACELAB than on a rocket, therefore the cost should be evaluated considering what flight time is indeed required. Dr. Brueckner expressed the optimist point of view that we are now in a learning stage and that the situation should normally improve and experiments tend to become cheaper.

In conclusion, the need for small/cheap short duration frequently flown experiments was clearly identified.

2. Are the available technical means sufficient ?

The discussion focused on platforms. Dr. H. Smith said that there are several arguments in favour of platforms provided the latter be kept simple. Pr. Pinkay said that the concept of platforms was lacking flexibility. The advantage of revisitable payloads was stressed in the discussion.

The Chairman concluded the Session in pointing out that the cost of space astronomy experiments should be kept low in as much as possible. The tendency of increasing costs makes space astronomy less and less popular and the experimental groups are faced with increasing difficulties in getting funds and flight opportunities frequent enough. He said that the whole astronomy community should be aware of this situation since space astronomy is now an indispensable segment of astronomy in general.

17 August 1979

BUSINESS MEETING Secretary : E. Peytreman

1. Commission 44 Newsletter

Four issues of the Newsletter have been published between 1976 and 1979. Y. Kondo who served as editor could not continue his task and the question of whether we should continue the edition was discussed. The Commission thought that the Newsletter was useful in conveying the information on International Space Astronomy Programmes and that this purpose at least should continue to be served.

The President asked volunteering editors but no one in the room thought he had the necessary administrative and financial support to take over the job. The question of whether the information Bulletin could be used was discussed. This solution was seen as offering the advantage of a larger audience. The President said that he would discuss the matter with the General Secretary.

The Commission members mandated the President and his successor to find a solution in the next months.

2. Relation with COSPAR

Dr. L. Peterson reported on the re-organization of COSPAR. He said that space astrophysics was now covered in Commission E itself subdivided into 2 subgroups. He informed the Commission that one concern was to avoid duplication of meetings and that there should not be any Commission E meeting in the year when the IAU General Assembly meets. He also outlined the major symposia of the next COSPAR meetings scheduled for June 1980 in Budapest.

The liaison between Commission 44 and the new Commission E was discussed and it was agreed that the duplication of activities should be avoided by for example, having common members in the organization Committees of both Commissions.

3. Name and re-organization of the Commission

The question of the renaming Commission 44 was discussed and the Commissi Commission agreed to adopt the new name "Space Astronomy" (Astronomie Spatiale). The suggestion of subdividing the Commission into subcommissions or working groups was discussed but not retained at that stage.

4. Organizing Committee

The President re-opened the discussion on the composition of the Organizing Committee. The Commission members agreed on a list of names which was then submitted to the General Secretary.

It was found that 4 members were also members of COSPAR Commission E.

5. Resolutions

The text of two resolutions was submitted to the Executive Committee.

The first resolution encouraged members of the IAU to support the concept of a one meter class Space Schmidt Telescope. The second expressed a sustained interest in the continuation of coordinated observations of the sun in the next years spanning the period of maximum activity.

6. Dr. Y. Kondo reported to the Commission members the outcome of a meeting of Commission 5 on nomenclature of new sky sources. This matter was found relevant to the activities of Commission 44 and those who had any suggestions were requested to convey them to Dr. C. Jaschek.

7. The Commission complained that it was involved in too many meetings. It was consequently agreed that this situation should be corrected for in the future by for example not accepting too many proposals of joint meetings and discussions and avoiding as much as possible duplications of topics between two consecutive general assemblies.

COMMISSION 45: STELLAR CLASSIFICATION
(CLASSIFICATION STELLAIRE)

Report of Meetings, 14, 15, 16, 17, 18 and 21 August 1979

PRESIDENT: B. Hauck SECRETARY: A. Maeder

I. JOINT MEETINGS

14 August: Commission representatives attended the meeting on Astronomical Nomenclature: Stellar, Galactic, Extragalactic Objects, held by Commission 5 and representatives of Commissions 8, 24, 25, 26, 27, 28, 29, 30, 34, 37, 40, 42 and 45.

16 August: A meeting on Space Astrometry was held with Commissions 24, 8, 20, 25, 26, 30, 33, 37 and 44.

17 August: Commission 45 held a joint meeting with Commission 29 on Am stars, the full report of the meeting being given below.

18 August: Meeting with Commissions 29 and 36 on Stellar Abundances

21 August: Meeting with Commissions 29 and 36 on Stellar Rotation.

These scientific meetings are outlined, in each case, in the report of the first listed Commission.

Commission 45 also cosponsored the Joint Discussion on Ultraviolet Astronomy held on 15 August.

17 August

II. ADMINISTRATIVE MEETING

1. Activity Report

The President reported on the two meetings organized in cosponsorship by Commission 45, namely IAU Symposium No. 80, "The HR Diagram" held in Washington from 2 to 5 November 1977 and dedicated to H.N. Russell, and IAU Colloquium No. 47 "Classification of the Future", held at Vatican City and dedicated to A. Secchi. A workshop on problems of calibration of multicolour photometric systems was also held in spring 1979 at Dudley Observatory and the proceedings have been published as Dudley Obs. Report No. 14.

The Commission Report published in the Transactions of the IAU, Vol. 17A gives an account of the Commission's activities; particular mention is also made of the new atlas of Drs. Morgan, Abt and Tapscott and of that of Dr. Keenan. The report was approved unanimously.

Dr. C. Jaschek also reported on the activities of the Working Group on Spectroscopic and Photometric Data, which publishes bi-annual lists of catalogues recently published or in preparation. The six lists published since the Grenoble Assembly contained a total of 150 catalogues and six atlases. The different astronomical data centres have prepared a common list of all catalogues existing on magnetic tape.

2. Name of the Commission

Dr. Wing (OSU) had addressed a proposal to the President for a change of name of the Commission. The Organizing Committee and members consulted all agreed with the fact that it was necessary to change the name of the Commission: "spectral classification" is related to a subject area, while "multi-colour band indices" is not on the same level. The proposals to change the name of the Commission and to adopt the name: STELLAR CLASSIFICATION (CLASSIFICATION STELLAIRE) were both accepted by a large majority.

3. Resolution

Commission 45 voted to support the proposal of Commission 24 and the addendum proposed by Commission 25 to have an IAU resolution on astrometric programmes.

4. Membership and Organizing Committee

Seventeen new members were proposed and accepted: Buser R. (Switzerland), Christy J. (USA), Clariá J.J. (Venezuela), Gerbaldi M. (France), Gurzadyan G. (USSR), Hartoog M.R. (USA), Hayes D.S. (USA), Heck A. (Belgium), Levato H. (Argentina), Mead J. (USA), Parsons S. (USA), Richer H.R. (Canada), Stock J. (Venezuela), Strobel A. (Poland), Warren W. (USA), Wyckoff S. (UK), Zdanavicius K. (USSR).

The new Organizing Committee was elected as follows: President: A. Slettebak (USA); Vice-President: V.L. Straizys (USSR); A.L. Ardeberg (Sweden), R.A. Bartaya (USSR), R.A. Bell (USA), R.F. Garrison (Canada), B. Hauck (Switzerland), M. Jaschek (France), A.G.D. Philip (USA).

Dr. C. Jaschek was reelected chairman of the Working Group on Spectroscopic and Photometric Data, the composition of which remains unmodified with the exception of the retirement of W.P. Bidelman.

5. Activity for the next years

A. Slettebak was willing to continue to inform members by sending circular letters and recommended members to send him any material. C. Jaschek would appreciate it if these letters contained the list of national meetings in the field of the Commission. M. McCarthy requested the same for summer schools. B. Hauck thought it best if everybody informed the President.

Future meetings: a) E.E. Mendoza had proposed a Symposium on IR Photometry and Spectroscopy to be held in 1981 at Mexico. This symposium will also be sponsored by Commissions 25 and 29. Commission 45 voted unanimously in favour of Dr. Mendoza's proposal.

b) R.F. Garrison proposed a Colloquium on MK Spectral Classification Criterion and Application to be held in June 1983 in Toronto. K. Nandy asked whether information from the UV would be incorporated, the reply being that the Scientific Committee would decide. The assembly agreed to support the colloquium.

c) M. McCarthy mentioned R.A. Bartaya's suggestion to have a meeting on automatic classification. P.C. Keenan pointed out that it could be very useful to discuss spectral classification in areas of the HR diagram, where classification breaks down. In the discussion which followed, W.P. Bidelman pointed out that the Commission could endorse a meeting on UV classification.

6. Various items

a) W.P. Bidelman, who was the Commission's representative at the session on Astronomical Nomenclature held on 14 August, reported in detail on the main recommendations and decisions taken during that session.

b) P.C. Keenan pointed out a small change in the designation of the brightest supergiants, which henceforth will be noted Ia $\overline{0}$ (instead of Ia 0 or Ia$^+$). B. Hauck commented on the difficulty of this representation for catalogues on magnetic tape.

c) B. Hauck reported that Commission 29 had envisaged the creation of a Working Group on Ap stars and suggested that this Working Group be common to both Commissions 29 and 45. Members elected by Commission 45 to this Working Group are: W.P. Bidelman, M. Jaschek, M. Gerbaldi, A. Heck, B. Hauck, E.E. Mendoza.

7. Discussion

a) J.R. Lesh reported on the applicability of MK standards in wavelength regions outside the photographic region and at dispersions appreciably different from 100 Å/mm. She emphasized some principles used in the "Revised MK Spectral Atlas for Stars Earlier than the Sun" by W.W. Morgan, H.A. Abt and J.W. Tapscott, published in 1978, continuing with the following important remarks:

"The message contained in this text is clear: if one wants to do MK classification in any wavelength region, one uses MK standards, because the system is defined by the standard stars. The process of transferring the MK grid to a new region may, of course, be a delicate one, and requires the same degree of thoughtfulness and thoroughness as any other pioneering operation in scientific research. [...]

There is, of course, always the possibility that someone may want to set up an entirely new classification system in another wavelength region (or even in the ordinary photographic region), and that the new system may be more useful for some purposes than the MK system itself. In that case, the classifier will choose whatever standards and/or criteria he wishes. The only caveat is that such an independent classification system should not be expressed in the MK notation (especially by using the Roman numerals for luminosity classes). Using the same notation for spectral types obtained by fundamentally different methods does a disservice both to the MK system and to the new system, and defeats one of the basic purposes of spectral classification itself: namely, to provide a large body of homogeneous data concerning the normal stars."

b) R. Nandy presented a report on Spectral Classification of Faint Stars, emphasizing the main points in this extension of spectral classification.

III. SCIENTIFIC MEETING

1. Joint Meeting with Commission 29 on Am Stars

K. Kodaira (Japan) reviewed the spectroscopic works devoted to the decomposition of Am binary systems since 1976 and discussed the interrelation between the Am property and the δ Scuti variability concerning the decomposed individual stars. Finally, the observations indicating the spectral inhomogeneities on the surface of Am stars were presented.

M. and C. Jaschek (France) reported, as they have shown in previous papers, that Am stars can be detected in the ultraviolet of the TDI spectra, in the wavelength region λλ1350-2750. Spectral type and luminosity may also be determined (see Cucchiaro et al., *Astron. Astrophys. Suppl.* 35, p. 75). However the spectral type determined with two different criteria gives two different types. This parallels the situation in the classic λλ3600-4800 region.

H. Levato (Argentina) reviewed the latest results contained in Abt's statistical study (*Astrophys. J.* 230, 485; 1979), which is primarily based on Abt and Levato's extensive observations of open clusters and associations. Among the main results, it was found that Am stars appear in clusters as young as $10^{5.7}$ y., that there is no simple dependence of the frequency of Am with age of the cluster. The decrease of rotational velocities with age and the evolutionary sequences leading to Am stars were discussed.

A.A. Boyarchuk (USSR) reported on a survey of the physical characteristics of the atmospheres of Am stars, in which the similarity of parameters and structures of atmospheres of Am and A stars has been shown and anomalies of chemical composition discussed. The deficiency in C, O, Mg, Ca and Sc along with the excess of heavier than Ni elements with respect to Fe has been pointed out and the excess of Fe with respect to H also noticed. The hypothesis of the origin of observed anomalies was discussed briefly.

R. Kurucz (USA) presented a review on enhanced abundance model atmospheres. He has computed model atmospheres for three and ten times scaled solar abundances for 5500 to 20000K. These are not exactly appropriate for Am stars but they are a considerable improvement over using solar abundance models in interpreting spectra. In particular they have steeper temperature gradients that can considerably affect derived abundances and microturbulent velocities. Colours will be forthcoming. The models give reasonable agreement with the spectrophotometry and Balmer line profiles of Sirius.

B. Hauck (Switzerland) reviewed the properties of Am stars in various photometric systems. He mentioned Kurtz's work on seven very late Fm stars. In general many properties of the Am stars may be investigated using these systems. For example, a correlation of the metallicity index Δm_2 with Δd, the luminosity parameter, at a given temperature, was found for δ Del stars but not for Am stars. It was concluded that some evolved Am stars may be pulsators.

E.E. Mendoza (Mexico) reported that multicolour photometry of Am stars, when compared with normal main sequence stars, shows colour deficiencies in $U-V$ and $B-V$ (0.11 and 0.06 mag. respectively). There is perhaps a small colour excess in $V-R$ and $V-I$ (0.01) but normal colours in $V-J$ and $V-K$. The oxygen (λ7775Å) line is noticeably weaker, and still weaker in Ap stars. The photometric reference index is α(16), which measures the total strength of the H_α line.

C. Burkhart (France) reported that for main sequence Am stars (the early Am excluded), the correlation between the photometric index δm_1 and the projected rotational velocity shows that the degree of metallicity does not disappear progressively with increasing velocity and it seemed that rotational velocity was not the only metallicity parameter.

C. Van't Veer (France) and collaborators have shown that UV photometry of A and Am stars using the UV fluxes given by S2/68 experiment, allows the segregation of Am stars from standards in correlation with metallicity, with possible theoretical model atmosphere interpretation.

A. Heck (Belgium) and collaborators have performed a statistical parallax calibration of the absolute magnitude of Am stars versus several photometric indices. They comment quite satisfactorily giving a mean absolute magnitude of about $1^{m}.2$ with a spread between $0^{m}.4$ to $2^{m}.0$.

COMMISSION 46: TEACHING OF ASTRONOMY (ENSEIGNEMENT DE L'ASTRONOMIE)

Report of Meetings held in Montreal

PRESIDENT: E. Kononovich

SECRETARIES: J. Pasachoff, A. Frankoni, D. DuPuy, D. Wentzel

Session I, 14 August 1979

I. REPORT OF COMMISSION AND NATIONAL REPORTS

The President's report on Commission activities 1976 through 1978 was approved. National reports were received from 26 countries. The reports, together with analysis of the answers on a commission questionnaire and a short list of publications on astronomy education by commission members, are available from E. Kononovich. The length of the reports generally conformed to the decisions made in Grenoble.

II. MEMBERSHIP

With a view to avoiding proliferation of membership (see Grenoble report), the following was approved:

President: D. G. Wentzel
Vice-Presidents: L. Houziaux and M. Rigutti
Organizing Committee: W. Buscombe, C. Iwaniszewska, J. Kleczek, E. V. Kononovich (ex officio), L. Mavridis, S. E. Okoye, B. F. Peery, A. Ringuelet.

National Representatives:

Argentina	Dr. A. E. Ringuelet	Israel	Dr. R. Steinitz
Australia	Dr. A. W. Rodgers	Italy	Prof. E. Proverbio
Austria	Dr. H. F. Haupt	Japan	Prof. N. Owaki
Belgium	Prof. L. Houziaux	Korea (Rep.)	Prof. Kyung Loh Yu
Brazil	Prof. S. Ferraz-Mello	Mexico	Dr. S. Torres-Peimbert
Bulgaria	Dr. N. S. Nikolov	Netherlands	Dr. H. Hubenet
Canada	Dr. D. L. DuPuy	New Zealand	Dr. N. A. Doughty
Chile	Dr. H. Moreno	Nigeria	Dr. S. E. Okoye
Czechoslovakia	Dr. J. Kleczek	Poland	Dr. C. Iwaniszewska
Denmark	Dr. H. E. Jøogensen	Portugal	Prof. J. Osorio
Finland	Dr. K. Lumme	Roumania	Prof. N. Dinulescu
France	Dr. L Gouguenheim	Rep. of S. Africa	
Germany DDR	Dr. H. Zimmermann		Prof. A. H. Jarrett
Germany FDR	Dr. H. Scheffler	Spain	Dr. Mª. A. Catala-Poch
Greece	Dr. L. Mavridis	Sweeden	Dr. Aa. Sandqvist
Hungary	Dr. M. Marik	Switzerland	Prof. B. Hauck
India	Dr. K. D. Abhyankar	Taiwan	Prof. Chun Shan Chen
Indonesia	Dr. B. Hidajat	Turkey	Dr. S. Hazer
Ireland	Dr. T. Kiang		
United Arab Rep.	Dr. A. S. Asaad	U.S.S.R.	Dr. E. V. Kononovich
United Kingdom	Dr. D. McNally	Vatican City	Dr. J. Casanovas
U.S.A.	Dr. J. M. Pasachoff	Yugoslavia	Prof. Dr. F. Dominko

Members: H.L. Andrillat, I. Atanasijevic, K.A. Barkhatova, V. Barocas, L. Bottinelli, H. Campins, S. Codina, J.M. Chamberlain, D. Clarke, E.A. Dibai, D.R. Fawell, M. Gerbaldi, N.P. Grushinski, A. Hayli, V.V. Ivanov, J.E. Kennedy, H. Lambrecht, J.C.D. Marsh, H.G. Miles, E.A. Muller, B. Onderlicka, W. Osborn, N.W. Ovenden, J. Percy, V.V. Porfirev, R.R. Robbins, A.E. Roy, T. Schmidt, B.M. Sevarlic, E.v.P. Smith, S.K. Trehan.

Consulting members: K.P. Addoli, M.L. Aguilar, J. Ebdon, F. Egger, D.M.J. Fubara, N.T. Jiwaji, D. Khaltar, T. Murtagh, F.N. Okeke, L.I. Onuora, K.A. Portzevskij, C. Roslund, E. Schmitter, J. Siroky, A.E. Troche Boggino, R.H. Wilkinson, D.V. Zaitschek.

We should make an attempt to find members from countries not represented on the commission. All interested persons are welcome at commission meetings and may request commission publications, such as the newsletter.

III. NEWSLETTER

Editor McNally reported that five newsletters have been published since the last General Assembly, starting in June 1977. They are published in June and December/January. The print run is about 300. McNally pointed to the developing controversy in the newsletter about Piaget. He hopes that more such controversy will develop.

Kononovich proposed that McNally continue as editor, a proposal that met with general approval. McNally thanked Wentzel and the University of Maryland for undertaking the production and mailing costs.

The members agreed with McNally that the newsletter is valuable. Schatzman said that the name "newsletter" is misleading, that it is more of a journal. He said "Sooner or later this will have to be turned into a journal." A journal on the teaching of physics is being launched in Europe. The "American Journal of Physics" already exists in the USA.

Excerpt from Discussion Paper: The prime aim of the Newsletter is to provide a means of communicating between members of the Commission and others interested in the teaching of astronomy. The newsletter is open to all IAU members to submit their ideas, suggestions and views concerning teaching of astronomy. The newsletter will remain interesting only if the contents become more international. The editor should actively seek guest editorials. Also, more contributions from commission 46 national representatives are greatly needed.

IV. RELATIONS WITH ICSU, UNESCO, COSTED AND OTHER BODIES

Report of L. Houziaux. The ICSU Committee for Teaching of Science met in Nijmegen (Netherlands) April 1978 and Paris, March 1979. It has issued a series of booklets "Learning Strategies in University Science", edited by D. McNally, since it appeared that university science teachers were less aware of many innovative schemes and ideas than their colleagues in other levels of education. Another series of 23 pamphlets, in preparation, concerns the cooperation between teachers of mathematics and teachers of other sciences. They are intended primarily for teaching of science at the school level. The Committee has also issued a Newsletter which reports on the activities of the teaching commissions of the various unions. Three have been issued; the second mentioned our newsletter. The Committee is also participating in the preparation of an international conference on Science and Society in order to show the contributions of science and technology to development in changing societies.

Sources: Learning Strategies in University Science, University College,

Cardiff Press, PO Box 78, Cardiff, United Kingdom; 9 pounds for all ten booklets.

Newsletter of ICSU Committee on Teaching of Science, also booklet "The Importance of Education and Training", 23 pamphlets when finished and Integrated Science Education Worldwide ($5, report on International Council of Associations for Science Education, Nijmegen, 1978): Dr. John L. Lewis, ICSU-CTS, Malvern College, Malvern, Worcestershire, United Kingdom.

COSTED is concerned mainly with problems specific to developing countries. It holds seminars mainly in South East Asia and provides grants for scientists for science education purposes.

Houziaux thinks that it is important for the IAU to be represented in the ICSU committees in order to promote the teaching of astronomy among the various disciplines and to obtain the support of UNESCO for the various activities of commission 46. (See ISYA, below).

President Kononovich felt that the IAU and Commission 46 can be of great value to the other organisations. He recommended that Dr. Houziaux remain our representative. He asked Houziaux to give more frequent reports on the relations to the other organisations.

Session II, 18 August 1979

V. TEACHING SESSIONS AT REGIONAL ASTRONOMY MEETINGS

The following recommendation was sent to the Executive Committee:

Whereas the IAU supports the organisation of regional astronomy meetings, Commission 46 further recommends that the IAU urge the holding of sessions on the teaching of astronomy at such conferences. The members of Commission 46 are available to assist.

Such teaching sessions may well be specific to the main conference topic.

VI. INTERNATIONAL SCHOOLS FOR YOUNG ASTRONOMERS

The report of the school in Nigeria was circulated. During 1976-1978 two schools were held, in Brazil and Nigeria. Both were very successful and fully justified the support given by the IAU.

The ISYA in La Laguna, Tenerife, Spain from 10 to 22 September 1979 will be supported by IAU funds and by a UNESCO contract through ICSU, each $4000, as well as by Spanish sources. There will by 36 participants from Spain, Portugal, Canary Islands and five Latin-American countries.

The 11th school is being prepared for 17 September to 10 October 1980 at Island Hvar, S. R. Croatia, Yugoslavia, with the topic of the physics of the solar system and the stars, for students of Mediterranean and Balkan countries. Funding includes contributions from IAU funds, from UNESCO and from the Scientific Council, S. R. Croatia.

Requests for future schools should go to D. Wentzel and J. Kleczek. Iraq, East Africa, French-speaking Africa, and Venezuela have been mentioned as possibilities.

Amendment of rules for ISYA: The rules for Commission 46 as approved at Grenoble were amended by the addition of a paragraph to rule 1. This rule now reads:

1.) The President, Vice-President(s), Secretary of ISYA and Past President will constitute a sub-committee of Commission 46 to regulate the organisation of the ISYA. The Secretary of the ISYA will be the Secretary of the sub-committee and will keep the sub-committee informed of all proposed ISYA, the details of their organisation and evaluation. The President of the Commission will inform the General Secretary of the Union of the programme for each proposed school only when approved by the sub-committee.

The sub-committee approves the preliminary programme, then the President of the Commission informs the General Secretary. The final version of the programme must take into account the Executive Committee recommendations together with those given by sub-committee members.

VII. ASTRONOMY EDUCATION MATERIALS

Only the Slavic-language portion was available at the meeting. It was prepared by E. Kononovich. Dr. B. Peery is currently writing the English-language material, based on a print-out of all English-language materials by the computer services of the U.S. Library of Congress (but which is too extensive and needs to be evaluated for our purposes). A telegram from Dr. L. Mavridis indicates the material in other languages is ready for distribution.

The collation of material suitable for AEM continues to be the most difficult aspect. Kononovich and Peery agreed to continue collating their parts. Prof. Houziaux is to investigate possible help through his government. If possible, all three parts of AEM are to be published together, in a single format, with a view to commercial publication of subsequent issues. They should no longer be called Addendum, since the original volume is out of print and largely outdated. Thus some permanent sources, for instance of visual aids, should be repeated in each edition.

D. Schatz mentioned that a good beginning list of astronomy education materials is included in the workbook "Effective Astronomy Teaching and Student Reasoning Ability" available from the Astronomical Society of the Pacific (address below), $3 for IAU members.

A. Fraknoi summarized the activities of the Astronomical Society of the Pacific (1290 24th Ave., San Francisco, CA 94122, USA). These include i) "Mercury" magazine, which has a regular feature on astronomy education, ii) an internationally syndicated weekly newspaper column on astronomy (500 words) available for newspapers everywhere, iii) a series of resource materials for teachers including bibliographies, iv) a mail order catalogue of educational materials.

The association of French teachers in Physics publishes "Les Cahiers Clairaut", which provide exercises, review papers, etc. in astronomy for science school teachers.

VIII. PROJECT CONTRATYPE

M. Gerbaldi reported on her pilot scheme, in the period 1975-76, of distributing slides from a collection of 56 slides, at cost, from the Institut d'Astrophysique, Paris. The greatest success in reaching the developing countries was obtained from an announcement in the Courier de l'UNESCO, June 1976, which yielded 37 letters from developing countries, mostly from South America (plus 25 from developed countries other than France). She suggests two problems must be overcome: i) While there are commercial slides available, teachers from developing countries do not have the foreign exchange. There exist "UNESCO coupons". ii) The range of slides is huge and of variable quality, so that a critical analysis is needed, which could appear in our newsletter and be announced in the UNESCO bulletin. M. Gerbaldi agrees to start the analysis.

M. Gerbaldi and L. Gouguenheim have assembled two sets of 24 slides each, on the subjects of gravitation and of spectral analysis. Each set comes with a booklet of 40 pages. The booklet gives not only a description of the slides but an analysis from a physical point of view, which they consider their most important contribution. The booklet is in French. Can UNESCO provide translations so that wider distribution is possible? Source of the slides:

Centre National de Documentation Pedagogique
Service de Vente des Publications de l'Education Nationale
13 rue du Four
75006 PARIS - FRANCE

Serie: Diatheque Sciences Physiques
Astrophysique I: la loi de la Gravitation dans l'Univers.
Astrophysique II: Connaissance des Astres par leur rayonnement.

Kononovich in the discussion paper on Project Contratype emphasized the need of a complete treatment of slides from both an educational point of view and practical usage in the classroom, possibly through special publications in one of the well known astronomy magazines.

IX. VISITING PROFESSORS PROJECT

Dr. Rigutti commented that little interest exists for the project in its present form of merely listing opportunities for visits.

A sub-committee of Kononovich, McNally, Wentzel and Okoye formulated the following recommendation to the IAU Executive Committee:

Commission 46 having regard to the need a) to introduce astronomy throughout the world, b) to strengthen developing and existing teaching in astronomy, and c) to enhance awareness of the place of the Earth in the solar system and the wide universe, proposes that the IAU establish an Annual Visiting Lectureship.

The lecturer should be an astronomer of scientific distinction. He/She should also be particularly gifted and interested in public appreciation of astronomical topics. Cognisance should also be taken of fluency in the appropriate language of the country to be visited. The programme for a typical visit might involve 3 public lectures and 1 scientific seminar, during the course of 1 month duration in the country. Where possible, the visiting lecturer should be based at a local university or institute of higher learning. The lectureship should be used to promote knowledge and understanding of astronomy, particularly in the countries having little or no formal astronomy.

The cost to the IAU should be limited to a) transportation, b) subsistence, and c) a small honorarium to the lecturer. It would be hoped that the subsistence would usually be met through local hospitality.

The proposal should be seen as falling within the auspices of Commission 46 and the Visiting Astronomers Program of Commission 38. Both commissions are to be consulted over choices of lecturer and venue.

It was suggested that perhaps a portion of funds for Commission 38 could be used for this purpose.

The proposal is to be considered also by the working group on education of astronomers from developing countries (see below).

X. EDUCATION OF ASTRONOMERS FROM DEVELOPING COUNTRIES

Dr. Sandqvist (Sweden) asked the commission to support the resolution to the IAU General Assembly by the Swedish National Committee, which is to encourage fellowships for astronomy students from developing countries. Dr. Rigutti (Italy) urged that this be supported by funds from IAU and possibly from UNESCO. S. Ferraz-Mello (Brazil) pointed out that the mere training of students without a supporting home institution might not succeed in fostering astronomy in a developing country. After some discussion, President Kononovich summarized that it was the sense of the meeting that the Commission approves in principle.

A working group on the education of astronomers from developing countries was formed (see below).

XI. FUTURE MEETINGS OF THE COMMISSION

Incoming President Wentzel thanked Dr. Kononovich for all his hard work as President and noted in particular the very informative collection of National Reports. Several countries now hold meetings and workshops to help teachers to include more and better astronomy in their classes. The topic of teaching astronomy to teachers may be a good one for a session at the next General Assembly.

Session of Working Group on Education of Astronomers from Developing Countries, 22 August 1979

The working group is to consist of S. Ferraz-Mello (Brazil) as chairman, B. Hidajat (Indonesia), J. Kleczek (Czechoslovakia), S. Okoye (Nigeria), Aa. Sandqvist (Sweden), and two members from Commission 38, J. Delhaye (France) and M. Kaftan (Iraq).

The working group is to discuss the training of astronomers in countries with little or no formal astronomy and by August 1980 produce a draft, with justification, of the one or two most practicable proposals that the IAU might support. The deliberations should include the Swedish proposal referred to Commission 46 by the IAU Executive Committee. The President of Commission 46 will seek comments upon the draft, so that the working group can present a revised proposal by December 1981.

Address for comments on these topics: Dr. S. Ferraz-Mello, Universidade de Sao Paulo Departamento de Astronomia, Caixa Postal, 30627, 01000 Sao Paulo S.P., Brazil.

Discussion amoung group members (with B. Wood, D.A. MacRae and D. Wentzel also present) stressed the great variety of needs that might apply to various countries. It was decided first to enquire among sympathetic individuals in a small number of countries as to their local possibilities for introducing astronomy in their countries.

Session III, 22 August 1979
Teaching Astronomy at the University Level

An all day meeting was held to discuss the teaching of astronomy at the university level. Attendance was approximately 100-125, with good representation from a broad range of countries.

Four invited speakers participated. First, G. Abell (U.S.A.) presented "An Overview of Astronomy Education" with an emphasis on the goals of teaching graduate students, e.g. preparing some graduate students expressly for teaching in small, non-PhD granting institutions. He also stressed the opportunities for astronomers to reach large numbers of non-science students, through radio/TV programs as well as university and community college courses. The next invited speaker was H. Eichhorn (U.S.A.) who spoke on "Teaching Astrometric Concepts." He emphasized teaching astrometry by tieing it in with astrophysics.

The third invited speaker was J. Chamberlain (U.S.A.), who spoke on "The Use of the Planetarium in Teaching University-Level Astronomy." He discussed non-credit astronomy courses taught by planetariums, and instruction versus entertainment in planetarium shows. The final invited speaker was D. Clark (Scotland), who spoke on "Teaching Observational Studies;" he illustrated his presentation with examples from his highly successful teaching program at Glasgow.

Fifteen contributed papers were presented by teachers from eight different countries. Among these, Sandqvist (Sweden) demonstrated his innovative umbrella/star chart, and he described an impressive program in teaching astronomy to the public in Sweden. Pierce (U.S.A.) described how teaching astronomy is a community college (virtually unknown in most other countries) differed from the more usual university course. Two representatives from the University of London (Doretsky and McNally) described a student spectrograph, and mathematics in teaching astronomy. Iwaniszewska (Poland) reported that astronomy is a required course in secondary schools in Poland, and Kononovich (U.S.S.R.) described secondary school astronomy in the U.S.S.R.

13 August 1979
Session With Canadian School Teachers

About 35 teachers attended, 50% from Ontario, 30% from Quebec; others from as far as Saskatchewan and British Columbia.

Opening remarks by Dr. Kononovich included an outline of the role of astronomy in the school curriculum of the USSR and a summary of "Tendency of Astronomical Education", results of an international survey (available with the National Reports, see item I.) Dr. J.R. Percy (University of Toronto) discussed the place of astronomy in high school curricula and the role of astronomy in the Canadian high school education. As a comparison, Miss L. Gouguenheim (Observatoire de Paris) presented a paper on teaching of astronomy in French schools. Dr. R.L. Bishop (Acadia University, Canada) introduced the role of the Royal Astronomical Society of Canada in teaching of astronomy.

Practical aspects were presented by Dr. R. Robbins (University of Texas, USA) on audiovisual materials, Dr. W. Osborn (Central Michigan University, USA) on observational activities for school teachers, with emphasis on observations with simple, inexpensive equipment, and Dr. J. Holzinger (Franklin and Marshall College, USA) on laboratory experiments in astronomy. Parallel sessions concerned telescopes for teaching in schools with specific examples (in English) and astronomy in the Quebec curriculum (in French). D. Levy (Canada) reported on teaching young children in summer camps, using color slides and taped commentary.

The meeting concluded with discussions based on the Piaget Workshop with applications in teaching of astronomy, as organized by D. Schatz (Pacific Science Center, USA).

COMMISSION 47: COSMOLOGY (COSMOLOGIE)

Report of Meetings 18 abd 21 August 1979

PRESISENT: I.D. Novikov

18 August 1979

BUSINESS MEETING

The President described the Commission activities since the last General Assembly. The Commission had co-sponsored (with Commission 28) IAU Symposium No. 79 ("The Large-Scale Structure of the Universe") held at Tallinn, Estonia, USSR, 12 - 16 September 1977. The Symposium was attended by more than 160 astronomers, and considered the current state of knowledge on the large-scale distribution of matter in the universe. The Proceedings have now been published by D. Reidel.

The Commission was also co-sponsoring (with Commissions 28 and 48) Symposium No. 92 ("Objects of High Redshift") to be held in Los Angeles, California, USA, 28 - 31 August 1979. Dr. Reidel will also publish the proceedings of that conference.

The President thanked certain members of the Organizing Committee and others who assisted in the preparation of the Draft Report. (These are G.O. Abell, M.S. Longair, S. Hayakawa, A.G. Doroshkevich, Ya.B.Zel'dovich, A.A. Starobinsky and Y. Ne'eman.) There followed the election of President, Vice President, and members of the Organizing Committee for the period 1979 to 1982. The results are as follows:

President: G.O. Abell
Vice-President: J. Audouze
Organizing Committee: G. de Vaucouleurs, J. Gunn, S. Hayakawa, M.S. Longair, I. Novikov, G. Setti, and G.A. Tammann.

The President announced that one Commission member, V.A. Ambartsumian, felt that he could no longer remain active, and had asked to resign from the Commission. A list of prospective new members was presented and discussed; Some additional names were suggested by Commission members present. After final presentation, screening, and seconding, these candidates were elected Commission members. Their names are included in the full list of Commission members, which will be found elsewhere in this volume.

SCIENTIFIC SESSION — New Observations Relevant to Cosmology (Chair: I. Novikov)

A. Penzias: Primordial Deuterium in the Universe.

J. Audouze: The Light Elements.

G. Smoot (read G.O. Abell): The Large-Scale Anisotropy of the Microwave Background.

Yu. Parijskij: The Small-Scale Anisotropy of the Microwave Background.

R.D. Davies:	Jodrell Bank Measurements of the 3 K Background.
I. Novikov:	Further Comments on the Anisotropy.
G. de Vaucouleurs:	Status of H and the Local Anisotropy.
B. Tinsley:	Recent Work on Galaxy Counts

21 August 1979

SCIENTIFIC SESSION — Theoretical Considerations in Cosmology
(Chair: B. Abell)

J. Wheeler:	Physics Near the Gates of Time.
I. Segal:	Chronometric Cosmology.
F. Ahmed:	Some Thoughts on Cosmology.
G. de Zotti:	Cosmic Distances from X-ray and Microwave Measurements of Clusters of Galaxies.

COMMISSION 49: THE INTERPLANETARY PLASMA AND THE HELIOSPHERE (PLASMA INTERPLANETAIRE ET L'HELIOSPHERE)

Report of Meetings, 14 and 20 August 1979

PRESIDENT: A. Hewish VICE-PRESIDENT: H. J. Fahr

14 August 1979

ADMINISTRATIVE SESSION

The President welcomed members of the Commission and recorded with sorrow the death of S. F. Smerd. The names of Officers were proposed and accepted by the Commission, and the list of new members was presented. The President thanked the three members responsible for preparing the Report. It was suggested that in future the report should be circulated to members before the meeting. It was also suggested that the authors should, in future, be invited to present the highlights of their reviews during the scientific meetings.

20 August 1979

SCIENTIFIC PRESENTATIONS

The following invited review papers were presented.

1. D. S. Intriligator: Observations of the distant solar wind.
2. J. L. Linsky: Properties of the nearby interstellar medium.
3. G. Morfill: Dust within the heliosphere.
4. H. J. Fahr: Plasma-gas interactions in the distant solar wind and solar wind modelling.
5. M. A. Raadu: Critical velocity effects in astrophysical plasmas.

The following communications were presented.

M. K. Bird: Faraday rotation of pulsar signals during solar occultation.
E. Petelski: Generalisation of Parker's heliospheric shock equations.

COMMISSION 50: IDENTIFICATION AND PROTECTION OF EXISTING OR POTENTIAL OBSERVATORY SITES (IDENTIFICATION ET PROTECTION DES SITES D'OBSERVATOIRES EXISTANTS OU POTENTIELS)

Report of Meetings, 17 August, 1979

PRESIDENT: R. Cayrel SECRETARY: J. Kovalevsky

I. ADMINISTRATIVE SESSION

1. Election of new officers, new organizing committee, membership

The meeting was called to order by the President at 9:15 a.m.

The Commission unanimously approved the proposal of the organizing committee for the following new officers: President, F.G. Smith; Vice-President, A.A. Hoag.

The composition of the new organizing committee, the replacement of Dr. Treanor (deceased) by Dr. J. Casanovas (Vatican state) and the cooptation of A.J. Fisher, President of the Committee of Road Lighting at the CIE, as consultant, were also unanimously approved. It was felt that other changes should require further steps through the national committees each of which has been asked, in the first place, to appoint one member at Commission 50.

It was agreed that the new president will, in due time, ask the national committees if they wish to update their initial choice.

2. Report on the activities of the Commission

The President called for comments or remarks from the floor on the document entitled "Report and Recommendation of Commission 50" prepared at the request of the 1976 General Assembly and issued in June, 1978.

Several directors of observatories have already made use of this report in their negotiations with local authorities to preserve observing conditions at their observatories and found it useful. However, it was pointed out that the document should not be thought of as a panacea, and that in each particular case decisions have to be made at the local, provincial or national level.

It was also indicated that in Recommendation 1 of the Report, the natural level of continuum radiation near 550 nm was correctly stated as 10 Rayleigh per nm, but translated afterwards into improper units. The number 2.10^{-11} stilb and 2.10^{-7} Nit should be read 2.10^{-11} stilb $Å^{-1}$ and 2.10^{-7} Nit $Å^{-1}$.

This removes the apparent discrepancy with the currently quoted sky brightness of 2.10^{-4} Nit for the total visible light, the effective band width of the eye response being of the order of 1000 Å (or 100 nm).

A second report has been drafted to deal in more detail with practical ways to implement light control near observing sites even in cases where some degradation beyond the ideal limits has already taken place.

It has been thought that a report jointly sponsored by the IAU and the CIE (Commission Internationale de l'Eclairage) would be of greater weight and a draft was distributed to the members of the Commission for comments to be received by R. Cayrel before the end of September, 1979.

3. ICSU-UNESCO Grant

It was indicated that a grant of $5,000.00 had been made available to the IAU from a UNESCO program in order to gather, edit, and publish the information contained in the two reports mentioned in the above paragraph.

4. Relations with COSPAR

COSPAR has established a panel called PEDAS (Potentially Environmentally Detrimental Activities in Space) which is to report to the United Nations Committee on Peaceful Uses of Outer Space. The advice of Commission 50 is requested on space activities which might interfer with astronomical observations. The President-elect will be in charge of pursuing the matter with the advice of the O.C.

5. Miscellaneous

It was agreed that Commission 50 will make sure that recommendations for the protection of radio-astronomy observations, even if studied in the first place by Commission 40, will be ultimately included in Commission 50's report.

II. SCIENTIFIC SESSION

The following presentations were made:

W. Howard: meeting of the NSF Committee on site testing and preservation

P. Edwards: site testing in New Zealand

M. Roth: a new site in Mexico, Teotepec

F.G. Smith: a new ordinance controlling La Palma (Canary Islands)

M.F. Walker: seeing monitoring: the polaris trail method

R.G. Bingham: seeing monitoring: wave-front analysis

Bill Howard reported on a meeting of the National Science Foundation *ad hoc* committee on site testing and preservation held in Tucson, Sept. 18-19, 1978. The committee was concerned by gathering more complete and systematic information on good existing or potential sites under U.S. control. A request was made for additional seeing measurements, sky brightness and meteorological conditions. However, it was felt that it was not very likely to find anything significantly better than already identified sites. A special request was expressed for the protection of Mauna Kea Observatory (first-ranked site) and for Junipero Serra, California, considered as prime sites for ground-based astronomy. The preservation of the quality of other astronomical existing sites was also encouraged.

Paul Edwards reported measurements of atmospheric extinction made in New Zealand. The proportion of the extinction due to aerosols has been determined and found to be 5% at the zenith.

M. Roth reported on a site study carried out by a group at the Instituto de Astronomia (UNAM - Mexico). The site is the mountain Teotepec (3,500 m) located 100° 08' W, 17° 27' N. The site was picked up from an infrared satellite survey.

It is fairly good during the dry season, partly useable during the rainy season, thanks to the fact that it is frequently above the cloud layer.

F.G. Smith reported on a clause protecting the site of the International Astronomical Center at La Palma (one of the Canary Islands). This clause is part of an intergovernmental agreement signed on May 26, 1979, by the governments of Spain, Denmark, Sweden, and U.K. It states that the Government of the Kingdom of Spain shall guarantee the protection of the research work and "shall maintain the astronomical qualities of the observatories and endeavor to adhere to the recommendations of the International Astronomical Union".

It is the first time that the recommendations of the IAU are playing a major role in a site protection ordinance.

M.F. Walker summarized the broadly used "polaris trail method" and indicated its advantages and limitations. As the method has been already described in the literature (pub. A.S.P. 82, 672 and 83, 401) we have not given more details in these proceedings.

R.G. Bingham presented some new ideas on wave-front measurements with the goal to obtain more rigorously statistical characteristics of the wave-front. However, the practical aspects remain a little bit of a problem.

A general discussion followed on seeing monitoring devices.

WORKING GROUP FOR PLANETARY SYSTEM NOMENCLATURE (NOMENCLATURE DU SYSTEME PLANETAIRE)
(Committee of the Executive Committee)

PRESIDENT: P.M. Millman

Establishing an international system of nomenclature for the topographic features on the surfaces of planetary bodies is an exercise that requires assistance from many sources and depends on the cooperation of many individuals. The Working Group for Planetary System Nomenclature (WGPSN) has five nomenclature task groups reporting to it, and these groups prepare the detailed lists of names for the consideration of the WGPSN, which in turn makes recommendations to the Executive Committee of the IAU. The names of those who have been directly involved in preparing this present report (approved at the 17th General Assembly of the IAU in Montreal, Canada, 1979) are listed in IAU Trans. Vol. 17A, pp. 113-114, 1979. The current membership in the nomenclature groups is listed in this volume, p. .

During the three years following the 16th General Assembly in Grenoble, France the WGPSN has held three meetings:-

Fourth Meeting - Washington, D.C., U.S.A., June 1 and 2, 1977;
Fifth Meeting - Innsbruck, Austria, June 2, 1978;
Sixth Meeting - Montreal, Canada, August 13 and 15, 1979.

In addition to the items reported in IAU Trans. Vol. 17A, pp. 113-114, 1979, the following nomenclature rules and terminology have been approved:-

Commemorative names of persons recently deceased shall not be assigned to topographical features on planetary bodies until after a minimum period of three years from the date of death.

The following temporary nomenclature shall be used to announce the discovery of new satellites - year/first letter of primary/arabic number - for example, Janus would have been originally announced as 1966 S 1. A name should be assigned to a satellite only after the orbit has been uniquely determined. Roman numerals are reserved for established members of satellite system, thus an alternative designation for Janus is S X.

Five Latin terms have been added to the 25 listed in IAU Trans. Vol. 16B, pp. 321-369, 1977, for use in nomenclature on the surfaces of planetary bodies (plurals are given in brackets)-

FLEXUS (Flexus)	a very low curvilinear ridge with a scalloped pattern
LINEA (Lineae)	a dark or bright elongate marking, may be curved or straight
MACULA (Maculae)	a dark spot, may be irregular
REGIO (Regiones)	a large area marked by reflectivity or colour distinctions from adjacent areas
TERRA (Terrae)	a rolling or rough upland area

In addition to the 30 approved Latin terms, referred to above, the following two terms have been generally used in specific cases-

OCEANUS	a very large, dark area on the moon
SOLITUDO	a classical, dark-hued albedo feature on Mercury.

LUNAR NOMENCLATURE

Table 1. 54 names, approved as assigned to features on the lunar surface. Positions have been given in degrees of latitude and longitude.

CRATERS			
ALDER, Kurt	1902-1958 German chemist	48.6S	177.4W
ANUCHIN, Demitrii N.	1843-1923 Russian geographer,anthropologist	49.0S	101.3E
ARYABHATA	476-c550 Indian astronomer, mathematician	06.2N	035.2E
BARKLA, Charles G.	1877-1944 British physicist	10.7S	067.2E
BLACKETT, Patrick M.S.	1897-1974 English physicist	37.3S	115.6W
BOK, Priscilla F.	1896-1975 American astronomer	20.2S	171.6W
BORN, Max	1882-1970 German physicist	06.0S	066.8E
BRONK, Detlev W.	1897-1975 American physiologist	26.1N	134.5W
CARREL, Alexis	1873-1944 French surgeon, physiologist	10.7N	026.7E
CARRILLO, Flores N.	1911-1967 Mexican engineer	02.2S	080.9E
CORI, Gerty T.R.	1896-1957 Czechoslovakian/American biochemist	50.6S	151.9W
DE MORAES, A.	1916-1970 Brazilian astronomer	49.4N	141.5E
DIDEROT, Denis	1713-1784 French philosopher	20.4S	121.5E
DOERFEL, Georg S.	1643-1688 German astronomer	69.1S	107.9W
FEDOROV, A.P.	1872-1920 Russian rocket engineer	28.2N	037.0W
FINSEN, Niels R.	1860-1904 Danish physician	42.0S	177.9W
GUILLAUME, Charles E.	1861-1938 Swiss physicist	45.4N	173.4W
HARGREAVES, Frederick J.	1891-1970 English Astronomer	02.2S	064.1E
HARKHEBI	circa 300 BC Egyptian astronomer	39.9N	099.8E
HEINRICH, Wladimir W.	1884-1965 Czechoslovakian astronomer	24.8N	015.4W
HOPMANN, J.	1890-1975 Austrian astronomer	50.8S	160.3E

CRATERS (Cont'd.)

KARRER, Paul	1889-1971 Russian/Swiss chemist	52.1S	141.8W
KEPINSKI, F.	1885-1966 Polish astronomer	28.6N	126.8E
LINDSAY, Eric M. (replaces Dollond C)	1907-1974 Irish astronomer	07.0S	013.0E
LIPPMANN, Gabriel	1845-1921 French physicist	55.5S	114.6W
LIPSKIY, Yu.N.	1909-1978 Soviet astronomer	02.2S	179.5W
MENZEL, Donald H.	1901-1976 American astronomer	03.4N	036.9E
OLIVIER, Charles P.	1884-1975 American astronomer	59.1N	136.5E
PIKEL'NER, S.B.	1924-1975 Soviet astronomer	47.9S	123.3E
PLANTÉ, Gaston	1834-1889 French physicist	10.2S	163.3E
POCZOBUTT, O.M.	1728-1810 Polish astronomer	57.5N	099.3W
RICHARDSON, Owen W.	1879-1959 English physicist	31.1N	100.3E
SABATIER, Paul	1854-1941 French chemist	13.2N	079.0E
SHIRAKATSI, Anania	c620-c685 Armenian geographer	12.2S	128.5E
SIKORSKY, Igor I.	1889-1972 Russian/American aeronautical engineer	66.1S	103.2E
STEARNS, Carl L.	1892-1972 American astronomer	34.8N	162.6E
TAMM, Igor	1895-1971 Soviet physicist	04.4S	146.4E
THEILER, Max	1899-1972 South African bacteriologist	13.4N	082.9E
TISELIUS, Arne	1902-1971 Swedish biochemist	07.0N	176.5E
TUCKER, Richard H.	1859-1952 American astronomer	05.6S	088.2E
VERTREGT, M.	1897-1973 Dutch chemist, space scientist	19.7S	170.9E
VIRCHOW, Rudolph L.K.	1821-1902 German pathologist	09.8N	083.7E
VIRTANEN, Artturi I.	1895-1973 Finnish biochemist	15.5N	176.7E
VON BEHRING, Emil A.	1854-1917 German bacteriologist	07.8S	071.8E

CRATERS (Cont'd.)

VON BÉKÉSY, Georg	1899-1972 Hungarian physicist	51.9N	126.8E
WALLACH, Otto	1847-1931 German chemist	04.9N	032.3E
WILDT, Rupert	1905-1976 German/American astronomer	09.0N	075.8E
ZWICKY, Fritz	1898-1974 Swiss astrophysicist	15.9S	167.6E

DORSA

MAWSON, Douglas	1882-1958 English/Australian geologist	09.0S 050.5E to 06.0S 054.0E
TETYAEV, Mikhail M.	1882-1956 Soviet geologist	16N 067E to 22N 062E

LACUS

PERSEVERANTIA	Lake of Persistence	07.5N to 08.0N 061.0E to 062.5E

MONTES

USOV, Mikhail A.	1883-1933 Soviet geologist	12.0N	063.2E
VINOGRADOV, Alexander P. (replaces Euler β)	1895-1975 Soviet geochemist and cosmochemist	22.4N	032.5W

SINUS

SUCCESSUS	Gulf of Success	00 to 03N 057E to 060E

CATENAE

For three crater chains the designations listed below are recommended for printing on current lunar maps. These provide a cross-reference to the earlier Soviet lunar maps; GDL standing for Gas Dynamics Laboratory, GIRD for Group for the Study of Reaction Motion, and RNII for Rocket Research Institute.

Catena LEUSCHNER (G D L)	16N	123W	to	02N	109W
Catena MICHELSON (G I R D)	03N	126W	to	05S	114W
Catena LUCRETIUS (R N I I)	04S	126W	to	10S	119W

Table 2. 95 first names of men and women, approved as assigned on an international basis to small features on large-scale map sheets of the lunar surface. No name on this list refers to any particular person. Locations are given in degrees, minutes, and in some cases seconds, of latitude and longitude. Where no Latin term is listed the feature is a crater.

Sheet 38B2/S1 (1:50,000)

	° ' "	° ' "
Aloha	29 45 20N	053 54 W
Rima Cleopatra	30 00 N	054 00 W

Sheet 39A1/S1 (1:50,000)

	° ' "	° ' "
Rocco	28 54 N	045 01 W
Ruth	28 42 N	045 05 W

Sheet 39A3/S1 (1:50,000)

	° ' "	° ' "
Ivan	26 51 N	043 16 W
Vera	26 19 N	043 44 W

Sheet 39B2/S1 (1:25,000)

	° ' "	° ' "
Isabel	28 10 N	034 05 W
Louise	28 28 N	034 13 W
Samir	28 29 N	034 18 W
Walter	28 02 N	033 50 W

Sheet 39B2/S2 (1:25,000)

	° ' "	° ' "
Boris	30 31 N	033 31 W
Gaston	30 53 N	033 58 W
Linda	30 41 N	033 23 W

Sheet 39B3/S1 (1:50,000)

	° ' "	° ' "
Courtney	25 08 N	030 49 W
Dorsum Thera	24 30 N	031 25 W
Catena Yuri	24 24 N	030 24 W
Rima Zahia	25 00 N	030 28 W

Sheet 39C2/S1 (1:50,000)

	° ' "	° ' "
Akis	20 02 N	031 46 W
Ango	20 29 N	032 21 W
Jehan	20 42 N	031 53 W
Natasha	20 00 N	031 19 W
Catena Pierre	19 46 N	031 52 W
Rosa	20 19 N	032 21 W
Rima Wan-yu	20 00 N	031 27 W

Sheet 40A1/S1 (1:50,000)

	° ' "	° ' "
Annegrit	29 26 N	025 39 W
Charles	29 53 N	026 23 W
Mavis	29 45 N	026 22 W

Sheet 40A4/S1 (1:10,000)

	° ' "	° ' "
Artemis	25 01 N	025 23 W
Felix	25 04 30N	025 24 30W
Verne	24 56 30N	025 24 30W

Sheet 41A3/S1 (1:50,000)

	° ' "	° ' "
Ann	25 06 N	000 03 W
Ian	25 43 N	000 24 W
Kathleen	25 20 N	000 49 W
Michael	25 03 N	000 12 E
Patricia	24 57 N	000 24 E
Rima Vladimir	25 10 N	000 45 W

Sheet 41B4/S1 (1:50,000)

	° ' "	° ' "
Julienne	26 02 N	003 07 E

Sheet 41B4/S2 (1:50,000)

	° ' "	° ' "
Carlos	24 55 N	002 17 E

Sheet 41B4/S3 (1:50,000)

	° ' "	° ' "
Béla	24 40 N	002 15 E
Jomo	24 25 N	002 25 E
Taizo	24 42 N	002 11 E

Sheet 41C3/S1 (1:10,000)

	° ' "	° ' "
Agnes	18 39 N	005 19 20E
Dag	18 42 N	005 15 20E
Ina	18 38 40N	005 17 30E
Osama	18 36 N	005 15 30E

Sheet 42A4/S1 (1:10,000)

	° ' "	° ' "
Christel	24 32 N	011 04 E
Yoshi	24 33 10N	010 58 40E

Sheet 42A4/S2 (1:10,000)

	° ' "	° ' "
Krishna	24 30 20N	011 15 50E
Manuel	24 27 20N	011 20 30E
Sung-mei	24 36 N	011 16 40E

Sheet 42C3/S1 (1:10,000)

	° ' "	° ' "
Isis	18 56 40N	027 28 40E
Mary	18 55 N	027 23 E
Robert	19 01 20N	027 26 30E

Sheet 42C3/S2 (1:10,000)

	° ' "	° ' "
Catena Brigitte	18 31 N	027 32 10E
Jerik	18 31 N	027 37 30E
Rima Marcello	18 36 N	027 42 30E
Osiris	18 38 20N	027 38 30E
Rima Reiko	18 32 40N	027 42 10E

Table 2 (Cont'd.)

Sheet 42C3/S3 (1:50,000)

	° ' "	° ' "
Rima Carmen	19 53 N	029 20 E
Ching-te	20 00 30N	029 57 30E
Rima Rudolf	19 43 N	029 37 E
Stella	19 54 N	029 45 E

Sheet 61A2/S1 (1:50,000)

	° ' "	° ' "
Diana	14 18 N	035 40 E
Mons Esam	14 33 N	035 44 E
Grace	14 13 N	035 54 E

Sheet 61D2/S1 (1:50,000)

	° ' "	° ' "
Donna	07 13 N	038 18 E

Sheet 65C1/S1 (1:50,000)

	° ' "	° ' "
Mons André	05 04 N	120 35 E
Mons Ardeshir	05 01 N	121 01 E
Mons Dieter	04 54 N	120 21 E
Mons Dilip	05 35 N	120 51 E
Mons Ganau	04 50 N	120 30 E
Sita	04 34 N	120 46 E

Sheet 65C1/S2 (1:50,000)

	° ' "	° ' "
Carol	08 28 N	122 20 E
Ewen	07 41 N	121 26 E
Kasper	08 17 N	122 05 E
Melissa	08 05 N	121 46 E
Romeo	07 30 N	122 38 E
Shahinaz	07 32 N	122 24 E

Sheet 77D1/S1 (1:10,000)

	° ' "	° ' "
Alan	10 56 30S	006 09 10W
Delia	10 55 10S	006 06 30W
Harold	10 53 S	006 03 W
Osman	10 58 40S	006 14 W
Priscilla	10 57 40S	006 11 40W
Susan	11 00 20S	006 17 10W

Sheet 77D3/S1 (1:50,000)

	° ' "	° ' "
Chang-ngo	12 41 S	002 09 W
José	12 41 S	001 39 W
Monira	12 33 S	001 42 30W
Ravi	12 31 S	001 57 W
Soraya	12 52 S	001 37 W

Sheet 100C1/S1 (1:50,000)

	° ' "	° ' "
Bawa	25 17 S	102 34 E
Edith	25 46 S	102 20 E
Fairouz	26 05 S	102 57 E
Karima	25 56 S	102 59 E
Rima Siegfried	25 55 S	102 45 E

Sheet 102A1/S1 (1:50,000)

	° ' "	° ' "
Kira	17 43 S	132 50 E

MERCURY NOMENCLATURE

Table 3. 68 names, approved as assigned to craters on the surface of Mercury. Locations are given in degrees of latitude and longitude, crater diameters in km. Biographical information is listed in Table 4.

Borealis Sheet

CRATERS			Diam.
ARISTOXENES	82 N	011 W	65
GOETHE	79.5N	044 W	340
PURCELL	81 N	148 W	80
VAN DIJCK	76.5N	165 W	100
SAIKAKU	73 N	177 W	80
TUNG YÜAN	73.5N	055 W	60
MANSART	73.5N	120 W	75
JÓKAI	72.5N	136 W	85
GAUGUIN	66.5N	097 W	75
TURGENEV	66 N	135 W	110

Victoria Sheet

CRATERS			Diam.
MONTEVERDI	64 N	077 W	130
RUBENS	59.5N	073.5W	180
STRAVINSKY	50.5N	073 W	170
VYĀSA	48.5N	080 W	275
SHOLOM-ALEJKHEM	51 N	076.5W	190
MONET	44 N	009.5W	250
DERZHAVIN	44.5N	035.5W	145
HUGO	39 N	047.5W	190
VELÁZQUEZ	37 N	054 W	120
AL-HAMADHANI	39 N	089.5W	170

Table 3 (Cont'd.)

CRATERS			Diam.	CRATERS			Diam.
VERDI	64.5N	169 W	150	HOLBEIN	35.5N	029 W	85
NIZĀMĪ	71.5N	165 W	70	KUAN HAN-CH'ING	29 N	053 W	155
MARTIAL	69 N	178 W	45	PRAXITELES	27 N	060 W	175
DESPRÉZ	81 N	092 W	40	WREN	24.5N	036 W	215
MYRON	71 N	080 W	50	GLUCK	37.5N	018.5W	85
				SOR JUANA	49 N	024 W	80

Shakespeare Sheet / Michelangelo Sheet

CRATERS (Shakespeare Sheet)			Diam.	CRATERS (Michelangelo Sheet)			Diam.
BOTTICELLI	64 N	110 W	120	IVES	32.5S	112 W	20
BRAHMS	58.5N	177 W	75	SAYAT-NOVA	27.5S	122.5W	125
AHMAD BABA	58.5N	127 W	115	USTAD ISA	31.5S	166 W	105
ZOLA	50.5N	178 W	60	SARMIENTO	28.5S	188.5W	115
STRINDBERG	54 N	136 W	165	SŪR DĀS	46.5S	094 W	100
MANSUR	47.5N	163 W	75	MICHELANGELO	44.5S	110 W	200
SHAKESPEARE	48.5N	151 W	350	SURIKOV	37 S	125 W	105
CHŎNG CH'ŎL	47 N	116 W	120	DELACROIX	44.5S	129.5W	135
NERVO	43 N	179 W	50	SHELLEY	47.5S	128.5W	145
VAN EYCK	43.5N	159 W	235	TAKAYOSHI	37 S	164 W	105
SCARLATTI	40.5N	099.5W	135	YAKOVLEV	40.5S	163.5W	100
BRONTË	39 N	126.5W	60	DOSTOEVSKIJ	44.5S	177 W	390
DEGAS	37.5N	127 W	45	LIANG K'AI	39.5S	183.5W	105
MARCH	31.5N	176 W	55	RIEMENSCHNEIDER	52.5S	100.5W	120
COUPERIN	30 N	152 W	75	ALENCAR	63.5S	104 W	85
HEINE	33 N	124.5W	65	HAWTHORNE	51 S	116 W	100
MUSSORGSKIJ	33 N	096.5W	115	VINCENTE	56.5S	143 W	85
				DOWLAND	53 S	180 W	80
				BARTÓK	29 S	135 W	80
				BASHŌ	32 S	170.5W	70

Table 4. Biographical information for Mercury crater names

AHMAD BABA al-Tinbukti (1556-1627) Sudanese writer
ALENCAR, Jose de (1829-1877) Brazilian novelist
AL-HAMADHANI (d. 1007) Arabic writer
ARISTOXENES (c. 150 BCE) Greek musical theorist
BARTÓK, Belá (1881-1945) Hungarian composer
BASHŌ, Matsuo (17th cen) Japanese writer
BOTTICELLI, Sandro (1445-1510) Italian painter
BRAHMS, Johannes (1883-1897) German composer
BRONTË, Charlotte (1816-1855) and Emily (1818-1848) English novelists
CHŎNG CH'ŎL (1536-1593) Korean poet

COUPERIN, Francois (1688-1733) French composer
DEGAS, Edgar (1834-1917) French painter
DELACROIX, Eugene (1798-1865) French painter
DERZHAVIN, Gavrila Romanovich (1743-1816) Russian poet

Table 4 (Cont'd.)

DESPRÉZ, Josquin (C.1450-1521) Flemish/French composer
DOSTOEVSKIJ, F.M. (1821-1881) Russian writer
DOWLAND, John (1562-1625) English composer
GAUGUIN, Paul (1848-1903) French painter
GLUCK, Christoph Willibalt (1714-1787) German composer
GOETHE, Johann Wolfgang von (1749-1832) German poet and dramatist

HAWTHORNE, Nathaniel (1804-1864) U.S. novelist
HEINE, Heinrich (1797-1856) German poet
HOLBEIN, Hans (c.1425-1524) and Hans (c.1497-1543) German painters
HUGO, Victor (1802-1885) French writer
IVES, Charles (1874-1954) U.S. composer
JÓKAI, Mór (1825-1904) Hungarian novelist
KUAN HAN-CH'ING (c.1241-c.1322) Chinese dramatist
LIANG K'AI (c1140-c1210) Chinese painter
MANSART, Jules Hardouin (c.1646-1708) French architect
MANSUR (18th cen?) Indian painter

MARCH, Ausias (1397-1459) Spanish poet
MARTIAL, Marcus Valerius (c.40-c.104) Roman poet
MICHELANGELO, Buonarotti (1475-1564) Italian painter, sculptor, architect
MONET, Claud (1840-1926) French painter
MONTEVERDI, Claudio (1567-1643) Italian composer
MUSSORGSKIJ, Modest (1839-1881) Russian composer
MYRON (5th cen BCE) Greek sculptor
NERVO, Amado (1870-1919) Mexican poet
NIZĀMĪ, Ilyās b. Yüsuf (1141-1203) Persian/Azerbaijan poet
PRAXITELES (4th cen BCE) Greek sculptor

PURCELL, Henry (c.1659-1695) English composer
RIEMENSCHNEIDER, Tilman (16th cen) German sculptor
RUBENS, Peter Paul (1577-1640) Flemish painter
SAIKAKU, Ihara (17th cen) Japanese novelist and poet
SARMIENTO, Domingo Faustino (1811-1888) Argentine writer
SAYAT-NOVA (1712-1795) Armenian/Georgian/Azerbaijan poet
SCARLATTI, Domenico (1685-1757) and Alessandro (1660-1725) Italian composers
SHAKESPEARE, William (1564-1616) English poet and dramatist
SHELLEY, Percy Bysshe (1792-1822) English poet
SHOLEM ALEICHEM (1859-1916) Ukranian (Yiddish) writer

SOR JUANA, Inés de la Cruz (1651-1695) Mexican writer
STRAVINSKY, Igor Fedorovich (1882-1971) Russian/French composer
STRINDBERG, August (1849-1912) Swedish poet and dramatist
SŪR DĀS (1483-1573) Indian poet
SURIKOV, Vassily (1848-1916) Russian painter
TAKAYOSHI (12th cen) Japanese painter
TURGENEV, I.S. (1822-1899) Russian writer
TUNG YÜAN (10th cen) Chinese painter
USTAD ISA (17th cen) Turkish/Persian architect
VAN DIJCK, Antoon (1599-1641) Flemish painter

VAN EYCK, Jan (d.1441) Flemish painter
VELÁZQUEZ, Diego (1599-1660) Spanish painter
VERDI, Giuseppe (1813-1901) Italian composer
VINCENTE, Gil (c.1465-c.1537) Portuguse dramatist
VYÄSA (2nd cen BCE) Indian writer
WREN, Christopher (1632-1723) English architect
YAKOVLEV, Barma and Posnik (16th cen) Russian architects
ZOLA, Emile (1840-1902) French novelist

VENUS NOMENCLATURE

Table 5. 14 names, approved as assigned to features in the surface of Venus. Locations are given in degrees of latitude and longitude and these coordinates are approximate only.

CRATER-LIKE FORMS			
EVE	Symbolizes the first woman	32 S	000
LISE MEITNER	1878-1968, Austrian physicist	55.5S	321.5E
SAPPHO	fl. c610-580 BC, Lesbos, Asia Minor lyric poetess	14 N	016 E
MONTES			
AKNA	Yucatan, Goddess of birth	68N	318E
FREYJA	Mother of Odin in Teutonic mythology	73N	335E
HATHOR	Egyptian Goddess of the sky	25S	323E
MAXWELL, James C.	1831-1879, British physicist	65N	004E
RHEA	Female Titan in Grecian mythology	32N	283E
THEIA	Female Titan in Grecian mythology	25N	281E
PLANUM			
LAKSHMI	Indian Goddess of fortune and prosperity	60N to 75N	315E to 345E
REGIONES			
ALPHA	First letter in the Greek alphabet	20S to 30S	000 to 010E
BETA	Second letter in the Greek alphabet	20N to 35N	280E to 290E
TERRAE			
APHRODITE	Grecian Goddess of fertility	10S to 05N	075E to 130E
ISHTAR	Babylonian Goddess of morning and evening	60N to 75N	315E to 025E

MARS NOMENCLATURE

Table 6. 61 names of small craters, approved as assigned on large-scale maps of the two Viking landing sites. Locations on Mars are given in degrees of latitude and longitude. Names are of small towns and stations of the space program, chosen on an international basis.

Viking 1, YORKTOWN sheet, 1:250,000

ALBANY	USA, New York	23.30 N	049.04 W
AMSTERDAM	Holland	23.26 N	047.03 W
ANNAPOLIS	USA, Md.	23.42 N	047.76 W
BORDEAUX	France	23.36 N	048.92 W
BREMERHAVEN	Germany	23.49 N	048.57 W

Table 6. (Cont'd.)

BRIDGETOWN	Barbados	22.15 N	047.10 W
BRISTOL	United Kingdom	22.35 N	046.95 W
CADIZ	Spain	23.38 N	049.05 W
CHARLESTON	USA, S. C.	22.87 N	047.82 W
COLÓN	Panama	23.01 N	047.09 W
FUNCHAL	Madeira Islands	23.21 N	049.46 W
KINGSTON	Jamaica	22.36 N	047.05 W
LA PAZ	Mexico	21.28 N	049.03 W
LISBOA	Portugal	21.49 N	047.59 W
NEW BERN	USA, N. C.	21.77 N	049.16 W
NEW HAVEN	USA, Conn.	22.30 N	049.28 W
NEWPORT	USA, R. I.	22.47 N	048.97 W
OKHOTSK	USSR	23.22 N	047.36 W
PHILADELPHIA	USA, Pa.	22.01 N	048.00 W
PORT-AU-PRINCE	Dom. Rep. (Hispaniola)	21.34 N	048.19 W
PORTSMOUTH	USA, N. H.	22.78 N	049.09 W
PRINCETON	USA, N. J.	21.93 N	049.12 W
SAVANNAH	USA, Ga.	22.27 N	047.80 W
Viking 2, CANBERRA sheet, 1:250,000			
BAYKONYR	USSR	46.60 N	227.23 W
CANAVERAL	USA	47.11 N	224.03 W
CANBERRA	Australia	47.50 N	227.25 W
EVPATORIYA	USSR	47.25 N	225.52 W
GOLDSTONE	USA	48.06 N	225.29 W
HAMAGUIR	Algeria	48.98 N	227.38 W
HOUSTON	USA	48.51 N	223.90 W
HSÜANCH'ENG	China	47.01 N	227.22 W
JOHANNESBURG	Union S. Africa	48.22 N	226.70 W
JODRELL	United Kingdom	47.77 N	227.60 W
KAGOSHIMA	Japan	47.60 N	224.12 W
KALININGRAD	USSR	48.78 N	224.89 W
KOUROU	French Guiana	47.02 N	227.12 W
MADRID	Spain	48.73 N	224.40 W
TSUKUBA	Japan	48.89 N	225.87 W
VOLGOGRAD	USSR	48.39 N	224.82 W
WALLOPS	USA	46.87 N	227.19 W
WOOMERA	Australia	48.36 N	227.28 W
Viking 2, UPTOPIA sheet, 1:1,000,000			
ACHAR	Uruguay	45.70 N	236.80 W
BHOR	India	42.00 N	225.50 W
BULHAR	Somalia	50.60 N	225.60 W
CHINCOTEAGUE	USA (Va.)	41.50 N	236.00 W
CORBY	U.K.	43.10 N	222.40 W
HĪT	Iraq	47.30 N	221.50 W
KEUL	USSR (Russian SFSR)	46.30 N	237.70 W
KUFRA	Libyan Arab Jamahiriya	40.60 N	239.70 W
KUMARA	New Zealand	43.30 N	231.40 W
LOJA	Ecuador	41.40 N	223.80 W

Table 6 (Cont'd.)

NAIN	Canada (Newfoundland)	41.70 N	233.20 W
NAJU	Korea	45.30 N	237.10 W
ORINDA	USA (Calif)	45.60 N	233.00 W
RAUB	Malaysia	42.60 N	224.90 W
RIMAC	Peru	45.20 N	223.80 W
RYNOK	USSR (Russian SFSR)	44.30 N	238.20 W
STON	Yugoslavia	47.20 N	237.40 W
TSAU	Botswana	49.80 N	238.90 W
UMATAC	USA (Guam)	42.70 N	222.80 W
VIVERO	Spain	49.20 N	241.10 W

Table 7. 28 names of features on the surface of Mars, approved as assigned on a Soviet map based on Mars 4 and 5 data. Locations are given in degrees of latitude and longitude.

CRATERS (Large)			
ARKAHNGELSKIY		41.0 S	024.0 W
BUNGE		34.0 S	049.0 W
SHATSKIY		33.0 S	015.0 W
SUMGIN		37.0 S	048.0 W
VINOGRADOV		20.0 S	037.5 W
VOEYKOV		32.0 S	075.0 W
CRATERS (Small)			
AKI	Japan	36.0 S	060.5 W
ALITUS	USSR	35.1 S	038.0 W
ANIAK	USA	32.5 S	069.0 W
DUBKI	USSR	35.0 S	055.0 W
FOROS	USSR	34.0 S	028.0 W
GARI	USSR	36.0 S	071.0 W
KUMAK	USSR	36.0 S	067.5 W
LABRIA	Brazil	35.5 S	048.0 W
LOS	USA	35.5 S	076.5 W
LUKI	USSR	30.0 S	037.0 W
MAGADI	Kenya	35.0 S	045.5 W
MENA	USSR	32.5 S	018.5 W
PULAWY	Poland	36.5 S	077.0 W
RUZA	USSR	34.5 S	052.5 W
TABOR	Czechoslovakia	36.0 S	064.0 W
ZILAIR	USSR	32.0 S	033.0 W
ZONGO	Zair	33.0 S	041.5 W
VIK	Iceland	36.0 S	064.0 W
VALLES			
OSUGA Valles		15.0 S	038.0 W
PROTVA Valles		33.5 S	059.0 W
RUNA Vallis		29.0 S	036.5 W
SURINDA Valles		28.5 S	035.0 W

Table 8. 36 names for features on the surface of Mars, approved as assigned on maps of the Geological Survey of the USA. Locations are given in degrees of latitude and longitude.

CRATERS			
BUTA	U.S.S.R.	23.5 S	032.2 W
GORI	U.S.S.R.	23.2 S	028.6 W
LĀR	Iran	26.1 S	028.8 W
CATENAE			
ACHERON		35 N 103 W to	42 N 097 W
PHLEGETHON		35 N 105 W to	40 N 095 W
CHAOS			
AROMATUM		02.0 S	042.5 W
DORSUM			
ZEA Dorsa		47 S 283 W to	53 S 277 W
FOSSA			
ACHERON Fossae		37 N 130 W to	39 N 138 W
MONS			
LIBYA Montes		00	270 W
PLANUM			
BOSPOROS		32 S	56 W
ICARIA		44 S	106 W
RUPES			
CEBERUS		06 N 186 W to	11 N 206 W
SCOPULUS			
ERIDANIA		47 S to 60 S	208 W to 235 W
TERRAE			
AONIA		55 S	090 W
ARABIA		25 N	320 W
MARGARITIFER		10 S	025 W
NOACHIS		45 S	000
PROMETHEI		55 S	270 W
TEMPE		35 N	075 W
CIMMERIA		45 S	210 W
MERIDIANI		05 S	000
SABAEA		15 S	325 W
SIRENUM		50 S	150 W
TYRRHENA		10 S	280 W
XANTHE		10 N	040 W

Table 8 (Cont'd.)

VALLES					
DAO Vallis	23 S	265 W	to	40 S	275 W
HARMAKHIS Vallis	38 S	265 W	to	43 S	275 W
MAJA Vallis	16 N	057 W	to	21 N	049 W
MAWRTH Vallis	19 N	057 W	to	27 N	052 W
REULL Vallis	39 S	260 W	to	48 S	270 W
AXIUS Valles	53 S	285 W	to	58 S	295 W
HIMERA Valles	21 S	022 W	to	23 S	023 W
LOCRAS Valles	07 S	311 W	to	11 S	313 W
PARANÁ Valles	23 S	012 W	to	26 S	009 W
RAVI Vallis	02 S	043 W	to	02 N	039 W
WARREGO Valles	42 S	095 W	to	44 S	091 W

THE OUTER SOLAR SYSTEM

The name CHIRON is approved, as suggested by C.T. Kowal, for the body found by him in heliocentric orbit between Saturn and Uranus. Kowal's general suggestion is also approved - that any minor planets with similar characteristics shall be called Centaurian, and named after Centaurs. These bodies can also receive numbers and be classified as minor planets, but the nomenclature must remain distinctive to separate these objects from the inner asteroids.

The following scheme has been adopted for naming features on the surfaces of satellites of the outer planets:-

(a) The source of names shall be the ancient myths, legends, epics, and folktales of all cultures that have these traditions.

(b) The Latin names for types of features shall be the same ones used in the inner solar system, with the addition of such classical terms as shall be necessary to describe new types of features without reference to their origins.

(c) Naming shall be done in an orderly fashion to associate types of features, regions, and even entire satellites with certain categories of names.

AMALTHEA NOMENCLATURE

Table 9. Four names approved for surface features on Amalthea, satellite of Jupiter. Approximate coordinates are given in degrees of latitude and longitude of a preliminary coordinate system, and may be in error by as much as 15 degrees.

BRIGHT SPOTS			
IDA	the nymph who cared for the infant Zeus	20 N	175 W
LYCTOS	the area in Crete where Zeus was raised	20 S	170 W
CRATERS			
GAEA	Mother Earth goddess who brought Zeus to Crete	80 S	090 W
PAN	son of Amalthea, the shepherds' god, made music on pipes of reed	55 N	035 W

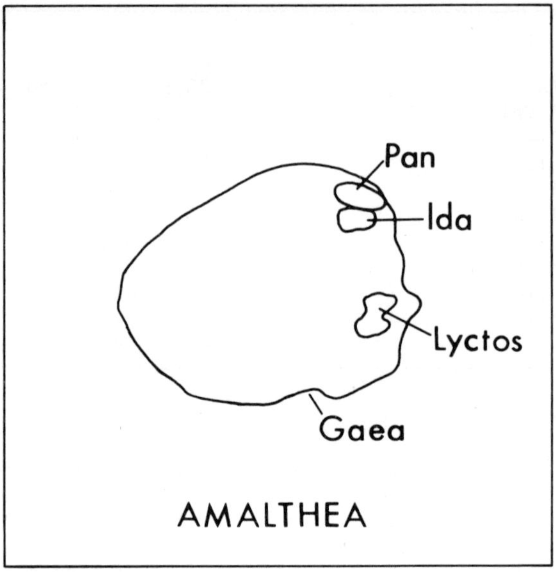

Identification diagram for the surface features on Amalthea.

Io NOMENCLATURE

Table 10. 66 names as approved for features on the surface of Io, satellite of Jupiter. Approximate positions are given in degrees of latitude and longitude with brackets indicating greater uncertainty.

ERUPTIVE CENTERS			
AMIRANI	Georgian god of fire	27 N	119 W
LOKI	Scandinavian mischievous demon god	19 N	305 W
MARDUK	Babylonian god of agriculture	28 S	210 W
MASUBI		45 S	053 W
MAUI	Maori hero who lifted up the sky	19 N	122 W
PELE	Hawaiian goddess of the volcano	19 S	257 W
PROMETHEUS	a Titan who stole fire from Zeus	03 S	153 W
SURT	Teutonic lord of the fire giants	45 N	337 W
VOLUND		22 N	177 W
CATENA			
MAZDA	Persian royal god	(10 S	312 W)
MONTES			
HAEMUS	place name connected with the Io myth	(70 S	044 W)
SILPIUM	" " " " " " "	62 S	282 W
PLANA			
DODONA	" " " " " "	(57 S	349 W)
NEMEA	" " " " " "	(70 S to 89 S	240 to 350 W)

Table 10 (Cont'd.)

REGIONES			
BACTRIA	place name connected with the Io myth	(45 S	125 W)
CHALYBES	" " " " " " "	(40 N to 65 N	060 to 110 W)
COLCHIS	" " " " " " "	(00 to 50 N	130 to 210 W)
LERNA	" " " " " " "	(55 S to 70 S	280 to 310 W)
MEDIA	" " " " " " "	(20 N to 30 N	000 to 100 W)
MYCENAE	" " " " " " "	(30 S to 55 S	130 to 180 W)
TARSUS	" " " " " " "	(20 S to 65 S	020 to 080 W)
THOLI			
APIS	name for Epaphus, son of Io and Zeus	(11 S	349 W)
INACHUS	river god, father of Io	16 S	349 W
PATERAE			
AMATERASU	Japanese sun goddess	38 N	307 W
ASHA	Persian god of fire	09 S	226 W
ĀTAR	son of the Persian god Mazda	30 N	279 W
ATEN	ancestor of human race, Egyptian mythology	(48 S	311 W)
BABBAR		40 S	272 W
BOCHICA	Columbian, Chibcha solar god	61 S	022 W
CREIDNE		(51 S	344 W)
CULANN	a Celtic hero	(20 S	162 W)
DAEDALUS	father of Icarus, escaped from the labyrinth by using wings	19 N	275 W
DAZHBOG	Salvonic sun god	54 N	302 W
EMAKONG		03 S	119 W
FUCHI		28 N	328 W
GALAI		11 S	289 W
GIBIL	Assyro-Babylonian god of fire, son of Anu	15 S	295 W
HENO		57 S	312 W
HEPHAESTUS	Grecian god of the forge	02 N	290 W
HIRUKO	Japanese solar god	(65 S	330 W)
HORUS	Egyptian solar god	10 S	340 W
INTI	Inca sun god	(68 S	349 W)
KANE		47 S	013 W
LOKI	Scandinavian mischievous demon god	13 N	310 W
MAASAW		40 S	341 W
MAFUIKE		(15 S	261 W)
MALIK		(33 S	127 W)
MANUA		35 N	322 W
MASAYA	Nicaraguan goddess of volcanos	(22 S	348 W)
MAUI	Maori hero who lifted up the sky	(20 S	123 W)
MIHR		16 S	306 W
NINA	Inca fire divinity	(40 S	165 W)
NUSKU	Assyro-Babylonian fire god	63 S	007 W
NYAMBE	Zambezi sun god	(00	344 W)
RA	Egyptian sun god	08 S	325 W
REIDEN		14 S	236 W
RUWA		(02 S	005 W)
SENGEN	Japanese god of Mt. Fuji	33 S	304 W
SHAKURU	Pawnee Indian sun god, Nebraska, USA	(24 N	267 W)
SHAMASH	Assyro-Babylonian sun god	(37 S	152 W)
TOHIL		(29 S	159 W)
SVAROG	the sky, Slavonic father of the gods	48 S	267 W

Table 10 (Cont'd.)

ÜLGEN		41 S	288 W
UTA	Assyro-Babylonian hero who survived the deluge	35 S	027 W
VAHAGN		(27 S	035 W)
VIRACOCHA	Inca god of rain and lake Titicaca	(62 S	282 W)

EUROPA NOMENCLATURE

Table 11. 18 names as approved for features on the surface of Europa, satellite of Jupiter. The coordinates are listed in degrees of latitude and longitude and are approximate only. Most of the coordinates give points of intersection between linear features and do not define the entire extent of the feature. The names are taken from classical mythology, in particular the Europa myth.

LINEAE				
ADONIS	son of Pheonix, nephew of Europa	33 S 123 W	to	75 S 113 W
AGENOR	Europa's father	45 S 210 W	to	30 S 174 W
ARGIOPE	a name for Europa's mother	29 S 219 W	to	06 S 182 W
ASTERIUS	Europa's husband	32 S 311 W	to	45 N 200 W
BELUS	Agenor's twin brother	03 S 215 W	to	35 N 105 W
CADMUS	brother of Europa	30 N 240 W	to	18 N 164 W
LIBYA	Agenor's mother	80 S 240 W	to	29 S 140 W
MINOS	son of Europa	48 S 218 W	to	34 S 149 W
PELORUS	from Dragon teeth, founder of Thebes	23 S 179 W	to	00 125 W
PHINEUS	brother of Europa	31 S 353 W	to	23 S 269 W
SARPEDON	son of Europa and Zeus	41 S 100 W	to	71 S 088 W
THASUS	brother of Europa	50 S 175 W	to	77 S 208 W
FLEXUS				
CILICIA	where Cilix searched for Europa	63 S 170 W	to	50 S 143 W
GORTYNA	place on Crete where Zeus brought Europa	53 S 171 W	to	40 S 149 W
SIDON	another name for Tyre	73 S 240 W	to	47 S 126 W
MACULAE				
THERA	island where Cadmus searched	48 S		182 W
THRACE	place where Cadmus searched	46 S		173 W
TYRE	where Zeus carried Europa	32 N		147 W

GANYMEDE NOMENCLATURE

Table 12. 53 names as approved for surface features on Ganymede, satellite of Jupiter. Approximate positions are given in degrees of latitude and longitude with brackets indicating greater uncertainty. Crater names are taken from the mythologies of Assyria, Babylon, Canaan and Egypt.

CRATERS			
ACHELOUS	Akkadian god of power, of heavens	62N	020W
ADAD	god of thunder and of the storm	57N	007W
ADAPA	lost his chance for immortality when he refused food offered by Ana	73N	040W
AMMURA	god of the "west"	34N	349W
ANU	river god, father of Ganymede	63N	349W

Table 12 (Cont'd.)

ASSHUR	Assyrian warrior	53N	338W
AYA	wife of Shamash	63N	332W
BAAL	Canaanite god	30N	336W
DANEL	mythical hero versed in art of divination	06S	026W
DIMENT	goddess of the dwelling place of the dead	25N	358W
ENLIL	god of the air, of hurricanes, of nature	52N	321W
ESHMUN	divinity of Sidon	14S	189W
ETANA	because dizzy and crashed to the Earth	69N	347W
GILGAMESH	Khumbuba the monster that guarded the mountain of cedars	(59S	122W)
GULA		64N	020W
HATHOR	protrectress of women, goddess of joy and love	(70S	265W)
ISIS	the name under which Io was worshipped in Egypt	(63S	198W)
KERET	legendery hero	15N	042W
KHUMBAM	Elamite god who created	24S	339W
KISHAR	female, terrestrial progenitor	70N	005W
MELKART	divinity of Tyre	04S	186W
MOR	spirit of the harvest	33N	334W
NABU	god of intellectual activity	47S	008W
NAMTAR	plague Demon	64S	347W
NIGIRSU		63S	321W
NUT	goddess of the sky	50S	270W
OSIRIS	god of the dead	(40S	160W)
RUTI	Byblos god	15N	310W
SAPAS		57N	046W
SEBEK		61N	004W
SIN	moon god	53N	003W
TANIT		57N	045W
TROS		09N	036W
ZAGAR	Sin's messenger brought dreams to men	58N	045W

SULCI

ANSHAR	celestial world home of Lakhmu and Lakhamu	32N to 15N	211W to 190W
APSU	primordial ocean	38S to 16S	242W to 218W
AQUARIUS	constellation of the water carrier	33N to 53N	032W to 003W
DARDANUS	promentary where Ganymede was abducted by eagles of Zeus	51S to 09N	043W to 036W
HARPAGIA	mount where Ganymede was abducted	05N to 23S	335W to 322W
KISHAR	terrestrial world, home of Lakhmu and Lakhamu	10S to 09N	229W to 211W
MASHU	where Ganymede and Phoebe are worshipped as raingivers	23N to 47N	182W to 242W
MYSIA	mount where Ganymede was abducted by eagles of Zeus	13S to 33N	002W to 334W
NUN	chaos, primordial ocean	48N	339W
PHILUS		33N to 62N	220W to 200W
PHRYGIA	kingdom where Ganymede was born	04N to 23N	034W to 321W
SICYON	where Ganymede and Hebe are worshipped as raingivers	30N to 43N	024W to 359W
TIAMAT	tumultuous sea from which everything was generated	19N to 07S	223W to 202W
URUK		53N to 10S	204W to 143W

Table 12 (Cont'd.)

REGIONES				
BARNARD	1857-1923	American astronomer	(05S to 37N	037W to 352W)
GALILEO	1564-1642	Italian astronomer	(01N to 65N	080W to 180W)
MARIUS	1570-1624	German astronomer	(75S to 58N	240W to 120W)
NICHOLSON	1891-1963	American astronomer	(45S to 00	061W to 318W)
PERRINE	1867-1951	American astronomer	(10N to 53N	075W to 350W)

CALLISTO NOMENCLATURE

Table 13. 89 names as approved for surface features on Callisto, satellite of Jupiter. Approximate positions are given in degrees of latitude and longitude with brackets indicating greater uncertainty. The names are taken primarily from the myths and folk-tales of northern cultures.

CATENA			
GIPUL	river (Norse)	64 N 074 W	to 71 N 034 W
LARGE-RINGED FEATURES			
ADLINDA	under the sea where souls are imprisoned after death (Eskimo)	58 S	018 W
ASGARD	where the gods dwell (Norse)	19 N	123 W
VALHALLA	Odinn's hall where dead heroes dwell (Norse)	18 N	054 W
CRATERS			
ADAL	son of Jarl and Erna (Norse)	76 N	082 W
ÄGRÖI		39 N	010 W
AKYCHA	Alaskan name for the sun (Eskimo)	73 N	315 W
ALFR	dwarf (Norse)	(09 S	222 W)
ALI	strongest of all men (Norse)	60 N	056 W
ANARR	dwarf (Norse)	48 N	002 W
ANINGAN	moon god (Eskimo)	53 N	008 W
ASKR	first man created from an ash tree (Norse)	52 N	323 W
BALKR	Ottar's ancestor (Norse)	30 N	011 W
BAVORR	dwarf (Norse)	50 N	019 W
BELI	giant (Norse)	64 N	082 W
BRAGI	skald-god (Norse)	76 N	072 W
BRAMI	Ottar's ancestor (Norse)	30 N	019 W
BRAN		(31 S	197 W)
BUGA	Tungu heaven god	22 N	326 W
BURI	dwarf (Norse)	(35 S	043 W)
BURR		34 N	115 W
DAG	Ottar's ancestor (Norse)	(61 N	077 W)
DANR	king (Norse)	64 N	078 W
DIA	Callisto's sister (Greek)	72 N	064 W
DRYOPS	son of Dia by Apollo (Greek)	79 N	019 W
DURINN	dwarf (Norse)	68 N	091 W
EGDIR	shepherd for the giants (Norse)	35 N	036 W
ERLIK		65 N	347 W
FADIR	farmer (Norse)	58 N	014 W
FILI	dwarf (Norse)	64 N	340 W
FINNR	dwarf (Norse)	17 N	004 W
FRODI	father of priestess Hledis (Norse)	68 N	142 W

Table 13 (Cont'd.)

FREKI	wolf (Norse)	78 N	358 W
FULLA	Frigg's maid (Norse)	72 N	105 W
FULNIR	son of Thrael and Thyr (Norse)	61 N	035 W
GERI	wolf (Norse)	65 N	343 W
GISL	steed on which the gods ride (Norse)	59 N	032 W
GLOI	dwarf (Norse)	41 N	241 W
GOLL	servant to the gods (Norse)	57 N	318 W
GONDUL	one of Odinn's maidens (Norse)	61 N	116 W
GRIMR		33 N	199 W
GUNNR	one of Odinn's maidens (Norse)	65 N	107 W
GYMIR		64 N	053 W
HABROK	hawk (Norse)	77 N	133 W
HAKI		25 N	314 W
HAR	dwarf (Norse)	02 S	358 W
HEPTI	dwarf (Norse)	65 N	022 W
HODR	Baldr's blind brother (Norse)	70 N	091 W
HOENIR	god who gave blood to first humans (Norse)	(36 S	261 W)
HOGNI	Ottar's ancestor (Norse)	10 S	005 W
IGALUK		07 N	314 W
IVARR	Ottar's ancestor (Norse)	07 S	321 W
JUMO		61 N	008 W
KARI	Ottar's ancestor (Norse)	48 N	106 W
KARL	Rigr's son (Norse)	56 N	328 W
LODURR	god who gave goodly colour to first humans (Norse)	47 S	268 W
LONI	dwarf (Norse)	08 S	205 W
LOSY		66 N	322 W
MERA	nymph of Artemis, seduced by Zeus (Greek)	65 N	076 W
MIMIR	giant (Norse)	34 N	054 W
MITSINA	Alaskan old man who perished while hunting on the ice (Eskimo)	57 N	097 W
MODI	son of Thorr and Sif (Norse)	67 N	122 W
NAMA	hero who built ark to save his family from the flood	57 N	329 W
NAR	dwarf (Norse)	02 N	048 W
NERIVIK	Alaskan name of Sedua (Eskimo)	14 S	051 W
NIDI	dwarf (Norse)	67 N	097 W
NORI	dwarf (Norse)	46 N	341 W
NUADA		(55 N	263 W)
OSKI	another name for Odinn (Norse)	(49 N	263 W)
OTTAR	a favorite of Freyja (Norse)	62 N	106 W
PEKKO		19 N	005 W
REGINN	dwarf (Norse)	(40 N	090 W)
RIGR	Heimdall in disguise (Norse)	(69 N	240 W)
SARAKKA		01 S	053 W
SEGINEK	the sun (Eskimo)	57 N	025 W
SHOLMO		55 N	015 W
SIGYN	Loki's wife (Norse)	37 N	027 W
SKOLL		55 N	314 W
SKULD	maiden, or norn, governing the fate of humans (Norse)	12 N	037 W
SUNDRI	dwarf (Norse)	56 N	139 W
SUMBUR		67 N	322 W
TINDR	Ottar's ancestor (Norse)	01 S	355 W

Table 13 (Cont'd.)

TORNARSUK	Greenland legendary hero (Eskimo)	28 N	129 W
TYN	war god, Odinn's son (Norse)	(69 N	228 W)
VALFODR	another name for Odinn (Norse)	01 S	239 W
VALI	Ottar's ancestor (Norse)	10 N	325 W
VESTRI	dwarf (Norse)	46 N	050 W
VITR	dwarf (Norse)	21 S	350 W
YMIR	giant from who Earth was created (Norse)	52 N	102 W

Ottawa, Ontario 3 April 1980

P A R T 4

A S T R O N O M E R ' S H A N D B O O K

ASTRONOMER'S HANDBOOK

IAU Transactions XIIC, 1967, was the first IAU Astronomer's Handbook. Subsequently, Volumes XIIIB, XIVB, XVB and XVIB have contained items for updating. Pending the reproduction of the whole "Handbook", the practice is continued here of indicating additions and changes to the former sections.

I. SHORT HISTORY OF THE INTERNATIONAL ASTRONOMICAL UNION

Table 2 (Volume XIIC *et req*) should be supplemented as follows:

Add: XVIIth General Assembly, 1979, Montreal, P.Q., Canada

Members of the Union, after the General Assembly	4 513
Participants (1163 Members)	1 965
Number of Adhering Countries	49
Number of Commissions	39
Volume of Transactions XVIIA (Parts 1, 2 and 3) pp.	xxi + 697
XVIIB	
Highlights of Astronomy Volume 5, 1979	

President:	M.K.V. Bappu	1979-1982
General Secretary:	P.A. Wayman	1979-1982
Assistant General Secretary:	R.M. West	1979-1982
Vice-Presidents:	M.W. Feast	1979-1985
	L. Kresak	1979-1985
	R. Wilson	1979-1985

II. ADMINISTRATION AND FINANCES OF THE UNION

A. Administration

The address of the General Secretary is:

Professor P.A. Wayman
IAU Secretariat
61, avenue de l'Observatoire
F-75014 Paris
France

The Secretariat consists of:

IAU Secretary:	Mme B. Manning
Typist/Assistant:	Ms. P. Smiley

with office address: IAU Secretariat, 61, avenue de l'Observatoire, F-75014 Paris, France.

The address of the Assistant General Secretary is:

Dr R.M. West
Sky Atlas Laboratory
ESO, c/o CERN
CH-1211 Genève 23
Switzerland

or, from September 1980:

Dr R.M. West
ESO
Karlsehwarzschild Str. 2
D. 8046 Garching.b. München
Germany, F.R.

(Enquiries concerning Symposia and Colloquia are directed to the Assistant General Secretary.)

B. FINANCES

The XVIIth General Assembly resolved that the unit of contribution payable by Adhering Countries to the IAU be 1 465.00 Swiss Francs for the years 1980, 1981 and 1982.

III. THE INTERNATIONAL SCIENTIFIC UNIONS

A. THE INTERNATIONAL COUNCIL OF SCIENTIFIC UNIONS

The ICSU Yearbook contains information for each year concerning IAU participation in ICSU Committees and scientific working groups, etc. This should be consulted for all current information on ICSU unions and committees.

The IAU is represented by the General Secretary, Professor P.A. Wayman, on the General Committee of ICSU.

B. IAU REPRESENTATION ON SCIENTIFIC AND SPECIAL COMMITTEES OF ICSU

1. Committee on Space Research (COSPAR): P.A. Wayman
2. ICSU Abstracting Board (ICSU-AB): W.D. Heintz
3. Federation of Astronomical and Geophysical Services (FAGS) H. Enslin and V. Bumba

   FAGS Services Representatives:

   a) Bureau International de l'Heure (BIH): A. Orte and H.M. Smith
   b) International Polar Motion Service (IPMS): K. Yokoyama
   c) International Ursigram and World Day Service (IUWDS): V. Bumba
4. Scientific Committee for Solar Terrestrial Physics (SCOSTEP): M.R. Kundu

C. IAU REPRESENTATION IN INTER-UNION COMMITTEES AND IN COMMISSIONS OF ICSU

1. Inter-Union Committee on Frequency Allocation for Radio Astronomy and Space Sciences (IUCAF): R. Schilizzi and G. Westerhout
2. Inter-Union Commission for Science Teaching (CTS): L. Houziaux
3. Committee for Data on Science and Technology (CODATA): B. Hauck
4. Inter-Union Commission on Spectroscopy (IUCS): B. Edlén, J.G. Phillips, M.J. Seaton
5. Committee on Science and Technology in Developing Countries (COSTED): L. Houziaux
6. Scientific Committee on Problems of the Environment (SCOPE): F. Graham Smith

D. IAU REPRESENTATION IN OTHER ORGANIZATIONS

1. Comité Consultatif pour la Définition de la Seconde (CCDS): W. Markowitz
2. Comité Consultatif pour la Définition du Mètre (CCDM): A.H. Cook

3. International Radio Consultative Committee (CCIR): L.H. Doherty, B.J. Robinson, H.M. Smith
4. European Physical Society Meetings Conference: R.M. West

IV. SERVICES AND FUNCTIONS OF THE IAU

This Section is to be found in full on pp.17-38 of *IAU Transactions* XIIC, 1966, with changes as follows:

IAU Transactions XIIIB, 1969, pp.235-236, §D.
" " XIVB, 1971, p. 253, ---
" " XVB, 1973, pp. 187-191, §D.
" " XVIB, 1977, p. 375, ---

The following change is now made:

§B.4 *The Minor Planet Center* (project supported by Commission 20)

Dr Paul Herget of the Cincinnati Observatory, on his retirement as Director on 30 June 1978, relinquished his direction of the Minor Planet Center. The Center was removed to the Smithsonian Institution Astrophysical Observatory, Cambridge, Massachussetts, U.S.A., where it operates under the direction of Dr Brian G. Marsden. The Assistant Director of the Center is Mr Conrad Bardwell, who has assisted with the work since 1958.

The Minor Planet Circulars are now published from Cambridge on a monthly basis. It is now practicable to include a certain amount of data on comets.

The file of observations from Cincinnati, consisting of 190,000 observations, is being maintained and continued at Cambridge.

§D. *The Central Bureau for Astronomical Telegrams,* Smithsonian Institution, Astrophysical Observatory.

The Bureau now operates a full service with communication equipment. Messages can be received at any time by TWX number 710-320-6842, answerback ASTROGRAM CAM.

V. PUBLICATIONS OF THE INTERNATIONAL ASTRONOMICAL UNION

Lists of IAU Publications have been published in the IAU Information Bulletin as follows:

No. 38, June 1977, pp. 18-23
No. 39, January 1978, p. 41
No. 40, June 1978, pp. 18-20
No. 41, January 1979, p. 32
No. 42, June 1979, pp. 56-58
No. 43, January 1980, pp. 49-50
No. 44, June 1980, pp. 26-27

A list also appears in the 1976-78 Report of the Executive Committee, Appendix III of this volume.

VI. SYMPOSIA OF THE INTERNATIONAL ASTRONOMICAL UNION

Details of IAU Symposia and other IAU scientific meetings are reported in successive issues of the IAU Information Bulletin and are also reported in the Report of the Executive Committee, Appendix III of this volume.

VII. EXECUTIVE COMMITTEE 1979-1982

President

Professor M.K.V. Bappu
Indian Institute of Astrophysics
Bangalore 560 034, India

Vice-Presidents

Dr M.W. Feast
South African Astronomical Observatory
P.O. Box 9, Observatory, Cape 7935
South Africa

Dr D.S. Heeschen
National Radio Astronomy Observatory
Edgemont Road, Charlottesville
Virginia 22901, U.S.A.

Academician E.K. Kharadze
Astrophysical Observatory
Mt. Canobili, Abastumani 383762
Georgia, U.S.S.R.

Dr L. Kresak
Astronomical Institute, Slovak Academy of Sciences
899 30 Bratislava, Czechoslovakia

Professor S. van den Bergh
Dominion Astrophysical Observatory
5071 West Saanich Road, R.R. 7
Victoria, B.C., V8V 3X3, Canada

Professor R. Wilson
Department of Physics and Astronomy
University College of London
Gower Street, London WC1E 6BT, U.K.

General Secretary

Professor P.A. Wayman
Dunsink Observatory
Castleknock, Co. Dublin, Ireland

Assistant General Secretary

Dr R.M. West
ESO c/o CERN
CH-1211 Geneva 23
Switzerland

from September 1980
ESO
Karlsehwarzschild Str. 2
D. 8046 Garching.b. München
Germany, F.R.

Advisers to the Executive Committee

Professor A. Blaauw
Sterrewacht, Postbus 9513
2300 RA, Leiden, Netherlands

Professor Edith A. Müller
Observatoire de Genève
CH-1290 Sauverny (GE)
Switzerland

VIII

UNION ASTRONOMIQUE INTERNATIONALE

STATUTS

I. DENOMINATION, BUTS ET DOMICILE

1. L'Union Astronomique Internationale (ci-après dénommée l'Union) est une organisation non-gouvernementale, qui a pour buts de:

   (a) développer l'astronomie par la coopération internationale,
   (b) encourager l'étude et le développement de l'astronomie sous tous ses aspects,
   (c) servir et sauvegarder les intérêts de l'astronomie.

2. L'Union a son siège légal à Bruxelles.

II. AFFILIATION DE L'UNION

3. L'Union adhère au Conseil International des Unions Scientifiques.

III. MEMBRES DE L'UNION

4. L'Union a pour membres:

   (a) des personnes morales (Pays adhérents)
   (b) des membres individuels (Membres).

IV. ORGANISATIONS AFFILIEES

5. L'Union peut accepter l'affiliation d'organisations internationales non-gouvernementales qui contribuent au développement de l'astronomie.

V. PAYS ADHERENTS

6. Les pays adhèrent à l'Union

   soit:
   (a) par l'intermédiaire de l'organisation par laquelle ils adhèrent au Conseil International des Unions Scientifiques, ou par l'intermédiaire d'un Comité National d'Astronomie approuvé par cette organisation,

   soit:
   (b) s'ils n'adhèrent pas au Conseil International des Unions Scientifiques, par l'intermédiaire d'un Comité National d'Astronomie reconnu par le Comité Exécutif de l'Union.

   (c) Les Organisations ou Comités mentionnés à l'article 6(a) et les Comités Nationaux d'Astronomie mentionnés à l'article 6(b) sont dénommés ci-après organismes adhérents.

7. L'adhésion d'un pays à l'Union est proposée par le Comité Exécutif et approuvée par l'Assemblée Générale; elle prend fin si le pays se retire de l'Union.

8. Les Pays Adhérents sont répartis en catégories. Le nombre des catégories est fixé par le Règlement. Un pays qui sollicite son adhésion indique la catégorie dans laquelle il désire être classé. La proposition peut être refusée par le Comité Exécutif si la catégorie est manifestement inadéquate.

INTERNATIONAL ASTRONOMICAL UNION

STATUTES

I. DENOMINATION, OBJECTS AND DOMICILE

1. The International Astronomical Union (referred to as the Union) is a non-governmental organization, whose objects are:

   (a) to develop astronomy through international co-operation,
   (b) to promote the study and development of astronomy in all aspects,
   (c) to further and safeguard the interests of astronomy.

2. The legal domicile of the Union is Brussels.

II. ADHERENCE TO THE UNION

3. The Union adheres to the International Council of Scientific Unions.

III. COMPOSITION OF THE UNION

4. The Union is composed of:

   (a) corporate members (Adhering Countries)
   (b) individual members (Members).

IV. AFFILIATED ORGANIZATIONS

5. The Union may admit the affiliation of international non-governmental organizations which contribute to the development of astronomy.

V. ADHERING COUNTRIES

6. Countries adhere to the Union

   either:
   (a) through the organization by which they adhere to the International Council of Scientific Unions, or through a National Committee of Astronomy approved by that organization,

   or
   (b) if they do not adhere to the International Council of Scientific Unions, through a National Committee of Astronomy recognized by the Executive Committee of the Union.

   (c) The Adhering Organizations and National Committees of Astronomy are referred to as adhering bodies.

7. Adherence of a country to the Union is approved, on the proposal of the Executive Committee, by the General Assembly; it terminates if the country withdraws from the Union.

8. Adhering Countries are classified in categories. The number of categories shall be specified in the By-laws.
   A country requesting adherence shall specify the category in which it desires to be classed. The specification may be declined by the Executive Committee if the category proposed is manifestly inadequate.

VI. MEMBRES

9. Les Membres sont admis dans l'Union par le Comité Exécutif, sur proposition de l'un des organismes adhérents mentionnés à l'article 6, en considération de leur activité dans une branche de l'astronomie.

VII. ASSEMBLEE GENERALE

10. (a) L'activité de l'Union est dirigée par l'Assemblée Générale des représentants des Pays Adhérents et des Membres. Chaque Pays Adhérent nomme un représentant autorisé à voter en son nom.
(b) L'Assemblée Générale rédige un Règlement qui précise les modalités d'application des Statuts.
(c) Elle nomme un Comité Exécutif chargé d'exécuter les décisions de l'Assemblée Générale, et d'administrer l'Union pendant la période séparant les réunions de deux Assemblées Générales ordinaires successives. Le Comité Exécutif rend compte de sa gestion à l'Assemblée Générale. L'Assemblée Générale, en acceptant le rapport du Comité Exécutif, le décharge de sa responsabilité.

11. (a) Sur les questions concernant l'administration de l'Union, sans implication budgétaire, le vote à l'Assemblée Générale a lieu par Pays Adhérent, chaque pays disposant d'une voix. Les Pays Adhérents qui ne sont pas à jour de leurs cotisations annuelles au 31 décembre de l'année précédant l'Assemblée Générale ne peuvent pas participer aux votes.
(b) Sur les questions engageant le budget de l'Union, le vote a lieu de même par Pays Adhérent, dans les conditions et avec les réserves prévues à l'article 11(a), le nombre de voix de chaque Pays Adhérent étant égal à l'indice de sa catégorie, définie conformément à l'article 8, augmenté d'une unité.
(c) Les Pays Adhérents peuvent voter par correspondance sur les questions figurant à l'ordre du jour de l'Assemblée Générale.
(d) Un scrutin n'est valable que si au moins deux tiers des Pays Adhérents disposant du droit de vote en vertu de l'article 11(a) y prennent part.

12. Sur les questions scientifiques n'engageant pas le budget de l'Union, les Membres de l'Union disposent chacun d'une voix.

13. Sur toutes les questions prévues aux articles 11 et 12, les décisions sont prises à la majorité absolue des suffrages. Cependant, une décision de modification des Statuts n'est valable que si elle a été prise à la majorité des deux tiers des voix des Pays Adhérents qui disposent du droit de vote en vertu de l'article 11(a).

14. Une proposition de modification des Statuts ne peut être discutée que si elle figure, en tant que telle, à l'ordre du jour de l'Assemblée Générale.

VII. COMITE EXECUTIF

15. Le Comité Exécutif se compose du Président de l'Union, de six Vice-Présidents, du Secrétaire Général et du Secrétaire Général Adjoint, élus par l'Assemblée Générale sur la proposition du Comité Spécial des Nominations.

IX. COMMISSIONS DE L'UNION

16. L'Assemblée Générale crée des Commissions en vue d'assurer la réalisation des buts qu'elle se propose.

VI. MEMBERS

9. Members are admitted to the Union by the Executive Committee, on the proposal of an adhering body referred to in article 6, with regard to their achievements in some branch of astronomy.

VII. GENERAL ASSEMBLY

10. (a) The work of the Union is directed by the General Assembly of representatives of Adhering Countries and of Members. Each Adhering Country appoints a representative authorized to vote in its name.
    (b) The General Assembly draws up By-laws governing the application of the Statutes.
    (c) It appoints an Executive Committee to implement the decisions of the General Assembly, and to direct the affairs of the Union in the interval between meetings of two successive ordinary General Assemblies. The Executive Committee reports to the General Assembly. The General Assembly, in accepting the report of the Executive Committee, discharges it of liability.

11. (a) On questions concerning the administration of the Union, not involving its budget, voting at the General Assembly is by Adhering Country, each country having one vote. Adhering Countries which have not paid their annual contributions up to 31 December of the year preceding the General Assembly may not participate in the voting.

    (b) On questions involving the budget of the Union, voting is similarly by Adhering Country, under the same conditions and with the same reservations as in article 11(a), the number of votes for each Adhering Country being one greater than the number of its category, as defined in article 8.
    (c) Adhering Countries may vote by correspondence on questions on the agenda for the General Assembly.
    (d) A vote is valid only if at least two thirds of the Adhering Countries having the right to vote by virtue of article 11(a) participate in it.

12. On scientific questions not involving the budget of the Union the Members of the Union each have one vote.

13. On all questions in articles 11 and 12, decisions are taken by an absolute majority of the votes cast. However, a decision to change the Statutes is only valid if taken with the approval of at least two thirds of the votes of the Adhering Countries having the right to vote by virtue of article 11(a).

14. A motion to change the Statutes can only be discussed if it appears, in specific terms, on the agenda for the General Assembly.

VIII. EXECUTIVE COMMITTEE

15. The Executive Committee consists of the President of the Union, six Vice-Presidents, the General Secretary and the Assistant General Secretary elected by the General Assembly on the proposal of a Special Nominating Committee.

IX. COMMISSIONS OF THE UNION

16. The General Assembly forms Commissions for such purposes as it may decide.

X. REPRESENTATION LEGALE DE L'UNION

17. Le Secrétaire Général est le représentant légal de l'Union.

XI. BUDGET ET COTISATIONS

18. (a) Pour chaque Assemblée Générale ordinaire, le Comité Exécutif prépare un projet de budget pour la période à courir jusqu'à l'Assemblée Générale ordinaire suivante, ainsi que les comptes de l'Union pour la période précédente. Il les soumet au Comité des Finances pour examen; ce Comité des Finances est composé de membres nommés par le organismes adhérents, à raison d'un membre par organisme, et il est approuvé par l'Assemblée Générale. Lors de sa première séance pendant l'Assemblée Générale, le Comité des Finances élit un Président parmi ses membres.

(b) Le Comité des Finances examine les comptes de l'Union pour voir si les dépenses engagées ont été conformes aux voeux émis lors de la précédente réunion de l'Assemblée Générale et il s'assure que le budget proposé vise à la poursuite de la politique de l'Assemblée Générale, telle qu'elle est interprétée par le Comité Exécutif. Il présente des rapports sur ces questions qu'il soumet à l'Assemblée Générale pour approbation des comptes, et pour décision sur le budget.

(c) Chaque Pays Adhérent verse annuellement à l'Union un nombre d'unités de cotisation qui est fonction de sa catégorie. Le nombre d'unités de cotisation pour chaque catégorie est fixé par le Règlement.

(d) Le montant de l'unité de cotisation est fixé par l'Assemblée Générale, sur la proposition du Comité Exécutif et avec l'avis du Comité des Finances.

(e) Le paiement des cotisations est à la charge des organismes adhérents. La responsabilité de chaque Pays Adhérent envers l'Union est limitée au moment des cotisations dues par ce pays à l'Union.

(f) Un Pays Adhérent qui cesse d'adhérer à l'Union renonce de ce fait à ses droits sur l'actif de l'Union.

XII. DISSOLUTION DE L'UNION

19. La décision de dissoudre l'Union n'est valable que si elle est prise à la majorité des trois quarts des voix des Pays Adhérents qui disposent du droit de vote en vertu de l'article 11(a).

XIII. DEVOLUTION DE L'AUTORITE EN CAS DE FORCE MAJEURE

20. Si, par suite d'événements indépendants de la volonté de l'Union, des circonstances apparaissent qui rendent impossible le respect des clauses de ces Statuts et du Règlement établi par l'Assemblée Générale, les organes et membres du Comité Exécutif de l'Union, dans l'ordre fixé ci-dessous, prendront toutes dispositions qu'ils jugeront nécessaires pour la continuation du fonctionnement de l'Union. Ces dispositions devront être soumises à une autorité supérieure dès que cela deviendra possible, jusqu'à ce qu'une Assemblée Générale extraordinaire puisse être réunie. L'autorité est dévolue dans l'ordre ci-dessous: l'Assemblée Générale; une Assemblée Générale extraordinaire; le Comité Exécutif, réuni ou par correspondance; le Président de l'Union; le Secrétaire Général; ou, à défaut de la possibilité de recourir à l'une de ces autorités ou de leur disponibilité, un des Vice-Présidents.

X. LEGAL REPRESENTATION OF THE UNION

17. The General Secretary is the legal representative of the Union.

XI. BUDGET AND DUES

18. (a) For each ordinary General Assembly the Executive Committee prepares a budget proposal covering the period to the next ordinary General Assembly, together with the accounts of the Union for the preceding period. It submits these to the Finance Committee for consideration; this Finance Committee consists of one member nominated by each adhering body and approved by the General Assembly. At its first meeting during the General Assembly, the Finance Committee elects a Chairman from among its members.

(b) The Finance Committee examines the accounts of the Union from the point of view of responsible expenditure within the intent of the previous General Assembly, and it considers whether the proposed budget is adequate to implement the policy of the General Assembly, as interpreted by the Executive Committee. It submits reports on these matters to the General Assembly for approval of the account and decision on the budget.

(c) Each Adhering Country pays annually to the Union a number of units of contribution according to its category. The number of units of contribution for each category shall be specified in the By-laws.

(d) The amount of the unit of contribution is determined by the General Assembly, on the proposal of the Executive Committee and with the advice of the Finance Committee.

(e) The payment of contributions is the responsibility of the adhering bodies. The liability of each Adhering Country in respect of the Union is limited to the amount of that country's dues to the Union.

(f) An Adhering Country that ceases to adhere to the Union resigns at the same time its rights to a share in the assets of the Union.

XII. DISSOLUTION OF THE UNION

19. The decision to dissolve the Union is only valid if taken with the approval of three quarters of the votes of the Adhering Countries having the right to vote by virtue of article 11(a).

XIII. EMERGENCY POWERS

20. If, through events outside the control of the Union, circumstances arise in which it is inpracticable to comply with the provisions of these Statutes and of the By-laws drawn up by the General Assembly, the organs and officers of the Union, in the order specified below, shall take such actions as they deem necessary for the continued operation of the Union. Such action shall be reported to a higher authority immediately this becomes practicable until such time as an extraordinary General Assembly can be convened. The following is the order of authority:
The General Assembly; an extraordinary General Assembly; the Executive Committee in meeting or by correspondence; the President of the Union; the General Secretary; or failing the practicability or availability of any of the above, one of the Vice-Presidents.

XIV. CLAUSES FINALES

21. Ces Statuts entrent en vigueur le 1er Septembre 1970.

22. Les présents Statuts sont publiés en versions française et anglaise. En cas d'incertitude, la version française fait seule autorité.

REGLEMENT

I. LES MEMBRES DE L'UNION

1. Les demandes d'adhésion des pays à l'Union Astronomique Internationale (ci-après dénommée l'Union) sont examinées par le Comité Exécutif et soumises à l'approbation de l'Assemblée Générale.

2. Les propositions de modifications de la liste des Membres sont, après examen attentif des suggestions des Présidents de Commissions, soumises pour avis au Comité des Nominations, composé d'un représentant de chaque Pays Adhérent désigné par l'organisme adhérent habilité, avant la décision du Comité Exécutif.

3. Les Commissions peuvent, avec l'approbation du Comité Exécutif, coopter des consultants qu'elles jugent en mesure d'apporter une contribution utile à leur travail. L'adhésion des consultants a pour terme le dernier jour de la première Assemblée Générale ordinaire qui suit leur admission, à moins qu'elle ne soit renouvelée.

4. Une organisation affiliée peut participer au travail de l'Union dans les conditions fixées par accord entre l'organisation et le Comité Exécutif.

II. L'ASSEMBLEE GENERALE

5. L'Union se réunit en Assemblée Générale ordinaire régulièrement une fois tous les trois ans. Si le lieu et la date de l'Assemblée Générale ordinaire n'ont pas été décidés lors de la précédente Assemblée Générale, ils sont fixés par le Comité Exécutif et communiqués aux organismes adhérents au moins six mois à l'avance.

6. Le Président peut convoquer, avec l'accord du Comité Exécutif, une Assemblée Générale extraordinaire. Il est tenu de le faire à la demande du tiers des Pays Adhérents.

7. L'Ordre du Jour de chaque Assemblée Générale ordinaire est arrêté par le Comité Exécutif et communiqué aux Organismes Adhérents au moins quatre mois avant le premier jour de la réunion. Il devra inclure la proposition du Comité Exécutif concernant le montant de l'unité de cotisation qui permet l'application de l'article 24.

8. (a) L'Ordre du Jour doit inclure toute motion ou proposition reçue par le Secrétaire Général au moins cinq mois avant le premier jour d'une Assemblée Générale ordinaire, qu'elle émane d'un organisme adhérent, d'une Commission de l'Union, ou d'une Commission mixte dans laquelle l'Union est représentée.

   (b) Une motion ou proposition concernant l'administration ou le budget de l'Union qui ne figure pas à l'Ordre du Jour, préparé par le Comité

XIV. FINAL CLAUSES

21. These Statutes enter into force on 1 September 1970.

22. The present Statutes are being published in French and English version. In case of doubt, the French version is the only authority.

BY-LAWS

I. MEMBERSHIP

1. Applications of countries for adherence to the International Astronomical Union (referred to as the Union) are examined by the Executive Committee and submitted to the General Assembly for approval.

2. Proposed changes in the list of Members are, with due regard to the suggestions of the Presidents of Commissions, submitted for advice to the Nominating Committee, consisting of one representative of each Adhering Country designated by the appropriate adhering body, before decision by the Executive Commitee.

3. Commissions may, with the approval of the Executive Committee, co-opt consultants whom they consider may contribute to their work. The adherence of consultants expires on the last day of the ordinary General Assembly next following their admission, unless renewed.

4. An affiliated organization may participate in the work of the Union as mutually agreed between the organization and the Executive Committee.

II. GENERAL ASSEMBLY

5. The Union meets in ordinary General Assembly, as a rule, once every three years. The place and date of the ordinary General Assembly, unless determined by the General Assembly at its previous meeting, shall be fixed by the Executive Committee and communicated to the adhering bodies at least six months beforehand.

6. The President, with the consent of the Executive Committee, may summon an extraordinary General Assembly. He must do so at the request of one third of the Adhering Countries.

7. The agenda of business for each ordinary General Assembly is determined by the Executive Committee and is communicated to the adhering bodies at least four months before the first day of the meeting. It shall include the proposal of the Executive Committee in regard to the unit of contribution as called for in article 24.

8. (a) Any motion or proposal received by the General Secretary at least five months before the first day of an ordinary General Assembly, whether from an adhering body, from a Commission of the Union, or from an Inter-Union Commission on which the Union is represented, must be placed on the agenda.

   (b) A motion or proposal concerning the administration or budget of the Union which does not appear on the agenda prepared by the Executive

Exécutif, ou tout amendement à une motion qui figure à l'Ordre du Jour, ne peut être discuté qu'avec l'accord prélable des deux tiers au moins des voix des Pays Adhérents représentés à l'Assemblée Générale et disposant du droit de vote en vertu de l'article 11(a) des Statuts.

9. S'il y a doute sur le caractère administratif ou scientifique d'une question donnant lieu à un vote, l'avis du Président est prépondérant.

10. En cas de partage égal des voix, le Président a voix prépondérante.

11. Le Président peut inviter des représentants d'autres organisations, des scientifiques et de jeunes astronomes à participer à l'Assemblée Générale. Avec l'accord du Comité Exécutif, il peut déléguer ce privilège au Secrétaire Général en ce qui concerne les représentants d'autres organisations, aux organismes adhérents en ce qui concerne les scientifiques et les jeunes astronomes.

III. LE COMITE SPECIAL DES NOMINATIONS

12. (a) Les propositions pour les élections du Président de l'Union, des six Vice-Présidents, du Secrétaire Général et du Secrétaire Général Adjoint sont soumises à l'Assemblée Générale par le Comité Spécial des Nominations. Ce Comité se compose du Président en fonction et du Président sortant, d'un membre proposé par le Comité Exécutif sortant et n'appartenant ni au Comité Exécutif actuel ni au Comité Exécutif précédent, et de quatre membres élus par le Comité des Nominations parmi 12 membres proposés par les Présidents de Commissions.
A l'exception du Président en fonction et du Président sortant, les membres actuels et les anciens membres du Comité Exécutif ne doivent pas faire partie du Comité Spécial des Nominations. Les membres du Comité Spécial des Nominations doivent tous appartenir à des pays différents.

(b) Le Secrétaire Général et le Secrétaire Général Adjoint participent au travail du Comité Spécial des Nominations à titre consultatif.

(c) Le Comité Spécial des Nominations est nommé par l'Assemblée Générale et est responsable directement devant elle. Il reste en fonction jusqu'à la fin de l'Assemblée Générale ordinaire qui suit immédiatement sa nomination, et il peut combler toute vacance survenant parmi ses membres.

IV. LE COMITE EXECUTIF ET SES MEMBRES

13. (a) Le Président de l'Union reste en fonction jusqu'à la fin de l'Assemblée Générale ordinaire qui suit immédiatement celle de son élection; les Vice-Présidents restent en fonction jusqu'à la fin de la deuxième Assemblée Générale ordinarie qui suit celle de leur élection. Ils ne pas rééligibles immédiatement pour les mêmes fonctions.

(b) Le Secrétaire Général et le Secrétaire Général Adjoint restent en fonction jusqu'à la fin de l'Assemblée Générale ordinaire qui suit immédiatement celle de leur élection. Normalement, le Secrétaire Général Adjoint succède au Secrétaire Général, mais l'un et l'autre peuvent être réélus aux mêmes fonctions pour une seconde période consécutive.

(c) Les élections ont lieu au cours de la dernière réunion de l'Assemblée Générale, les noms des candidates proposés ayant été annoncés au cours d'une réunion antérieure.

Committee, or any amendment to a motion that appears on the agenda, shall only be discussed with the prior approval of at least two thirds of the votes of Adhering Countries represented at the General Assembly and having the right to vote by virtue of Statute 11(a).

9. If there is doubt as to the administrative or scientific character of a question giving rise to a vote, the President determines the issue.

10. Where there is an equal division of votes, the President determines the issue.

11. The President may invite representatives of other organizations, scientists and young astronomers to participate in the General Assembly. Subject to the agreement of the Executive Committee he may delegate this privilege concerning representatives of other organizations to the General Secretary, and concerning scientists and young astronomers to the adhering bodies.

III. SPECIAL NOMINATING COMMITTEE

12. (a) Proposals for elections to the President of the Union, six Vice-Présidents, the General Secretary and the Assistant General Secretary are submitted to the General Assembly by the Special Nominating Committee. This consists of the President and past President of the Union, a member proposed by the retiring Executive Committee, and four members elected by the Nominating Committee from among twelve Members proposed by Presidents of Commissions. Other than the President and immediate past President, present and former members of the Executive Committee shall not serve on the Special Nominating Committee. No two members of the Special Nominating Committee shall belong to the same country.

(b) The General Secretary and the Assistant General Secretary participate in the work of the Special Nominating Committee in an advisory capacity.

(c) The Special Nominating Committee is appointed by the General Assembly to which it reports direct. It remains in office until the end of the ordinary General Assembly next following that of its appointment, and it may fill any vacancy occurring among its members.

IV. OFFICERS AND EXECUTIVE COMMITTEE

13. (a) The President of the Union remains in office until the end of the ordinary General Assembly next following that of his election; the Vice-Presidents remain in office until the end of the second ordinary General Assembly following that of their election. They may not be re-elected immediately to the same offices.

(b) The General Secretary and the Assistant General Secretary remain in office until the end of the ordinary General Assembly next following that of their election. Normally the Assistant General Secretary succeeds the General Secretary though both officers may be re-elected for another term.

(c) The election takes place at the last session of the General Assembly, the names of the candidates proposed having been announced at a previous session.

14. Le Président sortant et le Secrétaire Général sortant deviennent conseillers du Comité Exécutif jusqu'à la fin de l'Assemblée Générale ordinaire qui suit immédiatement celle de la fin de leur mandat. Ils participent au travail du Comité Exécutif et assistent à ses réunions sans droit de vote.

15. Le Comité Exécutif peut combler toute vacance survenant en son sein. Toute personne ainsi nommée reste en fonction jusqu'à l'Assemblée Générale ordinaire suivante.

16. Le Comité Exécutif peut rédiger et publier des Directives pour expliciter les Statuts et le Règlement.

17. Le Comité Exécutif nomme le représentant de l'Union qui doit siéger au sein du Comité Exécutif du Conseil International des Unions Scientifiques; si ce répresentant n'est pas déjà un membre élu du Comité Exécutif, il devient conseiller.

18. (a) Le Secrétaire Général est responsable auprès du Comité Exécutif des dépenses qu'il engage, qui ne doivent pas dépasser le montant des fonds mis à sa disposition.

    (b) Un bureau administratif, sous la direction du Secrétaire Général, est chargé de la correspondance, de la gestion des fonds de l'Union, et de la conservation des archives.

V. COMMISSIONS

19. (a) Les Commissions de l'Union poursuivent les buts scientifiques de l'Union par des moyens tels que l'étude de domaines particuliers de l'Astronomie, l'encouragement de recherches collectives et la discussion de questions relatives aux accords internationaux et à la standardisation.

    (b) Les Commissions de l'Union établissent des rapports sur les sujets qui leur ont été confiés.

20. Chaque Commission se compose de:

    (a) un Président et au moins un Vice-Président élus par l'Assemblée Générale sur la proposition du Comité Exécutif. Ils demeurent en fonction jusqu'à la fin de l'Assemblée Générale ordinaire qui suit immédiatement celle de leur élection. Ils ne sont pas normalement rééligibles.

    (b) un Comité d'Organisation, dont les membres sont désignés par la Commission sous réserve de l'approbation du Comité Exécutif. Le Comité d'Organisation assiste le Président et le(s) Vice-Président(s) dans leur tâche. Une Commission peut décider qu'elle n'a pas besoin de Comité d'Organisation.

    (c) des membres de l'Union, nommés par les Présidents, Vice-Président(s) et Comité d'Organisation, en considération de leurs spécialités; leur désignation est soumise à confirmation par le Comité Exécutif.

21. Entre deux Assemblées ordinaires, les Présidents de Commissions peuvent coopter, parmi les Membres de l'Union, de nouveaux membres des Comités d'Organisation et des Commissions elles-mêmes.

22. Les Commissions rédigent leur propre règlement. Les décisions sont prises, à l'intérieur des Commissions, par un vote de leurs membres.

14. The retiring President and the retiring General Secretary become advisers to the Executive Committee until the end of the ordinary General Assembly next following that of their retirement. They participate in the work of the Executive Committee and attend its meetings without voting right.

15. The Executive Committee may fill any vacancy occurring among its members. Any person so appointed remains in office until the next ordinary General Assembly.

16. The Executive Committe may draw up and publish Working Rules to implement the Statutes and By-laws.

17. The Executive Committee appoints the Union's representative on the Executive Committee of the International Council of Scientific Unions; if not already an elected member of the Executive Committee, this representative will become its adviser.

18. (a) The General Secretary is responsible to the Executive Committee for not incurring expenditure in excess of the funds at his disposal.

    (b) An administrative office, under the direction of the General Secretary, conducts the correspondence, administers the funds, and preserves the archives of the Union.

V. COMMISSIONS

19. (a) The Commissions of the Union shall pursue the scientific objects of the Union by activities such as the study of special branches of astronomy, the encouragement of collective investigations, and the discussion of questions relating to international agreements or to standardization.

    (b) The Commissions of the Union shall prepare reports on the work with which they are concerned.

20. Each Commission consists of:

    (a) A President and at least one Vice-President elected by the General Assembly on the proposal of the Executive Committee. They remain in office until the end of the ordinary General Assembly next following that of their election. They are not normally re-eligible.

    (b) An Organizing Committee, whose members are appointed by the Commission subject to the approval by the Executive Committee. The Organizing Committee assists the President and Vice-President(s) in their duties. A Commission may decide that it needs no Organizing Committee.

    (c) Members of the Union, appointed by the President, Vice-President(s) and the Organizing Committee, in consideration of their special interests; their appointment is subject to the confirmation by the Executive Committee.

21. Between two ordinary General Assemblies, Presidents of Commissions may co-opt, from among Members of the Union, new members to the Organizing Committees and to the Commissions themselves.

22. Commissions draw up their own rules. Decisions within Commissions are taken according to the vote of their members.

VI. ORGANISMES ADHERENTS

23. Le rôle des organismes adhérents est d'encourager et de coordonner, sur leurs territoires respectifs, l'étude des diverses branches de l'astronomie, particulièrement en ce qui concerne leurs besoins sur le plan international. Ils ont le droit de soumettre au Comité Exécutif des propositions pour discussion par l'Assemblée Générale.

VII. FINANCES

24. Chaque Pays Adhérent verse à l'Union une cotisation annuelle, qui est un multiple de l'unité de cotisation en fonction de sa catégorie, comme suit:

Catégories définies conformément à l'article 8 des Statuts:	1	2	3	4	5	6	7	8
Nombre respectif d'unités de cotisations:	1	2	4	6	10	14	20	30

25. Les ressources de l'Union sont consacrées à la poursuite de ses buts, y compris:

(a) les frais de publication et les dépenses administratives;

(b) l'encouragement des activités astronomiques qui nécessitent la coopération internationale.

(c) la cotisation due par l'Union au Conseil International des Unions Scientifiques.

26. Les ressources provenant de dons sont utilisées par l'Union en tenant compte des voeux exprimés par les donateurs.

VIII. PUBLICATIONS

27. L'Union a la propriété littéraire de tous les textes imprimés dans ses publications, sauf accord différent.

28. Les Membres de l'Union ont le droit de recevoir les publications de l'Union gratuitement ou à prix réduits, à la discrétion du Comité Exécutif qui décide en fonction de la situation financière de l'Union.

IX. CLAUSES FINALES

29. Ce règlement entre en vigueur le 1er septembre 1970.
Il peut être modifié avec l'approbation de la majorité absolue des voix des Pays Adhérents qui disposent du droit de vote en vertu de l'article 11(a) des Statuts.

30. Le présent règlement est publié en versions française et anglaise. En cas d'incertitude, la version française fait seule autorité.

VI. ADHERING BODIES

23. The functions of the adhering bodies are to promote and co-ordinate, in their respective territories, the study of the various branches of astronomy, more especially in relation to their international requirements. They are entitled to submit to the Executive Committee motions for discussion by the General Assembly.

VII. FINANCE

24. Each Adhering Country pays annually to the Union a number of units of contribution according to its category as follows:

Category as defined in Statute 8:	1	2	3	4	5	6	7	8
Number of units of contribution:	1	2	4	6	10	14	20	30

25. The income of the Union is to be devoted to its objects, included

(a) costs of publication and expenses of administration

(b) the promotion of astronomical enterprises requiring international co-operation

(c) the contribution due from the Union to the International Council of Scientific Unions.

26. Funds derived from donations are used by the Union in accordance with the wishes expressed by the donors.

VIII. PUBLICATIONS

27. The Union has the copyright to all materials printed in its publications, unless otherwise arranged.

28. Members of the Union are entitled to receive the publications of the Union free of charge or at reduced prices at the discretion of the Executive Committee taking due regard of the financial situation of the Union.

IX. FINAL CLAUSES

29. These By-laws enter into force on 1 September 1970. They can be changed with the approval of an absolute majority of the votes of the Adhering Countries having the right to vote by virtue of Statute 11(a).

30. The present By-laws are being published in French and English version. In case of doubt, the French version is the only authority.

IX

UNION ASTRONOMIQUE INTERNATIONALE

DIRECTIVES

Adoptées par le Comité Exécutif le 1er Septembre 1970

I. PUBLICATIONS

1. Les publications de l'Union Astronomique Internationale approuvées, dans le budget, par l'Assemblée Générale, sont préparées par le Bureau Administratif de l'Union.

2. Les Commissions de l'Union peuvent, avec l'approbation du Comité Exécutif, avoir leurs propres publications.

3. Le Comité Exécutif décide sur la proposition du Secrétaire Général, des modalités de distribution des publications de l'Union.

4. Les Membres de l'Union peuvent acquérir les publications de l'Union à un prix réduit.

II. APPARTENANCE A L'UNION

A. *Pays Adhérents*

6. Les demandes d'adhésion à l'Union formulées par les pays sont examinées par le Comité Exécutif compte tenu des points suivants:

(a) justesse du choix de la catégorie dans laquelle le pays souhaite être classé
(b) situation actuelle de l'astronomie dans le pays formulant la demande, et ses possibilités de développement
(c) mesure dans laquelle le futur organisme adhérent est représentatif de l'activité astronomique de son pays.

7. Les demandes proposant une contribution annuelle appropriée seront soumises pour décision à l'Assemblée Générale, avec la recommandation du Comité Exécutif.

B. *Membres*

8. Les personnes proposées pour devenir Membres de l'Union doivent en principe être choisies parmi des astronomes et des chercheurs dont les activités sont liées à l'astronomie, compte tenu de

(a) la qualité de leur œuvre scientifique
(b) la mesure dans laquelle leur activité scientifique implique des recherches astronomiques
(c) leur désir de contribuer à la poursuite des buts de l'Union.

9. Les jeunes astronomes doivent être considérés comme pouvant devenir Membres de l'Union dès qu'ils ont fait la preuve de leur capacité (en principe par une thèse de doctorat ou son équivalent) et de leur aptitude (quelques années d'activité fructueuse) à mener une recherche personnelle.

10. Pour les astronomes professionnels, leur contribution à l'astronomie peut consister soit en des recherches personnelles, soit en une collaboration assidue à des programmes importants d'observations.

INTERNATIONAL ASTRONOMICAL UNION

WORKING RULES

adopted by the Executive Committee on 1 September 1970

I. PUBLICATIONS

1. The publications of the International Astronomical Union approved, in the budget, by the General Assembly are prepared by the Administrative Office of the Union.

2. Commissions of the Union may, with the approval of the Executive Committee, issue their publications independently.

3. The distribution of publications of the Union is decided, on the proposal of the General Secretary, by the Executive Committee.

4. Members may purchase the publications of the Union at reduced prices.

II. MEMBERSHIP

A. *Adhering Countries*

6. Applications of countries for adherence to the Union are examined by the Executive Committee for

(a) the adequacy of the category in which the country wishes to be classified
(b) the present state and expected development of astronomy in the applying country
(c) the degree to which the prospective adhering body is representative of its country's astronomical activity.

7. Applications proposing an adequate annual contribution to the Union shall, with the recommendation of the Executive Committee, be submitted to the General Assembly for decision.

B. *Members*

8. Individuals proposed for Union membership should, as a rule, be chosen from among astronomers and scientists, whose activity is closely linked with astronomy taking into account

(a) the standard of their scientific achievement
(b) the extent to which their scientific activity involves research in astronomy
(c) their desire to assist in the fulfilment of the aims of the Union.

9. Young astronomers should be considered eligible for membership after they have shown their capability (as a rule Ph.D. or equivalent) of and experience (some years of successful activity) in conducting original research.

10. For full-time professional astronomers the achievement in astronomy may consist either of original research or of substantial contributions to major observational programmes.

11. Les autres personnes ne peuvent devenir Membres de l'Union que si certains de leurs travaux originaux concernent étroitement la recherche astronomique.

12. Huit mois avant une Assemblée Générale ordinaire, il sera demandé aux organismes adhérents de proposer de nouveaux Membres. Les propositions devront parvenir au Secrétaire Général au moins cinq mois avant la première session de l'Assemblée Générale. Les propositions reçues après cette date limite ne seront prises en considération que si des circonstances exceptionnelles justifient le retard.

13. Chaque proposition du nouveau Membre doit être présentée séparément et indiquer le nom, les prénoms et l'adresse postale du candidat (de préférence celle de son Institut ou Observatoire), ses date et lieu de naissance, l'Université devant laquelle il a soutenu sa thèse ou le diplôme équivalent, la date de soutenance, la situation actuelle du candidat, les titres et renseignements bibliographiques de deux ou trois de ses articles ou publications les plus significatifs et, s'il y a lieu, tous les renseignements susceptibles d'être pris en considération par le Comité des Nominations.

14. (a) Les Présidents de Commissions qui désirent suggérer de nouveaux Membres doivent adresser leurs suggestions au Secrétaire Général au moins cinq mois avant la première session d'une Assemblée Générale ordinaire. Les propositions devront fournir les mêmes renseignements que ceux mentionnés à l'article 13.
(b) Le Secrétaire Général fait part de ces suggestions aux organismes adhérents intéressés.

15. Le Secrétaire Général préparera deux listes pour le Comité des Nominations
(a) l'une contenant les noms des candidats proposés par les organismes adhérents
(b) l'autre contenant les noms des candidats proposés par les Présidents de Commissions, mais qui ne sont pas déjà inclus dans les propositions des organismes adhérents.

16. A partir des deux listes mentionnées à l'article 15, le Comité des Nominations prépare les propositions définitives de nouveaux Membres de l'Union.

17. Les organismes adhérents peuvent proposer la radiation de Membres ayant abandonné le domaine de l'astronomie pour d'autres activités, à moins qu'ils ne continuent à apporter une contribution à l'astronomie. Ces propositions doivent être portées à la connaissance du Secrétaire Général et du Membre concerné.

18. Le Secrétaire Général publiera la liste alphabétique des Membres de l'Union dans les Transactions de chaque Assemblée Générale ordinaire.

III. MEMBRES DES COMMISSIONS

19. Les membres des Commissions de l'Union sont cooptés par les Commissions. Cette procédure est régie par des règles établies par les Commissions elles-mêmes.

20. Les Commissions devraient choisir, ou approuver, la liste des membres de leurs commissions compte tenu de la spécialité de ces personnes, en particulier de leur activité scientifique dans le domaine de recherches de la Commission, et de leur contribution au travail de la Commission. Elles peuvent

(a) inviter des Membres de l'Union à devenir membres de la Commission
(b) radier les membres de la Commission qui n'ont pas contribué à son activité
(c) accepter ou refuser les demandes présentées par des Membres de l'Union, ou par des personnes proposées comme tels, en vue d'appartenir à la Commission
(d) suggérer l'élection comme Membres de l'Union de personnes n'y appartenant pas, ce qui leur permettrait alors de devenir membres de la Commission.

11. Others are eligible for membership only if they are making original contributions closely linked with astronomical research.

12. Eight months before an ordinary General Assembly, adhering bodies will be asked to propose new Members. The proposals should reach the General Secretary not later than five months before the first session of the General Assembly. Proposals received after the closing date will only be taken into consideration if the delay is justified by exceptional circumstances.

13. Each proposal shall be written separately. It should include the name, first names and postal address of the candidate, preferably that of his Institute or Observatory, his place and date of birth, the University and the year of his Ph.D. or equivalent title, his present occupation, titles and bibliographic data of two or three of his more important papers or publications, and details, if any, worthy to be considered by the Nominating Committee.

14. (a) Presidents of Union Commissions wishing to suggest new Members for admittance should address their suggestions to the General Secretary five months before the first session of an ordinary General Assembly. The proposals should contain particulars as in article 13.

(b) The General Secretary notifies the adhering bodies in questions of such suggestions.

15. The General Secretary shall prepare two lists for the Nominating Committee

(a) one containing the candidates proposed by the adhering bodies,
(b) the other containing those suggested by Presidents of Commissions, but not included among the proposals of adhering bodies.

16. The Nominating Committee prepares the final proposals for Union membership from the two lists as mentioned in article 15.

17. Adhering bodies should propose cancellation of Members who have left the field of astronomy for other interests, unless they continue to contribute to astronomy. Such proposals should be announced to the Member concerned and to the General Secretary.

18. The alphabetical list of Union Members will be published by the General Secretary in the Transactions of each ordinary General Assembly.

III. COMMISSION MEMBERSHIP

19. Members of Union Commissions are co-opted by Commissions. The rules governing the procedure of such co-option are drawn up by the Commissions themselves.

20. Commissions should choose, or approve of, Commission members taking into account their special interests, in particular their scientific activity in the appropriate fields of research and their contribution to the work of the Commission. They may

(a) invite Members to become members of their Commission
(b) remove members who have not contributed to the work of the Commission
(c) accept or reject applications for membership from existing or proposed Members
(d) suggest non-Members for election as Members, thus enabling them to become members of the Commission.

21. Les Membres de l'Union ne peuvent pas, en règle générale, appartenir à plus de trois Commissions.

22. Les Membres de l'Union peuvent demander à être admis dans une Commission en écrivant au Président de cette Commission. Ils ne devraient faire cette demande que si leur propre activité rentre dans le cadre des recherches de la Commission et s'ils sont décidés à contribuer au travail de la Commission.

23. Les membres des Commissions peuvent se retirer d'une Commission en écrivant à son Président.

24. En envoyant leurs propositions de nouveaux Membres, les organismes adhérents peuvent également suggérer le choix d'une Commission pour chaque candidat.

25. Le Secrétaire Général enregistera et analysera la liste des membres des Commissions; si cela est nécessaire, il tentera de trouver une solution aux anomalies évidentes.

26. Le Secrétaire Général publiera la liste des membres des Commissions dans les Transactions de chaque Assemblée Générale ordinaire.

IV. CONSULTANTS

27. Peuvent être élus Consultants des personnes qui ne sont pas astronomes, mais qui sont susceptibles de servir les intérêts de l'astronomie.

28. Les Commissions doivent en principe envoyer, pour approbation, leurs propositions de consultants au Secrétaire Général au moins cinq mois avant la première session d'une Assemblée Générale ordinaire.

29. Le Secrétaire Général préparera une liste des personnes proposées comme consultants et la soumettra pour approbation au Comité Exécutif.

30. Le Bureau Administratif établira une liste alphabétique des consultants.

31. Les consultants peuvent participer aux réunions de l'Union. Ils peuvent avoir droit de vote dans leurs Commissions respectives. Ils reçoivent gratuitement le Bulletin d'Information de l'Union.

21. Members may not, as a rule, be members of more than three Commissions.

22. Members may apply for Commission membership by writing to the President of the Commission concerned. Such applications should only be made if the Member is actively engaged in the appropriate field of research and is prepared to contribute to the work of the Commission.

23. Members of Commissions may resign from a Commission by writing to its President.

24. Adhering bodies, in sending in their proposals for new Members, may also suggest one Commission for each candidate.

25. The General Secretary will record and analyse the list of members of Commissions; if necessary he will try to resolve any outstanding anomalies.

26. The list of Commission members will be published by the General Secretary in the Transactions of each ordinary General Assembly.

IV. CONSULTANTS

27. Eligible as Consultants are non-astronomers in a position to further the interest in astronomy.

28. Proposals of Commissions for the approval of consultants should, as a rule, reach the General Secretary not later than five months before the first session of an ordinary General Assembly.

29. The General Secretary shall prepare a list of those proposed for admission as consultants and submit it to the Executive Committee for approval.

30. The Administrative Office will maintain an alphabetical list of consultants.

31. Consultants may participate in the meetings of the Union. They may have voting right in the respective Commission. They receive, free of charge, the Information Bulletin of the Union.

The following two articles were added to the Working Rules in August 1973:

V. SCIENTIFIC MEETINGS

32. The General Secretary shall publish rules for scientific meetings organized or sponsored by the Union.

VI. EXTERNAL CONTACTS

33. No dealings with third parties, attributable to the Union, shall be undertaken by any Member of the Union except on the authority of the General Secretary.

X

RULES FOR SCIENTIFIC MEETINGS

A. GENERAL

One of the essential tasks of the Union is to encourage circulation of ideas and to disseminate information internationally on current scientific results by organizing scientific meetings. Scientific Meetings are:

(a) Symposia
(b) Colloquia
(c) Regional Astronomy Meetings
(d) Commission Meetings
(e) Joint Discussions during General Assemblies

Except in the case (c), each meeting has to be sponsored by one or more Commissions. For Symposia there should always be more than one Commission involved, a principal *sponsoring* Commission with other *co-sponsoring* Commissions. In case (d), the Organizing Committee of the Commission is the Organizing Committee of the Meeting. In all other cases, a *Scientific Organizing Committee* is appointed by the Executive Committee with the advice of the proposer of the meeting.

The Chairman of the Scientific Organizing Committee (SOC) is appointed by the Executive Committee (EC). It is often the case that the President of the sponsoring Commission is proposed to the EC to be Chairman of the SOC, but it is not necessarily so. A proposer may propose himself as Chairman of SOC.

Symposia and Joint Discussions cover broader fields and have larger attendance, in general, than Colloquia. Proceedings of Symposia are published by the IAU in conjunction with a Publishing House. Whether or not the Proceedings of Colloquia are published, and in what form, is decided by the SOC of each Colloquium. Colloquia and Commission Meetings deal with narrower topics and are usually shorter. The IAU publishes short summaries of Colloquia in the Information Bulletin.

Regional Meetings are held at the proposal of a Host Organization in the country concerned. They should be confined to years when no IAU General Assembly is held. The EC can designate a meeting organized by a specific national or regional organization to be a Regional Astronomy Meeting.

Proposals for Symposia, Colloquia and Regional Meetings are addressed on the appropriate form to the Executive Committee through the Assistant General Secretary. Such proposals should be preceded by informal enquiries made at the earliest possible opportunity. *Enquiries* for Symposia and Colloquia should be directed to the Assistant General Secreary in the first instance; *in all other cases,* the first approach should be to the General Secretary. The formal proposal should be accompanied or preceded by letters of support from the Presidents of sponsoring and co-sponsoring Commissions. The following information is essential to a proposal:

1. A Title
2. Date and Place
3. Contact Address
4. Suggested composition of the SOC. No more than ten names should be included, one of the names being proposed as Chairman. There should be several countries represented and one institution should not be represented by more than one person unless some

other representation (e.g. a co-sponsoring Commission or relevant Working Group, etc.) is involved thereby. For Regional Meetings, the Assistant General Secretary is an ex-officio member of the SOC and in these cases more than ten names can be included, if desired.

5. Suggested Chairman and composition (or total number) of the proposed Local Organizing Committee (LOC).
6. The proposed number of participants.
7. The financial support expected, other than IAU support.
8. Whether or not the maximum IAU financial support is requested.
9. The proposed Editor, or Editors, of Proceedings, if these are to be published.
10. The proposed Scientific Programme, in outline.
11. A detailed account of why the meeting is useful or necessary at the time proposed and the relationship with other meetings.

Participation in Symposia and Colloquia is by invitation of the Chairman of the Scientific Organizing Committee. Invitations can be sought by suitably qualified scientists who are members of the IAU or who are resident in an adhering country. Restriction as to total number attending may be imposed by the Scientific Organizing Committee.

It is essential that there be no restriction based on race, colour, nationality, etc., imposed either by the organizers of Symposia and Colloquia or by authorities in the host country.

In the case of Regional Astronomy Meetings, invitation to attend is normally open to all interested astronomers of a particular geographical region for which invitations are primarily intended. The region will be outlined by the organizers in agreement with the Executive Committee.

B. CONTENTS OF MEETINGS

Commission Meetings held in the course of General Assemblies are of two kinds - (a) Business Meetings, including the organization of specific projects, and (b) Scientific Meetings on specific subjects. Commission Meetings can be organized jointly by two or more Commissions in order to deal with a topic of mutual interest; alternatively, a Joint Discussion, with an agreed Scientific Organizing Committee, can cover a topic of wider interest. Proposals for Meetings of Commissions and Joint Discussions during a General Assembly are addressed to the General Secretary, who decides the time-table for such proposals.

The agenda of scientific meetings may consist of:

(a) Invited Review Papers
(b) Invited Papers
(c) Contributed Papers
(d) Discussions, including Panel Discussions

The SOC first decides on the subjects, the general layout of the programme, and the mode of discussion, in accordance with suggestions that may be forthcoming from SOC members or from the Executive Committee. It can then proceed to decisions on the invited speakers, the acceptance or rejection of contributed papers, followed by the requirements for translation, recording of discussion, organization of the secretariat, etc., in agreement with the Local Organizing Committee.

C. TIME-TABLE AND OTHER COMMITMENTS

A timely presentation of proposals to the Executive Committee, taking into account the annual interval between EC meetings, is necessary if a satisfactory procedure of prior review is to be attained. The following types of notice are required:

(i) Preliminary notice must be received at the earliest possible date, so that the proposers can be informed of any similar proposals being made elsewhere. In the case of Regional Meetings, where the General Secretary receives proposals in the first instance, the outline proposal should normally be available 2-3 years in advance and contact maintained thereafter; in the case of all meetings ability to place *adequate prior notice* in the IAU Information Bulletin is a necessary service for IAU members generally. This requirement implies at least one year between the time the meeting is proposed on the official proposal form and the time of the meeting.

(ii) A firm proposal, still subject to approval and minor alteration by the Executive Committee, must be received in time for details to be investigated by the Assistant General Secretary in respect *inter alia*, of (a) composition of SOC (b) availability of IAU financial support (c) suitable nature and non-duplication of scientific content, *prior* to presentation for approval by the EC.

Fully documented proposals are submitted by the Assistant General Secretary to the Executive Committee. The EC either accepts or rejects the proposals. Since the Executive Committee normally meets annually in August and since it is desirable that discussion of critical points takes place at a formal meeting, rather than by correspondence, the actual dates for preparing proposals should take into account the dates of the EC meetings. March-April is a suitable time for making firm proposals for symposia and colloquia.

If the EC accepts a proposal it decides the title, place, date, Chairman of LOC, Chairman and members of SOC, financial support, Editor (in case of Symposia) and the scientific topics. Normally all these decisions are in accord with the firm proposals made by the organizers, but it is sometimes necessary to make changes in order to ensure the international character and proper scientific coverage and to avoid duplication with other meetings. If a meeting is proposed with less than one year's notice, it may be rejected because there is no time available for such changes to be made. However, *in the case of Colloquia only*, a proposal can be accepted by the EC at between eight and twelve months' notice if it is deemed to meet the normal IAU requirements without alteration, *provided always* that one announcement in the IAU Information Bulletin (prepared in May and November, and published in June and January), will give five months' notice to IAU members.

Presidents of related Commissions should be informed of the proposals by the organizers and invited to comment in their capacity as President.

The Chairman of SOC

(1) invites participants to a Symposium or Colloquium after the meeting is accepted by the Executive Committee;
(2) accepts or rejects requested invitations;
(3) invites contributions and sets a deadline for submission of abstracts
(4) informs the Assistant General Secretary as follows:
About 4 months before the meeting -
His recommendation for the distribution of the IAU financial allocation.

About 2 months before the meeting -
Detailed Programme, Choice of Participants, Local Organization.
About 1 month after the meeting -
He prepares a Report of the Meeting, suitable for inclusion in the Information Bulletin; Final Programme; List of actual Participants; Principal Abstracts.

The Assistant General Secretary should also be informed of any unexpected changes in the arrangements or programme of a meeting.

D. PARTICIPATION OF THE IAU IN JOINT SYMPOSIA

The IAU is willing to participate in Symposia, jointly with other Unions or Inter-Union Committees, Working Groups, etc. either as a principal sponsor or as a co-sponsor, or in a minor way.

As a principal sponsor, the IAU organizes a Symposium in accordance with its normal procedures, the proposal being made in consultation with other Unions. It invites the co-operation and participation of other Unions in the business of the Scientific Organizing Committee through representatives appointed by those Unions to the SOC. The IAU invites other Unions to discuss matters affecting the publication of Proceedings, but would normally arrange for publication of the Proceedings as one of the current IAU Symposium volumes.

The IAU would normally expect the other Unions to contribute financially towards the expenses of the Symposium.

As a co-sponsor, the IAU would expect to be invited to appoint a representative on the appropriate Organizing Committee, to be consulted through that representative on all important questions, and to be kept informed of decisions on other matters; the IAU would accept a leading role where called upon to do so, and would, in principle, contribute towards the expenses. The IAU would expect to be consulted in regard to publication of the Proceedings.

The IAU suggests that Symposia organized jointly by one or more Unions should always be announced in the following terms:
Organized by the IAU in co-operation with (e.g.) IUGG and COSPAR.
Organized by (e.g.) IUHPS in co-operation with the IAU.

It is to be noted that the IAU does not co-sponsor Symposia with National or Regional bodies, but only with International organizations.

In case of Colloquia or Regional Meetings, co-sponsorship can be proposed to the Executive Committee, who will decide upon acceptance or otherwise according to current resources and requirements, without specific restricition.

E. FINANCIAL CONSIDERATIONS

The Executive Committee allocates to each Symposium, Colloquium and Regional Meeting a limited sum of money which is meant as a 'catalyzer' to encourage other international organizations or national bodies to provide similar support. The maximum amount is specified in Swiss Francs, converted into other currencies, if necessary, according to current ICSU exchange rates. Colloquia receive smaller amounts than Symposia.

The IAU financial allocation is to be used principally for travel expenses of some of the participants. It can be used for partially supporting subsistence expenses of participants but in these cases should be limited to the nominal hotel or hostel charges rather than a *per diem* figure. If, following the conclusion of the meeting, some funds have not been allocated or have not been used, and the local organization requires further funds to meet its expenses, the remaining IAU allocation can *at the discretion of the General Secretary*, be used to contribute towards local expenses.

The Chairman of the SOC encourages every participant to apply for financial aid from his home institute or national organization. He gives notice to prospective participants that requests for IAU support can be submitted to him on the appropriate IAU Travel Grant form. If other forms of application are being used to allocate other monies, these forms can effectively replace the IAU Travel Grants forms. Having assessed the financial support available to applicants from other sources, the Chairman suggests to the Assistant General Secretary how the limited sum available should be distributed. This recommendation should reach the Assistant General Secretary 4 months before the meeting, so that prospective recipients can suitably arrange their affairs. The Assistant General Secretary will arrange payment to recipients either direct, or in the form of Bank Drafts to the Chairman, L.O.C. on an IAU account; arrangements to cash such drafts at a convenient place are the responsibility of the LOC.

IAU funds are intended to meet only a part of the travel and subsistence expenses of some participants; the Chairman of the SOC has to decide whether to help a moderate number of peopole with a small amount each, or to assist one or two principal participants appreciably. Either solution, or a compromise, is acceptable to the Executive Committee.

Expenses of the Secretariat of the Symposium, Colloquium or Regional Meeting are customarily met by organizations in the host country. Expenses of the Local Organizing Committee for receptions, excursions, buses, etc., are normally met either by the host municipality, national organization, or institute, or by means of participants' contributions in the form of registration fees or itemized charges, or by a combination of these sources. For Colloquia, registration fees and other charges should be kept to an absolute minimum.

F. LOCAL ORGANIZATION

The original proposal for a Symposium, Colloquium or Regional Meeting includes a proposed name for approval by the Executive Committee for the Chairman of the Local Organizing Committee. Even before such approval is finalized, this Chairman normally forms a Local Organizing Committee and informs the Chairman of the S.O.C. of its composition.

Under the personal guidance of its Chairman, the Local Organizing Committee takes care of the smooth running of the Meeting. It does not receive financial help from the IAU, the necessary expenses being met by local funds or by contributions from the participants.

The requirements of local organization are generally as follows:

1. Meeting Rooms suitable for the expected number of participants should be reserved.
2. Arrangements should be made for photocopying (Xeroxing) documents for participants.
3. Sufficient secretarial and technical assistance should be secured, with careful attention to the requirements of projection equipment, microphones, tape recorders, etc.

4. In conjunction with the requirements of the SOC, arrangements should be made to record verbal discussion. Reliance on tape recordings is often unsatisfactory and providing each contributor with a sheet of paper on which to record or summarize his remarks is advisable.
5. Information on accommodation (hostels, hotels, etc.) should be agreed with the Chairman of SOC and sent to prospective participants in good time. Block reservations are often advisable.
6. All participants should be asked to send their wishes as regards accommodation, excursions and social events to the Local Organizing Committee.
7. Receptions and excursions can be organized during the free period of the meeting, or just before or after the meeting. A guest programme is usually welcome.
8. Participants should be informed of the reservations made for them and how to reach their hotel or the meeting rooms on arrival.
9. The Local Organizing Committee should provide a Preliminary and a Final Programme, including useful auxiliary information, to be distributed to each participant at the appropriate time. A List of Participants, produced on about the second day of the meeting, is also extremely valuable if it corresponds closely with those actually present.

G. PUBLICATION OF SYMPOSIUM PROCEEDINGS

The IAU believes that the Proceedings of Symposia are of general interest for a considerable period of time and that early publication in uniform style to a high standard is desirable. Publication and distribution have therefore been entrusted to a commercial publishing house.

The main responsibility of the IAU as joint publisher is the maintenance of a high standard of scientific value, originality and accuracy. The commercial publishing house agrees upon a contract ensuring early publication and thereafter takes the financial responsibility.

The Executive Committee, by approving the choice of Editor or Editors, places the main burden of maintaining the required standard on one or two IAU members on the basis that they are familiar with the scientific matter of the Symposium and are persons with some experience in editorial tasks.

It is desirable that the Editor, or one of two joint Editors, should be from an English-speaking country.

The Editor is responsible for the scientific value and the appearance of the Proceedings. His main tasks are:

(a) to inform participants in ample time before the meeting in what general form their contributions should be submitted and what arrangements have been made with the publisher for receipt of camera-ready copy. The number of pages available to each contributor should be decided in good time.
(b) to inform the partcipants about IAU rules for publication of IAU Proceedings and to emphasize that any contributed papers that might be used must be refereed before acceptance in the published volume.
(c) to arrange with the Scientific Organizing Committee and the Local Organizing Committee the details for recording and reporting the scientific discussions that take place at meetings. Difficulties in this respect must not cause undue delay in preparing publication.
(d) to arrange with members of the SOC for the refereeing of any contributed paper if he is unable, as Editor, to referee such a paper himself.

In general, the Editor will not be able to allow time for substantial revision.

(e) to reduce the length of papers and discussion, to avoid duplication and to improve presentation.

(f) to check whether IAU rules have been followed in each contribution and to arrange for re-typing if necessary.

(g) to write the *Introduction, Table of Contents, etc.*, maintaining uniformity with recent IAU Symposium Proceedings.

(h) to maintain all necessary contact with the Publishing House, in accordance with current "Instructions for Editors" available from the Assistant General Secretary.

(i) to maintain close contact with the Assistant General Secretary on all matters affecting progress of publication arrangements, especially keeping him informed of the material sent to the publishers and of any unexpected delays or alterations.

As regards the *contents* of Proceedings volumes, duplication in respect of papers published elsewhere must be carefully avoided. Normally only *invited reviews* and *invited papers* should be included in Symposium volumes, and if material has been published elsewhere it should appear as abstract only. Detailed theoretical derivations, lists of observations, and comprehensive tabular data should be avoided.

Contributed papers should only be included in Symposium Proceedings if the material is of outstanding importance and interest. A refree's report should be obtained on every contributed paper that might be included. It is normal to publish abstracts only of contributed papers.

Normally, one introductory contribution is included, also a final summary of the Symposium, as provided for by the organizers.

In order to obtain a presentation that will indicate that the volumes of IAU Symposia form a series, it is requested that Editors adhere to the following recommendations:

Title Page The Title Page should include explicitly, above the title of Symposium and the date on which it was held, the words "IAU Symposium No. ...", followed, where appropriate, by the words "Organized by the IAU in co-operation with...", in which the list of organizations is limited to the Scientific Unions, the Scientific Committees, and Inter-Union Commissions of ICSU. Participation of UNESCO will be acknowledged by the following wording at the foot of the Title Page: "Published for the International Council of Scientific Unions with the Financial Assistance of UNESCO."

Table of Contents

Introduction The Editor's introduction should mention circumstances of the organization of the Symposium, and should list the supporting organizations and the members of the Scientific and Local Organizing Committees. It should express appreciation to those to whom it is due. The support of the IAU and other Unions, etc., should be recognized as well as that of other international, national, or local organizations.

Proceedings of the Symposium

Alphabetical Index of names and subject headings, with reference to the pagination arranged in the camera-ready copy by the publishers.

Symposium volumes should be published 6 - 8 months after the Symposium.

Participants obtain some free reprints of their contributions and other copies can be ordered. All IAU members can purchase Symposium volumes at reduced prices.

XI LIST OF COUNTRIES ADHERING TO THE UNION

The following is the list of 49 countries that adhere to the Union, giving the Adhering Organization, the year of admission, and the approximate number of individual members. (In the case of CHINA, the arrangement is subject to ratification by the next General Assembly of the Union.)

Country	*Adhering Organization*	*Year*	*Members*
ARAB REPUBLIC OF EGYPT	Academy of Scientific Research & Technology Dept. of Scientific Societies & International Scientific Unions 101 Kast El-Einy Street Cairo	1925	30
ARGENTINA	Comité Nacional de Astronomia Observatorio Astronómico "Félix Aguilar" Av. Benavídez 8175 Oeste 5407 Marquesado San Juan	1927	35
AUSTRALIA	Australian Academy of Sciences P.O. Box 783 Canberra City, ACT 2601	1939	95
AUSTRIA	Bundesministerium fuer Unterric und Kunst in Wien Minoritenplatz 5 A - 1014 Wien	1955	27
BELGIUM	Ministère de l'Education Nationale et de la Culture Française Cité Administrative de l'Etat rue Royale 204 B - 1010 Bruxelles	1920	78
BRAZIL	Conselho Nacional de Desenvolvimento Científico E Tecnológico - CNPq Av. W/3 Norte, Quadra 507/B Caixa Postal 11-1142 70.740 - Brasilia, D.F.	1961	23
BULGARIA	Bulgarian Academy of Sciences 7th November Street 1 1000 Sofia	1957	20
CANADA	National Research Council of Canada International Relations Ottawa K1A OR6	1920	144
CHILE	Universidad de Chile Observatorio Astronomico Nacional Casilla 36-D Santiago de Chile	1947	13

Country	*Adhering Organization*	*Year*	*Members*
CHINA	Chinese Astronomical Society Purple Mountain Observatory, Nanking	1935	50
CHINA	Astronomy Union located at Taipei, China Taipei Observatory, Taipei, Taiwan	1959	13
COLOMBIA	Universidad Nacional de Columbia Observatorio Astronomico Nacional Apartado Aereo 2584 Bogota D.E.	1967	2
CUBA	Academia de Ciencias de la Republica de Cuba Capitolio Nacional La Habana	1970	0
CZECHOSLOVAKIA	Czechoslovak Academy of Sciences Národní 3 Praha 1	1922	54
DENMARK	Det Kongelike Danske Videnskabernes Selskab The Royal Danish Academy of Sciences and Letters H.C. Andersens Boulevard 35 DK - 1553 Københavniv	1922	39
FINLAND	Académie des Sciences et Lettres Snellmaninkatu 9-11 Helsinki 17	1948	18
FRANCE	Académie des Sciences Institut de France 23, quai de Conti 75006 Paris	1920	350
GERMANY, D.R.	Akademie der Wissenschaften der DDR Der Generalsekretär Otto-Nuschke-Strasse 22/23 DDR - 108 Berlin	1951	42
GERMANY, F.R.	Rat Westdeutscher Sterwarten Institut f. Theoretische Physik u. Sternwarte der Universität Neue Universität, Haus N 2 c Olshausenstrasse 2300 Kiel	1951	227
GREECE	Academy of Athens Panepistimiou Street Athens	1920	47
HUNGARY	Hungarian Academy of Sciences Roosevelt tér 9 Budapest 5	1947	22
INDIA	Indian National Science Academy Bahadur Shah Zafar Marg New Delhi 110002	1964	78

Country	*Adhering Organization*	*Year*	*Members*
INDONESIA	National Institute for Sciences Tenku Cik Dithiro 43 Jakarta	1979	
IRAN	Institute of Geophysics Tehran University Amirabad Shomali Tehran	1969	3
IRAQ	Foundation of Scientific Research P.O. Box 255 Baghdad	1976	2
IRELAND	National Committee of Astronomy The Royal Irish Academy 19 Dawson Street Dublin 2	1947	19
ISRAEL	The Israel Academy of Sciences & Humanities P.O. Box 4040 Jerusalem 91040	1954	36
ITALY	Consiglio Nazionale delle Ricerche Piazale delle Scienze 7 00100 Rome	1921	185
JAPAN	Science Council of Japan 22-34 Roppongi, 7 chome Minato-ku Tokyo 106	1920	200
KOREA (DPR)	Academy of Sciences Pyogyang	1961	3
KOREA (REPUBLIC)	Korean National Astronomical Observatory Yoksam-Dong, Kangnam-ku Seoul, 134-03	1973	7
MEXICO	Instituto de Astronomia UNAM Apartado Postal 70-264 México D.F.	1921	28
NETHERLANDS	Koninklijke Nederlandse Akademie van Wetenschappen Kloveniersburgwal 29 Amsterdam - C	1922	122
NEW ZEALAND	The Royal Society of New Zealand 6 Kelburn Parade P.O. Box 12249 Wellington	1964	16
NORWAY	Det Norske Videnskaps-Akademi i Oslo Drammensveien 78 Oslo 2	1922	16

Country	*Adhering Organization*	*Year*	*Members*
POLAND	Polska Akademia Nauk Palac Kultury i Nauki Warsaw	1922	62
PORTUGAL	Secçáo Portuguesa das Uniões Internacionais Astronónica e Geodésica e Geofísica Praca da Estrela Lisboa 2	1924	14
ROUMANIA	Roumanian National Committee of Astronomy Astronomical Observatory Str. Cutitul de Argint 5 Bucarest 28	1928	18
SOUTH AFRICA	Council for Scientific & Industrial Research Information & Research Services P.O. Box 395 Pretoria 0001	1938	14
SPAIN	Comision Nacional de Astronomia Geografico y Cadastral General Ibanez 3 Madrid (3)	1922	36
SWEDEN	Kungl. Vetenskapsakademien Fack, S - 104 05 Stockholm 50	1925	56
SWITZERLAND	Schweizerische Naturforschende Gesellschaft Zentralsekretariat Laupenstrasse 10 3000 Bern	1923	45
TURKEY	Türk Astronomi Dernegi University Rasathanesi Beyazit, Istanbul	1961	31
UNITED KINGDOM	The Royal Society 6, Carlton House Terrace London SW1Y 5AG	1920	409
URUGUAY	Ministerio de Educacion y Cultura 8 de Octubre 3255 Montevideo	1970	0
U.S.A.	National Academy of Sciences Office of the Foreign Secretary 2101 Constitution Avenue, N.W. Washington, D.C. 20418	1920	1363
U.S.S.R.	Academy of Sciences of the USSR Leninskij Prospekt 14 Moscow 71	1935	410
VATICAN CITY STATE	Pontificia Academia delle Scienze Citta Del Vaticano	1932	5

Country	*Adhering Organization*	*Year*	*Members*
VENEZUELA	Asociación Venezolana de Astronomía Facultad de Ingenieria Departamento de Geodesia Universidad del Zulia Maracaibo	1953	5
YUGOSLAVIA	Savez društva matematičara, fizičara i astronoma Jugoslavije ul. Dr Ilije Duričića 4 Novi Sad	1935	26

A P P E N D I X

I. MEMBERSHIP OF COMMISSIONS

II. ALPHABETICAL LIST OF MEMBERS

III. REPORT OF THE EXECUTIVE COMMITTEE 1976-1978

IV. LETTERS EXCHANGED BETWEEN PROFESSOR A. BLAAUW AND PROFESSOR Y.C. CHANG

4. COMMISSION DES EPHEMERIDES (EPHEMERIDES)

PRESIDENT: A.M. Sinzi

VICE-PRESIDENT: T. Lederle

COMITE D'ORGANISATION: V.K. Abalakin, S. Aoki, R.L. Duncombe, J.H. Lieske, B. Morando, A. Orte, P.K. Seidelmann, G.A. Wilkins, B.D. Yallop.

MEMBRES: Arias-de Greiff J., Bandyopadhyay A., Bec-Borsenberger A., Bretagnon P., Chapront J., Deprit A., Dunham D.W., Emerson B., Fiala A.D., Fricke W., Fursenko M.A., Glebova N.I., Gondolatsch F., Haupt R.F., Henrard J., Janiczek P.M., Johnston K.J., King R.W., Jr., Kinoshita H., Klepczinski W.J., Kolaczek B., Kovalevsky J., Lahiri N.C., Morrison L.V., Mueller I.I., Oesterwinter C., O'Handley D.A., Reasenberg R.D., Rodriguez M., Schwan H., Shapiro I.I., Sochilina A.S., Standish E.M., Jr., Taylor G.E., Ting Y.T., Van Flandern T.C., Wackernagel H.B., Walter H.G., Williams J.G., Winkler G.M.R., Yamazaki A.

5. COMMISSION DE DOCUMENTATION ET DES DONNEES ASTRONOMIQUES (DOCUMENTATION AND ASTRONOMICAL DATA)

PRESIDENT: W.D. Heintz

VICE-PRESIDENT: G.A. Wilkins

COMITE D'ORGANISATION: B. Hauck, J. Kleczek, P. Lantos, S.A. Mitton, J.C. Pecker, L. Schmadel, J.R. Shakeshaft, I.S. Shcherbina-Samojlova.

MEMBRES: Abt H.A., Aoki S., Baker N., Beer A., Bidelman W., Bouska J., Davies R.J., Davis M.S., Dewhirst D.W., Dixon R.S., Dluzhnerskaya O.B., Duncombe R.L., Fleckenstein J.O., Fricke W., Garstang R.H., Griffin R.F., Heck A., Hirose H., Ivanov V.V., Jaschek C., Kadla Z., Lederlé T., Lequeux J., Lortet M.C., Lyngå G., Martynov D. Ya., Maxwell A., McNamara D.N., Mead J.M., Meadows A.J., Mein P., Moffat A.F.J., Ogorodnikov K.F., Pasinetti L.E., Philip A.G.D., Radlova L.N., Remy-Battiau L., Renson P.F., Sahade J., Schmidt K.H., Spite F., Sykes J.B., Terashita Y., Velghe A.G., Voigt H.H., Warren W.H., Jr., Weidemann V., Wempe J., Westerhout G., Worley C.E.,

6. COMMISSION DES TELEGRAMMES ASTRONOMIQUES (ASTRONOMICAL TELEGRAMS)

PRESIDENT: J. Hers

VICE-PRESIDENT: M.P. Candy

Sans Comité d'Organisation

MEMBRES: Biraud F., Cesco C.U., Cunningham L.E., Everhart E., Grindlay J.E., Kozai Y., Marsden B.G., Martynov D.Ya., Mrkos A., Pounds K.A., Roemer E., Rosino L., Simon P.

7. COMMISSION DE LA MECANIQUE CELESTE (CELESTIAL MECHANICS)

PRESIDENT: Y. Kozai

VICE-PRESIDENT: J. Kovalevsky

COMITE D'ORGANISATION: E.P. Aksenov, V.A. Brumberg, S. Ferraz-Mello, J. Hadjidemetrious, P.J. Message, F. Nahon, J. Schubart, P.K. Seidelman, V. Szebehely.

MEMBRES: Abalakin V.K., Aksnes K., Altavista C.A., Antonocopoulos G., Aoki S., Balimino G., Batrakov Y., Bec-Borsenberger A., Bettis D.G., Bohme S., Bozis G., Brookes C.J., Broucke R., Calame O., Candy M.P., Cefola P.J., Chapront J., Contopoulos G., Cook A.H., Counselman C.C., Cunningham L.E., Danby J.M.A., Davis M.S., Débarbat S., Demin V.G., Deprit A., Dormand J.R., Duboshin G.N., Duncombe R.L., Dvorak R., Everhart E., Eichhorn H.K., Erdi B., Fabre H., Farinella P., Fiala A., Froeschlé C., Galibina I.V., Gaposchkin E.M., Garfinkel B., Gaska S., Giacaglia G.E.O., Goldreich P., Goudas C.I., Grebenikov E.A., Greeberg R.J., Grouchinsky N.P., Hamid S.E., Heggie D.C., Hénon M., Henrard J., Herget P., Hori G., Janiczek P.M., Jeffreys W.H., Jupp A.H., Katsis D.N., Kaula W.M., Kholshevnikov V.K., King-Hele D.G., Kinoshita H., Kustaanheimo P., Lala P., Lazovic J.P., Lieske J.H., Lunquist C.A., Magnaradze N.C., Marchal C., Markellos V., Marsden B.G., Matas V., Meffroy J., Melbourne W.G., Merman G.A., Moller O., Morando B., Mulholland J.D., Musen P., Myachin V.F., Nacozy P.E., O'Handley D.A., Omarov T.B., Oesterwinter C., Orlov A.A., Peale S.J., Petrovskaya M.S., Pierce D.A., Popovic B., Robinson W.J., Roy A.E., Ryabov J.A., Sagnier J.L., Scholl H., Sconzo P., Sehnal L., Shapiro I.I., Sharaf S.G., Shteins K.A., Siegel C.L., Sinclair A.T., Siry J.W., Soultanov G.F., Standish E.M., Stellmacher I., Thiry Y., Vashkovyak S.N., Vilhena de Moraes R., Vinti J.P., Williams C.A., Yarov-Yarovoy M.S. Yoshilda J., Zare K.

8. COMMISSION DE L'ASTRONOMIE DE POSITION (POSITIONAL ASTRONOMY)

PRESIDENT: E. Høg

VICE-PRESIDENT: K.N. Tavatsherna

COMITE D'ORGANISATION: C.A. Anguita, G. Billaud, S. Débarbat, W. Fricke, J.A. Hughes, B.L. Klock, J.A. Lopez, I. Nikoloff, G. Teleki, R.H. Tucker, H. Yasuda.

MEMBRES: Argyracos J.G., Atkinson R. d'E., Bacchus P., Backer D.C., Bagildinski B.K., Bem J., Benevides-Soares P., Brouw W.N., Bykov M.F., Carestia R.A., Carrasco G., Chamberlain J.M., Chernega N.A., Chlietovsky F., Chollet F., Corbin T.E., Counselman C.C. (III), Dambara T., Dejaiffe R.J., De Vegt C., Dravskikh A.F., Duma D.P., Duncombe R.L., Eichhorn H.K., Federova R.T., Feissel M., Fogh Olsen H.-J., Gauss F.S., Gay J., Gliese W., Grudler P., Gubanov V.S., Gulyaev A.P., Haas J., Hansson N., Heintz W.D., Hurukawa K., Jackson P., Johnston K., Kharin A.S., Konin V.V., Kosin G.S., Kovalevsky J., Laclare F., Lacroute P., Larink J., Laustsen S., Lederlé T., Levy J.R., Loyola P., Manrique W.T., Marcus E., Melchior P., Mitic L.A., Miyamoto M., Murray C.A., Nefed'eva A.I., Nemiro A.A., Noel F., Okuda T., Osorio J.J.S.P., Petrov G.M., Pham-Van J., Pilowski K., Pinigin G.I., Podobed V.V., Polnitzky G., Polozhentsev D.D., Poma A., Proverbio E., Quijano L., Raimond E., Reiz A., Requieme Y., Rhynsburger R.W., Robertson W.H., Rousseau J.-M., Russell J.L., Rusu I., Sadzakov S., Sanchez M., Scheepmaker A.C., Schmeidler F., Schuler W., Schwan H., Sevarlic B.M., Sims K.P., Slaucitajs S., Smith C.A., Thomas D.V., Tsubokawa, I., Tuzi K., Von der Heide J.E.B., Walter H.G., Wood H.W., Woolsey E.G., Yamazaki A., Yatskiv Y.S., Zverev M.S.

9. COMMISSION DES INSTUMENTS ASTRONOMIQUES (ASTRONOMICAL INSTRUMENTS)

PRESIDENT: E.H. Richardson

VICE-PRESIDENT: W.C. Livingston

COMITE D'ORGANISATION: R. Albrecht, W.B. Burton, P. Connes, J. Davis, J.L. Heudier, I.M. Kopylov, A. Labeyrie, D. McMullan, A.B. Meinel, N.N. Mikhel'son, J. Ring, J. Rösch, N. Steshenko, G.A.H. Walker, M.F. Walker.

MEMBRES: Ables H.D., A'Hearn M.F., Atkinson R.d'E., Babcock H.W., Baum W.A., Beck N.G., Beer R., Bensammar S., Bhattacharyya J.C., Bingham R.G., Blitzstein W., Bolton C.T., Borgnino J., Brahde R., Brault J., Breckinridge J.B., Brück H., Campbell B., Carruthers G., Chapman G.A., Charvin P., Clarke D., Craine E.R., Crawford D.L., Currie D.G., Delbouille L., Dobronravin P.P., Dravins D., Duchesne M., Dunham T., Jr., Dunkelman L., Epstein G.L., Ewing M., Fellgett P.B., Fomenko A.F., Ford W.K., Jr., Gay J., Gehrz R.D., Gillingham P.R., Giovanelli R.G., Grigorjev V.M., Hallam K.L., Hammerschlag R., Harmer C.F.W., Harmer D.L., Hewitt A., Hickson P., Hilliard R., Hoekstra R., Honneycutt R., Hysom E.J., Jayarajan A.P., Jelley J.V., Jenkner H., Joseph R.D., Karpinskij V.N., Khokhlova V.L., Kipper T.A., Korovyakovskij Y.P., Kovatechev B.J., Lasker B.E., le Maitre G.R., Luud L.S., Maillard J.P., Martin F., Martins D.H., McGee J.D., Mertz L., Millikan A.G., Moroz V.I., Murray S.S., Nikonov V.B., Novikov S.B., Odgers G., Oliver J.P., Paxton H.J.B., Penny A.J., Petford A.D., Pope J.D., Pritchett C.J., Prokof'eva V.V., Purgathofer A. Th., Rakos K.D., Ramsay L.W., Reay N.K., Robinson L.B., Roddier C., Roddier F., Rusconi L., Rylov V.S., Sanchez Magro F., Santin P., Schöneich W., Schumann J.D., Shcheglov P.V., Schroeder D.J., Shakhbazyan Y.L., Smyth M.J., Texereau J., Tull R.C., Tyson J.A., Valnicek B., van Citters G.W., Veismann U.K., Viik T., Vrba R.J., Wayman P.A., Wells D.C., West R.M., Westphal J.A., Williams T.B., Wlérick G., Wolf R.E.A., Worden S.P., Wynne Ch. G., Zealey B.

10. COMMISSION DE L'ACTIVITE SOLAIRE (SOLAR ACTIVITY)

PRESIDENT: V. Bumba

VICE-PRESIDENT: E. Tandberg Hanssen

COMITE D'ORGANISATION: R. Bray, M. Dryer, O. Engvold, J. Harvey, R. Howard, G. Newkirk, Jr., M. Pick, D. Rust, V.E. Stepanov, H. Yoshimura, H. Zirin.

MEMBRES: Abrami A. Ahluwalia, H.S. Alissandrakis C., Altrock R.C., Altschuler M.D., Aly M.K.M., Ambroz P., Anderson K.A., Athay R.G., Avignon Y., Balli E., Banin V., Barrow C.H., Beckers J.M., Bell B., Benz A.O., Bhatnagar A., Billings D.E., Bonov A.D., Bougeret J.-L., Brown J.C., Bruckner G., Bruner E., Bruzek A., Cannon C.J., Carlquist P., Chapman G.A., Cheng ChungChieh, Chupp E.L., Cimino M., Coffey H.E., Coutrez R., Covington A.E., Crannel C.J., Culhane J.L., Datlowe D., de Groot T., de Jager C., Dennis B.R., Deubner F.L., Dezsö L., Dizer M., Dodson-Prince H., Dollfus A., Dubov E.E., Dulk G.A., Duncan R.A., Dunn R.B., Eddy J., Elste G.H., Erickson W., Falciani R., Feibelman W.A., Fokker A.D., Fortini T., Friedman H., Fritzova-Svestkova L., Gabriel A.H., Gaizauskas V., Gel'frejkh G.B., Gergely T.E., Gilman P.A., Giovanelli R.G., Gleissberg W., Gnevyshev M.N., Gneysheva R.S., Godoli G., Gokhale M.H., Gopasjuk S.I., Gurtovenko E.A., Hagan J.P., Hanasz J., Hansen R.T., Haug E., Hedeman E.R., Hildner E., Holzer T.E., Hyder C., Ioshpa B., Jager F.W., Jakimiec J., Jensen E., Jockers K., Jones H.P., Kaburaki O., Kai K., Kanno M., Kjeldseth M.O., Kleczek J., Koeckelenberg A., Kopecky M., Koutchmy S., Kozhevnikov N.I., Krat V.A., Krause F., Krivsky L., Kruger A., Kubota J., Kuklin G.V., Kundu M.R.,

Kunzel H., Kuperus M., Landecker P.B., Landman D.A., Leibacher J., Leroy J.-L., Lincoln J.V., Livshits M.A., Loughhead R.E., MacQueen R., Macris C.J., Makita M., Malitson H.H., Maltby P., Martres M.J., Matsura O.T., Mattig W., Maxwell A., McCabe M., McKenna Lawlor S.M.P., McLean D., Mein P., Melville J.M., Mendel'shtam S.L., Mergentaler J., Michalitsianos A.G., Michard R., Mogilevskij E.I., Mohler O.C., Moiseev I.G., Moreton G.E., Morozhenko N.N., Motta S., Muller R., Mustel E.R., Nagasawa S., Nakagawa Y., Namba O., Neupert W.M., Nishi K., Noyes R.W., Nussbaumer H., Obridko V.N., Ohki K., Orrall F.Q., Parkinson J.P., Parkinson W.H., Pedersen B.M., Picat J.-P., Piddington J.H., Pneuma G.W., Popovici C., Priest E., Raadu M.A., Rao P., Rayrole J., Razmadze T.S., Reeves E.M., Rigutti M., Roberts B., Roemer M., Romanchuk P.R., Rompolt B., Rösch J., Roselet J.P., Roxburgh I.W., Ruzdjak V., Saito K., Sakurai K., Sarabhai V.A., Sawyer C., Schatten K.H., Schluter A., Schmahl E.J., Schmieder B., Schmidt H.U., Schoolman S.A., Schroeter E.H., Semel M., Servajean R., Severny A.B., Shapley A.H., Shea M.A., Sheeley N.R., Jr., Shine R.A., Silberberg R., Sitnik G.F., Slonim E.M., Smith D.F., Smith E.v.P., Smith Henry J., Smith S., Smolkov G., Spicer D.S., Stellmacher G., Stenflo J.O., Stepanian N.N., Steshenko N.V., Stix M., Sturrock P.A., Suemoto Z., Svestka Z., Sykora J., Takakura T., Talon R., Tamenaga T., Tanaka H., Tanaka K., Teske R.G., Tifrea A., Tlamicha A., Topolova-Euzickova B., Trellis M., Tritakis B., Treumann R., Tuominen J., Uchida Y., Underwood J.H., Valnicek B., Van Allen J.A., Van Hoven G., Van't Veer F., Votjtech L., Vorpahl J., Vitinskij Ju. I., Vyalshin G.F., Waldmeier M., Wentzel D.G., Wiehr E., Wilcox J., Wild J.P., Wilson P., Wittmann A., Wöhl H., Wu Shi-tsan, Xanthakis J., Yakovkin N.A., Zappala R.A., Zelenka A., Zlobec P., Zwaan C.

12. COMMISSION DE LA RADIATION ET DE LA STRUCTURE DE L'ATMOSPHERE SOLAIRE (RADIATION AND STRUCTURE OF THE SOLAR ATMOSPHERE)

PRESIDENT: Y. Uchida

VICE-PRESIDENT: R.W. Noyes

COMITE D'ORGANISATION: M.K.V. Bappu, P. Delache, F.L. Deubner, A. Gabriel, S.I. Gopasyuk, J.W. Harvey, R. Howard, P.R. Wilson, C. Zwaan

MEMBRES: Acton L., Adam M.G., Aime C., Allen C.W., Altrock R.C., Altschuler M., Ansari G.M.R., Athay R.G., Battacharyya J.C., Beard D.B., Beckers J.M., Beckman J.E., Bel N., Benford G., Billings D.E., Bhatnagar A., Blackwell D.E., Blamont J.E., Bohm K.H., Bohm-Vitense E., Bonnet R., Brault J.W., Bray R.J., Brueckner G.E., Bruner E.C., Jr., Bruzek A., Bumba V., Cannon C., Ceppatelli G., Chapman G.A., Cheng C.C., Clark T.A., Cox A.N., Craig I.J.D., Cram L.E., Delbouille L., de Jager C., Dezsö L., Dravins D., Dubov E.E., Dumont S., Dunn R.B., Edmonds F.N., Elliot J., Elste G., Epstein G.L., Evans J.W., Falciani R., Fiala A., Fossat E., Frazier E.M., Friedman H., Gingerich O.J., Giovanelli R.G., Gnevyshev M.N., Godoli G., Gokdogan N., Goldberg L., Gomez M.T., Gordon C., Grevesse N., Gurtovenko E.A., Haupt R.F., Hiei E., Holweger H., Horton B.H., Hotinli M., House L., Jager F.W., Jefferies J.T., Jones H.P., Jordan C., Jordan S., Kalman B., Kanno M., Kato S., Kawaguchi I., Khetsuriani T.S., Kneer F., Kononovich E.V., Kopecky M., Kotov V.A., Koutchmy S., Kramer G.G., Krat V.A., Kubicela A., Kuklin G.V., Kuperus M., Laborde G., Labs D., Landman D.A., Leibacher J., Leighton R.B., Leroy J.-L., Linsky J.L., Livingston W.C., Locke J.L., Lopez Arroyo M., Loughhead R.E., Lust R., Makarova E.A., Marik M., Marilli E., Matsushima S., Mattig W., McKenna Lawlor S.M.P., Mein P., Melrose D.B., Mergentaler J., Mewe R., Michard R., Migeotte M.V., Mihalas D., Milkey R., Moore-Sitterly Ch., Mouradian Z., Müller E.A., Munro R.H., Namba O., Nevel L., Newkirk G.A.,

Nicolet M., Nikolsky G.M., Nishi K., Nordlund A., Orral F.Q., Oster L., Pajdusakova L., Pande C., Papathanasoglou D., Parkinson W.H., Pasachoff J.M., Pecker J.-C., Peyturaux R., Pierce A.K., Reeves E.M., Righini G., Rigutti M., Roberti G., Roddier F., Roland C., Rutten R.J., Samain D., Schmahl E.J., Schroter E.H., Seaton M.J., Semel M., Severny A.B., Shallis M.J., Shine R.A., Sivaraman K.R., Sitnik G.F., Skumanich A., Sobolev V.M., Souffrin P., Spicer D.S., Staude J., Stellmacher G., Stenflo J.O., Stepanov V.E., Stix M., Suemoto Z., Svestka Z., Swensson J.W., Tanaka K., Tandberg-Hanssen E., Teplitskaya R.B., Thomas R.N., Thomas J.H., Tousey R., Tsap T.T., Unno W., Uus U.K., Unsold A., van Hoven G., Vasil'eva G. Ya., Volonte S., Vukicevik-Karabin M., Waldmeier M., Warwick J.W., Wittmann A.D., Wlérick G., Wohl H., Worden S.P., Wyller A., Youssef N.H., Zelenka A., Zirin H., Zirker J.B.

14. COMMISSION DES DONNEES ATOMIQUES ET MOLECULAIRES (ATOMIC AND MOLECULAR DATA)

PRESIDENT: J.G. Philips

VICE-PRESIDENT: A.H. Gabriel

COMITE D'ORGANISATION: D.R. Johnson, S.L. Mandel'shtam, R.W. Nicholls, S. Sahal, K. Takayanagi, E. Trefftz, W. Wiese.

MEMBRES: Adelman S.J., Allen C.W., Andrew K.L., Artru M.-Chr., Baird K.M., Barnard A.J., Barrow R., Bates D.R., Bely O., Bely-Dubau F., Biemont E., Black J.H., Blaha M., Branscomb L.M., Brault J.W., Bromage G.E., Burgess A., Carbon D.F., Carroll P.K., Carver J.H., Cook A.H., Corliss C.H., Czyzak S.J., Dalgarno A., Delsemme A.H., Desesquelles J., Diercksen G.H.F., Dobronravin P.P., Douglas A.E., Dressler K., Dubau J., Dufay M., Edlen B., Engelhard E., Epstein G.L., Faucher P., Feautrier N., Felenbok P., Fink U., Flower D., Garstang R.H., Garton W.R.S., Glagolevskij Yu. V., Goldbach C., Grant I.P., Green L.C., Heddle D.W.O., Hefferlin R.A., Herman R., Herzberg G., Hesser J.E., House L.L., Huber M.C.E., Huebner W.F., Humphreys C.J., Jacquinot P., Johnson F.M., Joly F., Jordan C., Jordan H.L., Junkes J., Kato T., Kessler K.G., King R.B., Kingston A.E., Kipper T.A., Kohl J.L., Lagerquist A., Landman D.A., Lang J., Launay J.-M., Lawrence G.M., Laywer D., Lochte-Holtgreven W., Loulergue M., Lovas F.J., Lutz B.L., McWhirter R.W.P., Maillard J.P., Malinovsky M., Martin W.C., Mason H.E., Mel'nikov O.A., Mewe R., Migeotte M., Mohler O.C., Monfils A., Moore-Sitterly Ch., Nevin T.E., Newsom G.H., Nollez G., Nussbaumer H., Obi S., Oetken L., Oka T., Omont A., Orton G.S., Parkinson W.H., Peach G., Petrini D., Petropoulos B., Pfennig H., Prokof'ev V.K., Querci F., Richter J., Ross J.E., Roueff E., Schadee A., Schrijver J., Seaton M.J., Shore B.W., Smit J.A., Smith G., Smith P.L., Smith W.H., Somerville W.B., Sørensen N.G., Steenman-Clark L., Summers H.P., Swings J.P., Swings P., Tatum J.B., Tech J., Tousey R., van Regemorter H., van Rensbergen W., Varsavsky C.M., Völk H., Volonte S., Vujnovic V., Wares G.W., Weisheit J.C., Weniger S., Wilson R., Winnewisser G., Young L.G., Zeippen C.J., Zirin H.

15. COMMISSION POUR L'ETUDE PHYSIQUE DES COMETES, DES PETITES PLANETES ET DES METEORITES (PHYSICAL STUDY OF COMETS, MINOR PLANETS AND METEORITES)

PRESIDENT: B.D. Donn

VICE-PRESIDENT: B.J. Levin, C.R. Chapman

COMITE D'ORGANISATION: O.V. Dobrovolsky, M. Festou, L. Grossman, L. Kresak,

D.L. Matson, D.D. Morrison, N.B. Ritcher, Z. Sekanina, G.W. Wetherill.

MEMBRES: A'hearn F., Anders E., Andrienko A., Arpigny C., Beard B., Bertaud Ch., Beyer M., Biermann L., Blamont J.E., Bobrovnikoff N.T., Bouska J., Bowell E.L.G., Brandt J.C., Brecher A., Brunk W., Burlaga L.G., Burns J.A., Candy M.P., Ceplecha Z., Cherednichenko V.I., Cristescu C., Cruikshank D.P., Danks A., Delsemme A.H., Demenko A.A., Deutschman W.A., Dossin F., Douglas A.E. Dryer M., Ershkovich A.I., Everhart E., Eviatar A., Fechtig H., Feldman P., Gehrels T., Gérard E., Giese R.H., Greenberg J.M., Grudzinska S., Halliday I., Hanner M.A., Hapke B.W., Harris A.W., Harwitt M.O., Haser L., Haupt H., Herzberg G., Huebner W.F., Ibadinov K., Jackson W.M., Johnson T.V., Keller H.U., Knacke R.F., Kohoutek L., Konopleva V.P., Kowal C.T., Krinov E.L., Lancaster Brown P., Larson H.P., Liller W., Lillie C.F., Lüst R., Lyttleton R.A., McCord T.B., McCroskey R.E., Malaise D.J., Maran S.P., Marsden B.G., Martel-Chossat M.T., Matsuura O.T., Meisel D., Mendis D.A., Milet B., Miller F.D., Millman P., Moore E., Mrkos A., Neff J.S., Ness N.F., Newburn R., O'Dell C.R., O'Keefe J.A., Opik E.J., Perez-de-Tejada H.A., Pilcher C.B., Pittich E.M., Proisy P., Rahe J., Remy-Battiau L., Roemer E., Scaltriti F., Schmidt H.U., Shul'man L.M., Simonenko A.M., Sivaraman K.R., Snyder L.E., Swings P., Tomito K., Vanysek V., Veeder G.J., Veverka J.F., Vsekhsvjatskij S.K., Wallis M., Wasson J., Waterfield R.L., Wehinger P., Whipple F.L., Wilkening L.L., Wisniewski W., Wood J.A., Woszcyk A., Wyckoff S., Yavnel' A.A., Yeomans D., Zappala V., Zellner B.H.

16. COMMISSION POUR L'ETUDE PHYSIQUE DES PLANETES ET DES SATELLITES (PHYSICAL STUDY OF PLANETS AND SATELLITES)

PRESIDENT: B.A. Smith

VICE-PRESIDENT: V.G. Teifel'

COMITE D'ORGANISATION: E. Anders, J.E. Blamont, A. Brahic, M.S. Bobrov, W.M. Kaula, B.J. Levin, V.I. Moroz, D. Morrison, T.C. Owen, G.W. Wetherill, A. Wosczyk.

MEMBRES: Akabane T., Arthur D.W.G., Ashbrook J., Barrow C.H., Baum W.A., Beer R., Belton M.S., Bender P., Berge G.L., Bertaux J.L., Bondarenko L.N., Boyce P.B., Brecher A., Broadfoot A.L., Brunk W.E., Bullen K.E., Bystrov N.F., Calame O., Cameron W.S., Camichel H., Campbell D.B., Catalano S., Chamberlain J.W., Chapman C.R., Collinson E.H., Colombo G., Cook A.F. II, Connes J., Counselman C.C. III, Cruikshank D.P., Davies M.E., De Marcus W.C., De Mottoni y Palacios G., Dickel J.R., Dollfus A., Drake F.D., Dzhapiashvili V.P., El Baz F., Elliot J.L., Elston W.E., Eshleman V.R., Evans D.S., Fielder G., Fink U., Florenskij K.P., Galkin L.S., Gautier D., Gavrilov I.V., Geake J.E., Gehrels T., Geiss J., Giclas H.L., Gierasch P., Gold T., Goldreich P., Goldstein R.M., Goody R.M., Gorenstein P., Goudas C.L., Green J., Grossman L., Guerin P., Guest J.E., Gulkis S., Habibullin S.T., Hagfors T., Hall J.S., Halliday I., Hartmann W.K., Herzberg G., Hovenier J.W., Hubbard W.B., Hunten D.M., Ingrao C., Irvine W.M., Iwasaki K., Jeffreys H., Johnson T.V., Karandikar R.V., Kiladze R.I., Kislyuk V.S., Kondo Y., Kopal Z., Kozyrev N.A., Ksanfomaliti L.V., Kumar S.S., Kurt V.G., Kuzmin A.D., Larson H.P., Lejkin G.A., Link F., Lockwood G.W., Lutz B.L., McCord T.B., McElroy M.B., Martynov D. Ya., Masursky H., Matson D.L., Mayer C.H., Meadows A.J., Middlehurst B.M., Mikhail J.S., Millis R.L., Millman P.M., Miyamoto S., Moore P., Morozhenko A.V., Morrison L.V., Moutsoulas M., Mulholland J.D., Murphy R.E., Ness N.F., O'Keefe J.A., Opik E.J., Ottelet I., Petropoulous B., Pettengill G.H., Pollack J.B., Rosch J., Runcorn S.K., Safronov V., Sagan C., Sagitov M.U., Saissac J., Shapiro I.I., Shevchenko V.V., Shimizu T., Shoemaker E., Sinton W., Smith H.J.,

Smoluchowski R., Sonett C.P., Strom R.G., Strong J.D., Tombaugh C., Trafton L., Troitsky V.S., Tyler G.L., Jr., Urey H.C., Van Allen J.A., Van Flandern T.C., Veverka J., Walker R.M., Wallace L.V. Wasson J.T., Weimer T., Whitaker E.A., Wildey R.L., Williams I.P., Williams O.G., Wood J.A., Young A.T., Young L.G.

19. COMMISSION DE LA ROTATION DE LA TERRE (ROTATION OF THE EARTH)

PRESIDENT: P. Pâquet

VICE-PRESIDENT: Ya. S. Yatskiv

COMITE D'ORGANISATION: B. Elsmore, H. Enslin, C. Kakuta, B. Kolaczek, W. Klepczynski, V. Naumov, E. Proverbio, E. Silverberg, R. Vicente.

MEMBRES: Abraham H.J.M., Atkinson R.d'E., Barlier F., Bender P.L., Blinov N.S., Billaud G., Bonanomi J., Brosche P., Calame O., Cannon W., Currie D., Davies J.G., Débarbat S.V., Dejaiffe R., Djurovic D., Drâmbá C., Fedorov E.P., Feissel M., Fleckenstein J.O., Fliegel H.F., Gaignebet J., Gaposhkin E.M., Groten E., Guinot B., Hall R.G., Hellwig H., Hemmleb G., Hosoyama K., Hurukawa K., Iijima S., Jeffreys H., Kalmykov A.M., King R., Knowles S.B., Kostina L.D., Kozai Y., Lambeck K., Lederle T., Lefebvre M., Levallois J.J., Locke J.L., Markowitz Wm., McCarthy D.D., Meinig M., Melbourne W., Melchior P.J., Mietelski J., Mikhailov A.A., Milovanović V., Miyadi M., Moczko J., Morgan P., Morrison L.V., Mueller I.I., Mulholland J.D., Nicolini T., O'Hara N.P.G., Okamoto I., Okuda T., Ooe M., Opalski W., Orte A., Oterma L., Parijskij N.N., Pavlov N.N, Pilkington J.D.H., Popelar J., Popov N.A., Prodan Y.I., Randić L., Rochester M.G., Runcorn S.K., Sakharov V.I., Sanchez M., Sasao T., Sato K., Sekiguchi N., Sevarlic B.M., Sevilla M., Shapiro I.I., Slade M., Smith F.G., Smith H.M., Smylie D.E., Stanila G., Stoyko A., Sugawa C., Takagi S., Tapley B.D., Teleki G., Thomas D.V., Torao M., Tsao M., Tsubokawa I., Veis G., Wako Y., Ward W., Wilkins G., Williams J.G., Winkler G., Yokohama K., Yumi S.

20. COMMISSION DES POSITIONS ET DES MOUVEMENTS DES PETITES PLANETES, DES COMETES ET DES SATELLITES (POSITIONS AND MOTIONS OF MINOR PLANETS, COMETS AND SATELLITES)

PRESIDENT: G. Sitarski

VICE-PRESIDENT: E. Roemer

COMITE D'ORGANISATION: Yu. V. Batrakov, M.P. Candy, Y.C. Chang, F.K. Edmondson, Y. Kozai, L. Kresák, B.G. Marsden, J. Schubart, G.E. Taylor, P. Wild.

MEMBRES: Abalakin V.K., Aksnes K., Arend S.J.V., Babadzhanov P.B., Bennett J.C., Bielicki M., Borsenberger-Bec A., Bowell E.L.G., Bruwer J.A., Burns J.A., Calame O., Chernykh N.S., Cristescu C., Cunningham L.E., Debehogne H., Delsemme A.H., de Freitas Mourão R.R., de Pascual M., Dirikis M.A., Dunham D.W., Dvorak R., Elliot J.L., Evdokimov Yu. V., Everhart E., Ferraz-Mello S., Franklin F.A., Froeschlé C., Galivina I.V., Garcia F.L., Garfinkel B., Gehrels T., Giclas H.L., Gilmore A.C., Greenberg R.J., Harrington R.S., Harris A.W., Hasegawa I., Herget P., Hirose H., Hurnik H., Hurukawa K., Ianna P.A., Kazimirchak-Polonskaya E.I., Kiang T., Klemola A.R., Kohoutek L., Kovalevsky J., Kowal C.T., Kristensen L.K., Lieske J.H., Lindblad B.A., Lovas M., McCrosky R.E., Milet B., Millis R., Mintz-Blanco B., Morando B., Mulholland J.D., Nacozy P.E., Nakamura T., Oterma L., Pascu D., Pierce D.A., Pittich E.M., Popović B.,

Porter J.G., Protitch M.B., Quijano L., Rapaport M., Rickman H., Robertson W.H., Sagnier J.L., Schmadel L.D., Scholl H., Schrutka G., Schuster H.-E., Seidelmann P.K., Sekanina Z., Shelus P.J., Shor V.A., Shtejns K.A., Sinclair A.T., Strobel W., Sultanov G.F., Tomita Ko., Torres C., Vaghi S., Van Flandern T.C., van Houten C.J., van Houten-Groeneveld I., Vsekhsvyatskij S.K., Vu D.T., Wasserman L.H., West R.M., Whipple F.L., Williams J.G., Yakhontova-Samojlova N., Yeomans D.K., Zadunaisky P.E., Zappalá V., Ziolkowski K.

21. COMMISSION DE LA LUMIERE DU CIEL NOCTURNE (THE LIGHT OF THE NIGHT SKY)

PRESIDENT: H. Tanabe

VICE-PRESIDENT: R.H. Giese

COMITE D'ORGANISATION: R. Dumont, M. Harwit, C. Leinert, A.C. Levasseur-Regourd, K. Mattila, Y.L. Trutse, J.L. Weinberg.

MEMBRES: Anderson K., Angione R., Asaad A.S., Baggaley W.J., Banos C.J., Bates D.R., Blackwell D.E., Blamont J.E., Broadfoot A.L., Chamberlain J.W., Cook A.F., Dachs J., Divari N.B., Dufay M., Dunkelman L., Elsässer H., Fechtig H., Feldman P.A., Fishkova L.M., Fracassini M., Galperin Y.I., Greenberg J.M., Halliday I., Hanner M.S., Haug U., Henry R.C., Huruhata M., James J.F., Jarrett A.H., Kaplan J., Karandikar R.V., Karyagina Z.V., Kastler A., Knaflich H.B., Kordylewski K., Koutchmy S., Kulkarni P.V., Lamy P., Link F., Matsumoto T., Maucherat-Joubert M., Megrelishvili T.G., Misconi N.Y., Morgan D.H., Mukai S., Mukai T., Nawar S., Neuzil L., Ney E.P., Nicolet M., Pearse R.W.B., Pfleiderer J., Pitz E., Rapaport M., Roach F.E., Robley R., Roosen R.G., Rozhkovskij D.A., Sanchez-Martinez F., Saxena P.P., Schmidt T., Schuerman D.W., Sharov A.S., Soberman R.K., Sparrow J.G., Staude H.J., Toroshelidze T.I., Tyson J.A., Van Allen J.A., Van de Hulst H.C., Wallace L.V., Weill G., Weniger S., Witt A.N., Wolstencroft R.D.

22. COMMISSION DES METEORES ET DE LA POUSSIERE INTERPLANETAIRE (METEORS AND INTERPLANETARY DUST)

PRESIDENT: W.G. Elford

VICE-PRESIDENT: O.I. Belkovich

COMITE D'ORGANISATION: D.E. Brownlee, H. Fechtig, I. Halliday, J.A.M. McDonnell, J. Rajchl, Z. Sekanina, A.N. Simonenko, K. Tomita.

MEMBRES: Abbott W.N., Anders E., Babadzhanov P.B., Baggaley W.J., Beard D.B., Bibarsov R.S., Ceplecha Z., Clifton K.S., Cook A.F., Davies J.G., Davis J., Ellyett C.D., Fireman E.L., Giese R.H., Glass B.P., Guth V., Hajduk A., Hanner M.S., Harvey G.A., Hawkins G.S., Hemenway C.L., Hey J.S., Hirose H., Hodge P.W., Hong S.S., Hoppe J.A., Hughes D.W., Jacchia L.G., Jennison R.C., Jones J., Kaiser T.R., Kashscheev B.L., Katasev L.A., Keay C.S.L., Kizilirmak A., Kostylev K.V., Kramer E.N., Kresák L., Kresáková M., Krinov E.L., Kviz Z., Lebedinets V.N., Levin B.J., Lindblad, B.A., Lovell A.C.B., McCrosky R.E., McIntosh B.A., Millman P.M., Misconi N.Y., O'Keefe J.A., Opik E.J., Padevět V., Plavcová Z., Polnitzky G., Porubčan V., ReVelle D.O., Roosen R.G., Russell J.A., Schuerman D.W., Shao C.Y., Simek M., Soberman R.K., Southworth R.B., Štohl J., Verniani F.F., Weinberg J.L., Wetherill G.W., Whipple F.L., Wood J.A., Yavnel' A.A.

24. COMMISSION D'ASTROMETRIE PHOTOGRAPHIQUE (PHOTOGRAPHIC ASTRONOMY)

PRESIDENT: H.K. Eichhorn

VICE-PRESIDENT: W. Gliese

COMITE D'ORGANISATION: Ch. de Vegt, L.W. Frederick, G. Gatewood, R.S. Harrington, C.A. Murray, H.I. Potter, A. Upgren.

MEMBRES: Abhyankar K.D., Benedict G.F., Blaauw A., Bouigue R., Brosche P., Christy J.W., Churms J., Clube S.V.M., Connes P., Corbin T.E., Crézé M., Cudworth K., Dahn C.C., Delhaye J., Deutsch A.N., Dieckvoss W., Dommanget J., Elsmore B., Fallon F.W., Fatchikhin N.V., Firneis M.G., Fracastoro M.G., Franz O.G., Gallouet L., Gavrilov I.V., Giclas H.L., Goyal A.N., Haas J., Hanson R.B., Heintz W.D., Herget P., Hershey J.L., Heudier J.-L., Hill G., Hoffleit E.D., Ianna P.A., Jahreiss H., Jones B.F., Jones D.H.P., Kanaev I.I., Klemola A.R., Kock B.L., Kolchinskij I.G., Lacroute P., Lapushka K.K., Latypov A.A., Lavdovskij V.V., Le Poole R.S., Lippincott S.L., Lozinskij A.M., Lu P.K., Lutz T.E., Luyten W.J., Matiagin V.S., McAlister H.A., Ménessier M.O., Meurers J., Morgan W.W., Nicholson W., O'Connell J.K., Oja T., Onegina A.B., Podobed V.V., Polozhentsev D.D., Prochazka F.V., Quijano L., Rizvanov N.G., Robertson W.H., Roemer E., Russell J.L., Sanders W.L., Schilt J., Sims K., Smith C.A., Stock J.D., Strand K. Aa., Thomas D.V., Valbousquet A., van Altena W.F., van de Kamp P., Vasilevskis S., Wagman N.E., Wasserman L.H., Wesselink A.J.

25. COMMISSION DE PHOTOMETRIE ET DE POLARIMETRIE STELLAIRE (STELLAR PHOTOMETRY AND POLARIMETRY)

PRESIDENT: J.A. Graham

VICE-PRESIDENT: J. Tinbergen

COMITE D'ORGANISATION: A. Behr, M.S. Bessell, P.F. Chugajnov, G.V. Coyne, A. Feinstein, B. Hauck, P.W. Hill, A.R. Hyland, A.U. Landolt, T. Markkanen, M.F. McCarthy, F.G. Rufener, W. Schoneich.

MEMBRES: Ables H.D., Adelman S., Albrecht R., Argue A.N., Bahng J.D.R., Becker W., Bigay J.H., Blanco B.M., Bok B.J., Bookmyer B., Borgman J., Breger M., Brück H.A., Chis G., Cousins A.W.J., Crawford D.L., Dachs J., Dahn C.C., Edwards P., Eelsalu H.T., Fernie J.D., Fitch W.S., Gallouet L., Gehrels T., Gehrz R.D., Golay M., Goy G., Guttierrez-Moreno A., Hardie R.H., Hayes D., Hiltner W.A., Holmberg E.B., Irwin J.B., Jerzykiewicz M., Joshi S.C., Kron G.E., Kunkel W., Lasker B.M., Lockwood G.W., Marraco H.G., Masani A., Mayer P., McLean I.S., Mendoza V.E.E., Menzies J., Mianes P., Milone E., Mitchell R.I., Moffett T., Moreno H., Morris S.C., Muller A.B., Mumford G.S., Nikonov V.B., Notni P., Osawa K., Pel J.W., Perry C.L., Piirola Y., Robinson E.L., Roslund C., Rybka E., Rydgren A.E., Sarma M.B.K., Schmidt E.G., Schmidt H., Shakhovskoj, Shawl S.J., Smyth M.J., Steinlin U.W., Stock J., Stockman H.S., Stone R.P.S., Straizys V.L., Strohmeier W., Szkody P., Thomas J.A., Todoran I., Tolbert C.R., Ulrich B., Ureche V., Vardanian R.A., Velghe A.G., Vrba F.J., Wallenquist A.A.E., Walraven Th., Warren Jr. W.H., Weistrop D., Wesselink A., White N.M., Willstrop R.V., Wisniewski W., Yamashita Y., Young A.T.

26. COMMISSION DES ETOILES DOUBLES (DOUBLE STARS)

PRESIDENT: O.G. Franz

VICE-PRESIDENT: M. Fracastoro

COMITE D'ORGANISATION: R. Harrington, P.J. Kulikovsky, P. Muller, A. Proveda C.D. Scarfe, C.E. Worley.

MEMBRES: Abt H.A., Allen C., Arend S., Baize P., Batten A.H., Cester B., Couteau P., Da Silva A.S., Deutsch A.N., Dunham D.W., Djurkovic P.M., Dommanget J., Freitas Mourao R.R., Gatewood G., Geyer E.H., Hershey J.L., Heintz W.D., Hidajat B., Holden F., Johnson M.C., Kiselev A.A., Lippincott S.L., McAlister H.A., Meyer C., Morbey C.L., Popovic G.M., Rakos K.D., Russell J.L., Strand K.As., van de Kamp P., van Dessel E., Walker R.L., Weis E.W., Wieth-Knudsen N.P., Wilson R.H.

27. COMMISSION DES ETOILES VARIABLES (VARIABLE STARS)

PRESIDENT: J.D. Fernie

VICE-PRESIDENT: N.H. Baker

COMITE D'ORGANISATION: M. Breger, A.M. Cherepashschuk, A.N. Cox, N.R. Evans, M.W. Feast, H. Gursky, W.E. Kunkel, J. Smak, W. Wenzel.

MEMBRES: Aizenman M.L., Alania I.F., Andrews A.D., Arkhipova V.P., Ashbrook J., Bakos G.A., Barnes T.G., Bath G.T., Balazs-Detre J., Bateson F.M., Belserene E.P., Bertaud Ch., Bessell M.S., Bochonko D.R., Bond H., Bond H.E., Bopp B.V., Boulon J., Boyarchuk A.A., Boyarchuk M.E., Butler C.J., Butler D., Byrne P.B., Clement Coutts C.M., Cogan B.C., Cohen M., Chavira E., Chis G., Christy R.F., Chugainov P.F., De Groot M., de Kock R.P., Demers S., Deupree R.G., Dickens R.J., Dupuy D., Dziembowski W., Efremov Yu.N., Edwards P.J., Eskioglu A.N., Evans D.S., Feibelman W.A., Fitch W.S., Friedjung M., Frolov M.S., Gahm G., Gallagher J.S., Gaposchkin S., Gascoigne S.C.B., Gershberg R.E., Geyer E.H., Glagolevskij Yu.V., Godoli G., Gorbatsky V.G., Graham J., Grasdalen G.L., Grygar J., Guerrero G., Guinan E.F., Hackwell J.A., Hall D.S., Hansen C.J., Haro G., Heidmann N., Heiser A., Herbig G.H., Herr R.B., Hesser J.E., Heyden F.J., Hill A., Hoffleit N., Houk N., Huruhata M., Hutchings J.B., Iben I., Jarzebowski T., Jerzykiewicz M., Jones A.F., Karp A.H., Kholopov P.N., Kippenhahn R., Kopylov I.M., Kordylewski K., Kraft R.P., Kreiner J.M., Krzemiński W., Kubiak M., Kuhi L.V., Kukarkin B.V., Kumsishvili J.I., Kwee K.K., Landolt A.U., Laskarides P.G., Ledoux P., Lesh J.R., Leung K.C., Lockwood G.W., Lortet-Zuckermann M.C., Luud L.S., McNamara D.H., Madore B.F., Maffei P., Makarenko E.N., Mannino G., Martin W.L., Masuni A., Mattei J.A., Mavridis L.N., Mayall M.W., Miller W.J., Milone E.F., Milone L.A., Mirzoyan L.V., Moffett T.J., Morguleff N., Morrison N.D., Mumford G.S., Murdin P.G., Nather R.E., Neff J.S., Nicolov N.S., O'Connell D.J.K., Odgers G.J., Opolski A., Osaki Y., Osawa K., Oskanyan V., Papaloizou J., Parthasarathy M., Peltier L.C., Percy J.R., Petersen J.O., Piirola V., Piotrowski S.L., Plaut L., Popova M., Pringle J.E., Pskovskij Yu.P., Rakos K.D., Renson P., Richter G., Robinson E.L., Rodgers A.W., Rodomo M., Romano G., Romanov Yu.S., Rosino L., Sadik A.R., Sanyal A., Sarma M.B.K., Sawyer Hogg H.B., Schmidt E.G., Schoembs R., Schwartz P.R., Scuflaire R., Sherwood W.A., Shobbrook R.R., Sinvhal S.D., Smeyers P., Smith Harlan J., Smith M.A., Starrfield S., Stellingwerf R.F., Stepien K Sterken C.L., Stobie R.S., Strohmeier W., Swope H.H., Szeidle B., Szkody P., Takeuti M., Tammann G.A., Tempesti P., Terzan A., Tjin A Djie H.R.E., Torres C.A., Tremko J., Tsesevich V.P., Tutukov A.V., Usher P.D., Valtier J.C., van Agt S., van Genderen A.M., van Hoof A., Vogt N., Wachmann A.A., Walker E.N., Walker M.F., Wallerstein G., Walraven T., Warner B., Webbink R.F., Wehlau A., Wesselink A.J., Williamon R.M., Willson L.A., Wilson L.A., Wing R.F., Wisniewski W., Wood P.R., Wright F.W., Zuckerman B.

28. COMMISSION DES GALAXIES (GALAXIES)

PRESIDENT: B.E. Westerlund

VICE-PRESIDENT: V.C. Rubin

COMITE D'ORGANISATION: H.C. Arp, E.A. Dibaj, J.A. Graham, J. Heidmann, P.W. Hodge, B.E. Markarian, M. Peimbert, G.A. Tammann, M.H. Ulrich, P.C. van der Kruit.

MEMBRES: Abell G.O., Ables H.D., Aguero E.L., Alfvén H., Allen R.J., Ambartsumian V.A., Andrillat Y., Arakelian M.A., Ardeberg A.L., Athanassoula E.O., Azzopardi M., Babadzhaniane M.K., Bahcall J.N., Barbieri C., Barbon R., Baum W.A., Benedict G.F., Berman R.H., Bertola F., Bigay J.H., Binnev J.J., Boeshaar G.O., Bok B.J., Boksenberg A., Bolton J.G., Bondi H., Bonnor W.B., Börngen F., Bottinelli L., Braccesi A., Brecher K., Brosche P., Brück M.J., Bruston P., Burbidge E.M., Burbidge G.R., Byrd G.G., Cannon R.D., Capaccioli M., Carranza G.J., Carswell R.R., Casini C., Chamaraux P., Chincarini G.L., Contopoulos G., Corwin H.G., Courtès G., Danks A.C., Davidsen A.F., Davies R.D., Denisyuk E.K., de Vaucouleurs G., di Serego A.S., di Tullio G., d'Odorico S., Einasto J.E., Ekers R.D., Elvius A.M., Esipov V.F., Evans D.A., Fairall A.P., Feast M.W., Fehrenbach C., Field G.B., Fomenko A.F., Ford W.K., Fraser C.W., Freeman K.C., Gallagher J.S., Galleta G., Gascoigne S.C.B., Ghigo F.D., Goss W.M., Gottesman S.T., Goughenheim L., Grandi S.A., Grasdalen G.L., Griffiths R.R., Hara T., Hardy E., Harms R.J., Haro G., Heckmann O., Heeschen D.S., Hickson P., Hintzen P.M.N., Holmberg E.B., Hoyle F., Huchra J.P., Illingworth G.D., Johnson H.M., Jones Th. W., Jugaku J., Kafatos M., Kalinkov M., Kalloglian A.T., Karachentsev I.D., Kellerman K.I., Khachikian E.Y., Kiang T., King I.R., Kirshner R.P., Kodaira K., Kojoian G., Kopylov I.M., Kormendy J., Korovyakovskij Yu.P., Kumar C.K., Kundu M.R., Kustaanheimo P., Lafon J.P.J., Larson R.B., Lasker B.M., Lauberts A., Lausberg A., Layzer D., Lelièvre G., Lequex J., Lin C.C., Lindblad P.O., Lo K.-Y., Lovelace R.V.E., Low F.J., Lynden-Bell D., Lynds B.T., Lynds C.R., Lyuty V.M., Madore B.F., McClure R.H., McCrea W.H., MacGillivray H.T., McGimsey B.Q. Jr., Mackay C.D., McVittie G.C., Mark J.W.-K., Maurice E., Mayall N.U., Menon T.K., Miller H.R., Miller J.S., Miller R.H., Mills B.Y., Mirabel I.F., Morgan W.W., Murray S.S., Narlikar J.V., Neugebauer G., Neyman J., Noonan T.W., Oemler A. Jr., Oke J.B., Oleak H., Omer G.C., Oort J.H., Osman A.M., Osterbrock D.E., Owen F.N., Pachner J., Pacholcyk A.G., Page T.L., Peters W.L. III, Peterson Ch.J., Pismish de Recillas P., Popov V.S., Poveda A., Prendergast K.H., Press W.H., Pritchet Ch.J., Pronik I.I., Pronik V.I., Reaves G., Reddish V.C., Reinhardt M., Richer H.B., Richter N., Rickard L.J., Rindler W., Roberts M.S., Roberts W.W., Robinson I., Rood H.J., Rosino L., Rudnicki K., Sancisi R., Sanders R.H., Sarazin C.L., Sargent W.L.W., Saslaw W.C., Schmidt M., Schücking E.L., Schuster H.E., Schwarz U.J., Schweizer F., Sciama D.W., Scott E.L., Searle L., Segal I.E., Sersic J.L., Setti G., Shahbazian R.K., Shakeshaft J.R., Shane C.D., Shen B.S.P., Sherwood W.A., Shields G.A., Shostak G.S., Smith B.F., Smith H.E. Jr., Smith M.G., Spinrad H., Stein W.A., Stibbs D.W.N., Stone R.P.S., Strom R.G., Sulentic J.W., Takarada K., Takase B., Talbot R.J., Terebizh V. Yu., Terzian Y., Thompson L.A., Thonnard N., Thuan T.X., Tifft W.G., Tinsley B., Toomre A., Tovmassian H.M., Treder H.J., Turner E.L., Tyson J.A., van Albada T.S., van den Bergh S., van der Laan H., van Woerden H., Véron M.P., Visvanathan N., Vorontsov-Velyaminov B.A., Wakamatsu K., Warner J.W., Weedman D.W., Wehinger P.A., Welch G.A., Whitford A.E., Wielen R., Wild P., Williams R.E., Williams Th.B.,

Wills B.J., Wilson A.G., Wlérick G., Wood R., Wynn-Williams G., Zasov A.V., Zel'dovich Ya. B., Zinn R.J.

29. COMMISSION DES SPECTRES STELLAIRES (STELLAR SPECTRA)

PRESIDENT: W.K. Bonsack

VICE-PRESIDENT: J. Jugaku

COMITE D'ORGANISATION: Y. Andrillat, G. Cayrel-de-Strobel, C.R. Cowley, C.O.R. Jaschek, V. Khokhlova, J.P. Swings, R. Viotti, K.O. Wright.

MEMBRES: Abhyankar K.D., Abt H.A., Adelman S.J., Aller L.H., Andrillat H., Asaad A.S., Aslanov I.A., Bappu M.K.V., Berger J., Bertaud Ch., Bessel M.A., Bidelman W.P., Boesgaard A.M., Boggess A. III, Bouigue R., Boulon J., Boyarchuk A.A., Buscombe W., Catalano S., Cayrel R., Climenhaga J.L., Code A.D., Conti P.S., Cowley A.P., Divan L., Doazan V., Dobronravin P.P., Dolidze M.V., Dravins D., Dunham Jr. T., Edmonds Jr. F.N., Evans D.S., Faraggiana R., Feast M.W., Freire R., Friedjung M., Fringant A.M., Fujita Y., Garrison R.F., Gershberg R.E., Gilra D.P., Glagolevskij Ju.V., Glushneva I.N., Goldberg L., Gorbatskij V.G., Gratton L., Gray D.F., Greenstein J.L., Griffin R.E.M., Griffin R.F., Groth H.G., Guthrie B.N.G., Hack M., Haro G., Heintze J.R.W., Henize K.G., Herbig C.H., Herman R., Houziaux L., Hyland A.R., Jaschek M., Johnson H.R., Keenan P.C., King R.B., Kipper T.A., Kodaira K., Kogure T., Kopylov I.N., Kovachev R.J., Kraft R.P., Labs D., Lambert D.L., Larsson-Leander G., Locanthi D.D., Luud L.S., McNamara D.H., Maitzen H.M., Mel'nikov O.A., Milligan J.E., Moffat A.F.J., Morguleff N., Morton D.C., Mustel E.R., Myerscough V.P., Nicholls R.W., Nikitin A.A., Nishimura S., Oetken L., Oke J.B., Osawa K., Pagel B.E.J., Parson S.B., Pasinetti L.E., Peat D.W., Pedoussant A., Peery B.F., Perrin M.N., Peterson R.C., Pilachovski C.A., Plavec M., Preston G.W. III, Przybylski A., Querci F., Rego Fernandez M., Reimers D., Ringuelet A.E., Rodgers A.W., Sahade J., Sanwal N.B., Schild R.E., Seggewiss W., Shchegolev D.E., Sinnerstad U., Slettebak A., Smak J., Smith M., Smolinski J., Snow Th., Spite F., Spite M., Stalio R., Stawikowski A., Stecher T.P., Svolopoulos S.N., Swensson J.W., Swings P., Taffara S., Thompson G.I., Tsuji T., Underhill A.B., Utsumi K., Van't Veer C., Vilhu O., Voigt H.H., Vreux J.M., Wallerstein G.N., Waterworth M.D., Wehlau W.H., Wellmann P., Wempe J., Wehinger F., Weniger S., Weiss W.W., Williams P.M., Wilson O.C., Wilson R., Wing R.F., Wolff S.C., Wood H.J., Wyckoff S., Wyller A.A., Yamashita Y.

30. COMMISSION DES VITESSES RADIALES (RADIAL VELOCITIES)

PRESIDENT: M. Duflot

VICE-PRESIDENT: A.G.D. Philip

COMITE D'ORGANISATION: J. Andersen, A.H. Batten, E. Maurice, J. Sahade.

MEMBRES: Abt H.A., Barbier-Brossat M., Beardsley W.R., Bertiau F.C., Bolton C.T., Bouigue R., Boulon J., Crampton D., De Jonge J.K., De Vaucouleurs G., Edmondson F.K., Eelsalu H.T., Fehrenbach C., Fletcher J.M., Florsch A., Georgelin Y., Griffin R.F., Harding G.A., Heintze J.R.W., Hill G., Hube D.P., Imbert M., Karachentsev I.D., Kraft R.P., Martin N., Mayor M., Mel'nikov O.A., Morbey C.L., Nordström B., Oetken L., Pedoussaut A., Perry C.L., Preston G.W. III, Prévot L., Rebeirot E., Rubin V.C., Sanwal N.B., Scarfe C.D., Serkowski K., Willstrop R.V.

31. COMMISSION DE L'HEURE (TIME)

PRESIDENT: S. Iijima

VICE-PRESIDENT: G. Hemmleb

COMITE D'ORGANISATION: D.J. Belotserkovskij, H. Enslin, H.F. Fliegel, B. Guinot, D.D. McCarthy, A. Orte, P. Pâquet, E. Proverbio, A.M. Sinzi, G.A. Wilkins.

MEMBRES: Abele M.K., Afanaseva P.M., Benavente J., Bender P.L., Billaud G., Blinov N.S., Bonanomi J., Caprioli G., Chamberlain J.M., Costain C.C., Dominski I., Drâmbă C., Gaignebet J., Gökmen T., Hall R.G., Hellwig H., Hers J., Klepczynski W.J., Lederle T., Lieske J.H., Markowitz Wm., Melbourne W.G., Melchior P., Mikhailov A.A., Morgan P., Morrison L.V., Mueller I.I., Naumov V.A., Noel F., Opalski W., Parcelier P., Pavlov N.N., Penny C.J.A., Pensado J., Pilkington J., Pushkin S.B., Randić L., Schuler W., Scheglov V.P., Shiryaev A.V., Smith H.M., Smylie E., Stanila G., Starytsin G.V., Stoyko A., Takagi S., Tsao Mo., Tsuchiya A., Vicente R.O., Wackernagel H.B., Webrová L., Weith-Knudsen N.P., Winkler G.M.R., Yasuda H., Yumi S.

33. COMMISSION DE LA STRUCTURE ET DE LA DYNAMIQUE DU SYSTEME GALACTIQUE (THE STRUCTURE AND DYNAMICS OF THE GALACTIC SYSTEM)

PRESIDENT: G.G. Kuzmin

VICE-PRESIDENT: R. Wielen

COMITE D'ORGANISATION: W.B. Burton, D.L. Crawford, M. Fujimoto, W. Iwanowska, F.J. Kerr, E.K. Kharadze, P.O. Lindblad, D. Lynden-Bell.

MEMBRES: Aarseth S., Agekjan T.A., Altenhoff W.J., Ambartsumian V.A., Andrle P., Antonov V.A., Aoki S., Ardeberg A.L., Athanassoula E., Baldwin J.E., Barbanis B., Bash F.N., Becker W., Beer A., Berkhuijsen E.M., Bigay J.H., Binney J.J., Blaauw A., Blanco V.M., Bok B.J., Boulon J., Burke B.F., Carrasco L., Churchwell E.B., Clube S.V.M., Contopoulos G., Courtès G., Crampton D., Crézé M., Cudworth K.M., Cuperman S., Davies R.D., De Jong T., Delhaye J., Dickel H.R., Dickel J.R., Dieter N.H., Downes D., Drilling J.S., Dzigvashvili R.M., Edmondson F.K., Efremov Y.N., Einasto J., Elmegreen B.G., Elsässer H., Elvius T., Evangelidis E., Faber S.M., Feast M.W., Fehrenbach Ch., Feitzinger J.V., Fenkart R.P., Fitzgerald M.P., Freeman K.C., Fricke W., Genkin I.L., Georgelin Y.M., Georgelin Y.P., Gliese W., Goldreich P., Gomez A.E., Gordon M.A., Grayzeck E.J., Gyldenkerne K., Habing H.J., Haug U., Hayli A., Heckmann O., Heiles C.E., Hénon M., Herbst W., Hobbs R.W., Hohl F., Hori G., Hughes V.A., Hulsbosch A.N.M., Humphreys R.M., Hunter C., Ikeuchi S., Innanen K.A., Irwin J.B., Isobe S., Iwaniszewska-Lubienska C., Jackson P.D., Jahreiss H., Jaschek C.O.R., Johnson H.M., Jones D.H.P., Kaburaki M., Kalandadze N.B., Kalnajs A.J., Kato S., Kholopov P.N., King I.R., Kinman T.D., Klare G., Knapp G.R., Kolesnik L.N., Kormendy J., Kulsrud R., Kurth R., Kutuzov S.A., Lafon J.P., Larson R.B., Lecar M., Lin C.C., Lodén K., Lodén L.O., Lunel M., Luyten W.J., Lyngå G., MacConnell D.J., MacRae D.A., Maksumov M.N., Malmquist K.G., Mark J.W.K., Marochnik L., Martinet L., Mathewson D.S., Mavridis L.N., Mayor M., McCarthy M.F., Ménessier M.O., Mezger P.G., Miller R.H., Mirabel I.F., Mirzoyan L.V., Miyamoto M., Moffat A.F.J., Monnet G., Morgan W.W., Morris M.R., Münch G., Murray C.A., Muzzio J.C., Nahon F., Neckel Th., Nelson A.H., Niimi H., Ogorodnikov K.F., Oja T., Okuda H., Ollongren A., Oort J.H., Ostriker J.P., Ovenden M.W., Palmer P.E., Pauls T.A.,

Pavlovskaya E.D., Peimbert M., Perek L., Perry C.L., Pesch P., Philip A.G. Davis, Pilowski K., Pismis P., Plaut L., Price R.M., Priester W., Ramberg J.M., Riegel K.W., Roberts M.S., Roberts W.W., Rohlfs K., Roman N.G., Rubin V.C., Ruíz M.T., Rybicki G.B., Saar E., Sandqvist Aa., Sanduleak N., Schilt J., Schmidt H., Schmidt K.H., Schmidt M., Schmidt-Kaler Th., Seggewiss W., Shane W.W., Sharov A.S., Sher D., Shimizu T., Shu F., Shuter W.L.H., Simonson S.C., Slettebak A., Solomon P.M., Spiegel E.A., Stecker F.W., Steinlin U.W., Stephenson C.B., Stibbs D.W.N., Sturch C., Svolopoulos S.N., Szebehely V., Tammann G.A., Thè Pik Sin, Toomre A., Tosa M., Trefzger C.F., Tsioumis A., Turon-Lacarrieu P., Upgren A.R., van der Kruit P.C., Vanderlinden H.L., Vandervoort P.O., van Hoff A., van Woerden H., Varsavksy C.M., Velghe A.G., Venugopal V.R., Verschuur G.L., Vetesník M., Voroshilov V.I., Wayman P.A., Weaver H.F., Weistrop D., Westerhout G., Westerlund B.E., White R.E., Whiteoak J.B., Whittet D.C.B., Wielebinski R., Wilson Th. L., Woltjer L., Woodward P.R., Woolley Sir Richard, Yuan C.

34. COMMISSION DE LA MATIERE INTERSTELLAIRE ET DES NEBULEUSES PLANETAIRES (INTERSTELLAR MATTER AND PLANETARY NEBULAE)

PRESIDENT: V. Radhakrishnan

VICE-PRESIDENT: M. Peimbert

COMITE D'ORGANISATION: T. de Jong, B.D. Donn, G.B. Field, L.A. Higgs, E.B. Kostyakova, J. Lequex, U. Mebold, M. Morimoto, Y. Terzian, B. Zuckerman.

MEMBRES: Aannestad P., Acker A., Aitken D.K., Adabane K., Aller L.H., Altenhoff W., Andrew B.H., Andriesse C.D., Andrillat H., Andrillat Y., Arhipova V.P., Arny T.T., Avery L., Axford W.I., Baars J.W.M., Baker P., Baldwin J.E., Barlow M., Barrett A.H., Barnes A., Bash F.N., Baudry A., Becklin E., Beckman J., Bel N., Bergeron J., Berkhuijsen E.M., Bernat A., Biermann L., Bieging J., Birkle K., Black J., Blades J., Blair G., Bless R.C., Boeshaar G., Boggess A., Bohlin R.C., Bok B.J., Borgman J., Brand P.W.J.L., Braunsfurth E., Brown R., Bruck M.T., Bunner A., Burgess A., Burke B.F., Burton W.B., Bystrova N.V., Capriotti E.R., Carruthers R.G., Caswell J.L., Cesarsky C., Cesarsky D., Chopinet M., Churchwell W.B., Clark F., Code A.D., Cohen M., Collin-Souffrin S., Costero R., Courtès G., Cowie L., Cox D.P., Coyne G.V., Crovisier J., Cruvellier P., Cudaback D.D., Cugnon P., Czyzak S.J., Dahn C., Dalgarno A., Danks A.C., Davies R.D., de Boer K.S., Dibaj E.A., Dickel H.R., Disney M.J., D'Odorico S., Dopita M.A., Dorschner J., Downes D., Drapatz S., Dunham T., Dupree A., Dyson J.E., Elliott K., Elmegreen B., Elvius A.M., Emerson J., Encrenaz P., Esipov V.F., Falk S., Falle S.A.E.G., Faulkner D.J., Feibelman W., Felli M., Felten J.E., Flannery B., Flower D., Ford H., Friedemann C., Gardner F.F., Gaustad J.E., Gay J., Gehrels T., Georgelin Y., Gerard E., Gerola H., Gezari D., Gilra D.P., Giovannelli R., Goebel J., Goldreich P., Goldsmith D.W., Goldsmith P., Goldstein S.J., Goldsworthy F.A., Gordon C.P., Gordon M.A., Gosachinsky I.V., Goss W.M., Grasdalen G., Greenberg J.M., Grewing M., Guelin M., Gull T.R., Gurzadjan G.A., Habing H.J., Hackwell J., Hall J.S., Hardebeck E.G., Harrington J., Harten R.H., Harvey P., Hausiaux L., Haynes R., Hefele H., Heiles C.E., Helfer H.L., Henize K.G., Herbig G.H., Herzberg G., Hidajat B., Hiltner W.A., Hippelein H.H., Hjellming R.M., Hobbs L.M., Hoglund B., Hollenbach D., Hollis J., Hong S.S., Hua C.T., Hughes V.A., Hulsbosch A.N.M., Hummer D., Hutchings J.B., Irvine W.M., Isobe S., Jackson P., Jenkins E.B., Jennings R.B., Johnson F.M., Johnson H.M., Jones F., Johnston K.J., Jura M., Kafatos M., Kaftan-Kassin M.A., Kahn F.D., Kaifu N., Kaler J.B., Kamijo F., Kazes I., Kegel W.H., Kerr F.J., Kharadze E.K., Khromov G.S., Kimura H., Kirkpatrick R.C., Kirshner R., Knacke R., Knapp G.R.,

Knude J., Kohoutek L., Komesaroff M.M., Krishna Swamy K.S., Kumar C., Kundu M.R., Kutner M., Kwok S., Lada C., Lambrecht H., Langer W., Lasker B.M., Laurent C., Lee T.J., Le Squeren A.M., Leung C.M., Liller W., Lin C.C., Linke R., Lo K.Y., Loden L.O., Loren R., Lortet-Zuckerman M.C., Lovas F., Low F.J., Lucas R., Lutz B.L., Lynds B.T., Maihara T., Manchester R.N., Manfroid J., Mariagin V., Martin R., Mathews W.G., Mathewson D.S., Mathis J.S., Mattila K., McCray R.A., McCrea W.H., McGee R.X., McKee C., McKeith C., McLeod J., McNally D., Meaburn J., Mendez R., Menon T.K., Meszaros P., Mezger P.G., Michalitsioanos A., Michel K.W., Millar T., Miller J.S., Minn Y., Morgan D.H., Morris M., Mouschovias T.P., Morton D.C., Mufson S., Munch G., Myers P., Nandy K., Neugebauer G., Nussbaumer H., O'Dell C.R., Okuda H., Oort J.H., Osaki T., Osborne J., Osterbrock D.E., Ozernoy L.M., Palmer P.E., Panagia N., Pankonin V., Parker E.N., Pauls T., Pecker J.-C., Penston M.V., Penzias A., Perinotto M., Peters W., Petrosian V., Phillips T., Pishmish P., Poppel W., Pottasch S.R., Price R.M., Pronik I.I., Pskovskij Yu. P., Raimond E., Reynolds R.J., Richard L.J., Righini-Cohen G., Robbins R.R., Roberts W.W., Robinson B.J., Roger R., Rogers A.E.E., Rohlfs K., Rose W.K., Rosino L., Roxborough I., Rozhkovskij D.A., Rubin V.C., Sabano Y., Sabbadin F., Salpeter E.E., Sancisi R., Sandquist A., Sarazin C., Sato F., Sato S., Savage B., Savedoff M.P., Scalo J., Schalen C., Schatzman E., Scherb F., Sheuer P.A.G., Schmid-Burgk J., Schmidt K.H., Schmidt Th., Schmidt-Kaler Th., Schwartz P., Schwarz U.J., Seaton M.J., Seki M., Serkowski K., Shah G.A., Shane W.W., Shao C.Y., Shapiro M.M., Sharpless S.L., Shaver P.A., Shawl S.J., Shcheglov P.V., Sherwood W., Shields G., Shklovskij I.S., Shu F., Shuter W.L., Silberberg R., Silk J., Simons S., Sivan J.P., Skilling J., Smith B., Smith P., Snow T.P., Sobolev V.V., Sofia S., Solomon P.M., Somerville W.B., Sorensen S.A., Sorochenko R.L., Spitzer L., Stasinska G., Staude H., Stecher T.P., Steigman G., Strel'nitskij V., Strom R.G., Stromgren B., Sullivan W.T., Swings J.P., Takakubo K., Tamura S., Tarafdar S., Tenorio-Tagle G., Thaddeus P., The P.K., Thompson A.R., Thonnard N., Torres-Peimbert S., Townes C.H., Turner B.E., Turner K.C., Ulrich M.H., Unno W., van de Hulst H.C., van den Bout P., van der Laan H., van Woerden H., Vanysek V., Verschuur G.L., Vidal J.L., Vidal-Madjar A., Viner M., Visvanathan N., Vorontsov-Vel'yaminov B.A., Vrba F., Walker G.A.H., Wannier P., Weaver H.F., Webster B.L., Weiler K.W., Weisheit J.C., Wendker H.J., Wesselius P.R., Westerhout G., Weymann R.J., Whiteoak J.B., Whittet D., Wickramasinghe N.C., Williams D., Williams D.A., Williams I.P., Williams R.E., Willner S., Wilson R.W., Winnberg A., Witt A.N., Wolstencroft R.D., Woltjer L., Woodward P., Woolf N.J., Wright E., Wynn-Williams G., Yabushita S., York G., Yorke H., Zealey W., Zeilik M., Zimmerman H.

35. COMMISSION DE LA CONSTITUTION DES ETOILES (STELLAR CONSTITUTION)

PRESIDENT: R.J. Tayler

VICE-PRESIDENT: A.N. Cox

COMITE D'ORGANISATION: D.J. Faulkner, P. Giannone, I. Iben, R. Kippenhahn, B. Paczynski, G. Ruben, D. Sugimoto, A. Tutukov, J.P. Zahn.

MEMBRES: Aizenmann M., Anand S.P.S., Arai K., Arnett W.D., Appenzeller I., Audouze J., Ayashi C., Baglin A., Baym G., Beaudet G., Biermann L., Bludman S.A., Bodenheimer P., Böhm K.H., Bondi H., Boury A., Brownlee R.R., Burbidge G.R., Cameron A.G.W., Carson T.R., Castellani V., Castor J.I., Caughlan G.R., Chandrasekar S., Chevalier C., Chin H.Y., Chiosi C., Christensen-Dalsgaard J., Christy R.F., Cohen J.M., Cowan J.J., Cowling T.G., Cox J.P., D'Antona F., Davis C.G., Dearborn D.S.P., De Greve J.-P., Deinzer W., Demarque P.R., Despain K.H., Dcupree R.G., Dingens P.S.A., Djie T.A., Dluzhnevskaya O.B., Durisen R.H., Dziembowski W.A., Edwards A.C., Eggleton P.P., Eminzade T.A., Endal A.S., Epstein I., Ergma E.V., Ezer D., Faulkner J.,

Flannery B.P., Fontaine G., Forbes J.E., Fowler W.A., Frantzman Yu. L., Fricke K.J., Fujimoto M., Gabriel M., Goossens M., Gough D.O., Graham E., Hamada I., Hejlesen P.M., Hilf E.R.H., Hitotuyanagi Z., Hoshi R., Hoyle F., Imschennik V.S., Isaak G.R., Itoh N., James R.A., Kähler H., King D.S., Kochar R.K., Kolesnik I.G., Köster D., Kothari D.S., Kovetz A., Kozowski M., Krook M., Kumar S.S., Kushwaha R., Lamb D.Q. Jr., Lamb S.A., Larson R.B., Latour J., Lauterborn D., Lebovitz N.R., Ledoux P., Lépine J.R.D., McCrea W.H., Maeder A., Marx G., Masani A., Massevich A.G., Mazurek T.J., Mazzitelli I., Mestel L., Meyer-Hofmeister E., Michaud G., Monaghan J.J., Moore D.R., Morris S.C., Moss D.L., Möllenhoff C., Nadyozhin D.K., Nakano T., Nakazawa K., Nariai K., Narita S., Newman M.J., Nishida M., Nomoto K., Ohyama N., Okamoto I., Osaki Y., Ostriker J.P., Papaloizou J.C.B., Pines D., Plavec M., Popova M.D., Porfir'ev V.V., Poveda A., Prentice A.J.R., Provost J., Rädler K.-H., Ramadurai S., Reeves H., Reiz A., Renzini A., Rood R., Rosseland S., Rouse C.A., Roxburgh I., Sakashita S., Sakurai T., Salpeter E.E., Sato K., Savedoff M.P., Scalo J.M., Schatten K.H., Schatzman E., Schramm D.N., Schutz B.F., Schwartzschild M., Scuflaire R., Sears R.S. Seidov Z.F., Sengbusch K.V., Shaviv G., Shibata Y., Smeyers P., Smith R.C., Sobouti Y., Souffrin P., Sparks H., Spiegel E.A., Starrfield S.S., Stellingwerf R.F. Stibbs D.W.N., Strittmatter P.A., Strömgren B., Suda K., Sweet P.A., Sweigart A.V., Temesvary S., Thomas H.C., Tinsley B.M., Toomre J., Trimble-Weber V.L., Truran J.W., Tuominen I.V., Tuominen J.V., Uchida J., Ulrich R., Uus U., van der Borght R., van Horn H.M., van Riper K.A., Vardya M.S., Vauclair G., Vilas S.C., Vilhu O., Ward R.A., Weaver T.A., Webbink R.F., Weigert A., Weiss N.O., Wood P.R., Wright G.A.E., Yorke H.W., Zhevakin S.A., Ziolkowski J.

36. COMMISSION DE LA THEORIE DES ATMOSPHERES STELLAIRES (THE THEORY OF STELLAR ATMOSPHERES)

PRESIDENT: G. Traving

VICE-PRESIDENT: B. Gustafsson

COMITE D'ORGANISATION: H. Firsch, A.G. Hearn, W. Kalkofen, K. Kodaira, J.M. Marlborough, D. Mihalas, A.A. Sapar, M.J. Seaton, A.B. Underhill.

MEMBRES; Abhyankar K.D., Aller L.H., Altrock R.C., Arpigny C., Athay R.G., Auer L.H., Auman J.R., Avrett E.H., Baschek B., Bell R.A., Bernat A.P., Bertaud C., Biermann L., Blanco C., Bless R.C., Boesgaard A.M., Böhm K.H., Böhm-Vitense E., Bues I.D., Carbon D., Carson T.R., Cassinelli J., Castor J.I., Cayrel R., Cayrel de Strobel G., Clegg R.F.S., Conti P.S., Cowley C.R., Cram L.E., Cuny Y., Davis C.G., DeJager C., Delache Ph., Domke H., Dufton P.L., Dumont S., Edmonds F.N., Elste G.H., Evangelinis E., Faraggiana R., Finn G.D., Fischel D., Foy R., Frisch U., Gail H., Gebbie K.B., Gingerich O., Gökdogan N., Gordon Ch., Gough D., Grant I.P., Gray D.F., Greenstein J.L., Grevesse N., Grinin V.P., Groth H.G., Gussmann E.A., Hack M.H., Hardorp J., Heidmann N., Hekela J., Hitotuyangai Z., Holweger H., Holzer T.E., Hotinli M., House L.L., Houtgast J., Hummer D.G., Hunger K. Hutchings J.R., Ivanov V.V., Jefferies J.T., Johnson H.R., Kamp L.W., Kandel R.S., Karp A.H., Khokhlova V.L., Klein R.I., Kolesov A.K., Kontizas E., Krishna Swamy K.S., Kudritzki R.P., Kuhi L.V., Kumar S.S., Kurucz R.L., Kushwaha R.S., Lambert D.L., Leibacher J., Linsky J.L., Matsumoto M., Matsushima S., Miyamoto S., Michaud G., Mukai S., Müller E.A., Münch G., Mustel E.R., Mutschlecner J.P., Myerscough V.P., Nariai K., Neff J.S., Neven L., O'Mara B.J., Orrall F.Q., Oxenius J., Pagel B.E.J., Panek R.J., Pasinetti L., Pecker J.-C., Peraiah A., Peters G.J., Phillips J.G., Pottasch S.R., Praderie F., Ramsey L.W., Reimers D., Ross J.E.R., Rybicki G.B., Saito S., Sakhibullin N.A., Schmalberger D.C., Schmid-Burgk J., Scholz M., Schönberner D., Sedlmayr E., Shine R.A., Shipman H.L., Skumanich A., Simonneau E., Sitnik G.F., Soblev V.V.,

Sobouti Y., Souffrin P., Spiegel E.A., Spite F., Spite M., Stalio R., Stein R.F., Stepien K., Stibbs D.W.N., Strom S.E., Swihart T.L., Tarafdar S.P., Thomas R.N., Thompson R.I., Toomre J., Tsuji T., Ueno S., Uesugi A., Unno W., Unsöld A., van Regemorter H., van t'Veer C., van t'Veer F., Vardya M., Viik T., Wehrse R., Weidemann V., Wellmann P., Wickramsinghe D.T., Wilson P.R., Wright K.O., Wyller A.A., Yorke H.W., Zahn J.P., Zirker J.B., Zwaan C.

37. COMMISSION DES AMAS STELLAIRES ET DES ASSOCIATIONS (STAR CLUSTERS AND ASSOCIATIONS)

PRESIDENT: G. Lyngå

VICE-PRESIDENT: K.C. Freeman

COMITE D'ORGANISATION: T.A. Agekjan, B. Balazs, G.L.H. Harris, D.C. Heggie, S. van den Bergh.

MEMBRES: Aarseth S., Alcaino G., Allen C.P., Alksnis A.K., Artiukhina N.M., Barkhatova K.A., Becker W., Bijaoui A., Blaauw A., Bouvier P.B., Breger M., Burkhead M.S., Burnichon M.L., Butler D., Byrd G.G., Cannon R.D., Chun M.S., Clariá J.J., Cudworth K.M., Cuffey J., DeGraeve E., Demers S., Dickens R.J., Dluzhnevskaya Q.B., Efremov Yu. N., Einsato J., Elmegreen B.G., Feast M.W., Feinstein A., Fitzgerald M.P., Gascoigne S.C.B., Golay M., Grubissich C., Hanes D.A., Harris W.E., Hassan S.M., Hawarden T.G., Herbst W., Hénon M., Hesser J.E., Hills J.G., Iben I., Illingworth G.D., Janes K.A., Jones D.H.P., Kadla Z., Kamp L.W., Kholopov P.N., King I.R., Kontizas M., Kron G.E., Lada C.J., Larsson-Leander G., Laval A., Lavodovskij V.V., Liller M.W., Lodén L.O., Lohmann W., Lu P.K., Lynden-Bell D., Maeder A., Markkanen T., Marraco H.G., Martins D.H., Menon T.K., Menzies J.W., Meurers J., Moffat A.F.J., Morgan W.W., Murray C.A., Newell E.B., Osborn W., Penny A.J., Peterson C.J., Petrovskaya I.V., Philip A.D.G., Pilachowski C.A., Popova M.D., Poveda A., Rajamohan R., Reddish V.C., Rosino L., Rountree-Lesh J., Ruprecht J., Salukvadze G.M., Sanders W.L., Sawyer Hogg H.B., Schuster H.E., Sharov A., Sher D., Simoda M., Swope H.H., Terzan A., Thé P.S., Turner D.G., Upgen A., Van Altena A.F., Vanderlinden H.L., Veltman Yu. K., Vogt N., Von Hoerner S., Walker G.A.H., Walker M.F., Wallenquist Å.A.E., Warren W.H., Weaver H.F., Wehlau A., White R.E., Wielen R., Woolley R., Zinn R.J.

38. COMMISSION POUR L'ECHANGE DES ASTRONOMES (EXCHANGE OF ASTRONOMERS)

PRESIDENT: J. Delhaye

VICE-PRESIDENT: F.B. Wood

COMITE D'ORGANISATION: G.S. Kromov, D.A. MacRae, E.A. Müller, P.M. Routly, J. Sahade, F.C. Smith, H.H. Voigt, J.P. Wild.

MEMBRES: Abetti G., Al-Sabti A.W., Bok B.J., Godoli G., Haupt H., Keller G., Kourganoff V., Leung K.-C., Marik M., Miyamoto S., Okoye S.E., Opolski A., Page T.L., Rosseland S., Ruben G., Smith E.v.P., Stoy R.H., Swings P., Teleki G., Tolbert C.R., Wyatt S.P. Jr.

40. COMMISSION DE LA RADIOASTRONOMIE (RADIO ASTRONOMY)

PRÉSIDENT: G. Swarup

VICE-PRÉSIDENT: K.I. Kellermann

COMITÉ D'ORGANISATION: J.G. Ables, R.S. Booth, N.W. Broten, G.A. Dulk, R. Fanti, B. Höglund, N.S. Kardashev, A.T. Moffet, M. Morimoto, S. Okoye, H. van der Laan, L. Weliachew.

MEMBRES: Abrami A., Akabane K., Alexander J.K., Allen R.J., Aller H., Aller M.F., Altenhoff W.J., Andrew B.H., Argyle P.E., Assousa G.E., Avery L.W., Avignon Y., Baars J.W.M., Baart E.E., Backer D.C., Baldwin J.E., Barret A.H., Barrow C., Bates R.H.T., Benz A.O., Berge G.L., Berkhuijsen E., Bieging J.H., Biraud F., Blandford R.D., Blum E.J., Böhme A., Boischot A., Bolton J.G., Bracewell R.N., Braude S.Ya., Bridle A.H., Brouw W.N., Browne I.W.A., Burbidge G.R., Burke B.F., Caroubalos C., Carr T.D., Caswell J.L., Ceccarelli M., Christiansen W.N., Clark B.G., Clark D.H., Clark F.O., Cohen M.H., Cohen R.J., Cole T., Condon J.J., Conklin E.K., Conway R.G., Cooper B.F.C., Costain C.H., Coutrez R.A.J., Covington A.E., Croom D.I., Crovisier J., Cudaback D.D., Daintree E.G., Daishido T., Davies J.G., Davies R.D., Davis M.M., Davis R.J., de Groot T., de Jager G., Delannoy J., Denisse J.F., Dent W.A., Dickel H.R., Dickel J.R., Dieter N.H., Dixon R., Douglas J.N., Downes D., Downs G.S., Drake F.D., Dravskikh A.F., Dyson F.J., Ekers R.D., Elgaroy O., Ellis G.R.A., Elsmore B., Elwert G., Epstein E.E., Erickson W.C., Eriksen G.,Eshleman V.R., Evans K.D., Ewing M., Feix G., Feldman P.A., Felli M., Felten J.E., Field G.B., Findlay J.W., Fleischer R., Fokker A.D., Fomalont E.B., Fort D.N., Forster P.R., Frater R.H., Friedman H., Galt J., Gardner F.F., Gel'frejkh G.B., Gent H., Gergely T.E., Ghigo F.D., Gibson D.M., Ginzburg V.L., Gioia I.M., Gold T., Goldsmith P.F., Goldwire H.C., Goldstein S.J., Gonze R., Gopala Rao U.V., Gordon M.A., Gorgolewski S., Gosachinsky I.V., Goss W.M., Gower J.F.R., Gregory P.C., Grewing M., Guelin M., Gulkis S., Gull S.F., Hachenberg O., Haddock F.T., Hagen J.P., Hamilton P.A., Hanasz J., Hanbury Brown R., Harris D.E., Harten R.H., Hartz T.R., Haslam C.G.T., Haynes R.F., Hazard C., Heeschen D.S., Heidmann J., Heiles C., Hewish A., Hey J.S.,Higgs L.A., Hill E.R., Hills R.E., Hirabayashi H., Hjellming R.M., Hoang Binh D., Hobbs R.W., Högbom J.A., Hogg D.E., Hollis J.M., Hooghoudt B.G., Howard W.E., Huchtmeier W.K.-H., Hughes V.A., Hulsbosch A.N.M., Ikhsanova V.N., Jaffe W.J., Jauncey D.L., Jennison R.C., Johnson D.R., Johnston K., Joly F., Joshi M.N., Kaftan-Kassim M., Kahn F.D., Kai K., Kaifu N., Kakinuma T., Kapahi V.K., Kaufmann P., Kawabata Kin-Aki, Kazes I., Kenderdine S., Kerr F.J., Kesteven M.L., Kislyakov A.G., Ko Hsien-Ching, Kojoian G., Komesaroff M., Korolkov D.V., Kotelnicov V.A., Krauss J.D., Krishnan T., Kronberg P.P., Kuijpers J.M.E., Kundt W., Kundu M.R., Kuril'chik V.N., Kutner M.L., Kuz'min A.D., Lada C.J., Laffineur M., Lang K.R., Langer W.D., Lantos P., Large M.I., Leblanc Y., Lépine J.R.D., Lequeux J., Le Squeren A.M., Leung C.M., Lilley A.E., Linke R.A., Little A.G., Little L.Th., Lo K.-Y., Locke J.L., Longair M.S., Loren R.B., Lovell Sir Bernard, Lyne A.G., McAdam W.B., McCulloch P.M., McGee R.X., McKenna-Lawlor S., McLean D.J., MacLeod D., MacRae D.A., Macdonald G.H., Machalski J., Manchester R.N., Maran S.P., Martin R.N., Mattila K., Matveyenko L.I., Maxwell A., Mayer C.H., Mebold U., Meeks M.L., Menon T.K., Mezger P.G., Miley G.K., Mills B.Y., Milne D.K., Mirabel I.F., Moiseev I.G., Moran J., Morison I., Moriyama F., Morris D., Morris M.R., Muller C.A., Myers P., Nicolson G.D., Oort J.H., Osterbrock D.E., O'Sullivan J.D., Owen F.N., Pacholczyk A.G., Palmer H.P., Palmer P., Pankonin V.L., Papagiannis M.,

Parijskij Yu.N., Parker E.A., Parrish A., Pasachoff J.M., Pauliny-Toth I.I.K., Pauls T.A., Pedlar A., Penzias A.A., Peters W.L., Pettengill G.H., Phillips T.G., Pick-Gutman M., Ponsonby J.E.B., Pooley G.G., Preuss E., Price R.M., Priester W., Radhakrishnan V., Raimond E., Ramaty R., Rao A.P., Razin V.A., Readhead A.C.S., Reber G., Reid M.J., Ribes J.-C., Rickard L.J., Rickett B.J., Riley J.M., Roberts J.A., Roberts M.S., Robinson B.J., Roeder R.C., Roger R.S., Rogers A.E.E., Rogstad D.H., Rohlfs K., Rowson B., Rudnick L., Rydbeck O.E.H., Ryle Sir Martin, Ryzhkov N.F., Salomonovitch A.E., Salpeter E.E., Sanamjan V.A., Sastry Ch.V., Sato F., Scalise E.Jr., Scheuer P.A.G., Schilizzi R.T., Schmidt M., Schwartz P.R., Schwarz U.J., Scott J.S., Scott P.F., Seaquist E.R., Seielstad G.A., Shaffer D., Shakeshaft J.R., Sheridan K.V., Shimmins A.J., Shklovsky I.S., Sholomitsky G.B., Shuter W.L.H., Sieber W., Simon P., Slade M.A., Slee O.B., Slysh V.I., Smith A.G., Smith D.F., Smith F.G., Smith H.J., Soboleva N.S., Sofue Y., Sorochenko R.L., Spencer R.E., Sramek R., Stanley G.J., Stannard D., Steinberg J.L., Stewart P., Stone R.G., Strom R.G., Sullivan W.T. III, Swenson G.W.Jr., Tabara H., Takagi K., Takakura T., Tanaka H., Tanaka R., Tarter J.C., Terzian Y., Thomasson P., Thompson A.R., Tlamicha A., Tolbert C.R., Tomasi P., Tovmassian H.M., Townes C.H., Tritton K.P., Troitsky V.S., Tsuruta S., Turlo Z., Turner B.E., Turner K.C., Turtle A.J., Ulrich B.T., Demoulin Ulrich M.H., Vallée J.P., van de Hulst H.C., Vanden Bout P., van der Kruit P.C., van Nieuwkoop J., van Woerden H., Varsavsky C.M., Velusamy T., Venugopal V.R., Véron P., Verschuur G.L., Viner M.R., Wade C.M., Wall J.V., Walmsley M.C., Walsh D., Wannier P.G., Wardle J.F.C., Warner P.J., Warwick J.W., Weaver H.F., Wendker H.J., Westerhout G., Westfold K.C., Whiteoak J.B., Wickramasinghe N.C., Wielebinski R., Wild J.P., Wilkinson P.N., Williams D., Willis A.G., Wills B., Wills D., Wilson A.S., Wilson R.W., Wilson T.L., Wilson W.J., Wink J.E., Winnberg A., Winnewisser G., Witzel A., Woltjer L., Wright A.E., Zaitsev V.V., Zhelezniakov V.V., Zlobec P., Zuckerman B.M.

41. COMMISSION DE L'HISTOIRE DE L'ASTRONOMIE (HISTORY OF ASTRONOMY)

PRESIDENT: M.A. Hoskin

VICE-PRESIDENT: O. Pedersen

COMITE D'ORGANISATION: S.M.R. Ansari, J. Dobrzycki, J.A. Eddy, E. Forbes, P.G. Kulikovsky.

MEMBRES: Abetti G., Argyracos J.G., Ashbrook J., Beer A., Berendzen R., Brunet J.P., Cimino M., Dadic Z., Dekker E., Deeming T.J., Dewhirst D.W., Douglas A.V., Eelsalu H., Erpylev N.P., Evans D.S., Ferni D., Ferrari-D'Occhieppo D.K., Filliozat J., Firneis M.G., Fleckenstein J.O., Freiesleben H.C., de Freitas-Mourao R.R., Hawkins G.S., Hayli A., Heggie D., Herrmann D., Hirose H., Hysom E.J., Horsky Z., Howse H.D., Jackisch G., Kennedy J.E., Khromov G.S., Kiang T., King H.C., Kotsakis D., Lang K.R., Levy J.R., Link P., McKenna-Lawlor S.M.P., Meadows A.J., Merleau-Ponty J., Michel H., Moesgaard K.P., Nakayama S., Norlind J., North J., Omer G.C., Osterbrock D., Pelseneer J., Petri W., Pingree D., Pogo A., Ronan C.A., Rosen E., Rybka E., Rybka P., Shcheglov V.P., Shukla K.S., Sullivan W., Svolopoulos S.N., Swerdlow N., Taton R., Thoren V., Verdet J.P., Wattenberg D., Whitaker E.A., Whitrow G.J., Wright H., Yabuuti K., Yeomans D.K., Zagar F.

42. COMMISSION DES ÉTOILES DOUBLES SERRÉES (CLOSE BINARY STARS)

PRÉSIDENT: B. Warner

VICE-PRÉSIDENT: A.H. Batten

COMITÉ D'ORGANISATION: A.M. Cherepashchuk, M.G. Fracastoro, K. Gyldenkerne, M. Kitamura, R.H. Koch, Y. Kondo, G. Larsson-Leander, L.B. Lucy, S.D. Sinvhal, J. Smak, E.P.J. van den Heuvel, J.A.J. Whelan.

MEMBRES: Abhyankar K.D., Andersen J., Bappu M.K., Bath G.T., Beer A., Binnendijk L., Blitzstein W., Bolton C.T., Bookmyer B.B., Bopp B.W., Breinhorst R., Broglia P., Brownlee R.R., Budding E., Bunner A.N., Catalano S., Cester B., Chambliss R.C., Chen K.-Y., Chis G., Cillié G.G., Clausen J.V., Collins G.W., Cowley A.P., Cristaldi S., Dadaev A.N., De Grève J.-P., de Groot M.J.H., de Kort J., Doughty N.A., Duerbeck H.W., Eaton J.A., Ebbighausen E.G., Faulkner J., Ferrari d'Occhieppo K., Flannery B.P., Fracassini M., Frantsman Yu.L., Fredrick L.W., Fresa A., Friedjung M., Gaposchkin S., Garmany C.D., Geyer E.H., Giannone P., Gibson D.M., Grygar J., Guinan E.F., Gursky H., Hall D.S., Hammerschlag-Hensberge G., Harmanec P., Hazlehurst J., Herczeg T.J., Hilditch R.W., Hill G., Hjellming R.M., Horák T., Hube D.P., Hutchings J., Irwin J.B., Jaschek C., Joss P.C., Jurkevich I., Kawabata S., Kizilirmak A., Kopal Z., Kraft R.P., Krat V.A., Kreiner J.M., Kříž S., Kron K.G., Kruszewski A., Krzeminski W., Kumsiashvily M.I., Kviz Z., Kwee K.K., Lamb D.Q., Landolt A.U., Lavrov M.I., Leung K.-Ch., Linnell A.P., Magalashvily N.L., Mammano A., Marino B.F., Martynov D.Ya., Mattei J.A., Mauder H., Mayer P., McCluskey G.E., Merrill J.E., Meyer-Hofmeister E., Milone E.F., Morgan T.H., Mumford G.S., Nariai K., Nather R.E., Nelson B., O'Connell D.J.K., Oliver J.P., Ovenden M.V., Paczyński B., Peters G.J., Piirola V., Piotrowski S., Plavec M., Popper D.M., Purgathofer A., Rahe J., Rakos K.D., Refsdal S., Reuning E.G., Ritter H., Robinson E.L., Rodonò M., Rovithis P., Roxburgh I.W., Ruciński S., Sahade J., Sanwal N.B., Sanyal A., Scarfe C.D., Schmidt H., Schober H.J., Schöffel E.F., Seggewiss W., Semeniuk I., Serkowski K., Shakura N.I., Shapley M.B., Shaviv G., Shu F.H., Shulberg A.R., Smith R.C., Sobieski S., Solheim J.-E., Sparks W.M., Strohmeier W., Sugimoto D., Szafraniec R., Szkody P., Todoran I., Tremko J., Trimble V., Tsessevich V.P., Ureche V., van Hoof A., van Paradijs J., van't Veer F., Vetešnik M., Vilhu O., Walker W.S.G., Walter K., Webbink R.F., Wehlau W.H., Weigert A., Weiler E.J., Wellmann P., Wesselink A.J., Widorn T., Williamon R.M., Wilson R.E., Wood D.B., Wood F.B., Wright K.O., Yamasaki A., Ziółkowski J.

44. COMMISSION DE L'ASTRONOMIE A PARTIR DE L'ESPACE (ASTRONOMY FROM SPACE)

PRESIDENT: R.J. van Duinen

VICE-PRESIDENT: M. Oda

COMITE D'ORGANISATION: R.M. Bonnet, A.A. Boyarchuk, R.C. Catura, K. Fredga, Y. Kondon, L. Peterson, E. Peytreman, K. Pinkau, A.B. Underhill, B. Valnicek.

MEMBRES: Acton L.W., Alexander J.K., Aschenbach B., Benedict G.F., Blamont J.E., Bless R.C., Bogess A., Bohlin R.C., Boksenberg A., Boyd R.L.F., Brandt J.C., Brinkman A.C.H., Brueckner G.E., Bruner E.C., Bumba V., Burger M., Burton W.M., Carver J.H., Chubb T.A., Clark T.A., Code A.D., Courtès G.,

Grannell C.J., Culhane J.L., Davidsen A.F., Davis R.J., Dennis B.R., Dolan J.F., Dunkelman L., Dupree A., Faraggiana R., Fazio G.G., Fichter C.E., Fisher D.C., Friedman H., Gabriel A.H., Gezari D.Y., Giacconi R., Gilra D.P., Glaser H., Gold T., Goldberg L., Gondhalekar P., Greyber H.D., Griffiths R.E., Gursky H., Hack M., Haddock F., Hallam K.L., Harms R.J., Hartz T.R., Harvey Ch., Harvey P.M., Hayakawa S., Hearn A.C., Helmken H., Heise J., Henize K.G., Henoux J.C., Hinteregger H.E., Hoffman J.A., Houziaux L., de Jager C., Jamar C., Jechlee C., Jenkins E.B., Jordan C., Kramer G.G., Kraushaar W.L., Kurt V.G., Lamers H.J., Leckrone D.S., Lemaire P., Lindblad B.A., Lovell B., Lust P., Malaise D.J., Malitson H.M., Manara A., Mandelstham S.L., Maran S.P., Marov M.Y., Matsuoka M., McCracken K.G., McCluskey G.E., McWhirter R.W.P., Mehltretter J.P., Mewe R., Mikhailov A.A., Modisette S.P., Monfils A., Moos H.W., Morgan T.H., Morton D.C., Müller E.A., Ness N.F., Neupert W.H., Newkirk G.A., Novick R., Noyes R., Olthof M., Owens T., Pacini F., Papagiannis M., Parkinson J.H., Parkinson W.H., Prokofe'v V.D., Rahe J., Rao U.R., Rees M., Reeves E.M., Rense W.A., Righini-Cohen G., Roman N.G., Rosendhal J.D., Salomonovitch A.E., Sato H., Savage G., Schuerman D.W., Selvelli P., Severny A.B., Shivanandan K., Sholomitsky G.B., Simon P., Simon P.C., Smith H.J., Snow T.P., Sofia S., Speer R., Stecher T.P., Steinberg J.L., Stockman H.S., Stone R.G., Takakura T., Upson W.L., Van Beek H.F., Van der Hucht K.A., Van de Hulst H.C., Van Speybroeck L.P., Vidal-Madjar A., Walsh D., Warner J.W., Weiler E.J., Weinberg J.L., Wesselius P., Westphal J.A., Willner S.P., Wilson R., Wray J.D., Wu C.C., Yamashita K., Zombeck M.V.

45. COMMISSION DE CLASSIFICATION STELLAIRE (STELLAR CLASSIFICATION)

PRESIDENT: A. Slettebak

VICE-PRESIDENT: V.L. Straizys

COMITE D'ORGANISATION: A.L. Ardeberg, R.A. Bartaya, R.A. Bell, R.F. Garrison, B. Hauck, M. Jaschek, A.G.D. Philip.

MEMBRES: Albers H., Bahng J.D.R., Bappu M.K.V., Barbier-Brossat M., Beer A., Bidelman W.P., Blanco V.M., Boyce P.B., Buscombe W., Buser R., Cester B., Cherepashchuk A.M., Christy J., Claria J.J., Cowley A.P., Crampton D., Crawford D.L., Dessy J.L., Divan L., Duflot M., Elvius T., Evans D.S., Feast M.W., Fehrenbach Ch., Fracassini M., Gerbaldi M., Geyer E.H., Glagolevskij Ju. V., Golay M., Gurzadyan G., Gyldenkerne K., Hack M.H., Hallam K.L., Hartoog M.R., Hayes D.S., Heck A., Henize K.G., Herman R., Hill P.W., Hiltner W.A., Hoag A.A., Houk N., Humphreys R., Jaschek C., Keenan P.C., Kharadze E.K., Kron G.E., Levato H., Lodén K., Low F.J., Lyngå G., Lutz J., McCarthy M.F., McClure R.H., MacConnell D.J., McNamara D.H., Maeder A., Maehara H., Mavridis L.M., Mead J., Mendoza E.E., Morgan W.W., Morguleff N., Nandy K., Nikonov V.B., Notni P., Oja T., Osawa K., Osborn W., Parsons S., Pasinetti L.E., Perry C.L., Preston G.W. III, Reddish V.C., Richer H.R., Rountree-Lesh J., Rudkjøbing M., Sanduleak N., Sanwal N.B., Schild R.E., Schmidt-Kaler Th., Seitter W.C., Sharpless S.L., Sinnerstad U., Sinvhal S.D., Steinlin U.W., Stephenson C.B., Stock J., Strobel A., Strömgren B., Upgren A.R., Walborn N.R., Walker G.A.H., Warner B., Warren W., West R.M., Westerlund B., Wesselius P.R., Williams J.A., Wilson O.C., Wing R., Wyckoff S., Yoss K.M., Zdanavicius K.

46. COMMISSION POUR L'ENSEIGNEMENT DE L'ASTRONOMIE (THE TEACHING OF ASTRONOMY)

PRESIDENT: D.G. Wentzel

VICE-PRESIDENT: L.N. Houziaux and M. Rigutti

COMITE D'ORGANISATION: W. Buscombe, C. Ivaniszewska, J. Kleczek, E.V. Kononovich, L. Mavridis, S.E. Okoye, B.F. Peery, A. Ringuelet.

MEMBRES: Abhyankar K.D., Andrillat H.L., Asaad A.S., Atanasijevic I., Barkhatova K.A., Barocas V., Bottinelli L., Brahic A., Casanovas J., Catala-Poch Ma.A., Chamberlain J.M., Chun-Shan-Shen, Clarke D., Dibai E.A., Dinulescu N., Dominko F., Doughty N.A., Dy Puy D.L., Elgaröy O., Fawell D.R., Ferraz-Mello S., Gerbaldi M., Gouguenheim L., Grushinskiy N.P., Hayli A., Hauck B., Haupt H.F., Hazer S., Hidajat B., Ivanov V.V., Jarret A.H., Jørgensen H.E., Kennedy J.E., Kiang T., Kourganoff V., Kyung Loh Yu, Lambrecht H., Lumme K., Marik M., Marsh J.C.D., McNally D., Miles H.G., Moreno H., Müller E.A., Nikolov N.S., Onderlička B., Osborn W., Osorio J., Ovenden N.W., Owaki N., Pasachoff J.M., Percy J., Porfir'ev V.V., Proverbio E., Robbins R.R., Rodgers A.W., Roy A.E., Sandqvist Aa., Scheffler H., Schmidt T., Sevarlić B.M., Steinitz R., Torres-Peimbert S., Trehan S.K., Zimmermann H., Zwaan C.

47. COMMISSION DE LA COSMOLOGIE (COSMOLOGY)

PRESIDENT: G.O. Abell

VICE-PRESIDENT: J. Audouze

COMITE D'ORGANISATION: G. De Vaucouleurs, J. Gunn, S. Hayakawa, M.S. Longair, I. Novikov, G. Setti, G.A. Tammann.

MEMBRES: Aizu K., Alfvén H., Andrillat H., Baldwin J.E., Bardeen J.M., Barnothy J.M., Beckman J.E., Bel N., Bertola F., Bludman S.A., Bondi H., Bonnor W.B., Brecher K., Burbidge G.R., Calvani M., Cavaliere A., Chitre D.M., Cohen J.M., Condon J.J., Davidson W., Davis M.M., Demaret J., de Zotti G., Dicke R.H., Dyonisiou D., Doroshkevich A.G., Ehlers J., Ellis G.F.R., Faber S.M., Fall M.S., Falk S.W., Felten J.E., Field G.B., Florides P.S., Forman W.R., Fujimoto M., Fukui T., Godart O., Gold T., Goldsmith D.W., Goret P., Gratton L., Gregory S.A., Greyber H.D., Grishchuk L.P., Hara K., Hardy E., Harms R.J., Harrison E.R., Hawking S.W., Hayashi C., Heckmann O., Heidmann J., Heller M., Hoyle F., Huchra J.P., Icke V., Jauncey D.L., Jones B.J.T., Karachentsev I.D., Kasper U., Kato S., Kawabata K., Kormendy J., Kovetz A., Kozlovsky B.Z., Kustaanheimo P.E., Lasota J.P., Lausberg A., Layzer D., Liebscher D.E., McCrea W.H., McVittie G.C., Mészaros P., Misner C.W., Nariai H., Narlikar J.V., Ne'eman Y., Neyman J., Nishida M., Noerdlinger P.D., Noonan T.W., O'Connell R.F., Oemler A., Okoye S.E., Omer G.C., Omnes R., Oort J.H., Ozernoi L.M., Paal G., Pachner J., Palmer H.P., Peebles P.J.E., Penzias A., Persides S., Petrosian V., Press W.H., Rees M.J., Reeves H., Reinhardt M., Rindler W., Roberts D.H., Robinson I., Roeder R.C., Roxburgh I.W., Rubin V.C., Ruddy V., Rudnick L., Ryle M., Sapar A.A., Sargent W.L.W., Sato H., Savage A., Schatzman E., Scheuer P.A.G., Schmidt M., Schramm D.N., Schücking E., Sciama D.W., Scott E.L., Segal I.E., Seielstad G.A., Sersic J.L., Shaviv G., Shivanandan K., Simon R.L.E., Smith H.E., Spyrou N., Steigman G., Stewart J.M., Sunyaev R.A., Tartar J.C., Tauber G.E., Tayler R.J., Thompson L.A., Thuan X., Tifft W.G., Tinsky B.M., Treder H.J., Tully R.B., Turner E.L., Vaidya P.C., Van der Laan H., Vanysek V., Von Borzeszkowski H.H., Webster A.S., Wheeler J.A., Whitrow G.J., Will C.M., Wilson A.G., Zeldovich Ya. B., Zelmanov A.L., Zięba A., Zięba S.

48. COMMISSION DE L'ASTROPHYSIQUE DE GRANDE ENERGIE (HIGH ENERGY ASTROPHYSICS)

PRESIDENT: F. Pacini

VICE-PRESIDENT: R. Giacconi

COMITE D'ORGANISATION: J. Audouze, J. Culhane, K. Kellermann, E. Parker, E. Salpeter, I. Shklovski, J. Truemper.

MEMBRES: Abramowicz M., Adams D., Ahluwalia H.S., Alfven H., Alvarez L., Apparao K.M.V., Arons J., Aschenbach B., Asseo E., Avni Y., Axford W.I., Barnothy J., Baym G.A., Benford G., Bergeron J., Biswas S., Blandford R.D., Bludman S.A., Bonazzola S., Bonometto S., Boyd R.L.F., Brecher K., Bunner A.N., Burbidge G.R., Cameron A.G.W., Casse M., Catura R.C., Caughlan G.K., Cavaliere A., Chandrasekhar S., Chitre S.N., Chubb T.A., Chupp E.L., Clark G., Cohen J.M., Collin-Souffrin S., Condon J.J., Cowie L.L., Cowsik R., Cruise M., Dautcourt G., Davidsen A.F., Davidson W., Davis L., Jr., Davis M.M., Davison P., Debrunner H., De Felke F., De Graaf T., Dennis B.R., Dewitt B.S., De Young D.S., Dicke R.H., Disney M., Dolan J.F., Drake F.D., Duthie J.G., Edwards P.J., Eichler D., Fabian A.C., Fazio G., Felten J.E., Fenton K.B., Ferrari A., Fichtel C., Field G.B., Fisher P.C., Forman W.R., Fowler W.A., Friedman H., Garmire G.P., Ginzberg V.L., Gold T., Goldsmith D.W., Gratton L., Greisen K.I., Greyber H., Grewing M., Griffiths R.E., Grindlay J., Gunn J.E., Gursky H., Harwit M.O., Hawking S.W., Hayakawa S., Heise J., Henriksen R.N., Henry R.C., Hoang Binh D.Y., Hoffman J.A., Holloway N.J., Hoyle F., Ipser J., Israel W., Ito K., Jackson J.C., Jaffe W.J., Jelley J.V., Jokipii J.R., Jones F.C., Jones T.W., Joss P.C., Kafka P., Kahn F.D., Kellogg E., Kondo M., Kozlovsky B.Z., Kreisel E., Kristiansson K.G.B., Kulsrud R.M., Kundt W., Kurt V.G., Lamb D.Q., Jr., Lamb F., Lamb S.A., Lameron A.G.W., Lasher G.J., Lattimer J.M., Lea S.M., Longair M.S., Lovelace R.V.E., Lüst R., Lynden-Bell D., McBreen B.P., McCray R.A., Maccacaro T., Matsuoka M., Mazurek T.J., Mestel L., Meszáros P., Meyer F., Meyer J.-P., Michel F.C., Miller J.C., Miyamoto S., Morrison P., Ne'eman Y., Novick R., O'Connell R.F., Oda M., Okoye S.E., Ostriker J.P., Ozernoi L.N., Pacholczyk A.G., Page C.G., Palumbo G.G.C., Parkison J.H., Pauliny-Toth I., Perola C., Peterson B.A., Peterson L.E., Petrosian V., Piddington J.M., Pinkau K., Porter N.A., Pounds K.A., Preuss E., Quintana H., Ramadurai S., Rees M.J., Reeves H., Reinhart M., Rossi B., Sartori L., Saslaw W.C., Savedoff M.P., Scargle J.D., Scharlemann E.T., Schatten R.H., Schatzman E., Scheurer P.A.G., Schilizzi R., Schramm D.N., Sciama D.W., Scott J.S., Seielstad G., Setti G., Shaham J., Shapiro M.M., Shaviv G., Shields G.A., Shukre C.S., Silberberg R., Skilling J., Smith B.W., Smith F.G., Sofia S., Spitzer L., Steigman G., Stepanian A.A., Stockman H.S. Jr., Sturrock P.A., Sunyaev R.A., Taylor R.J., Thorne K.S., Trimble V.L., Truan J., Tsuruta S., Van Riper K.A., Vidal N., Weaver T.A., Webster A.S., Weisskopf M.C., Wentzel D.G., Westford K.C., Wheeler J.A., Will C.M., Wilson J.R., Wolstercroft R., Woltjer L., Zeldovich Ya.B., Zombeck M.V.

49. COMMISSION DU PLASMA INTERPLANETAIRE ET DE L'HELIOSPHERE (THE INTERPLANETARY PLASMA AND THE HELIOSPHERE)

PRESIDENT: H.J. Fahr

VICE-PRESIDENT: I.W. Roxburgh

COMITE D'ORGANISATION: S. Cuperman, A.Z. Dolginov, S. Grdzielski, H.U. Keller, F. Paresce, S.T. Suess, C.S. Weller.

MEMBRES: Ahluwalia H.S., Anderson K.A., Ansari S.M.R., Antonucci E., Barnes A., Barrow C.H., Barth C.A., Bertaux J.L., Blackwell D.E., Blum P.W., Bonnet R.M., Bowyer C.S., Brandt J.C., Burlaga L.F., Chamberlain J.W., Crifo F., de Jager C., Delache P., de Young D.S., Dryer M., Durney B., Dyson J.E., Fshleman V.R., Eviatar A., Feynman J., Field G.B., Gosling J.T., Harvey C.C., Heyvaerts J., Holzer T.E., Jokipii J.R., Kakinuma T., Kaplan S.A., Lafon J.P., Levy E., Lotova N.A., Lüst R., MacQueen R., Mangeney A., Matsura O.T., Mendis D.A. Mestel L., Michel F.C., Moussas X., Makagawa Y., Newkirk G.A., Parker E.N., Perkins F. Jr., Pflug K., Pneuman G., Raadu M.A., Readhead A.C.S., Reay N.K., Rickett B.J., Riddle A.C., Roach F.E., Sarris E.T., Sawyer C., Schatzman E., Scherb F., Schmidt H.U., Severny A.B., Shawhan S.D., Shea M.A., Smith D.F., Sonnett C.P., Stone R.G., Sturrock P.A., Tamaka H., Van Allen J.A., Wallis M.K., Watanabe T., Wilcox J.M., Wild J.P., Wu S.T.

50. COMMISSION DE LA PROTECTION DES SITES D'OBSERVATOIRE EXISTANT ET POTENTIELS (PROTECTION OF EXISTING AND POTENTIAL OBSERVATORY SITES)

PRESIDENT: F.G. Smith

VICE-PRESIDENT: A.A. Hoag

COMITE D'ORGANISATION: C. Anguita, A. Ardeberg, F. Bertola, V. Blanco, R. Cayrel, J.T. Jefferies, W. Mattig, B.E.J. Oke, M.F. Walker.

MEMBRES: Barreto L.M., Bhattacharyya J.C., Blaauw A., Boyarchuk A.A., Casanovas J., Crawford D.L., Dommanget J., Elsässer H., Gascoigne S.C.B., Göbel F., Gyldenkerne K., Kubicela A., Leibowitz E.M., Lewis B.M., Markannen T.K.J., Marx S., Mavridis L.N., Mendoza E., Osawa K., Osorio J., Sanchez F., Tremko J., van den Bergh S., Wayman P.A., Woolley Sir Richard, Woszczyk A.

GROUPE DE TRAVAIL POUR LA NOMENCLATURE DU SYSTEME PLANETAIRE (WORKING GROUP FOR PLANETARY SYSTEM NOMENCLATURE)

PRESIDENT : P.M. Millman

VICE-PRESIDENT : H. Masursky

MEMBRES : B. Yu. Levin, D. Morrison, T.C. Owen, G.H. Pettengill, B.A. Smith, V.G. Teijfel

TASK GROUPS :

1) *Lunar Nomenclature*

P.M. Millman, (Acting Chairman), A. Dollfus, F. El-Baz, K.P. Florenskij, Yu. N. Lipskij, H. Masursky, S.K. Runcorn, V.V. Shevchenko

2) *Mercury Nomenclature*

D. Morrison (Chairman), M.E. Davies, A. Dollfus, O.J. Gingerich, J.E. Guest, Yu. N. Lipskij, B.A. Smith

3) *Venus Nomenclature*

G.H. Pettengill (Chairman), R.M. Goldstein, M.Ya. Marov, H. Masursky

4) *Mars Nomenclature*

B.A. Smith (Chairman), A. Dollfus, M.Ya. Marov, H. Masursky, S. Miyamoto, A.V. Morozhenko, C. Sagan

5) *Outer Solar System Nomenclature*

T.C. Owen (Chairman), K. Aksnes, M.S. Bobrov, M.E. Davies, D. Gautier, B.A. Smith, V.G. Tejfel'.

II.

LISTE ALPHABETIQUE DES MEMBRES
ALPHABETICAL LIST OF MEMBERS

Cette liste est préparée d'une bande magnétique tenu par D. Reidel Publishing Company et soutenu en comparaison des archives du Secrétariat de l'UAI.

Les noms figurent à la première lettre du premier mot; par exemple, le nom De Jong se trouve à la lettre D.

Les noms des pays sont données sous leur forme anglaise.

Les membres de l'Union sont priés de bien vouloir informer le Secrétaire de l'UAI, 61 ave. de l'Observatoire, F-75014 Paris, France, de toutes les erreurs qu'ils pourraient relever dans la liste.

This list is prepared from a magnetic tape held by D. Reidel Publishing Company and collated with the records of the IAU Secretariat.

Names are given under the first letter of the first word; for example, the name De Jong is to be found under D.

The names of countries are given in the English form.

Please inform the Secretary of the IAU at 61 ave. de l'Observatoire, F-75014 Paris, France, of any errors that you may find in this list.

A'HEARN MICHAEL F DR
ASTRONOMY PROGRAM
UNIV OF MARYLAND
COLLEGE PK MD 20742
U.S.A.

AANESTAD PER ARNE DR
PHYSICS DEPT
ARIZONA STATE UNIV
TEMPE AZ 85281
U.S.A.

AARSETH SVERRE J DR
INST OF ASTRONOMY
MADINGLEY RD
CAMBRIDGE CB3 OHA
U.K.

ABALAKIN VICTOR K DR
INSTITUTE FOR THEORE
TICAL ASTRONOMY
10 KUTUZOV QUAY
192187 LENINGRAD
U.S.S.R.

ABBASOV ALIK R DR
SHEMAKHA ASTRO-
PHYSICAL OBSERVATORY
373243 SHEMAKHA AZER
U.S.S.R.

ABBOTT WILLIAM N DR
MICHALACOPOULOU 42
ATHENS 612
GREECE

ABDALA J DR
AV SOLANO
EDIFICIO ARAGUANEY
PLANTA BAJA NO 1
CHACAITO CARACAS 106
VENEZUELA

ABELE M K DR
LATVIAN STATE UNIVER
SITY ASTRONOMICAL
OBSERVATORY
226098 RIGA
U.S.S.R.

ABELL GEORGE O PROF
DEPT OF ASTRONOMY
UNIVERSITY OF
CALIFORNIA
LOS ANGELES CA 90024
U.S.A.

ABETTI G PROF
IST NAZ OTTICA
ARCETRI
LARGO ENRICO FERMI 7
I-50125 FIRENZE
ITALY

ABHYANKAR KRISHNA D PROF
DEPT OF ASTRONOMY
OSMANIA UNIVERSITY
HYDERABAD 500 007
INDIA

ABLES HAROLD D DR
FLAGSTAFF STATION
US NAVAL OBSERVATORY
P O BOX 1149
FLAGSTAFF AZ 86001
U.S.A.

ABLES HAROLD J G DR
CSIRO
DIV OF RADIOPHYSICS
P O BOX 76
EPPING NSW 2121
AUSTRALIA

ABOU-EL-ELLA MOHAMED S DR
HELWAN OBSERVATORY
HELWAN CAIRO
ARAB REP. OF EGYPT

ABRAHAM HENRY J M
7 CASTLEREAGH
CRESCENT
MACQUARIE ACT 2614
AUSTRALIA

ABRAHAM ZULEMA DR
CRAAM/CN/CNPQ
RUA PARA 277
01243 SAO PAULO SP
BRAZIL

ABRAMI ALBERTO PROF
OSSERVATORIO ASTRONO
MICO DI TRIESTE
I-34131 TRIESTE
ITALY

ABRAMOWICZ MAREK DR
N COPERNICUS ASTRON
CENTER
BARTYCKA 18
00-716 WARSAW
POLAND

ABREU DE BRITO E J C DR
OBSERVATORIO ASTRO-
NOMICO DE LISBOA
TAPADA
LISBON
PORTUGAL

ABT HELMUT A DR
KITT PEAK NATL OBS
BOX 26732
TUCSON AZ 85726
U.S.A.

ABU EL ATA NABIL DR
BUREAU D LONGITUDES
77 AVENUE DENFERT-
ROCHEREAU
F-75014 PARIS
FRANCE

ACKER AGNES DR
OBSERVATOIRE
11 RUE DE
L UNIVERSITE
F-67000 STRASBOURG
FRANCE

ACTON LOREN W DR
LOCKHEED PALO ALTO
RESEARCH LAB
3251 HANOVER ST
PALO ALTO CA 94304
U.S.A.

ADAM MADGE G DR
DEPT OF ASTROPHYSICS
SOUTH PARKS RD
OXFORD OX1 3RQ
U.K.

ADAMS A N MR
6549 N 35TH ROAD
ARLINGTON VA 22213
U.S.A.

ADAMS DAVID J DR
ASTRONOMY DEPT
THE UNIVERSITY
LEICESTER LE1 7RH
U.K.

ADAMS THOMAS F DR
G-7 MS 329 LOS ALA-
MOS SCIENTIFIC LAB
P O BOX 1663
LOS ALAMOS NM 87545
U.S.A.

ADEL ARTHUR F PROF EMER
NORTHERN ARIZONA
UNIVERSITY
BOX 5679
FLAGSTAFF AZ 86001
U.S.A.

ADEL SHARAF MOHAMED DR
DEPT OF ASTRONOMY
CAIRO UNIVERSITY
CAIRO
ARAB REP. OF EGYPT

ADELMAN SAUL J DR
DEPT OF PHYSICS
THE CITADEL
CHARLESTON SC 29409
U.S.A.

ADOLFSSON TORD DR
KRAGEHOLMSGATAN 12
S-216 19 MALMO
SWEDEN

AFANASJEVA PRASKOVYA M DR
PULKOVO OBSERVATORY
196140 LENINGRAD
U.S.S.R.

AGEKJAN TATEOS A PROF
UNIVERSITY OF
LENINGRAD
OBSERVATORY
199178 LENINGRAD
U.S.S.R.

AGRINIER BERNARD L MR
C E N SACLAY
B P NO 2
F-91190 GIF-YVETTE
FRANCE

AGUERO ESTELA L DR
OBS ASTRONOMICO
LAPRIDA NO 854
5000 CORDOBA
ARGENTINA

AHLUWALIA HARJIT SINGH DR
DEPT PHYSICS ASTRON
UNIV OF NEW MEXICO
800 YALE BLVD N E
ALBUQUERQUE NM 87131
U.S.A.

AHMED IMAM IBRAHIM
HELWAN OBSERVATORY
HELWAN NEAR CAIRO
ARAB REP. OF EGYPT

AHNERT P DR
ZENTRAL INST FUER
ASTROPHYSIK
STERNWARTE SONNEBERG
DDR-64 SONNEBERG
G.D.R.

AIAD A Z DR
CAIRO UNIVERSITY
FACULTY OF SCIENCE
ASTRONOMY DEPT
GEZA ORMAN
ARAB REP. OF EGYPT

AIKAWA TOSHIKI DR
ASTRONOMICAL INST
TOHOKU UNIVERSITY
AOBAYAMA SENDAI
JAPAN

AIKMAN GEORGE C L DR
DOMINION ASTROPH OBS
5071 W SAANICH ROAD
R R 7
VICTORIA BC V8X 3X3
CANADA

AIME C DR
DEPT D'ASTROPHYSIQUE
UNIVERSITE DE NICE
PARC VALROSE
F-06034 NICE CEDEX
FRANCE

AITKEN DAVID K DR
DEPT PHYS AND ASTRO
UNIV COLLEGE LONDON
GOWER STREET
LONDON WC1E 6BT
U.K.

AIZENMAN MORRIS L DR
DIV ASTRONOMICAL SC
NAT SC FOUNDATION
RM 615 1800 G ST NW
WASHINGTON DC 20550
U.S.A.

AIZU KO PROF
PHYSICS DEPT
RIKKYO UNIVERSITY
NISHI-IKEBUKURO 3
TOSHIMAKU TOKYO 171
JAPAN

AKABANE KENJI A PROF
TOKYO ASTRONOMICAL
OBSERVATORY
OSAWA MITAKA
TOKYO 181
JAPAN

AKABANE TOKUHIDE DR
HIDA OBSERVATORY
506-13 KAMITAKARA
GIFU
JAPAN

AKCAYLI MELEK M A DR
EGE UNIVERSITESI
FEN FAKUTESI
GOK BILIMLERI ENST
BORNOVA IZMIR
TURKEY

AKSENOV E P DR
STERNBERG STATE AS-
TRONOMICAL INSTITUTE
117234 MOSCOW
U.S.S.R.

AKSNES KAARE DR.
NORWEGIAN DEFENCE
RESEARCH ESTABLISHM
P O BOX 25
2007 KJELLER
NORWAY

AKYOL MUSTAFA U PROF
EGE UNIVERSITY
SCIENCE FACULTY
DEPT OF ASTRONOMY
BORNOVA IZMIR
TURKEY

AL-SABTI ABDUL ADIM DR
PHYSICS DEPT
SCIENCE COLLEGE
BAGHDAD UNIVERSITY
BAGHDAD
REP. OF IRAQ

ALANIJA I F DR
ABASTUMANI ASTROPHY-
SICAL OBSERVATORY
383762 ABASTUMANI
GEORGIA
U.S.S.R.

ALBANO J DR
OBS ASTRONOMICO
LA PLATA
ARGENTINA

ALBERS HENRY PROF
VASSAR COLLEGE
OBSERVATORY
POUGHKEEPSIE NY12601
U.S.A.

ALBRECHT RUDOLF DR
INS F ASTRONOMIE
UNIVERSITAET WIEN
TUERKENSCHANZSTR 17
A-1180 WIEN
AUSTRIA

ALDROVANDI RUBEN DR
INSTITUTO DI FISICA
TEORICA
RUA PAMPLONA 145
01405 SAO PAULO SP
BRAZIL

ALDROVANDI SUELI M V DR
INSTITUTO ASTRONOMI
CO E GEOFISICO
CP 30627
01000 SAO PAULO SP
BRAZIL

ALEXANDER JOHN B
ROYAL GREENWICH OBS
HAILSHAM
EAST SUSSEX BN27 1RP
U.K.

ALEXANDER JOSEPH K
CODE 693
GODDARD SPACE FLIGHT
CENTER
GREENBELT MD 20771
U.S.A.

ALFVEN HANNES PROF
DEPT OF PLASMA
PHYSICS
ROYAL INST OF TECHN
S-100 44 STOCKHOLM
SWEDEN

ALISSANDRAKIS C PH D
LAB OF ASTROPHYSICS
UNIV OF ATHENS
PANEPISTIMIOPOLIS
ATHENS 621
GREECE

ALKSNE ZENTA DR
RADIOASTROPHYSICAL
OBSERVATORY
TURGENEVA 19
226524 RIGA LATVIA
U.S.S.R.

ALKSNIS ANDREJS DR
RADIOASTROPHYSICAL
OBSERVATORY
226524 RIGA LATVIA
U.S.S.R.

ALLADIN SALEH MOHAMED DR
DEPT OF ASTRONOMY
OSMANIA UNIVERSITY
HYDERABAD 500 007
INDIA

ALLAMANDOLA LOUIS JOHN DR
STERREWACHT
HUYGENS LAB
POSTBUS 9513
2300 RA LEIDEN
THE NETHERLANDS

ALLEN C W PROF
MT STROMLO
OBSERVATORY
WODEN PO ACT 2606
AUSTRALIA

ALLEN CHRISTINE
INSTITUTO DE ASTRONO
MIA UNAM
APDO POSTAL 70-264
MEXICO 20 D F
MEXICO

ALLEN DAVID A DR
ANGLO AUSTRALIAN
OBSERVATORY
PO BOX 296
EPPING NSW 2121
AUSTRALIA

ALLEN RONALD J DR
KAPTEYN LABORATORIUM
POSTBUS 800
9700 AV GRONINGEN
THE NETHERLANDS

ALLER HUGH D DR
DEPT OF ASTRONOMY
UNIVERSITY MICHIGAN
953 PHYS-ASTRO BLDG
ANN ARBOR MI 48109
U.S.A.

ALLER LAWRENCE HUGH
ASTRONOMY DEPT
UNIVERSITY OF CALIF
MATH-SCIENCES BLDG
LOS ANGELES CA 90024
U.S.A.

ALLER MARGO F DR
ASTRONOMY DEPT
UNIVERSITY MICHIGAN
ANN ARBOR MI 48109
U.S.A.

ALLOIN DANIELLE DR
OBSERVATOIRE DE NICE
B P 252
F-06007 NICE CEDEX
FRANCE

ALMAR IVAN DR
SATELLITE GEODETIC
OBSERVATORY
POB 546
1373 BUDAPEST
HUNGARY

ALTAVISTA CARLOS A DR
OBS ASTRONOMICO
PASEO DEL BOSQUE
1900 LA PLATA
ARGENTINA

ALTENHOFF WILHELM J DR
MPI FUER
RADIOASTRONOMIE
AUF DEM HUEGEL 69
D-5300 BONN
GERMANY, F.R.

ALTROCK RICHARD C DR
AIR FORCE GEOPHYSICS
LABORATORY
SACRAMENTO PEAK.OBS
SUNSPOT NM 88349
U.S.A.

ALTSCHULER MARTIN D PROF
DEPT COMPUTER SC
STATE UNIVERSITY OF
NEW YORK AT BUFFALO
AMHERST NY 14226
U.S.A.

ALURKAR S K DR
PHYSICAL RES LAB
NAVRANGPURA
AHMEDABAD 380 009
INDIA

ALVAREZ HECTOR DR.
DEPTO DE ASTRONOMIA
UNIVERSIDAD DE CHILE
CASILLA 36-D
SANTIAGO
CHILE

ALVAREZ LUIS DR
LAWRENCE RADIATION
LABORATORY
UNIVERSITY OF CALIF
BERKELEY CA 94720
U.S.A.

ALY M K PROF DIRECTOR
RABITATUL AALAM AL-
ISLAMI
ISLAMIC OBSERVATORY
MECCA ALMUKARRAMA
SAUDI ARABIA

AMBARTSUMIAN V A PROF DR
BYURAKAN ASTROPHYSI
CAL OBSERVATORY
378433 ARMENIA
U.S.S.R.

AMBROZ PAVEL DR
ASTRONOMICAL INST OF
THE CZECHOSLOVAK
ACADEMY OF SCIENCES
25165 ONDREJOV
CZECHOSLOVAKIA

ANAND S P S DR
APPLIED RESEARCH AND
SYSTEMS
1998 GENERALS HWY
ANNAPOLIS MD 21401
U.S.A.

ANDERS EDWARD PROF
UNIVERSITY OF
CHICAGO
5640 S ELLIS AVE
CHICAGO IL 60637
U.S.A.

ANDERSEN JOHANNES LEKTOR
BRORFELDE OBS
COPENHAGEN UNIV
DK-4340 TOLLOSE
DENMARK

ANDERSON BRYAN DR
UNIV OF MANCHESTER
JODRELL BANK
MACCLESFIELD
CHESHIRE SK11 9DL
U.K.

ANDERSON CHRISTOPHER M DR
WASHBURN OBSERVATORY
UNIVERSITY OF
WISCONSIN
MADISON WI 53706
U.S.A.

ANDERSON KINSEY A PROF
SPACE SCIENCES LAB
UNIVERSITY OF CALIF
BERKELEY CA 94720
U.S.A.

ANDO HIROYASU DR
TOKYO ASTRONOMICAL
OBSERVATORY
OSAWA 2-21-1 MITAKA
TOKYO 181
JAPAN

ANDREW BRYAN H DR
HERZBERG INST OF
ASTROPHYSICS
NAT RES COUNCIL
OTTAWA ONT K1A 0R6
CANADA

ANDREW KENNETH L PROF
DEPT OF PHYSICS
PURDUE UNIVERSITY
W LAFAYETTE IN 47907
U.S.A.

ANDREWS DAVID A DR
ARMAGH OBSERVATORY
ARMAGH BT61 9DG
U.K.

ANDREWS DONALD H
DOMINION ASTRO OBS
NAT RES COUNC CANADA
5071 WEST SAANICH RD
VICTORIA BC V8X 3X3
CANADA

ANDREWS PETER J DR
ROYAL GREENWICH
OBSERVATORY
HAILSHAM
EAST SUSSEX BN27 1RP
U.K.

ANDRIENKO DMITRY A DR
VLADIMIRSKAYA 51 53
FLAT 75
252003 KIEV
U.S.S.R.

ANDRIESSE CORNELIS D DR
KAPTEYN STERREWACHT
MENSINGHEWEG 20
RODEN DR
THE NETHERLANDS

ANDRILLAT HENRI L PROF
LAB D'ASTRONOMIE
UNIV DES SCIENCES ET
TECHN DU LANGUEDOC
F-34060 MONTPELLIER
FRANCE

ANDRILLAT YVETTE DR
OBSERVATOIRE
DE HAUTE-PROVENCE
ST-MICHEL L OBS
F-04300 FORCALQUIER
FRANCE

ANDRLE P DR
ASTRONOMICAL INST
CZECH ACAD OF SC
BUDECSKA 6
12023 PRAHA 2
CZECHOSLOVAKIA

ANGEL J ROGER P PROF
STEWARD OBSERVATORY
UNIVERSITY OF
ARIZONA
TUCSON AZ 85721
U.S.A.

ANGELETTI LUCIO DR
OSSERVATORIO ASTRO-
NOMICO DI ROMA
I-00100 ROMA
ITALY

ANGIONE RONALD J DR
ASTRONOMY DEPT
SAN DIEGO STATE
UNIVERSITY
SAN DIEGO CA 92182
U.S.A.

ANGUITA C A DR
UNIVERSIDAD DE CHILE
OBS ASTRON NACIONAL
CASILLA 36-D
SANTIAGO
CHILE

ANILE ANGELO M
OSSERVATORIO
ASTROFISICO
CITTA UNIVERSITARIA
I-95123 CATANIA
ITALY

ANSARI S M RAZAULLAH DR
PHYSICS DEPT
ALIGARH MUSLIM
UNIVERSITY
ALIGARH UP 202001
INDIA

ANTONACOPOULOS GREG PROF
DEPT OF ASTRONOMY
UNIVERSITY OF PATRAS
PATRAS
GREECE

ANTONOV V A DR
LENINGRAD STATE UNI-
VERSITY ASTRONOMICAL
OBSERVATORY
199178 LENINGRAD
U.S.S.R.

ANTONUCCI ESTER DR
ISTITUTO DI FISICA
UNIVERSITA DI TORINO
C D AZEGLIO 46
I-10125 TORINO
ITALY

ANZER ULRICH DR
MPI FUER PHYSIK UND
ASTROPHYSIK
FOEHRINGER RING 6
D-8000 MUENCHEN 6
GERMANY, F.R.

AOKI SHINKO PROF
TOKYO ASTRONOMICAL
OBSERVATORY
MITAKA
TOKYO 181
JAPAN

APPARAO K M V DR
TATA INST FUND RES
HOMI BHABH ROAD
BOMBAY 400 005
INDIA

APPENZELLER IMMO PROF
LANDESSTERNWARTE
KOENIGSTUHL
D-6900 HEIDELBERG 1
GERMANY, F.R.

ARAI KENZO DR
DEPT OF PHYSICS
KUMAMOTO UNIVERSITY
KUMAMOTO 860
JAPAN

ARAKELIAN MARAT A DR
BYURAKAN ASTROPHYSI-
CAL OBSERVATORY
378433 BYURAKAN
ARMENIA
U.S.S.R.

ARDAVAN H DR
INST OF ASTRONOMY
MADINGLEY ROAD
CAMBRIDGE
U.K.

ARDEBERG ARNE L DR
LUND OBSERVATORY
SVANEGATAN 9
S-222 24 LUND
SWEDEN

AREND S DR
AVENUE DE SATURNE 11
B-1180 BRUXELLES
BELGIUM

ARGUE A NOEL MR
INST OF ASTRONOMY
THE OBSERVATORIES
MADINGLEY RD
CAMBRIDGE CB3 0HA
U.K.

ARGYLE P E DR
DOMINION RADIO
ASTROPHYSICAL OBS
BOX 248
PENTICTON BC V2A 6K3
CANADA

ARGYRAKOS JEAN PROF DR
UNIV TECHNIQUE
NATIONALE D'ATHENES
193 PATISSION ST
ATHENS 816
GREECE

ARIAS DE GREIF JORGE PROF
OBSERVATORIO
NACIONAL
APARTADO 2584
BOGOTA
COLOMBIA

ARKHIPOVA V P DR
STERNBERG STATE AS-
TRONOMICAL INSTITUTE
117234 MOSCOW
U.S.S.R.

ARNETT W DAVID PROF
ENRICO FERMI INST
933 E 56TH STREET
CHICAGO IL 60637
U.S.A.

ARNOULD MARCEL L DR
INST D'ASTRONOMIE
UNIV LIB BRUXELLES
AVE F D ROOSEVELT 50
B-1050 BRUXELLES
BELGIUM

ARNQUIST WARREN N DR
8127 DELGANY AVE
PLAYA DEL REY
CA 90291
U.S.A.

ARNY THOMAS T DR
DEPT PHYSICS ASTR
UNIVERSITY OF MA
GRC TOWER B
AMHERST MA 01002
U.S.A.

ARONS JONATHAN
DEPT OF ASTRONOMY
UNIVERSITY OF CALIF
BERKELEY CA 94720
U.S.A.

ARP HALTON DR
HALE OBSERVATORIES
813 SANTA BARBARA ST
PASADENA CA 91101
U.S.A.

ARPIGNY CLAUDE PROF
INST D ASTROPHYSIQUE
UNIVERSITE DE LIEGE
B-4200 COINTE OUGREE
BELGIUM

ARTHUR DAVID W G
US GEOLOGICAL SURVEY
FLAGSTAFF ARIZONA
FLAGSTAFF AZ 86001
U.S.A.

ARTRU M C DR
OBS DE MEUDON
F-92190 MEUDON
FRANCE

ASAAD ADLY S PROF
OBSERVATORY OF
HELWAN
HELWAN
ARAB REP. OF EGYPT

ASCHENBACH BERND PH D
HEITERWANGERSTR 54
D-8000 MUENCHEN 70
GERMANY, F.R.

ASHBROOK JOSEPH DR
16 SUMMER ST
WESTON MA 02193
U.S.A.

ASLAN ZEKI DR
ASTRONOMY DEPT
UNIV OF ANKARA
FEN FAKULTESI
ANKARA
TURKEY

ASLANOV I A DR
SHEMAKHA ASTROPHYSI-
CAL OBSERVATORY
373243 AZERBAIDZAN
U.S.S.R.

ASSEO ESTELLE DR
OBSERVATOIRE DE
MEUDON DAF
F-92190 MEUDON
FRANCE

ASSOUSA GEORGE ELIAS DR
DEPT TERR MAGNETISM
CARNEGIE INST WASH
5241 BROAD BR RD N W
WASHINGTON DC 20015
U.S.A.

ASTERIADIS GEORGIOS DR
DEPT OF GEODETIC
ASTRONOMY
UNIV THESSALONIKI
THESSALONIKI
GREECE

ATANASIJEVIC IVAN DR
FACULTY OF SCIENCES
6500 GL NIJMEGEN
THE NETHERLANDS

ATHANASSOULA E O DR
OBSERVATOIRE
DE BESANCON
41B AVE OBSERVATOIRE
F-25000 BESANCON
FRANCE

ATHAY R GRANT DR
HIGH ALTITUDE
OBSERVATORY
P O BOX 1470
BOULDER CO 80302
U.S.A.

AUDOUZE JEAN DR
INST D'ASTROPHYSIQUE
98 BIS BLVD ARAGO
F-75014 PARIS
FRANCE

AUMAN JASON R PROF
DEPT OF GEOPHYSICS
AND ASTRONOMY
UNIVERSITY OF BC
VANCOUVER BC V6T 1W5
CANADA

AVERY LORNE W DR
HERZBERG INST OF
ASTROPHYSICS
NAT RES COUNC CANADA
OTTAWA ONT K1A 0R6
CANADA

AVRETT EUGENE H DR
CENTER FOR
ASTROPHYSICS
60 GARDEN ST
CAMBRIDGE MA 02138
U.S.A.

AZZOPARDI M MR
OBS DE TOULOUSE
1 AV C FLAMMARION
F-31510 TOULOUSE
FRANCE

BABADZHANIANC MICHAIL DR
ASTRONOMICAL OBS
LENINGRAD UNIVERSITY
LENINGRAD
U.S.S.R.

BACCHUS PIERRE PROF
1 IMPASSE DE
L-OBSERVATOIRE
F-59000 LILLE
FRANCE

BAEHR ULRICH DR
QUINCKESTRASSE 44A
D-6900 HEIDELBERG
GERMANY, F.R.

BAGHOS BALEGH B DR
HELWAN OBSERVATORY
HELWAN CAIRO
ARAB REP. OF EGYPT

BAHCALL JOHN N PROF
INSTITUTE FOR
ADVANCED STUDY
PRINCETON NJ 08540
U.S.A.

BAIRD GEORGE A DR
PHYSICS DEPT
UNIVERSITY COLLEGE
BELFIELD
DUBLIN 4
IRELAND

ATKINSON ROBERT D E DR
SWAIN HALL WEST 319
ASTRONOMY DEPT
INDIANA UNIVERSITY
BLOOMINGTON IN 47401
U.S.A.

AUER LAWRENCE H DR
HIGH ALTITUDE
OBSERVATORY
P O BOX 1470
BOULDER CO 80302
U.S.A.

AUNER GERHARD DR
INST FUER ASTRONOMIE
UNIVERSITAETSSTR 4
A-6020 INNSBRUCK
AUSTRIA

AVIGNON YVETTE DR
OBSERVATOIRE
DE MEUDON
F-92190 MEUDON
FRANCE

AXFORD W IAN PROF
MPI FUER AERONOMIE
POSTFACH 20
D-3411 KATLENBURG-
LINDAU 3
GERMANY, F.R.

BAARS JAAP W M DR
MPI FUER
RADIOASTRONOMIE
AUF DEM HUEGEL 69
D-5300 BONN
GERMANY, F.R.

BABADZHANOV P B DR
ASTROPHYSICAL INSTI-
TUTE OF TADJIKIAN
ACADEMY OF SCIENCES
734670 DUSHANBE
U.S.S.R.

BACKER DONALD CH DR
601 CAMPBELL HALL
UNIV OF CALIFORNIA
BERKELEY CA 94720
U.S.A.

BAERENTZEN JORN
ELMEHOJVEJ 66
DK-8270 HOJBJERG
DENMARK

BAGILDINSKIJ BRONISLAV
K DR
PULKOVO OBSERVATORY
196140 LENINGRAD
U.S.S.R.

BAHNER KLAUS DR
LANDESSTERNWARTE
KOENIGSTUHL
D-6900 HEIDELBERG 1
GERMANY, F.R.

BAIRD KENNETH M DR
DIVISION OF PHYSICS
NAT RES COUNC CANADA
MONTREAL ROAD
OTTAWA ONT K1A 0R6
CANADA

AUBIER MONIQUE G DR
OBSERVATOIRE DE
MEUDON
F-92190 MEUDON
FRANCE

AULUCK F C PROF
DEPT OF PHYSICS AND
ASTROPHYSICS
UNIVERSITY OF DELHI
DELHI 110007
INDIA

AURIEMMA GIULIO DR
CNR ASTROFISICO
SPAZIALE C P 67
I-00044 FRASCATI
ROMA
ITALY

AVNI YORAM DR
WEIZMANN INSTITUTE
REHOVOT
ISRAEL

AYDIN CEMAL ASSOC PROF
AU FEN FAKULTESI
ASTRONOMI BOLUMU
BESEVLER ANKARA
TURKEY

BAART EDWARD E PROF
DEPT OF PHYSICS
RHODES UNIVERSITY
P O BOX 94
GRAHAMSTOWN 6140
SOUTH AFRICA

BABCOCK HORRACE W DR
HALE OBSERVATORIES
813 SANTA BARBARA ST
PASADENA CA 91101
U.S.A.

BAECK NICOLE A L DR
195 E-3 PLEIN
B-9218 LEDEBERG
BELGIUM

BAGGALEY WILLIAM JACK DR
PHYSICS DEPT
UNIV OF CANTERBURY
CHRISTCHURCH
NEW ZEALAND

BAGLIN ANNIE DR
OBSERVATOIRE DE NICE
BP 252
F-06007 NICE CEDEX
FRANCE

BAHNG JOHN D R PROF
DEARBORN OBSERVATORY
NORTHWESTERN
UNIVERSITY
EVANSTON IL 60201
U.S.A.

BAIZE PAUL DR
6 RUE DAUBIGNY
F-75017 PARIS
FRANCE

BAJAJA E DR
INSTITUTO ARGENTINO
DE RADIOASTRONOMIA
CASILLA NO 5
VILLA ELISA PR BS AS
ARGENTINA

BAKER JAMES GILBERT DR
7 GROVE ST
WINCHESTER MA 01890
U.S.A.

BAKER NORMAN H PROF
ASTRONOMY DEPT
COLUMBIA UNIVERSITY
PUPIN HALL 538 W 120
NEW YORK N Y 10027
U.S.A.

BAKOS GUSTAV A PROF
DEPT OF PHYSICS
UNIVERSITY WATERLOO
WATERLOO ONT N2L 3G1
CANADA

BALAZS BELA A DR
DEPT OF ASTRONOMY
EOTVOS UNIVERSITY
KUN BELA TER 2
H-1083 BUDAPEST
HUNGARY

BALAZS LAJOS G DR
KONKOLY OBSERVATORY
BOX 67
H-1525 BUDAPEST
HUNGARY

BALDINELLI LUIGI DR
VIA MONTEBELLO 4
I-40141 BOLOGNA
ITALY

BALDWIN JACK A DR
INST OF ASTRONOMY
MADINGLEY ROAD
CAMBRIDGE CB3 OHA
U.K.

BALDWIN JOHN E DR
CAVENDISH LABORATORY
MADINGLEY RD
CAMBRIDGE CB3 OHE
U.K.

BALDWIN RALPH B
3110 MANHATTAN LANE
S E
GRAND RAPIDS MI49506
U.S.A.

BALICK BRUCE PROF
ASTRONOMY DEPT FM-20
UNIV OF WASHINGTON
SEATTLE WA 98195
U.S.A.

BALKLAVS A E DR
RADIOASTROPHYSICAL
OBSERVATORY LATVIAN
ACADEMY OF SCIENCES
226026 RIGA
U.S.S.R.

BALL JOHN A DR
HARVARD COLLEGE
OBSERVATORY
HARVARD MA 01451
U.S.A.

BALLARIO M C PROF
OSSERVATORIO ASTRONO
MICO DI ARCETRI
VIA S LEONARDO
I-50100 FIRENZE
ITALY

BALLI EDIBE PROF
UNIVERSITE
RASATHANESI
ISTANBUL
TURKEY

BALMINO GEORGES C DR
CTR NAT D'ETUDES SPA
GROUP RECH GEOD SPAT
18 AVE EDOUARD BELIN
F-31400 TOULOUSE
FRANCE

BALONA LUIS ANTERO DR
S A ASTRON OBS
P O BOX 9
OBSERVATORY 7935
CAPE
SOUTH AFRICA

BANDERMANN L W DR
ROYAL OBSERVATORY
BLACKFORD HILL
EDINBURGH EH9 3HJ
U.K.

BANDYOPADHYAY A DR
INDIA METEO DEPT
INDIAN ASTRON EPH UN
P546 BK N NW ALIPORE
CALCUTTA 700 053
INDIA

BANIN V G DR
SIBIZMIR
P B 4
664697 IRKUTSK 33
U.S.S.R.

BANOS COSMAS J DR
OBSERVATOIRE NATIO-
NAL D ATHENES
ATHENS 306
GREECE

BANOS GEORGE J PROF
UNIVERSITY OF
IOANNINA
ASTRONOMY DEPT
IOANNINA
GREECE

BAPPU M K V DR
DIRECTOR
INDIAN INSTITUTE OF
ASTROPHYSICS
BANGALORE 560034
INDIA

BARANNE A DR
OBSERVATOIRE
DE MARSEILLE
2 PLACE LE VERRIER
F-13004 MARSEILLE
FRANCE

BARATTA G B DR
OSSERVATORIO ASTRO-
NOMICO DI ROMA
I-00100 ROMA
ITALY

BARBANIS BASIL PROF
DEPT OF ASTRONOMY
UNIVERSITY OF PATRAS
PATRAS
GREECE

BARBARO G DR
OSSERVATORIO ASTRON
VICOLO DELL
OSSERVATORIO
PADOVA
ITALY

BARBIER MADELEINE AST ADJ
OBSERVATOIRE
DE MARSEILLE
2 PLACE LE VERRIER
F-13004 MARSEILLE
FRANCE

BARBIERI CESARE PROF
INST DI ASTRONOMIA
UNIVERSITA DI PADOVA
I-35100 PADOVA
ITALY

BARBON ROBERTO PROF
OSSERVATORIO
ASTROFISICO
I-36012 ASIAGO VI
ITALY

BARCZA SZABOLCS DR
KONKOLY OBSERVATORY
P O BOX 67
H-1525 BUDAPEST
HUNGARY

BARDEEN JAMES M PROF
PHYSICS DEPT FM-15
UNIVERSITY OF
WASHINGTON
SEATTLE WA 98195
U.S.A.

BARKAT ZALMAN PROF
THEORETICAL PHYSICS
HEBREW UNIVERSITY
JERUSALEM
ISRAEL

BARKER EDWIN S DR
MCDONALD OBSERVATORY
FORT DAVIS TX 79734
U.S.A.

BARKHATOVA KLAUDIA PROF
STATE UNIVERSITY
LENIN ST
620083 SVERDLOVSK
U.S.S.R.

BARLAI KATALIN DR
KONKOLY OBSERVATORY
BOX 67
H-1525 BUDAPEST
HUNGARY

BARLETTI RAFFAELE ENG
OSSERVATORIO ARCETRI
LARGO FERMI 5
I-50125 FIRENZE
ITALY

BARLIER FRANCOIS E DR
CERGA
AV COPERNIC
F-06130 GRASSE
FRANCE

BARLOW MICHAEL J DR
ANGLO AUSTRALIAN OBS
P O BOX 296
EPPING NSW 2121
AUSTRALIA

BARNARD HANNES A J DR
PHYSICS DEPT
UNIVERSITY OF B C
2075 WESBROOK MALL
VANCOUVER V6T 1W6
CANADA

BARNES AARON DR
NASA AMES RES CTR
CODE 245-3
MOFFETT FLD CA 94035
U.S.A.

BARNES III THOMAS G DR
DEPT OF ASTRONOMY
UNIVERSITY OF TEXAS
AUSTIN TX 78712
U.S.A.

BARNOTHY JENO DR PROF
833 LINCOLN ST
EVANSTON IL 60201
U.S.A.

BAROCAS VINICIO DR
JEREMIAH HORROCKS
AND WILFRED HALL OBS
MOOR PARK
PRESTON PR1 6AD
U.K.

BARRETO LUIZ MUNIZ PROF
OBSERVATORIO
NACIONAL
RUA GEN BRUCE 586
20000 RIO DE JANEIRO
BRAZIL

BARRETT ALAN H PROF
ROOM 26-331
MASSACHUSETT INST
OF TECHNOLOGY
CAMBRIDGE MA 02139
U.S.A.

BARROW COLIN H DR
DEPT OF PHYSICS
UNIVERSITY OF THE
WEST INDIES
KINGSTON W I 7
JAMAICA

BARROW RICHARD F DR
PHYSICAL CHEMISTRY
LABORATORY
SOUTH PARKS RD
OXFORD OX1 3QZ
U.K.

BARRY DON C DR
DEPT OF ASTRONOMY
UNIVERSITY OF
SOUTHERN CALIFORNIA
LOS ANGELES CA 90007
U.S.A.

BARTAYA R A DR
ABASTUMANI ASTROPHY-
SICAL OBSERVATORY
383762 ABASTUMANI
GEORGIA
U.S.S.R.

BARTH CHARLES A PROF
LASP UNIVERSITY
OF COLORADO
BOULDER CO 80309
U.S.A.

BARTHOLDI PAUL DR
OBSERVATOIRE
DE GENEVE
CH-1290 SAUVERNY GE
SWITZERLAND

BARTOLINI CORRADO
OSSERVATORIO
ASTRONOMICO
VIA ZAMBONI 33
I-40126 BOLOGNA
ITALY

BASART JOHN P
NAT RADIO ASTRON OBS
P O BOX 0
SOCORRO NM 87801
U.S.A.

BASCHEK BODO PROF
INST F THEORETISCHE
ASTROPHYSIK
IM NEUENHEIMER F 294
D-6900 HEIDELBERG
GERMANY, F.R.

BASH FRANK N PROF
ASTRONOMY DEPT
UNIVERSITY OF TEXAS
AUSTIN TX 78712
U.S.A.

BASTIN J A PROF
PHYSICS DEPT
QUEEN MARY COLLEGE
MILE END ROAD
LONDON E1 4NS
U.K.

BASU BAIDYANATH READER
APPLIED MATHEMATICS
DEPARTMENT C U
92 A P C ROAD
CALCUTTA-700009
INDIA

BASU DIPAK DR
DEPT OF PHYSICS
UNIVERSITY OF THE
WEST INDIES
ST AUGUSTINE
TRINIDAD AND TOBAGO

BATES BRIAN DR
DEPT OF PURE AND
APPLIED PHYSICS
QUEENS UNIVERSITY
BELFAST BT7 1NN
U.K.

BATES DAVID R PROF
DEPT OF APPLIED MATH
AND THEORETICAL PHYS
QUEENS UNIVERSITY
BELFAST BT7 1NN
U.K.

BATES RICHARD HEATON T DR
ELECTRICAL ENG DEPT
UNIV OF CANTERBURY
CHRISTCHURCH
NEW ZEALAND

BATESON FRANK M OBE
P O BOX 3093
GREERTON TAURANGA
NEW ZEALAND

BATH GEOFFREY T DR
DEPT OF ASTROPHYSICS
UNIVERSITY OF OXFORD
OXFORD OX1 3RQ
U.K.

BATRAKOV J M DR
INSTITUTE OF THEORE-
TICAL ASTRONOMY
192187 LENINGRAD
U.S.S.R.

BATTANER EDUARDO DR
INSTITUTO DE ASTROFI
SICA DE ANDALUCIA
APDO 2144
GRANADA
SPAIN

BATTEN ALAN H DR
DOMINION ASTROPHYS
OBSERVATORY
5071 W SAANICH RD
VICTORIA BC V8X 3X3
CANADA

BATTISTINI PIERLUIGI DR
OSSERVATORIO
ASTRONOMICO
VIA ZAMBONI 33
I-40126 BOLOGNA
ITALY

BAUDRY ALAIN DR
OBSERVATOIRE
DE BORDEAUX
F-33270 FLOIRAC
FRANCE

BAUER CARL A DR
506 DAVEY
PENNA STATE UNIV
UNIV PK PA 16802
U.S.A.

BAUM WILLIAM A DR
LOWELL OBSERVATORY
P O BOX 1269
FLAGSTAFF AZ 86002
U.S.A.

BAUSTIAN W W MR
KITT PEAK NAT OBS
950 N CHERRY AVE
P O BOX 26732
TUCSON AZ 85726
U.S.A.

BAUTZ LAURA P DR
1325 18TH ST NW
APT 506
WASHINGTON DC 20036
U.S.A.

BAYM GORDON ALAN DR
DEPT OF PHYSICS
UNIV OF ILLINOIS
URBANA IL 61801
U.S.A.

BEALE JOHN S DR
ROYAL GREENWICH
OBSERVATORY
HAILSHAM
EAST SUSSEX BN27 1RP
U.K.

BEARD DAVID B DR
DEPT PHYSICS ASTRON
UNIV OF KANSAS
LAWRENCE KS 66045
U.S.A.

BEARDSLEY WALLACE R DR
DEPT PHYS AND ASTRON
UNIV OF PITTSBURGH
ALLEGHENY OBS B 7658
PITTSBURGH PA 15214
U.S.A.

BEAUDET GILLES DR
DEPT DE PHYSIQUE
UNIV DE MONTREAL
CP 6128
MONTREAL PQ H3C 3J7
CANADA

BEAVERS WILLET I DR
ERWIN W FICK OBS
IOWA STATE
UNIVERSITY
AMES IA 50010
U.S.A.

BEC-BORSENBERGER ANNICKDR
BUREAU LONGITUDES
77 AVE DENFERT-
ROCHEREAU
F-75014 PARIS
FRANCE

BECK H G
VEB CARL ZEISS
FORSCHUNGSZENTRUM
CARL ZEISS STR 1
DDR 69 JENA
G.D.R.

BECKER FRIEDRICH DR PROF
WEITLSTR 66-2109
D-8000 MUENCHEN 45
GERMANY, F.R.

BECKER ROBERT A DR
P O BOX 4609
CARMEL CA 93921
U.S.A.

BECKER WILHELM PROF
ASTRON INST
UNIVERSITAET BASEL
VENUSSTR 7
CH- 4102 BINNINGEN
SWITZERLAND

BECKERS JACQUES M DR
MULTIPLE MIRROR
TELESCOPE OBS
UNIV OF ARIZONA
TUCSON AZ 85718
U.S.A.

BECKLIN ERIC E DR
INST FOR ASTRONOMY
2680 WOODLAWN DR
HONOLULU HI 96822
U.S.A.

BECKMAN JOHN E DR
DEPT OF PHYSICS
QUEEN MARY COLLEGE
MILE END ROAD
LONDON E1 4NS
U.K.

BEEBE RETA FAYE DR
NEW MEXICO ST UNIV
DEPT OF ASTRONOMY
BOX 4500
LAS CRUCES NM 88003
U.S.A.

BEER ARTHUR DR
188 HUNTINGDON ROAD
CAMBRIDGE
U.K.

BEER REINHARD DR
JET PROPULSION LAB
CALIF INST OF TECH
4800 OAK GROVE DR
PASADENA CA 91103
U.S.A.

BEGGS DENIS W MR
INSTITUTE OF
ASTRONOMY
MADINGLEY ROAD
CAMBRIDGE CB3 OHA
U.K.

BEHR ALFRED PROF
HAMBURGER STERNWARTE
GOJENBERGSWEG 112
D-2050 HAMBURG 80
GERMANY, F.R.

BEINTEMA DOUWE A DR
SPACE RESEARCH DEPT
UNIV OF GRONINGEN
P O BOX 800
GRONINGEN
THE NETHERLANDS

BEKENSTEIN JACOB D DR
PHYSICS DEPT
BEN GURION UNIV
P O B 653
BEERSHEVA
ISRAEL

BEL NICOLE J DR
OBSERVATOIRE
DE MEUDON
F-92190 MEUDON
FRANCE

BELKOVICH O I DR
ENGELHARDT ASTRONOMI
CAL OBSERVATORY
422526 KAZAN
U.S.S.R.

BELL BARBARA DR
CENTER FOR
ASTROPHYSICS
60 GARDEN STREET
CAMBRIDGE MA 02138
U.S.A.

BELL KENNETH LLOYD DR
DEPT OF APPLIED MATH
AND THEORET PHYSICS
QUEEN'S UNIVERSITY
BELFAST BT7 1NN
U.K.

BELL MORLEY B
HERZBERG INST OF
ASTROPHYSICS
NAT RES COUNCIL
OTTAWA ONT K1A OR6
CANADA

BELL ROGER A DR
ASTRONOMY PROGRAM
UNIVERSITY OF
MARYLAND
COLLEGE PARK MD20742
U.S.A.

BELOTSERKOVSKIJ
DAVID J DR
GOSSTANDART USSR
LENINSKY PROSPECT 9
117049 MOSCOW
U.S.S.R.

BELTON MICHAEL J S DR
SOLAR SYSTEM PROG
KITT PEAK NAT OBS
950 N CHERRY AVE
TUCSON AZ 85726
U.S.A.

BELVEDERE GAETANO DR
IST DI ASTRONOMIA
CITTA UNIVERSITARIA
I-95125 CATANIA
ITALY

BELY OLEG DR
OBSERVATOIRE DE NICE
B P 252
F-06007 NICE CEDEX
FRANCE

BELY-DUBAU FRANCOISE
OBSERVATOIRE DE NICE
B P 252
F-06007 NICE CEDEX
FRANCE

BEM J DR
OBSERVATORIUM ASTR
UL KOPERNIKA 11
51-622 WROCLAW
POLAND

BENAVENTE JOSE
INST Y OBSERVATORIO
DE MARINA
SAN FERNANDO CADIZ
SPAIN

BENDER PETER L DR
JILA
UNIVERSITY OF
COLORADO
BOULDER CO 80309
U.S.A.

BENEDICT G F DR
DEPT OF ASTRONOMY
UNIV OF TEXAS
AUSTIN TX 78712
U.S.A.

BENEST D DR
OBS DE NICE
B P 252
F-06007 NICE CEDEX
FRANCE

BENEVIDES SOARES P DR
INSTITUTO ASTRONOMI-
CO E GEOFISICO
CAIXA POSTAL 30 627
SAO PAULO 00 SP
BRAZIL

BENFORD GREGORY DR
PHYSICS DEPT
UNIV OF CALIFORNIA
IRVINE CA 92717
U.S.A.

BENNETT JOHN CAISTER DR
90 MALAN STREET
RIVIERA
PRETORIA 0084
SOUTH AFRICA

BENSAMMAR S DR
OBS DE MEUDON
F-92190 MEUDON
FRANCE

BENVENUTI P DR
OSSERVATORIO ASTRO-
FISICO
I-36012 ASIAGO
ITALY

BENZ ARNOLD DR
RADIO ASTRONOMY
GROUP ETHZ
HOCHSTR 58
CH-8044 ZUERICH
SWITZERLAND

BERENDZEN RICHARD DR
PROVOST
THE AMERICAN
UNIVERSITY
WASHINGTON DC 20016
U.S.A.

BERG RICHARD A DR
DEPT PHYS AND ASTRON
UNIV OF ROCHESTER
419 SPACE SCI CTR
ROCHESTER NY 14627
U.S.A.

BERGE GLENN L DR
OWENS VALLEY RADIO
OBSERVATORY
CALTECH 102 24
PASADENA CA 91125
U.S.A.

BERGEAT JACQUES G DR
OBSERVATOIRE DE LYON
F-69230 SAINT-GENIS-
LAVAL
FRANCE

BERGER CHRISTIANE DR
CERGA
AVENUE COPERNIC
F-06130 GRASSE
FRANCE

BERGER JACQUES G DR
OBSERVATOIRE
DE PARIS
61 AV OBSERVATOIRE
F-75014 PARIS
FRANCE

BERGER XAVIER DR
CERGA
AVENUE COPERNIC
F-6130 GRASSE
FRANCE

BERGERON JACQUELINE A DR
ESC C-C CERN
CH-1211 GENEVE 23
SWITZERLAND

BERGSTRALH JAY T DR
JPL
M/X 183B - 365
4800 OAK GROVE DRIVE
PASADENA CA 91103
U.S.A.

BERKHUIJSEN ELLY M DR
MPI FUER
RADIOASTRONOMIE
AUF DEM HUEGEL 69
D-5300 BONN
GERMANY, F.R.

BERMAN ROBERT HIRAM DR
MIT
RM 36-227
77 MASSACHUSETTS AVE
CAMBRIDGE MA 02139
U.S.A.

BERNACCA P DR
OSSERVATORIO ASTRO-
FISICO
ASIAGO VICENZA
ITALY

BERNAT ANDREW PLOUS DR
KITT PEAK NAT OBS
950 NORTH CHERRY AVE
P O BOX 26732
TUCSON AZ 85726
U.S.A.

BERTAUD CHARLES DR
OBSERVATOIRE
DE MEUDON
F-92190 MEUDON
FRANCE

BERTAUX J L DR
SERVICE D AERONOMIE
B P NO 3
F-91370 VERRIERES-
LE-BUISSON
FRANCE

BERTHOMIEU GABRIELLE DR
OBSERVATOIRE DE NICE
BP 252
F-06007 NICE CEDEX
FRANCE

BERTIAU FLOR C PROF
WAVERSEBAAN 220
B-3030 HEVERLEE
BELGIUM

BERTIN GIUSEPPE DR
SCUOLA NORMALE SUP
PIAZZA DEI CAVALIERI
I-56100 PISA
ITALY

BERTOLA FRANCESCO PROF
OSSERVATORIO
ASTRONOMICO
I-35100 PADOVA
ITALY

BERTOUT CLAUDE PH D
LANDESSTERNWARTE
KOENIGSTUHL
D-69.. HEIDELBERG 1
GERMANY, F.R.

BESSELL MICHAEL S DR
MT STROMLO
OBSERVATORY
WODEN PO ACT 2606
AUSTRALIA

BETTIS DALE G PROF
TICOM
UNIVERSITY OF TEXAS
AT AUSTIN
AUSTIN TX 78712
U.S.A.

BEUERMANN KLAUS P PROF
PH BERLIN ABT 4
MALTESERSTR 74-100
D-1000 BERLIN 46
GERMANY, F.R.

BEYER MAX DR
JUSTUS-BRINCKMANN-
STRASSE 101
D-2050 HAMBURG 80
GERMANY, F.R.

BHATIA PREM K DR
DEPT OF MATHEMATICS
UNIVERSITY OF
JODHPUR
JODHPUR-342001
INDIA

BHATIA V B DR
DEPT OF PHYSICS AND
ASTROPHYSICS
DELHI UNIVERSITY
DELHI-110007
INDIA

BHATNAGAR ARVIND DR
VEDHSHALA UDAIPUR
SOLAR OBS
11 VIDYA MARG
UDAIPUR RAJAS 313001
INDIA

BHATNAGAR K B DR
ZAKIR HUSSAIN COLL
UNIV OF DELHI
AJMERI GATE
NEW DELHI 110 006
INDIA

BHATTACHARYYA J C PROF
INDIAN INSTITUTE OF
ASTROPHYSICS
BANGALORE 560034
INDIA

BHONSLE RAJARAM V PROF
PHYSICAL RESEARCH
LABORATORY
NAVRANGPURA
AHMEDABAD 380009
INDIA

BIANCHINI ANTONIO DR
OSSERVATORIO ASTRO-
FISICO
I-36012 ASIAGO
ITALY

BIBARSOV RAVIL'SH DR
ASTROPHYSICAL INST
ACAD OF SCS OF THE
TADJIK SSR
734670 DUSHANBE
U.S.S.R.

BICAK JIRI DR
DEPT OF MATH PHYSICS
CHARLES UNIVERSITY
POVLTAVSKA 8
180 00 PRAHA 8
CZECHOSLOVAKIA

BIDELMAN WILLIAM P PROF
WARNER AND SWASEY
OBSERVATORY
1975 TAYLOR RD
E CLEVELAND OH 44112
U.S.A.

BIEGING JOHN HAROLD DR
RADIO ASTRONOMY LAB
CAMPBELL HALL
UNIV OF CALIFORNIA
BERKELEY CA 94720
U.S.A.

BIELICKI MACIEJ DR
ASTRONOMICAL
OBSERVATORY
AL UJAZDOWSKIE 4
00-478 WARSZAWA
POLAND

BIEMONT EMILE DR
RUE A LECLEROS 39
B-4652 XHENDELESSE
HERVE
BELGIUM

BIERMANN LUDWIG PROF DR
MPI F ASTROPHYSIK
FOEHRINGER RING 6
D-8000 MUENCHEN 40
GERMANY, F.R.

BIERMANN PETER L DR
MPI FUR RADIOASTRON
AUF DEM HUEGEL 69
D-5300 BONN
GERMANY, F.R.

BIGAY J H DR
10 AVE DU GENERAL
DELESTRAINT
F-01000 BOURG-EN-
BRESSE
FRANCE

BIJAOUI ALBERT DR
OBSERVATOIRE DE NICE
B P 252
F-06007 NICE CEDEX
FRANCE

BILLAUD G J
CERGA
AVE COPERNIC
F-6130 GRASSE
FRANCE

BILLINGS DONALD E PROF
UNIVERSITY COLORADO
DEPT OF ASTRO-
GEOPHYSICS
BOULDER CO 80309
U.S.A.

BINGHAM RICHARD G DR
ROYAL GREENWICH
OBSERVATORY
HAILSHAM
EAST SUSSEX BN27 1RP
U.K.

BINNENDIJK L PROF DR
FLOWER AND COOK
OBSERVATORY
UNIV OF PENNSYLVANIA
PHILADELPHIA PA19104
U.S.A.

BINNEY JAMES J DR
DEPT THEOR PHYSICS
1 KEBLE ROAD
OXFORD OX1 3NP
U.K.

BIRAUD FRANCOIS DR
OBSERVATOIRE
DE MEUDON
F-92190 MEUDON
FRANCE

BIRKLE KURT PH D
MPI FUER ASTRONOMIE
KOENIGSTUHL
D-6900 HEIDELBERG 1
GERMANY, F.R.

BISIACCHI GIANFRANCO DR
INST DE ASTRONOMIA
UNAM
APDO POSTAL 70-264
MEXICO D F
MEXICO

BISNOVATYI-KOGAN G S DR
SPACE RESEARCH INSTI
TUTE USSR ACADEMY OF
SCIENCES
117810 MOSCOW
U.S.S.R.

BISWAS SUKUMAR DR
SOLAR PHYS GROUP
TATA INST FUND RES
HOMI BHABHA ROAD
BOMBAY 400 005
INDIA

BLAAUW A PROF DR
STERREWACHT
POSTBUS 9513
2300 RA LEIDEN
THE NETHERLANDS

BLACK JOHN HARRY DR
HARVARD COLLEGE OBS
CTR FOR ASTROPHYSICS
60 GARDEN ST
CAMBRIDGE MA 02138
U.S.A.

BLACKWELL DONALD E PROF
DEPT OF ASTROPHYSICS
SOUTH PARKS ROAD
OXFORD OX1 3RQ
U.K.

BLADES JOHN CH DR
ANGLO AUSTRALIAN OBS
P O BOX 296
EPPING NSW 2121
AUSTRALIA

BLAHA MILAN DR
DEPT OF PHYSICS AND
ASTRONOMY
UNIVERSITY MARYLAND
COLLEGE PARK MD20742
U.S.A.

BLAIR GUY NORMAN DR
ASTRONOMY PROGRAM
DEPT EARTH SPACE SCS
SUNY
STONY BROOK NY 11794
U.S.A.

BLAMONT JACQUES E PROF
SERVICE D AERONOMIE
DU CNRS B P 3
F-91370 VERRIERES
LE BUISSON
FRANCE

BLANCO CARLO DR
OSSERVATORIO
ASTROFISICO
CITTA UNIVERSITARIA
I-95125 CATANIA
ITALY

BLANCO VICTOR M DR
CERRO TOLOLO
INTERAMERICAN OBS
CASILLA 63 D
LA SERENA
CHILE

BLANDFORD R D DR
INST OF ASTRONOMY
MADINGLEY ROAD
CAMBRIDGE CB3 OHA
U.K.

BLASIUS KARL RICHARD DR
PLANETARY SC INST
283 S LAKE
SUITE 218
PASADENA CA 91101
U.S.A.

BLECHA ANDRE BORIS G DR
16 RUE ET-DUMONT
CH-1204 GENEVA
SWITZERLAND

BLEEKER JOHAN A M DR IR
HUYGENS LABORATORY
WASSENAARSEWEG 78
2300 RA LEIDEN
THE NETHERLANDS

BLESS ROBERT C PROF
ASTRONOMY DEPT
UNIVERSITY WISCONSIN
475 N CHARTER ST
MADISON WI 53706
U.S.A.

BLUDMAN SIDNEY A PROF
DEPT OF PHYSICS
UNIVERSITY OF
PENNSYLVANIA
PHILADELPHIA PA19104
U.S.A.

BOBROV M S DR
ASTRONOMICAL COUNCIL
USSR ACADEMY SCI
PYATNITSKAYA UL 48
109017 MOSCOW
U.S.S.R.

BODENHEIMER PETER PROF
LICK OBSERVATORY
UNIVERSITY OF CALIF
SANTA CRUZ CA 95064
U.S.A.

BOEHME A DR
HEINRICH HERTZ INST
SOLAR TERRESTR PHYS
TELEGRAFENBERG
DDR 15 POTSDAM
G.D.R.

BOERNGEN FREIMUT DR PH
ZENTR INST ASTROPHYS
K SCHWARZSCHILD OBS
DDR-6901 TAUTENBURG
G.D.R.

BOGGESS ALBERT DR
CODE 673
GODDARD SPACE
FLIGHT CENTER
GREENBELT MD 20771
U.S.A.

BOHLIN J DAVID DR
CODE ST
NASA HEADQUARTERS
WASHINGTON DC 20546
U.S.A.

BOISCHOT ANDRE DR
OBSERVATOIRE DE
MEUDON
F-92190 MEUDON
FRANCE

BOLCAL C DR
UNIV OBSERVATORY
UNIVERSITE
ISTANBUL
TURKEY

BOLTON C THOMAS PROF
DAVID DUNLAP OBS
P O BOX 360
RICHMOND HILL
ONT L4C 4Y6
CANADA

BONAZZOLA S DR
OBS DE MEUDON
F-92190 MEUDON
FRANCE

BLINOV N S DR
STERNBERG STATE AS-
TRONOMICAL INSTITUTE
117234 MOSCOW
U.S.S.R.

BLUM EMILE-JACQUES DR
IRAM
53 RUE DES MARTYRS
F-38026 GRENOBLE
FRANCE

BOBROVNIKOFF NICHOLAS T
DR
1623 VISALIA AVE
BERKELEY CA 94707
U.S.A.

BOEHM KARL-HEINZ PROF
ASTRONOMY DEPT
UNIVERSITY OF
WASHINGTON
SEATTLE WA 98195
U.S.A.

BOEHME SIEGFRIED DR
ASTRONOMISCHES
RECHEN INSTITUT
MOENCHHOFSTR 12-14
D-6900 HEIDELBERG
GERMANY, F.R.

BOESGAARD ANN M PROF
INST FOR ASTRONOMY
2680 WOODLAWN DR
HONOLULU HI 96822
U.S.A.

BOGGESS NANCY W DR
NASA HEADQUARTERS
WASHINGTON DC 20546
U.S.A.

BOHLIN RALPH C DR
CODE 672
GODDARD SPACE
FLIGHT CENTER
GREENBELT MD 20771
U.S.A.

BOK BART J PROF
STEWARD OBSERVATORY
UNIVERSITY ARIZONA
TUCSON AZ 85721
U.S.A.

BOLDT ELIHU DR
CODE 661
GODDARD SPACE FLIGHT
CENTER
GREENBELT MD 20771
U.S.A.

BOLTON JOHN G
ANRAO
BOX 276
PARKES NSW 2870
AUSTRALIA

BOND HOWARD E DR
DEPT PHYSICS AND AST
LOUISIANA STATE
UNIVERSITY
BATON ROUGE LA 70803
U.S.A.

BLITZSTEIN WILLIAM DR
DEPT OF ASTRONOMY
UNIV OF PENNSYLVANIA
D RITTENHOUSE LAB E1
PHILADELPHIA PA19174
U.S.A.

BLUM PETER PROF
INST F ASTROPHYSIK
UNIVERSITAET BONN
AUF DEM HUEGEL 71
D-5300 BONN 1
GERMANY, F.R.

BOCHONKO DONALD R DR
DEPT MATH ASTRONOMY
UNIV OF MANITOBA
WINNIPEG MB R3T 2N2
CANADA

BOEHM-VITENSE ERIKA PROF
ASTRONOMY DEPT
UNIVERSITY OF
WASHINGTON
SEATTLE WA 98195
U.S.A.

BOERNER GERHARD DR
MPI FUER PHYSIK UND
ASTROPHYSIK
FOEHRINGER RING 6
D-8000 MUENCHEN
GERMANY, F.R.

BOESHAAR GREGORY ORTH DR
DEPT OF PHYSICS
UNIV OF OREGON
EUGENE OR 97403
U.S.A.

BOGORODSKIJ A F DR
KIEV STATE UNIVERSI-
TY ASTRONOMICAL
OBSERVATORY
252053 KIEV
U.S.S.R.

BOHRMANN ALFRED PROF
SCHAERSTRASSE 23
D-2050 HAMBURG 80
GERMANY, F.R.

BOKSENBERG ALEC DR
DEPT OF PHYSICS
UNIV COLLEGE LONDON
GOWER STREET
LONDON WC1E 6BT
U.K.

BOLEY FORREST
WILDER LABORATORY
DARTMOUTH COLLEGE
HANOVER NH 03755
U.S.A.

BONANOMI J DR
OBSERVATOIRE
DE NEUCHATEL
CH-2000 NEUCHATEL
SWITZERLAND

BONDARENKO L N DR
STERNBERG STATE AS-
TRONOMICAL INSTITUTE
117234 MOSCOW
U.S.S.R.

BONDI HERMANN PROF SIR
DEPT OF ENERGY
THAMES HOUSE SOUTH
MILLBANK
LONDON SW1P 4QJ
U.K.

BONEV T MR
OBS POPULAIRE
STARA-ZAGORA
BULGARIA

BONIFAZI ANGELO DR
OSSERVATORIO ASTRO-
NOMICO
I-40100 BOLOGNA
ITALY

BONNET ROGER M DR
LPSP CNRS
BP 10
F-91370 VERRIERES
LE BUISSON
FRANCE

BONNOR W B PROF
QUEEN ELIZABETH
COLLEGE
CAMPDEN HILL ROAD
LONDON W8 7AH
U.K.

BONOMETTO SILVIO A DR
ISTITUTO DI FISICA
G GALILEI 8
VIA MARZOLO
I-35100 PADOVA
ITALY

BONOV ANGEL DR
ASTRONOMICAL
OBSERVATORY
P O BOX 36
1504 SOFIA 4
BULGARIA

BONSACK WALTER K PROF
INST FOR ASTRONOMY
2680 WOODLAWN DR
HONOLULU HI 96822
U.S.A.

BOOKMYER BEVERLY B DR
DEPT OF PHYSICS AND
ASTRONOMY
CLEMSON UNIVERSITY
CLEMSON SC 29631
U.S.A.

BOOTH ROY S DR
NUFFIELD RADIO ASTRO
NOMY LABORATORIES
JODRELL BANK
MACCLESFIELD SK119DL
U.K.

BOPP BERNARD W DR
DEPT PHYSICS ASTRON
UNIV OF TOLEDO
TOLEDO OH 43606
U.S.A.

BORGMAN JAN DR PROF
KAPTEYN OBSERVATORY
MENSINGHEWEG 20
RODEN DR
THE NETHERLANDS

BORGNINO J DR
DEPT D'ASTROPHYSIQUE
UNIVERSITE DE NICE
PARC VALROSE
F-06034 NICE CEDEX
FRANCE

BORRA ERMANNO F DR
DEPT DE PHYSIQUE
UNIVERSITE LAVAL
STE FOY
QUEBEC PQ G1K 7P4
CANADA

BOSMA PIETER B DR
SUBFAC NATUURKUNDE
VRIJE UNIVERSITEIT
DE BOELELAAN 1081
1007 MC AMSTERDAM
THE NETHERLANDS

BOSMAN-CRESPIN DENISE
BLVD D AVROY 68
BTE 093
B-4000 LIEGE
BELGIUM

BOTELHEIRO A DR
OBS ASTRONOMICO
LISBOA
PORTUGAL

BOTEZ ELVIRA DR
INSTITUT D'ENSEIGNE-
MENT SUPERIEUR
5800 SUCEAVA
ROUMANIA

BOTTINELLI LUCETTE DR
OBSERVATOIRE
DE MEUDON
RADIOASTRONOMIE
F-92190 MEUDON
FRANCE

BOUGERET J L DR
OBS DE MEUDON
F-92190 MEUDON
FRANCE

BOUIGUE ROGER PROF
LAB D'ASTRONOMIE
UNIVERSITE
PAUL SABATIER
F-31077 TOULOUSE CDX
FRANCE

BOULON JACQUES J DR
OBSERVATOIRE PARIS
61 AVENUE DE
L OBSERVATOIRE
F-75014 PARIS
FRANCE

BOURY ARSENE J DR
INST D ASTROPHYSIQUE
UNIVERSITE DE LIEGE
B-4200 COINTE OUGREE
BELGIUM

BOUSKA JIRI DR
DEPT OF ASTRONOMY
CHARLES UNIVERSITY
SVEDSKA 8
150 00 PRAHA
CZECHOSLOVAKIA

BOUVIER PIERRE PROF
OBSERVATOIRE
DE GENEVE
CH-1290 SAUVERNY GE
SWITZERLAND

BOWELL E L G DR
LOWELL OBS
P O BOX 1269
FLAGSTAFF AZ 86002
U.S.A.

BOWEN EDWARD G DR
5010 MAXWELL AVENUE
WEST RIVER MD 20881
U.S.A.

BOWYER C STUART PROF
UNIVERSITY
OF CALIFORNIA
ASTRONOMY DEPT
BERKELEY CA 94720
U.S.A.

BOYARCHUK A A DR
CRIMEAN ASTROPHYSI-
CAL OBSERVATORY USSR
ACAD OF SCI NAUCHNY
334413 CRIMEA
U.S.S.R.

BOYARCHUK MARGARITA E DR
CRIMEAN ASTROPHYS
OBSERVATORY
P/O NAUCHNYJ
334413 CRIMEA
U.S.S.R.

BOYCE PETER B DR
1604 HOBART ST
WASHINGTON DC 20009
U.S.A.

BOYD ROBERT L F PROF
MULLARD SPACE
SCIENCE LAB UCL
HOLMBURY ST MARY
DORKING RH5 6NS
U.K.

BOYTON P E DR
ASTRONOMY DEPT
UNIV OF WASHINGTON
SEATTLE WA 98195
U.S.A.

BOZIS GEORGE PROF
DEPT THEOR MECHANICS
UNIVERSITY
OF THESSALONIKI
THESSALONIKI
GREECE

BOZKURT SUKRU DR
EGE UNIVERSITY
OBSERVATORY
PK 21
BORNOVA IZMIR
TURKEY

BRACCESI ALESSANDRO PROF
OSSERVATORIO
ASTRONOMICO
VIA ZAMBONI 33
I-40126 BOLOGNA
ITALY

BRACEWELL RONALD N PROF
DURAND 329
STANFORD UNIVERSITY
STANFORD CA 94305
U.S.A.

BRAHIC ANDRE DR
OBS DE MEUDON
F-92190 MEUDON
FRANCE

BRANDIE GEORGE W DR
ENVIRONMENTAL ENG
DUPUIS HALL
QUEENS UNIVERSITY
KINGSTON ONT K7L 3N6
CANADA

BRANSON NICHOLAS J B A DR
GENERAL BOARD OF THE
FACULTIES
THE OLD SCHOOLS
CAMBRIDGE CB2 1TT
U.K.

BRAUNINGER H DR
MPI FUER PHYSIK UND
ASTROPHYSIK
INST EXTRATERR PHYS
D-8046 GARCHING
GERMANY, F.R.

BREALEY G A P ENG
DOMINION ASTROPHYS-
ICAL OBSERVATORY
5071 W SAANICH RD
VICTORIA BC V8X 3X3
CANADA

BRECKINRIDGE JAMES B DR
J P L CALTECH
MAIL STOP 183-301
4800 OAK GROVE DR
PASADENA CA 91103
U.S.A.

BREJCC IZABELLA I DR
PULKOVO OBSERVATORY
196140 LENINGRAD
U.S.S.R.

BRIEVA EDUARDO PROF
OBSERVATORIO
NACIONAL
APARTADO 2584
BOGOTA
COLOMBIA

BRINKMAN BERT C DR
SPACE RESEARCH LAB
BENELUXLAAN 21
UTRECHT
THE NETHERLANDS

BROGLIA PIETRO DR
OSSERVATORIO
ASTRONOMICO
I-22055 MERATE COMO
ITALY

BROSCHE PETER PROF
OBSERVATORIUM
HOHER LIST
D-5568 DAUN
GERMANY, F.R.

BRAES L L E DR
STERREWACHT
POSTBUS 9513
2300 RA LEIDEN
THE NETHERLANDS

BRANCH DAVID DR
DEPT OF PHYSICS AND
ASTRONOMY
UNIVERSITY OKLAHOMA
NORMAN OK 73019
U.S.A.

BRANDT JOHN C DR
CODE 680
GODDARD SPACE
FLIGHT CENTER
GREENBELT MD 20771
U.S.A.

BRAUDE SEMION YA PROF AG
INST RADIOPHY ELECTR
ACADEMY OF SCIENCES
THE UKRAINIAN SSR
310085 KHARKOV
U.S.S.R.

BRAUNSFURTH EDWARD PH D
ASTRONOMISCHES INST
RUHR UNIVERSITAET
POSTFACH 102148
D-4630 BOCHUM
GERMANY, F.R.

BRECHER AVIVA DR PROF
DEPT OF PHYSICS
WELLESLEY COLLEGE
WELLESLEY MA 02181
U.S.A.

BREGER MICHEL DR
DEPT OF ASTRONOMY
UNIVERSITY OF TEXAS
AUSTIN TX 78712
U.S.A.

BRETAGNON P DR
BUREAU D LONGITUDES
77 AVENUE DENFERT-
ROCHEREAU
F-75014 PARIS
FRANCE

BRIHAYE CHARLES C A DR
SCHEMERINGLAAN 14A
1900 OVERIJSE
BELGIUM

BROADFOOT A LYLE DR
3625 EAST AJO WAY
TUCSON AZ 85713
U.S.A.

BROMAGE GORDON E DR
ASTROPHYSICS RES DIV
S R C APPLETON LAB
CULHAM LAB
ABINGDON OXON
U.K.

BROSTERHUS E B F DR
DOMINION ASTRO-
PHYSICAL OBSERVATORY
5071 W SAANICH RDRR7
VICTORIA BC V8X 3X3
CANADA

BRAHDE ROLF DR
INSTITUTE OF THEORE-
TICAL ASTROPHYSICS
UNIVERSITY OF OSLO
BLINDERN OSLO 3
NORWAY

BRAND PETER W J L DR
DEPT OF ASTRONOMY
UNIVERSITY
OF EDINBURGH
EDINBURGH EH8 9YL
U.K.

BRANSCOMB L M DR
NATIONAL BUREAU OF
STANDARDS
WASHINGTON DC 20025
U.S.A.

BRAULT JAMES W DR
KITT PEAK NATIONAL
OBSERVATORY
PO BOX 26732
TUCSON AZ 85726
U.S.A.

BRAY ROBERT J DR
CSIRO NATIONAL MEASU
REMENT LABORATORY
PO BOX 218
LINDFIELD NSW 2070
AUSTRALIA

BRECHER KENNETH PROF
MASSACHUSETTS INSTI-
TUTE OF TECHNOLOGY
ROOM 6-201
CAMBRIDGE MA 02139
U.S.A.

BREINHORST ROBERT A DR
ASTRON INSTITUT
STERNWARTE
AUF DEM HUEGEL 71
D-5300 BONN 1
GERMANY, F.R.

BRIDLE ALAN H PROF
DEPT OF PHYSICS
QUEENS UNIVERSITY
KINGSTON ONT K7L 3N6
CANADA

BRINI D PROF
LABORATORIO TESRE
VIA CASTAGNOLI 1
I-40100 BOLOGNA
ITALY

BRODERICK JOHN DR
PHYSICS DEPT
VPI AND SU
BLACKSBURG VA 24061
U.S.A.

BROOKES CLIVE J DR
1 RYEFIELD CLOSE
WEST HAGLEY
STOURBRIDGE
W MIDLANDS DY9 0JS
U.K.

BROTEN NORMAN W
HERZBERG INST OF
ASTROPHYSICS
NAT RES COUNC CANADA
OTTAWA ONT K1A 0R6
CANADA

BROUCKE ROGER DR
7203 RUNNING ROPE
CIRCLE
AUSTIN TX 78731
U.S.A.

BROUW W N DR
RADIOSTERRENWACHT
OUDE HOOGEVEENSE
DIJK 4
DWINGELOO 7514
THE NETHERLANDS

BROWN HARRISON DR
DIV OF GEOLOGICAL
SCIENCES
CALTECH
PASADENA CA 91103
U.S.A.

BROWN JOHN C DR
DEPT OF ASTRONOMY
GLASGOW UNIVERSITY
GLASGOW G12 8QQ
U.K.

BROWN ROBERT L DR
NATIONAL RADIO
ASTRONOMY'OBS
BOX 2
GREEN BANK WV 24944
U.S.A.

BROWN RONALD D PROF
CHEMISTRY DEPT
MONASH UNIVERSITY
WELLINGTON ROAD
CLAYTON VICT 3168
AUSTRALIA

BROWNE IAN W A DR
NUFFIELD RADIO ASTRO
NOMY LABORATORIES
JODRELL BANK
MACCLESFIELD SK119DL
U.K.

BROWNLEE DONALD E PROF
DEPT OF ASTRONOMY
UNIV OF WASHINGTON
SEATTLE WA 98195
U.S.A.

BROWNLEE ROBERT R DR
G-DO MS 570
LOS ALAMOS SCI
LABORATORY
LOS ALAMOS NM 87544
U.S.A.

BRUCK HERMANN A PROF
CRAIGOWER
PENICUIK EH26 9LA
U.K.

BRUCK MARY T DR
ROYAL OBSERVATORY
EDINBURGH EH9 3HJ
U.K.

BRUECKNER GUENTER E DR
CODE 7142
NAVAL RESEARCH
LABORATORY
WASHINGTON DC 20375
U.S.A.

BRUMBERG VICTOR A DR
INSTITUTE FOR THEORE
TICAL ASTRONOMY
192187 LENINGRAD
U.S.S.R.

BRUNER EL C JR DR
LOCKHEED P A RES LAB
DEPT 52-12 BLDG 202
3251 HANOVER ST
PALO ALTO CA 94304
U.S.A.

BRUNET JEAN-PIERRE DR
OBSERVATOIRES
1 AVE C FLAMMARION
F-31500 TOULOUSE
FRANCE

BRUNK WILLIAM E DR
OFF SPACE SC CODE SL
NASA HEADQUARTERS
400 MARYLAND AVE S W
WASHINGTON DC 20546
U.S.A.

BRUSTON P DR
L P S P
B P NO 10
F-91370 VERRIERES-
LE-BUISSON
FRANCE

BRUWER J A
SOUTH AFRICAN ASTRO-
NOMICAL OBSERVATORY
P O BOX 3718
JOHANNESBURG 2000
SOUTH AFRICA

BRUZEK ANTON DR
KIEPENHEUER INSTITUT
FUER SONNENPHYSIK
SCHOENECKSTRASSE 6
D-7800 FREIBURG
GERMANY, F.R.

BUCHLER J ROBERT PROF
DEPT OF PHYSICS AND
ASTRONOMY
UNIVERSITY FLORIDA
GAINESVILLE FL 32611
U.S.A.

BUDDING EDWIN DR
ASTRONOMY DEPT
UNIVERSITY OF
MANCHESTER
MANCHESTER M13 9PL
U.K.

BUES IRMELA D DR
REMEIS STERNWARTE
STERNWARTSTRASSE 7
D-8600 BAMBERG
GERMANY, F.R.

BUFF JAMES S DR
DEPT PHYSICS ASTRON
DARTMOUTH COLLEGE
HANOVER NH 03755
U.S.A.

BUHL DAVID DR
INFR RADIO ASTRO BR
NASA GODDARD SPACE
FLIGHT CTR CODE 693
GREENBELT MD 20771
U.S.A.

BULLARD EDWARD SIR
IGPP A025
UNIVERSITY OF
CALIFORNIA
LA JOLLA CA 92093
U.S.A.

BULLEN K E PROF
DEPT OF APPLIED MATH
UNIVERSITY OF SYDNEY
SYDNEY NSW 2006
AUSTRALIA

BUMBA VACLAV DR
ASTRONOMICAL INST OF
THE CZECHOSLOVAK
ACADEMY OF SCIENCES
251 65 ONDREJOV
CZECHOSLOVAKIA

BUNNER A N DR
PERKIN-ELMER CORP
M S 846
100 WOOSTER HTS RD
DANBURY CT 06810
U.S.A.

BURBIDGE E MARGARET PROF
DEPT OF PHYSICS
UNIV OF CALIFORNIA
SAN DIEGO MCCO-11
LA JOLLA CA 92093
U.S.A.

BURBIDGE GEOFFREY R PROF
OFF OF THE DIRECTOR
KITT PEAK NAT OBS
P O BOX 26732
TUCSON AZ 85726
U.S.A.

BURCH CECIL R DR
PHYSICS DEPT
THE UNIVERSITY
BRISTOL 8
U.K.

BURGER J J DR IR
WERKGROEP KOSMISCHE
STRALING K ONNES LAB
NIEUWSTEEG 18
LEIDEN
THE NETHERLANDS

BURGER MARIJKE DR
SPARRENSTRAAT 58
1170 BRUSSEL
BELGIUM

BURGESS ALAN DR
DEPT OF APPLIED MATH
AND THEORETICAL PHYS
SILVER STREET
CAMBRIDGE CB3 9EW
U.K.

BURGESS DAVID D DR
BLACKETT LABORATORY
IMPERIAL COLLEGE OF
SCI AND TECHNOLOGY
LONDON SW7 2BZ
U.K.

BURKE BERNARD F DR
DEPT OF PHYSICS
M I T RM 26-335
CAMBRIDGE MA 02139
U.S.A.

BURKE J ANTHONY DR
DEPT OF PHYSICS
UNIVERSITY OF
VICTORIA
VICTORIA BC V8W 2Y2
CANADA

BURKHART C DR
OBS DE LYON
69230 ST-GENIS-LAVAL
FRANCE.

BURKHEAD MARTIN S
ASTRONOMY DEPT
INDIANA UNIVERSITY
SWAIN HALL WEST
BLOOMINGTON IN 47405
U.S.A.

BURKI GILBERT DR
OBS DE GENEVE
CH-1290 SAUVERNY
SWITZERLAND

BURLAGA L F DR
NASA GSFC
CODE 692
GREENBELT MD 20771
U.S.A.

BURNELL SUSAN J B DR
MULLARD SPACE SC LAB
UNIV COLLEGE LONDON
LONDON
U.K.

BURNS JOSEPH A PROF
CENTER RADIOPHYSICS
SPACE RESEARCH
CORNELL UNIVERSITY
ITHACA NY 14850
U.S.A.

BURSA MILAN DR
ASTRONOMICAL INST
CSAV
BUDECSKA 6
120 23 PRAHA 2
CZECHOSLOVAKIA

BURTON W BUTLER DR
DEPT OF ASTRONOMY
UNIV OF MINNESOTA
116 CHURCH ST S E
MINNEAPOLIS MN 55455
U.S.A.

BURTON WILLIAM M
ASTROPHYSICS RES DIV
CULHAM LABORATORY
ABINGDON
OXON OX14 3DB
U.K.

BUSCOMBE WILLIAM PROF
ASTRONOMY DEPT
NORTHWESTERN
UNIVERSITY
EVANSTON IL 60201
U.S.A.

BUSER ROLAND DR
ASTRONOMICAL INST
UNIV OF BASEL
VENUSSTRASSE 7
CH-4102 BINNINGEN
SWITZERLAND

BUTI B PROF
PHYSICAL RESEARCH
NAVRANGPURA
AHMEDABAD 380009
INDIA

BUTLER C JOHN DR
ARMAGH OBSERVATORY
ARMAGH BT61 9DG
U.K.

BUTLER DENNIS DR
ASTRONOMY DEPT
YALE UNIV
BOX 2023 YALE STAT
NEW HAVEN CT 06520
U.S.A.

BYARD P L DR
DEPT OF ASTRONOMY
OHIO STATE UNIV
174 W 18TH AVE
COLUMBUS OH 43210
U.S.A.

BYKOV M F DR
ASTRONOMICAL INST
UZBEKIAN ACADEMY OF
SCIENCES
700000 TASHKENT
U.S.S.R.

BYRD GENE G DR
DEPT PHYSICS ASTRON
UNIV OF ALABAMA
BOX 1921
UNIVERSITY AL 35486
U.S.A.

BYRNE PATRICK B DR
ARMAGH OBSERVATORY
ARMAGH BT61 9DG
U.K.

BYSTROV NIKOLAI F DR
PULKOVO OBSERVATORY
196140 LENINGRAD
U.S.S.R.

BYSTROVA NATALIJA V DR
SPECIAL ASTROPHYSI-
CAL OBSERVATORY
LENINGRAD BRANCH
196140 LENINGRAD
U.S.S.R.

CABRITA EZEQUIEL DR
OBSERVATORIO
DE LISBOA
TAPADA LISBOA
PORTUGAL

CACCIANI ALESSANDRO DR
OSSERVATORIO
ASTRONOMICO DI ROMA
I-00100 ROMA
ITALY

CACCIANI CARLA DR
IST DI ASTRONOMIA
I-40100 BOLOGNA
ITALY

CACCIN BRUNO
OSSERVATORIO ASTRONO
MICO DI CAPODIMONTE
VIA MOIARIELLO 16
I-80131 NAPOLI
ITALY

CADEZ ANDREJ DR
DEPT OF PHYSICS
UNIV OF LJUBLJANA
JADRANSKA 19
61000 LJUBLJANA
YUGOSLAVIA

CAHN JULIUS H ASSOC PROF
UNIVERSITY OF
ILLINOIS OBSERVATORY
URBANA IL 61801
U.S.A.

CALAMAI G PROF
OSSERVATORIO ASTRO
FISICO DI ARCETRI
FIRENZE
ITALY

CALAME ODILE DR
CERGA
AVENUE COPERNIC
F-06130 GRASSE
FRANCE

CALLEBAUT DIRK K DR
DEPT OF PHYSICS UIA
UNIVERSITEITSPLEIN 1
B 2610 WILRIJK-ANTW
BELGIUM

CALDI VITTORIA DR
CNR
ASTROFISICA SPAZIALE
C P 67 ROMA
I-00044 FRASCATI
ITALY

CALVANI MASSIMO DR
OSSERVATORIO ASTRO-
NOMICO
I-35100 PADOVA
ITALY

CALVO MANUEL
DEPTO DE ASTRONOMIA
UNIVERSIDAD ZARAGOZA
ZARAGOZA 9
SPAIN

CAMARENA BADIA V ADJ PROF
DPTO FIS TIERRA COS
FACULTAD DE CIENCIAS
UNIV DE ZARAGOZA
ZARAGOZA
SPAIN

CAMERON ALASTAIR G W PROF
HARVARD COLLEGE
OBSERVATORY
60 GARDEN STREET
CAMBRIDGE MA 02138
U.S.A.

CAMERON WINIFRED S MRS
CODE 601
GODDARD SPACE FLIGHT
CENTER
GREENBELT MD 20771
U.S.A.

CAMICHEL HENRI DR
OBSERVATOIRE
DE TOULOUSE
1 CAMILLE FLAMMARION
F-31500 TOULOUSE
FRANCE

CAMM G LESLIE DR
MATHEMATICS DEPT
THE UNIVERSITY
MANCHESTER M13 9PL
U.K.

CAMPBELL BRUCE DR
CFHT CORPORATION
BOX 1597
KAMUELA HI 96743
U.S.A.

CAMPBELL DONALD B
ARECIBO OBSERVATORY
BOX 995
ARECIBO PR 00612
U.S.A.

CAMPBELL JAMES W
ROYAL OBSERVATORY
BLACKFORD HILL
EDINBURGH EH9 3HJ
U.K.

CANAL RAMON M DR
DPTO FIS TIERRA COS
UNIVERSIDAD
DE BARCELONA
BARCELONA 14
SPAIN

CANAVAGGIA RENEE DR
OBSERVATOIRE
DE PARIS
61 AV L OBSERVATOIRE
F-75014 PARIS
FRANCE

CANDY MICHAEL P MR
PERTH OBSERVATORY
WALNUT ROAD
BICKLEY 6076
AUSTRALIA

CANFIELD RICHARD C DR
PHYSICS C-011 UCSD
LA JOLLA CA 92093
U.S.A.

CANIZARES CLAUDE R PROF
MIT
RM 37-501
CAMBRIDGE MA 02139
U.S.A.

CANNON C J DR
DEPT OF APPLIED
MATHEMATICS
UNIVERSITY OF SYDNEY
SYDNEY NSW 2006
AUSTRALIA

CANNON RUSSELL D DR
ROYAL OBSERVATORY
BLACKFORD HILL
EDINBURGH EH9 3HJ
U.K.

CANNON WAYNE H DR
DEPT OF PHYSICS
YORK UNIVERSITY
4700 KEELE STREET
DOWNSVIEW ONT M3J1P3
CANADA

CANTU ALBERTO M DR
OSSERVATORIO ARCETRI
LARGO FERMI 5
I-50125 FIRENZE
ITALY

CAPACCIOLI MASSIMO DR
ASTRONOMICAL OBS
UNIVERSITY OF PADOVA
I-35100 PADOVA
ITALY

CAPEN CHARLES F
LOWELL OBSERVATORY
BOX 1269
FLAGSTAFF AZ 86001
U.S.A.

CAPLAN JAMES
OBSERVATOIRE DE
MARSEILLE
PLACE LE VERRIER
F-13004 MARSEILLE
FRANCE

CAPRIOLI GIUSEPPE PROF
OSSERVATORIO ASTRONO
MICO DI ROMA
VIA TRIONFALE 204
ROMA
ITALY

CAPRIOTTI EUGENE R DR
DEPT OF ASTRONOMY
OHIO STATE UNIV
5058 A SMITH LAB
COLUMBUS OH 43210
U.S.A.

CAPUTO FILIPPINA DR
CNR ASTROFISICA
SPAZIALE C P 67
ROMA
I-00044 FRASCATI
ITALY

CARBON DUANE F DR
KITT PEAK NAT OBS
P O BOX 26732
TUCSON AZ 85726
U.S.A.

CARDONA OCTAVIO DR
INST NAC DE ASTROFIS
OPTYCA Y ELECTRONICA
APDO POSTAL 216
PUEBLA PUE
MEXICO

CARDUS ALMEDA J DR
OBSERVATORIO DEL
EBRO
ROQUETAS
TORTOSA BARCELONA
SPAIN

CARESTIA REINALDO A DR
OBS ASTRON F AGUILAR
AV BENAVIDEZ 8175
OESTE RIVADAVIA
5400 SAN JUAN
ARGENTINA

CARLETON NATHANIEL P DR
SMITHSONIAN ASTROPHY
SICAL OBSERVATORY
60 GARDEN ST
CAMBRIDGE MA 02138
U.S.A.

CARLQVIST PER A DR
DEPT PLASMA PHYSICS
ROYAL INST OF
TECHNOLOGY
S-10044 STOCKHOLM 70
SWEDEN

CARM LAWRENCE E DR
SACRAMENTO PEAK OBS
SUNSPOT NM 88349
U.S.A.

CAROFF LAWRENCE J
SPACE SC DIV NASA
AMES RES CTR MS245-3
MOFFETT FIELD
CA 94035
U.S.A.

CAROUBALOS C A PROF
LAB ELECTRONIC PHYS
UNIVERSITY OF ATHENS
104 SOLONOS STREET
ATHENS 144
GREECE

CARPENTER LLOYD DR
GODDARD SPACE FLIGHT
CENTER
CODE 921
GREENBELT MD 20771
U.S.A.

CARR THOMAS D PROF
DEPT OF PHYSICS AND
ASTRONOMY
UNIVERSITY FLORIDA
GAINESVILLE FL 32611
U.S.A.

CARRANZA G J DR
LAPRIDA 880
CORDOBA
ARGENTINA

CARRASCO GARRORENA R DR
CLARA DEL REY 14
MADRID 2
SPAIN

CARRASCO GUILLERMO DR
OBS ASTRON NATIONAL
UNIVERSIDAD DE CHILE
CASILLA 36-D
SANTIAGO
CHILE

CARRASCO LUIS DR
INSTITUTO DE
ASTRONOMIA
APDO POSTAL 70264
MEXICO 20 DF
MEXICO

CARROLL P KEVIN PROF
PHYSICS DEPT
UNIVERSITY COLLEGE
BELFIELD
DUBLIN 4
IRELAND

CARRUTHERS GEORGE R DR
SPACE SCIENCE DIV
US NAVAL RES LAB
CODE 7123
WASHINGTON DC 20375
U.S.A.

CARSON T R DR
UNIVERSITY
OBSERVATORY
BUCHANAN GARDENS
ST ANDREWS FIFE
U.K.

CARSWELL ROBERT F DR
INST OF ASTRONOMY
MADINGLEY ROAD
CAMBRIDGE CB3 OHA
U.K.

CARVER JOHN H PROF
PHYSICS DEPT
UNIVERSITY
OF ADELAIDE
ADELAIDE 5001
AUSTRALIA

CASANOVAS JUAN DR
VATICAN OBSERVATORY
I-00120 VATICAN CITY
VATICAN CITY STATE

CASINI CATERINA DR
ISTITUTO DI
ASTRONOMIA
VIA BRERA 28
I-20121 MILANO
ITALY

CASSATELLA ANGELO DR
LAB ASTROFISICA SPAZ
P C B 67
I-00044 FRASCATI
ITALY

CASSE M DR
SECT D'ASTROPHYSIQUE
CEN SACLAY
B P NO 2
F-91190 GIF-S-YVETTE
FRANCE

CASSINELLI JOSEPH P DR
ASTRONOMY DEPT
UNIVERSITY WISCONSIN
MADISON WI 53706
U.S.A.

CASTELLANI VITTORIO PROF
LAB ASTROFISICA
C P 67
I-00044 FRASCATI
ITALY

CASTELLI FIORELLA DR
OSSERVATORIO ASTRO-
NOMICO
VIA TIEPOLO 11
I-34131 TRIESTE
ITALY

CASTELLI JOHN P
AFGL-PHP
HANSCOM AFB MA 01731
U.S.A.

CASTOR JOHN I DR
JILA
UNIVERSITY COLORADO
BOULDER CO 80309
U.S.A.

CASWELL J L DR
CSIRO RADIOPHYSICS
P O BOX 76
EPPING NSW 2121
AUSTRALIA

CATALA ASUNCION P DR
DPTO FIS TIERRA COS
UNIVERSIDAD
DE BARCELONA
BARCELONA 28
SPAIN

CATALANO FRANCESCO A DR
OSSERVATORIO
ASTROFISICO
CITTA UNIVERSITARIA
I-95125 CATANIA
ITALY

CATALANO SANTO DR
IST DI ASTRONOMIA
CITTA UNIVERSITARIA
I-95125 CATANIA
ITALY

CATO TORGNY DR
RODDBAATSGATAN 3
S-421 71 VAESTRA
FROELUNDA
SWEDEN

CATURA RICHARD C DR
LOCKHEED RES LAB
DEPT 52 12 BLDG 202
3251 HANOVER ST
PALO ALTO CA 94304
U.S.A.

CAUGHLAN GEORGEANNE R
PROF
DEPT OF PHYSICS
MONTANA STATE UNIV
BOZEMAN MT 59715
U.S.A.

CAVALIERE ALFONSO G PROF
IST DI ASTRONOMIA
VIC OSSERVATORIO 5
I-35100 PADOVA
ITALY

CAYREL DE STROBEL GIUSADR
CANADA FRANCE HAWAII
TELESCOPE CORP
KAMUELA HI 96743
U.S.A.

CAYREL ROGER DR
CANADA FRANCE HAWAII
TELESCOPE CORP
KAMUELA HI 96743
U.S.A.

CAZENAVE ANNY DR
CNES-GRGS
18 AVE EDQUARD BELIN
F-31400 TOULOUSE
FRANCE

CAZZOLA P DR
OSSERVATORIO ASTRON
VICOLA DELL
OSSERVATORIO
PADOVA
ITALY

CECCARELLI M PROF
ISTITUTO DI FISICA
GENERALE
UNIV DI BOLOGNA
I-40100 BOLOGNA
ITALY

CEFOLA PAUL J DR
MAIL STATION 64
C S DRAPER LAB
555 TECHNOLOGY SQ
CAMBRIDGE MA 02139
U.S.A.

CELNIKIER LUDWIK DR
OBSERVATOIRE
DE MEUDON
F-92190 MEUDON
FRANCE

CEPLECHA ZDENEK DR
ASTRONOMICAL INST
CZECHO ACAD SCIENCES
251 65 ONDREJOV
OBSERVATORY
CZECHOSLOVAKIA

CEPPATELLI GUIDO DR
OSSERVATORIO ASTRO-
FISICO DI ARCETRI
I-50125 FIRENZE
ITALY

CESARSKY CATHERINE J DR
SEP-CEN SACLAY
B P NO 2
F-91190 GIF-YVETTE
FRANCE

CESARSKY DIEGO A DR
OBSERVATOIRE
DE MEUDON
RADIO ASTRONOMIE
F-92190 MEUDON
FRANCE

CESCO CARLOS DR
OBS ASTR F AGUILAR
BENAVIDEZ 8175 OESTE
5407 MARQUESADO
SAN JUAN
ARGENTINA

CESTER BRUNO PROF
OSSERVATORIO
ASTRONOMICO
VIA TIEPOLO 11
I-34131 TRIESTE
ITALY

CHAFFEE FREDERIC H DR
MT HOPKINS
OBSERVATORY
P O BOX 97
AMADO AZ 85640
U.S.A.

CHAISSON ERIC J PROF
HARVARD OBSERVATORY
CAMBRIDGE MA 02138
U.S.A.

CHAMARAUX P DR
CPS DE MEUDON
F-92190 MEUDON
FRANCE

CHAMBERLAIN JOSEPH M DR
ADLER PLANETARIUM
1300 S LAKE SHORE DR
CHICAGO IL 60201
U.S.A.

CHAMBERLAIN JOSEPH W PROF
DEPT SPACE PHYSICS
AND ASTRONOMY
RICE UNIVERSITY
HOUSTON TX 77001
U.S.A.

CHAMBLISS CARLSON R DR
KUTZTOWN STATE
COLLEGE
KUTZTOWN PA 19530
U.S.A.

CHANDRASEKHAR S PROF
933 E 56 STREET
CHICAGO IL 60637
U.S.A.

CHANMUGAN GABESAR PROF
DEPT OF PHYSICS
AND ASTRONOMY
LA STATE UNIVERSITY
BATON ROUGE LA 70803
U.S.A.

CHAPMAN CLARK R DR
PLANETARY SC INST
2030 E SPEEDWAY
SUITE 201
TUCSON AZ 85719
U.S.A.

CHAPMAN GARY A DR
DEPT PHYSICS ASTRON
CALIF STATE UNIV
18111 NORDHOFF ST
NORTHRIDGE CA 91324
U.S.A.

CHAPMAN ROBERT D DR
CODE 680
GODDARD SPACE FLIGHT
CENTER
GREENBELT MD 20771
U.S.A.

CHAPRONT J DR
BUREAU D LONGITUDES
77 AVE DENFERT-
ROCHEREAU
F-75014 PARIS
FRANCE

CHAPRONT-TOUZE M DR
BUREAU D LONGITUDES
77 AVENUE DENFERT-
ROCHEREAU
F-75014 PARIS
FRANCE

CHARVIN PIERRE DR
INAG
77 AVENUE DENFERT
ROCHEREAU
F-75014 PARIS
FRANCE

CHAU WAI Y PROF
PHYSICS DEPT
QUEEN S UNIVERSITY
KINGSTON ONT K7L 3N6
CANADA

CHAVIRA ENRIQUE SR
INST NAC ASTROPHYS
OPTICA Y ELECTRONICA
AP POSTAL 216 Y 51
PUEBLA PUE
MEXICO

CHEN KWAN-YU PROF
DEPT OF PHYSICS AND
ASTRONOMY
UNIVERSITY FLORIDA
GAINESVILLE FL 32611
U.S.A.

CHENG CHUNG-CHIEH DR
NAVAL RESEARCH LAB
CODE 7174C
WASHINGTON DC 20375
U.S.A.

CHEREDNICHENKO V I DR
KIEV POLYTECHNICAL
INSTITUTE
252056 KIEV
U.S.S.R.

CHEREPASHCHUK A M DR
STERNBERG STATE
ASTRONOMICAL INST
NOMICAL INSTITUTE
117234 MOSCOW
U.S.S.R.

CHERNEGA N A A DR
OBSERVATORY OF THE
KIEV UNIVERSITY
OBSERVATORNAYA 3
252053 KIEV
U.S.S.R.

CHERNYKH N S DR
CRIMEAN ASTRO-
PHYSICAL OBSERVATORY
NAUCHNY
334413 CRIMEA
U.S.S.R.

CHERTOPRUD V E DR
HYDROMETEOROLOGICAL
CENTER OF THE USSR
123376 MOSCOW
U.S.S.R.

CHEVALIER CLAUDE DR
OBSERVATOIRE
DE MEUDON
F-92190 MEUDON
FRANCE

CHEVALIER ROGER A DR
DEPT OF ASTRONOMY
UNIV OF VIRGINIA
P O BOX 3818
CHARLOTTESVL VA22903
U.S.A.

CHINCARINI GUIDO L DR
DEPT PHYS AND ASTRON
UNIV OF OKLAHOMA NEL
SON BD 440 W BROOKS
NORMAN OK 73069
U.S.A.

CHIOSI CESARE S DR
ISTITUTO DI
ASTRONOMIA
UNIVERSITA DI PADOVA
I-35100 PADOVA
ITALY

CHIS CHEORGHE PROF DR
STR REPUBLICII 109
3400 CLUJ NAPOCA
ROUMANIA

CHISTYAKOV V E DR
SOLAR STATION
PRIMORSKY KRAY
692533 GORNOTAEZHNOE
U.S.S.R.

CHITRE D M DR
RAMAN RESEARCH INST
P C HEBBAL
BANGALORE 560 006
INDIA

CHITRE SHASHIKUMAR M DR
TATA INSTITUTE
HOMI BHABHA ROAD
COLABA
BOMBAY 400005
INDIA

CHIU HONG-YEE DR
MUDD BLDG RM 828
COLUMBIA UNIV
NEW YORK NY 10027
U.S.A.

CHIUDERI CLAUDIO PROF
OSSERVATORIO ARCETRI
LARGO FERMI 5
I-50125 FIRENZE
ITALY

CHIUDERI-DRAGO F PROF
OSSERVATORIO ASTROFI
SICO DI ARCETRI
LARGO E FERMI 5
I-50125 FIRENZE
ITALY

CHLIETOVSKY FRANCA DR
OSSERVATORIO ASTRO-
NOMICO
BRERA
MILANO
ITALY

CHMIELEWSKI YVES DR
DEPT OF ASTRONOMY
UNIVERSITY OF TEXAS
R L MOORE HALL
AUSTIN TX 78712
U.S.A.

CHOI BON CHOL
ACADEMY OF SCIENCES
PYONGYANG ASTRONOMI-
CAL OBSERVATORY
PYONGYANG
DEM PEOPLE'S REP OF KOREA

CHOLLET FERNAND
OBSERVATOIRE PARIS
61 AVENUE DE
L OBSERVATOIRE
F-75014 PARIS
FRANCE

CHOPINET MARGUERITE DR
OBSERVATOIRE
DE BORDEAUX
F-33270 FLOIRAC
FRANCE

CHOU KYONG CHOL PROF
DEPT OF ASTRONOMY
AND METEOROLOGY
YOUNSEI UNIVERSITY
SEOUL 120
REPUBLIC OF KOREA

CHRISTENSEN-DALSGAARD J
INST D'ASTROPHYSIQUE
UNIVERSITE DE LIEGE
AVENUE DE COINTE 5
B-4200 COINTE-OUGREE
BELGIUM

CHRISTIANSEN WILBUR PROF
SCHOOL OF ELECTRICAL
ENGINEERING
UNIVERSITY OF SYDNEY
SYDNEY NSW 2006
AUSTRALIA

CHRISTY JAMES WALTER DR
EXPLORATORY DEV ST
U S NAVAL OBS
WASHINGTON DC 20390
U.S.A.

CHRISTY ROBERT F DR
PROVOST OFFICE
CALIFORNIA INSTITUTE
OF TECHNOLOGY
PASADENA CA 91125
U.S.A.

CHUBB TALBOT A DR
5023 N 38TH ST
ARLINGTON VA 22207
U.S.A.

CHUGAJNOV P F DR
CRIMEAN ASTROPHYSI-
CAL OBSERVATORY USSR
ACAD OF SCI NAUCHNIY
334413 CRIMEA
U.S.S.R.

CHUN MUN-SUK DR
DEPT OF ASTROPHYSICS
SOUTH PARKS ROAD
OXFORD OX1 3RQ
U.K.

CHUPP EDWARD L DR
PHYSICS DEPT
DEMERITT HALL
UNIV NEW HAMPSHIRE
DURHAM NH 03824
U.S.A.

CHURCHWELL EDWARD B DR
WASHBURN OBSERVATORY
UNIVERSITY WISCONSIN
475 N CHARTER ST
MADISON WI 53706
U.S.A.

CHURMS JOSEPH
S A ASTRONOMICAL OBS
P O BOX 9
OBSERVATORY
CAPE 7935
SOUTH AFRICA

CHUVAEV K K DR
CRIMEAN ASTROPHYSI-
CAL OBSERVATORY USSR
ACAD OF SCI NAUCHNIY
334413 CRIMEA
U.S.S.R.

CHVOJKOVA WOYK E DR
ASTRONOMICAL INST
CZECHO ACAD OF SC
BUDECSKA 6
12023 PRAHA 2
CZECHOSLOVAKIA

CIATTI FRANCO DR
OSSERVATORIO
ASTROFISICO
I-36012 ASIAGO VI
ITALY

CID PALACIOS RAFAEL PROF
FACULTAD DE CIENCIAS
CIUDAD UNIVERSITARIA
ZARAGOZA
SPAIN

CILLIE G G PROF
4 MINSERIE STREET
STELLENBOSCH 7600
SOUTH AFRICA

CIMINO MASSIMO A PROF
VIA A CADLOLO 19
I-00136 ROMA
ITALY

CLARIA J DR
ASOC VENEZOLANA
DE ASTRONOMIA
APT DE CORREOS 264
MERIDA
VENEZUELA

CLARK ALFRED JR PROF
DEPT MECH AND AERO
SCIENCES
UNIVERSITY ROCHESTER
ROCHESTER NY 14627
U.S.A.

CLARK BARRY G DR
NATIONAL RADIO
ASTRONOMY OBS
P O BOX O
SOCORRO NM 87801
U.S.A.

CLARK DAVID H DR
ROYAL GREENWICH
OBSERVATORY
HAILSHAM
EAST SUSSEX BN27 1RP
U.K.

CLARK FRANK OLIVER DR
DEPT PHYSICS ASTRON
UNIV OF KENTUCKY
LEXINGTON KY 40506
U.S.A.

CLARK GEORGE W PROF
MIT ROOM 37-611
CAMBRIDGE MA 02139
U.S.A.

CLARK THOMAS A DR
RADIO ASTRO BRANCH
NASA GODDARD SPACE
FLIGHT CTR CODE 693
GREENBELT MD 20771
U.S.A.

CLARK THOMAS ALAN DR
PHYSICS DEPT
UNIV OF CALGARY
2920-24 AVENUE N W
CALGARY ALB T2N 1N4
CANADA

CLARKE DAVID DR
DEPT OF ASTRONOMY
THE UNIVERSITY
GLASGOW G12 8QQ
U.K.

CLARKE THOMAS R DR
MCLAUGHLIN PLANETAR
ROYAL ONTARIO MUSEUM
100 QUEENS PK CRESC
TORONTO ONT M5S 2C6
CANADA

CLAUSEN JENS VIGGO LEKTOR
COPENHAGEN UNIVER-
SITY OBSERVATORY
BRORFELDE
DK-4340 TOLLOSE
DENMARK

CLAYTON DONALD D PROF
DEPT SPACE PHYSICS
AND ASTRONOMY
RICE UNIVERSITY
HOUSTON TX 77001
U.S.A.

CLEGG PETER E DR
QUEEN MARY COLLEGE
MILE END RD
LONDON E1 4NS
U.K.

CLEGG ROBIN F S DR
DEPT OF ASTRONOMY
UNIV OF TEXAS
AUSTIN TX 78712
U.S.A.

CLEMENT MAURICE J PROF
DEPT OF ASTRONOMY
UNIVERSITY OF
TORONTO
TORONTO ONT M5S 1A7
CANADA

CLIFTON KENNETH ST DR
NASA E S 64
MARSHALL SFC AL35812
U.S.A.

CLIMENHAGA JOHN L PROF
DEPT OF PHYSICS
UNIVERSITY
OF VICTORIA
VICTORIA BC V8W 2Y2
CANADA

CLUBE S V M DR
ROYAL OBSERVATORY
EDINBURGH EH9 3HJ
U.K.

COCHRAN WILLIAM DAVID DR
DEPT OF ASTRONOMY
UNIV OF TEXAS
AUSTIN TX 78712
U.S.A.

COCKE W JOHN PROF
STEWARD OBSERVATORY
UNIVERSITY ARIZONA
TUCSON AZ 85721
U.S.A.

CODE ARTHUR D
WASHBURN OBSERVATORY
UNIV OF WISCONSIN
MADISON WI 53706
U.S.A.

CODINA LANDABERRY S J DR
INSTITUTO ASTRONOMI-
CO I GEOFISICO USP
CAIXA POSTAL 30627
01000 SAO PAULO SP
BRAZIL

CODINA VIDAL J M DR
AV JOSE ANTONIO 679
BARCELONA 13
SPAIN

COELHO BALSA M C DR
OBS ASTRONOMICO
UNIV DE COIMBRA
COIMBRA
PORTUGAL

COFFEEN DAVID L DR
INST FOR SPACE STUDY
2880 BROADWAY
NEW YORK NY 10025
U.S.A.

COFFEY HELEN E DR
NOAA
NGSDC D63
BOULDER CO 80303
U.S.A.

COGAN BRUCE C DR
X-1 MAIL STOP 531
LOS ALAMOS SC LAB
LOS ALAMOS NM 87545
U.S.A.

COHEN JEFFREY M DR
PHYSICS DEPT
UNIV OF PENNSYLVANIA
PHILADELPHIA PA19174
U.S.A.

COHEN JUDITH DR
KITT PEAK NATIONAL
OBSERVATORY
P O BOX 26732
TUCSON AZ 85726
U.S.A.

COHEN LEON PROF
DEPT OF PHYSICS
HUNTER COLLEGE
695 PARK AVE
NEW YORK NY 10021
U.S.A.

COHEN MARSHALL H PROF
CALTECH 102-24
PASADENA CA 91125
U.S.A.

COHEN MARTIN DR
ASTRONOMY DEPT
UNIV OF CALIFORNIA
BERKELEY CA 94720
U.S.A.

COHEN RAYMOND J DR
NRAL
JODRELL BANK
MACCLESFIELD
CHESHIRE SK11 9DL
U.K.

COLBURN DAVID S DR
245-3 NASA
AMES RESEARCH CENTER
MOFFETT FIELD
CA 94035
U.S.A.

COLE TREVOR W DR
CSIRO RADIOPHYSICS
P O BOX 76
EPPING NSW 2121
AUSTRALIA

COLGATE STIRLING A DR
THEORETICAL DIVISION
LOS ALAMOS SCIENTI-
FIC LAB MS 210
LOS ALAMOS NM 87845
U.S.A.

COLLIN-SOUFFRIN SUZY DR
OBSERVATOIRE
DE MEUDON
F-92190 MEUDON
FRANCE

COLLINS GEORGE W II PROF
174 WEST 18TH AVE
COLUMBUS OH 43210
U.S.A.

COLLINSON EDWARD H
THE COPSE
CHURCH LANE PLAYFORD
IPSWICH
SUFFOLK IP6 9DR
U.K.

COLOMB FERNANDO R DR
INST ARGENTINO DE
RADIOASTRONOMIA
CAS DE CORREO NO 5
1894 VILLA ELISA BA
ARGENTINA

COLOMBO G PROF DR
IST MECCANICA APPL
UNIV DI PADOVA
VIA F MARZOLO 9
I-35100 PADOVA
ITALY

COMBES M MR
OBS DE MEUDON
F-92190 MEUDON
FRANCE

CONCONI PAOLO DR
OSSERVATORIO ASTRO-
NOMICO
BRERA
MILANO
ITALY

CONDON JAMES J DR
PHYSICS DEPT VA POLY
TECHN INST STATE UNI
325 ROBESON HALL
BLACKSBURG VA 24061
U.S.A.

CONKLIN EDWARD K
C O FORTH INC
815 MANHATTAN AVE
MANHATTAN BEACH
CA 90266
U.S.A.

CONNES JANINE DR
CIRCE
BAT 506
F-91440 ORSAY CEDEX
FRANCE

CONNES PIERRE DR
SERVICE D'ASTRONOMIE
B P 3
F-91370 VERRIERES LE
BUISSON
FRANCE

CONTI PETER S DR
JILA
UNIVERSITY COLORADO
BOULDER CO 80309
U.S.A.

CONTOPOULOS GEORGE PROF
ASTRONOMY DEPT
PANEPISTIMIOPOLIS
ATHENS 621
GREECE

CONWAY ROBIN G DR
NRAL
JODRELL BANK
CHESHIRE SK11 9DL
U.K.

COOK ALAN H PROF
CAVENDISH LABORATORY
MADINGLEY RD
CAMBRIDGE CB3 0HE
U.K.

COOK ALLAN F DR
CENTER FOR ASTROPHY-
SICS HCO-SAO
60 GARDEN ST
CAMBRIDGE MA 02138
U.S.A.

COOKE B A DR
X-RAY ASTRONOMY
GROUP PHYSICS DEPT
THE UNIVERSITY
LEICESTER LE1 7RH
U.K.

CORBIN THOMAS ELBERT DR
U S NAVAL OBS
TRANSIT CIRCLE DIV
WASHINGTON DC 20390
U.S.A.

CORLISS C H DR
NATIONAL BUREAU OF STANDARDS
WASHINGTON DC 20234
U.S.A.

CORNEJO ALEJANDRO A MSS
INAOE
APDO POSTAL 216 Y 51
PUEBLA PUE
MEXICO

CORWIN HAROLD G JR DR
DEPT OF ASTRONOMY
UNIV OF EDINBURGH
ROYAL OBS
EDINBURGH EH9 3HJ
U.K.

COSMOVICI CRISTIANO B DR
MPI FUER EXTRATER-RESTRISCHE PHYSIK
D-8046 GARCHING
GERMANY, F.R.

COSTAIN C C DR
DIV OF PHYSICS
NAT RES COUNCIL
OTTAWA K1A OR6
CANADA

COSTAIN CARMAN H DR
DOMINION RADIO
ASTROPHYSICAL OBS
BOX 248
PENTICTON BC V2A 6K3
CANADA

COSTERO RAFAEL MC
INST DE ASTRONOMIA
APDO POSTAL 70-264
MEXICO 20 D F
MEXICO

COUDERC PAUL DR
5 SQUARE DELAMBRE
F-75014 PARIS
FRANCE

COUNSELMAN CHARLES C PROF
MIT DEPT EARTH AND
PLANETARY SCIENCES
CAMBRIDGE MA 02139
U.S.A.

COUPINOT GERARD DR
OBSERVATOIRE
DE TOULOUSE
1 CAMILLE FLAMMARION
F-31500 TOULOUSE
FRANCE

COURTES GEORGES PROF
OBSERVATOIRE DE
MARSEILLE
2 PLACE LE VERRIER
F-13004 MARSEILLE
FRANCE

COUSINS A W J DR
S A ASTRONOMICAL OBS
P O BOX 9
OBSERVATORY
CAPE 7935
SOUTH AFRICA

COUTEAU PAUL PROF
OBSERVATOIRE DE NICE
B P 252
F-06007 NICE
FRANCE

COUTREZ RAYMOND A J PROF
UNIVERSITE LIBRE DE
BRUXELLES
50 AV F D ROOSEVELT
B-1050 BRUXELLES
BELGIUM

COUTTS-CLEMENT CHRISTINE M PROF
DEPT OF ASTRONOMY
UNIVERSITY TORONTO
TORONTO ONT M5S 1A7
CANADA

COVINGTON ARTHUR E
269 PLEASANT PARK RD
OTTAWA ONT K1H 5M7
CANADA

COWAN JOHN J DR
HARVARD SMITHS OBS
CTR FOR ASTROPHYSICS
60 GARDEN STREET
CAMBRIDGE MA 02138
U.S.A.

COWIE LENNOX LAUCHLAN DR
PRINCETON UNIV OBS
PEYTON HALL
PRINCETON NJ 08540
U.S.A.

COWLEY ANNE P DR
ASTRONOMY DEPT
UNIVERSITY
OF MICHIGAN
ANN ARBOR MI 48109
U.S.A.

COWLEY CHARLES R PROF
ASTRONOMY DEPT
UNIVERSITY
OF MICHIGAN
ANN ARBOR MI 48109
U.S.A.

COWLING THOMAS G PROF
19 HOLLIN GARDENS
HEADINGLY
LEEDS LS16 5NL
U.K.

COWSIK R DR
TATA INST FUND RES
HOMI BHABHA ROAD
BOMBAY 400 005
INDIA

COX ARTHUR N DR
LOS ALAMOS
SCIENTIFIC LAB
P O BOX 1663
LOS ALAMOS NM 87545
U.S.A.

COX DONALD P PROF
DEPT OF PHYSICS
UW MADISON
1150 UNIVERSITY AVE
MADISON WI 53706
U.S.A.

COX JOHN P PROF
JILA
UNIVERSITY OF
COLORADO
BOULDER CO 80309
U.S.A.

COYNE GEORGE V DR
LUNAR AND PLANETARY
LABORATORY
UNIVERSITY ARIZONA
TUCSON AZ 85721
U.S.A.

CRAIG IAN JONATHAN D DR
DEPT OF APPLIED MATH
UNIV OF WAIKATO
HAMILTON
NEW ZEALAND

CRAINE ERIC RICHARD DR
STEWARD OBS
UNIV OF ARIZONA
TUCSON AZ 85721
U.S.A.

CRAM LAWRENCE EDWARD DR
SACRAMENTO PEAK OBS
SUNSPOT NM 88349
U.S.A.

CRAMPTON DAVID DR
DOMINION ASTROPHYSI-CAL OBSERVATORY
5071 W SAANICH ROAD
VICTORIA BC V8X 3X3
CANADA

CRANNELL CAROL JO DR
NASA-GSFC
CODE 684
GREENBELT MD 20771
U.S.A.

CRAWFORD DAVID L
KITT PEAK NATIONAL
OBSERVATORY
BOX 26732
TUCSON AZ 85726
U.S.A.

CREZE MICHEL DR
OBSERVATOIRE DE
BESANCON
41 AV L OBSERVATOIRE
F-25000 BESANCON
FRANCE

CRIFO FRANCOISE DR
DAPHE
OBSERVATOIRE DE
MEUDON
F-92190 MEUDON
FRANCE

CRISTALDI SALVATORE DR
OSSERVATORIO
ASTROFISICO
CITTA UNIVERSITARIA
I-95125 CATANIA
ITALY

CRISTESCU CORNELIA G DR
OBSERVATOIRE
DE BUCAREST
75212 BUCAREST
ROUMANIA

CRCOM DAVID L DR
APPLETON LABORATORY
DITTCN PARK
SLOUGH BERKS SL3 9JX
U.K.

CROVISIER J DR
OBS DE MEUDON
F-92190 MEUDON
FRANCE

CRUIKSHANK DALE P DR
INST FOR ASTRONOMY
2680 WOODLAWN DR
HONOLULU HI 96822
U.S.A.

CRUISE ADRIAN MICHAEL DR
MULLARD SPACE SC LAB
HCLMBURY ST MARY
DCRKING
SURREY RH5 6NS
U.K.

CRUTCHER RICHARD M DR
UNIVERSITY OF
ILLINOIS OBSERVATORY
URBANA IL 61801
U.S.A.

CRUVELLIER PAUL E DR
LAB ASTRO SPATIALE
TRAVERSE DU SIPHON
LES TROIS LUCS
F-13012 MARSEILLE
FRANCE

CSADA IMRE K DR
KCNKCLY OBSERVATORY
BOX 67
H-1525 BUDAPEST
HUNGARY

CUDABACK DAVID D DR
RADIO ASTRONOMY LAB
UNIVERSITY
OF CALIFORNIA
BERKELEY CA 94720
U.S.A.

CUDWORTH KYLE MCCABE DR
YERKES OBS
WILLIAMS BAY WI53191
U.S.A.

CUFFEY J MR
N M STATE UNIV
DEPT OF EARTH SC AND
ASTRONOMY
UNIV PK NM 88001
U.S.A.

CUGNON PIERRE DR
OBSERVATOIRE ROYAL
DE BELGIQUE
AVE CIRCULAIRE 3
B-1180 BRUXELLES
BELGIUM

CUGUSI LEONINO DR
IST DI ASTRONOMIA
UNIVERSITA
CAGLIARI
ITALY

CULHANE LEONARD DR
MULLARD SPACE SC LAB
HCLMBURY ST MARY
DCRKING
SURREY RH5 6NS
U.K.

CULVER ROGER BRUCE DR
DEPT OF PHYSICS
COLORADO STATE UNIV
FT COLLINS CO 80523
U.S.A.

CUNNINGHAM LELAND E PROF
DEPT OF ASTRONOMY
UNIVERSITY OF
CALIFORNIA
BERKELEY CA 94720
U.S.A.

CUNY YVETTE J DR
CBSERVATOIRE DE
MEUDON
PLACE JANSSEN
F-92190 MEUDON
FRANCE

CUPERMAN SAMI PROF
DEPT OF PHYSICS AND
ASTRONOMY
TEL-AVIV UNIVERSITY
RAMAT-AVIV
ISRAEL

CURRIE DOUGLAS G DR
DEPT PHYSICS ASTRON
UNIV OF MARYLAND
COLLEGE PK MD 20742
U.S.A.

CZYZAK STANLEY J DR
ASTRONOMY DEPT
CHIC STATE UNIV
174 W 18TH ST
CCLUMBUS OH 43210
U.S.A.

D'ANTONA FRANCESCA DR
OSSERVATORIO ASTRO-
NOMICO
I-00100 ROMA
ITALY

D'ODORICO SANDRO DR
OSSERVATORIO
ASTROFISICO
I-36012 ASIAGO VI
ITALY

DA ROCHA VIEIRA E DR
INT DE FISICA
UNIV FEDERAL DA
RIO GRANDE DO SUL
90000 PORTO ALLEGRE
BRAZIL

DA SILVA A V C S
OBS ASTRONOMICO
UNIV SANTA CLARA
COIMBRA
PORTUGAL

DA SILVA I F PROF DR
FACULDADE DE CIENCIA
LISBON
PORTUGAL

DA SILVA LICIO DR
CBSERVATORIO NACION
R GEN BRUCE 586
S CRISTOVAO
20000 RIO DE JANEIRO
BRAZIL

DACHS JOACHIM PROF DR
ASTRONOMISCHES INST
RUHR-UNIVERSITAET
POSTFACH 102148
D-4630 BOCHUM 1
GERMANY, F.R.

DADAEV ALEKSANDR N DR
PULKOVO OBSERVATORY
196140 LENINGRAD
U.S.S.R.

DADIC ZARKO DR
INST SCIENCE HISTORY
YUGCS ACAD SC ART
DEMETROVA 18
41000 ZAGREB
YUGOSLAVIA

DAHN CONARD CURTIS DR
U S NAVAL OBS
P O BOX 1149
FLAGSTAFF AZ 86002
U.S.A.

DAIGNE GERARD G
OBS DE MEUDON
F-92190 MEUDON
FRANCE

DAINTREE EDWARD J DR
NUFFIELD RADIO ASTRO
NOMY LABS
JODRELL BANK
MACCLESFIELD SK119DL
U.K.

DAISHIDO TSUNEAKI DR
DEPT OF SCIENCE
SCHOOL OF EDUCATION
WASEDA UNIVERSITY
SHINJUKU-KU TOKYO
JAPAN

DALGARNO ALEXANDER PROF
HARVARD COLLEGE OBS
60 GARDEN STREET
CAMBRIDGE MA 02138
U.S.A.

DALLAPORTA N PROF
ISTITUTO ASTRONOMIA
UNIVERSITA
VICOLO OSSERVATORIO
I-35100 PADOVA
ITALY

DALTABUIT ENRIQUE DR
INSTITUTO DE
ASTRONOMIA UNAM
APDO POSTAL 70264
MEXICO DF
MEXICO

DAMBARA TAKESHI PROF
SHIZUOKA UNIVERSITY
OTANI SHIZUOKA 422
JAPAN

DANBY J M ANTHONY DR
DEPT OF MATHEMATICS
NORTH CAROLINA STATE
UNIVERSITY
RALEIGH NC 27607
U.S.A.

DANESE LUIGI DR
OSSERVATORIO ASTRO-
NOMICO
I-35100 PADOVA
ITALY

DANKS ANTHONY C R
ESO
SCHLEISSHEIMER ST 17
D-8046 GARCHING B M
GERMANY, F.R.

DANZIGER I JOHN DR
TELESCOPE PROJECT
EUROPEAN SOUTHERN OB
C/O CERN
CH-1211 GENEVA 23
SWITZERLAND

DATLOWE AYTON DR
LOCKHEED P A RES LAB
DEPT 52-12 BLDG 202
3251 HANOVER STREET
PALO ALTO CA 94304
U.S.A.

DAUBE-KURZEMNIECE I A DR
RADIOASTROPHYSICAL
OBSERVATORY LATVIAN
ACADEMY OF SCIENCES
226026 RIGA
U.S.S.R.

DAUTCOURT G DR
ZENTRINST ASTROPHYS
STERNWARTE BABELSB
R LUXEMBURG STR 17A
DDR-1502 POTSDAM
G.D.R.

DAVIDSEN ARTHUR FALNES DR
DEPT OF PHYSICS
JOHNS HOPKINS UNIV
CHARLES AND 34TH STS
BALTIMORE MD 21218
U.S.A.

DAVIDSON KRIS DR
SCHOOL PHYS AND ASTR
UNIV OF MINNESOTA
116 CHURCH ST S E
MINNEAPOLIS MN 55455
U.S.A.

DAVIDSON WILLIAM PROF
MATHEMATICS DEPT
UNIVERSITY OF OTAGO
P O BOX 56
DUNEDIN
NEW ZEALAND

DAVIES JOHN G DR
NRAL
JODRELL BANK
MACCLESFIELD
CHESHIRE SK11 9DC
U.K.

DAVIES MERTON E MR
1414 SAN REMO DRIVE
PACIFIC PALISADES
CA 90272
U.S.A.

DAVIES RODNEY D PROF
NUFFIELD RADIO
ASTRONOMY LABS
JODRELL BANK
MACCLESFIELD SK119DL
U.K.

DAVIS CECIL G JR
UNIVERSITY OF
CALIFORNIA LASL
BOX 1663 MS 420
LOS ALAMOS NM 87545
U.S.A.

DAVIS JOHN DR
SCHOOL OF PHYSICS
UNIVERSITY OF SYDNEY
SYDNEY NSW 2006
AUSTRALIA

DAVIS LEVERETT JR PROF
405-47 CALTECH
PASADENA CA 91125
U.S.A.

DAVIS MARC DR
DEPT OF ASTRONOMY
HARVARD UNIVERSITY
60 GARDEN ST
CAMBRIDGE MA 02138
U.S.A.

DAVIS MICHAEL M DR
ARECIBO OBSERVATORY
P O BOX 995
ARECIBO PR 00612
U.S.A.

DAVIS MORRIS S PROF
DEPT PHYS ASTRONOMY
UNIV NORTH CAROLINA
284 PHILLIPS H 039A
CHAPEL HILL NC 27514
U.S.A.

DAVIS RICHARD J DR
NUFFIELD RADIO
ASTRONOMY LABS
JODRELL BANK
MACCLESFIELD SK119DL
U.K.

DAVIS ROBERT J DR
CENTER FOR ASTRO
PHYSICS
60 GARDEN ST
CAMBRIDGE MA 02138
U.S.A.

DAWE JOHN ALAN DR
DEPT OF ASTRONOMY
THE UNIVERSITY
LEICESTER LE1 7RH
U.K.

DE BERGH C DR
GROUPE PLANETES
SECT D'ASTROPHYSIQUE
OBS DE MEUDON
F-92190 MEUDON
FRANCE

DE BIASE G A DR
OSS ASTRONOMICO
VIALE DEL PARCO
MELLINI 84
ROMA
ITALY

DE BOER K S DR
WASHBURN OBSERVATORY
475 N CHARTER STREET
MADISON WI 53706
U.S.A.

DE BRUYN ANTONIUS G DR
RADIOSTERREWACHT
POSTBUS 2
7990 AA DWINGELOO
THE NETHERLANDS

DE CASTRO A DR
OBSERVATORIO
ASTRONOMICO
ALFONSO XII 5
MADRID 7
SPAIN

DE FELICE FERNANDO DR
IST FISICA GALILEO
GALILEI
UNIVERSITA DI PADOVA
I-35100 PADOVA
ITALY

DE FREITAS PACHECO J A DR
OBSERVATORIO DE SAO
PAULO USP
CAIXA POSTAL 30627
SAO PAULO
BRAZIL

DE GRAAF T DR
INST VOOR FONETISCHE
WETENSCHAPPEN
GROTE ROZENSTRAAT 31
9712 TG GRONINGEN
THE NETHERLANDS

DE GRAAFF W DR
STERREKUNDIG INST
ZONNENBURG 2
3512 NL UTRECHT
THE NETHERLANDS

DE GRAAUW TH DR
SPACE SCIENCE DEPT
ESA
NOORDWIJK
THE NETHERLANDS

DE GREVE JEAN PIERRE DR
ASTROPHYSICAL INST
V U B PLEINLAAN 2
B-1050 BRUSSELS
BELGIUM

DE GROOT MART DR
ARMAGH OBSERVATORY
ARMAGH BT61 9DG
U.K.

DE GROOT T DR
STERREKUNDIG
INSTITUUT
ZONNENBURG 2
3512 NL UTRECHT
THE NETHERLANDS

DE JAGER CORNELIS PROF
THE ASTRONOMICAL
INSTITUTE
BENELUXLAAN 21
3527 HS UTRECHT
THE NETHERLANDS

DE JAGER G DR
DEPT OF PHYSICS
RHODES UNIVERSITY
P O BOX 94
GRAHAMSTOWN 6140
SOUTH AFRICA

DE JONG T DR
ASTRONOMICAL INST
ROETERSTRAAT 15
AMSTERDAM 1004
THE NETHERLANDS

DE JONGE J K DR
DEPT OF ASTRONOMY
UNIV OF PITTSBURGH
RIVERVIEW PARK
PITTSBURGH PA 15214
U.S.A.

DE KORT JULES J DR
HOUTLAAN 4
6500 GV NIJMEGEN
THE NETHERLANDS

DE KORTE P A J DR
COSMIC RAY WORKING
GROUP HUYGENS LAB
WASSENAARSEWEG 78
2300 RA LEIDEN
THE NETHERLANDS

DE LA HERRAN V JOSE ENG
INSTITUTO DE
ASTRONOMIA MEXICO
AP POSTAL 971
MEXICO 1 DF
MEXICO

DE LA NOE JEROME DR
OBS DE BORDEAUX
AV PIERRE SEMIROT
F-33270 FLOIRAC
FRANCE

DE LA REZA RAMIRO DR
OBSERVATORIO
NACIONAL
R GEN BRUCE 586 ZC08
20000 RIO DE JANEIRO
BRAZIL

DE LOORE CAMIEL PROF
ASTROPHYSICAL INST
VRIJE UNIVERSITEIT
VAN BRUSSEL
B-1050 BRUSSELS
BELGIUM

DE MARCUS WENDELL C PROF
UNIVERSITY OF
KENTUCKY-LEXINGTON
LEXINGTON KY 40506
U.S.A.

DE MOTTONI Y PALACIOS DR
PRIVAT OBSERVATORY
VIA ROSSELLI 15-23
I-16145 GENOVA
ITALY

DE PASCUAL MARTINEZ M DR
OBSERVATORIO
ASTRONOMICO
ALFONSO XII 5
MADRID 7
SPAIN

DE SABBATA V PROF DR
IST DI FISICA
UNIV DI BOLOGNA
VIA IRNERIO 46
I-40100 BOLOGNA
ITALY

DE VAUCOULEURS G PROF
DEPT OF ASTRONOMY
UNIVERSITY OF TEXAS
AUSTIN TX 78712
U.S.A.

DE VEGT CHR PROF DR
HAMBURGER STERNWARTE
GOJENBERGSWEG 112
D-2050 HAMBURG 80
GERMANY, F.R.

DE YOUNG DAVID S DR
NAT RADIO ASTR OBS
EDGEMONT RD
CHARLOTTESVILLE
VA 22901
U.S.A.

DE ZOTTI GIANFRANCO DR
OSSERVATORIO ASTRO-
NOMICO
I-35100 PADOVA
ITALY

DEARBORN DAVID S P DR
STEWARD OBS
UNIV OF ARIZONA
TUCSON AZ 85721
U.S.A.

DEBARBAT SUZANNE V DR
OBS DE PARIS
61 AV L'OBSERVATOIRE
F- 75014 PARIS
FRANCE

DEBEHOGNE HENRI DR SC
OBSERVATOIRE ROYAL
DE BELGIQUE
3 AVE CIRCULAIRE
B-1180 BRUXELLES
BELGIUM

DEBRUNNER HERMANN DR
PHYSIKALISCHES INST
UNIVERSITAET BERN
SIDLERSTRASSE 5
CH-3000 BERN
SWITZERLAND

DECAUX BERNARD
20 RUE BOIS LE VENT
F-75016 PARIS
FRANCE

DEEMING T J DR
DEPT OF ASTRONOMY
404 PHYSICS BLDG
UNIVERSITY OF TEXAS
AUSTIN TX 78712
U.S.A.

DEERENBERG AD J DR
COSMIC RAY WORKING
GROUP HUYGENS LAB
WASSENAARSEWEG 78
2300 RA LEIDEN
THE NETHERLANDS

DEHARVENG JEAN MICHEL DR
LAS TRAVERSE DU SIP-
HON LES TROIS LUCS
F-13012 MARSEILLE
FRANCE

DEINZER W PROF DR
UNIVERSITAETS-
STERNWARTE
GEISMARLANDSTR 11
D-3400 GOETTINGEN
GERMANY, F.R.

DEJAIFFE RENE J DR
OBSERVATOIRE ROYAL
DE BELGIQUE
AVE CIRCULAIRE 3
B-1180 BRUXELLES
BELGIUM

DEKKER E DR
STERREWACHT
POSTBUS 9513
2300 RA LEIDEN
THE NETHERLANDS

DELACHE PHILIPPE J DR
OBSERVATOIRE DE NICE
BP 252
F-06007 NICE CEDEX
FRANCE

DELANNOY JEAN DR
OBSERVATOIRE DE
BORDEAUX
AVE PR P SEMIROT
F-33270 FLOIRAC
FRANCE

DELBOUILLE LUC PROF
INST D ASTROPHYSIQUE
UNIVERSITE DE LIEGE
B-4200 COINTE OUGREE
BELGIUM

DELCROIX ANDRE J S DR
19A R E VANDERVELDE
B-7230 FRAMERIES
BELGIUM

DELHAYE JEAN PROF
OBSERVATOIRE PARIS
61 AVE DE
L OBSERVATOIRE
F-75014 PARIS
FRANCE

DELSEMME ARMAND H PROF DR
DEPT OF PHYSICS AND
ASTRONOMY
UNIVERSITY OF TOLEDO
TOLEDO OH 43606
U.S.A.

DEMARCQ JEAN L ING
OBSERVATOIRE DE NICE
B P 252
F-06007 NICE CEDEX
FRANCE

DEMARET JACQUES DR
158/033 AVENUE DE
L'OBSERVATOIRE
B-4000 LIEGE
BELGIUM

DEMARQUE P PROF
YALE UNIVERSITY OBS
BOX 2023
YALE STATION
NEW HAVEN CT 06520
U.S.A.

DEMENKO A A DR
KIEV STATE UNIVERSI-
TY ASTRONOMICAL OB-
SERVATORY
252053 KIEV
U.S.S.R.

DEMERS SERGE DR
DEPT DE PHYSIQUE
UNIV DE MONTREAL
CP 6128 SUCC A
MONTREAL PQ H3C 3J7
CANADA

DEMIN V G PROF DR
STERNBERG STATE AS-
TRONOMICAL INSTITUTE
117234 MOSCOW
U.S.S.R.

DENIS CARLO DR
INST D'ASTROPHYSIQUE
UNIVERSITE DE LIEGE
AVENUE DE COINTE 5
B-4200 COINTE-OUGREE
BELGIUM

DENISSE JEAN F PROF
OBSERVATOIRE
DE MEUDON
F-92190 MEUDON
FRANCE

DENISYUK EDVARD K DR
ASTROPHYSICAL INST
480068 ALMA ATA
U.S.S.R.

DENNIS BRIAN ROY DR
NASA-GSFC
CODE 684
GREENBELT MD 20771
U.S.A.

DENNISON E W DR
ELECTRONIC VISION CO
11526 SORRENTO
VALLEY ROAD
SAN DIEGO CA 92121
U.S.A.

DENNISON P A DR
DEPT OF PHYSICS
UNIV OF ADELAIDE
ADELAIDE SA 5001
AUSTRALIA

DENT WILLIAM A PROF
PHYSICS ASTRONOMY
DEPT GRC TOWER B
UNIVERSITY OF MASS
AMHERST MA 01002
U.S.A.

DEPRIT ANDRE PROF
CTR FOR APPLIED MATH
NATIONAL BUREAU OF
STANDARDS
WASHINGTON DC 20234
U.S.A.

DERMENDJIEV VLADIMIR DR
DEPT OF ASTRONOMY
BULGARIAN ACAD OF SC
7TH NOVEMBER STR 1
SOFIA
BULGARIA

DESESQUELLES JEAN DR
UNIVERSITE LYON I
CAMPUS LA DOUA
F-69621 VILLEURBANNE
FRANCE

DESPAIN KEITH HOWARD DR
DEPT OF ASTRONOMY
HAVERFORD COLLEGE
HAVERFORD PA 19041
U.S.A.

DEUBNER FRANZ-LUDWIG DR
INST F ASTRONOMIE U
ASTROPHYSIK DER UNIV
AM HUBLAND
D-8700 WUERZBURG
GERMANY, F.R.

DEUPREE ROBERT G DR
DEPT OF ASTRONOMY
BOSTON UNIV
725 COMMONWEALTH AVE
BOSTON MA 02215
U.S.A.

DEUTSCH ALEKSANDR N
PROF DR
PULKOVO OBSERVATORY
196140 LENINGRAD
U.S.S.R.

DEUTSCHMAN W.A. DR.
DEPT.PHYSICS ASTRON.
ROSE-HULMAN INST.
5500 WABASH AVE.
TERRE HAUTE IN 47803
U.S.A.

DEVINNEY E J DR
100 UNION AVE
DELANCO NJ 08075
U.S.A.

DEWHIRST DAVID W DR
INST OF ASTRONOMY
THE OBSERVATORIES
MADINGLEY RD
CAMBRIDGE CB3 0HA
U.K.

DEWITT B S DR
DEPT OF PHYSICS
UNIV OF TEXAS
AUSTIN TX 78712
U.S.A.

DEWITT JOHN H JR
3602 HOODS HILLS RD
NASHVILLE TN 37215
U.S.A.

DEWITT-MORETTE C PROF
DEPT OF ASTRONOMY
UNIVERSITY OF TEXAS
AUSTIN TX 78712
U.S.A.

DEZSO LORANT PROF
HELIOPHYSICAL
OBSERVATORY
H-4010 DEBRECEN
HUNGARY

DI SEREGO A SPERELLO DR
IST DI ASTRONOMIA
UNIV DI PADOVA
I-35100 PADOVA
ITALY

DI TULLIO GRAZIELLA DR
OSSERVATORIO ASTRO-
NOMICO
I-35100 PADOVA
ITALY

DIBAY E A DR
STERNBERG STATE AS-
TRONOMICAL INSTITUTE
117234 MOSCOW
U.S.S.R.

DICKE ROBERT H PROF
JOSEPH HENRY LABS
PHYSICS DEPT
PRINCETON UNIVERSITY
PRINCETON NJ 08540
U.S.A.

DICKEL HELENE R DR
UNIVERSITY OF
ILLINOIS OBSERVATORY
URBANA IL 61801
U.S.A.

DICKEL JOHN R ASSOC PROF
UNIVERSITY OF
ILLINOIS OBSERVATORY
URBANA IL 61801
U.S.A.

DICKENS ROBERT J DR
ROYAL GREENWICH
OBSERVATORY
HAILSHAM
EAST SUSSEX BN27 1RP
U.K.

DICKINSON DALE F DR
JPL
183-301
4800 OAK GROVE DRIVE
PASADENA CA 91103
U.S.A.

DIECKVOSS WILHELM KE PROF
HAMBURGER STERNWARTE
D-2050 HAMBURG 80
GERMANY, F.R.

DIERCKSEN GEERD H F PH D
MPI FUER PHYSIK UND
ASTROPHYSIK
FOEHRINGER RING 6
D-8000 MUENCHEN 40
GERMANY, F.R.

DIETER N H DR
RADIO ASTRONOMY LAB
UNIV OF CALIFORNIA
BERKELEY CA 94720
U.S.A.

DINESCU A DR
INST GEODESIE PHOTO-
GRAMMETRIE CARTOGRAP
1A BC DE LEXPOSITION
78334 BUCAREST
ROUMANIA

DINGENS P PROF DR
STERRENKUNDIG INST
RIJKSUNIVERSITEIT
KORTRIJKSE STWEG 523
B-9000 GENT
BELGIUM

DINULESCU NICOLAE I PROF
SOSEANA KISELEFF 13
72168 BUCURESTI
SECTOR 1
ROUMANIA

DIONYSIOU DEMETRIOS D DR
DEPT OF ASTRONOMY
UNIVERSITY OF ATHENS
ATHENS 621
GREECE

DIRIKIS M A DR
LATVIAN STATE UNIVER
SITY ASTRONOMICAL
OBSERVATORY
226098 RIGA
U.S.S.R.

DISNEY MICHAEL J DR
ROYAL GREENWICH
OBSERVATORY
HAILSHAM
EAST SUSSEX BN27 1RP
U.K.

DIVAN LUCIENNE DR
INSTITUT
ASTROPHYSIQUE
98 BIS BD ARAGO
F-75014 PARIS
FRANCE

DIVARI N B DR
ODESSA POLYTECHNICAL
INSTITUTE
270044 ODESSA
U.S.S.R.

DIXON ROBERTS S DR
OHIO STATE UNIVER-
SITY RADIO OBS
2015 NEIL AVE
COLUMBUS OH 43210
U.S.A.

DIZER MUAMMER PROF
KANDILLI OBSERVATORY
CENGELKOY
ISTANBUL
TURKEY

DJURKOVIC PETAR M PROF
OBSERVATOIRE
DE BELGRADE
VOLGINA 7
11050 BELGRADE
YUGOSLAVIA

DJUROVIC DRAGUTIN M DR
DEPT OF ASTRONOMY
FAC OF SCIENCES
STUDENTSKI TRG 16
11000 BEOGRAD
YUGOSLAVIA

DLUZHNEVSKAYA O B DR
ASTRONOMICAL COUNCIL
USSR ACAD OF SCI
PYATNITSKAYA UL 48
109017 MOSCOW
U.S.S.R.

DOAN H H DR
OBSERVATOIRE DE LYON
F-69230 SAINT-GENIS
LAVAL
FRANCE

DOAZAN VERA DR
OBSERVATOIRE
DE PARIS
61 AVE L OBSERVATOIR
F-75014 PARIS
FRANCE

DOBRITSCHEV V M MR
DEPT OF ASTRONOMY
7TH NOVEMBER ST 1
SOFIA
BULGARIA

DOBRONRAVIN PETER P DR
CRIMEAN ASTROPHYSI-
CAL OBSERVATORY
334413 NAUCHNYJ
CRIMEA
U.S.S.R.

DOBROVOLSKY OLEG V PROF
INST OF ASTROPHYSICS
734670 DUSHANBE
U.S.S.R.

DOBRZYCKI JERZY PROF
MEKSYKANSKA 3 M 30
03 951 WARSZAWA
POLAND

DODD RICHARD J DR
ROYAL OBSERVATORY
BLACKFORD HILL
EDINBURGH EH9 3HJ
U.K.

DOGAN NADIR PROF
A U FEN FAKULTESI
ANKARA
TURKEY

DOHERTY LORNE H DR
HERZBERG INST OF
ASTROPHYSICS
NAT RES COUNC CANADA
OTTAWA ONT K1A 0R6
CANADA

DOHERTY LOWELL R PROF
ASTRONOMY DEPT
UNIVERSITY WISCONSIN
475 N CHARTER ST
MADISON WI 53706
U.S.A.

DOKUCHAEVA O D DR
STERNBERG STATE AS-
TRONOMICAL INSTITUTE
117234 MOSCOW
U.S.S.R.

DOLAN JOSEPH F DR
CODE 681
NASA GODDARD SPACE
FLIGHT CENTER
GREENBELT MD 20771
U.S.A.

DOLGINOV ARKADY Z PROF DR
IOFFE PHYSICAL
TECHNICAL INSTITUTE
194021 LENINGRAD
U.S.S.R.

DOLIDZE M V DR
ABASTUMANI ASTROPHY-
SICAL OBSERVATORY
383762 ABASTUMANI
GEORGIA
U.S.S.R.

DOLLFUS AUDOUIN PROF
OBSERVATOIRE DE
MEUDON
F-92190 MEUDON
FRANCE

DOMINKO FRAN PROF DR
SARANOVICEVA 11
61000 LJUBLJANA
YUGOSLAVIA

DOMINSKI IRENEUSZ DR
BOROWIEC
63-120 KORNIK
POLAND

DOMKE HELMUT PH D
ZENTR INST ASTROPHYS
DER ADW DER DDR
ROSA-LUXEMBURGST 17A
DDR-1502 POTSDAM
G.D.R.

DOMMANGET J DR
OBSERVATOIRE ROYAL
DE BELGIQUE
AVE CIRCULAIRE 3
B-1180 BRUXELLES
BELGIUM

DONN BERTRAM D
GODDARD SPACE FLIGHT
CENTER
CODE 691
GREENBELT MD 20771
U.S.A.

DOPITA MICHAEL A DR
MT STROMLO
OBSERVATORY
PRIVATE BAG
WODEN PO ACT 2606
AUSTRALIA

DORMAND JOHN R DR
MATHEMATICS DEPT
TEESSIDE POLYTECHNIC
MIDDLESBROUGH
CLEVELAND TS1 3BA
U.K.

DOROSHKEVICH A G DR
INSTITUTE F APPLIED
MATHEMATICS
USSR ACADEMY OF SCI
125047 MOSCOW
U.S.S.R.

DORSCHNER JOHANN DR
UNIVERSITAETS
STERNWARTE JENA
SCHILLERGAESSCHEN 2
DDR-69 JENA
G.D.R.

DOS REIS M PROF
DIRECTOR OF THE
OBSERVATORY
COIMBRA
PORTUGAL

DOSCHEK GEORGE A DR
NAVAL RESEARCH
LABORATORY
CODE 7125.8
WASHINGTON DC 20375
U.S.A.

DOSSIN F DR
UNIVERSITE DE LIEGE
INST D ASTROPHYSIQUE
AVE DE COINTE 5
B-4200 COINTE OUCREE
BELGIUM

DOUGHTY NOEL A DR
PHYSICS DEPT
UNIVERSITY OF
CANTERBURY
CHRISTCHURCH 1
NEW ZEALAND

DOUGLAS A VIBERT DR
402 67 SYDENHAM ST
KINGSTON ONT K7L 3H2
CANADA

DOUGLAS ALEXANDER E DR
HERZBERG INSTITUTE
OF ASTROPHYSICS
NAT RESEARCH COUNCIL
OTTAWA ONT K1A 0R6
CANADA

DOUGLAS JAMES N PROF
DEPT OF ASTRONOMY
UNIV OF TEXAS
AUSTIN TX 78712
U.S.A.

DOWNES DENNIS D DR
MPI F RADIO-
ASTRONOMIE
AUF DEM HUEGEL 69
D-5300 BONN
GERMANY, F.R.

DOWNS GEORGE S DR
JPL
MAIL STOP 238/420
4800 OAK GROVE DRIVE
PASADENA CA 91103
U.S.A.

DRAKE FRANK D PROF
N A I C
402 SPACE SCI BLDG
CORNELL UNIVERSITY
ITHACA NY 14853
U.S.A.

DRAMBA C PROF
OBS ASTRONOMIQUE
5 CUTITUL DE ARGINT
P.O.BOX 28
75212 BUCAREST
ROUMANIA

DRAPATZ SIEGFRIED W DR
MPI F EXTRATERRESTRI
SCHE PHYSIK
D-8046 GARCHING
GERMANY, F.R.

DRAVINS DAINIS DR
LUND OBSERVATORY
SVANEGATAN 9
S-222 24 LUND
SWEDEN

DRAVSKIKH A F DR
SPECIAL ASTROPHYSI-
CAL OBS USSR ACAD SC
LENINGRAD BRANCH
196140 LENINGRAD
U.S.S.R.

DRESSLER KURT PROF
LAB PHYSIK CHEMIE
ETH ZENTRUM
CH-8092 ZUERICH
SWITZERLAND

DREVER RONALD W P DR
DEPT OF NATURAL PHIL
GLASGOW UNIVERSITY
GLASGOW G12 8QQ
U.K.

DRILLING JOHN S
DEPT OF PHYSICS AND
ASTRONOMY
LOUISIANA STATE UNIV
BATON ROUGE LA 70803
U.S.A.

DROFA VASILIY K DR
ASTRONOMICAL OBS
KIEV UNIVERSITY
252053 KIEV
U.S.S.R.

DRYER MURRAY DR
SPACE ENVIRONMENT
LAB
NOAA ERL
BOULDER CO 80302
U.S.A.

DUBAU J DR
OBS DE MEUDON
F-92190 MEUDON
FRANCE

DUBINSKIJ B A DR
INSTITUTE OF RADIO
AND ELECTRONICS USSR
ACADEMY OF SCIENCES
103907 MOSCOW
U.S.S.R.

DUBOSHIN G N PROF DR
STERNBERG STATE AS-
TRONOMICAL INSTITUTE
117234 MOSCOW
U.S.S.R.

DUBOV EMIL E PROF
WDC-B2
MOLODEZHNAYA 3
117296 MOSCOW
U.S.S.R.

DUCHESNE MAURICE DR
OBSERVATOIRE
DE PARIS
61 AVE OBSERVATOIRE
F-75014 PARIS
FRANCE

DUERBECK HILMAR W DR
OBS HOHER LIST
UNIV STERNWARTE BONN
D-5568 DAUN
GERMANY, F.R.

DUERST JOHANNES DR
SWISS FEDERAL OBS
SCHMELZBERGSTR 25
CH 8006 ZUERICH
SWITZERLAND

DUFAY MAURICE PROF
UNIVERSITE CLAUDE-
BERNARD-LYON I
43 BD DU 11 NOVEMBRE
F-69621 VILLEURBANNE
FRANCE

DUFLOT MARCELLE DR
OBSERVATOIRE DE
MARSEILLE
2 PLACE LE VERRIER
F-13004 MARSEILLE
FRANCE

DUFLOT RENEE DR
OBSERVATOIRE
DE MARSEILLE
F-13004 MARSEILLE
FRANCE

DUFTON PHILIP L DR
DEPT OF APPLIED MATH
QUEEN'S UNIVERSITY
BELFAST BT7 1NN
U.K.

DULEY WALTER W PROF
PHYSICS DEPT
YORK UNIVERSITY
DOWNSVIEW ONT M3J1P3
CANADA

DULK GEORGE A PROF
DEPT ASTRO-GEOPHYS
UNIV OF COLORADO
BOULDER CO 80302
U.S.A.

DUMA DMITRIJ P DR
MAIN ASTRONOMICAL OB
UKRAINIAN ACAD OF SC
252127 KIEV
U.S.S.R.

DUMONT RENE DR
OBSERVATOIRE
DE BORDEAUX
F-33270 FLOIRAC
FRANCE

DUMONT SIMONE DR
INST D'ASTROPHYSIQUE
98 BIS BLVD ARAGO
F-75014 PARIS
FRANCE

DUNCAN ROBERT A PROF
CSIRO
DIV OF RADIOPHYSICS
P O BOX 76
EPPING NSW 2121
AUSTRALIA

DUNCOMBE RAYNOR L DR
DEPT OF AEROSPACE
ENGINEERING
UNIVERSITY OF TEXAS
AUSTIN TX 78712
U.S.A.

DUNHAM DAVID W
COMPUTER SCS CORP
COLLESVILLE RD
SILVER SPG MD 20910
U.S.A.

DUNHAM THEODORE JR DR
P O BOX D
CHOCORUA NH 03817
U.S.A.

DUNKELMAN LAWRENCE
CODE 685 LAB F ASTRO
AND SOLAR PHYSICS
GODDARD SPACE FL CTR
GREENBELT MD 20771
U.S.A.

DUNN RICHARD B DR
SACRAMENTO PEAK
OBSERVATORY
SUNSPOT NM 88349
U.S.A.

DUPREE ANDREA K DR
CTR F ASTROPHYSICS
60 GARDEN ST
CAMBRIDGE MA 02138
U.S.A.

DUPUY DAVID L DR
DEPT OF ASTRONOMY
ST MARY S UNIVERSITY
HALIFAX NS B3H 3C3
CANADA

DURISEN RICHARD H DR
DEPT OF ASTRONOMY
SWAIN WEST 319
INDIANA UNIV
BLOOMINGTON IN 47401
U.S.A.

DURNEY BERNARD R DR
NATIONAL CTR FOR AT-
MOSPHERIC RESEARCH
P O BOX 3000
BOULDER CO 80303
U.S.A.

DURRANT CHRISTOPHER J DR
KIEPENHEUER INST
FUER SONNENPHYSIK
SCHOENECKSTR 6
D-7800 FREIBURG
GERMANY, F.R.

DUTHIE JOSEPH G PROF
UNIVERSITY ROCHESTER
DEPT OF PHYSICS AND
ASTRONOMY
ROCHESTER NY 14627
U.S.A.

DVORAK RUDOLF DR
ASTRONOMISCHES INST
DER UNIVERSITAET
A-8010 GRAZ
AUSTRIA

DWORETSKY MICHAEL M DR
DEPT OF PHYSICS AND
ASTRONOMY
UNIV COLLEGE LONDON
LONDON WC1E 6BT
U.K.

DYCK M DR
KITT PEAK NATIONAL
OBSERVATORY
P O BOX 26732
TUCSON AZ 85726
U.S.A.

DYER CHARLES CHESTER DR
PHYS SCS GR RM R4227
SCARBOROUGH COLLEGE
UNIV OF TORONTO
TORONTO ONT M1G 3S7
CANADA

DYER EDWARD R DR
NATIONAL ACADEMY OF
SCIENCES
2101 CONSTITUTION AV
WASHINGTON DC 20418
U.S.A.

DYSON F J DR
INSTITUTE FOR
ADVANCED STUDIES
PRINCETON NJ 08540
U.S.A.

DYSON JOHN E DR
ASTRONOMY DEPT
UNIVERSITY OF
MANCHESTER
MANCHESTER M13 9PL
U.K.

DZHAPIASHVILI V P DR
ABASTUMANI ASTROPHY-
SICAL OBSERVATORY
383762 ABASTUMANI
GEORGIA
U.S.S.R.

DZIEMBOWSKI W DR
ASTRONOMICAL CENTER
UL BARTYCKA 18
00-716 WARSAW
POLAND

DZIGVASHVILI R M DR
ABASTUMANI ASTROPHY-
SICAL OBSERVATORY
383762 ABASTUMANI
U.S.S.R.

EATON JOEL A DR
ASTRONOMY DEPT
PENNA STATE UNIV
525 DAVEY LAB
UNIV PK PA 16802
U.S.A.

EBBIGHAUSEN E G
PHYSICS DEPT
UNIVERSITY OF OREGON
EUGENE OR 97403
U.S.A.

ECCLES MICHAEL J DR
29 BUCKSTONE LOAN E
EDINBURGH 10
U.K.

EDDY JOHN A DR
HIGH ALTITUDE
OBSERVATORY
BOX 3000
BOULDER CO 80307
U.S.A.

EDLEN BENGT PROF
DEPT OF PHYSICS
SOELVEGATAN 14
S-223 62 LUND
SWEDEN

EDMONDS FRANK N JR DR
RLM 15.212
UNIVERSITY OF TEXAS
AUSTIN TX 78712
U.S.A.

EDMONDSON FRANK K PROF
GOETHE LINK OBS
319A SWAIN HALL WEST
INDIANA UNIVERSITY
BLOOMINGTON IN 47405
U.S.A.

EDWARDS ALAN CH DR
DEPT OF ASTROPHYSICS
SOUTH PARKS ROAD
OXFORD OX1 3RQ
U.K.

EDWARDS P J PROF
DEPT OF PHYSICS
UNIV OF OTAGO
P O BOX 56
DUNEDIN
NEW ZEALAND

EELSALU HEINO DR
TARTU OBSERVATORY
202-444 TORAVERE
ESTONIA
U.S.S.R.

EFREMOV YU I DR
INSTITUTE F APPLIED
MATHEMATICS
USSR ACADEMY OF SCI
125047 MOSCOW
U.S.S.R.

EFREMOV YURY N DR
STERNBERG
ASTRONOMICAL INST
117234 MOSCOW
U.S.S.R.

EGGLETON PETER P DR
INSTITUTE OF
ASTRONOMY
MADINGLEY RD
CAMBRIDGE CB3 OHE
U.K.

EHLERS JURGEN PROF
MPI FUER PHYSIK UND
ASTROPHYSIK MUENCHEN
D-8000 MUENCHEN 40
GERMANY, F.R.

EHRGMA E V DR
ASTRONOMICAL COUNCIL
USSR ACADEMY OF SCI
PYATNITSKAYA UL 48
109017 MOSCOW
U.S.S.R.

EICHHORN HEINRICH K DR
DEPT OF ASTRONOMY
UNIV OF FLORIDA
231 SPACE SC RES BLD
GAINESVILLE FL 32611
U.S.A.

EICHLER DAVID DR
ASTRONOMY PROGRAM
UNIV OF MARYLAND
COLLEGE PK MD 20742
U.S.A.

EINICKE O H LEKTOR
UNIVERSITY
OBSERVATORY
OSTERVOLDGADE 3
DK-1350 COPENHAGEN K
DENMARK

EINSATO JAAN DR
TARTU OBSERVATORY
202444 TORAVERE
ESTONIA
U.S.S.R.

EKERS RONALD J DR
KAPTEYN LABORATORIUM
POSTBUS 800
GRONINGEN
THE NETHERLANDS

EL-BAZ F DR
CTR EARTH AND PLANE-
TARY STUDIES
SMITHSONIAN INST.
WASHINGTON DC 20560
U.S.A.

EL-RAEY MOHAMED E DR
DEPT OF PHYSICS
FACULTY OF SCIENCE
UNIV OF ALEXANDRIA
ALEXANDRIA
ARAB REP. OF EGYPT

EL-SHAARAWY M B DR
HELWAN OBSERVATORY
HELWAN NEAR CAIRO
ARAB REP. OF EGYPT

EL-YASBERG P E PROF DR
SPACE RESEARCH
INSTITUTE
USSR ACADEMY OF SCI
117810 MOSCOW
U.S.S.R.

ELFORD W G DR
DEPT OF PHYSICS
UNIV OF ADELAIDE
ADELAIDE
AUSTRALIA

ELGAROY OYSTEIN PROF
INST THEOR ASTROPHYS
UNIVERSITY OF OSLO
P O BOX 1029
BLINDERN OSLO 3
NORWAY

ELLDER JOEL DR
ONSALA SPACE
OBSERVATORY
S-430 34 ONSALA
SWEDEN

ELLIOT HARRY PROF
BLACKETT LABORATORY
IMPERIAL COLLEGE
LONDON SW7 2BZ
U.K.

ELLIOTT IAN DR
DUNSINK OBSERVATORY
CASTLEKNOCK
CO DUBLIN
IRELAND

ELLIOTT JAMES L DR
328 SPACE SC BLDG
CORNELL UNIVERSITY
ITHACA NY 14853
U.S.A.

ELLIOTT KENNETH H DR
DEPT OF SPACE RES
UNIV OF BIRMINGHAM
P O BOX 363
BIRMINGHAM B15 2TT
U.K.

ELLIS G R A PROF
UNIV OF TASMANIA
P O BOX 2520
HOBART TASMANIA
AUSTRALIA

ELLIS GEORGE F R PROF
DEPT OF APPLIED MATH
UNIV OF CAPE TOWN
RONDEBOSCH CAPE 7700
SOUTH AFRICA

ELLYETT CLIFTON D PROF
DEPT OF PHYSICS
UNIVERSITY NEWCASTLE
NEWCASTLE NSW 2308
AUSTRALIA

ELMEGREEN BRUCE GORDON DR
DEPT OF ASTRONOMY
PUPIN LABS
COLUMBIA UNIV
NEW YORK NY 10027
U.S.A.

ELSAESSER HANS PROF
MPI FUER ASTRONOMIE
D-6900 HEIDELBERG-
KOENIGSTUHL
GERMANY, F.R.

ELSMORE BRUCE DR
CAVENDISH LABORATORY
MADINGLEY ROAD
CAMBRIDGE CB3 OHE
U.K.

ELST E DR
KONINGLIJKE STERREN-
WACHT VAN BELGIE
RINGLAAN 3
B-1180 BRUSSEL
BELGIUM

ELSTE GUNTHER H DR
DEPT OF ASTRONOMY
UNIVERSITY OF
MICHIGAN
ANN ARBOR MI 48109
U.S.A.

ELSTON WOLFGANG E PROF
DEPT OF GEOLOGY
UNIVERSITY OF
NEW MEXICO
ALBUQUERQUE NM 87131
U.S.A.

ELVIS MARTIN S DR
46 HOWARD ROAD
LEICESTER
U.K.

ELVIUS AINA M PROF
STOCKHOLM OBS
S-133 00
SALTSJOEBADEN
SWEDEN

ELVIUS TORD PROF
LUND OBSERVATORY
SVANEGATAN 9
S-222 24 LUND
SWEDEN

ELWERT GERHARD PROF
LEHRSTUHL F THEORE-
TISCHE ASTROPHYSIK
UNIV TUEBINGEN
D-7400 TUEBINGEN
GERMANY, F.R.

EMERSON BRIAN MR
ROYAL GREENWICH OBS
HERSTMONCEUX CASTLE
HAILSHAM
E SUSSEX BN27 1RP
U.K.

EMERSON JAMES P R
DEPT OF PHYSICS
QUEEN MARY COLLEGE
MILE END ROAD
LONDON E1 4NS
U.K.

EMINZADE T A DR
SHEMAKHA ASTROPHYSI-
CAL OBSERVATORY
373243 AZERBAIDZAN
U.S.S.R.

ENCRENAZ PIERRE J DR
ECOLE NORMALE
SUPERIEURE
24 RUE LHOMOND
F-75005 PARIS
FRANCE

ENCRENAZ T DR
GROUPE PLANETES
SECT D'ASTROPHYSIQUE
OBS DE MEUDON
F-92190 MEUDON
FRANCE

ENDAL ANDREW S DR
DEPT PHYSICS ASTRON
LOUISIANA STATE UNIV
BATON ROUGE LA 70803
U.S.A.

ENGELHARD E J G PROF DR
SACKRING 34
D-3300 BRAUNSCHWEIG
GERMANY, F.R.

ENGIN SEMANUR ASSOC PROF
DEPT OF ASTRONOMY
UNIVERSITY OF ANKARA
FEN FAKULTESI
ANKARA
TURKEY

ENGVOLD ODDBJOERN DR
INST THEOR ASTROPHYS
UNIVERSITY OF OSLO
P O BOX 1029
BLINDERN OSLO 3
NORWAY

ENOME SHINZO ASSOC PROF
RESEARCH INSTITUTE
OF ATMOSPHERICS
NAGOYA UNIVERSITY
TOYOKAWA 442
JAPAN

ENSLIN HEINZ DR
DEUTSCHES HYDROGRA-
PHISCHES INSTITUT
POSTFACH 220
D-2000 HAMBURG 4
GERMANY, F.R.

EPPS HARLAND WARREN PROF
DEPT OF ASTRONOMY
UNIVERSITY OF CALIF
MATH SCI RM 8983
LOS ANGELES CA 90024
U.S.A.

EPSTEIN EUGENE E DR
I A GETTING LABS
130-422 AERO CORP
BOX 92957
LOS ANGELES CA 90009
U.S.A.

EPSTEIN GABRIEL LEO DR
NASA-GSFC
CODE 682
GREENBELT MD 20771
U.S.A.

EPSTEIN ISADORE PROF
ASTRONOMY DEPT
COLUMBIA UNIVERSITY
NEW YORK NY 10027
U.S.A.

EPSTEIN RICHARD I DR
NORDITA
BLEGDAMSVEJ 17
DK-2100 COPENHAGEN O
DENMARK

ERDI B DR
ASTRONOMICAL DEPT
LORAND EOTVOS UNIV
KUN BELA TER 2
H-1083 BUDAPEST
HUNGARY

ERICKSON WILLIAM C DR
ASTRONOMY PROGRAM
UNIVERSITY MARYLAND
COLLEGE PARK
MD 20742
U.S.A.

ERIKSEN GUNNAR PROF
INST THEOR ASTROPHYS
UNIVERSITY OF OSLO
P O BOX 1029
BLINDERN OSLO 3
NORWAY

ERPYLEV N P DR
ASTRONOMICAL COUNCIL
USSR ACADEMY OF SCI
PYATNITSKAYA UL 48
109017 MOSCOW
U.S.S.R.

ERSHKOVICH ALEXANDER DR
58/1 HANKIN ST
RAANANA
ISRAEL

ERUSHEV N N DR
CRIMEAN ASTROPHYSI-
CAL OBSERVATORY USSR
ACAD OF SCI NAUCHNIY
334413 CRIMEA
U.S.S.R.

ESHLEMAN VON R PROF
DURAND 221
STANFORD UNIVERSITY
STANFORD CA 94305
U.S.A.

ESIPOV V F DR
STERNBERG STATE AS-
TRONOMICAL INSTITUTE
117234 MOSCOW
U.S.S.R.

ESKIOGLU A NIHAT
P K 14
MALTEPE ANKARA
TURKEY

ESPOSITO F PAUL PROF
DEPT OF PHYSICS 11
UNIVERSITY OF
CINCINNATI
CINCINNATI OH 45221
U.S.A.

EVANGELIDIS E DR
UNIVERSITY OF PATRAS
DEPT OF ASTRONOMY
PATRAS
GREECE

EVANS DAVID S PROF
DEPT OF ASTRONOMY
UNIVERSITY OF TEXAS
AUSTIN TX 78712
U.S.A.

EVANS J V DR
MIT LINCOLN LAB
MILLSTONE/HAYSTACK
OBSERVATORY
LEXINGTON MA 02173
U.S.A.

EVANS JOHN W DR
SACRAMENTO PEAK
OBSERVATORY
SUNSPOT NM 88349
U.S.A.

EVANS KENTON D DR
PHYSICS DEPT
THE UNIVERSITY
LEICESTER LE1 7RN
U.K.

EVANS NANCY R DR
183 BRIAR HILL AVE
TORONTO ONT M4R 1H8
CANADA

EVANS NEAL J II ASST PROF
DEPT OF ASTRONOMY
UNIVERSITY OF TEXAS
AUSTIN TX 78712
U.S.A.

EVANS ROGER G DR
RUTHERFORD
LABORATORY
DIDCOT OX11 0QX
U.K.

EVDOKIMOV YU V DR
ENGELHARDT ASTRONOMI
CAL OBSERVATORY
422526 KAZAN
U.S.S.R.

EVERHART EDGAR DR
PHYSICS DEPT AND
CHAMBERLIN OB
UNIVERSITY OF DENVER
DENVER CO 80208
U.S.A.

EVIATAR AHARON PROF
DEPT OF GEOPHYSICS
AND PLANETARY SC
TEL-AVIV UNIVERSITY
TEL-AVIV
ISRAEL

EWEN HAROLD I DR
60 BEAVER ROAD
WESTON MA 02193
U.S.A.

EWING MARTIN S
CALTECH 102-24
PASADENA CA 91125
U.S.A.

EZER DILHAN PROF
MIDDLE EAST TECH-
NICAL UNIVERSITY
ANKARA
TURKEY

FABER SANDRA M PROF
LICK OBSERVATORY
UNIVERSITY OF
CALIFORNIA
SANTA CRUZ CA 95064
U.S.A.

FABIAN ANDREW C DR
INSTITUTE OF
ASTRONOMY
MADINGLEY ROAD
CAMBRIDGE CB3 0HE
U.K.

FABRE HERVE DR
2 AVENUE MARECHAL
FOCH
F-06310 BEAULIEU MER
FRANCE

FAELTHAMMAR CARL GUNNE PR
DEPT PLASMA PHYSICS
ROYAL INST OF
TECHNOLOGY
S-100 44 STOCKHOLM70
SWEDEN

FAHLMAN GREGORY G DR
DEPT GEOPHYS AND AST
UNIVERSITY OF B C
2075 WESBROOK PLACE
VANCOUVER BC V6T 1W5
CANADA

FAHR HANS JOERG PROF DR
INST ASTROPHY EXTRA-
TERRESTR F UNIV BONN
AUF DEM HUEGEL 71
D-5300 BONN
GERMANY, F.R.

FAHY EDWARD F PROF
PHYSICS DEPT
UNIVERSITY COLLEGE
CORK
IRELAND

FAIRALL ANTHONY P DR
DEPT OF ASTRONOMY
UNIV OF CAPE TOWN
RONDEBOSCH 7700
CAPE
SOUTH AFRICA

FALCIANI ROBERTO DR
CAPODIMONTE ASTRO
OBSERVATORY
VIA MOIARIELLO 16
I-80131 NAPOLI
ITALY

FALK SYDNEY W JR DR
DEPT OF ASTRONOMY
UNIV OF TEXAS
AUSTIN TX 78712
U.S.A.

FALL S M DR
INST OF ASTRONOMY
MADINGLEY ROAD
CAMBRIDGE CB3 OHA
U.K.

FALLE SAMUEL A DR
UNIVERSITY OF LEEDS
LEEDS LS2 9JT
U.K.

FALLER JAMES E PROF
JILA NBS
UNIVERSITY COLORADO
BOULDER CO 80309
U.S.A.

FALLON FREDERICK W DR
DEPT OF ASTRONOMY
UNIV OF S FLORIDA
TAMPA FL 33620
U.S.A.

FANTI CARLA GIOVANNINI
LAB DI RADIOASTRONO-
MIA
VIA IRNERIO 46
I-40126 BOLOGNA
ITALY

FANTI R DR
PHYSICAL INST
UNIV OF BOLOGNA
VIA IRNERIO 46
I-40126 BOLOGNA
ITALY

FARAGGIANA ROSANNA PROF
OSSERVATORIO
ASTRONOMICO
VIA TIEPOLO 11
I-34131 TRIESTE
ITALY

FARINELLA PAOLO DR
OSSERVATORIO ASTRO-
NOMICO
BRERA
MILANO
ITALY

FATCHIKHIN NIKOLAJ V DR
PULKOVO OBSERVATORY
196140 LENINGRAD
U.S.S.R.

FAUCHER P DR
OBS DE NICE
B P 252
F-06007 NICE CEDEX
FRANCE

FAULKNER DONALD J DR
MT STROMLO
OBSERVATORY
WODEN PO ACT 2606
AUSTRALIA

FAULKNER JOHN PROF
UNIVERSITY OF
CALIFORNIA
SANTA CRUZ CA 95064
U.S.A.

FAWELL DEREK R DR
UNIVERSITY OF LONDON
OBSERVATORY
MILL HILL PARK
LONDON NW7 2QS
U.K.

FAY THEODORE D DR
MAIL STOP 19
TELEDYNE BROWN ENG
CUMMINGS RES PARK
HUNTSVILLE AL 35807
U.S.A.

FAZIO GIOVANNI G DR
CENTER FOR
ASTROPHYSICS HCO SAO
60 GARDEN ST
CAMBRIDGE MA 02138
U.S.A.

FEAST MICHAEL W DR
S A A O
P O BOX 9
OBSERVATORY
CAPE 7935
SOUTH AFRICA

FEAUTRIER NICOLE DR
OBSERVATOIRE
DE MEUDON
F-92190 MEUDON
FRANCE

FECHTIG HUGO DR
SANSERWEG 3
D-6906 LEIMEN
GERMANY, F.R.

FEDOROV E P DR
MAIN ASTRON OBS
UKRAINIAN ACAD OF SC
GOLOSEEVO
252127 KIEV
U.S.S.R.

FEDOROVA RIMMA T DR
ASTRONOMICAL OBS
327000 NIKOLAEV
U.S.S.R.

FEGAN DAVID J DR
PHYSICS DEPT
UNIV COLLEGE
BELFIELD
DUBLIN 4
IRELAND

FEHRENBACH CHARLES PROF
OBSERVATOIRE DE
HAUTE PROVENCE
ST MICHEL L OBS
F-04300 FORCALQUIER
FRANCE

FEIBELMAN WALTER A DR
NASA-GSFC
CODE 685
GREENBELT MD 20771
U.S.A.

FEINSTEIN A DR
OBS ASTRONOMICO
1900 LA PLATA
ARGENTINA

FEISSEL MARTINE
OBSERVATOIRE PARIS
61 AVE DE
L-OBSERVATOIRE
F-75014 PARIS
FRANCE

FEITZINGER JOHANNES V PHD
ASTRONOMISCHES INST
RUHR UNIVERSITAET
POSTFACH 102148
D-4630 BOCHUM
GERMANY, F.R.

FEIX GERHARD DR
RUHR UNIVERSITAET
DEPT XII
POSTFACH 102 148
D-4630 BOCHUM
GERMANY, F.R.

FEJES I TVAN DR
FOMI SATELLITE
GEODETIC OBSERVATORY
PO BOX 546
1373 BUDAPEST
HUNGARY

FELDMAN PAUL A DR
HERZBERG INST ASTROD
NAT RES COUNC CANADA
100 SUSSEX DR
OTTAWA ONT K1A 0R6
CANADA

FELDMAN PAUL DONALD DR
PHYSICS DEPT
JOHNS HOPKINS UNIV
BALTIMORE MD 21218
U.S.A.

FELDMAN U DR
DEPT OF PHYSICS AND
ASTRONOMY
TEL-AVIV UNIV
RAMAT-AVIV
ISRAEL

FELENBOK PAUL DR
OBS DE MEUDON
F-92190 MEUDON
FRANCE

FELLGETT PETER B PROF
DEPT OF CYBERNETICS
3 EARLY GATE
WHITEKNIGHTS
READING RG6 2AL
U.K.

FELLI MARCELLO DR
OSSERVATORIO ARCETRI
LARGO FERMI 5
I-50125 FIRENZE
ITALY

FELTEN JAMES E DR
CODE 660
GODDARD SPACE FL CTR
GREENBELT MD 20771
U.S.A.

FENKART ROLF P DR
ASTRONOMISCHES INST
UNIVERSITAET BASEL
VENUSSTR 7
CH-4102 BINNINGEN
SWITZERLAND

FENTON K B DR
DEPT OF PHYSICS
UNIV OF TASMANIA
HOBART TASM 7001
AUSTRALIA

FERNANDEZ-FIGUEROA M J DR
ASTROFIS FAC DE FIS
UNIV COMPLUTENSE
CIUDAD UNIVERSITARIA
MADRID 3
SPAIN

FERNIE J DONALD PROF
DAVID DUNLAP OBS
P O BOX 360
RICHMOND HILL
ONT L4C 4Y6
CANADA

FERRARI ATTILIO DR
ISTITUTO DI FISICA
GENERALE DELL.UNIV.
CR M D AZEGLIO 46
I-10125 TORINO
ITALY

FERRARI D OCCHIEPPO K DR
OESTERREICHISCHE AK
DER WISSENSCHAFTEN
DR-I-SEIPEL-PLATZ2
A-1110 WIEN
AUSTRIA

FERRAZ-MELLO S PROF DR
UNIV DE SAO PAULO
DEPT ASTRONOMIA
CAIXA POSTAL 30627
01000-SAO PAULO SP
BRAZIL

FESTOU M DR
SERVICE D'AERONOMIE
CNRS B P NO 3
F-91370 VERRIERES
LE BUISSON
FRANCE

FEYNMAN JOAN DR
DEPT OF PHYSICS
BOSTON COLLEGE
CHESTNUT HL MA 02167
U.S.A.

FIALA ALAN D DR
U S NAVAL OBS
WASHINGTON DC 20390
U.S.A.

FICARRA A MR
LAB NAZIONALE DI
RADIOASTRONOMIA
VIA IRNERIO 46
I-40126 BOLOGNA
ITALY

FICHTEL CARL E DR
GODDARD SPACE FLIGHT
CENTER
CODE 662
GREENBELT MD 20771
U.S.A.

FIELD GEORGE B PROF
CENTER FOR
ASTROPHYSICS
60 GARDEN ST
CAMBRIDGE MA 02138
U.S.A.

FIELDER GILBERT DR
LUNAR AND PLANETARY
UNIT E S DEPT
LANCASTER UNIVERSITY
LANCASTER LA1 4YR
U.K.

FILLIOZAT JEAN PROF
35R FRANCOIS ROLLAND
F-94130 NOGENT-MARNE
FRANCE

FINDLAY JOHN W DR
NAT RADIO ASTR OBS
EDGEMONT RD
CHARLOTTESVILLE
VA 22901
U.S.A.

FINK U DR
LUNAR AND PLANETARY
LABORATORY
UNIVERSITY ARIZONA
TUCSON AZ 85721
U.S.A.

FINN G D DR
INST FOR ASTRONOMY
2525 CORREA ROAD
HONOLULU HI 96822
U.S.A.

FINZI ARRIGO DR
DEPT OF MATHEMATICS
TECHNION I I T
HAIFA
ISRAEL

FIREMAN EDWARD L
ASTROPHYSICAL OBS
SMITHSONIAN INST
60 GARDEN ST
CAMBRIDGE MA 02138
U.S.A.

FIRMANI CLAUDIO A PROF
INSTITUTO DE
ASTRONOMIA UNAM
A P 70-264
MEXICO 20 DF
MEXICO

FIRNEIS MARIA G DR
INST FUR ASTRONOMIE
TUERKENSCHANZSTR 17
A-1180 WIEN
AUSTRIA

FIROR JOHN W DR
NATIONAL CTR FOR AT-
MOSPHERIC RESEARCH
P O BOX 3000
BOULDER CO 80703
U.S.A.

FISCHEL DAVID DR
CODE 671
GODDARD SPACE FLIGHT
CENTER
GREENBELT MD 20771
U.S.A.

FISCHER P LEANDER PROF
FREYUNG 6
A-1010 WIEN
AUSTRIA

FISCHER STANISLAV DR
ASTRONOMICAL INST
CSAV
BUDECSKA 6
120 23 PRAHA 2
CZECHOSLOVAKIA

FISHER J RICHARD
NRAO
P O BOX 2
GREEN BANK WV 24944
U.S.A.

FISHER PHILIP C
2401 SHARON OAKS DR
MENLO PARK CA 94025
U.S.A.

FISHER RICHARD R DR
HIGH ALTITUDE OBS
P O BOX 3000
BOULDER CO 80307
U.S.A.

FISHKOVA L M PROF
ABASTUMANI ASTROPHY-
SICAL OBSERVATORY
383762 ABASTUMANI
GEORGIA
U.S.S.R.

FITCH WALTER S DR
STEWARD OBSERVATORY
UNIV OF ARIZONA
TUCSON AZ 85721
U.S.A.

FITZGERALD M PIM PROF
DEPT OF PHYSICS
UNIVERSITY WATERLOO
WATERLOO ONT N2L 3G1
CANADA

FIX JOHN D DR
DEPT OF PHYSICS AND
ASTRONOMY
UNIVERSITY OF IOWA
IOWA CITY IA 52240
U.S.A.

FLANNERY BRIAN PAUL DR
CTR FOR ASTROPHYSICS
60 GARDEN STREET
CAMBRIDGE MA 02138
U.S.A.

FLECKENSTEIN J O PROF DR
MATHEMATISCHES INST
UNIVERSITAET
RHEINSPRUNG 21
CH-4051 BASEL
SWITZERLAND

FLEISCHER ROBERT DR
1733 CHURCH ST N W
WASHINGTON DC 20036
U.S.A.

FLETCHER J MURRAY
DOMINION ASTROPHYSI-
CAL OBSERVATORY
5071 W SAANICH RR7
VICTORIA BC V8X 3X3
CANADA

FLIEGEL HENRY F
JPL
4800 OAK GROVE DRIVE
PASADENA CA 91103
U.S.A.

FLORENSKY KYRILL P DR
VERNADSKY INST OF
GEOCHEMISTRY
47 A VOROBJOVSKOJE
117334 MOSCOW
U.S.S.R.

FLORENTIN-NIELSEN R M SC
COPENHAGEN UNIVERSI-
TY OBSERVATORY
BRORFELDE
DK-4340 TOLLOSE
DENMARK

FLORIDES PETROS S PROF
SCHOOL OF MATHEMA-
TICS TRINITY COLLEGE
DUBLIN 2
IRELAND

FLORSCH ALPHONSE DR
OBSERVATOIRE
DE STRASBOURG
11 RUE L UNIVERSITE
F-67000 STRASBOURG
FRANCE

FLOWER DAVID R DR
DEPT OF PHYSICS
UNIVERSITY OF DURHAM
DURHAM DH1 3LE
U.K.

FOGARTY WILLIAM G DR
2511 PLOVER RD
WISCONSIN RAPIDS
WI 54494
U.S.A.

FOGH OLSEN H J
COPENHAGEN UNIV
OBSERVATORY
BRORFELDE
DK-4340 TOLLOSE
DENMARK

FOKKER AAD D DR
STERREWACHT
SONNENBORGH
ZONNENBURG 2
UTRECHT
THE NETHERLANDS

FOMALONT EDWARD B DR
NATIONAL RADIO
ASTRONOMY OBS
P O BOX O
SOCORRO NM 87801
U.S.A.

FOMENKO ALEXANDER F DR
SPECIAL ASTROPHYS OB
NIZHNIJ ARKHYZ
357140 STAVROPOLSKIJ
KRAJ
U.S.S.R.

FONTAINE GILLES DR
DEPT OF PHYSICS
UNIV OF MONTREAL
P O BOX 6128
MONTREAL PQ H3C 3J7
CANADA

FORBES ERIC G DR
HISTORY DEPT
EDINBURGH UNIVERSITY
50 GEORGE SQUARE
EDINBURGH EH8 9JY
U.K.

FORBES J E DR
P O BOX 88120
INDIANAPOLIS IN46208
U.S.A.

FORD HOLLAND C ASSOC PROF
DEPT OF ASTRONOMY
LOS ANGELES CA 90024
U.S.A.

FORD W K JR DR
TERRESTRIAL MAGNET
CARNEGIE INSTITUTION
5241 BROAD BR RD N W
WASHINGTON DC 20015
U.S.A.

FORMAN WILLIAM RICHARD DR
SMITHSONIAN OBS
60 GARDEN STREET
CAMBRIDGE MA 02138
U.S.A.

FORT BERNARD P DR
OBSERVATOIRE
DE MEUDON
F-92190 MEUDON
FRANCE

FORT DAVID NORMAN DR
733 LONSDALE ROAD
OTTAWA K1K 0J9
CANADA

FORTI GIUSEPPE DR
OSSERVATORIO ARCETRI
LARGO FERMI 5
I-50125 FIRENZE
ITALY

FORTINI TERESA DR
OSSERVATORIO
ASTRONOMICO
I-00136 ROMA
ITALY

FOSBURY ROBERT A E DR
TELESCOPE PROJECT
EUROPEAN SOUTHERN OB
C/O CERN
CH-1211 GENEVA 23
SWITZERLAND

FOSSAT ERIC DR
OBSERVATOIRE DE NICE
BP 252
F-06007 NICE CEDEX
FRANCE

FOUKAL PETER V DR
CENTER FOR
ASTROPHYSICS
60 GARDEN ST
CAMBRIDGE MA 02138
U.S.A.

FOWLER WILLIAM A PROF
CALTECH 106 38
PASADENA CA 91125
U.S.A.

FOX KENNETH DR
DEPT PHYSICS ASTRON
UNIV OF TENNESSEE
MEMPHIS TN 38103
U.S.A.

FOX W E MR
40 WINDSOR ROAD
NEWARK NOTTINGHAMS
U.K.

FOY RENAUD DR
OBSERVATOIRE
DE MEUDON
F-92190 MEUDON
FRANCE

FRACASSINI MASSIMO PROF
OSSERVATORIO
ASTRONOMICO DI BRERA
VIA BRERA 28
I-20121 MILANO
ITALY

FRACASTORO MARIO G PROF
OSSERVATORIO ASTRONO
MICO DI TORINO
PINO TORINESE
I-10025 TORINO
ITALY

FRANDSEN SOEREN PROF
INST OF ASTRONOMY
UNIVERSITY OF AARHUS
DK-8000 AARHUS C
DENMARK

FRANKLIN FRED A DR
PLANETARY SC DIV
CTR F ASTROPHYSICS
60 GARDEN ST
CAMBRIDGE MA 02138
U.S.A.

FRANTSMAN YU L DR
RADIOASTROPHYSICAL
OBSERVATORY LATVIAN
ACADEMY OF SCIENCES
226026 RIGA
U.S.S.R.

FRANZ OTTO G DR
LOWELL OBSERVATORY
P O BOX 1269
FLAGSTAFF AZ 86002
U.S.A.

FRASER C W DR
UNIV OBSERVATORY
BUCHANAN GARDENS
ST ANDREWS
FIFE KY16 9LZ
U.K.

FRATER ROBERT H DR
UNIVERSITY OF SYDNEY
SYDNEY N S W 2006
AUSTRALIA

FRAZIER EDWARD N DR
AEROSPACE
CORPORATION
P O BOX 92957
LOS ANGELES CA 90009
U.S.A.

FREDGA KERSTIN PROF
SWEDISH BOARD FOR
SPACE ACTIVITIES
TRITONVAEGEN 27
S-171 54 SOLNA
SWEDEN

FREDRICK LAURENCE W PROF
LEANDER MCCORMICK
OBSERVATORY BOX 3818
CHARLOTTESVILLE
VA 22903
U.S.A.

FREEMAN KENNETH C DR
MT STROMLO
OBSERVATORY
WODEN PO ACT 2606
AUSTRALIA

FREIESLEBEN H C DR
FLORENTINER STR 20
APP 6120
D-7000 STUTTGART 75
RIEDENBERG
GERMANY, F.R.

FREIRE FERRERO R G DR
DEPT ASTROPHYS FOND
OBS DE PARIS
5 PLACE JANSEN
F-92190 MEUDON
FRANCE

FREITAS MOURAO R R DR
OBS NACIONAL
R GENERAL BRUCE 586
SAO CRISTOVAO
RIO DE JANEIRO
BRAZIL

FRESA A PROF
VIA PORTO 23
I-84100 SALERNO
ITALY

FRESNEAU A DR
OBS DE STRASBOURG
11 R DE L'UNIVERSITE
F-67000 STRASBOURG
FRANCE

FRICKE KLAUS DR
UNIVERSITAETSSTERNW
UNIV GOETTINGEN
GEISMARLANDSTR 11
D-3400 GOETTINGEN
GERMANY, F.R.

FRICKE WALTER PROF DR
DIRECTOR ASTRONOMI-
SCHES RECHENINSTITUT
12-14 MOENCHHOFSTR
D-6900 HEIDELBERG
GERMANY, F.R.

FRIEDEMANN CHRISTIAN DR
UNIVERSITAETS
STERNWARTE JENA
SCHILLERGAESSCHEN 2
DDR-69 JENA
G.D.R.

FRIEDJUNG MICHAEL DR
INST D ASTROPHYSIQUE
98 BIS BLVD ARAGO
F-75014 PARIS
FRANCE

FRIEDLANDER MICHAEL PROF
DEPT OF PHYSICS
WASHINGTON
UNIVERSITY
ST LOUIS MO 63130
U.S.A.

FRIEDMAN HERBERT DR
US NAVAL RESEARCH
LAB CODE 7100
WASHINGTON DC 20375
U.S.A.

FRINGANT ANNE-MARIE DR
OBS DE PARIS
BAT B
61 AV L'OBSERVATOIRE
F-75014 PARIS
FRANCE

FRISCH HELENE DR
OBSERVATOIRE DE NICE
BP 252
F-06007 NICE-CEDEX
FRANCE

FRISCH URIEL DR
OBSERVATOIRE DE NICE
B P 252
F-06007 NICE-CEDEX
FRANCE

FRITZOVA-SVESTKA L DR
147 DOPPERSTRAAT
BUNSCHOTEN
THE NETHERLANDS

FROESCHLE CHRISTIANE D DR
OBSERVATOIRE DE NICE
BP 252
F-06007 NICE CEDEX
FRANCE

FROESCHLE CLAUDE DR
OBSERVATOIRE DE NICE
B P 252
F-06007 NICE-CEDEX
FRANCE

FROGEL JAY A DR
CERRO TOLOLO INTER-
AMERIC OBSERVATORY
CASILLA 63 D
LA SERENA
CHILE

FROLOV M S DR
ASTRONOMICAL COUNCIL
USSR ACADEMY OF SCI
PYATNITSKAYA UL 48
109017 MOSCOW
U.S.S.R.

FROST KENNETH J DR
NASA GSFC
CODE 684
GREENBELT MD 20771
U.S.A.

FRYE GLENN M PROF
PHYS DEPT ROCK BLDG
CASE WESTERN RESERVE
UNIVERSITY
CLEVELAND OH 44106
U.S.A.

FUCHS JOSEPH PROF DR
GOLDSCHLAGSTR 185-9
A-1140 WIEN
AUSTRIA

FUJIMOTO MASAYUKI DR
NIIGATA UNIVERSITY
1-7-2 SAIJYO-MACHI
JYOETSU
NIIGATA 943
JAPAN

FUJIMOTO MITSUAKI DR
DEPT OF PHYSICS
NAGOYA UNIVERSITY
NAGOYA 464
JAPAN

FUJITA YOSHIO PROF
DEPT OF ASTRONOMY
UNIVERSITY OF TOKYO
BUNKYO KU
TOKYO 113
JAPAN

FUKUI TAKAO DR
DEPT OF LIBERAL ARTS
DOKKYO UNIVERSITY
SAKAE-MACHI 600
SOKA SAITAMA
JAPAN

FULCHIGNONI MARCELLO PROF
LABORATORIO DI
ASTROFISICA SPAZIALE
E.N.R.REP PLANETOL
I-00044 FRASCATI
ITALY

FURENLID K INGEMAR DR
KITT PEAK NATIONAL
OBSERVATORY
P O BOX 26732
TUCSON AZ 85726
U.S.A.

FURSENKO M A DR
INSTITUTE OF THEORE-
TICAL ASTRONOMY
192187 LENINGRAD
U.S.S.R.

GABRIEL ALAN H
ARD
CULHAM LABORATORY
ABINGDON OXFORDSHIRE
U.K.

GABRIEL MAURICE R DR
INST D ASTROPHYSIQUE
UNIVERSITE DE LIEGE
B-4200 COINTE OUGREE
BELGIUM

GAHM GOESTA F DR
STOCKHOLM
OBSERVATORY
S-133 00
SALTSJOEBADEN
SWEDEN

GAIGNEBET J DR
CERGA
AVE COPERNIC
F-06130 GRASSE
FRANCE

GAIL H P DR
INST THEOR ASTROPHYS
DER UNIVERSITAET
IM NEUENHEIMER F 294
D-6900 HEIDELBERG 1
GERMANY, F.R.

GAIZAUSKAS VICTOR DR
HERZBERG INST OF
ASTROPHYSICS
NAT RES COUNC CANADA
OTTAWA ONT K1A OR6
CANADA

GALAL A A DR
HELWAN OBSERVATORY
HELWAN NEAR CAIRO
ARAB REP. OF EGYPT

GALEOTTI PIERO PROF
LABORATORIO DI
COSMO-GEOFISICA
CORSO FIUME 4
I-10133 TORINO
ITALY

GALIBINA I V DR
INSTITUTE OF THEORE-
TICAL ASTRONOMY
192187 LENINGRAD
U.S.S.R.

GALKIN LEONID S A DR
CRIMEAN ASTROPHYSI-
CAL OBSERVATORY
334413 PO NAUCHNY
CRIMEA
U.S.S.R.

GALLAGHER III JOHN S DR
OBSERVATORY
UNIVERSITY OF
ILLINOIS
URBANA IL 61801
U.S.A.

GALLET R M
964 7TH STREET
BOULDER CO 80302
U.S.A.

GALLETTA GIUSSEPPE DR
OSSERVATORIO ASTRO-
NOMICO
I-35100 PADOVA
ITALY

GALLOUET LOUIS DR
OBSERVATOIRE
DE PARIS
61 AVE OBSERVATOIRE
F-75014 PARIS
FRANCE

GALPERIN YU I DR
SPACE RESEARCH
INSTITUTE
USSR ACADEMY OF SCI
117810 MOSCOW
U.S.S.R.

GALT JOHN A DR
DOMINION RADIO
ASTROPHYSICAL OBS
BOX 248
PENTICTON BC V2A 6K3
CANADA

GAMALELDIN ABDULLA I DR
HELWAN OBSERVATORY
HELWAN-CAIRO
ARAB REP. OF EGYPT

GAMMELGAARD PETER PROF
HANS BROGESVEJ 8
DK-8220 BRABAND
DENMARK

GAPOSCHKIN E DR
SMITHSONIAN ASTRO
OBSERVATORY
60 GARDEN ST
CAMBRIDGE MA 02138
U.S.A.

GAPOSHKIN SERGEI ILLAR DR
HARVARD COLLEGE
OBSERVATORY
60 GARDEN ST
CAMBRIDGE MA 02138
U.S.A.

GARDNER FRANCIS F DR
CSIRO DIVISION OF
RADIOPHYSICS
P O 76
EPPING NSW 2121
AUSTRALIA

GARFINKEL BORIS DR
YALE UNIVERSITY
OBSERVATORY
NEW HAVEN CT 06520
U.S.A.

GARLICK GEORGE F DR
267 SOUTH BELOIT AVE
LOS ANGELES CA 90049
U.S.A.

GARMANY CATHERINE D DR
JILA
UNIV OF COLORADO
BOULDER CO 80309
U.S.A.

GARMIRE GORDON P PROF
320-47 CALTECH
PASADENA CA 91125
U.S.A.

GARNIER ROBERT ING
OBSERVATOIRE DE LYON
F-69230 ST-GENIS-
LAVAL
FRANCE

GARRISON ROBERT F PROF
DAVID DUNLAP OBS
P O BOX 360
RICHMOND HILL
ONT L4C 4Y6
CANADA

GARSTANG ROY H PROF
JILA UNIVERSITY
COLORADO
BOULDER CO 80309
U.S.A.

GARTON W R S PROF
BLACKETT LABORATORY
IMPERIAL COLLEGE
LONDON SW7 2BZ
U.K.

GARZOLI S L DR
M T DE ALVEAR 2412
4 H
BUENOS AIRES
ARGENTINA

GASCOIGNE S C B DR
MT STROMLO
OBSERVATORY
WODEN PO ACT 2606
AUSTRALIA

GASKA STANISLAW DR
INSTITUTE ASTRONOMY
UL CHOPINA 12-18
87-100 TORUN
POLAND

GATEWOOD GEORGE DIRECTOR
ALLEGHENY OBS
OBSERVATORY STATION
PITTSBURG PA 15214
U.S.A.

GAUSS F STEPHEN
US NAVAL OBSERVATORY
WASHINGTON DC 20390
U.S.A.

GAUSTAD JOHN E PROF
DEPT OF ASTRONOMY
UNIVERSITY OF
CALIFORNIA
BERKELEY CA 94720
U.S.A.

GAUTIER DANIEL
OBSERVATOIRE
DE MEUDON
F-92190 MEUDON
FRANCE

GAVIOLA ENRIQUE PROF DR
INSTITUTO BALSEIRO
8400 BARILOCHE
ARGENTINA

GAVRILOV I V DR
MAIN ASTRONOMICAL
OBS OF THE UKRAINIAN
ACAD OF SCIENCES
252127 KIEV
U.S.S.R.

GAY JEAN DR
CERGA
AVE COPERNIC
F-06130 GRASSE
FRANCE

GEAKE JOHN E DR
PHYSICS DEPT
U M I S T
MANCHESTER M60 1QD
U.K.

GEBBIE KATHARINE B DR
JILA
UNIVERSITY OF
COLORADO
BOULDER CO 80302
U.S.A.

GEHRELS TOM PROF
LUNAR LABORATORY
UNIVERSITY OF
ARIZONA
TUCSON AZ 85721
U.S.A.

GEHREN THOMAS PH D
MPI FUER ASTRONOMIE
KOENIGSTUHL
D-6900 HEIDELBERG 1
GERMANY, F.R.

GEHRZ ROBERT DOUGLAS DR
DEPT PHYSICS ASTRON
UNIV OF WYOMING
UNIV STA BOX 3905
LARAMIE WY 82071
U.S.A.

GEISS JOHANNES PROF
PHYS INST
UNIVERSITAET BERN
SIDLERSTR 5
CH-3012 BERN
SWITZERLAND

GELFREJKH GEORGIJ B DR
PULKOVO OBSERVATORY
196140 LENINGRAD
U.S.S.R.

GENKIN IGOR L PROF DR
PHYSICS FACULTY
KAZAKH STATE UNIV
KOMSOMOLSKAYA 96
480012 ALMA ATA
U.S.S.R.

GENT HUBERT MR
35 ALBERT ROAD NORTH
MALVERN
WORCS WR14 2TP
U.K.

GEORGELIN YVON P DR
OBSERVATOIRE
DE MARSEILLE
2 PLACE LE VERRIER
F-13004 MARSEILLE
FRANCE

GEORGELIN YVONNE M DR
OBS DE MARSEILLE
2 PLACE LE VERRIER
F-13004 MARSEILLE
THE NETHERLANDS

GERARD ERIC DR
OBSERVATOIRE DE
MEUDON DEPT DE RADIO
ASTRONOMIE
F-92190 MEUDON
FRANCE

GERBAL D DR
OBS DE MEUDON
F-92190 MEUDON
FRANCE

GERBALDI MICHELE DR
INST D ASTROPHYSIQUE
98 BIS BD ARAGO
F-75014 PARIS
FRANCE

GERGELY TOMAS ESTEBAN DR
ASTRONOMY PROGRAM
UNIV OF MARYLAND
COLLEGE PK MD 20742
U.S.A.

GERLEI OTTO
HELIOPHYSICAL
OBSERVATORY
4010 DEBRECEN
HUNGARY

GEROLA HUMBERTO DR
IBM THOMAS J.WATSON
RESEARCH CENTER
P O BOX 218
YORKTWN HTS NY 10598
U.S.A.

GERSHBERG R E DR
CRIMEAN ASTROPHYSI-
CAL OBSERVATORY
NAUCHNY
334413 CRIMEA
U.S.S.R.

GETMANTSEV G G DR
RADIOPHYSICAL
RESEARCH INSTITUTE
603600 GORKIJ
U.S.S.R.

GEYER EDWARD H PROF DR
OBSERVATORIUM
HOHER LIST
D-5568 DAUN EIFEL
GERMANY, F.R.

GEZARI DANIEL YSA DR
NASA-GSFC
CODE 693 2
GREENBELT MD 20771
U.S.A.

GHIGO FRANCIS D DR
DEPT OF ASTRONOMY
UNIV OF MINNESOTA
116 CHURCH ST SE
MINNEAPOLIS MN 55455
U.S.A.

GHOBROS ROSHDY AZER DR
HELWAN OBSERVATORY
HELWAN NR CAIRO
ARAB REP. OF EGYPT

GIACAGLIA G E O PROF
ESCOLA POLITECHNICA
U S P
CAIXA POSTAL 8174
SAO PAULO
BRAZIL

GIACCONI RICCARDO PROF
DEPT OF ASTRONOMY
HARVARD UNIVERSITY
60 GARDEN ST
CAMBRIDGE MA 02138
U.S.A.

GIACHETTI RICCARDO DR
VIA BOLOGNESE 98
I-50100 FIRENZE
ITALY

GIANNONE PIETRO PROF
OSSERVATORIO
ASTRONOMICO
V D PARCO MELLINI 84
I-00136 ROMA
ITALY

GIANNUZZI MARIA A DR
PIAZZA GRAMSCI 4
I-00041 ALBANO
LAZIALE
ITALY

GIBSON DAVID MICHAEL DR
PHYSICS DEPT
NM INST MINING TECHN
CAMPUS STATION
SOCORRO NM 87801
U.S.A.

GICLAS H L MR
LOWELL OBSERVATORY
BOX 1269
FLAGSTAFF AZ 86002
U.S.A.

GIERASCH PETER J DR
318 SPACE SC BLDG
CORNELL UNIVERSITY
ITHACA NY 14853
U.S.A.

GIESE RICHARD H PROF
RUHR UNIVERSITAET
BEREICH EXTRATERRE-
STRISCHE PHYSIK
D-4630 BOCHUM
GERMANY, F.R.

GIESEKING FRANK DR
OBS HOHER LIST
UNIV STERNWARTE BONN
D-5568 DAUN
GERMANY, F.R.

GIETZEN JOSEPH W
ROYAL GREENWICH
OBSERVATORY
HAILSHAM
EAST SUSSEX BN27 1RP
U.K.

GILLINGHAM PETER DR
ANGLO AUSTRALIAN OBS
P O BOX 76
EPPING N S W 2121
AUSTRALIA

GILMAN PETER A DR
ADVANCED STUDY PROGR
NAT CTR FOR ATM RES
BOX 3000
BOULDER CO 80307
U.S.A.

GILMORE ALAN C MR
101 HAPPY VALLEY RD
WELLINGTON 2
NEW ZEALAND

GILRA DAYA P DR
KAPTEYN ASTRONOMICAL
INSTITUTE
UNIVERSITY GRONINGEN
GRONINGEN
THE NETHERLANDS

GINGERICH OWEN PROF
CTR FOR ASTROPHYSICS
CAMBRIDGE MA 02138
U.S.A.

GINZBURG VITALY L PROF
P N LEBEDEV PHY INST
LENINSKY PROSPECT 53
117924 GSP
MOSCOW B 333
U.S.S.R.

GICIA ISABELLA M DR
LAB RADIOASTRONOMIA
VIA IRNERIO 46
I-40126 BOLOGNA
ITALY

GIOVANELLI RONALD G DR
NATIONAL MEASUREMENT
LABORATORY
PO BOX 218
LINDFIELD NSW 2070
AUSTRALIA

GIOVANNELLI FRANCO DR
LAB ASTROFISICO SPAZ
P O B 67
I-00044 FRASCATI
ITALY

GIOVANNELLI RICCARDO DR
OSSERVATORIO ASTRO-
FISICO DI ARCETRI
I-50125 FIRENZE
ITALY

GIOVANNINI FONTI C
OSS ASTRONOMICO
VIA ZAMBONI 33
BOLOGNA
ITALY

GLAGOLEVSKIJ JU V DR
SPECIAL ASTROPHYSI-
CAL OBSERVATORY
USSR ACADEMY OF SCI
357140 N ARKHYZ
U.S.S.R.

GLASER HAROLD DR
NASA
WASHINGTON DC 20546
U.S.A.

GLASPEY JOHN W DR
DEPT DE PHYSIQUE
UNIV DE MONTREAL
CP 6128 SUCC A
MONTREAL QUE H3C 3J7
CANADA

GLASS BILLY PRICE DR
DEPT OF GEOLOGY
UNIV OF DELAWARE
NEWARK DE 19711
U.S.A.

GLASS IAN S DR
S A ASTRONOMICAL OBS
P O BOX 9
OBSERVATORY CP 7935
SOUTH AFRICA

GLASSGOLD ALFRED E PROF
NEW YORK UNIVERSITY
PHYSICS DEPT
NEW YORK NY 10003
U.S.A.

GLEBOCKI ROBERT DR
INSTITUTE OF PHYSICS
UL WITA STWOSZA 57
80-308 GDANSK
POLAND

GLEBOVA NINA I DR
INST THEORET ASTRON
USSR ACAD OF SCS
10 KUTUZOV QUAY
192187 LENINGRAD
U.S.S.R.

GLEDHILL JOHN A PROF
DEPT OF PHYSICS
RHODES UNIVERSITY
P O BOX 94
GRAHAMSTOWN 6140
SOUTH AFRICA

GLEISSBERG WOLFGANG
PROF DR
BUCHENWEG 12
D-6374 OBERURSEL
GERMANY, F.R.

GLENCROSS WILLIAM M DR
DEPT OF PHYSICS-ASTR
UNIVERSITY COLLEGE
GOWER STREET
LONDON WC1E 6BT
U.K.

GLIESE WILHELM DR
ASTRONOMISCHES
RECHEN INSTITUT
MOENCHHOFSTR 12 14
D-6901 HEIDELBERG
GERMANY, F.R.

GLUSHNEVA I N DR
STERNBERG STATE AS-
TRONOMICAL INSTITUTE
117234 MOSCOW
U.S.S.R.

GNEVYSHEV MSTISLAV N DR
PULKOVO OBSERVATORY
196140 LENINGRAD
U.S.S.R.

GNEVYSHEVA RAISA S DR
PULKOVO OBSERVATORY
196140 LENINGRAD
U.S.S.R.

GODART ODON PROF
RUE DE CHATEAU 96
B-1288 ROUSVAL
BELGIUM

GODOLI GIOVANNI PROF
OSSERVATORIO ASTRO-
FISICO DI ARCETRI
LARGO ENRICO FERMI 5
I-50125 FIRENZE
ITALY

GODWIN J G DR
UNIV OBSERVATORY
SOUTH PARKS ROAD
OXFORD OX1 3RQ
U.K.

GOEBEL ERNST DR
INST F ASTRONOMIE
UNIVERSITAET WIEN
TUERKENSCHANZSTR 17
A-1180 WIEN
AUSTRIA

GOEBEL JOHN H DR
SPACE SCIENCE DIV
ASTROPHYS EXP BRANCH
NASA-AMES RES CTR
MOFFETT FLD CA 94035
U.S.A.

GOKDOGAN NUZHET PROF
UNIVERSITY
OBSERVATORY
UNIV ISTANBUL
ISTANBUL
TURKEY

GOKHALE MORESHWAR HARI DR
INDIAN INSTITUTE OF
ASTROPHYSICS
BANGALORE 560034
INDIA

GOKMEN TARIK ASSOC PROF
UNIVERSITY
OBSERVATORY
34 ISTANBUL
TURKEY

GOLAY MARCEL PROF
OBSERVATOIRE
DE GENEVE
CH-1290 SAUVERNY GE
SWITZERLAND

GOLD THOMAS PROF
SPACE SCIENCE BLDG
CORNELL UNIVERSITY
ITHACA NY 14853
U.S.A.

GOLDBACH C MME
INST D'ASTROPHYSIQUE
98 BIS BD ARAGO
F-75014 PARIS
FRANCE

GOLDBERG LEO PROF
KITT PEAK NATIONAL
OBSERVATORY
P O BOX 26732
TUCSON AZ 85726
U.S.A.

GOLDREICH P DR
CALTECH
PASADENA CA 91109
U.S.A.

GOLDSMITH DONALD W. DR.
2153 RUSSELL STREET
BERKELEY CA 94705
U.S.A.

GOLDSMITH PAUL F DR
DEPT PHYSICS ASTRON
GRC TOWER B RM 626
UNIV OF MASS
AMHERST MA 01003
U.S.A.

GOLDSMITH S. DR.
DEPT.OF PHYSICS AND
ASTRONOMY
TEL AVIV UNIVERSITY
TEL AVIV
ISRAEL

GOLDSTEIN RICHARD M DR
JET PROPULSION
LABORATORY
4800 OAK GROVE DRIVE
PASADENA CA 91011
U.S.A.

GOLDSTEIN SAMUEL J PROF
UNIVERSITY VIRGINIA
PO BOX 3818
CHARLOTTESVILLE
VA 22903
U.S.A.

GOLDSWORTHY FREDERICK A
PROF
SCHOOL MATHEMATICS
UNIVERSITY OF LEEDS
LEEDS LS2 9JT
U.K.

GOLDWIRE HENRY C JR
UNIVERSITY OF
CALIFORNIA LASL
BOX 1663 MS 420
LOS ALAMOS NM 87545
U.S.A.

GOLLNOW H DR
MT STROMLO OBS
CANBERRA ACT
AUSTRALIA

GOLUB LEON DR
HARVARD COLLEGE
OBSERVATORY
60 GARDEN ST
CAMBRIDGE MA 02138
U.S.A.

GOMES ALERCIO M DR
R GAVIAO PEIXOTO 13
AP 1401
ICARAI 24000
NITEROJ ERJ
BRAZIL

GOMEZ GONZALEZ J DR
PASEO IMPERIAL 29 6H
MADRID 5
SPAIN

GOMEZ MARIA THERESA DR
OSSERVATORIO ASTRO-
NOMICO
NAPOLI
ITALY

GONDHALEKAR PRABHAKAR DR
14 REDWOOD CLOSE
SOUTHMOOR
OXON OX13 5DH
U.K.

GONDOLATSCH FRIERICH PROF
ASTRONOMISCHES
RECHENINSTITUUT
MOENCHHOFSTR 12-14
D-6900 HEIDELBERG
GERMANY, F.R.

GONZALEZ G
INST NAC DE ASTROFIS
OPT Y ELECTRONICA
APT POSTALES 216 51
PUEBLA PUE
MEXICO

GONZE ROGER F J IR
OBSERVATOIRE ROYAL
DE BELGIQUE
AVE CIRCULAIRE 3
B-1180 BRUXELLES
BELGIUM

GOODY R M
CEPP
PIERCE HALL
29 OXFORD STREET
CAMBRIDGE MA 02138
U.S.A.

GOOSSENS MARCEL DR
LOZENHOEKSTRAAT 127
B-2850 KEERBERGEN
BELGIUM

GOPALA RAO U V MR
ASTROPHYSICAL OBS
KODAIKANAL 3
INDIA

GOPASYUK S I DR
CRIMEAN ASTROPHYSI-
CAL OBSERVATORY USSR
ACAD OF SCI NAUCHNIY
334413 CRIMEA
U.S.S.R.

GORBATSKY VITALIJ G PROF
LENINGRAD UNIVERSITY
OBSERVATORY
199178 LENINGRAD
U.S.S.R.

GORDON CHARLOTTE PROF
5 SQUARE PORT-ROYAL
F-75013 PARIS
FRANCE

GORDON COURTNEY P PROF
HAMPSHIRE COLLEGE
AMHERST MA 01002
U.S.A.

GORDON I M DR
INSTITUTE OF RADIO
PHYSICS AND
ELECTRONICS
310085 KHARKOV
U.S.S.R.

GORDON KURTISS J PROF
HAMPSHIRE COLLEGE
AMHERST MA 01002
U.S.A.

GORDON MARK A DR
NAT RADIO ASTRONOMY
OBS SUITE 100
2010 N FORBES BLVD
TUCSON AZ 85705
U.S.A.

GORENSTEIN PAUL DR
CENTER FOR ASTRO-
PHYSICS
60 GARDEN STREET
CAMBRIDGE MA 02138
U.S.A.

GORET P DR
SECT D'ASTROPHYSIQUE
CEN SACLAY
B P NO 2
F-91190 GIF-S-YVETTE
FRANCE

GOSLING JOHN T DR
P4- MS436
LOS ALAMOS SCIENTI-
FIC LABORATORY
LOS ALAMOS NM 87545
U.S.A.

GOTT III J RICHARD
ASST PROF
DEPT ASTROPHYS SC
PRINCETON UNIVERSITY
PRINCETON NJ 08540
U.S.A.

GOTTLIEB KURT
46 JENNINGS STREET
CURTIN ACT 2605
AUSTRALIA

GOUGUENHEIM LUCIENNE
OBSERVATOIRE
DE MEUDON
RADIOASTRONOMIE
F-92190 MEUDON
FRANCE

GOWER J F R DR
1615 MCTAVISH ROAD
R R 2
SIDNEY B C V8L 3S1
CANADA

GRABOSKE HAROLD C JR
LAWRENCE LIVERMORE
LABORATORY
P O BOX 808
LIVERMORE CA 94550
U.S.A.

GRAHAM ERIC DR
HIGH ALTITUDE OBS
NATL CENTER ATMO RES
PO BOX 3000
BOULDER CO 80307
U.S.A.

GRAINGER J F DR
PHYSICS DEPT
UMIST
MANCHESTER M60 1QD
U.K.

GRASDALEN GARY L DR
DEPT PHYSICS ASTRON
UNIV OF WYOMING
P O BOX 3905 UN STA
LARAMIE WY 82071
U.S.A.

GRAYZECK EDWIN J DR
PHYSICS DEPT
UNIV OF NEVADA
4505 S MARYLAND PKY
LAS VEGAS NV 89154
U.S.A.

GREEN LOUIS C PROF
HAVERFORD COLLEGE
HAVERFORD PA 19041
U.S.A.

GORGOLEWSKI STANISLAW DR
INST OF ASTRONOMY
NIC COPERNICUS UNIV
CHOPINA 12-18
87100 TORUN
POLAND

GOSS W MILLER DR
KAPTEYN ASTRONOMICAL
INSTITUTE
POSTBUS 800
9700 AV GRONINGEN
THE NETHERLANDS

GOTTESMAN STEPHEN T DR
PHYSICS AND
ASTRONOMY
UNIVERSITY FLORIDA
GAINESVILLE FL 32611
U.S.A.

GOUDAS CONSTANTINE L PROF
DEPT OF MECHANICS
UNIVERSITY OF PATRAS
PATRAS
GREECE

GOULD ROBERT J PROF
PHYSICS DEPT
UCSD
LA JOLLA CA 92093
U.S.A.

GOY GERALD DR
OBSERVATOIRE
DE GENEVE
CH-1290 SAUVERNY
SWITZERLAND

GRABOWSKI BOLESLAW DR
INSTITUTE OF PHYSICS
UL OLESKA 48
45951 OPOLE
POLAND

GRAHAM JOHN A DR
CERRO TOLOLO
INTERAMERICAN OBS
CASILLA 63 D
LA SERENA
CHILE

GRANDI STEVEN ALDRIDGE DR
DEPT OF ASTRONOMY
MATH SCS BLDG
UNIV OF CALIFORNIA
LOS ANGELES CA 90024
U.S.A.

GRATTON LIVIO PROF
LABORATORIO DI
ASTROFISICA SPAZIALE
CAS POST 67
I-00044 FRASCATI
ITALY

GREBENIKOV E A PROF DR
INSTITUTE OF THEORE-
TICAL AND EXPERIMEN-
TAL PHYSICS
117259 MOSCOW
U.S.S.R.

GREEN ROBIN M DR
DEPT OF ASTRONOMY
GLASGOW UNIVERSITY
GLASGOW G12 8QQ
U.K.

GOSACHINSKIJ I V DR
LENINGRAD BRANCH OF
SAO
196140 LENINGRAD
U.S.S.R.

GOSSNER SIMONE D MRS
415 E 50TH STREET
NEW YORK NY 10022
U.S.A.

GOTTLIEB CARL A DR
GODDARD INST FOR
SPACE STUDIES
2880 BROAD WAY
NEW YORK NY 10025
U.S.A.

GOUGH DOUGLAS O DR
INST OF ASTRONOMY
MADINGLEY RD
CAMBRIDGE CB3 0HA
U.K.

GOWER ANN C DR
UNIVERSITY OF VICTO-
RIA OBSERVATORY
VICTORIA BC V8W 2Y2
CANADA

GOYAL A N DR
DEPT OF MATHEMATICS
UNIV OF RAJASTHAN
JAIPUR
INDIA

GRADSZTAJN E DR
INSTITUT DE PHYSIQUE
NUCLEAIRE D-ORSAY
F-91190 GIF-YVETTE
FRANCE

GRAHL BERND H DR
MPI FUER RADIOASTRO-
NOMIE
AUF DEM HUEGEL 69
D-5300 BONN
GERMANY, F.R.

GRANT IAN P DR
PEMBROKE COLLEGE
OXFORD OX1 1DW
U.K.

GRAY DAVID F PROF
DEPT OF ASTRONOMY
UNIVERSITY OF
WESTERN ONTARIO
LONDON ONT N6A 5B9
CANADA

GREEN JACK PROF
DEPT OF GEOLOGY
CALIF STATE UNIVER-
SITY-LONG BEACH
LONG BEACH CA 90840
U.S.A.

GREENBERG J MAYO DR
HUYGENS LABORATORY
UNIVERSITY OF LEIDEN
WASSENAARSEWEG 78
2300 RA LEIDEN
THE NETHERLANDS

GREENBERG RICHARD DR
PLANETARY SCIENCE
INSTITUTE
2030 E SPEEDWAY
TUCSON AZ 85719
U.S.A.

GREENSTEIN GEORGE PROF
AMHERST COLLEGE
ASTRONOMY DEPT
AMHERST MA 01002
U.S.A.

GREENSTEIN J L PROF
HALE OBSERVATORIES
CALIF INST OF TECHN
1201 E CALIFORNIA ST
PASADENA CA 91125
U.S.A.

GREGORY PHILIP C DR
PHYSICS DEPT
UNIVERSITY OF B C
2075 WESBROOK MALL
VANCOUVER BC V6T 1W5
CANADA

GREGORY STEPHEN ALBERT DR
PHYSICS DEPT
BOWLING GREEN STATE
UNIVERSITY
BOWLING GRN OH 43403
U.S.A.

GREGUL A YA DR
OBSERVATORY OF THE
KIEV UNIVERSITY
OBSERVATORNAYA 3
252053 KIEV
U.S.S.R.

GREISEN KENNETH I PROF
DEPT OF ASTRONOMY
228 SPACE SC BLDG
CORNELL UNIVERSITY
ITHACA NY 14853
U.S.A.

GRENON MICHEL DR
OBS DE GENEVE
CH-1290 SAUVERNY
SWITZERLAND

GREVESSE N DR
INSTITUT
D ASTROPHYSIQUE
B-4200 COINTE-OUGREE
BELGIUM

GREWING MICHAEL PROF
ASTRONOMISCHES INST
UNIVERSITAET
WALDHAUSERSTR 64
D-7400 TUEBINGEN
GERMANY, F.R.

GREYBER HOWARD D DR
10123 FALLS ROAD
POTOMAC MD 20854
U.S.A.

GRIFFIN RITA E M DR
THE OBSERVATORIES
MADINGLEY ROAD
CAMBRIDGE CB3 0HA
U.K.

GRIFFIN ROGER F DR
THE OBSERVATORIES
MADINGLEY ROAD
CAMBRIDGE CB3 0HA
U.K.

GRIFFITH JOHN S PROF
DEPT OF MATH SC
LAKEHEAD UNIVERSITY
THUNDER BAY
ONT P7B 5E1
CANADA

GRIFFITHS RICHARD E DR
HARVARD SMITHSONIAN
CTR FOR ASTROPHYSICS
60 GARDEN STREET
CAMBRIDGE MA 02138
U.S.A.

GRIGOREVSKY VITALI M PROF
TERESHKOVA ST 39-2
FLAT 95
270072 ODESSA
U.S.S.R.

GRIGORJEV VICTOR M DR
SIBERIAN INST OF TER
MAGN,IONOSPH,RAD W P
P BOX 4
664697 IRKUTSK
U.S.S.R.

GRINDLAY JONATHAN E DR
HARVARD SMITHSONIAN
CENTER FOR ASTROPHYS
60 GARDEN ST
CAMBRIDGE MA 02138
U.S.A.

GRININ V P DR
CRIMEAN ASTROPHYSI-
CAL OBSERVATORY USSR
ACAD OF SCI NAUCHNIY
334413 CRIMEA
U.S.S.R.

GRISHCHUK L P DR
STERNBERG STATE AS-
TRONOMICAL INSTITUTE
117234 MOSCOW
U.S.S.R.

GROSBOL PREBEN JOHNSON DR
COPENHAGEN UNIV
OBSERVATORY
OSTER VOLDGADE 3
DK-1350 COPENHAGEN-K
DENMARK

GROSS PETER G PROF
WARNER SWASEY
OBSERVATORY
1975 TAYLOR RD
CLEVELAND OH 44112
U.S.A.

GROSSMAN ALLEN S PROF
ERWIN FICK
OBSERVATORY
IOWA ST UNIVERSITY
AMES IA 50011
U.S.A.

GROSSMAN LAWRENCE PROF
DEPT GEOPHYSICAL SC
UNIVERSITY CHICAGO
5734 SOUTH ELLIS AVE
CHICAGO IL 60637
U.S.A.

GROSSMANN-DOERTH U DR
KIEPENHEUER INST
FUER SONNENPHYSIK
SCHOENECKSTR 6
D-7800 FREIBURG
GERMANY, F.R.

GROTEN ERWIN PROF
INST F PHYSIKALISCHE
GEODAESIE
PETERSENSTR 13
D-6100 DARMSTADT
GERMANY, F.R.

GROTH EDWARD J III
PHYSICS DEPT
JADWIN HALL
PRINCETON UNIVERSITY
PRINCETON NJ 08540
U.S.A.

GROTH HANS G PROF DR
INST FUER ASTRONOMIE
UND ASTROPHYSIK
SCHEINERSTR 1
D-8000 MUENCHEN 80
GERMANY, F.R.

GRUBISSICH C PROF DR
VIA AOSTA 34/5
I-35100 PADOVA
ITALY

GRUDLER PIERRE
C E R G A
AVE COPERNIC
F-06130 GRASSE
FRANCE

GRUDZINSKA STEFANIA DR
INST OF ASTRONOMY
UL CHOPINA 12-18
MIKOLAJA KOPERNIKA
87100 TORUN
POLAND

GRUEFF GAVRIL DR
LABORATORIO DI RADIO
ASTRONOMIA C N R
VIA IRNERIO 46
I-40126 BOLOGNA
ITALY

GRUSHINSKIJ N P PROF DR
STERNBERG STATE AS-
TRONOMICAL INSTITUTE
117234 MOSCOW
U.S.S.R.

GRYGAR JIRI DR
ASTRONOMICAL INST
CZECHOSLOVAK ACAD SC
OBSERVATORY
251 65 ONDREJOV
CZECHOSLOVAKIA

GRZEDZIELSKI STANISLAW DR
CENTER FOR SPACE RES
UL BARTYCKA 18
00-716 WARSAW
POLAND

GUARNIERI ADRIANO DR
OSSERVATORIO
ASTRONOMICO
VIA ZAMBONI 33
I-40126 BOLOGNA
ITALY

GUBANOV VADIM S DR
PULKOVO OBSERVATORY
196140 LENINGRAD
U.S.S.R.

GUDUR N DR
EGE UNIVERSITY
OBSERVATORY
P.K.21 BORNOVA
IZMIR
TURKEY

GUELIN MICHEL DR
OBSERVATOIRE DE
MEUDON
F-92190 MEUDON
FRANCE

GUENTZEL-LINGNER U DR
ASTRONOMISCHES-
RECHENINSTITUT
MOENCHHOFSTR 12-14
D-69.. HEIDELBERG
GERMANY, F.R.

GUERIN PIERRE DR
INST D ASTROPHYSIQUE
98 BIS BLVD ARAGO
F-75014 PARIS
FRANCE

GUERRERO GIANANTONIO DR
OSSERVATORIO ASTRO-
NOMICO
MERATE
MILANO
ITALY

GUEST JOHN E DR
UNIVERSITY OF LONDON
OBSERVATORY
MILL HILL PARK
LONDON NW7 2QS
U.K.

GUIBERT JEAN DR
OBSERVATOIRE
DE MEUDON
F-92190 MEUDON
FRANCE

GUIDICE DONALD A DR
A F GEOPHYSICS
LABORATORY
HANSCOM AFB MA 01731
U.S.A.

GUINAN EDWARD FRANCIS DR
DEPT OF ASTRONOMY
VILLANOVA UNIV
VILLANOVA PA 19085
U.S.A.

GUINOT BERNARD R PROF
BUREAU INTERNATIONAL
DE L HEURE
61 AV L OBSERVATOIRE
F-75014 PARIS
FRANCE

GULKIS SAMUEL DR
JET PROPULSION
LABORATORY
4800 OAK GROVE DRIVE
PASADENA CA 91103
U.S.A.

GULL STEPHEN F DR
ST JOHN'S COLLEGE
CAMBRIDGE
U.K.

GULL THEODRE R DR
CODE 683
GODDARD
SPACE FLIGHT CENTER
GREENBELT MD 20771
U.S.A.

GULMEN OMUR DR
EGE UNIVERSITY
OBSERVATORY
PK 21
BORNOVA IZMIR
TURKEY

GULYAEV A P DR
STERNBERG ASTRO INST
UNIVERSITETSKIJ
PROSPEKT 13
117234 MOSCOW
U.S.S.R.

GULYAEV RUDOLF A DR
IZMIRAN
AKADEMGORODOK
142092 MOSCOW REGION
U.S.S.R.

GUNN JAMES E PROF
CALTECH
105-24
PASADENA CA 91125
U.S.A.

GURM HARDEN S DR
DEPT OF ASTRONOMY
AND SPACE SCIENCES
PUNJABI UNIVERSITY
PATIALA 147002
INDIA

GURSHTEIN A A DR
SPACE RESEARCH
INSTITUTE
USSR ACADEMY OF SCI
117810 MOSCOW
U.S.S.R.

GURSKY HERBERT DR
CENTER FOR
ASTROPHYSICS
60 GARDEN STREET
CAMBRIDGE MA 02138
U.S.A.

GURTOVENKO E A DR
MAIN ASTRONOMICAL
OBS UKRAINIAN ACAD
OF SCIENCES
252127 KIEV
U.S.S.R.

GURZADIAN G A PROF DR
BYURAKAN ASTROPHYSI-
CAL OBSERVATORY
378433 ARMENIA
U.S.S.R.

GUSEINOV O KH DR
SHEMAKHA ASTROPHYSI-
CAL OBSERVATORY
373243 AZERBAIDZAN
U.S.S.R.

GUSEJNOV R EH DR
SHEMAKHA ASTROPHYSI-
CAL OBSERVATORY
373243 AZERBAIDZAN
U.S.S.R.

GUSSMANN E A DR
ZENTRALINST ASTROPHY
ASTROPHYSIKALIS OBS
TELEGRAFENBERG
DDR-15 POTSDAM
G.D.R.

GUSTAFSSON BENGT DR
ASTRONOMICAL
OBSERVATORY
BOX 515
S-751 20 UPPSALA
SWEDEN

GUTH VLADIMIR DR PROF
ASTRONOMICAL INST
CZECHOSLOVAK ACAD SC
251 65 ONDREJOV
OBSERVATORY
CZECHOSLOVAKIA

GUTHRIE BRUCE N G DR
ROYAL OBSERVATORY
BLACKFORD HILL
EDINBURGH
EDINBURGH EH9 3HJ
U.K.

GUTIERREZ-MORENO A DR MRS
DEP DE ASTRONOMIA
UNIVERSIDAD DE CHILE
CASILLA 36-D
SANTIAGO
CHILE

GYLDENKERNE KJELD DR
COPENHAGEN UNIVER-
SITY OBSERVATORY
BRORFELDE
DK-4340 TOLLOSE
DENMARK

HAAS J PROF DR
LIPSCHITZ STR 1
D-5300 BONN
GERMANY, F.R.

HABIBULLIN SH T PROF DR
ENGELGARDT ASTRONO-
MICAL OBSERVATORY
422526 KAZAN
U.S.S.R.

HABING H J DR
STERREWACHT
POSTBUS 9513
2300 RA LEIDEN
THE NETHERLANDS

HACHENBERG OTTO PROF DR
RADIOASTRON INST DER
UNIV BONN
AUF DEM HUEGEL 71
D-5300 BONN
GERMANY, F.R.

HACK MARGHERITA PROF
OSSERVATORIO
ASTRONOMICO
VIA TIEPOLO 11
I-34131 TRIESTE
ITALY

HACKWELL JOHN A DR
DEPT PHYSICS ASTRON
UNIV OF WYOMING
LARAMIE WY 82070
U.S.A.

HACYAN SHAHEN DR.
GROUPE D ASTROPHYSI-
QUE RELATIVISTE
OBS DE PARIS
F-92190 MEUDON
FRANCE

HADDOCK FRED T DR
UNIV OF MICHIGAN
937 PHYSICS-ASTRONO-
MY BLDG
ANN ARBOR MI 48104
U.S.A.

HADJIDEMETRIOU JOHN DPROF
DEPT OF THEORETICAL
MECHANICS
UNIV OF THESSALONIKI
THESSALONIKI
GREECE

HAEFNER REINHOLD DR
UNIV STERNWARTE
SCHEINERSTRASSE 1
D-8000 MUENCHEN 80
GERMANY, F.R.

HAERENDEL G DR
MPI FUER PHYSIK UND
ASTROPHYSIK
INST EXTRATERR PHYS
D-8046 GARCHING
GERMANY, F.R.

HAGEN JOHN P
613 W PARK AVE
STATE COLLEGE
PA 16801
U.S.A.

HAGEN-THORN V A DR
LENINGRAD UNIVERSITY
10 LINIJA 33
199178 LENINGRAD
U.S.S.R.

HAGFORS T DR
MIT LINCOLN LAB
MILLSTON/HAYSTACK
LEXINGTON MA 02173
U.S.A.

HAJDUK ANTON DR
ASTRONOMICAL INST
SAV
DUBRAVSKA CESTA A/11
89930 BRATISLAVA
CZECHOSLOVAKIA

HALL DONALD N DR
KITT PEAK NATIONAL
OBSERVATORY
P O 26732
TUCSON AZ 85726
U.S.A.

HALL DOUGLAS S DR
DYER OBSERVATORY
VANDERBILT
UNIVERSITY
NASHVILLE TN 37235
U.S.A.

HALL JOHN S DR
STAR ROUTE 1
BOX 97
SEDONA AZ 86336
U.S.A.

HALL R GLENN DR
US NAVAL OBSERVATORY
WASHINGTON DC 20390
U.S.A.

HALLAM KENNETH L DR
CODE 685 LAB FOR
ASTRON SOLAR PHYSICS
GSFC
GREENBELT MD 20771
U.S.A.

HALLIDAY I DR
HERZBERG INSTITUTE
OF ASTROPHYSICS
NATIONAL RES COUNCIL
OTTAWA ONT K1A 0R6
CANADA

HAMADA TETSUO PROF
DEPT OF PHYSICS
IBARAKI UNIVERSITY
310 MITO
JAPAN

HAMDY M A M DR
HELWAN OBSERVATORY
HELWAN
ARAB REP. OF EGYPT

HAMEEN ANTILLA KAARLE A
PROF DR
ASTRONOMY DEPT
UNIVERSITY OF OULU
SF-90100 OULU 10
FINLAND

HAMID S EL DIN DR
DEPT OF ASTRONOMY
FACULTY OF SCIENCE
UNIVERSITY FOUAD
GIZA CAIRO
ARAB REP. OF EGYPT

HAMILTON P A DR
UNIV OF TASMANIA
HOBART TASMANIA
AUSTRALIA

HAMMERSCHLAG ROBERT H
DR IR
ASTRONOMICAL INST
SERVAAS BOLWERK 13
UTRECHT
THE NETHERLANDS

HAMMERSCHLAG-HENSBERGE
G DR
ASTRONOMICAL INST
ROETERSTRAAT 15
1018 WB AMSTERDAM
THE NETHERLANDS

HAMZAOGLU ESAT E H DR
EGE UNIVERSITESI
FEN FAKULTESI
GOK BILIMLERI ENST
BORNOVA IZMIR
TURKEY

HANASZ JAN DR
ASTRONOMICAL CENTER
UL CHOPINA 12/18
NICAL CENTRE P A N
87100-TORUN
POLAND

HANBURY BROWN R PROF
SCHOOL OF PHYSICS
UNIVERSITY OF SYDNEY
SYDNEY NSW 2006
AUSTRALIA

HANES DAVID A DR
INST OF ASTRONOMY
UNIV OF CAMBRIDGE
CAMBRIDGE
U.K.

HANNER MARTHA S DR
JPL
MS 183-301
PASADENA CA 91103
U.S.A.

HANSEN CARL J PROF
JILA
UNIVERSITY OF
COLORADO
BOULDER CO 80309
U.S.A.

HANSEN LEIF LECTURER
COPENHAGEN UNIVERSI-
TY OBSERVATORY
OSTERVOLDGADE 3
DK-1350 COPENHAGEN K
DENMARK

HANSEN RICHARD T MR
HIGH ALTITUDE
OBSERVATORY
P O BOX 3000
BOULDER CO 80303
U.S.A.

HANSON ROBERT B DR
LICK OBSERVATORY
UNIV OF CALIFORNIA
SANTA CRUZ CA 95064
U.S.A.

HANSSON NILS DR
LUND OBSERVATORY
SVANEGATAN 9
S-222 24 LUND
SWEDEN

HAPKE BRUCE W DR
DEPT EARTH PLANET SC
321 OLD ENGIN HALL
UNIV OF PITTSBURGH
PITTSBURGH PA 15235
U.S.A.

HARA KEN NOSUKE DR
SENDAI-SHIRITSU-JOSH
SENIOR HIGH SCHOOL
KASHIWAGI 3-3-1
SENDAI 980
JAPAN

HARA TETSUYA DR
DEPT OF PHYSICS
KYOTO SANGYO UNIV
KYOTO 603
JAPAN

HARDEBECK ELLEN G DR
3106 TUMBLEWEED RD
BISHOP CA 93514
U.S.A.

HARDIE R PROF
DYER OBSERVATORY
VANDERBILT
UNIVERSITY
NASHVILLE TN 37235
U.S.A.

HARDING GEORGE A
ROYAL GREENWICH
OBSERVATORY
HAILSHAM
EAST SUSSEX BN27 1RP
U.K.

HARDORP JOHANNES PROF
DEPT EARTH SPACE SC
STATE UNIVERSITY
STONY BROOK NY 11794
U.S.A.

HARDY EDUARDO DR PH
DEPT DE PHYSIQUE
UNIVERSITE LAVAL
FAC DES SCS GENIE
QUEBEC G1K 7P4
CANADA

HARMANEC PETR DR
ASTRONOMICAL INST
CZECHOSLOVAK ACAD
OF SCIENCES
251 65 ONDREJOV
CZECHOSLOVAKIA

HARMER CHARLES F W MR
ROYAL GREENWICH OBS
HERSTMONCEUX CASTLE
HAILSHAM
E SUSSEX BN27 1RP
U.K.

HARMER DIANNE L MRS
ROYAL GREENWICH OBS
HERSTMONCEUX CASTLE
HAILSHAM
E SUSSEX BN27 1RP
U.K.

HARMS RICHARD JAMES DR
PHYSICS DEPT B-019
UNIV OF CALIFORNIA
SAN DIEGO
LA JOLLA CA 92093
U.S.A.

HARO GUILLERMO DR
INST NAC ASTROFISICA
OPTICA Y ELECTRONICA
A P 216 Y 51
PUEBLA PUE
MEXICO

HARRINGTON J PATRICK DR
ASTRONOMY PROGRAM
UNIVERSITY MARYLAND
COLLEGE PK MD 20742
U.S.A.

HARRINGTON ROBERT S DR
U S NAVAL
OBSERVATORY
WASHINGTON DC 20390
U.S.A.

HARRIS ALAN WILLIAM DR
JPL
MS 264-720
4800 OAK GROVE DRIVE
PASADENA CA 91103
U.S.A.

HARRIS DANIEL E DR
DRAO
P O BOX 248
PENTICTON BC V2A 6K3
CANADA

HARRIS GRETCHEN L H DR PH
DEPT OF PHYSICS
UNIV OF WATERLOO
WATERLOO ONT
CANADA

HARRIS WILLIAM E DR
DEPT OF PHYSICS
MCMASTER UNIV
HAMILTON ONT L8S 4M1
CANADA

HARRISON EDWARD R PROF
DEPT OF PHYSICS AND
ASTRONOMY
UNIVERSITY OF MASS
AMHERST MA 01003
U.S.A.

HARROWER GEORGE A DR
LAKEHEAD UNIVERSITY
THUNDER BAY
ONT P7B 5E1
CANADA

HART MICHAEL H DR
PHYSICS DEPT
TRINITY UNIV
SAN ANTONIO TX 78284
U.S.A.

HARTEN RONALD H DR
RADIOSTERRENWACHT
DWINGELOO 7514
THE NETHERLANDS

HARTL HERBERT DR
INST F ASTRONOMIE
UNIVERSITAETSSTR 4
A-6020 INNSBRUCK
AUSTRIA

HARTLEY KENNETH F DR
ROYAL GREENWICH
OBSERVATORY
HAILSHAM
EAST SUSSEX BN27 1RP
U.K.

HARTNER WILLY R PROF
UNIVERSITAET
D-6000 FRANKFURT
GERMANY, F.R.

HARTOOG MARK RICHARD DR
LICK OBS
UNIV OF CALIFORNIA
SANTA CRUZ CA 95064
U.S.A.

HARTWICK F DAVID A DR
UNIVERSITY VICTORIA
DEPT OF PHYSICS
VICTORIA BC V8W 2Y2
CANADA

HARTZ THEODORE R DR
COMMUNICATIONS
RESEARCH CENTER
POB 11490 STATION H
OTTAWA ONT K2H 8S2
CANADA

HARVEY CHRISTOPHER C DR
OBSERVATOIRE
DE MEUDON
F-92190 MEUDON
FRANCE

HARVEY GALE A DR
INST F GESCHICHTE D
NATURWISSENSCHAFTEN
GOETHE UNIVERSITAET
6000 FRANKFURT
GERMANY, F.R.

HARVEY J W DR
KITT PEAK NATIONAL
OBSERVATORY
950 N CHERRY AVE
TUCSON AZ 85726
U.S.A.

HARVEY PAUL MICHAEL DR
STEWARD OBS
UNIV OF ARIZONA
TUCSON AZ 85721
U.S.A.

HARWIT MARTIN PROF
SPACE SCIENCE BLDG
CORNELL UNIVERSITY
ITHACA NY 14853
U.S.A.

HASEGAWA HIROICHI DR
DEPT OF PHYSICS
KYOTO UNIVERSITY
SAKYO-KU
KYOTO 606
JAPAN

HASEGAWA ICHIRO DR
2-3-11 SAIDAIJINO-
GAMI
NARA 631
JAPAN

HASER LEO N K DR
MPI FUER EXTRATERRE-
STRISCHE PHYSIK
D-8046 GARCHING
GERMANY, F.R.

HASLAM C GLYN T DR
MPI F RADIO
ASTRONOMIE
AUF DEM HUEGEL 69
D-5300 BONN
GERMANY, F.R.

HASSAN S M DR
HELWAN INST OF ASTRO
NOMY AND GEOPHYSICS
HELWAN NEAR CAIRO
ARAB REP. OF EGYPT

HATTORI AKIRA A PROF
HIDA OBSERVATORY
KAMITAKARA
GIFU KEN 506 13
JAPAN

HAUCK BERNARD PROF
INST D ASTRONOMIE
UNIVERSITE LAUSANNE
CH-1290 CHAVANNES
DES BOIS
SWITZERLAND

HAUG EBERHARD DOZ DR
MOZARTSTRASSE 20
D-7418 METZINGEN
GERMANY, F.R.

HAUG ULRICH PROF
HAMBURGER STERNWARTE
GOJENBERGSWEG 112
D-2050 HAMBURG 80
GERMANY, F.R.

HAUGE OIVIND DR
INSTITUTE OF THEORE-
TICAL ASTROPHYSICS
UNIVERSITY OF OSLO
BLINDERN OSLO 3
NORWAY

HAUPT HERMANN F PROF
ASTRONOMISCHES INST
DER UNIVERSITAET
UNIVERSTAETSPLATZ 5
A-8010 GRAZ
AUSTRIA

HAUPT RALPH F
3170 ADDERLEY COURT
SILVER SPRING
MD 20906
U.S.A.

HAUPT WOLFGANG DR
GIRONDELLE 105
D-4630 BOCHUM
GERMANY, F.R.

HAVLEN ROBERT J DR
NAT RADIO ASTRON OBS
EDGEMONT ROAD
CHARLOTTESVL VA22901
U.S.A.

HAVNES OVE DR
AURORAL OBSERVATORY
UNIVERSITY OF TROMSO
N-9000 TROMSO
NORWAY

HAWARDEN T G DR
UK 48 INCH SCHMIDT
TELESCOPE UNIT
PRIVATE BAG
COONABARAN NSW 2857
AUSTRALIA

HAWKING STEPHEN W PROF
DEPT APPLIED MATHS
AND THEOR PHYSICS
SILVER STREET
CAMBRIDGE CB3 9EW
U.K.

HAWKINS G S DR
16 GARDEN ROAD
WELLESLEY MA 02181
U.S.A.

HAYAKAWA SACHIO PROF
DEPT OF PHYSICS
NAGOYA UNIVERSITY
FUROKO CHIKUSAKU
NAGOYA 464
JAPAN

HAYASHI CHUSHIRO PROF
PHYSICS DEPT
FACULTY OF SCIENCE
KYOTO UNIVERSITY
KYOTO 606
JAPAN

HAYES DONALD S DR
1807 E WAVERLY ST
TUCSON AZ 85719
U.S.A.

HAYLI AVRAM PROF
OBSERVATOIRE DE LYON
F-69230 ST GENIS
LAVAL
FRANCE

HAYMES ROBERT C PROF
DEPT OF SPACE PHY-
SICS AND ASTRONOMY
RICE UNIVERSITY
HOUSTON TX 77001
U.S.A.

HAYNES RAYMOND F PROF
CSIRO
DIV OF RADIOPHYSICS
P O BOX 76
EPPING N S W 2121
AUSTRALIA

HAZARD CYRIL DR
INST OF ASTRONOMY
MADINGLEY ROAD
CAMBRIDGE CB3 OHE
U.K.

HAZER S DR
FACULTY OF SCIENCE
DEPT OF ASTRONOMY
P K 21
BORNOVA IZMIR
TURKEY

HAZLEHURST J DR
ASTRONOMY CENTRE
UNIVERSITY OF SUSSEX
FALMER
BRIGHTON BN1 9QH
U.K.

HEAP SARA R DR
CODE 672 GODDARD
SPACE FLIGHT CENTER
GREENBELT MD 20771
U.S.A.

HEARN ANTHONY G DR
STERREKUNDIG INST
SERVAAS BOLWERK 13
UTRECHT
THE NETHERLANDS

HEARNSHAW JOHN B DR
DEPT OF PHYSICS
UNIVERSITY OF
CANTERBURY
CHRISTCHURCH
NEW ZEALAND

HECK ANDRE DR
VILLAFRANCA SATEL-
LITE TRACKING STAT
APARTADO 54065
MADRID
SPAIN

HECKMANN OTTO PROF DR
WOHNSTIFT GOETTINGEN
APP B 1403
CHARLOTTENB.STR 29
D-3400 GOETTINGEN
GERMANY, F.R.

HEDDLE DOUGLAS W O PROF
DEPT OF PHYSICS
ROYAL HOLLOWAY
COLLEGE
EGHAM SURREY TW200EX
U.K.

HEDEMAN E RUTH
MCMATH HULBERT
OBSERVATORY
895N LAKE ANGELUS RD
PONTIAC MI 48055
U.S.A.

HEESCHEN DAVID S DR DIR
NAT RADIO ASTR OBS
EDGEMONT RD
CHARLOTTESVILLE
VA 22901
U.S.A.

HEFELE HERBERT PH D
MPI FUER ASTRONOMIE
KOENIGSTUHL
D-6900 HEIDELBERG 1
GERMANY, F.R.

HEFFERLIN RAY A PROF
PHYSICS DEPT
S MISSISSIPPI COLL
DRAWER H
COLLEGEDALE TN 37315
U.S.A.

HEGGIE DOUGLAS C DR
UNIV OF EDINBURGH
EDINBURGH EH9 3JZ
U.K.

HEGYI DENNIS J ASS PROF
RANDALL LABORATORY
UNIVERSITY OF
MICHIGAN
ANN ARBOR MI 48109
U.S.A.

HEIDMANN JEAN DR
OBS DE MEUDON
F-92190 MEUDON
FRANCE

HEIDMANN NICOLE DR
INST D ASTROPHYSIQUE
98 BIS BLVD ARAGO
F-75014 PARIS
FRANCE

HEILES CARL PROF
ASTRONOMY DEPT
UNIVERSITY OF
CALIFORNIA
BERKELEY CA 94720
U.S.A.

HEILESEN POUL MARTIN DR
BOULEVARDEN 14 2 16
DK-2800 LYNGBY
DENMARK

HEINTZ WULFF D DR
DEPT OF ASTRONOMY
SWARTHMORE COLLEGE
SWARTHMORE PA 19081
U.S.A.

HEINTZE J R W DR
STERREWACHT
SONNENBORGH
SERVAAS BOLWERK 13
UTRECHT
THE NETHERLANDS

HEISE JOHN DR
SPACE RESEARCH LAB
BENELUXLAAN 21
3527 HS UTRECHT
THE NETHERLANDS

HEISER ARNOLD M DR
DYER OBSERVATORY
VANDERBILT UNIV
BOX 1803-STA B
NASHVILLE TN 37235
U.S.A.

HEKELA JAN DR
ASTRONOMICAL INST
CZECHOSLOVAK ACAD
OF SCIENCES
251 65 ONDREJOV
CZECHOSLOVAKIA

HELALI YHYA E DR
HELWAN OBSERVATORY
HELWAN-CAIRO
ARAB REP. OF EGYPT

HELFER H LAWRENCE PROF
DEPT OF PHYSICS AND
ASTRONOMY
UNIVERSITY ROCHESTER
ROCHESTER NY 14627
U.S.A.

HELLER MICHAEL DR
ASTRONOMICAL OBS
UL ORLA 171
30-244 KRAKOW
POLAND

HELLWIG HELMUT WILHELM DR
NAT BUREAU STANDARDS
TIME & FREQ DIV
325 BROADWAY RM 2031
BOULDER CO 80303
U.S.A.

HELMKEN HENRY F DR
SMITHSONIAN
ASTROPHYSICAL OBS
60 GARDEN ST
CAMBRIDGE MA 02138
U.S.A.

HELT BODIL E
UNIVERSITY
OBSERVATORY
OSTER VOLDGADE 3
DK-1350 COPENHAGEN
DENMARK

HEMENWAY CURTIS L PROF
DUDLEY OBS PLAZA 7
1202 TROY
SCHENECTADY RD
LATHAM NY 12110
U.S.A.

HEMMLEB GERHARD DR
ZENTRALINSTITUT
PHYSIK DER ERDE
TELEGRAFENBERG A 17
DDR-15 POTSDAM
G.D.R.

HENIZE KARL G ASTRONAUT
CODE CP
JOHNSON SPACE CENTER
HOUSTON TX 77058
U.S.A.

HENON MICHEL C DR
OBSERVATOIRE DE NICE
B P 252
F-06007 NICE CEDEX
FRANCE

HENOUX JEAN-CLAUDE DR
OBSERVATOIRE
DE MEUDON
F-92190 MEUDON
FRANCE

HENRARD JACQUES PROF
FAC UNIV DE NAMUR
61 RUE DE BRUXELLES
B-5000 NAMUR
BELGIUM

HENRIKSEN RICHARD N DR
ASTRONOMY GROUP
DEPT OF PHYSICS
QUEENS UNIVERSITY
KINGSTON ONT K7L 3N6
CANADA

HENRY RICHARD C. PROF.
PHYSICS DEPARTMENT
JOHNS HOPKINS UNIV.
BALTIMORE MD 21218
U.S.A.

HERBIG GEORGE H DR
LICK OBSERVATORY
UNIVERSITY OF
CALIFORNIA
SANTA CRUZ CA 95064
U.S.A.

HERBST ERIC DR
DEPT OF CHEMISTRY
CLG WILLIAM & MARY
WILLIAMSBG VA 23185
U.S.A.

HERBST WILLIAM DR
ASTRONOMY DEPT
WESLEYAN UNIV
MIDDLETOWN CT 06457
U.S.A.

HERCZEG TIBOR J PROF DR
DEPT OF PHYSICS AND
ASTRONOMY
UNIVERSITY OKLAHOMA
NORMAN OK 73069
U.S.A.

HERGET PAUL DR
3522 HERSCHEL VIEW
CINCINNATI OH 45208
U.S.A.

HERMAN RENEE DR
7 BIS RUE TRUDON
F-92160 ANTONY
FRANCE

HERR RICHARD B DR
PHYSICS DEPT
UNIVERSITY DELAWARE
NEWARK DE 19711
U.S.A.

HERRMANN DIETER DR
ARCHENHOLD OBS
AND PUBLIC OBS
1193 BERLIN-TREPTOW
G.D.R.

HERS JAN
P.O.BOX 48
6573 SEDGEFIELD
SOUTH AFRICA

HERSHEY JOHN L DR
SPROUL OBSERVATORY
SWARTHMORE COLLEGE
SWARTHMORE PA 19081
U.S.A.

HERZBERG GERHARD DR
HERZBERG INSTITUTE
OF ASTROPHYSICS
NAT RES COUNCIL
OTTAWA ONT K1A OR6
CANADA

HESSER JAMES E DR
DOMINION ASTROPHYSI-
CAL OBSERVATORY
5071 W SANNICH RD
VICTORIA BC V8X 3X3
CANADA

HEUDIER J L MR
CERGA
AVENUE COPERNIC
F-06130 GRASSE
FRANCE

HEWISH ANTONY PROF
CAVENDISH LABORATORY
MADINGLEY RD
CAMBRIDGE CB3 OHE
U.K.

HEWITT A V DR
US NAVAL OBSERVATORY
P O BOX 1149
FLAGSTAFF AZ 86001
U.S.A.

HEY JAMES STANLEY DR
4 SHORTLANDS CLOSE
WILLINGDON
EASTBOURNE
SUSSEX BN22 0JE
U.K.

HEYDEN FRANCIS J SJ DR
MANILA OBSERVATORY
P O BOX 1231
MANILA
REPUBLIC OF PHILIPPINES

HEYVAERTS JEAN DR
OBSERVATOIRE
DE MEUDON
F-92190 MEUDON
FRANCE

HIBBS ALBERT R MGR PLANS
JET PROPULSION
LABORATORY
4800 OAK GROVE DRIVE
PASADENA CA 91103
U.S.A.

HICKSON PAUL DR
DEPT GEOPHYS ASTRON
UNIVERSITY OF B C
VANCOUVER BC V6T 1W5
CANADA

HIDAJAT BAMBANG PROF DR
BOSSCHA OBSERVATORY
LEMBANG JAVA
INDONESIA

HIDALGO MIGUEL A DR
FACULTAD DE CIENCIAS
CIUDAD UNIVERSITARIA
ZARAGOZA
SPAIN

HIDE RAYMOND PROF
GEOPHYSICAL FLUID
DYNAMICS LABORATORY
METEOROLOGICAL OFF
BRACKNELL BERKSHIRE
U.K.

HIEI EIJIRO DR
TOKYO ASTRONOMICAL
OBSERVATORY
OSAWA MITAKA
TOKYO 181
JAPAN

HIGGS LLOYD A DR
HERZBERG INSTITUTE
OF ASTROPHYSICS
NAT RES COUNC CANADA
OTTAWA ONT K1A 0R6
CANADA

HILDITCH RONALD W DR
UNIVERSITY
OBSERVATORY
BUCHANAN GARDENS
ST ANDREWS KY16 9LZ
U.K.

HILDNER ERNEST DR
HIGH ALTITUDE
OBSERVATORY
P O BOX 3000
BOUDER CO 80307
U.S.A.

HILF EBERHARD R H PH D
PAUL-WAGNERSTR 56
D-6100 DARMSTADT
GERMANY, F.R.

HILL GRAHAM DR
DOMINION ASTRO OBS
NAT RES COUNC CANADA
5071 WEST SAANICH RD
VICTORIA BC V8X 3X3
CANADA

HILL HENRY ALLEN DR
DEPT OF PHYSICS
BLDG 81
UNIV OF ARIZONA
TUCSON AZ 85721
U.S.A.

HILL PHILIP W DR
UNIVERSITY
OBSERVATORY
ST ANDREWS
FIFE KY16 9LZ
U.K.

HILLEBRANDT WOLFGANG PH D
MPI FUER PHYSIK UND
ASTROPHYSIK
FOEHRINGER RING 6
D-8000 MUENCHEN 40
GERMANY, F.R.

HILLIARD R DR
STEWARD OBSERVATORY
UNIV OF ARIZONA
TUCSON AZ 85721
U.S.A.

HILLS JACK G DR
DEPT ASTRON ASTROPHY
MICHIGAN STATE UNIV
PHYS-ASTROPHYS BLDG
E LANSING MI 48824
U.S.A.

HILLS RICHARD E DR
CAVENDISH LABORATORY
MADINGLEY RD
CAMBRIDGE CB3 0HE
U.K.

HILTNER W ALBERT PROF
DEPT OF ASTRONOMY
UNIVERSITY MICHIGAN
ANN ARBOR MI 48109
U.S.A.

HINTEREGGER HANS E DR
AIR FORCE GEOPHYSICS
LABORATORY
HANSCOM AFB MA 01731
U.S.A.

HINTZEN PAUL MICHAEL N DR
NASA-GSFC
CODE 681
GREENBELT MD 20771
U.S.A.

HIPPELEIN HANS H DR
MPI FUER ASTRONOMIE
D-6900 HEIDELBERG
GERMANY, F.R.

HIRABAYASHI HISASHI DR
TOKYO ASTRONOMICAL
OBSERVATORY
OSAWA MITAKA
TOKYO 181
JAPAN

HIRAI MASANORI DR
FUKUOKA UNIVERSITY
OF EDUCATION
729 MUNAKATA
FUKUOKA 811-41
JAPAN

HIRAYAMA TADASHI DR
TOKYO ASTRONOMICAL
OBSERVATORY
OSAWA MITAKA
TOKYO 181
JAPAN

HIROSE HIDEO PROF
TOKYO ASTRONOMICAL
OBSERVATORY
OSAWA 2-CHOME MITAKA
TOKYO 181
JAPAN

HIRST WILLIAM P
1 CLIFFORD CRESCENT
BERGVLIET CP 7800
SOUTH AFRICA

HIRTH WOLFGANG ERNST PH D
THEODOR-HEUSSTR 18
D-5354 WEILERSWIST
GERMANY, F.R.

HITOTSUYANAGI JUICHI PROF
DEPT OF ASTRONOMY
TOHOKU UNIVERSITY
ARAMAKI SENDAI 980
JAPAN

HJALMARSON AKE G DR
ONSALA SPACE
OBSERVATORY
S-430 34 ONSALA
SWEDEN

HJELLMING ROBERT M DR
NATIONAL RADIO ASTRO
NOMY OBSERVATORY
P O BOX 0
SOCORRO NM 87801
U.S.A.

HOAG ARTHUR A DR
LOWELL OBSERVATORY
BOX 1269
FLAGSTAFF AZ 86002
U.S.A.

HOANG BINH DY DR
OBSERVATOIRE
DE MEUDON
F-92190 MEUDON
FRANCE

HOBBS LEWIS M DR
YERKES OBSERVATORY
UNIVERSITY OF
CHICAGO
WILLIAMS BAY WI53191
U.S.A.

HOBBS R W DR
NASA-GODDARD SPACE
FLIGHT CENTER
CODE 685
GREENBELT MD 20771
U.S.A.

HODGE PAUL W PROF
ASTRONOMY FM20
UNIVERSITY OF
WASHINGTON
SEATTLE WA 98195
U.S.A.

HOEG ERIK LEKTOR
COPENHAGEN OBS
BRORFELDE
DK-4340 TOLLOSE
DENMARK

HOEGBOM JAN A DR
STOCKHOLM
OBSERVATORY
S-133 00
SALTSJOEBADEN
SWEDEN

HOEGLUND BERTIL PROF
ONSALA SPACE
OBSERVATORY
S-430 34 ONSALA
SWEDEN

HOEKSTRA ROEL DR
SPACE RESEARCH
LABORATORY
UTRECHT
THE NETHERLANDS

HOEY MICHAEL J DR
PHYSICS DEPT
UNIVERSITY COLLEGE
BELFIELD
DUBLIN 4
IRELAND

HOFFLEIT E DORRIT DR
DEPT OF ASTRONOMY
BOX 2023
YALE STATION
NEW HAVEN CT 06520
U.S.A.

HOFFMAN JEFFREY ALAN DR
NASA-JSC
CODE CB-4
HOUSTON TX 77058
U.S.A.

HOGG DAVID E DR
NAT RADIO ASTR OBS
EDGEMONT RD
CHARLOTTESVILLE
VA 22901
U.S.A.

HOHL FRANK DR
NASA MS-160
LANGLEY RESEARCH
CENTER
HAMPTON VA 23665
U.S.A.

HOLDEN FRANK
70 CANNOCK ROAD
STAFFORD ST17 0QQ
U.K.

HOLLENBACH DAVID JOHN DR
ASTRONOMY DEPT
UNIV OF CALIFORNIA
BERKELEY CA 94720
U.S.A.

HOLLIS JAN MICHAEL DR
NAT RADIO ASTRON OBS
2010 NORTH FORBES BD
TUCSON AZ 85705
U.S.A.

HOLLOWAY NIGEL J DR
ASTRONOMY CENTRE
UNIV OF SUSSEX
FALMER
BRIGHTON BN1 9QH
U.K.

HOLMBERG ERIK B PROF
ENELIDEN 2
S-433 00 PARTILLE
SWEDEN

HOLWEGER HARTMUT PROF DR
INST THEOR PHY STERN
NEUE UNIV PHYS ZENTR
OLSHAUSENST GEBN 61C
D-2300 KIEL 1
GERMANY, F.R.

HOLZER THOMAS EDWARD DR
HIGH ALTITUDE OBS
NCAR
P O BOX 3000
BOULDER CO 80307
U.S.A.

HONEYCUTT R KENT PROF
ASTRONOMY DEPT
SWAIN HALL WEST
INDIANA UNIVERSITY
BLOOMINGTON IN 47405
U.S.A.

HONG SEUNG SOO DR
DEPT OF ASTRONOMY
COLLEGE NATURAL SCS
SEOUL NATIONAL UNIV
SEOUL 150
REPUBLIC OF KOREA

HOOGHOUDT B G IR
PRINSENLAAN 10
2341 KT OEGSTGEEST
THE NETHERLANDS

HOPPE J A PROF DR
SONNENBERGSTR 12
DDR 69 JENA
G.D.R.

HORAK TOMAS B DR
GEOFYZIKA NP
JECNA 29A
61200 BRNO
CZECHOSLOVAKIA

HORAK ZDENEK PROF DR
VIETNAMSKA 2
16000 PRAHA 6
CZECHOSLOVAKIA

HORI GENICHIRO PROF
DEPT OF ASTRONOMY
UNIVERSITY OF TOKYO
BUNKYO TOKYO 113
JAPAN

HOROWITZ GERALD PROF
RACAH INST OF PHYS
HEBREW UNIVERSITY
JERUSALAM
ISRAEL

HORSKY JAN DR
DEPT OF THEOR PHYS
PURKYNE UNIVERSITY
KOTLARSKA 2
611 37 BRNO
CZECHOSLOVAKIA

HORSKY Z DR
ASTRONOMICAL INST
CSAV
BUDECSKA 6
120 23 PRAHA
CZECHOSLOVAKIA

HORTON BRIAN H DR
DEPT OF PHYSICS
UNIV OF ADELAIDE
GPO BOX 498
ADELAIDE S A 5001
AUSTRALIA

HOSHI REIUN DR
DEPT OF PHYSICS
RIKKYO UNIVERSITY
NISHIIKEBUKURO 3-CH
TOSHIMA TOKYO 171
JAPAN

HOSKIN MICHAEL A DR
CHURCHILL COLLEGE
CAMBRIDGE CB3 ODS
U.K.

HOSKING ROGER J PROF
MATHEMATICS DEPT
UNIVERSITY WAIKATO
HAMILTON
NEW ZEALAND

HOSOKAWA YOSHIMASA H PROF
DEPT OF PHYSICS
YAMAGATA UNIVERSITY
KOSHIRAKAWA 1 CHOME
YAMAGATA CITY 990
JAPAN

HOSOYAMA KENNOSHUKE DR
INTERNATIONAL LATITU
DE OBSERVATORY
HOSHIGAOKA 2-CHOME
MIZUSAWA IWATE 023
JAPAN

HOTINLI METIN DR
UNIVERSITE
RASATHANESI
UNIVERSITE ISTANBUL
TURKEY

HOUCK JAMES R
ASTRONOMY DEPT
CORNELL UNIVERSITY
220 SPACE SCIENCE BD
ITHACA NY 14853
U.S.A.

HOUK NANCY DR
DEPT OF ASTRONOMY
UNIV OF MICHIGAN
1045 PHYS-ASTRO BLDG
ANN ARBOR MI 48109
U.S.A.

HOUSE FRANKLIN C DR
ESA-ESOC
ROBERT-BOSCH-STR 5
D-6100 DARMSTADT
GERMANY, F.R.

HOUSE LEWIS L DR
HIGH ALTITUDE OBS
NAT CTR ATMOSPH RES
P O BOX 3000
BOULDER CO 80303
U.S.A.

HOUTGAST JAKOB DR
STERREWACHT
ZONNENBURG 2
3512 NK UTRECHT
THE NETHERLANDS

HOUZIAUX L PROF
INST D'ASTROPHYSIQUE
5 AV DE COINTE
B-4200 COINTE-OUGREE
BELGIUM

HOVENIER J W DR
FREE UNIVERSITY
DEPT PHYSICS/ASTRON
DE BOELELAAN
1081 HV AMSTERDAM
THE NETHERLANDS

HOWARD III WILLIAM E DR
DIV ASTRONOMICAL SC
NAT SCIENCE FOUND
1800 G STREET NW
WASHINGTON DC 20550
U.S.A.

HOWARD ROBERT F DR
HALE OBSERVATORIES
813 SANTA BARBARA ST
PASADENA CA 91101
U.S.A.

HOWARD W MICHAEL DR
LAWRENCE LIVERMORE
LABORATORY L 71
LIVERMORE CA 94550
U.S.A.

HOWSE H DEREK
NATIONAL MARITIME
MUSEUM GREENWICH
LONDON SE10 9NF
U.K.

HOYLE FRED SIR
COCKLEY MOOR
DOCKRAY
PENRITH
CUMBRIA CA11 CLG
U.K.

HOYNG PETER DR
SPACE RESEARCH LAB
BENELUXLAAN 21
UTRECHT
THE NETHERLANDS

HUA CHON TRUNG DR
VILLA LA MAJORICA
LES EMEUS
F-13190 ALLAUCH
FRANCE

HUBBARD WILLIAM B PROF
DEPT OF PLANETARY
SCIENCES
UNIVERSITY ARIZONA
TUCSON AZ 85721
U.S.A.

HUBE DOUGLAS P DR
DEPT OF PHYSICS
UNIVERSITY ALBERTA
EDMONTON ALTA T6G2J1
CANADA

HUBENET HENRI DR
STERRENWACHT
SONNENBORGH
ZONNENBURG 2
UTRECHT 2500
THE NETHERLANDS

HUBER MARTIN C DR
ATOM & ASTROPHYSICS
GROUP
ETH ZENTRUM
CH 8092 ZUERICH
SWITZERLAND

HUBERT-DELPLACE
ANNE-MARIE DR
OBSERVATOIRE
DE MEUDON
F-92190 MEUDON
FRANCE

HUCHRA JOHN PETER DR
CTR FOR ASTROPHYSICS
60 GARDEN STREET
CAMBRIDGE MA 02138
U.S.A.

HUCHTMEIER WALTER K DR
MPI FUER
RADIOASTRONOMIE
AUF DEM HUEGEL 69
D-5300 BONN 1
GERMANY, F.R.

HUDSON HUGH S DR
PHYSICS C-011
UCSD
LA JOLLA CA 92093
U.S.A.

HUEBNER WALTER F DR
T-4
LOS ALAMOS SCIENTI-
FIC LABORATORY
LOS ALAMOS NM 87545
U.S.A.

HUGHES DAVID W DR
DEPT OF PHYSICS
THE UNIVERSITY
SHEFFIELD S3 7RH
U.K.

HUGHES JAMES A DR
US NAVAL OBSERVATORY
WASHINGTON DC 20390
U.S.A.

HUGHES VICTOR A PROF
DEPT OF PHYSICS
QUEENS UNIVERSITY
KINGSTON ONT K7L 3N6
CANADA

HUGUENIN G RICHARD
DEPT PHYS AND ASTRON
UNIV OF MASS
GRADUATE RES CTR
AMHERST MA 01002
U.S.A.

HULSBOSCH A N M DR
STERRENKUNDIG INST
KATH UNIVERSITEIT
NIJMEGEN
THE NETHERLANDS

HUMMER DAVID G DR
JOINT INST F LABORA-
TORY ASTROPHYSICS
BOULDER CO 80309
U.S.A.

HUMPHREYS CURTIS JUDSON
DR
6065 RIVERSIDE AVE
RIVERSIDE CA 92506
U.S.A.

HUMPHREYS ROBERTA M PROF
ASTRONOMY DEPT
UNIVERSITY OF
MINNESOTA
MINNEAPOLIS MN 55455
U.S.A.

HUMPHRIES C M DR
ROYAL OBSERVATORY
BLACKFORD HILL
EDINBURGH EH9 3HJ
U.K.

HUNDHAUSEN A DR
HIGH ALTITUDE OBS
BOULDER CO 80302
U.S.A.

HUNGER KURT PROF
INST THEOR PHY STERN
NEUE UNIV PHYS ZENTR
OLSHAUSENST CB N 61C
D-2300 KIEL 1
GERMANY, F.R.

HUNSTEAD RICHARD W DR
SCHOOL OF PHYSICS
UNIVERSITY OF SYDNEY
SYDNEY N S W 2006
AUSTRALIA

HUNT GARRY E DR
LAB PLANETARY ATMOSP
DEPT PHYS ASTRONOMY
GOWER STREET
LONDON WC1E 6BT
U.K.

HUNT ROLAND DR
DEPT OF MATHEMATICS
UNIV OF STRATHCLYDE
26 RICHMOND STREET
GLASGOW G1 1XH
U.K.

HUNTEN DONALD M PROF
LUNAR AND PLANETARY
LABORATORY
UNIVERSITY ARIZONA
TUCSON AZ 85721
U.S.A.

HURNIK HIERONIM PROF
OBSERWATORIUM
ASTRONOMICZNE UAM
SLONECZNA 36
60-286 POZNAN
POLAND

HUTCHEON RICHARD J DR
X-RAY ASTRONOMY GR
PHYSICS DEPT
UNIV OF LEICESTER
LEICESTER LE1 7RH
U.K.

HYLAND A R HARRY DR
MT STROMLO
OBSERVATORY
WODEN PO ACT 2606
AUSTRALIA

HYUN JONG-JUNE PROF
SEOUL NAT UNIVERSITY
SINLIM-DONG
KWANAK-KU
SEOUL 151
REPUBLIC OF KOREA

IBADINOV KHURSANDKUL DR
INST OF ASTROPHYSICS
ACAD OF SCS OF THE
TADJIK SSR
734670 DUSHANBE
U.S.S.R.

IBEN ICKO JR PROF
ASTRONOMY DEPT
UNIVERSITY OF
ILLINOIS
URBANA IL 61801
U.S.A.

IDLIS G M DR
INSTITUTE F HISTORY
OF SCI AND TECHNOLO-
GY USSR ACAD OF SCI
103012 MOSCOW
U.S.S.R.

IKEUCHI SATORU DR
DEPT OF PHYSICS
HOKKAIDO UNIVERSITY
SAPPORO 060
JAPAN

ILL MARTON J DR
TOTH KALMAN U 19
OBSERVATORY
6501 BAJA
HUNGARY

IMBERT MAURICE DR
OBSERVATOIRE
DE MARSEILLE
2 PLACE LE VERRIER
F-13004 MARSEILLE
FRANCE

INNANEN KIMMO A PROF
CROSS PHYSICS DEPT
YORK UNIVERSITY
4700 KEELE STREET
DOWNSVIEW ONT M3J1P3
CANADA

HUNTER CHRISTOPHER PROF
MATHEMATICS DEPT
FLORIDA STATE
UNIVERSITY
TALLAHASSEE FL 32306
U.S.A.

HURUHATA MASAAKI PROF
TOKYO ASTRONOMICAL
OBSERVATORY
OSAWA 2-CHOME
MITAKA TOKYO 181
JAPAN

HUTCHINGS JOHN B DR
DOMINION ASTRO OBS
NAT RES COUNC CANADA
5071 WEST SAANICH RD
VICTORIA BC V8X 3X3
CANADA

HYNEK J ALLEN PROF
DEARBORN OBSERVATORY
NORTHWESTERN
UNIVERSITY
EVANSTON IL 60201
U.S.A.

IANNA PHILIP A
PO BOX 3818
CHARLOTTESVILLE
VA 22903
U.S.A.

IBANOGLU C DR
EGE UNIVERSITY
OBSERVATORY
BORNOVA IZMIR
TURKEY

ICHIMARU SETSUO DR
DEPT OF PHYSICS
UNIVERSITY OF TOKYO
BUNKYO-KU
TOKYO 113
JAPAN

IIJIMA SHIGETAKA PROF
TOKYO ASTRONOMICAL
OBSERVATORY
OSAWA MITAKA TOK 181
JAPAN

IKHSANOV ROBERT N DR
PULKOVO OBSERVATORY
196140 LENINGRAD
U.S.S.R.

ILLES ALMAR ERZSEBET DR
KONKOLY OBSERVATORY
BOX 67
1525 BUDAPEST
HUNGARY

IMSHENNIK V S DR
INST THEOR EXP PHYS
CHEREMUSHKINSKAJA 25
117259 MOSCOW
U.S.S.R.

INOUE TAKESHI PROF
KYOTO SANGYO
UNIVERSITY
KYOTO 603
JAPAN

HUNTER JAMES H DR
5641 DEL PRADO DR
APT 237
TAMPA FL 33617
U.S.A.

HURUKAWA KIICHIRO DR
TOKYO ASTRONOMICAL
OBSERVATORY
OSAWA MITAKA
181 TOKYO
JAPAN

HYDER C L DR
NASA SOLAR GROUP
UNIV OF NEW MEXICO
800 YALE BLVD N E
ALBUQUERQUE NM 87106
U.S.A.

HYSOM E J
65 DALKEITH RD
HARPENDEN
HERTS AL5 5PP
U.K.

IANNINI GUALBERTO DR
OBSERVATORIO ASTRONO
MICO DE CORDOBA
LAPRIDA 854
5000 CORDOBA
ARGENTINA

IBBETSON PETER AARON DR
WISE OBSERVATORY
TEL AVIV UNIVERSITY
TEL AVIV
ISRAEL

ICKE V DR
DEPT OF ASTRONOMY
UNIV OF MINNESOTA
116 CHURCH ST S E
MINNEAPOLIS MN 55455
U.S.A.

IKEDA TETSURO DR
1-18 TENMONDAI DORI
MIZUSAWA IWATE 023
JAPAN

IKHSANOVA VERA N DR
PULKOVO OBSERVATORY
196140 LENINGRAD
U.S.S.R.

ILLINGWORTH GARTH D DR
KITT PEAK NAT OBS
P O BOX 26732
TUCSON AZ 85726
U.S.A.

INGRAO HECTOR C
CAMBRIDGE SYST CORP
545 TECHNOLOGY SQ
LOWER LEVEL
CAMBRIDGE MA 02139
U.S.A.

IOANNISIANI B K DR
MAIN ASTRONOMICAL
OBSERVATORY PULKOVO
196140 LENINGRAD
U.S.S.R.

ICSHPA B A DR
INSTITUTE OF TERRE-
STRIAL MAGNETISM AND
IONOSPHERE
142092 AKADEMGORODOK
U.S.S.R.

IPSER JAMES R PROF
LASR
933 E 56TH ST
CHICAGO IL 60637
U.S.A.

IRELAND JOHN G DR
C-O 13 GORDON RD
BELVEDERE
KENT DA17 6EA
U.K.

IRIARTE B MR
INST NAC DE ASTROFIS
OPT Y ELECTRONICA
APT POSTALES 216 51
PUEBLA PUE
MEXICO

IRVINE WILLIAM M PROF
ASTRONOMY GRC
TOWER B
UNIVERSITY MASS
AMHERST MA 01003
U.S.A.

IRWIN JOHN B PROF
2744 TYNDALL AVE
TUCSON AZ 85719
U.S.A.

ISAAK GEORGE R DR
DEPT OF PHYSICS
UNIV OF BIRMINGHAM
P O BOX 363
BIRMINGHAM B15 2TT
U.K.

ISHIDA KEIICHI DR
TOKYO ASTRONOMICAL
OBSERVATORY
2-21-1 OSAWA MITAKA
JAPAN

ISHIZAWA TOSHISUKE A PROF
DEPT OF ASTRONOMY
UNIVERSITY OF KYOTO
KYOTO 606
JAPAN

ISOBE SHUZO DR
TOKYO ASTRONOMICAL
OBSERVATORY
181 MITAKA TOKYO
JAPAN

ISRAEL WERNER PROF
PHYSICS DEPT
UNIVERSITY OF
ALBERTA
EDMONTON ALTA T6G2E1
CANADA

ISSA ALI DR
HELWAN OBSERVATORY
HELWAN-CAIRO
ARAB REP. OF EGYPT

ISSERSTEDT JOERG DR
INSTITUT ASTRONOMIE
UND ASTROPHYSIK
AM HUBLAND
D-7800 WUERZBURG
GERMANY, F.R.

ITO KENSUKE A PROF
RIKKYO UNIVERSITY
NISHI-IKEBUKURO
TOKYO 171
JAPAN

ITOH NAOKI DR
DEPT OF PHYSICS
SOPHIA UNIVERSITY
7 KIOICHO CHIYODA-KU
TOKYO 102
JAPAN

IVANCHUK VICTOR I DR
KIEV UNIVERSITY
OBSERVATORNAYA 3
252053 KIEV
U.S.S.R.

IVANOV G R DR
DEPT OF ASTRONOMY
ANTON IVANOV ST 5
P O 120
1126 SOFIA
BULGARIA

IVANOV VSEVOLOD V DR
ASTRONOMICAL OBS
LENINGRAD UNIVERSITY
10TH LINIJA 33
199178 LENINGRAD
U.S.S.R.

IVANOV-KHOLODNY G S DR
INSTITUTE OF APPLIED
GEOPHYSICS USSR ACA-
DEMY OF SCIENCES
107150 MOSCOW
U.S.S.R.

IVES JOHN CH DR
MULLARD SPACE SC LAB
UNIV COLLEGE LONDON
LONDON
U.K.

IWANISZEWSKA CECYLIA DR
INST OF ASTRONOMY
N COPERNICUS UNIV
CHOPINA 12 18
87-100 TORUN
POLAND

IWANOWSKA WILHELMINA DR
INST OF ASTRONOMY
CHOPINA 12 18
87-100 TORUN
POLAND

IWASAKI KYOSUKE DR
KWASAN OBSERVATORY
YAMASHINA KYOTO 607
JAPAN

IZVEKOV V A DR
INST OF THEORETICAL
ASTRONOMY
10 KUTUZOV QUAY
192187 LENINGRAD
U.S.S.R.

JAAKKOLA TOIVO S
OBSERVATORY
TAHTITORNINMAKI
SF-00130 HELSINKI
FINLAND

JACCHIA LUIGI G DR
CENTER FOR
ASTROPHYSICS
60 GARDEN ST
CAMBRIDGE MA 02138
U.S.A.

JACKISCH G DR
ZENTRALINSTITUT FUER
ASTROPHYSIK
STERNWARTE SONNEBERG
DDR-64 SONNEBERG
G.D.R.

JACKSON C MR
OBS AUSTRAL YALE -
COLUMBIA
C DE C 263
SAN JUAN
ARGENTINA

JACKSON J C DR
16 THE PARK
NEWARK
NOTTINHAMS NG24 1SO
U.K.

JACKSON PAUL DR
INST FUER ASTRONOMY
DER UNIV WIEN
TUERKENSCHANZSTR 17
A-1180 WIEN
AUSTRIA

JACKSON PETER DOUGLAS DR
ASTRONOMY PROGRAM
UNIV OF MARYLAND
COLLEGE PK MD 20742
U.S.A.

JACKSON WILLIAM M DR
DEPT OF CHEMISTRY
HOWARD UNIVERSITY
WASHINGTON DC 20059
U.S.A.

JACOBS KENNETH C DR
CENTER 344
BELL LABORATORIES
600 MOUNTAIN AVENUE
MURRAY HILL NJ 07974
U.S.A.

JACOBSEN THEODOR S PROF
6205 17TH AVE NE
SEATTLE WA 98115
U.S.A.

JACQUINOT P DR
LAB AIME COTTON
BAT 505
UNIVERSITE PARIS SUD
F-91405 ORSAY CEDEX
FRANCE

JAEGER FRIEDRICH W PROF
ZENTRALINSTITUT FUER
SOLARTERR PHYSIK
DDR-15 POTSDAM
G.D.R.

JAFFE WALTER JOSEPH DR
NAT RADIO ASTRONOMY
OBSERVATORY
EDGEMONT ROAD
CHARLOTTESVL VA22901
U.S.A.

JAHREISS HARTMUT DR
ASTRONOMISCHES
RECHEN INSTITUT
MOENCHHOFSTR 12-14
D-6900 HEIDELBERG 1
GERMANY, F.R.

JAKIMIEC JERZY DR
ASTRONOMICAL INST
KOPERNIKA 11
51 622 WROCLAW
POLAND

JAMAR CLAUDE A J DR
8 RUE DE BATTY
B-4200 COINTE-OUGREE
BELGIUM

JAMES JOHN F MR
SCHUSTER LABORATORY
THE UNIVERSITY
MANCHESTER M13 9PL
U.K.

JAMES RICHARD A DR
DEPT OF ASTRONOMY
THE UNIVERSITY
MANCHESTER M13 9PL
U.K.

JAMESON RICHARD F DR
ASTRONOMY DEPT
LEICESTER UNIVERSITY
LEICESTER LE1 7RH
U.K.

JANES KENNETH A DR
ASTRONOMY DEPT
BOSTON UNIVERSITY
725 COMMONWEALTH AVE
BOSTON MA 02215
U.S.A.

JANICZEK PAUL M DR
US NAVAL OBSERVATORY
WASHINGTON DC 20390
U.S.A.

JANKOVICS ISTVAN DR
KONKOLY OBSERVATORY
BOX 67
1525 BUDAPEST
HUNGARY

JARNEFELT GUSTAF J PROF
OHJAAJANTIE 3 B 18
SF-00400 HELSINKI 40
FINLAND

JARRETT ALAN H PROF
BOYDEN OBSERVATORY
P O BOX 334
BLOEMFONTEIN 9300
SOUTH AFRICA

JARZEBOWSKI TADEUSZ DR
ASTRONOMICAL
INSTITUTE
KOPERNIKA 11
51-622 WROCLAW
POLAND

JASCHEK CARLOS O R PROF
OBSERVATOIRE
11 R DE L UNIVERSITE
F-67000 STRASBOURG
FRANCE

JASCHEK MERCEDES DR
OBSERVATOIRE
11 R DE L UNIVERSITE
F-67000 STRASBOURG
FRANCE

JASTROW ROBERT
INST F SPACE STUDIES
2880 BROADWAY
NEW YORK NY 10025
U.S.A.

JAUNCEY DAVID L DR
CSIRO DIVISION OF
RADIOPHYSICS
PO BOX 76
EPPING NSW 2121
AUSTRALIA

JAVET PIERRE PROF
AV DE BEAUMONT 36
CH-1012 LAUSANNE
SWITZERLAND

JAYARAJAN A P MR
ASTROPHYSICAL OBS
KODAIKANAL 3
INDIA

JEFFERIES JOHN T DR
INST FOR ASTRONOMY
2680 WOODLAWN DR
HONOLULU HI 96822
U.S.A.

JEFFERS STANLEY DR
CRESS PHYSICS DEPT
YORK UNIVERSITY
4700 KEELE ST
DOWNSVIEW M3J 1P3
CANADA

JEFFREYS HAROLD PROF SIR
160 HUNTINGDON ROAD
CAMBRIDGE CB3 0LD
U.K.

JEFFREYS WILLIAM H DR
ASTRONOMY DEPT
UNIVERSITY OF TEXAS
AUSTIN TX 78712
U.S.A.

JELLEY JOHN V PHD FRAS
NP-DIV BLDG 8 AERE
HARWELL OXON OX110RA
U.K.

JENKINS EDWARD B DR
PRINCETON UNIVERSITY
OBSERVATORY
PRINCETON NJ 08540
U.S.A.

JENKINS L F MS
YALE UNIV OBS
BOX 2023
YALE STATION
NEW HAVEN CT 06520
U.S.A.

JENKNER HELMUT DR
UNIV-STERNWARTE
TUERKENSCHANZSTR 17
A-1180 WIEN
AUSTRIA

JENNER DAVID C PROF
DEPT OF ASTRONOMY
UNIV.OF WASHINGTON
SEATTLE, WA 98195
U.S.A.

JENNINGS R E DR
DEPT PHYSICS AND AST
UNIVERSITY COLLEGE
GOWER STREET
LONDON WC1E 6BT
U.K.

JENNISON ROGER C PROF
ELECTRONICS LAB
UNIVERSITY OF KENT
CANTERBURY CT2 7NT
U.K.

JENSCH A
PESTALOZZISTR 9
DDR 69 JENA
G.D.R.

JENSEN EBERHART PROF
INSTITUTE OF THEORE-
TICAL ASTROPHYSICS
UNIVERSITY OF OSLO
BLINDERN OSLO 3
NORWAY

JERZYKIEWICZ MIKOLAJ DR
ASTRONOMICAL INST
KOPERNIKA 11
51-622 WROCLAW
POLAND

JOCKERS KLAUS DR
MPI FUER AERONOMIE
POSTFACH 20
D-3411 KATLENBURG-L
GERMANY, F.R.

JOHANSEN KAREN T LEKTOR
IGELSO
DK-4300 HOLBAEK
DENMARK

JOHNSON DONALD R DR
716 TIFFANY COURT
GAITHERSBURG MD20760
U.S.A.

JOHNSON FRED M PROF DR
DEPT PHYS AND ASTRON
CALIF STATE UNIV
2308 E AMERIGE AVE
FULLERTON CA 92634
U.S.A.

JOHNSON HOLLIS R PROF
ASTRONOMY DEPT
INDIANA UNIVERSITY
BLOOMINGTON IN 47405
U.S.A.

JOHNSON HUGH M DR
ORGN 52 12 BLDG 202
3251 HANOVER ST
PALO ALTO CA 94304
U.S.A.

JOHNSON MARTIN DR
PHYSICS DEPT
UNIVERSITY
BIRMINGHAM B15 2TT
U.K.

JOHNSON TORRENCE V DR
JET PROPULSION
LABORATORY
4800 OAK GROVE DRIVE
PASADENA CA 91103
U.S.A.

JOHNSTON KENNETH
CODE 7134
NAVAL RESEARCH
LABORATORY
WASHINGTON DC 20375
U.S.A.

JOKIPII J R PROF
DEPT OF PLANETARY SC
UNIVERSITY OF
ARIZONA
TUCSON AZ 85721
U.S.A.

JOLY FRANCOIS DR
UNIVERSITE DE
BORDEAUX I
40 RUE LAMARTINE
F-33400 TALENCE
FRANCE

JONES ALBERT F MR
31 RANUI ROAD
STOKE
NELSON
NEW ZEALAND

JONES BERNARD J T DR
INST OF ASTRONOMY
MADINGLEY ROAD
CAMBRIDGE
U.K.

JONES BURTON DR
LICK OBSERVATORY
UNIVERSITY
OF CALIFORNIA
SANTA CRUZ CA 95064
U.S.A.

JONES DEREK H P DR
ROYAL GREENWICH
OBSERVATORY
HAILSHAM
EAST SUSSEX BN27 1RP
U.K.

JONES ERIC M
LOS ALAMOS SCIENTI-
FIC LABORATORY
LOS ALAMOS NM 87545
U.S.A.

JONES FRANK CULVER DR
NASA-GSFC
CODE 660
GREENBELT MD 20771
U.S.A.

JONES HARRISON PRICE DR
KITT PEAK NAT OBS
SOLAR STATION
900 N CHERRY AVE
TUCSON AZ 85726
U.S.A.

JONES JAMES DR
DEPT OF PHYSICS
UNIV WESTERN ONTARIO
LONDON ONT N6A 5B9
CANADA

JONES JANET E DR
INST OF ASTRONOMY
MADINGLEY ROAD
CAMBRIDGE
U.K.

JONES THOMAS WALTER DR
DEPT OF ASTRONOMY
UNIV OF MINNESOTA
116 CHURCH ST SE
MINNEAPOLIS MN 55455
U.S.A.

JORDAN CAROLE DR
DEPT THEORETICAL PHY
1 KEBLE ROAD
OXFORD UNIVERSITY
OXFORD OX1 3NP
U.K.

JORDAN H L DR DIREKTOR
INST PLASMAPHYSIK
KERNFORSCHUNGSANLAGE
JUELICH GMBH PF 365
D-5170 JUELICH 1
GERMANY, F.R.

JORDAN STUART D DR
LAB ASTRO SOLAR PHYS
NASA-GODDARD SPACE
FLIGHT CTR CODE 682
GREENBELT MD 20771
U.S.A.

JORGENSEN HENNING E LEKT
UNIVERSITY
OBSERVATORY
OSTERVOLDGADE 3
DK-1350 COPENHAGEN K
DENMARK

JOSEPH J H DR
DEPT OF GEOPHYSICS
AND PLANETARY SC
TEL-AVIV UNIVERSITY
RAMAT AVIV
ISRAEL

JOSEPH ROBERT D DR
BLACKETT LAB
ASTRONOMY GROUP
IMPERIAL COLLEGE
LONDON SW7 2BZ
U.K.

JOSHI MOHAN N DR
RADIO ASTRONOMY
CENTRE OF TIFR
POST BOX 8
OOTACAMUND 643001
INDIA

JOSHI SURESH CHANDRA DR
UP STATE OBSERVATORY
MANORA PEAK
NAINI TAL 263129
INDIA

JOSS PAUL CHRISTOPHER DR
MIT
ROOM 6-203
CAMBRIDGE MA 02139
U.S.A.

JUGAKU JUN DR
TOKYO ASTRONOMICAL
OBSERVATORY
OSAWA MITAKA
TOKYO 181
JAPAN

JUNG JEAN DR
AMBASS FRANCE WASH
FRENCH SC MISSION
2011 EYE STREET NW
WASHINGTON DC 20006
U.S.A.

JUNKES JOSEF DR
12 HOHENBRUNNERSTR
D-8014 NEUBIBERG
GERMANY, F.R.

JUPP ALAN A DR
APPL MATH THEOR PHYS
UNIVERSITY LIVERPOOL
P O BOX 147
LIVERPOOL L69 3BX
U.K.

JURA MICHAEL DR
DEPT OF ASTRONOMY
UCLA
LOS ANGELES CA 90024
U.S.A.

JURKEVICH IGOR DR
3130 PORT WAY
ANNAPOLIS MD 21403
U.S.A.

KABURAKI MASAKI PROF
4-24-9 KICHIJYOJI
MINAMI
MUSASHINO TOKYO 180
JAPAN

KABURAKI OSAMU DR
ASTRONOMICAL INST
TOHOKU UNIVERSITY
SENDAI 980
JAPAN

KADLA ZDENKA I DR
PULKOVO OBSERVATORY
196140 LENINGRAD
U.S.S.R.

KAEHLER HELMUTH DR
HAMBURGER STERNWARTE
GOJENBERGSWEG 112
D-2050 HAMBURG 80
GERMANY, F.R.

KAFATOS MINEAS DR
PHYSICS DEPT
GEORGE MASON UNIV
FAIRFAX VA 22030
U.S.A.

KAFKA PETER
MPI FUER PHYSIK UND
ASTROPHYSIK
8000 MUENCHEN 40
GERMANY, F.R.

KAFTAN MAY A DR
ASTRONOMICAL OBS
FOUND OF SCIENTIFIC
RESEARCH P O BOX 255
JADIRIYAH BAGHDAD
REP. OF IRAQ

KAHLER STEPHEN W DR
AMERICAN SCIENCE AND
ENGINEERING INC
955 MASSACHUSETTS AV
CAMBRIDGE MA 02139
U.S.A.

KAHN FRANZ D PROF
DEPT OF ASTRONOMY
THE UNIVERSITY
MANCHESTER M13 9PL
U.K.

KAI KEIZO DR
NOBEYAMA SOLAR RADIO
OBSERVATORY
NOBEYAMA MINAMIMAKI
NAGANO 384-13
JAPAN

KAIFU NORIO DR
TOKYO ASTRONOMICAL
OBSERVATORY
OSAWA MITAKA
TOKYO 181
JAPAN

KAISER T R PROF
DEPT OF PHYSICS
THE UNIVERSITY
SHEFFIELD S3 7RH
U.K.

KAKINUMA TAKAKIYO T PROF
RESEARCH INSTITUTE
OF ATMOSPHERICS
NAGOYA UNIVERSITY
TOYOKAWA AICHI 442
JAPAN

KAKUTA CHUICHI DR
INTERNATIONAL LATITU
DE OBSERVATORY
HOSHIGAOKA 2-CHOME
MIZUSAWA IWATE 023
JAPAN

KALANDADZE N B DR
ABASTUMANY ASTROPHY-
SICAL OBSERVATORY
383762 ABASTUMANY
U.S.S.R.

KALER JAMES B PROF
UNIVERSITY OF
ILLINOIS OBSERVATORY
URBANA IL 61801
U.S.A.

KALINKOV MARIN P DR
DEPT OF ASTRONOMY
BULGARIAN AC OF SC
7 NOVEMBER ST 1
1000 SOFIA
BULGARIA

KALKOFEN W DR
SMITHSONIAN ASTROPHY
SICAL OBSERVATORY
60 GARDEN ST
CAMBRIDGE MA 02138
U.S.A.

KALLOGLIAN ARSEN T DR
BYURAKAN ASTROPHYSI-
CAL OBSERVATORY
378433 BYURAKAN
ARMENIA
U.S.S.R.

KALMAN BELA DR
HELIOPHYSICAL OBS
P O BOX 30
4010 DEBRECEN
HUNGARY

KALMYKOV A M DR
ASTRONOMICAL INSTITU
TE UZBEKIAN ACADEMY
OF SCIENCES
700000 TASHKENT
U.S.S.R.

KALNAJS AGRIS J DR
MT STROMLO
OBSERVATORY
WODEN PO ACT 2606
AUSTRALIA

KAMEL OSMAN M DR
FACULTY OF SCIENCES
ASTRONOMY DEPT
CAIRO UNIVERSITY
GIZA
ARAB REP. OF EGYPT

KAMIJO FUMIO PROF DR
DEPT OF ASTRONOMY
UNIVERSITY OF TOKYO
YAYOI BUNKYO KU
TOKYO 113
JAPAN

KAMINISHI KEISUKE PROF
FACULTY OF SCIENCE
KUMAMOTO UNIVERSITY
KUROKAMI 5 CHOME
860 KUMAMOTO
JAPAN

KAMP LUCAS WILLEM DR
DEPT OF ASTRONOMY
BOSTON UNIV
725 COMMONWEALTH AVE
BOSTON MA 02215
U.S.A.

KAMPER KARL W DR
DAVID DUNLAP
OBSERVATORY
RICHMOND HILL
ONT L4C 4Y6
CANADA

KANAEV IVAN I DR
PULKOVO OBSERVATORY
196140 LENINGRAD
U.S.S.R.

KANDEL ROBERT S DR
OBSERVATOIRE
DE MEUDON
F-92190 MEUDON
FRANCE

KANE SHARAD R DR
SPACE SCIENCES LAB
UNIVERSITY OF CALI-
FORNIA-BERKELEY
BERKELEY CA 94720
U.S.A.

KANEKO NOBORU DR
DEPT OF PHYSICS
FACULTY OF SCIENCE
HOKKAIDO UNIVERSITY
060 SAPPORO
JAPAN

KANNO MITSUO DR
HIDA OBSERVATORY
KAMITAKARA
GIFU-KEN 506 13
JAPAN

KANYO SANDOR DR
KONKOLY OBSERVATORY
BOX 67
1525 BUDAPEST
HUNGARY

KAPAHI V K DR
TIFR CENTRE
INDIAN INSTITUTE OF
SCIENCE
BANGALORE 560 012
INDIA

KAPLAN J DR
DEPT OF PHYSICS
UCLA
LOS ANGELES CA 90024
U.S.A.

KAPLAN L MR
JPL
4800 OAK GROVE DRIVE
PASADENA CA 91103
U.S.A.

KARACHENTSEV I D DR
SPECIAL ASTROPHYSI-
CAL OBSERVATORY
USSR ACADEMY OF SCI
357140 N ARKHYZ
U.S.S.R.

KARANDIKAR R V PROF
DEPT OF ASTRONOMY
OSMANIA UNIVERSITY
HYDERABAD 500 007
INDIA

KARDASHEV N S DR
SPACE RESEARCH
INSTITUTE
USSR ACADEMY OF SCI
117810 MOSCOW
U.S.S.R.

KARP ALAN HERSH DR
IBM PALO ALTO SC CTR
1530 PAGE MILL RD
PALO ALTO CA 94304
U.S.A.

KARPINSKIJ VADIM N DR
PULKOVO OBSERVATORY
196140 LENINGRAD
U.S.S.R.

KARYGINA ZOYA V DR
ASTROPHYSICAL INST
480068 ALMA-ATA
U.S.S.R.

KASHSCHEEV B L PROF DR
KHARKOV INSTITUTE
FOR RADIOELECTRONICS
310059 KHARKOV
U.S.S.R.

KASPER U DR
ZNTRINST ASTROPHYSIK
STERNW BABELSBERG
R LUXEMBURG STR 17A
DDR-1502 POTSDAM
G.D.R.

KASTLER ALFRED PROF
ECOLE NORMALE
SUPERIEURE
45 RUE D ULM
F-75005 PARIS
FRANCE

KATASEV L A DR
ASTRONOMICAL COUNCIL
USSR ACADEMY OF SCI
PYATNITSKAYA UL 48
109017 MOSCOW
U.S.S.R.

KATGERT PETER DR
STERREWACHT
POSTBUS 9513
2300 RA LEIDEN
THE NETHERLANDS

KATGERT-MERKELIJN J K DR
STERREWACHT
HUYGENS LAB
WASSENAARSEWEG 78
2300 RA LEIDEN
THE NETHERLANDS

KATO SHOJI PROF
DEPT OF ASTRONOMY
UNIVERSITY OF KYOTO
KITASHIRAKAWA OIWAKE
SAKYOKU KYOTO 606
JAPAN

KATO TAKAKO DR
INST OF PLASMA PHYS
NAGOYA UNIVERSITY
RURO-CHO CHIKUSA-KU
NAGOYA 464
JAPAN

KATSIS DEM DR
12 RUE VARNIS
NEA SMYRNE
ATHENS
GREECE

KATZ JOSEPH DR
THE ROUND HOUSE
CORNER OF HABICURIM
AND PLUGAT HAKOTEL S
JERUSALEM
ISRAEL

KAUFMAN MICHELE DR
PHYSICS DEPT
OHIO ST UNIVERSITY
174 W 18TH AVE
COLUMBUS OH 43210
U.S.A.

KAUFMANN JENS PETER DR
INST F ASTROPHYSIK
T U
ERNST REUTER PLATZ 7
D-1000 BERLIN 10
GERMANY, F.R.

KAUFMANN PIERRE PROF
CRAAM ON CNPQ
RUA CEARA NO 290
01243 SAO PAULO SP
BRAZIL

KAULA WILLIAM M PROF
INST OF GEOPHYSICS
UNIVERSITY
OF CALIFORNIA
LOS ANGELES CA 90024
U.S.A.

KAWABATA KINAKI PROF
DEPT OF PHYSICS
NAGOYA UNIVERSITY
FUROCHO CHIKUSAKU
NAGOYA 464
JAPAN

KAWABATA SHUSAKU PROF
KYOTO GAKUEN
UNIVERSITY
SOGABE MACHI KAMEOKA
KYOTO 621
JAPAN

KAWAGUCHI ICHIRO PROF
DEPT OF ASTRONOMY
FACULTY OF SCIENCE
UNIVERSITY OF KYOTO
606 KYOTO
JAPAN

KAWATA YOSHIYUKI DR
KANAZAWA INSTITUTE
OF TECHNOLOGY
NONOICHO
KANAZAWA MINAMI 921
JAPAN

KAZANTIS PANAYOTIS DR
DEPT OF ASTRONOMY
UNIV OF GLASGOW
GLASGOW G12 8QQ
U.K.

KAZES ILYA DR
OBSERVATOIRE
DE MEUDON
F-92190 MEUDON
FRANCE

KAZIMIRCHAK-POLONSKAYA
I DR
INSTITUTE OF THEORE-
TICAL ASTRONOMY
192187 LENINGRAD
U.S.S.R.

KEAY COLIN S L PROF
PHYSICS DEPT
NEWCASTLE UNIVERSITY
NEWCASTLE NSW 2308
AUSTRALIA

KEENAN PHILIP C PROF EMER
PERKINS OBSERVATORY
BOX 449
DELAWARE OH 43015
U.S.A.

KEGEL WILHELM H PROF
INST THEOR PHYSIK
UNIV FRANKFURT
R-MAYER-STR 8-10
D-6000 FRANKFURT/M 1
GERMANY, F.R.

KELLER CHARLES F
UNIVERSITY OF
CALIFORNIA LASL
BOX 1663 MS 420
LOS ALAMOS NM 87545
U.S.A.

KELLER GEOFFREY
410 HIGHGATE AVE
WORTHINGTON OH 43085
U.S.A.

KELLER H UWE DR
MPI FUER AERONOMIE
D-3411 KATLENBURG
LINDAU 3
GERMANY, F.R.

KELLER HANS U DR
OBSERVATORY AND
PLANETARIUM
PCB 161 NECKARST 47
D-7000 STUTTGART
GERMANY, F.R.

KELLERMANN KENNETH I DR
NATIONAL RADIO ASTRO
NOMY OBSERVATORY
P O BOX 2
GREEN BANK WV 24944
U.S.A.

KELLOGG EDWIN M DR
100 ROBBINS ROAD
WATERTOWN MA 02172
U.S.A.

KELLOGG WILLIAM W DR
NATIONAL CTR FOR
ATMOSPHERIC RESEARCH
BOULDER CO 80302
U.S.A.

KENDERDINE SIDNEY DR
MULLARD RADIO ASTRO-
NOMY LABORATORY
MADINGLEY ROAD
CAMBRIDGE CB3 OHE
U.K.

KENNEDY JOHN E PROF
UNIV OF SASKATCHEWAN
236 ARTS BLDG
SASKATOON
SASKATCHEWAN S7N OWO
CANADA

KERES KH P PROF DR
IAPHA ESTONIAN ACADE
MY OF SCIENCES
TORAVERE
202444 TARTU
U.S.S.R.

KERR FRANK J DR
ASTRONOMY PROGRAM
UNIVERSITY MARYLAND
COLLEGE PK MD 20742
U.S.A.

KERR ROY P PROF
UNIV OF CANTERBURY
PRIVATE BAG
CHRISTCHURCH
NEW ZEALAND

KESSLER KARL G DR
CTR ABS PHYS QUANTIT
NAT BUREAU STANDARDS
B-164 PHYSICS BLDG
WASHINGTON DC 20234
U.S.A.

KESTEVEN MICHAEL J L DR
QUEEN S UNIVERSITY
KINGSTON ONT K7L 3N6
CANADA

KHACHIKIAN E Y DR
BYURAKAN ASTROPHYSI-
CAL OBSERVATORY
378433 ARMENIA
U.S.S.R.

KHARADZE E K PROF
ABASTUMANY ASTROPHY-
SICAL OBSERVATORY
383762 ABASTUMANY
GEORGIA
U.S.S.R.

KHARE BISHUN N DR
306 SPACE SC BLDG
CORNELL UNIVERSITY
ITHACA NY 14853
U.S.A.

KHARIN A S DR
MAIN ASTRON OBS
UKRAINIAN ACAD OF SC
GOLOSEEVO
252127 KIEV
U.S.S.R.

KHARITONOV ANDREJ V DR
ASTROPHYSICAL INST
480068 ALMA-ATA
U.S.S.R.

KHETSURIANI T S DR
ABASTUMANY ASTROPHY-
SICAL OBSERVATORY
383762 ABASTUMANY
U.S.S.R.

KHOKHLOVA V L DR
ASTRONOMICAL COUNCIL
USSR ACADEMY OF SCI
PYATNITSKAYA UL 48
109017 MOSCOW
U.S.S.R.

KHOLOPOV P N DR
STERNBERG STATE AS-
TRONOMICAL INSTITUTE
117234 MOSCOW
U.S.S.R.

KHOLSHEVNIKOV K V DR
LENINGRAD STATE UNI-
VERSITY ASTRONOMICAL
OBSERVATORY
199178 LENINGRAD
U.S.S.R.

KHROMOV G S DR
ASTRONOMICAL COUNCIL
USSR ACADEMY OF SCI
PYATNITSKAYA UL 48
109017 MOSCOW
U.S.S.R.

KIANG TAO PROF
DUNSINK OBSERVATORY
CASTLEKNOCK
CO DUBLIN
IRELAND

KIASATPOOR AHMAD PROF
PHYSICS DEPT
UNIVERSITY OF
ISFAHAN
DANESHGAH E ISFAHAN
IRAN

KIBBLEWHITE EDWARD J DR
INSTITUTE OF
ASTRONOMY
MADINGLEY RD
CAMBRIDGE CB3 OHE
U.K.

KIKUCHI SADAEMON PROF
ASTRONOMICAL OBS
TOHOKU UNIVERSITY
ARAMAKI
SENDAI 980
JAPAN

KILADZE R I DR
ABASTUMANI ASTROPHY-
CAL OBSERVATORY
383762 ABASTUMANY
U.S.S.R.

KILAR BOGDAN DR
FACULTY OF GEODESY
LJUBLJANA UNIVERSITY
JAMOVA 2
61000 LJUBLJANA
YUGOSLAVIA

KILKENNY DAVID DR
UNIV OBSERVATORY
BUCHANAN GARDENS
ST ANDREWS
FIFE KY16 9LZ
U.K.

KIM YONG HEK DR
ACADEMY OF SCIENCES
PYONGYANG ASTRONOMI-
CAL OBSERVATORY
PYONGYANG
DEM PEOPLE'S REP OF KOREA

KIMURA HIROSHI DR
PURPLE MOUNTAIN OBS
ACADEMIA SINICA
NANKING
PEOPLE'S REPUBLIC CHINA

KING ANDREW R DR
ASTRONOMY DEPT
UNIV OF LEICESTER
LEICESTER
U.K.

KING DAVID S PROF
DEPT OF PHYSICS AND
ASTRONOMY
UNIVERSITY N MEXICO
ALBUQUERQUE NM 87131
U.S.A.

KING HENRY C DR
MCLAUGHLIN PLANETAR
ROYAL ONTARIO MUSEUM
100 QUEENS PARK
TORONTO ONT M5S 2C6
CANADA

KING IVAN R PROF
ASTRONOMY DEPT
UNIVERSITY OF
CALIFORNIA
BERKELEY CA 94720
U.S.A.

KING R B DR
P O BOX 725
MEDOCINO CA 95460
U.S.A.

KING ROBERT WILSON JR DR
DEPT EARTH PLANET SC
MIT 54-627
CAMBRIDGE MA 02139
U.S.A.

KING-HELE DESMOND G MR
ROYAL AIRCRAFT EST
FARNBOROUGH HANTS
U.K.

KINGSTON ARTHUR E DR
DEPT OF APPLIED MATH
AND THEORET PHYSICS
QUEEN'S UNIVERSITY
BELFAST BT7 1NN
U.K.

KINMAN THOMAS D DR
KITT PEAK NATIONAL
OBSERVATORY
P O BOX 26732
TUCSON AZ 85726
U.S.A.

KINOSHITA HIROSHI DR
TOKYO ASTRONOMICAL
OBSERVATORY
OSAWA MITAKA
181 TOKYO
JAPAN

KIPPENHAHN RUDOLF PROF
MPI FUER PHYSIK UND
ASTROPHYSIK
K SCHWARZSCHILD ST 1
D-8046 GARCHING B M
GERMANY, F.R.

KIPPER AKSEL PROF
IAPHA
ESTONIAN ACAD OF SC
TORAVERE
20244 TARTU
U.S.S.R.

KIPPER TONUA DR
IAPHA
ESTONIAN ACAD OF SC
TORAVERE
20244 TARTU
U.S.S.R.

KIRAL ADNAN PROF
UNIV OBSERVATORY
BEYAZIT
ISTANBUL
TURKEY

KIRBIYIK HALIL DR
MIDDLE EAST TECHNI-
CAL UNIVERSITY
PHYSICS DEPT
ANKARA
TURKEY

KIRKPATRICK RONALD C DR
MS 220
LOS ALAMOS SCIENTI-
FIC LABORATORY
LOS ALAMOS NM 87545
U.S.A.

KIRSHNER ROBERT PAUL DR
DEPT OF ASTRONOMY
UNIV OF MICHIGAN
ANN ARBOR MI 48103
U.S.A.

KISELEV ALEXEJ A DR
PULKOVO OBSERVATORY
196140 LENINGRAD
U.S.S.R.

KISLYAKOV ALBERT G DR
APPLIED PHYSICS INST
ACADEMY OF SCIENCE
ULYANOV STREET 46
603600 GORKY
U.S.S.R.

KISLYUK V S DR
MAIN ASTRON OBS
UKRAINIAN ACAD OF SC
GOLOSEEVO
252127 KIEV
U.S.S.R.

KISSELL KENNETH E DR
AF AVIONICS LABORA-
TORY RW
W-P AFB OH 45433
U.S.A.

KITAMURA M PROF
TOKYO ASTRONOMICAL
OBSERVATORY
MITAKA TOKYO 181
JAPAN

KITAMURA SEIICHI DR
SHIGA UNIVERSITY
OTSU SHIGA 520
JAPAN

KITCHIN CHRISTOPHER R DR
HATFIELD POLYTECHNIC
OBSERVATORY
BAYFORDBURY
HERTFORD HERTS
U.K.

KIZILIRMAK ABDULLAH PROF
EGE UNIVERSITY
SCIENCE FACULTY
ASTRONOMY DEPT.
BORNOVA IZMIR
TURKEY

KJAERGAARD PER DR
COPENHAGEN UNIVERSI-
TY OBSERVATORY
OSTER VOLDGADE 3
DK-1350 COPENHAGEN
DENMARK

KJELDSETH MOE OLAF DR
INSTITUTE OF THEORE-
TICAL ASTROPHYSICS
UNIVERSITY OF OSLO
BLINDERN OSLO 3
NORWAY

KLARE GERHARD DR
LANDESSTERNWARTE
KOENIGSTUHL
D-6900 HEIDELBERG 1
GERMANY, F.R.

KLARMANN JOSEPH PROF
DEPT OF PHYSICS
WASHINGTON
UNIVERSITY
ST LOUIS MO 63130
U.S.A.

KLECZEK JOSIP DR
ASTRONOMICAL
INSTITUTE
25165 ONDREJOV
CZECHOSLOVAKIA

KLEIN MICHAEL J DR
SPACE SCIENCES DIV
JET PROPULSION LAB
BLDG 183B-365
PASADENA CA 91103
U.S.A.

KLEIN RICHARD I DR
KITT PEAK NAT OBS
P O BOX 26732
TUCSON AZ 85726
U.S.A.

KLEINMANN DOUGLAS E DR
CENTER FOR ASTROPHYS
60 GARDEN ST
CAMBRIDGE MA 02138
U.S.A.

KLEMOLA ARNOLD R DR
LICK OBSERVATORY
UCSC
SANTA CRUZ CA 95064
U.S.A.

KLEMPERER W K DR
DIV ASTRONOMICAL SC
NAT SCIENCE FOUND
1800 G STREET N W
WASHINGTON DC 20550
U.S.A.

KLEPCZYNSKI WILLIAM J DR
US NAVAL OBSERVATORY
WASHINGTON DC 20390
U.S.A.

KLIMISHIN I A DR
PEDAGOGIC INSTITUTE
284000 IVANO-
FRANKOV SK
U.S.S.R.

KLINGLESMITH DANIEL A DR
GSFC
CODE 671
GREENBELT MD 20771
U.S.A.

KLIORE ARVYDAS JOSEPH DR
JPL
4800 OAK GROVE DRIVE
PASADENA CA 91103
U.S.A.

KLOCK B L DR
US NAVAL OBSERVATORY
WASHINGTON DC 20390
U.S.A.

KNACK ROGER F DR
DEPT EARTH SPACE SCS
SUNY
STONY BROOK NY 11794
U.S.A.

KNAFLICH H B MS
BOEING COMPANY
AEROSPACE GROUP
M S 88-21 POB 3999
SEATTLE WA 98124
U.S.A.

KNAPP GILLIAN R DR
OWENS VALLEY RADIO
OBSERVATORY
CALIF INST OF TECHN
PASADENA CA 91125
U.S.A.

KNEER F DR
KIEPENHEUER INST
FUER SONNENPHYSIK
SCHOENECKSTR 6
D-7800 FREIBURG-BR
GERMANY, F.R.

KNIFFEN DONALD A DR
NASA-GSFC
CODE 662
GREENBELT MD 20771
U.S.A.

KNOWLES STEPHEN H DR
CODE 7134
NAVAL RESEARCH
LABORATORY
WASHINGTON DC 20375
U.S.A.

KNUDE JENS KIRKESKOV DR
COPENHAGEN UNIV
OBSERVATORY
OSTER VOLDGADE 3
DK-1350 COPENHAGEN-K
DENMARK

KO HSIEN C PROF
DEPT OF ELECTRICAL
ENGINEERING
OHIO ST UNIVERSITY
COLUMBUS OH 43210
U.S.A.

KOBAYASHI EISUKE DR
DEPT OF EARTH SCS
SC ED INST OSAKA PR
4-15 KARITA-CHO SUMI
YOSHI-KU OSAKA 558
JAPAN

KOBRIN M M PROF DR
INSTITUTE OF APPLIED
PHYSICS
603600 GORKIJ
U.S.S.R.

KOCER D DR
UNIV OBSERVATORY
UNIVERSITY
ISTANBUL
TURKEY

KOCH LYDIE DR
DPH EP ES BAT 28
CEN SACLAY
F P 2
F-91190 GIF S YVETTE
FRANCE

KOCH ROBERT H DR
DEPT OF ASTRONOMY
UNIV OF PENNA DRL
33RD AND WALNUT STS
PHILADELPHIA PA19104
U.S.A.

KOCHAR R K DR
INDIA INSTITUTE OF
ASTROPHYSICS
BANGALORE 560 034
INDIA

KOCHAROV GRANT E PROF
PHYSICAL
TECHN INSTITUTE
ACAD SCIENCES USSR
194021 LENINGRAD
U.S.S.R.

KODAIRA KEIICHI A PROF
DEPT OF ASTRONOMY
FACULTY OF SCIENCE
UNIVERSITY OF TOKYO
113 TOKYO BUNKYO-KU
JAPAN

KOECKELENBERGH ANDRE DR
OBSERVATOIRE ROYAL
DE BELGIQUE
AVE CIRCULAIRE 3
B-1180 BRUXELLES
BELGIUM

KOEHLER H PROF DR
SAUERBRUCHSTRASSE 6
D-7920 HEIDENHEIM
A-D BRENZ
GERMANY, F.R.

KOEHLER JAMES A PROF
PHYSICS DEPT
UNIV OF SASKATOON
SASKATOON S7N OWO
CANADA

KOESTER DETLEV DR
INST THEOR PHY STERN
NEUE UNIV PHYS ZENTR
OLSHAUSENST GEBN 61C
D-2300 KIEL 1
GERMANY, F.R.

KOGURE TOMOKAZU DR
DEPT OF ASTRONOMY
FACULTY OF SCIENCE
UNIVERSITY OF KYOTO
KYOTO 606
JAPAN

KOHL JOHN L DR
CENTER FOR
ASTROPHYSICS
60 GARDEN ST
CAMBRIDGE MA 02138
U.S.A.

KOHOUTEK LUBOS DR
HAMBURGER STERNWARTE
GOJENSBERGSWEG
D-2050 HAMBURG 80
GERMANY, F.R.

KOJOIAN GABRIEL DR
DEPT OF PHYSICS
UNIV OF WISCONSIN
EAU CLAIRE WI 54701
U.S.A.

KOKURIN YURIJ L DR
LEBEDEV PHYS INST
USSR ACAD OF SCS
LENINSKY PROSPECT 53
117924 MOSCOW
U.S.S.R.

KOLACZEK BARBARA DR
PLAN. GEODESY DEPT
UL BARTYCKA 18
00-716 WARSAW
POLAND

KOLCHINSKIJ I G DR
MAIN ASTRONOMICAL
OBS UKRAINIAN ACAD
OF SCIENCES
252127 KIEV
U.S.S.R.

KOLESNIK L N DR
MAIN ASTRONOMICAL
OBS UKRAINIAN ACAD
OF SCIENCES
252127 KIEV
U.S.S.R.

KOLESOV A K DR
LENINGRAD STATE UNI-
VERSITY ASTRONOMICAL
OBSERVATORY
199178 LENINGRAD
U.S.S.R.

KOLLBERG ERIK L DR
RES LAB ELECTRONICS
CHALMERS UNIVERSITY
OF TECHNOLOGY FACK
S-402 20 GOETEBORG
SWEDEN

KOMAROV N S DR
ODESSA STATE UNIVER-
SITY ASTRONIMICAL
OBSERVATORY
270014 ODESSA
U.S.S.R.

KOMESAROFF MAX M D SC
CSIRO DIVISION OF
RADIOPHYSICS
BOX 76
EPPING NSW 2121
AUSTRALIA

KONDO MASAAKI DR
FAC GENERAL EDUC
UNIVERSITY OF TOKYO
KOMABA MEGURO
TOKYO 153
JAPAN

KONDO YOJI DR
CODE 685
GODDARD SPACE FL CTR
GREENBELT MD 20771
U.S.A.

KONIN V V DR
NIKOLAEV BRANCH OF
THE MAIN ASTRONOMI-
CAL OBSERVATORY
327000 NIKOLAEV
U.S.S.R.

KONONOVICH EDWARD V DR
STERNBERG ASTRONOMI-
CAL INSTITUTE
117234 MOSCOW
U.S.S.R.

KONOPLEVA V P DR
MAIN ASTRONOMICAL
OBS UKRAINIAN ACAD
OF SCIENCES
252127 KIEV
U.S.S.R.

KONTIZAS EVANGELOS PH D
NAT OBS OF ATHENS
ASTRONOMICAL INST
THISSION
ATHENS 306
GREECE

KONTIZAS MARY PH D
DEPT OF ASTRONOMY
UNIV OF ATHENS
PANEPISTIMIOPOLIS
ATHENS 621
GREECE

KOPAL ZDENEK PROF
DEPT OF ASTRONOMY
UNIVERSITY OF
MANCHESTER
MANCHESTER M13 9PL
U.K.

KOPECKY MILOSLAV DR
ASTRONOMICAL INST
CZECHOSLOVAK ACADEMY
OF SCIENCES
251 65 ONDREJOV OBS
CZECHOSLOVAKIA

KOPP ROGER A DR
LOS ALAMOS SCIENTI-
FIC LABORATORY
M S 531
LOS ALAMOS NM 87545
U.S.A.

KOPYLOV I M DR
SPECIAL ASTROPHYSI-
CAL OBSERVATORY
USSR ACADEMY OF SCI
357140 N ARKHYZ
U.S.S.R.

KORCHAK A A DR
INSTITUTE OF TERRE-
STRIAL MAGNETISM AND
IONOSPHERE
142092 ACADEMGORODOK
U.S.S.R.

KORDYLEWSKI KAZIMIERZ DR.
SKLODOWSKIEJ 6
31-025 KRAKOW
POLAND

KORMENDY JOHN DR
DOMINION ASTROPHYS-
ICAL OBSERVATORY
5071 W SAANICH ROAD
VICTORIA BC V8X 3X3
CANADA

KOROLKOV DMITRY V PROF DR
PULKOVO OBSERVATORY
196140 LENINGRAD
U.S.S.R.

KOROVYAKOVSKIJ YURIJ P DR
SPECIAL ASTROPHYS OB
NIZHNIJ ARKHYZ
357140 STAVROPOLSKIJ
KRAJ
U.S.S.R.

KOSIN GENNADIJ S DR
PULKOVO OBSERVATORY
196140 LENINGRAD
U.S.S.R.

KOSTINA LIDIJA D DR
PULKOVO OBSERVATORY
196140 LENINGRAD
U.S.S.R.

KOSTYAKOVA ELENA B DR
STERNBERG STATE
ASTRONOMICAL INST
117234 MOSCOW
U.S.S.R.

KOSTYLEV K V DR
ENGELHARDT ASTRONO-
MICAL OBSERVATORY
422526 KAZAN
U.S.S.R.

KOTELNIKOV V A ACAD
INSTITUTE OF RADIO
AND ELECTRONICS
USSR ACADEMY OF SCI
103907 MOSCOW
U.S.S.R.

KOTHARI D.S. DR.
DEPT.OF PHYSICS
UNIVERSITY OF DELHI
NEW DELHI
INDIA

KOTOV V A DR
CRIMEAN ASTROPHYSI-
CAL OBSERVATORY USSR
ACAD OF SCI NAUCHNIY
334413 CRIMEA
U.S.S.R.

KOTSAKIS DEMETRIUS PROF
HIPPOCRATES STR 189
ATHENS-708
GREECE

KOURGANOFF VLADIMIR PROF
20 AVE PAUL APPELL
F-75014 PARIS
FRANCE

KOUTCHMY SERGE DR
INSTITUT D ASTROPHY-
SIQUE CNRS
98 BIS BD ARAGO
F-75014 PARIS
FRANCE

KOVACHEV B J DR
ACADEMIE BULGARE
SECTION ASTRONOMIE
RUE 7 NOVEMBER NO 1
SOFIA
BULGARIA

KOVACS AGNES DR
HELIOPHYSICAL OBS
P O BOX 30
H-4010 DEBRECEN
HUNGARY

KOVAL I K DR
MAIN ASTRONOMICAL
OBSERVATORY
GOLOSEEVO
252127 KIEV
U.S.S.R.

KOVALEVSKY JEAN DR
CERGA
AV COPERNIC
F-06130 GRASSE
FRANCE

KOVAR N S DR
PHYSICS DEPT
UNIV OF HOUSTON
HOUSTON TX 77004
U.S.A.

KOVAR ROBERT P DR
9666 E ORCHARD DR
ENGLEWOOD CO 80111
U.S.A.

KOVETZ ATTAY PROF
DEPT OF PHYSICS AND
ASTRONOMY
TEL AVIV UNIVERSITY
TEL-AVIV
ISRAEL

KOWAL CHARLES THOMAS DR
DEPT OF ASTROPHYSICS
CALTECH
PASADENA CA 91125
U.S.A.

KOYAMA SHIN PROF DR
KAGAWA UNIVERSITY
SAIWAI CHO
TAKAMATSU 760
JAPAN

KOZAI YOSHIHIDE PROF
TOKYO ASTRONOMICAL
OBSERVATORY
OSAWA MITAKA
TOKYO 181
JAPAN

KOZHEVNIKOV N I DR
STERNBERG STATE AS-
TRONOMICAL INSTITUTE
117234 MOSCOW
U.S.S.R.

KOZIEL KAROL PROF DR
UL 22 LIPCA 16
43-460 WISLA
POLAND

KOZLOVSKY B Z DR
DEPT OF PHYSICS AND
ASTRONOMY
TEL-AVIV UNIVERSITY
RAMAT AVIV
ISRAEL

KOZLOWSKI MACIEJ DR
ASTRONOMICAL
OBSERVATORY
AL UJAZDOWSKIE 4
00-478 WARSZAWA
POLAND

KOZYREV NIKOLAJ A DR
PULKOVO OBSERVATORY
196140 LENINGRAD
U.S.S.R.

KRAEMER GERHARD DR
ERLENWEG 24
D-7400 TUEBINGEN
GERMANY, F.R.

KRAFT ROBERT P PROF
LICK OBSERVATORY
UNIVERSITY
OF CALIFORNIA
SANTA CRUZ CA 95064
U.S.A.

KRAICHEVA Z DR
DEPT OF ASTRONOMY
BULGARIAN AC OF SCS
7TH NOVEMBER ST 1
1000 SOFIA
BULGARIA

KRAMER KH N DR
ODESSA STATE UNIVER-
SITY ASTRONOMICAL
OBSERVATORY
270014 ODESSA
U.S.S.R.

KRANJC A DR
OSS ASTRONOMICO
I-40100 BOLOGNA
ITALY

KRASINSKY GEORGE A DR
INST THEORET ASTRON
USSR ACAD OF SCS
10 KUTUZOV QUAY
192187 LENINGRAD
U.S.S.R.

KRAT VLADIMIR A PROF DR
PULKOVO OBSERVATORY
196140 LENINGRAD
U.S.S.R.

KRAUS JOHN D PROF
RADIO OBSERVATORY
OHIO ST UNIVERSITY
2015 NEIL AVE
COLUMBUS OH 43210
U.S.A.

KRAUSE F DR
ZNTRINST ASTROPHYSIK
ASTROPHYS OBS
TELEGRAFENBERG
DDR-15 POTSDAM
G.D.R.

KRAUSHAAR WILLIAM L PROF
DEPT OF PHYSICS
UNIV OF WISCONSIN
1150 UNIVERSITY AVE
MADISON WI 53706
U.S.A.

KREINER JERZY MAREK DR
UL SPOLDZIELCOW 8/52
30-682 KRAKOW 47
POLAND

KREISEL E DR
ZNTRINST ASTROPHYSIK
STERNW BABELSBERG
ROSA LUXEMBURGST 17A
DDR-1502 POTSDAM
G.D.R.

KREMPEC JANINA DR
N.COPERNICUS AST CTR
ASTROPHYSICAL LAB
CHOPINA 12/18
87-100 TORUN
POLAND

KRESAK LUBOR DR
ASTRONOMICAL INST
SAV
DUBRAVSKA CESTA 11/A
89930 BRATISLAVA
CZECHOSLOVAKIA

KRESAKOVA MARGITA DR
ASTRONOMICAL INST
SAV
DUBRAVSKA CESTA 11/A
89930 BRATISLAVA
CZECHOSLOVAKIA

KRIEGER ALLEN S DR
AMERICAN SCIENCE AND
ENGINEERING
955 MASSACHUSETTS AV
CAMBRIDGE MA 02139
U.S.A.

KRINOV EUGEN L PROF
METEORITE COMMITTEE
ACADEMY OF SCIENCES
UL M ULIANOVOI 3
117313 MOSCOW
U.S.S.R.

KRISHNA SWAMY K S DR
ASTROPHYSICS GROUP
TATA INSTITUTE
COLABA
BOMBAY 400005
INDIA

KRISHNAN THIRUVENKATA DR
132-A LLOYDS ROAD
GOPALAPURAM
MADRAS 600086
INDIA

KRISTENSEN LEIF KAHL DR
INST OF PHYSICS
UNIV OF AARHUS
NY MUNKGADE
DK-8000 AARHUS C
DENMARK

KRISTENSON HENRIK DR
SLAANBAERSVAEGEN 9
S-381 00 KALMAR
SWEDEN

KRISTIAN JEROME DR
HALE OBSERVATORIES
813 SANTA BARBARA ST
PASADENA CA 91101
U.S.A.

KRISTIANSSON KRISTER
G B PROF
DEPT OF PHYSICS
SOELVEGATAN 14
S-223 62 LUND
SWEDEN

KRIVSKY L DR
ASTRONOMICAL
INSTITUTE
25165 ONDREJOV
CZECHOSLOVAKIA

KRIZ SVATOPLUK DR
DEPT.OF ASTRONOMY
CHARLES UNIVERSITY
SVEDSKA 8
150 000 PRAGUE
CZECHOSLOVAKIA

KROGDAHL W S DR
DEPT OF PHYSICS AND
ASTRONOMY
UNIVERSITY KENTUCKY
LEXINGTON KY 40506
U.S.A.

KRON GERALD E DR
PINECREST OBS
416 N BERTRAND ST
FLAGSTAFF AZ 86001
U.S.A.

KRON KATHERINE GORDON
PINECREST OBS
416 N BERTRAND ST
FLAGSTAFF AZ 86001
U.S.A.

KRONBERG PHILIPP DR
UNIVERSITY
OF TORONTO
SCARBOROUGH COLLEGE
TORONTO M1C 1A4
CANADA

KROOK M DR
HARVARD COLLEGE OBS
CAMBRIDGE MA 02138
U.S.A.

KRUEGER ALBRECHT DR
ZENTRALINSTITUT FUER
SOLAR-TERR.PHYSIK
RUDOWER CHAUSSEE 5
DDR-1199 BERLIN
G.D.R.

KRUEGER E PROF
VIA MAURO MACCHI 65
I-20100 MILANO
ITALY

KRUSZEWSKI ANDRZEJ PROF
ASTRONOMICAL
OBSERVATORY
AL UJAZDOWSKIE 4
00-478 WARSZAWA
POLAND

KRZEMINSKI W DR
COPERNICUS ASTRONO-
MICAL CENTER
UL BARTYCKA 18
00-716 WARSAW
POLAND

KSANFOMALITI L V DR
SPACE RESEARCH INSTI
TUTE USSR ACADEMY OF
SCIENCES
117810 MOSCOW
U.S.S.R.

KUBIAK MARCIN A DR
WARSAW UNIVERSITY
OBSERVATORY
AL UJAZDOWSKIE 4
00-478 WARSAW
POLAND

KUBICELA ALEKSANDAR DR
ASTRONOMSKA OPSERV
VOLGINA 7
11050 BEOGRAD
YUGOSLAVIA

KUBOTA JUN DR
KASAN OBSERVATORY
YAMASHINA
KYOTO 607
JAPAN

KUDRITZKI ROLF-PETER PH D
INST F THEOR PHYSIK
UND UNIV STERNWARTE
OLSHAUSENSTRASSE
D-2300 KIEL
GERMANY, F.R.

KUEHNE CHRISTOPH F DR
CARL ZEISS COMPANY
POSTFACH 35/36
D-7082 OBERKOCHEN
GERMANY, F.R.

KUHI LEONARD V PROF
ASTRONOMY DEPT
UNIVERSITY OF CALIF
BERKELEY CA 94720
U.S.A.

KUIJPERS JAN DR
STERREKUNDIG INST
SERVAAS BOLWERK 13
UTRECHT
THE NETHERLANDS

KUIPER THOMAS B H DR
JET PROPULSION LABO-
RATORY 183B-365
PASADENA CA 91103
U.S.A.

KUKLIN G V DR
SIBIZMIR
P B 4
664697 IRKUTSK 33
U.S.S.R.

KULIKOV K A PROF DR
STERNBERG STATE AS-
TRONOMICAL INSTITUTE
117234 MOSCOW
U.S.S.R.

KULIKOVSKIJ P G DR
STERNBERG STATE AS-
TRONOMICAL INSTITUTE
117234 MOSCOW
U.S.S.R.

KULKARNI PRABHAKAR V PROF
PHYSICAL RESEARCH
LABORATORY
AHMEDABAD 380 009
INDIA

KULSRUD R M DR
ASTROPHYSICAL SCS
PRINCETON UNIVERSITY
PRINCETON NJ 08540
U.S.A.

KUMAJGORODSKAYA RAISA DR
SPECIAL ASTROPHYS OB
NIZHNIJ ARKHYZ
357140 STAVROPOLSKIJ
KRAJ
U.S.S.R.

KUMAR CIDAMBI KRISHNA DR
DEPT PHYSICS ASTRON
HOWARD UNIV
WASHINGTON DC 20059
U.S.A.

KUMAR S S PROF
DEPT OF ASTRONOMY
UNIV OF VA POB 3818
CHARLOTTESVILLE
VA 22903
U.S.A.

KUMSIASHVILY MZIA I DR
ABASTUMANI ASTRO-
PHYSICAL OBSERVATORY
383762 GEORGIA
U.S.S.R.

KUMSISHVILI J I DR
ABASTUMANY ASTROPHY-
SICAL OBSERVATORY
383762 ABASTUMANY
GEORGIA
U.S.S.R.

KUNCHEV PETER DR
DEPT OF ASTRONOMY
FACULTY OF PHYSICS
ANTON IVANOV STR 5
1126 SOFIA
BULGARIA

KUNDT WOLFGANG PROF DR
INST F ASTROPHYSIK U
EXTRATERR FORSCHUNG
AUF DEM HUEGEL 71
D-5300 BONN 1
GERMANY, F.R.

KUNDU MUKUL R DR
ASTRONOMY PROGRAM
UNIVERSITY MARYLAND
COLLEGE PARK
MD 20742
U.S.A.

KUNKEL WILLIAM E DR
OBS NACIONAL
RUA GENERAL BRUCE
586 ZC 08
RIO DE JANEIRO GB
BRAZIL

KUNZEL HORST
ZENTRALINSTITUT FUER
SOLARTERRESTRISCHE
PHYSIK
DDR-15 POTSDAM
G.D.R.

KUPERUS MAX PROF DR
ASTRONOMICAL
INSTITUTE
ZONNENBURG 2
UTRECHT
THE NETHERLANDS

KUPPERIAN JAMES E DR
GODDARD SPACE FLIGHT
CENTER
CODE 410
GREENBELT MD 20771
U.S.A.

KURIL-CHIK V N DR
STERNBERG STATE AS-
TRONOMICAL INSTITUTE
117234 MOSCOW
U.S.S.R.

KUROCHKA L N DR
KIEV STATE UNIVER-
SITY ASTRONOMICAL
OBSERVATORY
252053 KIEV
U.S.S.R.

KURT V G DR
SPACE RESEARCH INSTI
TUTE USSR ACADEMY OF
SCIENCES
117810 MOSCOW
U.S.S.R.

KURTH R DR
PROF OF MATHEMATICS
GEORGIA INST OF
TECHNOLOGY
ATLANTA GA 30332
U.S.A.

KURUCZ ROBERT L DR
CTR FOR ASTROPHYSICS
60 GARDEN ST
CAMBRIDGE MA 02138
U.S.A.

KUSHWAHA R S PROF
DEPT OF MATHEMATICS
UNIV OF JODHPUR
JODHPUR RAJ
INDIA

KUSTAANHEIMO PAUL E PROF
DIA E 451
DANMARKS TEKNISKE
HOJSKOLE
DK-2800 LYNGBY
DENMARK

KUTNER MARC LESLIE DR
PHYSICS DEPT
RENSSELAER POLYTECH
INST
TROY NY 12181
U.S.A.

KUTTER G SIEGFRIED DR
THE EVERGREEN STATE
COLLEGE
OLYMPIA WA 98505
U.S.A.

KUTUZOV S A DR
LENINGRAD STATE UNI-
VERSITY ASTRONOMICAL
OBSERVATORY
199178 LENINGRAD
U.S.S.R.

KUZMENKO K N DR
ASTRONOMICAL OBSERVA
TORY KHARKOV STATE
UNIVERSITY
310077 KHARKOV
U.S.S.R.

KUZMIN ARKADII D
PROF DR SC
LEBEDEV PHYS INST
ACADEMY SC
117924 MOSCOW
U.S.S.R.

KUZMIN GRIGORI G PROF
IAPHA ESTONIAN
ACADEMY OF SCIENCES
TORAVERE
202444 TARTU
U.S.S.R.

KVIZ ZDENEK DR
UNIVERSITY OF
NEW SOUTH WALES
P O BOX 334
BROKEN HILL NSW 2880
AUSTRALIA

KWEE K K DR
STERREWACHT
POSTBUS 9513
2300 RA LEIDEN
THE NETHERLANDS

KWOK SUN DR
HERZBERG INST ASTRO
NAT RES COUNCIL CAN
100 SUSSEX DRIVE
OTTAWA ONT K1A 0R6
CANADA

LABEYRIE ANTOINE DR
CERGA
F-06460 ST VALLIER
FRANCE

LABEYRIE JACQUES DR
CENTRE DES FAIBLES
RADIOACTIVITES
LAB MIXTE CNRS-CEA
91190 GIF SUR YVETTE
FRANCE

LABORDE GEORGES DR
OBSERVATOIRE
DE MEUDON
F-92190 MEUDON
FRANCE

LABRUM NORMAN R MR
CSIRO DIVISION OF
RADIOPHYSICS
PO BOX 76
EPPING NSW 2121
AUSTRALIA

LABS DIETRICH PROF
LANDESSTERNWARTE
KOENIGSTUHL
D-6900 HEIDELBERG 1
GERMANY, F.R.

LACLARE F MR
CERGA
AVENUE COPERNIC
F-06130 GRASSE
FRANCE

LACROUTE PIERRE A PROF
2 RUE D ALISE
F-21000 DIJON
FRANCE

LADA CHARLES JOSEPH DR
STEWARD OBS
UNIV OF ARIZONA
TUCSON AZ 85721
U.S.A.

LAFFINEUR MARIUS MR
21 BLVD BRUNE
F-75014 PARIS
FRANCE

LAFON JEAN-PIERRE J DR
OBSERVATOIRE
DE MEUDON
F-92190 MEUDON
FRANCE

LAGERQVIST ALBIN PROF
INSTITUTE OF PHYSICS
VANADISVAEGEN 9
S-113 46 STOCKHOLM
SWEDEN

LAHIRI N C
INDIAN ASTRON EPH UN
INDIAN MET DEPT
P-546 BLOCK N
NEW ALIPORE CALCUTTA
INDIA

LAHULLA J FORNICS DR
OBSERVATORIO
ASTRONOMICO
ALFONSO XII 5
MADRID 7
SPAIN

LALA PETR DR
ASTRONOMICAL INST
CSAV
251 65 ONDREJOV
CZECHOSLOVAKIA

LAMB DONALD QUINCY JR DR
CTR FOR SPACE RES
BLDG 37-555
MIT
CAMBRIDGE MA 02139
U.S.A.

LAMB FREDERICK K PROF
PHYSICS DEPT
UNIVERSITY OF
ILLINOIS
URBANA IL 61801
U.S.A.

LAMB SUSAN ANN DR
DEPT OF PHYSICS
UNIV OF MISSOURI
8001 NATURAL BRG RD
ST LOUIS MO 63121
U.S.A.

LAMBECK KURT PROF
AUSTRALIA NAT UNIVER
SITY RESEARCH SCHOOL
EARTH SC PO BOX 4
CANBERRA 2600
AUSTRALIA

LAMBERT DAVID L PROF
DEPT OF ASTRONOMY
UNIVERSITY OF TEXAS
R L MOORE HALL
AUSTIN TX 78712
U.S.A.

LAMBRECHT HERMANN PROF DR
SCHILLERGAESSCHEN 2
DDR-69 JENA
G.D.R.

LAMERS H J G L M DR
LAB VOOR RUIMTE-
ONDERZOEK
BENELUXLAAN 21
UTRECHT
THE NETHERLANDS

LAMLA ERICH E DR
ASTRONOMISCHES INST
STERNW UNIV BONN
AUF DEM HUEGEL 71
D-5300 BONN
GERMANY, F.R.

LAMY P DR
LAB D'ASTRON SPAT
TRAVERSE DU SIPHON
LES-TROIS-LUCS
F-13012 MARSEILLE
FRANCE

LANCASTER BROWN PETER
41 LINDEN CLOSE
ALDEBURGH SUFFOLK
U.K.

LANDE KENNETH PROF
PHYSICS DEPT
UNIVERSITY OF
PENNSYLVANIA
PHILADELPHIA PA19104
U.S.A.

LANDECKER PETER BRUCE DR
AEROSPACE CORP
SPACE SCIENCES LAB
P O BOX 92957
LOS ANGELES CA 90009
U.S.A.

LANDECKER THOMAS L DR
DOMINION RADIO ASTRO
PHYSICAL OBSERVATORY
BOX 248
PENTICTON BC V2A 6K3
CANADA

LANDI DEGL INNOCENTI
EGIDIO DR
OSSERVATORIO ARCETRI
LARGO FERMI 5
I-50125 FIRENZE
ITALY

LANDI-DESSY J DR
OBSERVATORIO NACIO-
NAL ARGENTINO
CORDOBA
ARGENTINA

LANDINI MASSIMO PROF
OSSERVATORIO ARCETRI
LARGO FERMI
I-50125 FIRENZE
ITALY

LANDMAN DONALD ALAN DR
INST FOR ASTRONOMY
HALEAKALA OBS
P O BOX 157
KULA MAUI HI 96790
U.S.A.

LANDOLT ARLO U PROF
DEPT PHYS AND ASTRON
LOUISIANA STATE
UNIVERSITY
BATON ROUGE LA 70803
U.S.A.

LANDSTREET JOHN D PROF
DEPT OF ASTRONOMY
UNIVERSITY OF
WESTERN ONTARIO
LONDON ONT N6A 5B9
CANADA

LANG JAMES DR
ASTROPHYSICS RES DIV
CULHAM LABORATORY
ABINGDON
OXON OX14 3DB
U.K.

LANG KENNETH R ASST PROF
DEPT OF PHYSICS
TUFTS UNIVERSITY
MEDFORD MA 02155
U.S.A.

LANGER WILLIAM DAVID DR
DEPT PHYSICS ASTRON
GRC-B
UNIV OF MASS
AMHERST MA 01003
U.S.A.

LANTOS PIERRE
OBS DE MEUDON
F-92190 MEUDON
FRANCE

LAPAZ LINCOLN DR
INST OF METEORITICS
UNIV OF NEW MEXICO
CAMPUS PO BOX 23
ALBUQUERQUE NM 87131
U.S.A.

LAPOINTE S M DR
UNIVERSITE DE QUEBEC
2875 BOUL LAURIER
STE-FOY P Q G1V 2M3
CANADA

LAPUSHKA K K DR
LATVIAN STATE UNIVER
SITY ASTRONOMICAL
OBSERVATORY
226098 RIGA
U.S.S.R.

LAQUES PIERRE DR
OBSERVATOIRE
DU PIC DU MIDI
F-65200 BAGNERES
DE BIGORRE
FRANCE

LARGE MICHAEL I DR
SCHOOL OF PHYSICS
UNIVERSITY OF SYDNEY
SYDNEY NSW 2006
AUSTRALIA

LARI C DR
LAB NAZIONALE DI
RADIOASTRONOMIA
VIA IRNERIO 46
I-40126 BOLOGNA
ITALY

LARINK JOHANNES PROF DR
HAMBURGER STERNWARTE
BERGEDORF
GOJENBERGSWEG 78
D-2050 HAMBURG 80
GERMANY, F.R.

LARSON HAROLD P DR
LUNAR AND PLANETARY
LABORATORY
UNIVERSITY ARIZONA
TUCSON AZ 85721
U.S.A.

LARSON RICHARD B PROF
YALE UNIVERSITY
OBSERVATORY BOX 2023
YALE STATION
NEW HAVEN CT 06520
U.S.A.

LARSSON LEANDER GUNNAR
PROF
LUND OBSERVATORY
SVANEGATAN 9
S-222 24 LUND
SWEDEN

LASHER GORDON JEWETT DR
IBM WATSON RES CTR
YORKTWN HTS NY 10598
U.S.A.

LASKARIDES PAUL G ASSPROF
DEPT OF ASTRONOMY
UNIVERSITY OF ATHENS
PANEPISTIMIOPOLIS
ATHENS 621
GREECE

LASKER BARRY M DR
CERRO TOLOLO
INTERAMERICAN OBS
CASILLA 63 D
LA SERENA
CHILE

LASOTA JAN PIOTR DR
COPERNICUS ASTRONO-
MICAL CENTER
UL BARTYCKA 18
00 716 WARSAW
POLAND

LATHAM DAVID W DR
CTR F ASTROPHYSICS
60 GARDEN ST
CAMBRIDGE MA 02138
U.S.A.

LATOUR JEAN J
JILA
UNIV OF COLORADO
BOULDER CO 80309
FRANCE

LATTIMER JAMES M DR
DEPT OF ASTRONOMY
UNIV OF ILLINOIS
URBANA IL 61801
U.S.A.

LATYPOV A A DR
ASTRONOMICAL INST
UZBEKIAN ACAD OF SC
700000 TASHKENT
U.S.S.R.

LAUBERTS ANDRIS DR
ASTRONOMICAL
OBSERVATORY
BOX 515
S-751 20 UPPSALA
SWEDEN

LAUNAY J M DR
OBS DE MEUDON
F-92190 MEUDON
FRANCE

LAURENT BERTIL E PROF
INSTITUTE FOR
THEORETICAL PHYSICS
VANADISVAEGEN 9
S-113 46 STOCKHOLM
SWEDEN

LAURENT C DR
L P S P
B P NO 10
F-91370 VERRIERES
LE-BUISSON
FRANCE

LAUSBERG ANDRE DR
RUE STE JULIENNE 57
B-4621 RETINNE
BELGIUM

LAUSTSEN SVEND DR
AARHUS UNIVERSITY
DK-8000 AARHUS C
DENMARK

LAUTERBORN D PROF DR
HAMBURGER STERNWARTE
D-2050 HAMBURG 80
GERMANY, F.R.

LAUTMAN D A DR
SMITHSONIAN ASTRO-
PHYSICAL
60 GARDEN STREET
CAMBRIDGE MA 02138
U.S.A.

LAVAL ANNIE DR
OBSERVATOIRE DE
MARSEILLE
2 PLACE LE VERRIER
F-13004 MARSEILLE
FRANCE

LAVDOVSKIJ VICTOR V DR
PULKOVO OBSERVATORY
196140 LENINGRAD
U.S.S.R.

LAVROV M I DR
ENGELHARDT ASTRONO-
MICAL OBSERVATORY
422526 KAZAN
U.S.S.R.

LAVRUKHINA A K PROF DR
INSTITUTE OF GEOCHE-
MISTRY AND ANALYTIC
CHEMISTRY
117334 MOSCOW
U.S.S.R.

LAWRENCE G M DR
LASP
SPACE SCIENCE BLDG
UNIV OF COLORADO
BOULDER CO 80302
U.S.A.

LAYZER DAVID PROF
HARVARD COLLEGE
OBSERVATORY
60 GARDEN ST
CAMBRIDGE MA 02138
U.S.A.

LAZOVIC JOVAN P PROF
DEPT OF ASTRONOMY
FACULTY OF SCIENCES
STUDENTSKI TRG 16
11000 BEOGRAD
YUGOSLAVIA

LE CONTEL JEAN M
OBSERVATOIRE DE NICE
LE MONT GROS
B P 252
F-06007 NICE CEDEX
FRANCE

LE POOLE RUDOLF S DR
STERREWACHT
POSTBUS 9513
2300 RA LEIDEN
THE NETHERLANDS

LE SQUEREN ANNE-MARIE DR
OBSERVATOIRE
DE MEUDON
F-92190 MEUDON
FRANCE

LEA SUSAN MAUREEN DR
ASTRONOMY DEPT
UNIV OF CALIFORNIA
BERKELEY CA 94708
U.S.A.

LEBEDINETS V N DR
ASTRONOMICAL COUNCIL
USSR ACADEMY OF SCI
PYATNITSKAYA UL 48
109017 MOSCOW
U.S.S.R.

LEBLANC YOLANDE DR
OBSERVATOIRE
DE MEUDON
F-92190 MEUDON
FRANCE

LEBOVITZ NORMAN R PROF
MATHEMATICS DEPT
UNIV OF CHICAGO
5734 S UNIV AVE
CHICAGO IL 60637
U.S.A.

LECAR MYRON DR
CTR FOR ASTROPHYSICS
60 GARDEN ST
CAMBRIDGE MA 02138
U.S.A.

LECKRONE DAVID S DR
GODDARD SPACE
FLIGHT CENTER
CODE 671
GREENBELT MD 20771
U.S.A.

LEDERLE TRUDPERT DR
ASTRONOMISCHES
RECHEN INSTITUT
MOENCHHOFSTR 12 14
D-6900 HEIDELBERG 1
GERMANY, F.R.

LEDOUX P J PROF
INST D ASTROPHYSIQUE
UNIVERSITE DE LIEGE
B-4200 COINTE OUGREE
BELGIUM

LEE P D DR
LOUISIANA STATE UNIV
BATON ROUGE LA 70803
U.S.A.

LEE SEE-WOO DR
DEPT EARTH SCIENCES
KYUNGPOOK NATIONAL
UNIVERSITY
TAEGU CITY
REPUBLIC OF KOREA

LEE TERENCE J DR
U.K.INFRARED
TELESCOPE UNIT
900 LEILANI STREET
HILO HI 96720
U.S.A.

LEER EGIL PROF
AURORAL OBSERVATORY
UNIVERSITY OF TROMSO
P O BOX 953
N-9001 TROMSO
NORWAY

LEFEBVRE MICHEL DR
CNES-GRGS
18 AVE E BELIN
F-31055 TOULOUSE CED
FRANCE

LEFEVRE JEAN DR
OBSERVATOIRE DE NICE
B P 252
F-06007 NICE CEDEX
FRANCE

LEGG THOMAS H DR
HERZBERG INSTITUTE
OF ASTROPHYSICS
NAT RES COUNC CANADA
OTTAWA ONT K1A 0R6
CANADA

LEHNERT BO P PROF
DEPT PLASMA PHYSICS
ROYAL INSTITUTE OF
TECHNOLOGY
S-10044 STOCKHOLM 70
SWEDEN

LEIBACHER JOHN DR
SPACE ASTRONONOMY GR
LOCKHEED 52-12/202
3251 HANOVER ST
PALO ALTO CA 94304
U.S.A.

LEIBOWITZ ELIA M DR
DEPT OF PHYSICS
AND ASTRONOMY
TEL AVIV UNIVERSITY
TEL AVIV
ISRAEL

LEIGHTON R B PROF
HALE OBSERVATORIES
CALIF INST OF TECHN
1201 E CALIFORNIA ST
PASADENA CA 91125
U.S.A.

LEIKIN G A DR
ASTRONOMICAL COUNCIL
USSR ACAD OF SCI
PYATNITSKAYA UL 48
109017 MOSCOW
U.S.S.R.

LEINERT CHRISTOPH DR
MPI F ASTRONOMIE
KOENIGSTUHL
D-6900 HEIDELBERG 1
GERMANY, F.R.

LELIEVRE G DR
OBS DE MEUDON
F-92190 MEUDON
FRANCE

LEMAIRE PHILIPPE DR
L P S P CNRS
B P 10
F-91370 VERRIERES
LE BUISSON
FRANCE

LEMAITRE GERARD R DR
OBSERVATOIRE
DE MARSEILLE
2 PLACE LE VERRIER
F-13004 MARSEILLE
FRANCE

LEMKE DIETRICH DR
MPI FUER ASTRONOMIE
KOENIGSTUHL
D-6900 HEIDELBERG 1
GERMANY, F.R.

LENA PIERRE J PROF
OBSERVATOIRE
DE MEUDON
F-92190 MEUDON
FRANCE

LEORAT JACQUES DR
OBSERVATOIRE
DE MEUDON
F-92190 MEUDON
FRANCE

LEPINE JACQUES R D DR
CRAAM/ON/CNPQ
RUA CEARA 290
SAO PAULO
BRAZIL

LEQUEUX JAMES DR
OBSERVATOIRE DE
PARIS-MEUDON
F-92190 MEUDON
FRANCE

LEROY JEAN LOUIS
OBSERVATOIRE
DE PIC DE MIDI
F-65200 BAGNERES
DE BIGORRE
FRANCE

LESTER JOHN B DR
ASTRONOMY DEPT
ERINDALE COLLEGE
UNIV OF TORONTO
MISSISSAUGA L5L 1C6
CANADA

LETFUS VOJTECH DR
ASTRONOMICAL INST
CZECH ACAD SCIENCES
251 65 ONDREJOV
OBSERVATORY
CZECHOSLOVAKIA

LEUNG CHUN MING DR
DEPT OF PHYSICS
RENSSELAER POLYTECHN
INSTITUTE
TROY NY 12181
U.S.A.

LEUNG KAM CHING PROF
BEHLEN OBSERVATORY
DEPT PHYSICS ASTRON
UNIV OF NEBRASKA
LINCOLN NB 68588
U.S.A.

LEVALLOIS J J DR
INSTITUT GEOGRAPHI-
QUE NATIONAL
2 AVE PASTEUR
F-94160 SAINT-MANDE
FRANCE

LEVASSEUR-REGOURD
ANNY-CHANTAL DR
SERV AERO CNRS BP 2
F-91370 VERRIERES
LE BUISSON
FRANCE

LEVATO ORLANDO H DR
CALLE 38 NO 463
1900 LA PLATA
ARGENTINA

LEVIN B J DR
ASTRONOMICAL COUNCIL
USSR ACADEMY OF SCI
PYATNITSKAYA UL 48
109017 MOSCOW
U.S.S.R.

LEVINE RANDOLPH H DR
CTR F ASTROPHYSICS
60 GARDEN ST
CAMBRIDGE MA 02138
U.S.A.

LEVY EUGENE H DR
DEPT PLANETARY SCI
LUNAR AND PLANET LAB
UNIVERSITY ARIZONA
TUCSON AZ 85721
U.S.A.

LEVY JAQUES R DR
OBSERVATOIRE
DE PARIS
61 AV L OBSERVATOIRE
F-75014 PARIS
FRANCE

LEWIN WALTER H G PROF
MIT 37-627
CAMBRIDGE MA 02139
U.S.A.

LEWIS BRIAN MURRAY DR
CARTER OBSERVATORY
P O BOX 2909
WELLINGTON
NEW ZEALAND

LI J Y
PYONGYANG ASTRONOMI-
CAL OBS
ACADEMY OF SCIENCES
PYONGYANG
DEM PEOPLE'S REP OF KOREA

LI NED C DR
CALIFORNIA UNIV
6531 WITHWORTH RD
LOS ANGELES CA 90035
U.S.A.

LIDDELL U MR
NASA LUNAR AND PLAN-
ETARY PROGRAM OFF OF
SPACE SC AND APPL
WASHINGTON DC 20546
U.S.A.

LIEBSCHER D E DR
ZNTRINST ASTROPHYSIK
STERNW BABELSBERG
ROSA LUXEMBURGST 17A
DDR-1502 POTSDAM
G.D.R.

LIESKE JAY H DR
CALIF INST TECHN
JET PROPULSION LAB
4800 OAK GROVE DRIVE
PASADENA CA 91103
U.S.A.

LILLER MARTHA H DR
HARVARD COLLEGE
OBSERVATORY
CAMBRIDGE MA 02138
U.S.A.

LILLER WILLIAM PROF
CTR F ASTROPHYSICS
60 GARDEN ST
CAMBRIDGE MA 02138
U.S.A.

LILLEY EDWARD A PROF
HARVARD COLLEGE
OBSERVATORY
60 GARDEN ST
CAMBRIDGE MA 02138
U.S.A.

LILLIE C F DR
DEPT PHYSICS ASTRON
UNIV OF COLORADO
BOULDER CO 80302
U.S.A.

LIN CHIA C PROF
MIT
DEPT OF MATHEMATICS
77 MASSACHUSETTS AVE
CAMBRIDGE MA 02139
U.S.A.

LIN DOUGLAS E DR
INST OF ASTRONOMY
MADINGLEY ROAD
CAMBRIDGE CB3 0HA
U.K.

LINCOLN J VIRGINIA MISS
WDC-A FOR SOLAR-
TERRESTRIAL PHYSICS
NOAA D63
BOULDER CO 80303
U.S.A.

LINDBLAD BERTIL A DR
LUND OBSERVATORY
SVANEGATAN 9
S-222 24 LUND
SWEDEN

LINDBLAD PER OLOF PROF
STOCKHOLM
OBSERVATORY
S-133 00
SALTSJOEBADEN
SWEDEN

LINFOOT E H DR
THE OBSERVATORIES
MADINGLEY ROAD
CAMBRIDGE CB3 0HA
U.K.

LING CHIH-BING DR
INST OF MATHEMATICS
ACADEMIA SINICA
P O BOX NO 143
TAIPEI
TAIWAN

LINGENFELTER RICHARD E
PROF
DEPT OF ASTRONOMY
UNIV OF CALIFORNIA
LOS ANGELES CA 90024
U.S.A.

LINK FRANCOIS PROF
INSTITUT
D ASTROPHYSIQUE
98 BIS BLVD ARAGO
F-75014 PARIS
FRANCE

LINKE RICHARD ALAN DR
BELL LABS
CRAWFORD HILL LAB
HOLMDEL NJ 07733
U.S.A.

LINNELL ALBERT P PROF
DEPT OF ASTRONOMY
AND ASTROPHYSICS
MICHIGAN STATE UNIV
E LANSING MI 48824
U.S.A.

LINNIK V P PROF DR
MAIN ASTRONOMICAL
OBSERVATORY PULKOVO
196140 LENINGRAD
U.S.S.R.

LINSKY JEFFREY L DR
JILA
UNIVERSITY OF
COLORADO
BOULDER CO 80309
U.S.A.

LIPOVETSKY V A
SPECIAL ASTROPHYS OB
ZELENCUK
357140 STAVROPOLSKIJ
KRAJ
U.S.S.R.

LIPPINCOTT SARAH LEE DR
SPROUL OBSERVATORY
SWARTHMORE COLLEGE
SWARTHMORE PA 19081
U.S.A.

LITTLE ALEC G ASSOC PROF
PHYSICS DEPT
UNIVERSITY OF SYDNEY
SYDNEY NSW 2006
AUSTRALIA

LITTLE C GORDON
CHIEF DIV 87
NAT BUREAU STANDARDS
RM 3001 RADIO BLDG
BOULDER CO 80302
U.S.A.

LITTLE LESLIE T DR
ELECTRONICS LAB
UNIVERSITY OF KENT
CANTERBURY
KENT CT2 7NT
U.K.

LITVAK M DR
SMITHSONIAN ASTRO-
PHYSICAL OBS
CAMBRIDGE MA 02138
U.S.A.

LIU SOU-YANG DR
COMPUTER SCIENCES CO
SYSTEM SCIENCES DIV
8728 COLESVILLE ROAD
SILVER SPG MD 20910
U.S.A.

LIVINGSTON WILLIAM C
KITT PEAK NATIONAL
OBSERVATORY
P O BOX 26732
TUCSON AZ 85726
U.S.A.

LIVSHITS M A DR
INSTITUTE OF TERRE-
STRIAL MAGNETISM AND
IONOSPHERE
142092 AKADEMGORODOK
U.S.S.R.

LLOYD-EVANS THOMAS DR
S A ASTRONOMICAL OBS
P O BOX 9
OBSERVATORY
CAPE 7935
SOUTH AFRICA

LOCHTE-HOLTGREVEN W PROF
INSTITUT FUER EXPERI
MENTALPHYSIK DER
UNIVERSITAET
D-2300 KIEL
GERMANY, F.R.

LODEN KERSTIN R DR
STOCKHOLM
OBSERVATORY
S-133 00
SALTSJOEBADEN
SWEDEN

LONGAIR M S DR
ASTRON ROYAL FOR
SCOTLAND
ROYAL OBS
EDINBURGH EH9 3JH
U.K.

LOPEZ JOSE A ING
OBS ASTRONOMICO
FELIX AGUILAR
AV BENAVIDEZ 8175 OE
5407 MARQUESADO S J
ARGENTINA

LORTET MARIE CLAIRE
D A F
OBSERVATOIRE
DE MEUDON
F-92190 MEUDON
FRANCE

LOUGHHEAD RALPH E DR
NATIONAL MEASUREMENT
LABORATORY
BRADFIELD RD
LINDFIELD NSW 2070
AUSTRALIA

LOVAS FRANCIS JOHN DR
MOLECULAR SPECTR
DIV 545
NAT BUREAU STANDARDS
WASHINGTON DC 20234
U.S.A.

LOVELL SIR BERNARD PROF
JODRELL BANK
CHESHIRE SK11 9PL
U.K.

LOYOLA PATRICIO DR
OBS ASTRON NACIONAL
UNIVERSIDAD DE CHILE
CASILLA 36-D
SANTIAGO
CHILE

LU PHILLIP K DR
DEPT PHYSICS ASTRON
WESTERN CONN ST COL
181 WHITE ST
DANBURY CT 06810
U.S.A.

LUCCHIN F
IST DI FISICA
G GALILEI
VIA MARZOLO 8
I-35100 PADOVA
ITALY

LO KWOK-YUNG DR
OWENS VLY RADIO OBS
CALTECH 102-24
PASADENA CA 91125
U.S.A.

LOCKE JACK L DR
HERZBERG INSTITUTE
OF ASTROPHYSICS
NAT RES COUNC CANADA
OTTAWA ONT K1A OR6
CANADA

LODEN LARS OLOF PROF
ASTRONOMICAL OBSERVA
TORY BOX 515
S-751 20 UPPSALA
SWEDEN

LONGMORE ANDREW J
63 MOAT VIEW
ROSLIN MIDLOTHIAN
U.K.

LOPEZ-ARROYO M
OBSERVATORIO
ASTRONOMICO
ALFONSO X11 5
MADRID 7
SPAIN

LOSCO LUCETTE DR
FACULTE DES SCIENCES
BESANCON
F-25030 BESANCON
CEDEX
FRANCE

LOUISE RAYMOND DR
OBSERVATOIRE
DE MARSEILLE
2 PLACE LE VERRIER
F-13004 MARSEILLE
FRANCE

LOVAS MIKLOS
KONKOLY OBSERVATORY
BOX 67
1525 BUDAPEST
HUNGARY

LOW FRANK J DR
4940 CALLE BARRIL
TUCSON AZ 85718
U.S.A.

LOZINSKAYA T A DR
STERNBERG STATE AS-
TRONOMICAL INSTITUTE
117234 MOSCOW
U.S.S.R.

LUB JAN DR
C/O ESO
SCHLEISSHEIMERSTR 17
D-8046 GARCHING B M
GERMANY, F.R.

LUCKE PETER B DR
OBS DE GENEVE
CH-1290 SAUVERNY GE
SWITZERLAND

LOCANTHI DOROTHY DAVIS DR
2180 PINECREST DR
ALTADENA CA 91001
U.S.A.

LOCKWOOD G WESLEY DR
LOWELL OBSERVATORY
P O BOX 1269
FLAGSTAFF AZ 86002
U.S.A.

LOHMANN WERNER PROF
ASTRONOMISCHES
RECHEN INSTITUT
MOENCHHOFSTR 12 14
D-6900 HEIDELBERG
GERMANY, F.R.

LOPEZ GARCIA ZUELMA L DR
OBS ASTRONOMICO
FELIX AGUILAR
AV BENAVIDEZ 8175 OE
5407 MARQUESADO S J
ARGENTINA

LOREN ROBERT BRUCE DR
MILLIMETER WAVE OBS
MCDONALD OBS
UNIV OF TEXAS
FT DAVIS TX 79734
U.S.A.

LOTOVA N A DR
PHYSICAL INSTITUTE
USSR ACADEMY OF
SCIENCES
117924 MOSCOW
U.S.S.R.

LOULERGUE M DR
OBS DE MEUDON
F-92190 MEUDON
FRANCE

LOVELACE RICHARD V E DR
SPACE SCIENCES BLDG
CORNELL UNIV
ITHACA NY 14853
U.S.A.

LOWE ROBERT P DR
DEPT OF PHYSICS
UNIV WESTERN ONTARIO
LONDON ONT N6A 3K7
CANADA

LOZINSKIJ A M DR
ASTRONOMICAL COUNCIL
USSR ACADEMY OF SCI
PYATNITSKAYA UL 48
109017 MOSCOW
U.S.S.R.

LUCAS R DR
ECOLE NORMALE SUP
24 RUE LHOMOND
F-75005 PARIS
FRANCE

LUCY LEON B PROF
DEPT OF ASTRONOMY
COLUMBIA UNIVERSITY
NEW YORK NY 10027
U.S.A.

LUEST REIMAR PROF
MAX PLANCK SOCIETY
POSTFACH 647
D-8000 MUENCHEN 1
GERMANY, F.R.

LUEST RHEA DR
MPI FUER PHYSIK UND
ASTROPHYSIK
FOEHRINGER RING 6
D-8000 MUENCHEN 40
GERMANY, F.R.

LUKACEVIC ILIJA S DR
FACULTY OF SCIENCES
DEPT OF MECHANICS
STUDENTSKI TRG 16
11000 BEOGRAD
YUGOSLAVIA

LUMME KARI A DR
OBSERVATORY
TAHTITORNINMAKI
SF-00130 HELSINKI 13
FINLAND

LUNDQUIST CHARLES A DR
SPACE SCIENCES
LABORATORY
CODE ES51
HUNTSVILLE AL 35812
U.S.A.

LUNEL MADELEINE DR
OBSERVATOIRE
DE LYON
F-69230 SAINT GENIS
LAVAL
FRANCE

LUNGU NICOLAIE DR
INST POLITEHNIC
CAT DE MATEMATICA
STR EMIL ISAC 15
3400 CLUJ-NAPOCA
ROUMANIA

LUTZ BARRY L DR
LOWELL OBSERVATORY
P O BOX 1269
FLAGSTAFF AZ 86002
U.S.A.

LUTZ JULIE H DR
DEPT OF MATHEMATICS
WASHINGTON STATE
UNIVERSITY
PULLMAN WA 99164
U.S.A.

LUTZ THOMAS E DR
DEPT OF PURE AND
APPLIED MATHEMATICS
WASHINGTON ST UNIV
PULLMAN WA 99164
U.S.A.

LUUD LAURI DR
TARTU OBSERVATORY
202444 ESTONIA
U.S.S.R.

LUYTEN WILLEM J PROF
SPACE SCIENCE CENTER
UNIVERSITY OF
MINNESOTA
MINNEAPOLIS MN 55455
U.S.A.

LYNDEN-BELL DONALD PROF
INST OF ASTRONOMY
MADINGLEY ROAD
CAMBRIDGE CB3 0HA
U.K.

LYNDS BEVERLY T DR
KITT PEAK NATIONAL
OBSERVATORY
P O BOX 26732
TUCSON AZ 85726
U.S.A.

LYNDS ROGER C DR
KITT PEAK NATIONAL
OBSERVATORY
P O BOX 26732
TUCSON AZ 85726
U.S.A.

LYNE ANDREW G DR
NRAL
JODRELL BANK
MACCLESFIELD
CHESHIRE SK11 9PL
U.K.

LYNGA GOSTA DR
LUND OBSERVATORY
SVANEGATAN 9
S-222 24 LUND
SWEDEN

LYTTKENS EJNAR DR
SKOLGATAN 33 B
S-752 21 UPPSALA
SWEDEN

LYTTLETON RAYMOND A PROF
ST JOHNS COLLEGE
CAMBRIDGE
U.K.

LYUTY VICTOR M DR
CRIMEAN STATION OF
STERNBERG INSTITUTE
NAUCHNYJ
334413 CRIMEA
U.S.S.R.

MACCACARO TOMMASO DR
LAB RADIOASTRONOMIA
VIA IRNERIO 46
I-40126 BOLOGNA
ITALY

MACCHETTO FERDINANDO DR
EUROPEAN SP AGENCY
SPACE SCIENCE DEPT
ESTEC
NOORDWIJK
THE NETHERLANDS

MACCONNELL DARRELL J DR
CENTRO DE INVEST DE
ASTRONOMIA
APARTADO 264
MERIDA
VENEZUELA

MACDONALD GEOFFREY H DR
ELECTRONICS LAB
UNIVERSITY OF KENT
CANTERBURY
KENT CT2 7NT
U.K.

MACGILLIVRAY HARVEY T DR
ROYAL OBSERVATORY
BLACKFORD HILL
EDINBURGH EH9 3HJ
U.K.

MACHADO JOSE M A B DR
TECHNICAL UNIVERSITY
OF LISBON
AV DA IGREJA 17 1 D
LISBOA 5
PORTUGAL

MACHALSKI JERZY DR
UL MAZOWIECKA 36/33
30-019 KRAKOW
POLAND

MACKAY CRAIG D DR
INSTITUTE ASTRONOMY
UNIVERSITY CAMBRIDGE
MADINGLEY RD
CAMBRIDGE CB3 0HA
U.K.

MACLEOD JOHN M DR
HERZBERG INSTITUTE
OF ASTROPHYSICS
NAT RES COUNC CANADA
OTTAWA ONT K1A 0R6
CANADA

MACQUEEN ROBERT M DR
HIGH ALTITUDE
OBSERVATORY
P O BOX 3000
BOULDER CO 80307
U.S.A.

MACRAE DONALD A PROF
DAVID DUNLAP OBS
P O BOX 360
RICHMOND HILL
ONT L4C 4Y6
CANADA

MACRIS CONSTANTIN J PROF
RSAAM ACADEMY ATHENS
ANAGNOSTOPOULOU 14
ATHENS 136
GREECE

MACY WILLIAM WRAY DR
INST FOR ASTRONOMY
UNIV OF HAWAII
2800 WOODLAWN DRIVE
HONOLULU HI 96822
U.S.A.

MADDISON RONALD CH DR
2 CHURCH PLANTATION
KEELE PARK
KEELE STAFFS
U.K.

MADORE BARRY FRANCIS DR
DAVID DUNLAP OBS
UNIV OF TORONTO
RICHMOND HL L4C 4Y6
CANADA

MAEDER ANDRE DR
OBSERVATOIRE
DE GENEVE
CH-1290 SAUVERNY GE
SWITZERLAND

MAEHARA HIDEO DR
TOKYO ASTRONOMICAL
OBSERVATORY
OSAWA MITAKA
TOKYO 181
JAPAN

MAETZLER CHRISTIAN DR
INST APPLIED PHYSICS
SIDLERSTRASSE 5
CH-3012 BERN
SWITZERLAND

MAFFEI PAOLO PROF
OSSERVATORIO
ASTROFISICO
I-95125 CATANIA
ITALY

MAGALASHVILI N L DR
ABASTUMANI ASTROPHY-
SICAL OBSERVATORY
383762 ABASTUMANI
GEORGIA
U.S.S.R.

MAGALHAES ANTONIO A S DR
OBS ASTRON DE BARROS
UNIV DO PORTO
MONTE DA VIRGEM
4400 VILA NOVA GAIA
PORTUGAL

MAGNAN CHRISTIAN DR
LAT
98 BIS BLVD ARAGO
F-75014 PARIS
FRANCE

MAGNARADZE N G DR
TBILISI STATE
UNIVERSITY
380043 TBILISI
U.S.S.R.

MAGUN ANDREAS DR
INST APPLIED PHYSICS
SIDLERSTRASSE 5
CH-3012 BERN
SWITZERLAND

MAHDY HAMED A DR
HELWAN OBSERVATORY
HELWAN-CAIRO
ARAB REP. OF EGYPT

MAHMOUD FAROUK M A B DR
HELWAN OBSERVATORY
HELWAN CAIRO
ARAB REP. OF EGYPT

MAHRA H S DR
UTTAR PRADESH ST OBS
MANORA PEAK
NAINITAL 263 129
INDIA

MAIHARA TOSHINORI DR
DEPT OF PHYSICS
KYOTO UNIVERSITY
SAKYOKU KYOTO 606
JAPAN

MAILLARD JEAN-PIERRE DR
OBSERVATOIRE
DE MEUDON
F-92190 MEUDON
FRANCE

MAITZEN HANS M DR
UNIVERSITAETS-
STERNWARTE
TUERKENSCHANZSTR 17
A-1180 WIEN
AUSTRIA

MAKARENKO EKATERINA N DR
ASTRONOMICAL OBS
PARK SHEVCHENKO
270014 ODESSA
U.S.S.R.

MAKAROVA ELENA A DR
STERNBERG ASTRONOMI-
CAL INSTITUTE
117234 MOSCOW
U.S.S.R.

MAKITA MITSUGU DR
TOKYO ASTRONOMICAL
OBSERVATORY
OSAWA MITAKA
TOKYO 181
JAPAN

MALACARA DANIEL
INAOE
AP POSTAL 216 Y51
TONANTZINTLA
PUEBLA PUE
MEXICO

MALAISE DANIEL J DR
INSTITUT
D ASTROPHYSIQUE
B-4200 COINTE-OUGREE
BELGIUM

MALAKPUR IRADJ DR
INST OF GEOPHYSICS
TEHRAN UNIVERSITY
AMIRABAD SHOMALI
TEHRAN
IRAN

MALARODA STELLA M DR
OBS ASTRONOMICO
PASEO DEL BOSQUE
1900 LA PLATA
ARGENTINA

MALIN DAVID F DR
ANGLO-AUSTRALIAN OBS
P O BOX 296
EPPING N S W 2121
AUSTRALIA

MALINOVSKY MONIQUE DR
L P S P
B P NO 10
F-91370 VERRIERES LE
BUISSON
FRANCE

MALITSON HARRIET H MS
NASA GODDARD SPACE
FLIGHT CENTER
CODE 601
GREENBELT MD 20771
U.S.A.

MALLIA EDWARD A DR
DEPT OF ASTROPHYSICS
SOUTH PARKS ROAD
OXFORD OX1 3RQ
U.K.

MALLIK D C V DR
INDIAN INSTITUTE OF
ASTROPHYSICS
BANGALORE 560 034
INDIA

MALMQUIST K GUNNAR PROF
ST OLOFSGATAN 10 A
S-753 21 UPPSALA
SWEDEN

MALTBY PER PROF
INST THEOR PHYSICS
UNIVERSITY OF OSLO
P O BOX 1029
BLINDERN OSLO 3
NORWAY

MALVILLE J MCKIM PROF
DEPT ASTROGEOPHYSICS
UNIVERSITY OF
COLORADO
BOULDER CO 80302
U.S.A.

MAMEDBEJLI G D DR
SHEMAKHA ASTROPHYSI-
CAL OBSERVATORY
373243 AZERBAIDZAN
U.S.S.R.

MAMMANO A DR
OSS ASTROFISICO
I-36012 ASIAGO VIC
ITALY

MANARA ALESSANDRO A DR
OSSERVATORIO ASTRONO
MICO DI MILANO
VIA BRERA 28
I-20121 MILANO
ITALY

MANCHESTER RICHARD N DR
CSIRO RADIOPHYSICS
P O BOX 76
EPPING NSW 2121
AUSTRALIA

MANCUSO-SANTI DR
CAPODIMONTE ASTRONO-
MICAL OBSERVATORY
VIA MOIARIELLO 16
I-80131 NAPOLI
ITALY

MANDELSTAM S L PROF
INST OF SPECTROSCOPY
USSR ACADEMY OF SC
142092 AKADEMIGORO-
DOK MOSCOW REGION
U.S.S.R.

MANFROID JEAN DR
INST D'ASTROPHYSIQUE
UNIVERSITE DE LIEGE
AVENUE DE COINTE 5
B-4200 COINTE-OUGREE
BELGIUM

MANGENEY ANDRE DR
OBSERVATOIRE DE
MEUDON
F-92190
FRANCE

MANNINO GIUSEPPE PROF
ISTITUTO MATEMATICO
VIA CAMPI 181
I-41100 MODENA
ITALY

MANRIQUE WALTER T PROF
OBS ASTRONOMICO
FELIX AGUILAR
AV BENAVIDEZ 8175 OE
5407 MARQUESADO S J
ARGENTINA

MANSFIELD VICTOR N PROF
COLGATE UNIVERSITY
HAMILTON NY 13346
U.S.A.

MARAN STEPHEN P DR
NASA-GSFC CODE 680
GREENBELT MD 20771
U.S.A.

MARASCHI LAURA DR
ISTITUTO DI FISICA
VIA CELORIA 16
I-20133 MILANO
ITALY

MARCHAL C DR
DEPT D'ETUDES DE
SYNTHESE
ONERA
F-92320 CHATILLON
FRANCE

MARCUS ELLA PROF
OBSERVATOIRE
ASTRONOMIQUE
5 R CUTITUL ARGINT
BUCAREST 75212
ROUMANIA

MARGON BRUCE H PROF
ASTRONOMY DEPT
UNIVERSITY OF CALIF
LOS ANGELES CA 90024
U.S.A.

MARGONI R
OSS ASTROFISICO
I-36012 ASIAGO VIC
ITALY

MARIK MIKLOS DR.
ASTRONOMICAL DEPT.
EOTVOS UNIVERSITY
KUN BELA TER 2
H-1083 BUDAPEST
HUNGARY

MARILLI ETTORE DR
OSS ASTROFISICO
CITTA UNIVERSITARIA
I-95125 CATANIA
ITALY

MARINO BRIAN F ENG
BOX 72009
NORTHCOTE
AUCKLAND 9
NEW ZEALAND

MARK JAMES WAI-KEE DR
RES LAB ELECTRONICS
MASS INST TECHNOLOGY
BLDG 26 RM 215
CAMBRIDGE MA 02139
U.S.A.

MARKARIAN B E DR
BYURAKAN ASTROPHYSI-
CAL OBSERVATORY
378433 ARMENIA
U.S.S.R.

MARKELLOS VASSILIS V DR
ASTRONOMY DEPT
UNIV OF GLASGOW
GLASGOW G12 8QQ
U.K.

MARKKANEN TAPIO
OBSERVATORY
TAHTITORNINMAKI
SF-00130 HELSINKI 13
FINLAND

MARKOWITZ WILLIAM DR
2800 E SUNRISE BLVD
FT LAUDERDL FL 33304
U.S.A.

MARLBOROUGH J M PROF
DEPT OF ASTRONOMY
UNIVERSITY OF
WESTERN ONTARIO
LONDON ONT N6A 5B9
CANADA

MAROCHNIK L S PROF DR
USSR ACAD OF SCS
SPACE RESEARCH INST
PROFSOJOSNA A 84/32
117810 MOSCOW
U.S.S.R.

MAROV MIKHAIL YA DR
INST OF APPLIED MATH
USSR ACAD OF SCS
MIUSSKAYA SQ 4
MOSCOW A-45
U.S.S.R.

MARQUES DOS SANTOS P PROF
INST ASTRONOMICO E
GEOFISICO USP
CAIXA POSTAL 30627
01000 SAO PAULO SP
BRAZIL

MARQUES MANUEL N DR
OBSERVATORIO
ASTRONOMICO
TAPADA DA AJUDA
LISBOA 3
PORTUGAL

MARRACO HUGO G DR
OBS ASTRONOMICO
PASEO DEL BOSQUE
1900 LA PLATA
ARGENTINA

MARSDEN BRIAN G DR
SMITHSONIAN ASTROPHY
SICAL OBSERVATORY
CAMBRIDGE MA 02138
U.S.A.

MARSDEN PHILIP L PROF
DEPT OF PHYSICS
UNIVERSITY OF LEEDS
LEEDS LS2 9JT
U.K.

MARSH JULIAN C D
HATFIELD POLYTECHNIC
OBSERVATORY
BAYFORDBURY NR HERT-
FORD HERTS SG13 8LD
U.K.

MARSOGLU S DR
UNIV OBSERVATORY
UNIVERSITE
ISTANBUL
TURKEY

MARTEL MARIE-THERESE DR
OBSERVATOIRE
DE LYON
F-69230 SAINT GENIS
LAVAL
FRANCE

MARTIN DEREK H PROF
QUEEN MARY COLLEGE
MILE END ROAD
LONDON E1 4NS
U.K.

MARTIN F DR
DEPT D'ASTROPHYSIQUE
UNIV DE NICE
PARC VALROSE
F-06034 NICE CEDEX
FRANCE

MARTIN NICOLE DR
OBSERVATOIRE
DE MARSEILLE
2 PLACE LE VERRIER
F-13004 MARSEILLE
FRANCE

MARTIN PETER G PROF
DEPT OF ASTRONOMY
UNIVERSITY TORONTO
TORONTO ONT M5S 1A7
CANADA

MARTIN ROBERT N DR
MPI FUER RADIOASTRON
AUF DEM HUEGEL 69
D-5300 BONN
GERMANY, F.R.

MARTIN WILLIAM C DR
A167 PHYSICS
NATIONAL BUREAU OF
STANDARDS
WASHINGTON DC 20234
U.S.A.

MARTIN WILLIAM L DR
ROYAL GREENWICH OBS
HERSTMONCEUX CASTLE
HAILSHAM
E SUSSEX BN27 1RP
U.K.

MARTIN-LORON M DR
HERMANOS MIRALLES 14
MADRID 1
SPAIN

MARTINET LOUIS PROF
OBSERVATOIRE
DE GENEVE
CH-1290 SAUVERNY GE
SWITZERLAND

MARTINEZ M DR
INST DE ASTRONOMIA
UNAM
APT POSTAL 70-264
MEXICO DF
MEXICO

MARTINI ALDO DR
CNR ASTROFISICA
SPAZIALE C P 67
I-00044 FRASCATI
ROMA
ITALY

MARTINS DONALD HENRY DR
DEPT PHYSICS ASTRON
UNIV OF GEORGIA
ATHENS GA 30602
U.S.A.

MARTRES MARIE-JOSEPHE
OBSERVATOIRE DE
PARIS-MEUDON
F-92190 MEUDON
FRANCE

MARTYNOV D YA PROF DR
STERNBERG STATE AS-
TRONOMICAL INSTITUTE
117234 MOSCOW
U.S.S.R.

MARVIN URSULA B DR
CENTER FOR
ASTROPHYSICS
60 GARDEN ST
CAMBRIDGE MA 02138
U.S.A.

MARX GYORGY PROF.
EOTVOS UNIVERSITY
PUSKIN U 5
H-1088 BUDAPEST
HUNGARY

MARX S DR
ZNTRINST ASTROPHYSIK
KARL SCHWARZSCHILD
OBSERVATORIUM
DDR-6901 TAUTENBURG
G.D.R.

MASANI A PROF
OSS ASTRONOMICO DI
BRERA
VIA BRERA 28
I-20100 MILANO
ITALY

MASLOWSKI JOZEF DR
ASTRONOMICAL
OBSERVATORY
UL ORLA 171
30-244 KRAKOW
POLAND

MASNOU FRANCOISE DR
28 ALLEE GAMBAUDERIE
F-91190 GIF S YVETTE
FRANCE

MASNOU J L DR
SECT D'ASTROPHYSIQUE
CEN SACLAY
B P NO 2
F-91120 GIF-S-YVETTE
FRANCE

MASON HELEN E DR
DEPT APPLIED MATH
AND THEORET PHYSICS
SILVER STREET
CAMBRIDGE
U.K.

MASSEVICH A G DR
ASTRONOMICAL COUNCIL
USSR ACADEMY OF SCI
PYATNITSKAYA UL 48
109017 MOSCOW
U.S.S.R.

MASURSKY HAROLD PROF
US GEOLOGICAL SURVEY
BR OF ASTRO STUDIES
2255 NORTH GEMINI DR
FLAGSTAFF AZ 86001
U.S.A.

MATAS VLADIMIR R DR
ASTRONOMISCHES
RECHEN INSTITUT
MOENCHHOFSTR 12-14
D-6900 HEIDELBERG
GERMANY, F.R.

MATHEWS WILLIAM G PROF
LICK OBSERVATORY
UNIVERSITY OF CALIF
SANTA CRUZ CA 95064
U.S.A.

MATHEWSON DONALD S DR
MT STROMLO
OBSERVATORY
WODEN PO ACT 2606
AUSTRALIA

MATHIS JOHN S PROF
DEPT OF ASTRONOMY
UNIVERSITY OF
WISCONSIN
MADISON WI 53706
U.S.A.

MATIAGIN VALERY S DR
ASTROPHYSICAL INST
480068 ALMA-ATA
U.S.S.R.

MATSON DENNIS L DR
JPL 183-501
4800 OAK GROVE DRIVE
PASADENA CA 91103
U.S.A.

MATSUDA TAKUYA PROF
DEPT AERONAUTIC ENG
KYOTO UNIVERSITY
YOSHIDAHONMACHI
SAKYOKU KYOTO 606
JAPAN

MATSUMOTO MASAMICHI PROF
FACULTY OF
ENGINEERING
GIFU UNIVERSITY
504 KAGAMIGAHARA
JAPAN

MATSUMOTO TOSHIO DR
DEPT OF PHYSICS
NAGOYA UNIVERSITY
FUROCHO CHIKUSAKU
NAGOYA 464
JAPAN

MATSUOKA MASARU DR
INST SPACE AERON SC
UNIVERSITY OF TOKYO
KOMABA MEGURO
TOKYO 153
JAPAN

MATSUSHIMA SATOSHI DR
DEPT ASTRO PENNA ST
UNIV 525 DAVEY LAB
UNIVERSITY PARK
PA 16802
U.S.A.

MATTEI JANET AKYUZ DR
AAVSO
187 CONCORD AVE
CAMBRIDGE MA 02138
U.S.A.

MATTHEWS THOMAS A DR
ASTRONOMY PROGRAM
UNIVERSITY MARYLAND
COLLEGE PK MD 20742
U.S.A.

MATTIG W PROF DR
KIEPENHEUER INST
FUER SONNENPHYSIK
SCHOENECKSTRASSE 6
D-7800 FREIBURG-BR
GERMANY, F.R.

MATTILA KALEVI DR
OBSERVATORY
TAHTITORNINMAKI
SF-00130 HELSINKI 13
FINLAND

MATVEYENKO L I DR
SPACE RESEARCH
INSTITUTE
USSR ACADEMY OF SCI
117810 MOSCOW
U.S.S.R.

MATZNER RICHARD A PROF
PHYSICS DEPT
UNIVERSITY OF TEXAS
AUSTIN TX 78712
U.S.A.

MAUCHERAT-JOUBERT MME
LAB D'ASTRON SPAT
TRAVERSE DU SIPHON
LES TROIS LUCS
F-13012 MARSEILLE
FRANCE

MAUDER HORST PROF DR
ASTRONOMISCHES INST
WALDHAUSER STR 64
D-7400 TUEBINGEN
GERMANY, F.R.

MAURICE ERIC N
OBSERVATOIRE
DE MARSEILLE
2 PLACE LE VERRIER
F-13004 MARSEILLE
FRANCE

MAVRIDIS L N PROF
DEPT GEO ASTRONOMY
UNIVERSITY
OF THESSALONIKI
THESSALONIKI
GREECE

MAX CLAIRE E DR
L-477 LAWRENCE
LIVERMORE LABORATORY
P O BOX 808
LIVERMORE CA 94550
U.S.A.

MAXWELL ALAN DR
HARVARD RADIO
ASTRONOMY STATION
FORT DAVIS TX 79734
U.S.A.

MAY J ING
OBS RADIOASTRONOMICO
UNIV DE CHILE
CASILLA 68
MAIPU
CHILE

MAYALL MARGARET W
5 SPARKS ST
CAMBRIDGE MA 02138
U.S.A.

MAYALL NICHOLAS U ASTRON
7206 E CAMINO VEC'NO
TUCSON AZ 85715
U.S.A.

MAYER CORNELL H
SPACE SCIENCE DIV
NAVAL RESEARCH LAB
CODE 7130
WASHINGTON DC 20375
U.S.A.

MAYER PAVEL DR
ASTRONOMY DEPT
CHARLES UNIVERSITY
SVEDSKA 8
15000 PRAHA 5
CZECHOSLOVAKIA

MAYER URSULA DR
ASTRONOMISCHES INST
DER UNIVERSITAET
WALDHAUSERSTR 64
D-7400 TUEBINGEN
GERMANY, F.R.

MAYFIELD EARLE B DR
5536 MICHELLE DR
TORRANCE CA 90503
U.S.A.

MAYOR MICHEL DR
OBSERVATOIRE
DE GENEVE
CH-1290 SAUVERNY GE
SWITZERLAND

MAZUREK THADDEUS JOHN DR
DEPT OF ASTRONOMY
UNIV OF TEXAS
AUSTIN TX 78712
U.S.A.

MAZZITELLI ITALO DR
LAB ASTROFIZICA SPAZ
P O B 67
I-00044 FRASCATI
ITALY

MAZZUCCONI FABRIZIO DR
OSSERVATORIO ARCETRI
LARGO FERMI 5
I-50125 FIRENZE
ITALY

MCADAM W BRUCE DR
SCHOOL OF PHYSICS
UNIVERSITY OF SYDNEY
SYDNEY NSW 2006
AUSTRALIA

MCALISTER HAROLD A DR
DEPT OF PHYSICS
GEORGIA STATE UNIV
UNIVERSITY PLAZA
ATLANTA GA 30303
U.S.A.

MCBREEN BRIAN PHILIP DR
PHYSICS DEPT
UNIV COLLEGE
BELFIELD
DUBLIN 4
IRELAND

MCCABE MARIE K MS
INST FOR ASTRONOMY
2680 WOODLAWN DR
HONOLULU HI 96822
U.S.A.

MCCARROLL RONALD PROF
LAB D ASTROPHYSIQUE
UNIVERSITE
DE BORDEAUX I
F-33405 TALENCE
FRANCE

MCCARTHY DENNIS D DR
US NAVAL OBSERVATORY
WASHINGTON DC 20390
U.S.A.

MCCARTHY MARTIN F DR
SPECOLA VATICANA
CITTA DEL VATICANO
VATICAN CITY STATE

MCCLAIN EDWARD F
4133 MAPLE RD
MORNINGSIDE MD 20023
U.S.A.

MCCLINTOCK JEFFREY E DR
MASSACHUSETTS INSTI-
TUTE OF TECHNOLOGY
BLDG 37-521
CAMBRIDGE MA 02139
U.S.A.

MCCLURE ROBERT D PROF
DOMINION ASTROPHYSIC
OBSERVATORY
5071 SAANICH ROAD
VICTORIA BC V8X 3X3
CANADA

MCCLUSKEY GEORGE E JR DR
DIV OF ASTRONOMY
DEPT OF MATHEMATICS
LEHIGH UNIVERSITY
BETHLEHEM PA 18015
U.S.A.

MCCORD T B DR
INST F ASTRONOMY
2680 WOODLAWN DR
HONOLULU HI 96822
U.S.A.

MCCRACKEN KENNETH G DR
CSIRO
MINERAL PHYSICS SECT
P O BOX 136
NORTH RYDE NSW 2113
AUSTRALIA

MCCRAY RICHARD DR
JILA
UNIVERSITY OF
COLORADO
BOULDER CO 80309
U.S.A.

MCCREA WILLIAM H PROF
ASTRONOMY CENTRE
SUSSEX UNIVERSITY
BRIGHTON BN1 9QH
U.K.

MCCROSKY RICHARD E DR
CENTER FOR
ASTROPHYSICS
CAMBRIDGE MA 02138
U.S.A.

MCCULLOCH P M DR
DEPT OF PHYSICS
UNIV OF TASMANIA
HOBART TASMANIA
AUSTRALIA

MCCUTCHEON WILLIAM H PROF
DEPT OF PHYSICS
UNIVERSITY OF B C
2075 WESBROOK MALL
VANCOUVER BC V6T 1W5
CANADA

MCDONALD FRANK B DR
CODE 660
NASA GSFC
GREENBELT MD 20771
U.S.A.

MCDONALD J K PETRIE DR
DOMINION ASTROPHYSI-
CAL OBSERVATORY
5071 W SAANICH ROAD
VICTORIA BC V8X 3X3
CANADA

MCDONNELL J A M DR
SPACE SCIENCE LAB
UNIVERSITY OF KENT
CANTERBURY
KENT CT2 7NT
U.K.

MCDONOUGH THOMAS R DR
1299 CORDOVA RM 206
PASADENA CA 91106
U.S.A.

MCELROY M B DR
KITT PEAK NAT OBS
950 N CHERRY AVE
TUCSON AZ 85726
U.S.A.

MCGEE JAMES D PROF
PHYSICS DEPT
IMPERIAL COLLEGE
LONDON SW7 2BZ
U.K.

MCGEE RICHARD X DR
CSIRO DIVISION OF
RADIOPHYSICS
PO BOX 76
EPPING NSW 2121
AUSTRALIA

MCGIMSEY BEN Q JR DR
DEPT OF PHYSICS
GEORGIA STATE UNIV
UNIVERSITY PLAZA
ATLANTA GA 30303
U.S.A.

MCINTOSH BRUCE A DR
HERZBERG INSTITUTE
OF ASTROPHYSICS
NAT RES COUNCIL CAN
OTTAWA ONT K1A 0R6
CANADA

MCKEE CHRISTOPHER F PROF
PHYSICS DEPT
UNIVERSITY OF CALIF
BERKELEY CA 94720
U.S.A.

MCKEITH CONAL D DR
17 ADELAIDE PARK
BELFAST BT9 6FX
U.K.

MCKEITH NIALL ENDA DR
ST PATRICK'S COLLEGE
MAYNOOTH CO KILDARE
IRELAND

MCKENNA LAWLOR SUSAN
M P DR
ST PATRICKS COLLEGE
MAYNOOTH CO KILDARE
IRELAND

MCKINLEY DONALD W R DR
1889 FAIRMEADOW CR
OTTAWA ONT
CANADA

MCLAREN ROBERT A DR
DEPT OF ASTRONOMY
UNIV OF TORONTO
TORONTO ONT M5S 1A7
CANADA

MCLEAN DONALD J DR
DIV OF RADIOPHYSICS
CSIRO
P O BOX 76
EPPING NSW 2121
AUSTRALIA

MCLEAN IAN S DR
LUNAR AND PLANET LAB
UNIV OF ARIZONA
TUCSON AZ 85721
U.S.A.

MCMULLAN DENNIS DR
ROYAL GREENWICH
OBSERVATORY
HAILSHAM
EAST SUSSEX BN27 1RP
U.K.

MCNALLY DEREK DR
UNIVERSITY OF LONDON
OBSERVATORY
MILL HILL PARK
LONDON NW7 2QS
U.K.

MCNAMARA DELBERT H DR
DEPT OF PHYSICS AND
ASTRONOMY
BRIGHAM YOUNG UNIV
PROVO UT 84602
U.S.A.

MCVITTIE GEORGE C PROF
74 OLD DOVER ROAD
CANTERBURY
KENT CT1 3AY
U.K.

MCWHIRTER R W PETER DR
ASTROPHYSICS RES DIV
CULHAM LAB
ABINGDON
OXON OX14 3DB
U.K.

MEABURN J DR
DEPT OF ASTRONOMY
THE UNIVERSITY
MANCHESTER M13 9PL
U.K.

MEAD JAYLEE MONTAGUE DR
NASA-GSFC
CODE 680
GREENBELT MD 20771
U.S.A.

MEADOWS A JACK PROF
ASTRONOMY AND HISTO-
RY OF SCIENCE DEPT
UNIVERSITY LEICESTER
LEICESTER LE1 7RH
U.K.

MEBOLD U DR
MPI FUER RADIO-
ASTRONOMIE
AUF DEM HEUGEL 69
D-5300 BONN 1
GERMANY, F.R.

MEEKS M LITTLETON DR
LINCOLN LABORATORY
P O BOX 73
LEXINGTON MA 02173
U.S.A.

MEFFROY JEAN F PROF DR
INST MATHEMATIQUES
UNIV DES SCIENCES ET
TECHN DU LANGUEDOC
F-34060 MONTPELLIER
FRANCE

MEGESSIER CLAUDE DR
OBSERVATOIRE PARIS
61 AVE OBSERVATOIRE
F-75014 PARIS
FRANCE

MEGRELISHVILI T G PROF
ABASTUMANI ASTROPHY-
SICAL OBSERVATORY
383762 ABASTUMANI
GEORGIA
U.S.S.R.

MEHLTRETTER J P DR
KIEPENHEUER INST
FUER SONNENPHYSIK
SCHOENECKSTR 6
D-7800 FREIBURG-BR
GERMANY, F.R.

MEIN PIERRE
OBSERVATOIRE
DE MEUDON
5 PLACE JANSSEN
F-92190 MEUDON
FRANCE

MEINEL ADEN B PROF
OPTICAL SCIENCES CTR
UNIVERSITY OF
ARIZONA
TUCSON AZ 85721
U.S.A.

MEINIG MANFRED DR
ZENTRALINSTITUT
PHYSIK DER ERDE
TELEGRAFENBERG 17 A
DDR-15 POTSDAM
G.D.R.

MEISEL DAVID D DR
DEPT OF PHYSICS AND
ASTRONOMY
STATE UNIV COLLEGE
GENESEO NY 14454
U.S.A.

MEKLER YURI DR
DEPT OF GEOPHYSICS
TEL-AVIV UNIVERSITY
TEL-AVIV
ISRAEL

MELBOURNE WILLIAM G DR
JPL
4800 OAK GROVE DRIVE
PASADENA CA 91103
U.S.A.

MELCHIOR PAUL J PROF
OBSERVATOIRE ROYAL
DE BELGIQUE
AVENUE CIRCULAIRE 3
B-1180 BRUXELLES
BELGIUM

MELNIKOV OLEG A PROF DR
PULKOVO OBSERVATORY
196140 LENINGRAD
U.S.S.R.

MELROSE DONALD B DR
DEPT OF THEORETICAL
PHYSICS
UNIVERSITY OF SYDNEY
SYDNEY N S W 2006
AUSTRALIA

MEN' A V DR
INSTITUTE OF RADIO-
PHYSICS AND ELECTRO-
NICS
310085 KHARKOV
U.S.S.R.

MENDEZ MANUEL DR
INST DE ASTRONOMIA
UNAM
APT POSTAL 70-264
MEXICO DF
MEXICO

MENDEZ ROBERTO H DR
INST ASTRON FIS ESP
CASILLA 67
SUCURSAL 28
1428 BUENOS AIRES
ARGENTINA

MENDIS D ASOKA DR
APIS UCSD
LA JOLLA CA 92093
U.S.A.

MENDOZA V EUGENIO E DR
CIDA
APARTADO 264
MERIDA
VENEZUELA

MENEGUZZI MAURICE M DR
SEC ETUDES SPATIALES
DPEH SACLAY
BP 2
F-91190 GIF S YVETTE
FRANCE

MENNESSIER MARIE-ODILE DR
LABORATOIRE
D ASTRONOMIE
U S T L
F-34060 MONTPELLIER
FRANCE

MENON T K PROF
DEPT GEOPHYS/ASTRON
UNIVERSITY OF
BRITISH COLUMBIA
VANCOUVER BC V6T 1W5
CANADA

MENZIES JOHN W DR
SAAO
PO BOX 9
OBSERVATORY
CAPE TOWN 7935
SOUTH AFRICA

MERGENTALER JAN PROF
ASTRONOMICAL INST
KOPERNIKA 19
51-617 WROCLAW
POLAND

MERLEAU-PONTY J PROF
5 RUE DU GENERAL
DE CASTELNAU
F-75015 PARIS
FRANCE

MERMAN G A DR
INST OF THEORETICAL
ASTRONOMY
192187 LENINGRAD
U.S.S.R.

MERMAN NATALIA V DR
PULKOVO OBSERVATORY
196140 LENINGRAD
U.S.S.R.

MERRILL JOHN E DR
7836 HOSBROOK RD
CINCINNATI OH 45243
U.S.A.

MERTON GERALD MC MA PHD
17 HOLYWELL
OXFORD OX1 3SA
U.K.

MERTZ LAWRENCE N DR
287 FAIRFIELD ROAD
PALO ALTO CA 94306
U.S.A.

MERZANIDES CONSTANTINOS
DR
DEPT OF ASTRONOMY
UNIVERSITY OF
THESSALONIKI
GREECE

MESSAGE PHILIP J DR
DEPT APPLIED MATHS
THEORETICAL PHYSICS
THE UNIVERSITY
LIVERPOOL L69 3BX
U.K.

MESTEL LEON PROF
ASTRONOMY CENTRE
UNIVERSITY OF SUSSEX
BRIGHTON BN1 9QH
U.K.

MESZAROS PETER DR
MPI FUER PHYSIK UND
ASTROPHYSIK
FOEHRINGER RING 6
D-8000 MUENCHEN 40
GERMANY, F.R.

METZ K DR
INSTITUT F ASTRONO-
MIE UND ASTROPHYSIK
SCHEINERSTR 1
D-8000 MUENCHEN 80
GERMANY, F.R.

MEURERS JOSEPH PROF DR
UNIVERSITAETS-
STERNWARTE
TUERKENSCHANZSTR 17
A-1180 WIEN
AUSTRIA

MEWE R DR
LAB VOOR RUIMTE-
ONDERZOEK
BENELUXLAAN 21
UTRECHT
THE NETHERLANDS

MEYER CLAUDE DR
C E R G A
AVENUE COPERNIC
F-06130 GRASSE
FRANCE

MEYER FRIEDRICH DR
MPI FUER PHYSIK UND
ASTROPHYSIK
FOEHRINGER RING 6
D-8000 MUENCHEN 40
GERMANY, F.R.

MEYER J P DR
SECT D'ASTROPHYSIQUE
CEN SACLAY
B P NO 2
F-91190 GIF-S-YVETTE
FRANCE

MEYER-HOFMEISTER E DR
MPI FUER PHYSIK UND
ASTROPHYSIK
FOEHRINGER RING 6
D-8000 MUENCHEN 40
GERMANY, F.R.

MEZGER PETER G PROF
MPI FUER RADIO-
ASTRONOMIE
AUF DEM HUEGEL 69
D-5300 BONN 1
GERMANY, F.R.

MIANES PIERRE DR
OBSERVATOIRE DE
TOULOUSE
AV C FLAMMARION 1
F-31500 TOULOUSE
FRANCE

MICHALITSIANOS ANDREW
NASA-GSFC
CODE 685
GREENBELT MD 20771
U.S.A.

MICHARD RAYMOND DR
OBS DE PARIS
61 AV L OBSERVATOIRE
F-75014 PARIS
FRANCE

MICHAUD GEORGES J DR
250 DU FINISTERE
ST-LAMBERT J4S 1P5
CANADA

MICHEL F CURTIS PROF
RICE UNIVERSITY
HOUSTON TX 77001
U.S.A.

MICHEL H MR
54 RUE TEN BOSCH
B-1050 BRUXELLES
BELGIUM

MICHEL K W PROF
INST PHYSIK CHEMIE
UNIVERSITAET
STUTTGART
D-7000 STUTTGART 80
GERMANY, F.R.

MICZAIKA G R DR
TRW SYSTEMS R5-2291
1 SPACE PARK
REDONDO BCH CA 90278
U.S.A.

MIDDLEHURST BARBARA M MS
LUNAR AND PLANETARY
INSTITUTE
3303 NASA ROAD 1
HOUSTON TX 77058
U.S.A.

MIETELSKI JAN S DR
AL MICKIEWICZA 13-5
31-120 KRAKOW
POLAND

MIGEOTTE MARCEL V PROF
INSTITUT
D ASTROPHYSIQUE
B-4200 COINTE-OUGREE
BELGIUM

MIHAILA IERONIM M DR
BUCHAREST UNIVERSITY
ACADEMIEI 14
70109 BUCHAREST
ROUMANIA

MIHALAS DIMITRI DR
SACRAMENTO PEAK OBS
SUNSPOT NM 88349
U.S.A.

MIKESELL ALFRED H MR
2109 EAST COPPER ST
TUCSON AZ 85719
U.S.A.

MIKHAEL FAHMY I PROF DR
AIN SHAMS UNIVERSITY
CAIRO
ARAB REP. OF EGYPT

MIKHAIL JOSEPH SIDKY DR
HELWAN OBSERVATORY
CAIRO
ARAB REP. OF EGYPT

MIKHAILOV ALEXANDRA PROF
PULKOVO OBSERVATORY
196140 LENINGRAD
U.S.S.R.

MIKHELSON NIKOLAJ N DR
PULKOVO OBSERVATORY
196140 LENINGRAD
U.S.S.R.

MILANO LEOPOLDO DR
CAPODIMONTE ASTRONO-
MICAL OBSERVATORY
VIA MOIARIELLO 16
I-80131 NAPOLI
ITALY

MILES HOWARD G MR
21 BABBACOMBE RD
COVENTRY CV3 5PE
U.K.

MILET BERNARD L DR
OBSERVATOIRE DE NICE
B P 252
F-06007 NICE CEDEX
FRANCE

MILEY G K DR
STERREWACHT
POSTBUS 9513
2300 RA LEIDEN
THE NETHERLANDS

MILKEY ROBERT W DR
KITT PEAK NATIONAL
OBSERVATORY
BOX 26732 MS 18
TUCSON AZ 85726
U.S.A.

MILLAR THOMAS J DR
DEPT THEORET PHYSICS
OXFORD UNIV
1 KEBLE ROAD
OXFORD OX1 3NP
U.K.

MILLER FREEMAN D PROF
UNIVERSITY MICHIGAN
DEPT OF ASTRONOMY
DENNISON BLDG
ANN ARBOR MI 48109
U.S.A.

MILLER HUGH R PROF
DEPT OF PHYSICS
GEORGIA STATE
UNIVERSITY
ATLANTA GA 30303
U.S.A.

MILLER JOHN C DR
DEPT OF ASTROPHYSICS
UNIV OF OXFORD
SOUTH PARKS ROAD
OXFORD
U.K.

MILLER JOSEPH S PROF
LICK OBSERVATORY
UNIVERSITY OF CALIF
SANTA CRUZ CA 95060
U.S.A.

MILLER RICHARD H DR
ASTRONOMY DEPT
UNIV OF CHICAGO
5640 ELLIS AVE
CHICAGO IL 60637
U.S.A.

MILLIGAN J E
ASTROPHYSICS BRANCH
GODDARD SPACE FL CTR
GREENBELT MD 20771
U.S.A.

MILLIKAN ALLAN G DR
RESEARCH LABS B-59
EASTMAN KODAK CO
ROCHESTER NY 14650
U.S.A.

MILLIS ROBERT L DR
LOWELL OBSERVATORY
P O BOX 1269
FLAGSTAFF AZ 86001
U.S.A.

MILLMAN PETER M DR
HERZBERG INSTITUTE
OF ASTROPHYSICS
NAT RES COUNC CANADA
OTTAWA ONT K1A 0R6
CANADA

MILLS ALLAN A DR
DEPT OF ASTRONOMY
THE UNIVERSITY
LEICESTER LE1 7RH
U.K.

MILLS BERNARD Y PROF
SCHOOL OF PHYSICS
UNIVERSITY OF SYDNEY
SYDNEY NSW 2006
AUSTRALIA

MILNE K DR
CSIRO
DIV OF RADIOPHYSICS
P O BOX 76
EPPING NSW 2121
AUSTRALIA

MILONE EUGENE F DR
PHYSICS DEPT
UNIVERSITY
OF CALGARY
CALGARY ALTA T2N 1N4
CANADA

MILONE LUIS A DR
OBSERVATORIO
DE CORDOBA
LAPRIDA 854
5000 CORDOBA
ARGENTINA

MILOVANOVIC VLADETA DR
INSTITUT ZA GEODE-
ZIJU BEOGRAD
BUL REVOLUCIJE 73
11000 BEOGRAD
YUGOSLAVIA

MININ I N DR
LENINGRAD STATE UNI-
VERSITY ASTRONOMICAL
OBSERVATORY
199178 LENINGRAD
U.S.S.R.

MINN YOUNG KEY DR
KOREAN NATIONAL
OBSERVATORY
KANGNAM-KU
SEOUL 134-03
REPUBLIC OF KOREA

MINNET HARRY C MR
CSIRO DIVISION OF
RADIOPHYSICS
P O BOX 76
EPPING NSW 2121
AUSTRALIA

MINTZ BLANCO MA B MR
CERRO TOLOLO INTER-
AMERICAN OBS
CASILLA 63-D
LA SERENA
CHILE

MIRABEL IGOR FELIX DR
DEPT OF PHYSICS
FAC OF NATURAL SCS
UNIV OF PUERTO RICO
RIO PIEDRAS PR 00931
U.S.A.

MIRZOYAN L N DR PROF
BYURAKAN ASTROPHYSI-
CAL OBSERVATORY
378433 ARMENIA
U.S.S.R.

MISCONI NEBIL YOUSIF DR
SPACE ASTRONOMY LAB
SUNY
EXECUTIVE PARK EAST
ALBANY NY 12203
U.S.A.

MISNER CHARLES W PROF
DEPT OF PHYSICS AND
ASTRONOMY
UNIVERSITY MARYLAND
COLLEGE PARK MD20742
U.S.A.

MISSANA MARCO DR
VIA BERGAMO
I-22058 OSNAGO COMO
ITALY

MISSANA NATALE PROF
VIA PUCCINI N.3
I-10025 PINO
TORINESE
ITALY

MITALAS ROMAS ASSOC PROF
DEPT OF ASTRONOMY
UNIVERSITY OF
WESTERN ONTARIO
LONDON ONT N6A 5B9
CANADA

MITCHELL GEORGE F DR
DEPT OF ASTRONOMY
SAINT MARY S
UNIVERSITY
HALIFAX NS B3H 3C3
CANADA

MITCHELL R MR
C/O CICESE
AV CASTELUM NO 898
ENSENADA
BAJA CALIFORNIA
MEXICO

MITCHELL WALTER E JR
ASTRONOMY DEPT
OHIO STATE UNIV
174 W 18TH AVE
COLUMBUS OH 43210
U.S.A.

MITIC LJUBISA A DR
OBSERVATOIRE
DE BELGRADE
VOLGINA 7
11050 BELGRADE
YUGOSLAVIA

MITRA A P DR
ASST DIRECTOR
NATIONAL PHYSICAL
LABORATORY
NEW DELHI
INDIA

MITROFANOVA LYUDMILA A DR
PULKOVO OBSERVATORY
196140 LENINGRAD
U.S.S.R.

MITTON SIMON DR
CAMBRIDGE UNIV PRESS
P O BOX 110
CAMBRIDGE CB2 3RL
U.K.

MIYADI MASASI DR
SENGAWA 3-6-11
CHOFU-SHI 182
JAPAN

MIYAMOTO MASANORI DR
TOKYO ASTRONOMICAL
OBSERVATORY
OSAWA MITAKA
TOKYO 181
JAPAN

MIYAMOTO SIGINORI PROF
DEPT OF PHYSICS
OSAKA UNIVERSITY
MACHIKANEYAMA-CHO
TOYONAKA OSAKA 560
JAPAN

MIYAMOTO SYOTARO PROF DR
KASAN OBSERVATORY
YAMA HIGASHIYAMA
KYOTO 607
JAPAN

MOCZKO JANUSZ DR
BOROWIEC
63-120 KORNIK
POLAND

MODALI SARMA B DR
6404 FAIRBORN TERR
NEW CARROLLTON
MD 20784
U.S.A.

MODISETTE JERRY L PROF
HOUSTON BAPTIST
UNIVERSITY
7502 FONDREN AVE
HOUSTON TX 77074
U.S.A.

MOELLENHOFF CLAUS DR
LANDESSTERNWARTE
KOENIGSTUHL
D-6900 HEIDELBERG 1
GERMANY, F.R.

MOELLER OLE PROF
INST OF ASTRONOMY
UNIVERSITY OF AARHUS
DK-8000 AARHUS C
DENMARK

MOERDIJK WILLY G DR
STERRENKUNDIG INST
KORTRIJKSE
STEENWEG 523
B-9000 GENT
BELGIUM

MOESGAARD KRISTIAN P PH D
BYGADEN 1
TORRILD
DK-8300 ODDER
DENMARK

MOFFAT ANTHONY F J DR
DEPT DE PHYSIQUE
UNIVERSITE
DE MONTREAL
MONTREAL P Q H3C 3J7
CANADA

MOFFET ALAN T PROF
OWENS VALLEY RADIO
OBSERVATORY
CALTECH 102 24
PASADENA CA 91125
U.S.A.

MOFFETT THOMAS J PROF
DEPT OF PHYSICS
PURDUE UNIVERSITY
WEST LAFAYETTE
IN 47907
U.S.A.

MOGILEVSKIJ EH I DR
INSTITUTE OF TERRE-
STRIAL MAGNETISM AND
IONOSPHERE
142092 AKADEMGORODOK
U.S.S.R.

MOHLER ORREN C PROF
405 AWIXA RD
ANN ARBOR MI 48104
U.S.A.

MOISEEV I G DR
CRIMEAN ASTROPHYSI-
CAL OBSERVATORY USSR
ACAD OF SCI NAUCHNIY
334413 CRIMEA
U.S.S.R.

MOLCHANOV A P PROF
LENINGRAD STATE UNI-
VERSITY ASTRONOMICAL
OBSERVATORY
199178 LENINGRAD
U.S.S.R.

MOLES MARIANO J DR
JOSE ANTONIO 5
BINACED HUESCA
SPAIN

MOLNAR MICHAEL R PROF
UNIVERSITY OF TOLEDO
RITTER OBSERVATORY
TOLEDO OH 43606
U.S.A.

MONAGHAN JOSEPH J DR
MATHEMATICS DEPT
MONASH UNIVERSITY
CLAYTON VICT 3168
AUSTRALIA

MONFILS ANDRE G PROF
INSTITUT
D ASTROPHYSIQUE
B-4200 COINTE-LIEGE
BELGIUM

MONNET GUY J DR
OBSERVATOIRE DE LYON
AVENUE CHARLES ANDRE
F-69230 ST GENIS
LAVAL
FRANCE

MONSIGNORI FOSSI BRUNA DR
OSSERVATORIO ARCETRI
LARGO FERMI 5
I-50125 FIRENZE
ITALY

MONTES C DR
OBS DE NICE
B P 252
F-06007 NICE CEDEX
FRANCE

MONTMERLE T DR
SECT D'ASTROPHYSIQUE
CEN SACLAY
B P NO 2
F-91190 GIF-S-YVETTE
FRANCE

MOOK DELO E PROF
DEPT OF PHYSICS AND
ASTRONOMY
DARTMOUTH COLLEGE
HANOVER NH 03755
U.S.A.

MOORE D R DR
DEPT APPLIED MATHS
AND THEORETICAL PHYS
SILVER STREET
CAMBRIDGE CB3 9EW
U.K.

MOORE ELLIOTT P PROF
JOINT OBSERVATORY
F COMETARY RESEARCH
CAMPUS STATION
SOCORRO NM 87801
U.S.A.

MOORE PATRICK DR OBE
FARTHINGS
39 WEST STREET
SELSEY SUSSEX
U.K.

MOORE RONALD L DR
SOLAR ASTRONOMY
264 33
CALIF INST TECHNOLOG
PASADENA CA 91125
U.S.A.

MOORE-SITTERLY CH MRS DR
NATIONAL BUREAU OF
STANDARDS 151-00
WASHINGTON DC 20234
U.S.A.

MOOS HENRY WARREN DR
DEPT OF PHYSICS
JOHNS HOPKINS UNIV
BALTIMORE MD 21218
U.S.A.

MORAN JAMES M DR
CENTER FOR
ASTROPHYSICS
60 GARDEN ST
CAMBRIDGE MA 02138
U.S.A.

MORANDO BRUNO L DR
BUREAU D LONGITUDES
77 AVE DENFERT
ROCHEREAU
F-75014 PARIS
FRANCE

MORBEY CHRISTOPHER L
DOMINION ASTROPHYSI-
CAL OBSERVATORY
5071W SAANICH RD RR7
VICTORIA BC V8X 3X3
CANADA

MOREELS G DR
OBS DE BESANCON
41 BIS AVENUE DE
L'OBSERVATOIRE
F-25000 BESANCON
FRANCE

MORENO HUGO PROF
DEP DE ASTRONOMIA
UNIVERSIDAD DE CHILE
CASILLA 36-D
SANTIAGO
CHILE

MORETON G E
15-5 THE ESPLANADE
BALMORAL BCH NSW2088
AUSTRALIA

MORGAN DAVID H DR
ROYAL OBSERVATORY
BLACKFORD HILL
EDINBURGH EH9 3HJ
U.K.

MORGAN PETER DR
DIV OF NATIONAL
MAPPING
P O BOX 548
QUEANBEYAN NSW 2620
AUSTRALIA

MORGAN THOMAS H DR
HOUSTON BAPTIST
UNIVERSITY
7502 FONDREN AVE
HOUSTON TX 77074
U.S.A.

MORGAN W W PROF
YERKES OBSERVATORY
WILLIAMS BAY WI53191
U.S.A.

MORGULEFF NINA ING
INST D ASTROPHYSIQUE
98 BIS BD ARAGO
F-75014 PARIS
FRANCE

MORIMOTO MASAKI DR
TOKYO ASTRONOMICAL
OBSERVATORY
OSAWA MITAKA
TOKYO 181
JAPAN

MORISON IAN MR
NRAL
JODRELL BANK
CHESHIRE SK11 9DL
U.K.

MORIYAMA FUMIO DR
TOKYO ASTRONOMICAL
OBSERVATORY
OSAWA MITAKA
TOKYO 181
JAPAN

MOROZ V I PROF DR
SPACE RESEARCH
INSTITUTE
USSR ACADEMY OF SCI
117810 MOSCOW
U.S.S.R.

MOROZHENKO A V DR
MAIN ASTRON OBS
UKRAINIAN ACAD OF SC
GOLOSEEVO
252127 KIEV
U.S.S.R.

MOROZHENKO N N DR
MAIN ASTRON OBS
UKRAINIAN ACAD OF SC
GOLOSEEVO
252127 KIEV
U.S.S.R.

MORRIS DAVID DR
MPI FUER RADIOASTRON
AUF DEM HUEGEL 69
D-5300 BONN 1
GERMANY, F.R.

MORRIS MARK ROOT DR
DEPT OF ASTRONOMY
COLUMBIA UNIV
NEW YORK NY 10027
U.S.A.

MORRIS S C DR
DOMINION ASTROPHYSI-
CAL OBSERVATORY
5071W SAANICH RD RR7
VICTORIA BC V8X 3X3
CANADA

MORRISON DAVID PROF
CODE SL
NASA HEADQUARTERS
WASHINGTON DC 20546
U.S.A.

MORRISON LESLIE V
ROYAL GREENWICH
OBSERVATORY
HAILSHAM
EAST SUSSEX BN27 1RP
U.K.

MORRISON NANCY DUNLAP DR
DEPT PHYSICS ASTRON
UNIV OF TOLEDO
2801 W BANCROFT ST
TOLEDO OH 43606
U.S.A.

MORRISON PHILIP PROF
PHYSICS MIT 6-205
CAMBRIDGE MA 02139
U.S.A.

MORTON DONALD C DR
ANGLO AUSTRALIAN
OBSERVATORY
P O BOX 296
EPPING N S W 2121
AUSTRALIA

MORTON G A DR
1122 SKYCREST DRIVE
APT 6
WALNUT CREEK CA94595
U.S.A.

MOSS DAVID L DR
MATHEMATICS DEPT
MANCHESTER
UNIVERSITY
MANCHESTER M13 9PL
U.K.

MOTTA SANTO DR
OSS ASTROFISICO
CITTA UNIVERSITARIA
I-95125 CATANIA
ITALY

MOTZ LLOYD PROF
DEPT OF ASTRONOMY
COLUMBIA UNIVERSITY
PUPIN HALL BOX 57
NEW YORK NY 10027
U.S.A.

MOURADIAN ZADIG M DR
OBSERVATOIRE DE
PARIS-MEUDON
F-92190 MEUDON
FRANCE

MOUSCHOVIAS TELEMACHOS CH
DEPT OF ASTRONOMY
UNIVERSITY ILLINOIS
URBANA IL 61801
U.S.A.

MOUSSAS XENOPHON PH D
20 SKYLITSI STREET
ATHENS 707
GREECE

MOUTSOULAS MICHAEL PROF
ASTRONOMY DEPT
UNIVERSITY OF ATHENS
PANEPISTIMIOPOLIS
ATHENS 621
GREECE

MOVAHED REZA DR
P O BOX 6
BABOLSAR
IRAN

MRKOS ANTONIN DR
DEPT OF ASTRONOMY
CHARLES UNIVERSITY
SVEDSKA 8
150 00 PRAHA 5
CZECHOSLOVAKIA

MUELLER EDITH A PROF
OBSERVATOIRE
DE GENEVE
CH-1290 SAUVERNY GE
SWITZERLAND

MUELLER HELMUT O PROF
HERZOGENMUEHLESTR 4
CH-8051 ZUERICH
SWITZERLAND

MUELLER IVAN I PROF
GEODETIC SCIENCE
OHIO ST UNIVERSITY
1958 NEIL AVE
COLUMBUS OH 43210
U.S.A.

MUELLER R PROF DR
NEUBEUERER STRASSE 1
D-8201 NUSSDORF-INN
GERMANY, F.R.

MUENCH GUIDO PROF
MPI FUER ASTRONOMIE
KOENIGSTUHL
D-6900 HEIDELBERG 1
GERMANY, F.R.

MUFSON STUART LEE DR
ASTRONOMY DEPT
319 SWAIN WEST
INDIANA UNIV
BLOOMINGTON IN 47401
U.S.A.

MUKAI SONOYO DR
KANAZAWA INSTITUTE
OF TECHNOLOGY
ENKOJIHONMACHI
KANAZAWA 921
JAPAN

MUKAI TADASHI DR
KANAZAWA INSTITUTE
OF TECHNOLOGY
ENKOJIHONMACHI
KANAZAWA 921
JAPAN

MULDERS GERARD F W
4519 EVERETT ST
KENSINGTON MD 20795
U.S.A.

MULHOLLAND J DERRAL DR
DEPT OF ASTRONOMY
UNIVERSITY OF TEXAS
AUSTIN TX 78712
U.S.A.

MULLALY RICHARD F DR
SCHOOL OF ELECTRICAL
ENGINEERING
UNIV OF SYDNEY
SYDNEY NSW 2006
AUSTRALIA

MULLAN DERMOTT J DR
BARTOL RESEARCH
FOUNDATION
UNIV OF DELAWARE
NEWARK DE 19711
U.S.A.

MULLER A B DR
ESO
SCHLEISSHEIMERSTR 17
D-8046 GARCHING BEI
MUENCHEN
GERMANY, F.R.

MULLER C A PROF ING
ODINKSVELD 8
DELDEN
THE NETHERLANDS

MULLER PAUL PROF
CERGA
AV COPERNIC
F-06130 GRASSE
FRANCE

MULLER R DR
OBS DU PIC-DU-MIDI
F-65200 BAGNERES-DE-
BIGORRE
FRANCE

MUMFORD GEORGE S PROF
BALLOU HALL
TUFTS UNIVERSITY
MEDFORD MA 02155
U.S.A.

MUNRO RICHARD H DR
HIGH ALTITUDE
OBSERVATORY
P O BOX 3000
BOULDER CO 80307
U.S.A.

MURDIN PAUL G DR
ROYAL GREENWICH
OBSERVATORY
HAILSHAM
EAST SUSSEX BN27 1RP
U.K.

MURDOCH HUGH S DR
ASTROPHYSICS DEPT
UNIVERSITY OF SYDNEY
SYDNEY 2006
AUSTRALIA

MURPHY ROBERT E DR
CODE SL-4
NASA HEADQUARTERS
WASHINGTON DC 20546
U.S.A.

MURRAY C ANDREW
ROYAL GREENWICH
OBSERVATORY
HAILSHAM
EAST SUSSEX BN27 1RP
U.K.

MURRAY JOHN B DR
UNIV OF LONDON
OBSERVATORY
MILL HILL PARK
LONDON NW7 2QS
U.K.

MURRAY STEPHEN S DR
HARVARD-SMITHSONIAN
CTR FOR ASTROPHYSICS
60 GARDEN ST
CAMBRIDGE MA 02138
U.S.A.

MUSEN PETER DR
NASA GSFC
GREENBELT MD 20801
U.S.A.

MUSMAN STEVEN DR
SACRAMENTO PEAK
OBSERVATORY
SUNSPOT NM 88349
U.S.A.

MUSTEL E R PROF DR
ASTRONOMICAL COUNCIL
USSR ACAD OF SCI
PYATNITSKAYA UL 48
109017 MOSCOW
U.S.S.R.

MUTSCHLECNER J PAUL DR
ASTRONOMY DEPT
SWAIN HALL W
INDIANA UNIVERSITY
BLOOMINGTON IN 47405
U.S.A.

MUZZIO JUAN C DR
OBSERVATORIO
ASTRONOMICO
PASEO DEL BOSQUE
1900 LA PLATA
ARGENTINA

MYACHIN V F DR
INSTITUTE OF THEORE-
TICAL ASTRONOMY
192187 LENINGRAD
U.S.S.R.

MYERS PHILIP C PROF
PHYSICS DEPT
MIT
BLDG 26 RM 349
CAMBRIDGE MA 02139
U.S.A.

MYERSCOUGH VALERIE P DR
DEPT OF APPLIED MATH
QUEEN MARY COLLEGE
MILE END ROAD
LONDON E1 4NS
U.K.

NACOZY PAUL E PROF
AEROSPACE ENGR AND
ENGR MECHANICS
UNIVERSITY OF TEXAS
AUSTIN TX 78712
U.S.A.

NADOLSCHI V PROF DR
COM.ARDEOANI
OF TESCANI JUD
BACAU
ROUMANIA

NADYOZHIN D K DR
INST THEOR EXP PHYS
CHEREMUSHKINSKAJA 25
117259 MOSCOW
U.S.S.R.

NAGASAWA SHINGO PROF
TOKYO SCIENCE UNIV
2641 YAMAZAKI
HIGASHI KAMEYAMA
NODASHI 278
JAPAN

NAGIRNER DMITRIJ I DR
LENINGRAD UNIVERSITY
ASTRONOMICAL OBS
199178 LENINGRAD
U.S.S.R.

NAGNIBEDA VALERY G DR
LENINGRAD UNIVERSITY
ASTRONOMICAL
OBSERVATORY
199178 LENINGRAD
U.S.S.R.

NAHON FERNAND PROF
25 AVE DE L-EUROPE
F-92310 SEVRES
FRANCE

NAKAGAWA NAOYA DR
UNIV OF ELECTRO-
COMMUNICATIONS
CHOFU-SHI
TOKYO 181
JAPAN

NAKAGAWA YOSHINARI DR
HIGH ALTITUDE
OBSERVATORY
P O BOX 3000
BOULDER CO 80307
U.S.A.

NAKAMURA TSUKO DR
DODAIRA STATION
TOKYO ASTRON OBS
TOKIGAWA
SAITAMA 355-05
JAPAN

NAKANO SABURO DR
KOBINATO 1-21-7
BUNKYO KU
TOKYO 112
JAPAN

NAKANO TAKENORI DR
DEPT OF PHYSICS
KYOTO UNIVERSITY
KITASHIRAKAWA OIWAKE
SAKYOKU KYOTO 606
JAPAN

NAKAYAMA SHIGERU DR
FACULTY GENERAL EDUC
UNIVERSITY OF TOKYO
KOMABA MEGURO
TOKYO 153
JAPAN

NAKAZAWA KIYOSHI DR
DEPT OF PHYSICS
KYOTO UNIVERSITY
KITASHIRAKAWA OIWAKE
SAKYOKU KYOTO 606
JAPAN

NAMBA OSAMU DR
ASTRONOMICAL INST
ZONNENBURG 2
3512 NL UTRECHT
THE NETHERLANDS

NANDY KASHINATT DR
ROYAL OBSERVATORY
EDINBURGH
BLACKFORD EH9 3HJ
U.K.

NAPIER WILLIAM M DR
ROYAL OBSERVATORY
BLACKFORD HILL
EDINBURGH EH9 3HJ
U.K.

NAQVI A M DR
6274 DEL ROSA AVE
APT 11
S BERNARDINO CA92404
U.S.A.

NARANAN S PROF
TATA INSTITUTE OF
FUNDAMENTAL RESEARCH
HOMI BHABHA ROAD
BOMBAY 400005
INDIA

NARAYANA J V
REG METEOROLOGICAL
OFFICE
4 COLLEGE ROAD
MADRAS 600006
INDIA

NARIAI HIDEICHI
RESEARCH INST FOR
THEORETICAL PHYSICS
HIROSHIMA UNIVERSITY
725 TAKEHARA
JAPAN

NARIAI KYOJI DR
TOKYO ASTRONOMICAL
OBSERVATORY
MITAKA
TOKYO 181
JAPAN

NARITA SHINJI DR
DOSHISHA UNIVERSITY
KYOTO 602
JAPAN

NARLIKAR JAYANT V PROF
KING S COLLEGE
CAMBRIDGE
U.K.

NATALI GIULIANO DR
LAB ASTROFISICO SPAZ
P O BOX 67
I-00044 FRASCATI
ITALY

NATHER R EDWARD
DEPT OF ASTRONOMY
UNIVERSITY OF TEXAS
AUSTIN TX 78712
U.S.A.

NATTA ANTONELLA DR
CNR ASTROFISICA
SPAZIALE C P 67
I-00044 FRASCATI
ROMA
ITALY

NAUMOV VITALIJ A DR
PULKOVO OBSERVATORY
196140 LENINGRAD
U.S.S.R.

NAWAR SAMIR DR
HELWAN OBSERVATORY
HELWAN
ARAB REP. OF EGYPT

NECKEL HEINZ DR
HAMBURGER STERNWARTE
D-2050 HAMBURG 80
GERMANY, F.R.

NECKEL TH DR
MPI FUER ASTRONOMIE
D-6900 HEIDELBERG-
KOENIGSTUHL
GERMANY, F.R.

NEE TSU-WEI DR
INST OF PHYSICS AND
ASTRONOMY
NAT CENTRAL UNIV
CHUNG-LI
TAIWAN

NEEMAN YUVAL PROF
DEPT OF PHYSICS AND
ASTRONOMY
TEL AVIV UNIVERSITY
TEL AVIV
ISRAEL

NEFEDEVA ANTONINA I DR
ENGELHARDT ASTRONO-
MICAL OBSERVATORY
OBSERVATORY STATION
422526 KAZAN
U.S.S.R.

NEFF JOHN S ASSOC PROF
SPACE SCIE DIVISION
JET PROPULSION LAB
4800 OAK GROVE DRIVE
PASADENA CA 91103
U.S.A.

NELSON ALISTAIR H DR
DEPT OF APPLIED MATH
AND ASTRONOMY
UNIV COLLEGE
CARDIFF
U.K.

NELSON BURT DR
DEPT OF ASTRONOMY
SAN DIEGO STATE
UNIVERSITY
SAN DIEGO CA 92115
U.S.A.

NEMIRO ANDREJ A DR PROF
PULKOVO OBSERVATORY
196140 LENINGRAD
U.S.S.R.

NESS NORMAN F DR
NASA GSFC CODE 690
GREENBELT MD 20771
U.S.A.

NEUGEBAUER GERRY DR
PHYSICS DEPARTMENT
CALIF INST OF TECHN
320 DOWNS
PASADENA CA 91125
U.S.A.

NEUPERT WERNER M DR
NASA GODDARD SPACE
FLIGHT CENTER
CODE 682
GREENBELT MD 20771
U.S.A.

NEUZIL LUDEK DR
ASTRONOMICAL INST
CZECH ACAD SCIENCES
OBSERVATORY
251-65 ONDREJOV
CZECHOSLOVAKIA

NEVEN LUC DR
OBSERVATOIRE ROYAL
AVENUE CIRCULAIRE 3
B-1180 BRUXELLES
BELGIUM

NEVIN THOMAS E PROF
PHYSICS DEPT
UNIVERSITY COLLEGE
BELFIELD
DUBLIN 4
IRELAND

NEWELL EDWARD B DR
MT STROMLO AND
SIDING SPRING OBS
PRIVATE BAG
WODEN A C T 2606
AUSTRALIA

NEWKIRK GORDON A JR DR
HIGH ALTITUDE
OBSERVATORY NCAR
P O BOX 3000
BOULDER CO 80307
U.S.A.

NEWMAN MICHAEL JOHN DR
T-DOT MS 288
LOS ALAMOS SC LAB
P O BOX 1663
LOS ALAMOS NM 87545
U.S.A.

NEWSON GERALD H ASST PROF
ASTRONOMY DEPT
OHIO STATE UNIV
174 W 18TH AVE
COLUMBUS OH 43210
U.S.A.

NEWTON ROBERT R DR
JOHNS HOPKINS UNIV
APPLIED PHYSICS LAB
JOHNS HOPKINS RD
LAUREL MD 20810
U.S.A.

NEY EDWARD P PROF
DEPT OF ASTRONOMY
TATE LAB OF PHYSICS
UNIV OF MINNESOTA
MINNEAPOLIS MN 55455
U.S.A.

NEYMAN JERRY PROF
STATISTICAL LAB
UNIV OF CALIFORNIA
367 EVANS HALL
BERKELEY CA 94720
U.S.A.

NHA II-SEONG DR
66 SINCHONDONG
SEODAEMOONKU
SEOUL 120
REPUBLIC OF KOREA

NIARCHOS PANAYIOTIS PH D
DEPT OF ASTRONOMY
UNIV OF ATHENS
PANEPISTIMIOPOLIS
ATHENS 621
GREECE

NICHOLLS RALPH W PROF
CRESS DEPT OF PHYS
YORK UNIVERSITY
4700 KEELE ST
DOWNSVIEW ONT M3J1P3
CANADA

NICHOLSON WILLIAM
ROYAL GREENWICH
OBSERVATORY
HAILSHAM
EAST SUSSEX BN27 1RP
U.K.

NICOLET MARCEL PROF
UNIVERSITE
DE BRUXELLES
30 AVE DEN DOORN
B-1180 BRUXELLES
BELGIUM

NICOLINI T PROF
VIA E NICOLARDI 221
PALAZZO CUSAN
COLLI AMINEI
I-80131 NAPOLI
ITALY

NICOLOV N S DR
UNIVERSITE DE SOFIA
FACULTE PHYSIQUE
SOFIA
BULGARIA

NICOLSON GEORGE D DR
RADIO ASTRONOMY
OBSERVATORY
P O BOX 3718
JOHANNESBURG 2000
SOUTH AFRICA

NIELL ARTHUR E DR
JPL
FHL-D 430
4800 OAK GROVE DR
PASADENA CA 91103
U.S.A.

NIEMELA VIRPI S DR
CALLE 51 ESQ 11
1894 VILLA ELISA BA
ARGENTINA

NIEUWENHUIJZEN HANS DR
ASTRONOMICAL INST
ZONNENBURG 2
UTRECHT
THE NETHERLANDS

NIIMI HIDEYUKI DR
FACULTY ENGINEERING
UNIVERSITY OF KYOTO
YOSHIDAHONMACHI
SAKYOKU KYOTO 606
JAPAN

NIKITIN A A DR
LENINGRAD STATE UNI-
VERSITY ASTRONOMICAL
OBSERVATORY
199178 LENINGRAD
U.S.S.R.

NIKOLOFF IVAN DR
PERTH OBSERVATORY
BICKLEY WA 6076
AUSTRALIA

NIKOLOV ANDREJ DR
P O BOX 402
1000 SOFIA
BULGARIA

NIKOLSKY G M PROF DR
INSTITUTE OF TERRE-
STRIAL MAGNETISM AND
IONOSPHERE
142092 AKADEMGORODOK
U.S.S.R.

NIKONOV V B DR
CRIMEAN ASTROPHYSI-
CAL OBSERVATORY USSR
ACAD OF SCI NAUCHNIY
334413 CRIMEA
U.S.S.R.

NILSON PETER DR
ASTRONOMICAL
OBSERVATORY
BOX 515
S-751 20 UPPSALA
SWEDEN

NILSSON CARL DR
SMITHSONIAN ASTRO-
PHYSICS OBSERVATORY
60 GARDEN STREET
CAMBRIDGE MA 02138
U.S.A.

NININGER H H
BOX 420
SEDONA AZ 86336
U.S.A.

NISHI KEIZO DR
TOKYO ASTRONOMICAL
OBSERVATORY
OSAWA MITAKA
TOKYO 181
JAPAN

NISHIDA MINORU PROF
DEPT OF PHYSICS
KYOTO UNIVERSITY
KITASHIRAKAWA OIWAKE
SAKYOKU KYOTO 606
JAPAN

NISHIMURA JUN DR
INST SPACE AERON SC
UNIVERSITY OF TOKYO
KOMABA MEGURO-KU
TOKYO
JAPAN

NISHIMURA SHIRO DR
TOKYO ASTRONOMICAL
OBSERVATORY
OSAWA MITAKA
TOKYO 181
JAPAN

NISSEN POUL E PROF
INST OF ASTRONOMY
UNIVERSITY OF AARHUS
LANGELANDSKADE
DK-8000 AARHUS C
DENMARK

NOBILI L DR
OSS ASTRONOMICO
VICOLO DELL OSSERVA-
TORIO
PADOVA
ITALY

NOCI GIANCARLO PROF
OSSERVATORIO ARCETRI
LARGO FERMI 5
I-50125 FIRENZE
ITALY

NOEL FERNANDO
DPTO DE ASTRONOMIA
UNIVERSIDAD DE CHILE
CASILLA 36D
SANTIAGO
CHILE

NOELS ARLETTE DR
50 AVENUE DE LA PAIX
BOITE 063
B-4030 GRIVEGNEE
BELGIUM

NOERDLINGER PETER D.PROF.
NASA AMES RES CTR
245-3
MOFFETT FLD CA 94035
U.S.A.

NOLLEZ G DR
INST D'ASTROPHYSIQUE
98 BIS BD ARAGO
F-75014 PARIS
FRANCE

NOMOTO KENICHI
DEPT OF PHYSICS
IBARAKI UNIVERSITY
BUNKYO MITO 310
JAPAN

NOONAN THOMAS W PROF
PHYSICS DEPT
SUNY
BROCKPORT NY 14420
U.S.A.

NORDH H LENNART DR
STOCKHOLM
OBSERVATORY
S-133 00
SALTSJOEBADEN
SWEDEN

NORDLUND AKE DR
NORDITA
BLEGDAMSVEJ 17
DK-2100 COPENHAGEN O
DENMARK

NORDSTROM BIRGITTA DR
COPENHAGEN
OBSERVATORY
BRORFELDE
DK-4340 TOLLOSE
DENMARK

NORLIND WILHELM DR
SVANEGATAN 16
S-222 24 LUND
SWEDEN

NORLUND N E PROF
MALMOGADE 6
DK-2100 COPENHAGEN O
DENMARK

NORMAN COLIN DR
HUYGENS LAB
STERREWACHT
POSTBUS 9513
2300 RA LEIDEN
THE NETHERLANDS

NORRIS JOHN DR
MT STROMLO
OBSERVATORY
WODEN PO ACT 2606
AUSTRALIA

NORTH JOHN D PROF
28 CHALFONT ROAD
OXFORD
U.K.

NOSKOV BORIS N DR
STERNBERG STATE
ASTRONOMICAL INST
117234 MOSCOW
U.S.S.R.

NOTNI P DR
ZNTRINST ASTROPHYSIK
STERNW BABELSBERG
ROSA LUXEMBURGST 17A
DDR-1502 POTSDAM
G.D.R.

NOTUKI MASAO DR
4-36-7 NARITAHIGASHI
SUGINAMI-KU
TOKYO 166
JAPAN

NOVICK ROBERT
DEPT OF PHYSICS
COLUMBIA UNIVERSITY
538 W 120 ST
NEW YORK NY 10027
U.S.A.

NOVIKOV I D DR
SPACE RESEARH
INSTITUTE
USSR ACADEMY OF SCI
117810 MOSCOW
U.S.S.R.

NOVIKOV SERGEJ B DR
STERNBERG STATE
ASTRONOMICAL INST
117234 MOSCOW
U.S.S.R.

NOVOSELOV V S PROF DR
LENINGRAD STATE UNI-
VERSITY ASTRONOMICAL
OBSERVATORY
199178 LENINGRAD
U.S.S.R.

NOYES ROBERT W PROF
CTR F ASTROPHYSICS
60 GARDEN ST
CAMBRIDGE MA 02138
U.S.A.

NUSSBAUMER HARRY DR
ATOMIC AND ASTROPHYS
GROUP ETHZ
HUTTENSTR 34
CH-8092 ZUERICH
SWITZERLAND

O'CONNELL D J K J DR
BORGO S SPIRITO 5
I-00193 ROMA
ITALY

O'CONNELL ROBERT F PROF
LOUISIANA STATE
UNIVERSITY
BATON ROUGE LA 70803
U.S.A.

O'CONNELL ROBERT WEST DR
ASTRO DEPT UNIV VA
PO BOX 3818 UNIV STA
CHARLOTTESVILLE
VA 22903
U.S.A.

O'CONNOR SEAMUS L DR
PHYSICS DEPT
UNIVERSITY COLLEGE
DUBLIN 4
IRELAND

O'DELL STEPHEN L
PHYSICS DEPT
V P I AND S U
BLACKSBURG VA 24061
U.S.A.

O'HANDLEY DOUGLAS A DR
JET PROPULSION
LABORATORY
BLDG 169 422
PASADENA CA 91103
U.S.A.

O'HORA NATHY P J
ROYAL GREENWICH
OBSERVATORY
HAILSHAM
EAST SUSSEX BN27 1RP
U.K.

O'KEEFE JOHN A DR
GODDARD SPACE FLIGHT
CENTER CODE 680
GREENBELT MD 20771
U.S.A.

O'LEARY BRIAN T
DEPT OF PHYSICS
PRINCETON UNIVERSITY
PRINCETON NJ 08540
U.S.A.

O'MARA BERNARD J PROF
DEPT OF PHYSICS
UNIV OF QUEENSLAND
ST LUCIA
BRISBANE QLD 4067
AUSTRALIA

O'SULLIVAN JOHN DAVID DR
RADIOSTERREWACHT
POSTBUS 2
7990 AA DWINGELOO
THE NETHERLANDS

OBASHEV SAKEN O DR
ASTROPHYSICAL
INSTITUTE
480068 ALMA-ATA
U.S.S.R.

OBERLERCHNER DOROTHEA DR
UNIV-STERNWARTE
TUERKENSCHANZSTR 17
A-1180 WIEN
AUSTRIA

OBI SHINYA PROF
FAC OF GENERAL ED.
UNIV OF TOKYO
KOMABA MEGURO
TOKYO 153
JAPAN

OBREGON DIAZ OCTAVIO J DR
UAM IZTAPALAPA
PO BOX 55 534
MEXICO 13 DF
MEXICO

OBRIDKO VLADIMIR A DR
IZMIRAN
ACADEMGORODOK
142092 MOSCOW REGION
U.S.S.R.

OCCHIONERO FRANCO PROF
CNR ASTROFISICA
SPAZIALE C P 67
I-00044 FRASCATI
ROMA
ITALY

ODA MINORU PROF
INST SPACE AERON SC
UNIVERSITY OF TOKYO
KOMABA MEGURO KU
TOKYO 153
JAPAN

ODELL CHARLES R DR
DS30 MSFC
HUNTSVILLE AL 35812
U.S.A.

ODGERS GRAHAM J DR
DOMINION ASTRO OBS
NAT RES COUNC CANADA
5071 WEST SAANICH RD
VICTORIA BC V8X 3X3
CANADA

OEHMAN YNGVE PROF
THULEHEM 53
S-223 67 LUND
SWEDEN

OEMLER AUGUSTUS JR DR
YALE UNIV OBS
BOX 2023
YALE STATION
NEW HAVEN CT 06520
U.S.A.

OESTERWINTER CLAUS
COMMANDER
NAVAL SURFACE WEA-
PONS CENTER DK 10
DAHLGREN VA 22448
U.S.A.

OETKEN L DR
ZNTRINST ASTROPHYSIK
ASTROPHY OBSERVATORI
TELEGRAFENBERG
DDR-15 POTSDAM
G.D.R.

OGORODNIKOV KYRILL P PROF
DEPT OF ASTRONOMY
LENINGRAD UNIVERSITY
199164 LENINGRAD
U.S.S.R.

OHKI KENICHIRO DR
TOKYO ASTRONOMICAL
OBSERVATORY
OSAWA
MITAKA TOKYO 181
JAPAN

OHRING GEORGE PROF
DEPT OF GEOPHYSICS
PLANETARY SCIENCES
TEL AVIV
ISRAEL

OHYAMA NOBORU PROF
FAC OF ENGINEERING
SHIZUOKA UNIVERSITY
3 CHOME JYOHOKU
HAMAMATSU 432
JAPAN

OJA TARMO PROF
KVISTABERG
OBSERVATORY
S-190 51 BRO
SWEDEN

OKA TAKESHI DR
HERZBERG INSTITUTE
OF ASTROPHYSICS
NRCC
OTTAWA ONT K1A OR6
CANADA

OKAMOTO ISAO DR
INTERNATIONAL LATITU
DE OBSERVATORY
HOSHIGAOKA MIZUSAWA
IWATE 023
JAPAN

OKE J BEVERLY PROF
HALE OBSERVATORIES
CALIFORNIA INSTITUTE
OF TECHNOLOGY
PASADENA CA 91125
U.S.A.

OKI TOSHIO PROF
FACULTY OF EDUCATION
FUKUSHIMA UNIVERSITY
HAMADACHO
FUKUSHIMA 960
JAPAN

OKOYE SAMUEL E PROF
DEPT OF PHYSICS
UNIVERSITY OF
NIGERIA
NSUKKA
NIGERIA

OKUDA HARUYUKI DR PROF
DEPT OF PHYSICS
KYOTO UNIVERSITY
KITASHIRAKAWA SAKYO
KYOTO 606
JAPAN

OKUDA TOYOZO DR
5-11-9 HANAZONO
CHIBA 281
JAPAN

OLEAK H DR
ZNTRINST ASTROPHYSIK
STERNW BABELSBERG
ROSA LUXEMBURGST 17A
DDR-1502 POTSDAM
G.D.R.

OLIVER JOHN PARKER DR
DEPT PHYSICS ASTRON
UNIV OF FLORIDA
GAINESVILLE FL 32611
U.S.A.

OLLONGREN A DR
DEPT OF MATHEMATICS
WASSENAARSEWEG 80
POSTBOX 9512
2300 RA LEIDEN
THE NETHERLANDS

OLOFSSON S GOERAN DR
STOCKHOLM
OBSERVATORY
S-133 00
SALTSJOEBADEN
SWEDEN

OLSEN ERIK H
COPENHAGEN
OBSERVATORY
BRORFELDE
DK-4340 TOLLOSE
DENMARK

OLSEN KENNETH H DR
LOS ALAMOS SCIENTI-
FIC LABORATORY
BOX 1663 MS 676
LOS ALAMOS NM 87544
U.S.A.

OLSON EDWARD C PROF
OBSERVATORY
UNIVERSITY OF
ILLINOIS
URBANA IL 62801
U.S.A.

OLTHOF HINDERICUS DR
ESA
8-10 RUE MARIO NIKIS
F-75738 PARIS 15
FRANCE

OMAROV TUKEN B PROF
ASTROPHYSICAL
INSTITUTE
480068 ALMA ATA
U.S.S.R.

OMER GUY C JR PROF
1080 SW 11TH TERRACE
GAINESVILLE FL 32601
U.S.A.

OMNES ROLAND PROF
LPTHE BAT 211
UNIVERSITE DE
PARIS-SUD
F-91405 ORSAY
FRANCE

OMONT ALAIN PROF
OBSERVATOIRE
DE MEUDON
F-92190 MEUDON
FRANCE

ONDERLICKA BEDRICH DR
DEPT OF ASTRONOMY
PURKYNE UNIVERSITY
KOTLARSKA 2
61137 BRNO
CZECHOSLOVAKIA

ONEGINA A B DR
MAIN ASTRON OBS
UKRAINIAN AC OF SCS
GOLOSEEVO
252127 KIEV
U.S.S.R.

ONO YORO PROF
DEPT OF PHYSICS
UNIVERSITY HOKKAIDO
KITAHACHIJYO NISHI 8
SAPPORO HOKKAIDO 063
JAPAN

OOE MASATSUGU DR
INTERNATIONAL LATITU
DE OBSERVATORY
HOSHIGAOKA MIZUSAWA
IWATE 023
JAPAN

OORT JAN H PROF
PRES KENNEDYLAAN 169
OEGSTGEEST
THE NETHERLANDS

OPALSKI WIESLAW PROF
ELEKTORALNA 4-6 M4
00-139 WARSAW
POLAND

OPHER REUVEN PROF
PHYSICS DEPT
TECHNION-ISRAEL INST
OF TECHNOLOGY
HAIFA 32000
ISRAEL

OPIK ERNST J DR
ARMAGH OBSERVATORY
ARMAGH BT61 9DG
U.K.

OPOLSKI ANTONI PROF
ASTRONOMICAL
OBSERVATORY
KOPERNIKA 11
51-622 WROCLAW
POLAND

OPROIU TIBERIU DR
STR BUCIUM 25 BLOC
R-5 SC 1 ET 11 AP 10
3400 CLUJ-NAPOCA
ROUMANIA

ORLIN HYMAN DR
NAT ACADEMY SCIENCES
2101 CONSTITUTION
AVE N W
WASHINGTON DC 20418
U.S.A.

ORLOV A A DR
STERNBERG STATE AS-
TRONOMICAL INSTITUTE
117234 MOSCOW
U.S.S.R.

ORLOV MIKHAIL DR
MAIN ASTRONOMICAL OB
252127 KIEV
U.S.S.R.

ORRALL FRANK Q PROF
INSTITUTE FOR
ASTRONOMY
2680 WOODLAWN DRIVE
HONOLULU HI 96822
U.S.A.

ORTE ALBERTO
ISTI Y OBSERVATORIO
DE MARINA
SAN FERNANDO CADIZ
SPAIN

ORTON GLENN S DR
M S 183B-365
JPL
4800 OAK GROVE DRIVE
PASADENA CA 91103
U.S.A.

ORUS JUAN J PROF
DPTO FIS TIERRA COSM
UNIVERSIDAD
DE BARCELONA
BARCELONA 28
SPAIN

OSAKI TORU DR
RYUKOKU UNIVERSITY
FUKAKUSA TSUKAMOTO
FUSHIMIKU KYOTO 612
JAPAN

OSAKI YOJI DR
DEPT OF ASTRONOMY
UNIV OF TOKYO
YAYOI BUNKYO
TOKYO 113
JAPAN

OSAWA KIYOTERU DR
TOKYO ASTRONOMICAL
OBSERVATORY
OSAWA
MITAKA TOKYO 181
JAPAN

OSBORN WAYNE DR
PHYSICS DEPT
CENTRAL MICHIGAN
UNIVERSITY
MT PLEASANT MI 48859
U.S.A.

OSBORNE JOHN L DR
DEPT OF PHYSICS
UNIV OF DURHAM
SOUTH ROAD
DURHAM DH1 3LE
U.K.

OSKANYAN V S DR
BYURAKAN ASTROPHYSI-
CAL OBSERVATORY
378433 ARMENIA
U.S.S.R.

OSMAN ANAS MOHAMED DR
HELWAN OBSERVATORY
HELWAN
ARAB REP. OF EGYPT

OSMER PATRICK S DR
CERRO TOLOLO INTER-
AMERICAN OBSERVATORY
CASILLA 63 D
LA SERENA
CHILE

OSORIO JOSE J S P DR
OBS ASTRONOMICO
UNIV DO PORTO
MONTE DA VIRGEM
VILA NOVA DE GAIA
PORTUGAL

OSTER LUDWIG F PROF DR
JILA
UNIV OF COLORADO
BOULDER CO 80309
U.S.A.

OSTERBROCK DONALD E PROF
LICK OBSERVATORY
UNIVERSITY
OF CALIFORNIA
SANTA CRUZ CA 95064
U.S.A.

OSTRIKER JEREMIAH P PROF
PRINCETON UNIVERSITY
OBSERVATORY
PEYTON HALL
PRINCETON NJ 08540
U.S.A.

OTERMA LIISI PROF
SIRKKALANKATU 31
SF-20700 TURKU
FINLAND

OTTELET I J DR
INST D ASTROPHYSIQUE
UNIVERSITE DE LIEGE
B-4200 COINTE OUGREE
BELGIUM

OVENDEN MICHAEL W PROF
DEPT OF GEOPHYSICS
AND ASTRONOMY
UNIVERSITY OF B C
VANCOUVER BC V6T 1W5
CANADA

OWAKI NAOAKI DR
TOKYO GAKUGEI
UNIVERSITY
NUKUIKITAMACHI
KOGANEI TOKYO 184
JAPAN

OWEN FRAZER NELSON DR
NATIONAL RADIO ASTR
OBSERVATORY
EDGEMONT RD
CHARLOTTESVL VA22901
U.S.A.

OWEN TOBIAS C PROF
DEPT OF EARTH AND
SPACE SCIENCES
STATE UNIVERSITY NY
STONY BROOK NY 11794
U.S.A.

OWREN LEIF DR
FYSISK INSTITUTT
UNIVERSITETET BERGEN
ALLEGATEN 53 55
N-5014 BERGEN U
NORWAY

OXENIUS JOACHIM DR
UNIV LIBRE BRUXELLES
CP 231
CAMPUS PLAINE ULB
B-1050 BRUSSELS
BELGIUM

OZEMRE KAMURAN DOCENT DR
UNIVERSITY
OBSERVATORY
UNIVERSITE
ISTANBUL
TURKEY

OZERNOY LEONID M PROF
PHYSICAL INSTITUTE
USSR ACADEMIE OF
SCIENCE
117924 MOSCOW
U.S.S.R.

OZSVATH I PROF
EARTH AND PLANET SCS
GRADUATE RES CTR
P O BOX 30365
DALLAS TX 75230
U.S.A.

PAAL GYORGY DR
KONKOLY OBSERVATORY
BOX 67
1525 BUDAPEST
HUNGARY

PACHNER JAROSLAV PROF
DEPT OF PHYSICS AND
ASTRONOMY
UNIVERSITY OF REGINA
REGINA SK S4S 0A2
CANADA

PACHOLCZYK ANDRZEJ G PROF
STEWARD OBSERVATORY
UNIVERSITY ARIZONA
TUCSON AZ 85721
U.S.A.

PACINI FRANCO PROF
ISTITUTO ASTRONOMIA
LARGO E FERMI 5
I-50125 FIRENZE
ITALY

PACZYNSKI B PROF
COPERNICUS ASTRONOMI
-CAL CENTER
UL BARTYCKA
00 716 WARSAW
POLAND

PADEVET VLADIMIR DR
ASTRONOMICAL INST
CZECH ACADEMY SC
251 65 ONDREJOV
OBSERVATORY
CZECHOSLOVAKIA

PADRIELLI L
LAB DI RADIOASTRON
VIA IRNERIO 46
I-40126 BOLOGNA
ITALY

PAGE CLIVE G DR
PHYSICS DEPT
UNIV OF LEICESTER
LEICESTER LE1 7RH
U.K.

PAGE THORNTON L DR
18639 POINT LOOKOUT
DRIVE
HOUSTON TX 77058
U.S.A.

PAGEL BERNARD E J PROF
ROYAL GREENWICH
OBSERVATORY
HAILSHAM
EAST SUSSEX BN27 1RP
U.K.

PAL ARPAD PROF DR
STR RAKOCZI 72
3400 CLUJ NAPOCA
ROUMANIA

PALLAVICINI ROBERTO DR
OSSERVATORIO ARCETRI
LARGO FERMI 5
I-50125 FIRENZE
ITALY

PALMEIRA RICARDO A R DR
INPE CP 515
12200 SJOSE D CAMPOS
SAO PAULO
BRAZIL

PALMER HENRY P DR
NRAL
JODRELL BANK
CHESHIRE SK11 9DL
U.K.

PALMER PATRICK E PROF
DEPT OF ASTRONOMY
UNIVERSITY CHICAGO
1100 E 58TH ST
CHICAGO IL 60637
U.S.A.

PALUMBO GIORGIO G C DR
TE-S R E LAB CNR
VIA DE CASTAGNOLI 1
I-40100 BOLOGNA
ITALY

PAMJATN'IKH A A DR
ASTONOMICAL COUNCIL
USSR ACAD OF SCS
48 PYATNITSKAYA ST
109017 MOSCOW
U.S.S.R.

PANAGIA NINO DR
LABORATORIO DI
RADIOASTRONOMIA
VIA IRNERIO 46
I-40126 BOLOGNA
ITALY

PANDE MAHESH CHANDRA DR
UP STATE OBSERVATORY
MANORA PEAK
NAINI TAL 263129
INDIA

PANEK ROBERT J DR
DEPT OF ASTRONOMY
525 DAVEY LAB
UNIV OF PENNA
UNIV PK PA 16802
U.S.A.

PANKONIN VERNON LEE DR
DIV ASTRON SCIENCES
NAT SCIENCE FOUND
1800 G ST SW
WASHINGTON DC 20550
U.S.A.

PANOV K DR
DEPT OF ASTRONOMY
BULGARIAN AC OF SCS
7TH NOVEMBER ST 1
1000 SOFIA
BULGARIA

PAOLICCHI PAOLO DR
OSSERVATORIO ASTRO-
NOMICO
MERATE
MILANO
ITALY

PAPAGIANNIS MICHAEL D PRO
DEPT OF ASTRONOMY
BOSTON UNIVERSITY
725 COMMONWEALTH AVE
BOSTON MA 02215
U.S.A.

PAPALIOLIOS C DR
SMITHSONIAN ASTRO-
PHYSICAL OBS
CAMBRIDGE MA 02138
U.S.A.

PAPALOIZOU JOHN C B DR
INST OF ASTRONOMY
MADINGLEY ROAD
CAMBRIDGE CB3 0HA
U.K.

PAPATHANASOGLOU D DR
DEPT OF ASTRONOMY
UNIV OF ATHENS
PANEPISTIMIOPOLIS
ATHENS 621
GREECE

PAQUET PAUL EG DR
OBSERVATOIRE ROYAL
DE BELGIQUE
AVE CIRCULAIRE 3
B-1180 BRUXELLES
BELGIUM

PARCELIER PIERRE DR
OBSERVATOIRE
DE PARIS
61 AV L OBSERVATOIRE
F-75014 PARIS
FRANCE

PARESCE FRANCESCO DR
SPACE SCIENCES
LABORATORY
UNIVERSITY OF CALIF
BERKELEY CA 94720
U.S.A.

PARIJSKIJ N N DR
INSTITUTE OF PHYSICS
OF THE EARTH
USSR ACAD OF SCI
123810 MOSCOW
U.S.S.R.

PARIJSKIJ YU N DR
SPECIAL ASTROPHYS-
OBS USSR ACAD OF SCI
LENINGRAD BRANCH
196140 LENINGRAD
U.S.S.R.

PARKER EDWARD A DR
ELECTRONICS
LABORATORIES
THE UNIVERSITY
CANTERBURY CT2 7NT
U.K.

PARKER EUGENE N
LAB ASTROPHYS SP RES
UNIV OF CHICAGO
933 E 56TH ST
CHICAGO IL 60637
U.S.A.

PARKER ROBERT A R
CODE CB
NASA-JSC
HOUSTON TX 77058
U.S.A.

PARKINSON JOHN H DR
MULLARD SPACE SC LAB
UNIV COLLEGE LONDON
HOLMBURY ST MARY
DORKING SURREYRH56NS
U.K.

PARKINSON T DR
KITT PEAK NAT OBS
950 N CHERRY AVE
TUCSON AZ 85726
U.S.A.

PARKINSON WILLIAM H
HARVARD COLLEGE
OBSERVATORY
60 GARDEN ST
CAMBRIDGE MA 02138
U.S.A.

PARRISH ALLAN DR.
STATE UNIV.OF N.YORK
C/O 6191 GRC U MASS.
AMHERST MA 01003
U.S.A.

PARSAMYAN E S DR
BYURAKAN ASTROPHYSI-
CAL OBSERVATORY
378433 ARMENIA
U.S.S.R.

PARSONS SIDNEY B DR
DEPT OF ASTRONOMY
UNIVERSITY OF TEXAS
AUSTIN TX 78712
U.S.A.

PARTHASARATHY M DR
INDIAN INSTITUTE OF
ASTROPHYSICS
BANGALORE 560 034
INDIA

PARTRIDGE ROBERT B PROF
HAVERFORD COLLEGE
HAVERFORD PA 19041
U.S.A.

PASACHOFF JAY M PROF
WILLIAMS COLLEGE
HOPKINS OBSERVATORY
WILLIAMSTOWN MA01267
U.S.A.

PASCOAL ANTONIO J B SCI
OBS ASTRONOMICO
PROF MANUEL D BARROS
MONTE DA VIRGEM
VILA NOVA DE GAIA
PORTUGAL

PASCU DAN DR
US NAVAL OBSERVATORY
WASHINGTON DC 20390
U.S.A.

PASINETTI LAURA E PROF
OSSERVATORIO DI
MILANO MERATE
VIA E BIANCHI 46
I-22055 MERATE
ITALY

PASTORIZA MIRIANI G DR
INSTITUTO DE FISICA
UNIVERSIDADE FEDERAL
DO RIO GRANDE DO SUL
90000 PORTO ALEGRE
BRAZIL

PATERNO LUCIO DR
OSSERVATORIO
ASTROFISICO
CITTA UNIVERSITARIA
I-95125 CATANIA
ITALY

PATHRIA RAJ K PROF
DEPT OF PHYSICS
UNIVERSITY WATERLOO
WATERLOO ONT N2L 3G1
CANADA

PAULINY TOTH IVAN K K DR
MPI FUER RADIOASTRO-
NOMIE
AUF DEM HUEGEL 69
D-5300 BONN
GERMANY, F.R.

PAULS THOMAS ALBERT DR
MPI FUER RADIOASTRON
AUF DEM HUEGEL 69
D-5300 BONN 1
GERMANY, F.R.

PAVLOV NIKOLAJ N PROF DR
PULKOVO OBSERVATORY
196140 LENINGRAD
U.S.S.R.

PAVLOVSKAYA E D DR
STERNBERG STATE AS-
TRONOMICAL INSTITUTE
117234 MOSCOW
U.S.S.R.

PAXTON HAROLD J B R
ROYAL GREENWICH OBS
HERSTMONCEUX CASTLE
HAILSHAM
E SUSSEX BN27 1RP
U.K.

PEACH G DR
DEPT PHYSICS ASTRO
UNIVERSITY COLLEGE
GOWER STREET
LONDON WC1E 6BT
U.K.

PEACH JOHN V DR
DEPT OF ASTROPHYSICS
SOUTH PARKS ROAD
OXFORD OX1 3RQ
U.K.

PEALE STANTON J PROF
DEPT OF PHYSICS
UNIVERSITY OF
CALIFORNIA
SN BARBARA CA 93106
U.S.A.

PEARSE REGINALD W B DR
60 GREENLANDS ROAD
STAINES MDX TW18 4LR
U.K.

PEAT D W DR
THE OBSERVATORIES
MADINGLEY ROAD
CAMBRIDGE CB3 OHA
U.K.

PECKER J C PROF
LAT
98 BIS BLVD ARAGO
F-75014 PARIS
FRANCE

PEDERSEN B DR
OBS DE MEUDON
F-92190 MEUDON
FRANCE

PEDERSEN HOLGER DR
E S O
CASILLA 16317
SANTIAGO 9
CHILE

PEDERSEN OLAF PROF
HISTORY SCIENCE DEPT
UNIVERSITY OF AARHUS
NY MUNKEGADE
DK-8000 AARHUS C
DENMARK

PEDLAR ALAN DR
NRAL
JODRELL BANK
CHESHIRE SK11 9DL
U.K.

PEDOUSSAUT ANDRE
OBS DE TOULOUSE
1 AVE FLAMMARION
F-31500 TOULOUSE
FRANCE

PEEBLES P JAMES E
JOSEPH HENRY LABS
JADWIN HALL
PRINCETON NJ 08540
U.S.A.

PEERY BENJAMIN F PROF
DEPT OF PHYSICS AND
ASTRONOMY
HOWARD UNIVERSITY
WASHINGTON DC 20059
U.S.A.

PEIMBERT MANUEL DR
INST DE ASTRONOMIA
APDO POSTAL 70264
MEXICO 20 DF
MEXICO

PEKERIS CHAIM LEIB PROF
DEPT APPLIED MATHS
WEIZMANN INSTITUTE
OF SCIENCE
REHOVOT
ISRAEL

PEL J W DR
KAPTEIJN STERREWACHT
RODEN 8120
THE NETHERLANDS

PELLAS PAUL DR
61 RUE BUFFON
F-75005 PARIS
FRANCE

PELSENEER JEAN KON PROF
UNIV LIBRE DE BRUX
76 AV DES GRENADIERS
BOX 6
B-1150 BRUSSELS
BELGIUM

PELTIER LESLIE C
327 S BREDEICK ST
DELPHOS OH 45833
U.S.A.

PENNY ALAN JOHN DR
S A ASTRON OBS
P O BOX 9
OBSERVATORY 7935
CAPE
SOUTH AFRICA

PENNY C J A MISS
ROYAL GREENWICH
OBSERVATORY
HAILSHAM
EAST SUSSEX BN27 1RP
U.K.

PENSADO J DR
OBSERVATORIO
ASTRONOMICO
ALFONSO X11 5
MADRID 7
SPAIN

PENSTON MICHAEL V DR
VILLAFRANCA SATELLIT
TRACKING STATION
PO BOX APT 54065
MADRID
SPAIN

PENZIAS ARNO A DR
BELL LABORATORIES
BOX 400
HOLMDEL NJ 07733
U.S.A.

PERAIAH ANNAMANENI DR
INDIAN INSTITUTE OF
ASTROPHYSICS
BANGALORE 560034
INDIA

PERCY JOHN R PROF
ERINDALE COLLEGE
MISSISAUGA
ONT L5L 1C6
CANADA

PERDANG JEAN M DR
INSTITUT
D'ASTROPHYSIQUE
B-4200 COINTE-OUGREE
BELGIUM

PEREK LUBOS DR
DEPT OF PHYS C 011
UNIV OF CALIFORNIA
SAN DIEGO PO BOX 109
LA JOLLA CA 92037
U.S.A.

PEREZ-DE-TEJADA H A DR
INST DE ASTRONOMIA
UNAM
AV GASTELUM ENSENADA
BAJA CALIFORNIA
MEXICO

PEREZ-PERAZA J DR
INST DE ASTRONOMIA
UNAM
APT POSTAL 70-264
MEXICO DF
MEXICO

PERINOTTO MARIO DR
OSSERVATORIO ASTRO-
FISICO ARCETRI
I-50125 FIRENZE
ITALY

PERKINS FRANCIS W DR
PLASMA PHYSICS
LABORATORY
P O BOX 451
PRINCETON NJ 08540
U.S.A.

PEROLA G C DR
ISTITUTO FISICA
VIA CELORIA 16
I-20133 MILANO
ITALY

PERRIN MARIE-NOELLE DR
OBSERVATOIRE
DE PARIS
61 AVE OBSERVATOIRE
F-75014 PARIS
FRANCE

PERRY CHARLES L DR
DEPT OF PHYS/ASTRON
LOUISIANA STATE UNIV
BATON ROUGE LA 70803
U.S.A.

PERRY JUDITH J DR
MPI FUER PHYSIK UND
ASTROPHYSIK
FOEHRINGER RING 6
D-8000 MUENCHEN 40
GERMANY, F.R.

PERSIDES SOTIRIOS C
ASST PROF
DEPT OF ASTRONOMY
UNIV OF THESSALONIKI
THESSALONIKI
GREECE

PESCH PETER DR
WARNER AND SWASEY
OBSERVATORY
1975 TAYLOR RD
E CLEVELAND OH 44112
U.S.A.

PETERS GERALDINE JOAN DR
DEPT OF ASTRONOMY
UNIV SOUTHERN CALIF
LOS ANGELES CA 90024
U.S.A.

PETERS WILLIAM L III DR
ASTRONOMY DEPT
UNIV OF TEXAS
AUSTIN TX 78712
U.S.A.

PETERSON BRUCE A DR
ANGLO AUSTRALIAN
OBSERVATORY
P O BOX 296
EPPING NSW 2121
AUSTRALIA

PETERSON CHARLES JOHN DR
DEPT OF PHYSICS
223 PHYSICS BLDG
UNIV OF MISSOURI
COLUMBIA MO 65211
U.S.A.

PETERSON LAURENCE E PROF
DEPT OF PHYSICS C011
UNIVERSITY OF CALIF-
AT SAN DIEGO
LA JOLLA CA 92093
U.S.A.

PETERSON RUTH CAROL DR
607 MARION PLACE
PALO ALTO CA 94301
U.S.A.

PETFORD A DAVID DR
DEPT OF ASTROPHYSICS
SOUTH PARKS ROAD
OXFORD OX1 3RQ
U.K.

PETRI WINFRIED PROF DR
UNTERLEITEN 2
POSTFACH 106
8162 SCHLIERSEE OBB
GERMANY, F.R.

PETRINI DANIEL DR
OBSERVATOIRE DE NICE
B P 252
F-06007 NICE CEDEX
FRANCE

PETROPOULOS BASIL CH DR
RES CTR ASTRO AP MAT
ACADEMY OF ATHENS
14 ANAGNOSTOPOULOU
ATHENS 136
GREECE

PETROSIAN VAHE PROF
INSTITUTE FOR PLASMA
RESEARCH
STANFORD UNIVERSITY
STANFORD CA 94305
U.S.A.

PETROV G I PROF DR
SPACE RESEARCH
INSTITUTE
USSR ACADEMY OF SCI
117810 MOSCOW
U.S.S.R.

PETROV G M DR
NIKOLAEV DEPT OF THE
MAIN ASTRONOMICAL
OBSERVATORY
327000 NIKOLAEV
U.S.S.R.

PETROV NIKOLAI
ASTRONOMICAL OBS
VARNA
BULGARIA

PETROVSKAYA M S DR
INSTITUTE OF THEORE-
TICAL ASTRONOMY
192187 LENINGRAD
U.S.S.R.

PETTENGILL GORDON H PROF
RM 54-620
MASS INSTITUTE OF
TECHNOLOGY
CAMBRIDGE MA 02139
U.S.A.

PETTIT E DR
MT WILSON AND
PALOMAR OBS
813 ST BARBARA ST
PASASENA CA 91106
U.S.A.

PEYTREMANN ERIC DR
ESA
8 10 RUE MARIO NIKIS
F-75738 PARIS CDX 15
FRANCE

PEYTURAUX ROGER H PROF
INST D ASTROPHYSIQUE
98B BLVD ARAGO
F-75014 PARIS
FRANCE

PFENNIG HANS H DR
MPI F PHYSIK UND
ASTROPHYSIK
FOEHRINGER RING 6
D-8000 MUENCHEN 40
GERMANY, F.R.

PFLEIDERER JORG PROF
INST F ASTRONOMIE
UNIVERSITAETSSTR 4
A-6020 INNSBRUCK
AUSTRIA

PFLUG KLAUS DR
ZENTRALINSTITUT FUER
SOLARTERR PHYSIK
DDR-15 POTSDAM-
TELEGRAFENBERG
G.D.R.

PHAM-VAN J MME
CERGA
AVENUE COPERNIC
F-06130 GRASSE
FRANCE

PHILIP A G DAVIS
DUDLEY OBSERVATORY
1202 TROY-
SCHENECTADY RD
LATHAM NY 12110
U.S.A.

PHILLIPS JOHN G PROF
ASTRONOMY DEPT
UNIV OF CALIFORNIA
601 CAMPBELL HALL
BERKELEY CA 94720
U.S.A.

PHILLIPS THOMAS GOULD DR
BELL TELEPHONE LABS
MURRAY HILL NJ 07476
U.S.A.

PICAT J P DR
OBS DE MEUDON
F-92190 MEUDON
FRANCE

PICK MONIQUE DR
DASOP
OBSERVATOIRE
DE MEUDON
F-92190 MEUDON
FRANCE

PIDDINGTON JACK H RES FEL
NATIONAL MEASUREMENT
LABORATORY P O B 218
LINDFIELD
SYDNEY 2070
AUSTRALIA

PIERCE A KEITH DR
KITT PEAK NATIONAL
OBSERVATORY
PO 26732
TUCSON AZ 85726
U.S.A.

PIERCE DAVID ALLEN
PHYSICAL SCS DIV
EL CAMINO COLLEGE
TORRANCE CA 90506
U.S.A.

PIIROLA VILPPU E DR
OBSERVATORY
TAHTITORNINMAKI
SF-00130 HELSINKI 13
FINLAND

PILACHOWSKI CATHERINE DR
ASTRONOMY DEPT FM-20
UNIV OF WASHINGTON
SEATTLE WA 98195
U.S.A.

PILCHER CARL BERNARD DR
INST FOR ASTRONOMY
UNIV OF HAWAII
2680 WOODLAWN DRIVE
HONOLULU HI 96822
U.S.A.

PILKINGTON JOHN D H DR
ROYAL GREENWICH
OBSERVATORY
HAILSHAM
EAST SUSSEX BN27 1RP
U.K.

PILOWSKI K PROF DR
GEODAETISCHES INST
TECHN UNIVERSITAET
NIENBURGER STR 1
D-3000 HANNOVER
GERMANY, F.R.

PINEAULT SERGE DR
DEPT GEOPHYS ASTRON
UNIVERSITY OF B C
VANCOUVER BC V6T 1W5
CANADA

PINES DAVID PROF
DEPT OF PHYSICS
UNIVERSITY ILLINOIS
URBANA IL 61801
U.S.A.

PINGREE DAVID PROF
BROWN UNIV
PROVIDENCE RI 02912
U.S.A.

PINIGIN GENNADIJ I DR
MAIN ASTRONOMICAL OB
USSR ACAD OF SCS
PULKOVO M-140
196140 LENINGRAD
U.S.S.R.

PINKAU K PROF
MPI FUER EXTRATERRE
STRISCHE PHYSIK
D-8046 GARCHING
GERMANY, F.R.

PINTO GIROLAMO PROF
OSSERVATORIO
ASTRONOMICO
I-35100 PADOVA
ITALY

PIOTROWSKI STEFAN PROF
OBSERWATORIUM
ASTRONOMICZNE UW
AL UJAZDOWSKIE 4
00478 WARSZAWA
POLAND

PIPHER JUDITH L
PHYSICS AND ASTRONO-
MY DEPT
UNIV OF ROCHESTER
ROCHESTER NY 14627
U.S.A.

PISMIS DE RECILLAS PARIS
PROF
INST DE ASTRONOMIA
APDO POSTAL 70264
MEXICO 20 D F
MEXICO

PITTICH EDUARD M DR
ASTRONOMICAL INSTI-
TUTE SAV
DUBRAVSKA CESTA A 11
89930 BRATISLAVA
CZECHOSLOVAKIA

PITZ ECKHART DR
MPI-ASTRONOMY
KOENIGSTUHL
D-6900 HEIDELBERG
GERMANY, F.R.

PIZZELLA G DR
IST DI FISICA
G MARCONI
UNIVERSITA DI ROMA
ROMA
ITALY

PLAKIDIS STAVROS PROF
CRONOU 1 VAS PAVLOU
1 STR
PALEA PENTELI ATHENS
GREECE

PLASKETT HARRY HEMLEY
PROF
48 BLENHEIM DRIVE
OXFORD OX2 8DQ
U.K.

PLASSARD J DR
KSARA OBSERVATORY
KSARA
LEBANON

PLAUT LUCAS PROF
KAPTEYN LABORATORIUM
POSTBUS 800
9700 RV GRONINGEN
THE NETHERLANDS

PLAVCOVA Z DR
DEPT OF ASTRONOMY
UCLA
405 HILGARD AVE
LOS ANGELES CA 90024
U.S.A.

PLAVEC MIROSLAV PROF
DEPT OF ASTRONOMY
UNIVERSITY OF
CALIFORNIA MS 8979
LOS ANGELES CA 90024
U.S.A.

PNEUMAN GERALD W
HIGH ALTITUDE
OBSERVATORY
BOULDER CO 80302
U.S.A.

PODOBED V V DR
STERNBERG STATE AS-
TRONOMICAL INSTITUTE
117234 MOSCOW
U.S.S.R.

POEPPEL WOLFGANG G L DR
INSTITUTO ARGENTINO
DE RADIOASTRONOMIA
CASILLA DE CORREO 5
1894 VILLA ELISA BA
ARGENTINA

POGO ALEXANDER DR
HALE OBSERVATORIES
813 SANTA BARBARA ST
PASADENA CA 91101
U.S.A.

POHL ECKHARD DR
STERNWARTE NUERNBERG
LUETZOWSTR 10
D-8500 NUERNBERG
GERMANY, F.R.

POLAND ARTHUR I DR
HIGH ALTITUDE
OBSERVATORY
P O BOX 3000
BOULDER CO 80307
U.S.A.

POLETTO GIANNINA PROF
OSSERVATORIO ARCETRI
LARGO FERMI 5
I-50125 FIRENZE
ITALY

POLLACK JAMES B DR
SPACE SC DIV MP245-3
NASA AMES RES CTR
MOFFETT FIELD
CA 94035
U.S.A.

POLNITZKY GERHARD DR
INST FUER ASTRONOMIE
UNIV WIEN
TUERKENSCHANZSTR 17
A-1180 WIEN
AUSTRIA

POLOZHENTSEV DIMITRIJ DDR
PULKOVO OBSERVATORY
196140 LENINGRAD
U.S.S.R.

POLUPAN P N DR
KIEV STATE UNIVERSI-
TY ASTRONOMICAL OB-
SERVATORY
252053 KIEV
U.S.S.R.

POMA ANGELO DR
INT ASTRONOMICAL
STATION
CARLOFORTE
CAGLIARI
ITALY

PONSONBY JOHN E B DR
NRAL
JODRELL BANK
CHESHIRE SK11 9DL
U.K.

POOLEY GUY DR
CAVENDISH LABORATORY
MADINGLEY RD
CAMBRIDGE CB3 OHE
U.K.

POPE JOHN D
ROYAL GREENWICH
OBSERVATORY
HAILSHAM
EAST SUSSEX BN27 1RP
U.K.

POPELAR J DR
GRAVITY AND GEOD DIV
EARTH PHYSICS BRANCH
3 OBSERVATORY CRESC
OTTAWA ONT K1A OE4
CANADA

POPOV N A DR
GRAVIMETRICAL OBSER-
VATORY OF UKRAINIAN
ACADEMY OF SCIENCES
314029 POLTAVA
U.S.S.R.

POPOV VICTOR S DR
MAIN ASTRONOMICAL OB
USSR ACAD OF SCS
PULKOVO M-140
196140 LENINGRAD
U.S.S.R.

POPOVA MALINA D DR
DEPT OF ASTRONOMY
BULGARIAN ACAD SCI
7 NOVEMBER 1
SOFIA 1000
BULGARIA

POPOVIC BOZIDAR PROF DR
FACULTY OF TRAFFIC
TAKOVSKA 34
11000 BEOGRAD
YUGOSLAVIA

POPOVIC GEORGIJE DR
ASTRONOMICAL OBS
VOLGINA 7
11050 BEOGRAD
YUGOSLAVIA

POPPER DANIEL M PROF
DEPT OF ASTRONOMY
UNIVERSITY OF
CALIFORNIA
LOS ANGELES CA 90024
U.S.A.

PORFIR EV V V DR
KRUPSKAJA PEDAGOGIC
INSTITUTE
107846 MOSCOW
U.S.S.R.

PORTER JOHN G DR
WHITESTONES
HEMPSTEAD LANE
HAILSHAM
SUSSEX BN27 3AA
U.K.

PORTER NEIL A PROF
PHYSICS DEPT
UNIVERSITY COLLEGE
BELFIELD
DUBLIN 4
IRELAND

PORUBCAN VLADIMIR DR
ASTRONOMICAL INSTI-
TUTE SAV
DUBRAVSKA CESTA A 11
89930 BRATISLAVA
CZECHOSLOVAKIA

POTTASCH STUART R PROF
KAPTEYN LABORATORIUM
POSTBUS 800
GRONINGEN
THE NETHERLANDS

POTTER HEINO I DR
PULKOVO OBSERVATORY
196140 LENINGRAD
U.S.S.R.

POULAKOS CONSTANTINE DR
RESEARCH CTR ASTRONO
MY AND APPLIED MATHS
ACADEMY OF ATHENS
ATHENS 136
GREECE

POUMEYROL FERNAND MR
OBSERVATOIRE
DE BORDEAUX
F-33270 FLOIRAC
FRANCE

POUNDS KENNETH A PROF
DEPT OF PHYSICS
UNIVERSITY
LEICESTER LE1 7RH
U.K.

POUQUET A DR
OBS DE NICE
B P 252
F-06007 NICE CEDEX
FRANCE

POVEDA ARCADIO DR
INST DE ASTRONOMIA
UNIV NAC AUTO DE MEX
APDO POSTAL 70-264
MEXICO 20 DF
MEXICO

PRADERIE F DR
DESPA
OBSERVATOIRE MEUDON
F-92190 MEUDON
FRANCE

PRADHAN DR
NIZAMIAH OBS
HYDERABAD
INDIA

PRATAP R DR
PHYSICAL RESEARCH
LABORATORY
NAVRANGPURA
AHMEDABAD 380009
INDIA

PREITE-MARTINEZ ANDREA DR
LABORATORIO DI
ASTROFISICA SPAZIALE
CP 67
I-00044 FRASCATI
ITALY

PRENDERGAST KEVIN H PROF
DEPT OF ASTRONOMY
COLUMBIA UNIVERSITY
538 W 120 ST
NEW YORK NY 10027
U.S.A.

PRENTICE ANDREW J R DR
DEPT OF MATHEMATICS
MONASH UNIVERSITY
CLAYTON VICT 3168
AUSTRALIA

PRESS WILLIAM H DR
HARVARD COLLEGE OBS
60 GARDEN ST
CAMBRIDGE MA 02138
U.S.A.

PRESTON GEORGE W DR
HALE OBSERVATORIES
813 SANTA BARBARA ST
PASADENA CA 91101
U.S.A.

PREUSS EUGEN DR
MPI FUER RADIO-
ASTRONOMIE
AUF DEM HUEGEL 69
D-5300 BONN 1
GERMANY, F.R.

PREVOT BURNICHON MARIE
LOUISE DR
INST D ASTROPHYSIQUE
98 BIS BLVD ARAGO
F-75014 PARIS
FRANCE

PREVOT LOUIS DR
OBSERVATOIRE DE
MARSEILLE
2 PLACE LE VERRIER
F-13004 MARSEILLE
FRANCE

PRICE MICHAEL J. DR.
SCIENCE APPLICATIONS
5055 E BROADWAY
SUITE A-214
TUCSON AZ 85711
U.S.A.

PRICE R MARCUS DR
DIV ASTRONOMICAL SCS
NAT SCIENCE FOUND
1800 G STREET N W
WASHINGTON DC 20550
U.S.A.

PRIEST E R DR
APPLIED MATHS DEPT
THE UNIVERSITY
ST ANDREWS FIFE
KY16 9SS
U.K.

PRIESTER WOLFGANG PROF
INST F ASTROPHYSIK
AUF DEM HUEGEL 71
D-5300 BONN
GERMANY, F.R.

PRINCE HELEN DODSON PROF
4800 FILLMORE AVE
ALEXANDRIA VA 22311
U.S.A.

PRINGLE JAMES E DR
INST OF ASTRONOMY
MADINGLEY ROAD
CAMBRIDGE CB3 OHA
U.K.

PRITCHET CHRISTOPHER J DR
DEPT GEOPHYS ASTRON
UNIVERSITY OF B C
VANCOUVER BC V6T 1W5
CANADA

PROBSTEIN R F DR
DEPT OF MECHANICAL
ENGINEERING
MIT
CAMBRIDGE MA 02139
U.S.A.

PROCHAZKA FRANZ V DR
UNIVERSITAETS-
STERNWARTE
TUERKENSCHANZSTR 17
A-1180 WIEN
AUSTRIA

PRODAN Y I DR
STERNBERG STATE AS-
TRONOMICAL INSTITUTE
117234 MOSCOW
U.S.S.R.

PROISY PAUL E DR
OBSERVATOIRE DE LYON
F-69230 ST GENIS
LAVAL
FRANCE

PROKAKIS THEODORE J DR
OBSERVATOIRE
D ATHENES
ATHENS 306
GREECE

PROKOF'EV VLADIMIR K PROF
CRIMEAN ASTROPHYSI-
CAL OBSERVATORY
PO NAUCHNY
334413 CRIMEA
U.S.S.R.

PROKOF'EVA IRINA A DR
PULKOVO OBSERVATORY
196140 LENINGRAD
U.S.S.R.

PROKOF'EVA VALENTINA V DR
CRIMEAN ASTROPHYS OB
NAUCHNYJ
334413 CRIMEA
U.S.S.R.

PRONIK I I DR
CRIMEAN ASTROPHY OBS
ACAD SCIENCES USSR
334413 P O NAUCHNY
CRIMEA
U.S.S.R.

PRONIK V I DR
CRIMEAN ASTROPHY OBS
ACAD SCIENCES USSR
334413 P O NAUCHNY
CRIMEA
U.S.S.R.

PROTHEROE WILLIAM M PROF
DEPT OF ASTRONOMY
OHIO STATE UNIV
174 W 18TH AVE
COLUMBUS OH 43210
U.S.A.

PROTICH MILORAD B
ASTRONOMICAL OBSERVA
TORY OF BELGRADE
VOLGINA 7
11050 BELGRADE
YUGOSLAVIA

PROVERBIO EDOARDO PROF
IST DI ASTRONOMIA
VIA OSPEDALE 72
I-09100 CAGLIARI
ITALY

PROVOST J DR
OBS DE NICE
B P 252
F-06007 NICE CEDEX
FRANCE

PRYCE MAURICE H L DR
DEPT OF PHYSICS
UNIVERSITY OF B C
2075 WESBROOK MALL
VANCOUVER BC V6T 1W5
CANADA

PRZYBYLSKI ANTONI DR
MOUNT STROMLO
OBSERVATORY
WODEN A C T 2606
AUSTRALIA

PSKOVSKIJ JU P DR
STERNBERG STATE AS-
TRONOMICAL INSTITUTE
117234 MOSCOW
U.S.S.R.

PUCILLO MAURO DR
OSSERVATORIO
ASTRONOMICO
VIA TIEPOLO 11
I-34131 TRIESTE
ITALY

PUGET JEAN-LOUP DR
INSTITUT
D ASTROPHYSIQUE
98 BIS BLVD ARAGO
F-75014 PARIS
FRANCE

PUNETHA LALIT MOHAN DR
UP STATE OBSERVATORY
MANORA PEAK
NAINI TAL 263129
INDIA

PURGATHOFER ALOIS PROF DR
INST FUER ASTRONOMIE
UNIV WIEN
TUERKENSCHANZSTR 17
A-1180 WIEN
AUSTRIA

PURTON CHRISTOPHER R DR
CRESS DEPT OF PHYS
YORK UNIVERSITY
4700 KEELE STR
DOWNSVIEW ONT M3J1P3
CANADA

PUSHKIN SERGEY B DR
TIME AN FREQUENCY
SERVICE
GOSSTANDARD USSR
117049 MOSCOW
U.S.S.R.

PYE JOHN P DR
8 BEAUMONT WALK
LEICESTER LE4 0PQ
U.K.

PYPER D M DR
DEPT OF PHYSICS AND
ASTRONOMY
TEL-AVIV UNIVERSITY
TEL-AVIV
ISRAEL

QUAST G R
OBS NACIONAL
R GENERAL BRUCE 586
SAO CRISTOVAO
20000 RIO DE JANEIRO
BRAZIL

QUENBY JOHN J DR
BLACKETT LAB
IMPERIAL COLLEGE
PRINCE CONSORT ROAD
LONDON SW7 2BZ
U.K.

QUERCI FRANCOIS R DR
EQUIPE ETOILES
FROIDES
OBS DE MEUDON
F-92190 MEUDON
FRANCE

QUERCI MONIQUE DR
EQUIPE ETOILES
FROIDES
OBS DE MEUDON
F-92190 MEUDON
FRANCE

QUIJANO LUIS
INSTITUTO Y OBSERVA-
TORIO DE MARINA
SAN FERNANDO CADIZ
SPAIN

QUINTANA HERNAN DR
NAT RADIO ASTRO OBS
EDGEMONT RD
CHARLOTTESVILLE
VA 22901
U.S.A.

QUINTANA JOSE M DR
INSTITUTO DE ASTROFI
SICA DE ANDALUCIA
GRANADA
SPAIN

QUIRK WILLIAM J DR
24798 JO MARY CT
HAYWARD CA 94541
U.S.A.

QVIST BERTIL PROF
MATHEMATICS DEPT
ABO AKADEMI
SF-20500 ABO 50
FINLAND

RAADU MICHAEL A DR
DEPT PLASMA PHYSICS
ROYAL INSTITUTE OF
TECHNOLOGY
S-10044 STOCKHOLM 70
SWEDEN

RACHKOVSKY D N DR
CRIMEAN ASTROPHYSI-
CAL OBSERVATORY USSR
ACAD OF SCI NAUCHNIY
334413 CRIMEA
U.S.S.R.

RACINE RENE DR
DEPT DE PHYSIQUE
UNIVERSITE DE
MONTREAL BP 6128
MONTREAL PQ H3C 3J7
CANADA

RACKHAM THOMAS W DR
JODRELL BANK
CHESHIRE SK 11 9PL
U.K.

RADHAKRISHNAN V PROF
DIRECTOR
RAMAN RESEARCH
INSTITUTE
BANGALORE 560 006
INDIA

RADLOVA L N DR
INSTITUTE OF SCIENCE
AND TECHNICS INFORMA
TION
125219 MOSCOW
U.S.S.R.

RAGHAVAN NIRUPAMA DR
2133 INDIAN INSTITU-
TE OF TECHNOLOGY
CAMPUS
NEW DELHI 110 029
INDIA

RAIMOND ERNST DR
NETHERL FOUNDATION
FOR RADIOASTRONOMY
RADIOSTERRENWACHT
7991 AC DWINGELOO
THE NETHERLANDS

RAJCHL JAROSLAV DR
ASTRONOMICAL INST
CZECHOS ACAD SC
OBSERVATORY
251 65 ONDREJOV
CZECHOSLOVAKIA

RAKOS KARL D PROF
INST FUER ASTRONOMIE
UNIV WIEN
TURKENSCHANZSTR 17
A-1180 WIEN
AUSTRIA

RAMATY REUVEN DR
LAB HIGH ENERGY ASTR
NASA-GODDARD SPACE
FLIGHT CTR CODE 660
GREENBELT MD 20771
U.S.A.

RANDIC LEO PROF DR
GUNDULICEVA 54
ZAGREB
YUGOSLAVIA

RAO A P DR
RADIO ASTRONOMY CTR
P O BOX NO 8
OOTACAMUND 643 001
INDIA

RAPLEY CHRISTOPHER G DR
MULLARD SPACE SC LAB
UNIV COLLEGE LONDON
LONDON
U.K.

RAZIN V A DR
RADIOPHYSICAL
RESEARCH INSTITUTE
603600 GORKIJ
U.S.S.R.

REAVES GIBSON PROF
UNIVERSITY OF
SOUTHERN CALIFORNIA
LOS ANGELES CA 90007
U.S.A.

REBER GROTE DR
GENERAL DELIVERY
BOTHWELL TASM 7411
AUSTRALIA

RAEDLER K H DR
ZNTRINST ASTROPHYSIK
ASTROP OBSERVATORIUM
TELEGRAFENBERG
DDR-15 POTSDAM
G.D.R.

RAHE JURGEN PROF
UNIV ERL-NUERNBERG
ASTRONOMICAL INST
STERNWARTSTR 7
D-8600 BAMBERG
GERMANY, F.R.

RAINE DEREK J DR
DEPT OF ASTRONOMY
UNIVERSITY OF
LEICESTER
LEICESTER LEI 7RH
U.K.

RAJU P K DR
INDIAN INSTITUTE OF
ASTROPHYSICS
BANGALORE 560034
INDIA

RAKSHIT H PROF
BENGAL ENGINEERING
COLLEGE
SIBPORE
HOWRAH
INDIA

RAMBERG JOERAN M PROF
GENVAEGEN 4
S-133 00
SALTSJOEBADEN
SWEDEN

RANK DAVID M PROF
LICK OBSERVATORY
UNIVERSITY OF CALIF
SANTA CRUZ CA 95064
U.S.A.

RAO RAMACHANDRA V PROF
ISRO SATELLITE
CENTER PEENYA
BANGALORE 560058
INDIA

RASMUSEN H Q DR
VAERSLEVGAAREN
FRANKERUPVEI 40
DK-4400 KALUNDBORG
DENMARK

READHEAD ANTHONY C S DR
RADIO ASTRONOMY DEPT
CALIFORNIA INSTITUTE
OF TECHNOLOGY
PASADENA CA 91125
U.S.A.

REAY NEWRICK K DR
ASTRONOMY GROUP
BLACKETT LABORATORY
IMPERIAL COLLEGE
LONDON SW7 2BZ
U.K.

RECILLAS-CRUZ ELSA DR
INST DE ASTRONOMIA
UNAM
APDO POSTAL 70-264
MEXICO D F
MEXICO

RAFANELLI PIERO DR
OSSERVATORIO ASTRO-
NOMICO
I-35100 PADOVA
ITALY

RAIKOVA DONKA DR
DEPT OF ASTRONOMY
BULGARIAN AC OF SCS
7TH NOVEMBER ST 1
1000 SOFIA
BULGARIA

RAJAMOHAN R DR
INDIAN INSTITUTE OF
ASTROPHYSICS
BANGALORE 560 034
INDIA

RAKAVY GIDEON PROF
HEBREW UNIVERSITY
EINSTEIN INST PHYS
JERUSALEM
ISRAEL

RAMADURAI S DR
TATA INST FUND RES
HOMI BHABHA ROAD
BOMBAY 400 005
INDIA

RAMSEY LAWRENCE W DR
DEPT OF ASTRONOMY
525 DAVEY LAB
UNIV OF PENNA
UNIV PK PA 16802
U.S.A.

RANKIN JOANNA M DR
DEPT OF HISTORY
CORNELL UNIVERSITY
MCGRAW HALL
ITHACA NY 14853
U.S.A.

RAPAPORT M DR
OBS DE BORDEAUX
F-33270 FLOIRAC
FRANCE

RAYROLE JEAN R DR
OBSERVATOIRE
DE MEUDON
F-92190 MEUDON
FRANCE

REASENBERG ROBERT D DR
MIT
54-612
CAMBRIDGE MA 02139
U.S.A.

REBEIROT EDITH DR
OBS DE MARSEILLE
2 PLACE LE VERRIER
F-13004 MARSEILLE
FRANCE

REDDISH VINCENT C PROF
ROYAL OBSERVATORY
EDINBURGH EH12 9AJ
U.K.

REES MARTIN J PROF
INST OF ASTRONOMY
MADINGLEY RD
CAMBRIDGE CB3 OHE
U.K.

REEVES EDMOND M DR
HIGH ALTITUDE OBS.
P O BOX 3000
BOULDER CO 80307
U.S.A.

REEVES HUBERT PROF
SEP-SES BAT 28
CEN SACLAY
B P 2
F-91190 GIF S YVETTE
FRANCE

REFSDAL S PROF DR
HAMBURGER STERNWARTE
GOJENBERGSWEG 112
D-2050 HAMBURG 80
GERMANY, F.R.

REGO FERNANDES M. DR.
ASTROFISICA
FACULTA FISICA
UNIV.COMPLUTENSE
MADRID 3
SPAIN

REID JOHN H DR
1007 RESEDA DRIVE
HOUSTON TX 77062
U.S.A.

REID MARK JONATHAN DR
NAT RADIO ASTRON OBS
EDGEMONT RD
CHARLOTTESVL VA22901
U.S.A.

REIMERS DIETER DR
INST THEO PHY STERNW
NEUE UNIV PHYSIK ZTM
OLSHAUSENST GB N 61C
D-2300 KIEL 1
GERMANY, F.R.

REINHARDT MICHAEL PROF
ASTRONOMISCHES INST
RUHR UNIVERSITAET
POSTFACH 102148
D-4630 BOCHUM 1
GERMANY, F.R.

REINISCH G DR
OBS DE NICE
B P 252
F-06007 NICE CEDEX
FRANCE

REIZ ANDERS PROF
UNIVERSITY
OBSERVATORY
OSTERVOLDGADE 3
DK-1350 COPENHAGEN K
DENMARK

REMY BATTIAU LILIANE G A
DR
INST D ASTROPHYSIQUE
UNIVERSITY DE LIEGE
B-4200 COINTE OUGREE
BELGIUM

RENSE WILLIAM A DR
DEPT PHYS AND ASTROP
UNIV OF COLORADO
DUANE PHYSICAL LABS
BOULDER CO 80302
U.S.A.

RENSON P F M DR
INSTITUT
D ASTROPHYSIQUE
B-4200 COINTE-OUGREE
BELGIUM

RENZINI ALVIO PROF
OSSERVATORIO
ASTRONOMICO
VIA ZAMBONI 33
I-40126 BOLOGNA
ITALY

REQUIEME YVES DR
OBSERVATOIRE
DE BORDEAUX
F-33270 FLOIRAC
FRANCE

REUNING ERNEST G DR
DEPT OF PHYSICS AND
ASTRONOMY
UNIV OF GEORGIA
ATHENS GA 30602
U.S.A.

REVELLE DOUGLAS ORSON DR
DEPT OF PHYSICS
C U BOX 6010
NORTHERN ARIZ UNIV
FLAGSTAFF AZ 86011
U.S.A.

REYNOLDS JOHN H PROF
DEPT OF PHYSICS
UNIVERSITY OF CALIF
BERKELEY CA 94720
U.S.A.

REYNOLDS RONALD J DR
SPACE PHYSICS GROUP
1150 UNIVERSITY AVE
MADISON WI 53706
U.S.A.

RHYNSBURGER ROBERT W DR
US NAVAL OBSERVATORY
WASHINGTON DC 20390
U.S.A.

RIBES ELIZABETH DR
OBSERVATOIRE
DE MEUDON
F-92190 MEUDON
FRANCE

RIBES JEAN-CLAUDE DR
INAG
77 AVE DENFERT-
ROCHEREAU
F-75014 PARIS
FRANCE

RICE D A
COAST AND GEODETIC
SURVEY
ROCKVILLE MD 20852
U.S.A.

RICE JOHN B DR
DEPT PHYSICS AND
ASTRONOMY
BRANDON UNIVERSITY
BRANDON MAN R7A 6A9
CANADA

RICHARDSON E HARVEY DR
DOMINION ASTROPHYSI-
CAL OBSERVATORY
5071 W SAANICH RD
VICTORIA BC V8X 3X3
CANADA

RICHARDSON R S
GRIFFITH OBS
P O BOX 27787
LOS FELIX STATION
LOS ANGELES CA 90027
U.S.A.

RICHER HARVEY B DR
DEPT GEOPHYS-ASTRON
UNIVERSITY OF BC
2075 WESBROOK PLACE
VANCOUVER BC V6T 1W5
CANADA

RICHTER G A DR
ZENTRALINSTITUT FUER
ASTROPHYSIK
STERNWARTE SONNEBERG
DDR-64 SONNEBERG
G.D.R.

RICHTER JOHANNES PROF
INST F EXPER PHYSIK
PHYSIKZENTRUM
OLSHAUSENSTRASSE
D-2300 KIEL
GERMANY, F.R.

RICHTER N B PROF
POST ULLA
BEI WELZEL
DDR-5301 WEIMAR
G.D.R.

RICKARD JAMES JOSEPH DR
CLARK LAKE RADIO OBS
P O BOX 626
BORREGO SPRINGS
CO 92004
U.S.A.

RICKARD LEE J DR
NAT RADIO ASTRON OBS
EDGEMONT RD
CHARLOTTESVL VA22901
U.S.A.

RICKER GEORGE R DR
MIT
CTR FOR SPACE RES
77 MASS AV RM 37-527
CAMBRIDGE MA 02139
U.S.A.

RICKETT BARNABY JAMES DR
DEPT APPLIED PHYSICS
AND INFORMATION SC
UNIV CALIF SAN DIEGO
LA JOLLA CA 92093
U.S.A.

RICKMAN HANS DR
ASTRONOMISKA OBS
BOX 515
S-75120 UPPSALA
SWEDEN

RIDDLE ANTHONY C DR
DEPT ASTRO-GEOPHYS
UNIVERSITY OF
COLORADO
BOULDER CO 80302
U.S.A.

RIEGEL KURT W DR
DEPT OF ASTRONOMY
UCLA
LOS ANGELES CA 90024
U.S.A.

RIEU N G DR
OBSERVATOIRE
DE MEUDON
F-92190 MEUDON
FRANCE

RIGHINI ALBERTO PROF
OSSERVATORIO ASTRO-
FISICO ARCETRI
LARGO E FERMI 5
I-50125 FIRENZE
ITALY

RIGHINI-COHEN GIOVANNA DR
DEPT EARTH SPACE SCS
SUNY
STONY BROOK NY 11794
U.S.A.

RIGUTTI MARIO PROF
OSSERVATORIO ASTRONO
MICO DI CAPODIMONTE
MOIARIELLO 16
I-80131 NAPOLI
ITALY

RIIHIMAA JORMA J DR
AARNE KARJALAINEN
OBSERVATORY
UNIVERSITY OF OULU
SF-90100 OULU 10
FINLAND

RILEY JULIA M DR
MULLARD RADIO ASRON
LABORATORY
MADINGLEY ROAD
CAMBRIDGE
U.K.

RINDLER WOLFGANG PROF
UNIVERSITY OF TEXAS
BOX 688
RICHARDSON TX 75080
U.S.A.

RINEHART JOHN S PROF
HYPERDYNAMICS
P O BOX 392
SANTA FE NM 87501
U.S.A.

RING JAMES PROF
THE BLACKETT
LABORATORY
IMPERIAL COLLEGE
LONDON SW7 2BZ
U.K.

RINGNES TRULS S DR
INST THEOR ASTROPHYS
UNIVERSITY OF OSLO
P O BOX 1029
BLINDERN OSLO 3
NORWAY

RINGUELET ADELA E DR
49 342
1900 LA PLATA
ARGENTINA

RITTER HANS DR
MPI FUER PHYSIK UND
ASTROPHYSIK
FOEHRINGER RING 6
D-8000 MUENCHEN
GERMANY, F.R.

RIZVANOV NAUFAL G DR
ENGELHARDT OBS
422526 KASAN
U.S.S.R.

ROACH FRANKLIN E
2969 KALAKAUA AVE
APT 605
HONOLULU HI 96815
U.S.A.

ROARK TERRY P PROF
DEPT OF ASTRONOMY
OHIO ST UNIVERSITY
174 W 18TH AVE
COLUMBUS OH 43210
U.S.A.

ROBBINS R ROBERT PROF
UNIVERSITY OF TEXAS
ASTRONOMY DEPT
AUSTIN TX 78712
U.S.A.

ROBE H A G DR
INSTITUT
D ASTROPHYSIQUE
B-4200 COINTE-OUGREE
BELGIUM

ROBERTI GIUSEPPE DR
OSSERVATORIO ASTRO-
NOMICO
NAPOLI
ITALY

ROBERTS BERNARD DR
DEPT APPLIED MATH
UNIV OF ST ANDREWS
ST ANDREWS FIFE
KY16 8QP
U.K.

ROBERTS DAVID HALL DR
RES LAB ELECTRONICS
26-321
MIT
CAMBRIDGE MA 02139
U.S.A.

ROBERTS JAMES A DR
CSIRO DIVISION OF
RADIOPHYSICS
P O BOX 76
EPPING NSW 2121
AUSTRALIA

ROBERTS MORTON S DR
NAT RADIO ASTRO OBS
EDGEMONT RD
CHARLOTTESVILLE
VA 22901
U.S.A.

ROBERTS WALTER ORR DR
ASPEN INST HUM STUD
1229 UNIVERSITY AVE
2ND FLOOR
BOULDER CO 80302
U.S.A.

ROBERTS WILLIAM W JR PROF
DPT APPL MATH COM SC
THORNTON HALL UNIVVA
CHARLOTTESVILLE
VA 22901
U.S.A.

ROBERTSON WILLIAM H BSC
SYDNEY OBSERVATORY
OBSERVATORY PARK
SYDNEY NSW 2000
AUSTRALIA

ROBINSON BRIAN J DR
CSIRO RADIOPHYSICS
LABORATORY
P O BOX 76
EPPING NSW 2121
AUSTRALIA

ROBINSON EDWARD LEWIS DR
DEPT OF ASTRONOMY
UNIV OF TEXAS
AUSTIN TX 78712
U.S.A.

ROBINSON I PROF
UNIVERSITY OF TEXAS
BOX 688
MAIL STATION BE 32
RICHARDSON TX 75080
U.S.A.

ROBINSON LLOYD B DR
LICK OBSERVATORY
UNIVERSITY OF CALIF
SANTA CRUZ CA 95064
U.S.A.

ROBINSON WILLIAM J DR
DEPT OF MATHEMATICS
THE UNIVERSITY
BRADFORD BD7 1DP
U.K.

ROBLEY R DR
OBSERVATOIRE
DU PIC DU MIDI
1 AVE C FLAMMARION
F-31500 TOULOUSE
FRANCE

ROBSON IAN E DR
DIV PHYS AND ASTRON
PRESTON POLYTECHNIC
PRESTON PR1 2TQ
U.K.

ROCCA-VOLMERANGE B DR
LAB RENE BERNAS
BAT 108
F-91406 ORSAY
FRANCE

ROCHESTER MICHAEL G PROF
PHYSICS DEPT
MEMORIAL UNIVERSITY
ST JOHNS NFLD A1C5S7
CANADA

RODDIER C DR
DEPT D'ASTROPHYSIQUE
UNIVERSITE DE NICE
PARC VALRORE
F-06034 NICE CEDEX
FRANCE

RODDIER FRANCOIS PROF
UNIVERSITE DE NICE
DEPT ASTROPHYSIQUE
PARC VALROSE
F-06034 NICE CEDEX
FRANCE

RODGERS ALEX W DR
MT STROMLO
OBSERVATORY
WODEN PO ACT 2606
AUSTRALIA

RODMAN RICHARD B DR
65 LOCUST AVE
LEXINGTON MA 02173
U.S.A.

RODONO MARCELLO DR
OSSERVATORIO
ASTROFISICO
CITTA UNIVERSITARIA
I-95125 CATANIA
ITALY

RODRIGUEZ LUIS F
INST DE ASTRONOMIA
UNAM
APDO POSTAL 70-264
MEXICO D F
MEXICO

ROEDER ROBERT C PROF
DEPT OF ASTRONOMY
SCARBOROUGH COLLEGE
1265 MILITARY TRAIL
WEST HILL ONT M1C1A4
CANADA

ROEMER ELIZABETH PROF
LUNAR AND PLANETARY
LABORATORY
UNIVERSITY ARIZONA
TUCSON AZ 85721
U.S.A.

ROEMER MAX PROF
INST F ASTROPHYSIK
EXTRATERR FORSCHUNG
AUF DEM HUEGEL 71
D-5300 BONN 1
GERMANY, F.R.

ROENNAENG BERNT O DR
ONSALA SPACE
OBSERVATORY
S-430 34 ONSALA
SWEDEN

ROGER ROBERTS S DR
DOMINION RADIO ASTRO
PHYSICAL OBSERVATORY
BOX 248
PENTICTON BC V2A 6K3
CANADA

ROGERS ALAN E E DR
HAYSTACK OBSERVATORY
WESTFORD MA 01886
U.S.A.

ROGERSON JOHN B PROF
PRINCETON UNIVERSITY
DEPT ASTROPHYS SCS
PEYTON HALL
PRINCETON NJ 08540
U.S.A.

ROGSTAD DAVID H DR
MAIL CODE 102-24
CALIF INST TECHN
1201 E CALIF BLVD
PASADENA CA 91125
U.S.A.

ROHLFS K PROF DR
RUHR UNIV BOCHUM
INST FUR ASTROPHYS
POSTFACH 102 148
D-4630 BOCHUM 1
GERMANY, F.R.

ROLAND GINETTE DR
INST D ASTROPHYSIQUE
UNIVERSITE DE LIEGE
B-4200 COINTE OUGREE
BELGIUM

ROLLAND ANGEL DR
JUPITER 2
HUETOR VEGA
GRANADA
SPAIN

ROMAN NANCY G DR
ASTROPHYS DIV NASA
CODE SC-7
WASHINGTON DC 20546
U.S.A.

ROMANA ANTONIO DR
OBSERVATORIO DEL
EBRO
ROQUETAS
TORTOSA BARCELONA
SPAIN

ROMANCHUK P R DR
KIEV STATE UNIVERSI-
TY ASTRONOMICAL
OBSERVATORY
252053 KIEV
U.S.S.R.

ROMANO GIULIANO PROF
VIALE S ANTONIO
DA PADOVA 7
I-31100 TREVISO
ITALY

ROMANOV YURI S DR
ODESSA ASTRONOMICAL
OBSERVATORY
SHEVCHENKO PARK
270014 ODESSA
U.S.S.R.

ROMPOLT BOHDAN DR
ASTRONOMICAL INST
UL KOPERNIKA 11
51 622 WROCLAW
POLAND

RONAN COLIN A
13 ACORN AVE
BAR HILL
CAMBRIDGE CB3 8DT
U.K.

ROOD HERBERT J
52 ELIZABETH ST
SOUTH BOUND BROOK
NJ 08880
U.S.A.

ROOD ROBERT T DR
BOX 3818
UNIVERSITY STATION
CHARLOTTESVILLE
VA 22903
U.S.A.

ROOSEN ROBERT G DR
LAB FOR SOLAR PHYS
UNIV OF NEW MEXICO
ALBUQUERQUE NM 87131
U.S.A.

ROSADO MARGARITA DR
INST DE ASTRONOMIA
UNAM
APDO POSTAL 70-264
MEXICO D F
MEXICO

ROSCH JEAN PROF DIR
OBSERVATOIRES DU PIC
DU MIDI ET DE
TOULOUSE
F-65200 BAGNERES
FRANCE

ROSE WILLIAM K DR
ASTRONOMY PROGRAM
UNIVERSITY MARYLAND
COLLEGE PK MD 20742
U.S.A.

ROSEN E DR
DEPT OF HISTORY
CITY COLLEGE OF THE
CITY UNIV NEW YORK
NEW YORK NY 10031
U.S.A.

ROSENBERG J DR
STERRENWACHT
SERVAAS BOLWERK 13
UTRECHT
THE NETHERLANDS

ROSENDHAL JEFFEREY D DR
NASA HEADQUARTERS
CODE SA
WASHINGTON DC 20546
U.S.A.

ROSINO LEONIDA PROF
ASTRONOMICAL OBSERVA
TORY PADOVA
PADOVA
ITALY

ROSLUND CURT DR
DEPT OF ASTRONOMY
CHALMERS UNIVERSITY
OF TECHNOLOGY FACK
S-402 20 GOETEBORG
SWEDEN

ROSS DENNIS K PROF
PHYSICS DEPT
IOWA ST UNIVERSITY
AMES IA 50010
U.S.A.

ROSS JOHN E R DR
PHYSICS DEPT
UNIV OF QUEENSLAND
ST LUCIA
BRISBANE QLD 4067
AUSTRALIA

ROSSELAND SVEIN PROF
INST THEORET ASTROPH
UNIVERSITY OF OSLO
P O BOX 1029
BLINDERN OSLO 3
NORWAY

ROSSI BRUNO B PROF
MASS INSTITUTE OF
TECHNOLOGY
RM 37-667
CAMBRIDGE MA 02139
U.S.A.

ROSTAS FRANCOIS DR
OBSERVATOIRE
DE MEUDON
F-92190 MEUDON
FRANCE

ROTH MARIA-LUISE PH D
HAMBURGER STERNWARTE
D-205 HAMBURG 80
GERMANY, F.R.

ROTH MIGUEL R DR
INST DE ASTRONOMIA
UNAM
APDO POSTAL 70-264
MEXICO D F
MEXICO

ROTS ARNOLD H DR
NAT RADIO ASTRON OBS
P O BOX O
SOCORRO NM 87801
U.S.A.

ROTTENBERG J A DR
2911 BAYVIEW AVE
SUITE 110C
WILLOWDALE ONTM2K1E8
CANADA

ROUEFF EVELYNE M A DR
OBSERVATOIRE MEUDON
DEPT D ASTROPHYSIQUE
FDTLE
F-92190 MEUDON
FRANCE

ROUNTREE LESH JANET DR
LABORATORY ASTRONOMY
SOLAR PHYS CODE 681
GODDARD SP FLIGHT CT
GREENBELT MD 20771
U.S.A.

ROUSE CARL A DR
627 15TH ST
DEL MAR CA 92014
U.S.A.

ROUSSEAU JEAN-MICHEL MR
OBSERVATOIRE
DE BORDEAUX
F-33270 FLOIRAC
FRANCE

ROUSSEAU JEANINE DR
OBSERVATOIRE DE LYON
F-69230 ST GENIS
LAVAL
FRANCE

ROUTLEDGE DAVID DR
ELECTRICAL ENGG DEPT
UNIVERSITY OF
ALBERTA
EDMONTON T6G 2J1
CANADA

ROUTLY PAUL M DR
US NAVAL OBSERVATORY
34TH AND MASS AVE NW
WASHINGTON DC 20390
U.S.A.

ROVITHIS PETER EMM DR
NATIONAL OBSERVATORY
OF ATHENS
ATHENS 306
GREECE

ROVITHIS-LIVANIOU H PH D
ASTRONOMY DEPT
UNIV OF ATHENS
PANEPISTIMIOPOLIS
ATHENS 621
GREECE

ROWAN-ROBINSON MICHAEL DR
DEPT OF APPLIED MATH
QUEEN MARY COLLEGE
MILE END RD
LONDON E1 4NS
U.K.

ROWSON BARRIE DR
NRAL
JODRELL BANK
MACCLESFIELD
CHESHIRE SK11 9DL
U.K.

ROXBURGH IAN W PROF
DEPT APPLIED MATHS
QUEEN MARY COLLEGE
MILE END ROAD
LONDON E1
U.K.

ROY ARCHIE E PROF
DEPT OF ASTRONOMY
GLASGOW UNIVERSITY
GLASGOW G12 8QQ
U.K.

ROY JEAN-RENE
DEPT DE PHYSIQUE
UNIVERSITE LAVAL
CITE UNIVERSITAIRE
QUEBEC G1K 7P4
CANADA

ROZELOT JEAN P
150 BLVD DIDEROT
IMMEUBLE F 128
F-75012 PARIS
FRANCE

ROZHKOVSKIJ DIMITRIJ A
PROF
ASTROPHYSICAL
INSTITUTE
48068 ALMA ATA
U.S.S.R.

RUBASHEV BORIS M DR
PULKOVO OBSERVATORY
196140 LENINGRAD
U.S.S.R.

RUBEN G PROF DR
ZNTRINST ASTROPHYSIK
ASTROPHYS OBS
TELEGRAFENBERG
DDR-15 POTSDAM
G.D.R.

RUBIN VERA C DR
DEPT TERRESTRIAL MAG
CARNEGIE INST WASH
5241 BROAD BR RD N W
WASHINGTON DC 20015
U.S.A.

RUCINSKI SLAWOMIR M DR
ASTRONOMICAL
OBSERVATORY
AL UJAZDOWSKIE 4
00 478 WARSZAWA
POLAND

RUDDY VINCENT P DR
REGIONAL TECHNICAL
COLLEGE
ROSSA AVE
CORK
IRELAND

RUDERMAN MALVIN A
PHYSICS DEPT
COLUMBIA UNIVERSITY
NEW YORK NY 10027
U.S.A.

RUDKJOBING MOGENS PROF
ASTRONOMISK INSTITUT
LANGELANDSGADE
DK-8000 AARHUS C
DENMARK

RUDNICK LAWRENCE DR
NAT RADIO ASTRON LAB
EDGEMONT RD
CHARLOTTESVL VA22901
U.S.A.

RUDNICKI KONRAD PROF
ASTRONOMICAL OBS
UL ORLA 171
30 244 KRAKOW
POLAND

RUFENER FREDY G PROF
OBSERVATOIRE
DE GENEVE
CH-1290 SAUVERNY GE
SWITZERLAND

RUGGE HUGO R DR
THE AEROSPACE
CORPORATION
P O BOX 92957
LOS ANGELES CA 90009
U.S.A.

RUIZ MARIA TERESA DR
OBS ASTRON NACIONAL
UNIVERSIDAD DE CHILE
CASILLA 36-D
SANTIAGO
CHILE

RULE BRUCE H STAFF RETIRE
HALE OBSERVATORIES
2205 MONTE VISTA ST
PASADENA CA 91107
U.S.A.

RUMSEY NORMAN J
PHYSICS AND ENGINEE-
RING LABORATORY
D S I R
LOWER HUTT
NEW ZEALAND

RUNCORN S K PROF
SCHOOL OF PHYSICS
THE UNIVERSITY
NEWCASTLE UPON TYNE
NE1 7RU
U.K.

RUPRECHT JAROSLAV DR
ASTRONOMICAL INST OF
CZECHOSLOVAK ACAD SC
BUDECSKA 6
120 23 PRAHA 2
CZECHOSLOVAKIA

RUSCONI LUIGIA DR
VIA GHIRLANDAIO 22/5
I-34100 TRIESTE
ITALY

RUSKOL EUGENIA L DR
OJSCHMIDT INSTITUTE
OF PHYSICS OF EARTH
AC SCI USSR
123810 MOSCOW
U.S.S.R.

RUSSELL JANE L DR
ALLEGHENY OBS
OBS STATION
PITTSBURGH PA 15214
U.S.A.

RUSSELL JOHN A PROF
DEPT OF ASTRONOMY
UNIV OF SOUTHERN CAL
UNIVERSITY PARK
LOS ANGELES CA 90007
U.S.A.

RUSSEV RUSCHO DR
DEPT OF ASTRONOMY
ANTON IVANOV ST 5
1126 SOFIA
BULGARIA

RUSSO GUIDO DR
OSSERVATORIO ASTRO-
NOMICO CAPODIMONTE
VIA MOIARELLO 16
I-80131 NAPOLI
ITALY

RUST DAVID M DR
AMERICAN SCIENCE AND
ENGINEERING CO
955 MASS AVE
CAMBRIDGE MA 02139
U.S.A.

RUSU I DR
ASTRONOMICAL OBS
5 CUTITUL DE ARGINT
BUCAREST 28
ROUMANIA

RUSU L DR
ASTRONOMICAL OBS
5 CUTITUL DE ARGINT
BUCAREST 28
ROUMANIA

RUTTEN ROBERT JELLE DR
OBS SONNENBORGH
ZONNENBURG 2
3512 NL UTRECHT
THE NETHERLANDS

RUZDJAK VLADIMIR DR
INST OF PHYSICS
UNIVERSITY
P O BOX 304
41001 ZAGREB
YUGOSLAVIA

RUZICKOVA TOPOLOVA
BLAZENA DR
ASTRONOMICAL INST
CZECHOSLOVAK ACAD SC
251 65 ONDREJOV OBS
CZECHOSLOVAKIA

RYABOV YU A PROF DR
MATHEMATICS DEPT OF
MADI
125319 MOSCOW
U.S.S.R.

RYBICKI GEORGE B DR
HARVARD SMITHSONIAN
CTR F ASTROPHYSICS
60 GARDEN ST
CAMBRIDGE MA 02138
U.S.A.

RYBKA EUGENIUSZ PROF DR
UL SZOPENA 23
30047 KRAKOW
POLAND

RYBKA PRZEMYSLAW DR
INSTITUTE OF HISTORY
OF SCIENCE
NOWY SWIAT 72
00-330 WARSZAWA
POLAND

RYDBECK OLOF E H PROF
ONSALA SPACE
OBSERVATORY
S-430 34 ONSALA
SWEDEN

RYDGREN ALFRED ERIC JR DR
PHYSICS DEPT
RENSSELAER POLYTECHN
INSTITUTE
TROY NY 12180
U.S.A.

RYLE MARTIN PROF SIR
CAVENDISH LABORATORY
MADINGLEY ROAD
CAMBRIDGE CB3 OHE
U.K.

RYLOV V S DR
SPECIAL ASTROPHYSI-
CAL OBSERVATORY
USSR ACADEMY OF SCI
357140 N ARKHYZ
U.S.S.R.

RYTER CHARLES E DR
CEN SACLAY
DPH EP A
BP 2
F-91190 GIF S YVETTE
FRANCE

RYZHKOV N F DR
SPECIAL ASTROPHYS-
OBS USSR ACADEMY SCI
LENINGRAD BRANCH
196140 LENINGRAD
U.S.S.R.

RZHIGA O N DR
INSTITUTE OF RADIO
AND ELECTRONICS
USSR ACADEMY OF SCI
103907 MOSCOW
U.S.S.R.

SAAR ENN DR
OBSERVATORY 5-19
202444 TORAVERE
ESTONIA
U.S.S.R.

SABANO YUTAKA DR
ASTRONOMICAL INST
TOHOKU UNIVERSITY
ARAMAKI
SENDAI 980
JAPAN

SABBADIN FRANCO DR
OSSERVATORIO ASTRO-
FISICO
I-36012 ASIAGO
ITALY

SACK NOAM DR
DEPT OF THEORETICAL
PHYSICS
HEBREW UNIVERSITY
JERUSALEM
ISRAEL

SACKMANN I JULIANA DR
KELLOGG RADIATION
LABORATORY
CALIF INST OF TECHN
PASADENA CA 91125
U.S.A.

SADEH D DR
DEPT OF PHYSICS AND
ASTRONOMY
TEL AVIV UNIVERSITY
TEL AVIV
ISRAEL

SADLER DONALD H DR
8 COLLINGTON RISE
BEXHILL ON SEA
SUSSEX TN39 3RT
U.K.

SADZAKOV SOFIJA DR
ASTRONOMICAL OBS
VOLGINA 7
11050 BEOGRAD
YUGOSLAVIA

SAFRONOV VICTOR S DR
INSTITUTE OF PHYSICS
OF THE EARTH
B GRUZINSKAYA 10
123242 MOSCOW
U.S.S.R.

SAGAN CARL DR
302 SPACE SCIENCES
CORNELL UNIVERSITY
ITHACA NY 14853
U.S.A.

SAGGION ANTONIO PROF
ISTITUTO DI FISICA
G GALILEI
VIA MARZOLO 8
I-35100 PADOVA
ITALY

SAGITOV M U DR
STERNBERG STATE AS-
TRONOMICAL INSTITUTE
117234 MOSCOW
U.S.S.R.

SAGNIER JEAN LOUIS MR
BUREAU D LONGITUDES
77 AVENUE DENFERT-
ROCHEREAU
F-75014 PARIS
FRANCE

SAHADE JORGE PROF
49 342
1900 LA PLATA
ARGENTINA

SAHAL-BRECHOT SYLVIE DR
OBSERVATOIRE DE
PARIS-MEUDON
F-92190 MEUDON
FRANCE

SAISSAC JOSEPH DR
OBSERVATOIRE DU
PIC-DU-MIDI-TOULOUSE
F-65200 BAGNERES-DE-
BIGORRE
FRANCE

SAITO KUNIJI PROF
TOKYO ASTRONOMICAL
OBSERVATORY
OSAWA MITAKA
TOKYO 181
JAPAN

SAITO MAMORU DR
TOKYO ASTRONOMICAL
OBSERVATORY
OSAWA MITAKA
TOKYO 181
JAPAN

SAITO SUMISABURO DR
KASAN OBSERVATORY
YAMASH HIGASHIYAMA
KYOTO 607
JAPAN

SAKASHITA SHIRO PROF
DEPT OF PHYSICS
HOKKAIDO UNIVERSITY
KITAJYUJO NISHI 8
SAPPORO HOKKAIDO 063
JAPAN

SAKHAROV VLADIMIR I DR
PULKOVO OBSERVATORY
196140 LENINGRAD
U.S.S.R.

SAKHIBULLIN NAIL A DR
DEPT OF ASTRONOMY
KAZAN STATE UNIV
420008 KAZAN
U.S.S.R.

SAKURAI KUNITOMO PROF
DEPT OF PHYSICS
KANAGAWA UNIVERSITY
KANAGAWAKU
YOKOHAMA 252
JAPAN

SAKURAI TAKEO T PROF
FACULTY OF
ENGINEERING
KYOTO UNIVERSITY
SAKYO-KU KYOTO 606
JAPAN

SALETIC DUSAN
ASTRONOMICAL
OBSERVATORY
VOLGINA 7
11050 BEOGRAD
YUGOSLAVIA

SALISBURY J W DR
AFCRL
BLDG 1121
L G HANSCOM FIELD
BEDFORD MA 01730
U.S.A.

SALOMONOVICH A E DR
PHYSICAL INSTITUTE
USSR ACADEMY OF SCI
117924 MOSCOW
U.S.S.R.

SALPETER EDWIN E PROF
NEWMAN LAB OF NS
CORNELL UNIVERSITY
ITHACA NY 14853
U.S.A.

SALUKVADZE G N DR
ABASTUMANY ASTROPHY-
SICAL OBSERVATORY
383762 ABASTUMANY
U.S.S.R.

SAMAHA A H M PROF
5 WADI EL NILE ST
MAADI CAIRO
ARAB REP. OF EGYPT

SAMAIN D DR
LPSP
B P NO 10
F-91370 VERRIERES LE
BUISSON
FRANCE

SAMPSON DOUGLAS H PROF
DEPT OF ASTRONOMY
PENN STATE UNIV
525 DAVEY LAB
UNIV PARK PA 16802
U.S.A.

SANAMIAN V A DR
BYURAKAN ASTROPHYSI-
CAL OBSERVATORY
378433 ARMENIA
U.S.S.R.

SANCHEZ FRANCISCO PROF
INSTITUTO ASTROFISI-
CA CANARIAS
UNIVERSIDAD LAGUNA
LA LAGUNA TENERIFE
SPAIN

SANCHEZ MAGRO C DR
IAC INSTITUTO DE
ASTROFISICA CANARIAS
UNIV DE LA LAGUNA
TENERIFE
SPAIN

SANCHEZ MANUEL
INSTITUTO Y OBSERVA-
TORIO DE MARINA
SAN FERNANDO CADIZ
SPAIN

SANCISI RENZO DR
KAPTEYN LABORATORIUM
POSTBUS 800
9700 RV GRONINGEN
THE NETHERLANDS

SANDAKOVA E V DR
KIEV STATE UNIVERSI-
TY ASTRONOMICAL
OBSERVATORY
252053 KIEV
U.S.S.R.

SANDELL GORAN HANS L DR
STOCKHOLM OBS
S-13300 SALTSJOBADEN
SWEDEN

SANDERS R DR
KAPTEYN LABORATORY
P C BOX 800
9700 RV GRONINGEN
THE NETHERLANDS

SANDERS W L PROF
NEW MEXICO STATE
UNIVERSITY
BOX 4500
LAS CRUCES NM 88003
U.S.A.

SANDFORD MAXWELL T II
LOS ALAMOS SCIENTI-
FIC LABORATORY
LOS ALAMOS NM 87545
U.S.A.

SANDQVIST AAGE DR
STOCKHOLM
OBSERVATORY
S-133 00
SALTSJOEBADEN
SWEDEN

SANDULEAK N DR
WARNER AND SWASEY
OBSERVATORY
1975 TAYLOR RD
E CLEVELAND OH 44112
U.S.A.

SANFORD P W MR
MULLARD SPACE SC LAB
HOLMBURY ST MARY
DORKING
SURREY RH5 6NS
U.K.

SANTIN PAOLO DR
VIA DUDOVICH 3
I-34100 TRIESTE
ITALY

SANWAL N B DR
DEPT OF ASTRONOMY
OSMANIA UNIVERSITY
HYDERABAD 500 007
INDIA

SANYAL ASHIT DR
DEPT OF ASTRONOMY
PENNA STATE UNIV
UNIV PK PA 16802
U.S.A.

SAPAR ARVED DR
IAPHA ESTONIAN AC OF
SCIENCE TORAVERE
202444 TARTU
U.S.S.R.

SARAZIN CRAIG L DR
DEPT OF ASTRONOMY
UNIV OF VIRGINIA
POB 3818 UNIV STA
CHARLOTTESVL VA22903
U.S.A.

SAREYAN J P DR
OBS DE NICE
B P 252
F-06007 NICE CEDEX
FRANCE

SARGENT WALLACE L.W. DR.
HALE OBSERVATORIES
CALTECH
1201 E CALIFORNIA ST
PASADENA CA 91125
U.S.A.

SARMA M B K PROF
DEPT OF ASTRONOMY
OSMANIA UNIVERSITY
HYDERABAD 500 007
INDIA

SARRIS ELEFTHERIOS PH D
NAT OBS OF ATHENS
ASTRONOMICAL INST
ATHENS 306
GREECE

SARRIS EMMANUEL T PH D
DEMOCRITUS UNIV OF
THRACE
XANTHI
GREECE

SARTORI LEO PROF
BEHLEN LABORATORY OF
PHYSICS
UNIVERSITY NEBRASKA
LINCOLN NB 68588
U.S.A.

SASAO TETSUO DR
INTERNATIONAL
LATITUDE OBSERVATORY
OF MIZUSAWA
MIZUSAWA IWATE 023
JAPAN

SASLAW WILLIAM C PROF
ASTRONOMY DEPT
UNIVERSITY VIRGINIA
BOX 3818 UNIVSTATION
CHARLOTTESV VA 22903
U.S.A.

SASTRY CH V
ASTROPHYSICAL
OBSERVATORY
KODAIKANAL 3
INDIA

SATO FUMIO DR
CHIBA PREFECTURE
EDUCATION CENTER
10-1 KATSURAGI 2-CHO
CHIBA 280
JAPAN

SATO FUMITAKA PROF
INSTITUTE OF FUNDA-
MENTAL PHYSICS
UNIVERSITY OF KYOTO
SAKYOKU KYOTO 606
JAPAN

SATO KATSUHIKO DR
RESEARCH INST FOR
FUNDAMENTAL PHYSICS
KYOTO UNIVERSITY
SAKYOKU KYOTO 606
JAPAN

SATO KOICHI DR
INTERNATIONAL
LATITUDE OBSERVATORY
OF MIZUSAWA
MIZUSAWA IWATE 023
JAPAN

SATO NAONOBU DR
AKITA UNIVERSITY
AKITA 010
JAPAN

SATO SHUJI DR
DEPT OF PHYSICS
UNIVERSITY OF KYOTO
KITASHIRAKAWA
SAKYOKU KYOTO 606
JAPAN

SATO YUZO DR
OSAWA 4-8-19
MITAKA
TOKYO 181
JAPAN

SAUVAL JACQUES DR
OBS ROYAL DE BELG.
3 AVE CIRCULAIRE
B-1180 BRUXELLES
BELGIUM

SAVAGE ANN DR
ROYAL OBSERVATORY
BLACKFORD HILL
EDINBURGH EH9 3HJ
U.K.

SAVAGE BLAIR D DR
DEPT OF ASTRONOMY
UNIVERSITY WISCONSIN
475 N.CHARTER STR
MADISON WI 53706
U.S.A.

SAVEDOFF MALCOLM P PROF
PHYS AND ASTRON DEPT
UNIVERSITY ROCHESTER
BAUSCH AND LOMB BLDG
ROCHESTER NY 14627
U.S.A.

SAWYER CONSTANCE S DR
NOAA PMEL
3711 15TH AVE NE
SEATTLE WA 98105
U.S.A.

SAWYER HOGG HELEN B DR
DAVID DUNLAP OBS
P O BOX 360
RICHMOND HILL
ONT L4C 4Y6
CANADA

SAXENA P P DR
UTTAR PRADESH
STATE OBSERVATORY
MANORA PEAK
NAINI TAL
INDIA

SCALISE JR EUGENIO DR
CRAAM-ON-CNPQ
RUA PARA 277
01243 SAO PAULO
BRAZIL

SCALO JOHNMICHAEL DR
DEPT OF ASTRONOMY
UNIV OF TEXAS
AUSTIN TX 78712
U.S.A.

SCALTRITI FRANCO DR
OSSERVATORIO ASTRO-
NOMICO
PINO TORINESE
TORINO
ITALY

SCARFE COLIN D DR
DEPT OF PHYSICS
UNIV OF VICTORIA
P O BOX 1700
VICTORIA BC V8W 2Y2
CANADA

SCARGLE JEFFREY D DR
NASA-AMES RESEARCH
CENTER MS245-3
MOFFETT FIELD
CA 94035
U.S.A.

SCARROTT STANLEY M DR
PHYSICS DEPT
UNIV OF DURHAM
SOUTH ROAD
DURHAM DH1 3LE
U.K.

SCHADEE AERT DR
STERREWACHT
SONNENBORGH
UTRECHT 2501
THE NETHERLANDS

SCHAIFERS KARL DR
LANDESSTERNWARTE
KOENIGSTUHL
D-6900 HEIDELBERG 1
GERMANY, F.R.

SCHALEN CARL PROF
LUND OBSERVATORY
SVANEGATAN 9
S-222 24 LUND
SWEDEN

SCHANDA ERWIN PROF
INST OF APPLIED
PHYSICS
SIDLERSTR 5
CH-3012 BERN
SWITZERLAND

SCHARLEMANN ERNST T DR
DEPT OF ASTRONOMY
COLUMBIA UNIV
BOX 98 PUPIN
NEW YORK NY 10027
U.S.A.

SCHATTEN K H DR
PHYSICS DEPT
VICTORIA UNIVERSITY
PRIVATE BAG
WELLINGTON
NEW ZEALAND

SCHATZMAN EVRY PROF
OBSERVATOIRE DE NICE
BP 252
F-06007 NICE CEDEX
FRANCE

SCHEEPMAKER ANTON DR
COSM RAY WORK GROUP
HUYGENS LABORATORY
WASSENAARSEWEG 78
2300 RA LEIDEN
THE NETHERLANDS

SCHEFFLER HELMUT PROF
LANDESSTERNWARTE
KOENIGSTUHL
D-6900 HEIDELBERG 1
GERMANY, F.R.

SCHEIDECKER JEAN-PAUL DR
OBSERVATOIRE DE NICE
BP 252
F-06007 NICE CEDEX
FRANCE

SCHERB FRANK PROF
PHYSICS DEPT
UNIVERSITY WISCONSIN
MADISON WI 53706
U.S.A.

SCHERRER PHILIP H DR
INST OF PLASMA RES
STANFORD UNIV
VIA CRESPI
STANFORD CA 94305
U.S.A.

SCHEUER PETER A G DR
CAVENDISH LABORATORY
MADINGLEY RD
CAMBRIDGE CB3 OHE
U.K.

SCHILD RUDOLPH E DR
CTR F ASTROPHYSICS
60 GARDEN STREET
CAMBRIDGE MA 02138
U.S.A.

SCHILIZZI RICHARD T DR
RADIOSTERRENWACHT
DWINGELOO
THE NETHERLANDS

SCHILLER KARL PROF DR
PIRSCH-WEG 6
D-6072 DREIEICH-
BUCHSCHLAG
GERMANY, F.R.

SCHILT JAN DR
188 HILLSIDE AVE
LEONIA NJ 07605
U.S.A.

SCHKODROV V G DR
DEPT OF ASTRONOMY
7TH NOVEMBER ST 1
SOFIA
BULGARIA

SCHLESINGER BARRY M DR
SYSTEMS AND APPLIED
SCIENCES CORPORATION
6811 KENILWORTH AVE
RIVERDALE MD 20840
U.S.A.

SCHLOSSER WOLFHARD PROF
ASTRONOMISCHES
INSTITUT
POSTFACH 102148
D-4630 BOCHUM
GERMANY, F.R.

SCHLUETER A PROF DR
MPI F PLASMAPHYSIK
D-8046 GARCHING
GERMANY, F.R.

SCHLUETER DIETER PROF
INST THEO PHY STERNW
NEUE UNIV PHYSIK ZTM
OLSHAUSENST GB N 61C
D-2300 KIEL 1
GERMANY, F.R.

SCHMADEL LUTZ D DR
ASTRONOMISCHES
RECHEN-INSTITUT
MOENCHHOFSTR 12-14
D-6900 HEIDELBERG 1
GERMANY, F.R.

SCHMAHL EDWARD J DR
ASTRONOMY PROGRAM
UNIV OF MARYLAND
COLLEGE PK MD 20742
U.S.A.

SCHMAHL GUENTER PROF
UNIVERSITAETS-
STERNWARTE
GEISMARLANDSTR 11
D-3400 GOETTINGEN
GERMANY, F.R.

SCHMALBERGER DONALD C DR
DUDLEY OBSERVATORY
1202 TROY-
SCHENECTADY RD
LATHAM NY 12110
U.S.A.

SCHMEIDLER F PROF DR
INSTITUT F ASTRONO-
MIE UND ASTROPHYSIK
SCHEINERSTR 1
D-8000 MUENCHEN 80
GERMANY, F.R.

SCHMID-BURGK J DR
MPI RADIOASTRONOMIE
AUF DEM HUEGEL 69
D-5300 BONN 1
GERMANY, F.R.

SCHMIDT EDWARD G
DEPT OF PHYSICS
UNIVERSITY OF
NEBRASKA
LINCOLN NB 68588
U.S.A.

SCHMIDT H U DR
MPI FUER PHYSIK UND
ASTROPHYSIK
FOEHRINGER RING 6
D-8000 MUENCHEN 40
GERMANY, F.R.

SCHMIDT HANS PROF
UNIVERSITAETS-
STERNWARTE BONN
AUF DEM HUEGEL 71
D-5300 BONN 1
GERMANY, F.R.

SCHMIDT K H DR
ZNTRINST ASTROPHYSIK
STERNW BABELSBERG
ROSA LUXEMBURGST 17A
DDR-1502 POTSDAM
G.D.R.

SCHMIDT MAARTEN PROF
HALE OBSERVATORIES
813 SANTA BARBARA ST
PASADENA CA 91101
U.S.A.

SCHMIDT THOMAS DR
MPI FUER ASTRONOMIE
KOENIGSTUHL
D-6900 HEIDELBERG
GERMANY, F.R.

SCHMIDT-KALER TH PROF
STEINHUEGEL 105
D-5810 WITTEN
GERMANY, F.R.

SCHMIEDER B DR
OBS DE MEUDON
F-92190 MEUDON
FRANCE

SCHMITTER Y MARTIN DEL
CAMPO E F DR
INST ASTRONOMY UNAM
APT POSTAL 70-264
MEXICO D F
MEXICO

SCHNELL ANNELISE DR
INST FUER ASTRONOMIE
UNIVERSITAT WIEN
TUERKENSCHANZST 17
A-1180 WIEN
AUSTRIA

SCHNOPPER HERBERT W DR
CTR FOR ASTROPHYSICS
60 GARDEN ST
CAMBRIDGE MA 02138
U.S.A.

SCHNUR G F O
ESO ADMINISTRATION
ALONSO CORDOBA 3107
VITACURA CAS 11 P
SANTIAGO DE CHILE
CHILE

SCHOBER HANS J DR
ASTRONOMISCHES INST
DER UNIVERSITAET
UNIVERSITAETSPLATZ 5
A-8010 GRAZ
AUSTRIA

SCHOEFFEL EBERHARD F DR
ASTRONOMISCHES INST
RUHR UNIVERSITAET
POSTFACH 10 21 48
D-4630 BOCHUM 1
GERMANY, F.R.

SCHOEMBS ROLF DR
INST FUER ASTRONOMIE
ASTROPHYSIK
SCHEINERSTR 1
D-8000 MUENCHEN 81
GERMANY, F.R.

SCHOENBERNER DETLEF PH D
INST F THEOR PHYSIK
UND UNIV STERNWARTE
OLSHAUSENSTRASSE
D-2300 KIEL
GERMANY, F.R.

SCHOENEICH W DR
ZNTRINST ASTROPHYSIK
ASTROP OBSERVATORIUM
TELEGRAFENBERG
DDR-15 POTSDAM
G.D.R.

SCHOENFELDER VOLKER DR
MPI F EXTRATERRE-
STRISCHE PHYSIK
D-8046 GARCHING
GERMANY, F.R.

SCHOLL HANS DR
ASTRONOMISCHES
RECHEN-INSTITUT
MOENCHHOFSTR 12-14
D-6900 HEIDELBERG
GERMANY, F.R.

SCHOLZ GERHARD DR
ZENTR INST ASTROPHYS
DER ADW DER DDR
ROSA-LUXEMBURGST 17A
DDR-1502 POTSDAM
G.D.R.

SCHOLZ M PROF
INST F THEORETISCHE
ASTROPHYS DER UNIV
NEUENHEIMER FELD 294
D-6900 HEIDELBERG
GERMANY, F.R.

SCHOMBERT JOHN L MR
21920 DAVIS MILL RD
GERMANTOWN MD 20767
U.S.A.

SCHOOLMAN STEPHEN A DR
LOCKHEED PA RES LAB
3251 HANOVER ST
PALO ALTO CA 94304
U.S.A.

SCHRAMM DAVID N PROF
UNIVERSITY CHICAGO
EFI-LASR-205
933 E 56TH ST
CHICAGO IL 60637
U.S.A.

SCHRIJVER J DR
LABORATORIUM VOOR
RUIMTEONDERZOEK
BENELUXLAAN 21
3527 HS UTRECHT
THE NETHERLANDS

SCHROEDER DANIEL J PROF
DEPT PHYS AND ASTRON
BELOIT COLLEGE
BELOIT WI 53511
U.S.A.

SCHROEDER ROLF DR
HAMBURGER STERNWARTE
GOJENBERGSWEG 112
D-2050 HAMBURG 80
GERMANY, F.R.

SCHROETER EGON H PROF
KIEPENHEUER INST
FUER SONNENPHYSIK
SCHOENECKSTRASSE 6
D-7800 FREIBURG I BR
GERMANY, F.R.

SCHROLL ALFRED DR
SONNENOBSERVATORIUM
KANZELHOEHE
A-9520 SATTENDORF
AUSTRIA

SCHRUEFER EBERHARD DR
INST F ASTROPHYSIK
UNIVERSITAET BONN
AUF DEM HUEGEL 71
D-5300 BONN 1
GERMANY, F.R.

SCHRUTKA RECHTENSTAMM
GUNTRAM PROF DR
COTTAGEGASSE 56
A-1190 WIEN
AUSTRIA

SCHUBART JOACHIM DR
ASTRONOMISCHES
RECHEN INSTITUT
MOENCHHOFSTR 12-14
D-6900 HEIDELBERG
GERMANY, F.R.

SCHUECKING E L DR
DEPT OF PHYSICS
NEW YORK UNIVERSITY
NEW YORK NY 10012
U.S.A.

SCHUERER MAX PROF
ASTRONOMISCHES INST
UNIVERSITAET BERN
SIDLERSTR 5
CH-3012 BERN
SWITZERLAND

SCHUERMAN DONALD W DR
SPACE ASTRONOMY LAB
SUNY
EXECUTIVE PK EAST
ALBANY NY 12203
U.S.A.

SCHULER WALTER DR
STERNWARTE DER
KANTONSSCHULE
CH-4500 SOLOTHURN
SWITZERLAND

SCHULTE D H DR
ITEK CORPORATION
10 MAGUIRE ROAD
LEXINGTON MA 02173
U.S.A.

SCHUMANN JOERG DIETER DR
OBS HOHER LIST
UNIV STERNWARTE BONN
D-5568 DAUN
GERMANY, F.R.

SCHUSTER HANS-EMIL PH D
EUROPEAN SOUTHERN OB
CAS 16317 CORREO 9
ALONSO CORDOVA 3107
SANTIAGO
CHILE

SCHUTZ BERNARD F DR
APPLIED MATHEMATICS
AND ASTRONOMY
UNIVERSITY COLLEGE
CARDIFF CF1 1XL
U.K.

SCHWAN HEINER DR
ASTRONOMISCHES
RECHEN INSTITUT
MOENCHHOFSTR 12-14
D-6900 HEIDELBERG
GERMANY, F.R.

SCHWARTZ DANIEL A DR
CENTER FOR ASTROPHYS
60 GARDEN ST
CAMBRIDGE MA 02138
U.S.A.

SCHWARTZ PHILIP R DR
NAVAL RESEARCH LAB
CODE 7138
WASHINGTON DC 20375
U.S.A.

SCHWARTZ RICHARD D
ASST PROF DEPT PHYS
UNIVERSITY MISSOURI
8001 NAT BRIDGE RD
ST LOUIS MO 63121
U.S.A.

SCHWARTZ ROLF PH D
MPI FUER
RADIOASTRONOMIE
AUF DEM HUEGEL 69
D-5300 BONN 1
GERMANY, F.R.

SCHWARZ ULRICH J DR
KAPTEYN LABORATORIUM
POSTBUS 800
GRONINGEN
THE NETHERLANDS

SCHWARZSCHILD MARTIN PROF
PRINCETON UNIVERSITY
OBSERVATORY
PEYTON HALL
PRINCETON NJ 08540
U.S.A.

SCHWEIZER FRANCOIS DR
CERRO TOLOLO
INTER-AMERICAN OBS
CASILLA 63-D
LA SERENA
CHILE

SCIAMA DENNIS W DR
DEPT OF ASTROPHYSICS
SOUTH PARKS ROAD
OXFORD OX1 3RQ
U.K.

SCONZO PASQUALE DR
29 OLD MYSTIC ST
ARLINGTON MA 02174
U.S.A.

SCOTT ELIZABETH L PROF
STATISTICS LAB
UNIVERSITY OF CALIF
367 EVANS HALL
BERKELEY CA 94720
U.S.A.

SCOTT JOHN S DR
ASTRONOMY PROGRAM
UNIV OF MARYLAND
COLLEGE PK MD 20742
U.S.A.

SCOTT PAUL F DR
CAVENDISH LABORATORY
MADINGLEY ROAD
CAMBRIDGE CB3 0HE
U.K.

SCOVILLE NICHOLAS Z
ASTRO-PHYSICS DEPT
UNIVERSITY OF MASS
GRADUATE RES CENTER
AMHERST MA 01003
U.S.A.

SCUFLAIRE RICHARD DR
INST D'ASTROPHYSIQUE
UNIVERSITE DE LIEGE
5 AVENUE DE COINTE
B-4200 COINTE-OUGREE
BELGIUM

SEAQUIST ERNEST R PROF
DEPT OF ASTRONOMY
UNIVERSITY
TORONTO ONT M5S 1A7
CANADA

SEARLE L DR
HALE OBSERVATORIES
813 SANTA BARBARA ST
PASADENA CA 91101
U.S.A.

SEARS RICHARD LANGLEY DR
DEPT OF ASTRONOMY
UNIV OF MICHIGAN
PHYSICS-ASTRO BLDG
ANN ARBOR MI 48109
U.S.A.

SEATON MICHAEL J PROF
DEPT PHYS AND ASTRON
UNIV COLLEGE LONDON
GOWER ST
LONDON WC1E 6BT
U.K.

SEDDON HARRY MR
ROYAL OBSERVATORY
EDINBURGH EH9 3HJ
U.K.

SEDLMAYER ERWIN DR
INSTITUT FUER THEORE
TISCHE ASTROPHYSIK
NEUENHEIMER FELD 294
D-6900 HEIDELBERG
GERMANY, F.R.

SEDMAK GIORGIO DR
OSSERVATORIO ASTRONO
MICO DI TRIESTE
VIA TIEPOLO 11
I-34131 TRIESTE
ITALY

SEEGER CHARLES LOUIS III
SAN FRANCISCO STATE
UNIVERSITY
473 JAMES ROAD
PALO ALTO CA 94306
U.S.A.

SEEGER PHILIP A DR
LOS ALAMOS SCIENTI-
FIC LABORATORY
LOS ALAMOS NM 87545
U.S.A.

SEGAL IRVING E DR
MIT
2-224
CAMBRIDGE MA 02139
U.S.A.

SEGALUVITZ ALEXANDER DR
11 HABANIM ST
KEFAR SAVA
ISRAEL

SEGGEWISS WILHELM DR
OBSERVATORIUM
HOHER LIST
D-5568 DAUN EIFEL
GERMANY, F.R.

SEHNAL LADISLAV DR
ASTRONOMICAL INST
CSAV
25165 ONDREJOV
CZECHOSLOVAKIA

SEIDELMANN P KENNETH DR
US NAVAL OBSERVATORY
WASHINGTON DC 20390
U.S.A.

SEIDOV ZAKIR F DR
SHEMAKHA ASTRO-
PHYSICAL OBSERVATORY
373243 SHEMAKHA AZER
U.S.S.R.

SEIELSTAD GEORGE A
OWENS VALLEY RADIO
OBSERVATORY
PO BOX 387
BIG PINE CA 93513
U.S.A.

SEITTER WALTRAUT C PROF
ASTRONOMISHES INST
STEINFURTER STR 107
D-4400 MUENSTER/W
GERMANY, F.R.

SEKANINA ZDENEK DR
SMITHSONIAN ASTROPHY
SICAL OBSERVATORY
CAMBRIDGE MA 02138
U.S.A.

SEKI MUENZU DR
ASTRONOMICAL INST
TOHOKU UNIVERSITY
KAWAUCHI
SENDAI 980
JAPAN

SEKIGUCHI NAOSUKE DR
TOKYO ASTRONOMICAL
OBSERVATORY
OSAWA MITAKA
TOKYO 181
JAPAN

SELVELLI PIERLUIGI DR
OSS ASTRONOMICO
VIA TIEPOLO 11
I-34131 TRIESTE
ITALY

SEMEL MEIR DR
OBSERVATOIRE
DE MEUDON
F-92190 MEUDON
FRANCE

SEMENIUK IRENA DR
ASTRONOMICAL OBS
UL UJAZDOWSKIE 4
00478 WARZAWA
POLAND

SEN S N DR
INDIAN ASSN FOR THE
CULTIVATION OF SC
JADAVPUR
INDIA

SENGBUSCH KURT V DR
MPI FUER PHYSIK UND
ASTROPHYSIK
FOEHRINGER RING 40
D-8000 MUENCHEN 40
GERMANY, F.R.

SERKOWSKI KRYSZTOF DR
LUNAR AND PLANETARY
LABORATORY
UNIVERSITY ARIZONA
TUCSON AZ 85721
U.S.A.

SERRANO ALFONSO DR
INST DE ASTRONOMIA
UNAM
APDO POSTAL 70-264
MEXICO D F
MEXICO

SERSIC J L DR
OBSERVATORIO
ASTRONOMICO
5000 CORDOBA
ARGENTINA

SERVAJEAN ROGER DR
OBSERVATOIRE
DE MEUDON
F-92190 MEUDON
FRANCE

SETTI GIANCARLO PROF
LABORATORIO RADIOAST
ISTITUTO DI FISICA
VIA IRNERIO
I-40126 BOLOGNA
ITALY

SEVARLIC BRANISLAV M PROF
DEPT OF ASTRONOMY
UNIVERSITY BEOGRAD
VOLGINA 7
11050 BEOGRAD
YUGOSLAVIA

SEVERNYJ A B PROF DR
CRIMEAN ASTROPHYSI-
CAL OBSERVATORY USSR
ACAD OF SCI NAUCHNIY
334413 CRIMEA
U.S.S.R.

SEVILLA MIGUEL J DR
ALMANSA 76
MADRID 20
SPAIN

SHAFFER DAVID B DR
CODE 945
NASA/GFSC
GREENBELT MD 20771
U.S.A.

SHAH GHANSHYAM A DR
INDIAN INST ASTROPHY
SARJAPUR RD
KORAMANGALA
BANGALORE 560 034
INDIA

SHAHAM JACOB PROF
RACAH INSTITUTE OF
PHYSICS
HEBREW UNIVERSITY
JERUSALEM 91000
ISRAEL

SHAHBAZIAN ROMELIA K DR
BYURAKAN ASTRO-
PHYSICAL OBSERVATORY
378433 BYURAKAN ARM
U.S.S.R.

SHAKESHAFT JOHN R DR
CAVENDISH LABORATORY
MADINGLEY ROAD
CAMBRIDGE CB3 OHE
U.K.

SHAKHBAZYAN YURIJ L DR
PULKOVO OBSERVATORY
196140 LENINGRAD
U.S.S.R.

SHAKHOVSKOJ N M DR
CRIMEAN ASTROPHYSI-
CAL OBSERVATORY USSR
ACAD OF SCI NAUCHNIY
334413 CRIMEA
U.S.S.R.

SHAKURA N I DR
STERNBERG STATE AS-
TRONOMICAL INSTITUTE
117234 MOSCOW
U.S.S.R.

SHALLIS MICHAEL J DR
DEPT OF ASTROPHYSICS
SOUTH PARKS ROAD
OXFORD OX1 3RQ
U.K.

SHALTOUT MESALAM A M DR
HELWAN OBSERVATORY
HELWAN
ARAB REP. OF EGYPT

SHANE C D ASTRONOMER
BOX 582
SANTA CRUZ CA 95061
U.S.A.

SHANE W W DR
STERREWACHT
POSTBUS 9513
2300 RA LEIDEN
THE NETHERLANDS

SHAO CHING-YUAN
HARVARD COLLEGE
OBSERVATORY
60 GARDEN STR
CAMBRIDGE MA 02138
U.S.A.

SHAPIRO IRWIN I PROF
54-620 MIT
CAMBRIDGE MA 02139
U.S.A.

SHAPIRO MAURICE M
LAB FOR COSMIC RAY
PHYSICS CODE 7020
NAVAL RESEARCH LAB
WASHINGTON DC 20375
U.S.A.

SHAPIRO STUART L
ASST PROF
CRSR
CORNELL UNIVERSITY
ITHACA NY 14853
U.S.A.

SHAPLEY ALAN H
NOAA
BOULDER CO 80302
U.S.A.

SHAPLEY MARTHA B MRS
4170 PINON DR
BOULDER CO 80303
U.S.A.

SHARAF SH G DR
INSTITUTE OF THEORE-
TICAL ASTRONOMY
192187 LENINGRAD
U.S.S.R.

SHAROV A S DR
STERNBERG STATE AS-
TRONOMICAL INSTITUTE
117234 MOSCOW
U.S.S.R.

SHARPLESS STEWART PROF
DEPT OF PHYSICS AND
ASTRONOMY
UNIVERSITY ROCHESTER
ROCHESTER NY 14627
U.S.A.

SHAVER PETER A DR
KAPTEYN LABORATORIUM
POSTBUS 800
GRONINGEN
THE NETHERLANDS

SHAVIV GIORA DR
DEPT OF PHYSICS
TEL-AVIV UNIVERSITY
RAMAT-AVIV
TEL-AVIV
ISRAEL

SHAW J SCOTT DR
DEPT OF PHYSICS AND
ASTRONOMY
UNIVERSITY GEORGIA
ATHENS GA 30602
U.S.A.

SHAW JOHN H PROF
174 W 18TH AVE
COLUMBUS OH 43210
U.S.A.

SHAW R WILLIAM PROF
CORNEL UNIVERSITY
105 HALCYON HILL
ITHACA NY 14850
U.S.A.

SHAWHAN STANLEY D DR
DEPT PHYSICS ASTRON
UNIV OF IOWA
IOWA CITY IA 52242
U.S.A.

SHAWL STEPHEN J DR
DEPT OF PHYSICS AND
ASTRONOMY
UNIVERSITY OF KANSAS
LAWRENCE KS 66045
U.S.A.

SHCHEGLOV P V DR
STERNBERG STATE AS-
TRONOMICAL INSTITUTE
117234 MOSCOW
U.S.S.R.

SHCHEGLOV V P PROF DR
ASTRONOMICAL INSTITU
TE UZBEKIAN ACADEMY
OF SCIENCES
700000 TASHKENT
U.S.S.R.

SHCHEGOLEV DIMITRIJ E DR
PULKOVO OBSERVATORY
196140 LENINGRAD
U.S.S.R.

SHCHERBINA-SAMOJLOVA I S
INSTITUTE OF SCIENCE
AND TECHNICS INFORMA
TION
125219 MOSCOW
U.S.S.R.

SHEA MARGARET A DR
AIR FORCE GEOPHYSICS
LAB PHC
HANSCOM AFB MA 01732
U.S.A.

SHEELEY NEIL R DR
CODE 7140NS
NAVAL RESEARCH LAB
WASHINGTON DC 20375
U.S.A.

SHELUS PETER J DR
ASTRONOMY DEPT
UNIVERSITY OF TEXAS
R L M 16-316
AUSTIN TX 78712
U.S.A.

SHEN BENJAMIN S P PROF
DEPT ASTRONOMY E1
UNIVERSITY OF
PENNSYLVANIA
PHILADELPHIA PA19104
U.S.A.

SHEN CHUN-SHAN PRESIDENT
ASTRO SOCIETY CHINA
DEAN SCHOOL OF SC
NAT TSING-HUA UNIV
HSIN-CHU 300
TAIWAN

SHEN ZEE DR
DEPT OF MATHEMATICS
NATIONAL TAIWAN UNIV
TAIPEI
TAIWAN

SHER DAVID DR
2837 MINTO DR APT 2
CINCINNATI OH 45208
U.S.A.

SHERIDAN K V DR
C S I R O
DIV OF RADIOPHYSICS
P O BOX 76
EPPING NSW 2121
AUSTRALIA

SHERWOOD WILLIAM A DR
MPI FUER RADIO
ASTRONOMIE
AUF DEM HUEGEL 69
D-5300 BONN
GERMANY, F.R.

SHEVCHENKO V V DR
STERNBERG STATE AS-
TRONOMICAL INSTITUTE
117234 MOSCOW
U.S.S.R.

SHIBATA YUKIO DR
RESEARCH INST SCIEN-
TIFIC MEASUREMENTS
TOHOKU UNIVERSITY
SENDAI 980
JAPAN

SHIELDS GREGORY A DR
DEPT OF ASTRONOMY
RLM 15 212
UNIV OF TEXAS
AUSTIN TX 78712
U.S.A.

SHIM WOON-TAIK PROF
236-53 SINDAN-DONG
JOONG-KU
SEOUL 100
REPUBLIC OF KOREA

SHIMIZU MIKIO ASST PROF
INST SPACE AERON SC
UNIVERSITY OF TOKYO
KOMABA MEGURO KU
TOKYO 153
JAPAN

SHIMIZU TSUTOMU PROF
DEPT OF ASTRONOMY
UNIVERSITY OF KYOTO
SAKYOKU KYOTO 606
JAPAN

SHIMMINS A J
AUSTRALIAN NAT RADIO
ASTRONOMY OBS
P O BOX 276
PARKES NSW 2870
AUSTRALIA

SHIMODA MAHIRO DR
TOKYO GAKUGEI
UNIVERSITY
NUKUIKITAMACHI
KOGANEI TOKYO 184
JAPAN

SHINE RICHARD A DR
NASA GSFC
CODE 685
GREENBELT MD 20771
U.S.A.

SHIPMAN HENRY L DR
PHYSICS DEPT
UNIV OF DELAWARE
NEWARK DE 19711
U.S.A.

SHIRYAEV A V DR
LENINGRAD STATE UNI-
VERSITY ASTRONOMICAL
OBSERVATORY
199178 LENINGRAD
U.S.S.R.

SHIVANANDAN KANDIAH DR
NAVAL RESEARCH LAB
CODE 7122 1
WASHINGTON DC 20375
U.S.A.

SHKLOVSKY I S DR
SPACE RESEARCH
INSTITUTE
USSR ACADEMY OF SCI
117810 MOSCOW
U.S.S.R.

SHOBBROOK ROBERT R DR
CHATTERTON DEPT OF
ASTRONOMY
UNIVERSITY OF SYDNEY
SYDNEY NSW 2006
AUSTRALIA

SHOEMAKER E
US GEOLOGICAL SURVEY
MENLO PARK CA 94025
U.S.A.

SHOLOMITSKY G B DR
SPACE RESEARCH
INSTITUTE
USSR ACADEMY OF SC
117810 MOSCOW
U.S.S.R.

SHOR VIKTOR A DR
INST THEOR ASTRONOMY
NABEREZHNAYA
KUTUZOVA 10
192187 LENINGRAD
U.S.S.R.

SHORE BRUCE W
LAWRENCE LIVERMORE
LABORATORY
LIVERMORE CA 94550
U.S.A.

SHOSTAK G S DR
KAPTEYN ASTRON INST
POSTBUS 800
9700 AV GRONINGEN
THE NETHERLANDS

SHTEINS K A DR
LATVIAN STATE UNIVER
SITY ASTRONOMICAL
OBSERVATORY
226098 RIGA
U.S.S.R.

SHU FRANK H PROF
ASTRONOMY DEPT
UNIV OF CALIFORNIA
CAMPBELL HALL
BERKELEY CA 94720
U.S.A.

SHUKLA K
DEPT OF MATHEMATICS
AND ASTRONOMY
LUCKNOW UNIVERSITY
LUCKNOW U P
INDIA

SHUKRE C S DR
RAMAN RESEARCH INST
P O HEBBAL
BANGALORE 560 006
INDIA

SHUL MAN L M DR
MAIN ASTRONOMICAL OB
SERVATORY UKRAINIAN
ACADEMY OF SCIENCES
252127 KIEV
U.S.S.R.

SHUL'BERG A M DR
ODESSA STATE UNIVER-
SITY ASTRONOMICAL
OBSERVATORY
270014 ODESSA
U.S.S.R.

SHULOV O S DR
LENINGRAD STATE UNI-
VERSITY ASTRONOMICAL
OBSERVATORY
199178 LENINGRAD
U.S.S.R.

SHUTER WILLIAM L H DR
DEPT OF PHYSICS
UNIVERSITY OF B C
2075 WESTBROOK MALL
VANCOUVER BC V6T 1W5
CANADA

SIBILLE FRANCOIS
OBSERVATOIRE DE LYON
F-69230 ST-GENIS
LAVAL
FRANCE

SIDA DERRICK W PROF
DEPT OF PHYSICS
CARLETON UNIVERSITY
OTTAWA K1S 5B6
CANADA

SIDLICHOVSKY MILOS DR
ASTRONOMICAL INST
CSAV
BUDECSKA 6
120 23 PRAHA 2
CZECHOSLOVAKIA

SIEBER WOLFGANG PH D
MPI FUER
RADIOASTRONOMIE
AUF DEM HUEGEL 69
D-5300 BONN 1
GERMANY, F.R.

SIEGEL C L
UNIVERSITAET
GOETTINGEN
BUNSENSTR 3-5
D-3400 GOETTINGEN
GERMANY, F.R.

SILBERBERG REIN DR
NAVAL RESEARCH LAB
CODE 7028
WASHINGTON DC 20375
U.S.A.

SILK JOSEPH I PROF
ASTRONOMY DEPT
UNIV OF CALIFORNIA
BERKELEY CA 94720
U.S.A.

SILVERBERG ERIC C DR
MCDONALD OBSERVATORY
UNIVERSITY OF TEXAS
BOX 1337
FORT DAVIS TX 79734
U.S.A.

SIM MARY E DR
ROYAL OBSERVATORY
BLACKFORD HILL
EDINBURGH EH9 3HJ
U.K.

SIMEK MILOS DR
ASTRONOMICAL INST
CZECHO ACAD SCIENCES
OBSERVATORY
251 65 ONDREJOV
CZECHOSLOVAKIA

SIMKIN SUSAN M DR
MT STROMLO OBSERVA-
TORY
CANBERRA ACT 2606
AUSTRALIA

SIMMONEAU E DR
INST D'ASTROPHYSIQUE
98 BIS BD ARAGO
F-75014 PARIS
FRANCE

SIMO CHARLES DR
PJE MARIMON 11
BARCELONA 21
SPAIN

SIMON GEORGE W DR
SACRAMENTO PEAK
OBSERVATORY
SUNSPOT NM 88349
U.S.A.

SIMON J L MR
BUREAU D LONGITUDES
77 AVENUE DENFERT-
ROCHEREAU
F-75014 PARIS
FRANCE

SIMON MICHAL PROF
DEPT EARTH SPACE SC
STATE UNIVERSITY OF
NEW YORK
STONY BROOK NY 11794
U.S.A.

SIMON NORMAN R ASSOC PROF
BEHLEN LAB PHYSICS
UNIVERSITY NEBRASKA
LINCOLN NB 68588
U.S.A.

SIMON PAUL A DR
OBSERVATOIRE
DE MEUDON
F-92190 MEUDON
FRANCE

SIMON PAUL C DR
AV MANOIR D'ANJOU 14
B-1150 BRUXELLES
BELGIUM

SIMON RENE L E PROF
INSTITUT
D ASTROPHYSIQUE
B-4200 COINTE-LIEGE
BELGIUM

SIMONENKO ALLA N DR
ASTRON COUNCIL
USSR ACADEMY OF SCS
PYATNITSKAYA 48
109017 MOSCOW
U.S.S.R.

SIMONS STUART DR
170 HOLMLEIGH ROAD
LONDON N16 5PY
U.K.

SIMONSON S CHRISTIAN DR
1061 RUSSELL AVE
LOS ALTOS CA 94022
U.S.A.

SIMOVLJEVITCH JOVAN L DR
DEPT OF ASTRONOMY
FACULTY OF SCIENCES
STUDENTSKI TRG 16
11000 BEOGRAD
YUGOSLAVIA

SIMS KENNETH P DR
SYDNEY OBSERVATORY
SYDNEY N S W 2000
AUSTRALIA

SINCLAIR ANDREW T DR
ROYAL GREENWICH
OBSERVATORY
HAILSHAM
EAST SUSSEX BN27 1RP
U.K.

SINNERSTAD ULF E PROF
STOCKHOLMS
OBSERVATORIUM
S-13300 SALTSJOBADEN
SWEDEN

SINTON WILLIAM M
UNIV OF HAWAII
INST FOR ASTRONOMY
2680 WOODLAWN DR
HONOLULU HI 96822
U.S.A.

SINVHAL SHAMBHU DAYAL DR.
KUMAUN UNIVERSITY
NAINI TAL 263001
INDIA

SINZI AKIRA M DR
HYDROGRAPHIC DEPT
TSUKIJI 5 CHUO KU
TOKYO 104
JAPAN

SIRY JOSEPH W
4438 42ND ST N W
WASHINGTON DC 20016
U.S.A.

SISSON GEORGE M MR
PLANETREES WALL
HEXHAM NE46 4EQ
U.K.

SISTERO ROBERTO F DR
OBSERVATORIO
ASTRONOMICO
LAPRIDA 854
5000 CORDOBA
ARGENTINA

SITARSKI GRZEGORZ DR
CTR FOR SPACE RES
UL BARTYCKA 18
00716 WARSAW
POLAND

SITNIK G F DR
STERNBERG STATE AS-
TRONOMICAL INSTITUTE
117234 MOSCOW
U.S.S.R.

SIVAN J P DR
LAB D'ASTRON SPAT
TRAVERSE DU SIPHON
LES TROIS LUCS
F-13012 MARSEILLE
FRANCE

SIVARAMAN K R DR
INDIAN INSTITUTE OF
ASTROPHYSICS
BANGALORE 560034
INDIA

SJOGREN WILLIAM L DR
JPL
MS 264-738
4800 OAK GROVE DRIVE
PASADENA CA 91103
U.S.A.

SKALAFURIS ANGELO
MATH RESEARCH CTR
NAVAL RESEARCH LAB
C7840
WASHINGTON DC 20375
U.S.A.

SKILLING JOHN DR
DEPT APPLIED
MATHS AND THEO PHYS
SILVER STREET
CAMBRIDGE CB3 9EW
U.K.

SKUMANICH ANDRE PROF
HIGH ALTITUDE
OBSERVATORY
PO BOX 3000
BOULDER CO 80307
U.S.A.

SLADE MARTIN A III DR
JPL
264-720
4800 OAK GROVE DRIVE
PASADENA CA 91103
U.S.A.

SLAUCITAJS SERGEJS
PROF DR
OBSERVATORIO ASTRONO
MICO DE LA PLATA
1900 LA PLATA
ARGENTINA

SLEBARSKI TADEUSZ DR
UNIVERSITY OBS
BUCHANAN GARDENS
ST ANDREWS FIFE
KY16 9LZ
U.K.

SLEE O B DR
RADIOPHYSICS
CSIRO
BOX 76
EPPING NSW 2121
AUSTRALIA

SLETTEBAK ARNE PROF
PERKINS OBSERVATORY
DELAWARE OH 43015
U.S.A.

SLONIM E M DR
ASTRONOMICAL INSTITU
TE UZBEKIAN ACADEMY
OF SCIENCES
700000 TASHKENT
U.S.S.R.

SLYSH V I DR
SPACE RESEARCH
INSTITUTE
USSR ACADEMY OF SCI
117810 MOSCOW
U.S.S.R.

SMAK JOSEPH I PROF
COPERNICUS ASTRONOMI
CAL CENTER
UL BARTYCKA 18
00 716 WARSAW
POLAND

SMEYERS PAUL PROF
ASTRONOMISCH INST
NAAMSESTRAAT 61
B-3000 LEUVEN
BELGIUM

SMILEY CHARLES H DR
28 MONTAGUE ST
PROVIDENCE RI 02906
U.S.A.

SMIT J A PROF
PHYSICAL LAB
UNIVERSITY
SORBONNELAAN 4
UTRECHT UITHOF
THE NETHERLANDS

SMITH ALEX G PROF
DEPT PHYSICS AND
ASTRONOMY
231 SPACE SC BLDG
GAINESVILLE FL 32611
U.S.A.

SMITH ANDREW M DR
CODE 672
GODDARD SPACE FLIGHT
CENTER
GREENBELT MD 20771
U.S.A.

SMITH BARHAM W DR
G-6 MS 665
LASL
LOS ALAMOS NM 87544
U.S.A.

SMITH BRADFORD A PROF
DEPT OF PLANETARY
SCIENCES
UNIVERSITY ARIZONA
TUCSON AZ 85721
U.S.A.

SMITH BRUCE F DR
THEORET & PLAN STUD
BR 245-3
NASA-AMES RES CTR
MOFFETT FLD CA 94035
U.S.A.

SMITH CLAYTON A JR DR
U S NAVAL OBS
WASHINGTON DC 20390
U.S.A.

SMITH DEAN F DR
DEPT OF ASTRO-
GEOPHYSICS
UNIV OF COLORADO
BOULDER CO 80309
U.S.A.

SMITH ELSKE V P DR
ASTRONOMY PROGRAM
UNIVERSITY MARYLAND
COLLEGE PK MD 20742
U.S.A.

SMITH F GRAHAM PROF
ROYAL GREENWICH
OBSERVATORY
HAILSHAM
EAST SUSSEX BN27 1RP
U.K.

SMITH GEOFFREY DR
DEPT OF ASTROPHYSICS
SOUTH PARKS ROAD
OXFORD OX1 3RQ
U.K.

SMITH HARDING E JR DR
PHYSICS DEPT C-011
UNIV OF CALIFORNIA
SAN DIEGO
LA JOLLA CA 92093
U.S.A.

SMITH HARLAN J PROF
ASTRONOMY DEPT
UNIVERSITY OF TEXAS
RL MOORE HALL 15,206
AUSTIN TX 78712
U.S.A.

SMITH HAYWOOD C DR
DEPT OF ASTRONOMY
UNIVERSITY OF
SOUTH FLORIDA
TAMPA FL 33620
U.S.A.

SMITH HENRY J DR
NASA HEADQUARTERS
CODE SG
WASHINGTON DC 20546
MEXICO

SMITH HUMPHRY M
23 NORMANDALE
BEXHILL ON SEA
EAST SUSSEX TN39 3LU
U.K.

SMITH LINDSEY F DR
DEPT OF PHYSICS
UNIV OF WOLLONGONG
P O BOX 1144
WOLLONGONG NSW 2500
AUSTRALIA

SMITH MALCOLM G DR
ANGLO AUSTRALIAN
OBSERVATORY
P O BOX 296
EPPING N S W 2121
AUSTRALIA

SMITH MYRON A ASST PROF
DEPT OF ASTRONOMY
UNIVERSITY OF TEXAS
AUSTIN TX 78712
U.S.A.

SMITH PETER L DR
CTR FOR ASTROPHYSICS
60 GARDEN ST
CAMBRIDGE MA 02138
U.S.A.

SMITH ROBERT CONNON DR
ASTRONOMY CENTRE
UNIVERSITY OF SUSSEX
BRIGHTON BN1 9QH
U.K.

SMITH WM HAYDEN PROF
MCDONNELL CENTER FOR
SPACE SCIENCES
WASHINGTON UNIVERSIT
ST LOUIS MO 63130
U.S.A.

SMOLINSKI JAN DR
COPERNICUS ASTRONO-
MICAL CENTER
CHOPINA 12-18
87-100 TORUN
POLAND

SMOLKOV G YA DR
SIBIZMIR
P B 4
664697 IRKUTSK 33
U.S.S.R.

SMOLUCHOWSKI ROMAN PROF
ASTRONOMY DEPT
UNIVERSITY OF TEXAS
RLM-15-212
AUSTIN TX 78712
U.S.A.

SMRIGLIO FILIPPO PROF
OSSERVATORIO
ASTRONOMICO
VIA PARCO MELLINI 84
I-00136 ROMA
ITALY

SMYLIE DOUGLAS E DR
DEPT OF PHYSICS
YORK UNIVERSITY
4700 KEELE STREET
DOWNSVIEW ONT M3J1P3
CANADA

SMYTH MICHAEL J DR
DEPT OF ASTRONOMY
ROYAL OBSERVATORY
EDINBURGH EH9 3HJ
U.K.

SNOW THEODORE P PROF
DEPT OF PHYSICS AND
ASTROPHYSICS
UNIVERSITY COLORADO
BOULDER CO 80309
U.S.A.

SNYDER LEWIS E
DEPT OF ASTRONOMY
UNIV OF ILLINOIS
URBANA IL 61801
U.S.A.

SOBERMAN ROBERT K DR
FRANKLIN RES CENTER
THE FRANKLIN INST
20+B.FRANKLIN PARKW
PHILADELPHIA PA19103
U.S.A.

SOBIESKI STANLEY DR
CODE 673
GODDARD SPACE FLIGHT
CENTER
GREENBELT MD 20771
U.S.A.

SOBOLEV V V DR
LENINGRAD STATE UNIV
ASTRONOMICAL OBS
199178 LENINGRAD
U.S.S.R.

SOBOLEV VLADISLAV M DR
MAIN ASTRONOMICAL
OBSERVATORY PULKOVO
196140 LENINGRAD
U.S.S.R.

SOBOLEVA N S DR
SPECIAL ASTROPHYS
OBS USSR ACAD OF SCI
LENINGRAD BRANCH
196140 LENINGRAD
U.S.S.R.

SOBOUTI YOUSEF PROF
BIRUNI OBSERVATORY
AND PHYSICS DEPT
UNIVERSITY
SHIRAZ
IRAN

SOCHILINA ALLA S DR
INSTITUTE FOR THEORE
TICAL ASTRONOMY
10 KUTUZOV QUAY
192187 LENINGRAD
U.S.S.R.

SODERBLOM LARRY DR
BR ASTROGEOLOGIC ST
US GEOLOGICAL SURVEY
2555 N GEMINI DRIVE
FLAGSTAFF AZ 86001
U.S.A.

SOFIA SABATINO PROF
GODDARD SPACE FLIGHT
CENTER
CODE 681
GREENBELT MD 20771
U.S.A.

SOFUE YOSHIAKI DR
DEPT OF PHYSICS
UNIVERSITY OF NAGOYA
FUROCHO CHIKUSAKU
NAGOYA 464
JAPAN

SOIFER BARUCH T DR
320-47 DOWNS LAB
PHYSICS DEPT
CALTECH
PASADENA CA 91125
U.S.A.

SOLC IVAN DR
ASTRON INST CSAV
GROUP OF OPTICS
DVORAKOWA 298
511 01 TURNOV
CZECHOSLOVAKIA

SOLF JOSEF DR
MPI FUER ASTRONOMIE
KOENIGSTUHL
D-6900 HEIDELBERG 1
GERMANY, F.R.

SOLHEIM JAN ERIK
IMR P O BOX 953
N-9001 TROMSO
NORWAY

SOLOMON PHILIP M DR
ASTROPHYSICS PROGRAM.
DEPT EARTH SPACE SCS
SUNY AT STONY BROOK
STONY BROOK NY 11794
U.S.A.

SOMERVILLE WILLIAM B DR
UNIVERSITY COLLEGE
LONDON WC1E 6BT
U.K.

SONETT CHARLES P PROF
DEPT PLANETARY SCI
UNIVERSITY OF
ARIZONA
TUCSON AZ 85721
U.S.A.

SORENSEN GUNNAR DR
INSTITUTE OF PHYSICS
LANGELANDSGADE
DK-8000 AARHUS C
DENMARK

SORENSEN SOREN-AKSEL DR
UNIV OF LONDON
OBSERVATORY
MILL HILL PARK
LONDON NW7 2QS
U.K.

SOROCHENKO R L DR
PHYSICAL INSTITUTE
USSR ACADEMY OF
SCIENCES
117924 MOSCOW
U.S.S.R.

SORU-ESCAUT IRINA MRS
OBSERVATOIRE
DE MEUDON
F-92190 MEUDON
FRANCE

SOTIROVSKI PASCAL DR
OBSERVATOIRE
DE MEUDON
F-92190 MEUDON
FRANCE

SOUFFRIN PIERRE B DR
OBSERVATOIRE DE NICE
B P 252
F-06007 NICE CEDEX
FRANCE

SOULIE GUY
OBS DE L'UNIVERSITE
DE BORDEAUX
F-33270 FLOIRAC
FRANCE

SOUTHWORTH R B DR
HARVARD COLLEGE OBS
60 GARDEN STREET
CAMBRIDGE MA 02138
U.S.A.

SPADA GIANFRANCO DR
TE S R E
CNR
VIA DE CASTAGNOLI 1
I-40126 BOLOGNA
ITALY

SPAENHAUER ANDREAS MARTIN
ASTRON INST BASEL
VENUSSTRASSE 7
CH-4102 BINNINGEN
SWITZERLAND

SPARKS WARREN M DR
CODE 671
GODDARD SPACE FLIGHT
CENTER
GREENBELT MD 20771
U.S.A.

SPARROW JAMES G DR
AERONAUTICAL RES
LABORATORIES
BOX 4331
MELBOURNE 3001
AUSTRALIA

SPEER R J DR
DEPT OF PHYSICS
IMPERIAL COLLEGE
LONDON SW7 2BZ
U.K.

SPENCER RALPH E DR
NUFFIELD RADIO ASTRO
NOMY LABS
JODRELL BANK MACCLES
FIELD CHES SK11 9DL
U.K.

SPICER DANIEL SHIELDS DR
INST FOR PHYSICAL SC
AND TECHNOLOGY
UNIV OF MARYLAND
COLLEGE PK MD 20742
U.S.A.

SPIEGEL E DR
COURANT INST OF
MATHEMATICAL SCS
251 MERCER STREET
NEW YORK NY 10003
U.S.A.

SPINRAD HYRON PROF
DEPT OF ASTRONOMY
UNIVERSITY
OF CALIFORNIA
BERKELEY CA 94720
U.S.A.

SPITE FRANCOIS M DR
OBSERVATOIRE
DE PARIS
F-92190 MEUDON
FRANCE

SPITE MONIQUE DR
OBSERVATOIRE
DE PARIS
F-92190 MEUDON
FRANCE

SPITHAS ELEFTERIOS N DR
DEPT OF ASTRONOMY
UNIVERSITY OF ATHENS
PANEPISTIMIOPOLIS
ATHENS 621
GREECE

SPITZER LYMAN JR DR
PRINCETON UNIVERSITY
OBSERVATORY
PEYTON HALL
PRINCETON NJ 08540
U.S.A.

SPOELSTRA T A TH DR
STERREWACHT
POSTBUS 9513
2300 RA LEIDEN
THE NETHERLANDS

SPYROU NICOLAOS DR
DEPT OF ASTRONOMY
UNIVERSITY OF
THESSALONIKI
THESSALONIKI
GREECE

SRAMEK RICHARD A DR
NAT RADIO ASTRON OBS
P O BOX 0
SOCORRO NM 87801
U.S.A.

SREEKANTAN B V DR
TATA INSTITUTE OF
FUNDAMENTAL RESEARCH
HOMI BHABHA ROAD
BOMBAY 400005
INDIA

SREENIVASAN S RANGA PROF
PHYSICS DEPT
THE UNIVERSITY
2920-24 AVE N W
CALGARY ALTA T2N 1N4
CANADA

STABELL ROLF DR
INSTITUTE OF THEORE-
TICAL ASTROPHYSICS
UNIVERSITY OF OSLO
BLINDERN OSLO 3
NORWAY

STAHR-CARPENTER M DR
LEANDER MCCORMICK OB
BOX 3818
UNIVERSITY STATION
CHARLOTTESVL VA22903
U.S.A.

STALIO ROBERTO DR
OSSERVATORIO ASTRONO
MICO DI TRIESTE
VIA TREPOLO 11
I-34131 TRIESTE
ITALY

STANDISH E MYLES DR
JET PROPULSION
LABORATORY
PASADENA CA 91103
U.S.A.

STANILA GEORGE DR
ASTRONOMICAL OBS
5 CUTITUL DE ARGINT
BUCAREST 28
ROUMANIA

STANKEVICH K S DR
RADIOPHYSICAL
RESEARCH INSTITUTE
603600 GORKIJ
U.S.S.R.

STANLEY G J
CALTECH
PASADENA CA 91109
U.S.A.

STANNARD DAVID DR
JODRELL BANK
MACCLESFIELD
CHESHIRE SK11 9DL
U.K.

STARRFIELD SUMNER PROF
DEPT OF PHYSICS
ARIZONA STATE
UNIVERSITY
TEMPE AZ 85281
U.S.A.

STARYTSIN GENNADIJ V DR
PULKOVO OBSERVATORY
196140 LENINGRAD
U.S.S.R.

STASINSKA G DR
OBS DE MEUDON
F-92190 MEUDON
FRANCE

STAUBERT RUDIGER DR
ASTRONOMISCHES INST.
DER UNIV TUEBINGEN
WALDHAUSERSTR 64
D-7400 TUEBINGEN
GERMANY, F.R.

STAUDE HANS JAKOB PH D
MPI FUER ASTRONOMIE
KOENIGSTUHL
D-69 HEIDELBERG 1
GERMANY, F.R.

STAUDE JUERGEN DR
ZENTRALINST FUER
SOLAR-TERR PHYSIK
TELEGRAFENBERG
DDR-15 POTSDAM
G.D.R.

STAWIKOWSKI ANTONI DR
ASTRONOMICAL CENTER
UL CHOPINA 12-18
87-100 TORUN
POLAND

STEARNS C L DR
VAN VLECK OBS
MIDDLETOWN CT 06457
U.S.A.

STECHER THEODORE P
CODE 672
GODDARD SPACE FLIGHT
CENTER
GREENBELT MD 20771
U.S.A.

STECKER FLOYD W DR
LAB FOR HIGH ENERGY
ASTROPHYS 660 NASA
GODDARD SPACE CENTER
GREENBELT MD 20771
U.S.A.

STEENMAN-CLARK L DR
OBS DE NICE
B P 252
F-06007 NICE CEDEX
FRANCE

STEFANOVITCH-GOMEZ A E DR
OBSERVATOIRE
DE MEUDON
F-92190 MEUDON
FRANCE

STEIGER W R PROF
HAWAII INST GEOPHYS
UNIV OF HAWAII
2525 CORREA ROAD
HONOLULU HI 96822
U.S.A.

STEIGMAN GARY PROF
BARTOL RESEARCH
FOUNDATION
UNIV OF DELAWARE
NEWARK DE 19711
U.S.A.

STEIN ROBERT F ASSOC PROF
ASTRONOMY DEPT
MICHIGAN STATE
UNIVERSITY
EAST LANSING MI48824
U.S.A.

STEIN WAYNE A PROF
SCHOOL OF PHYSICS
AND ASTRONOMY
UNIVERSITY MINNESOTA
MINNEAPOLIS MN 55455
U.S.A.

STEINBERG JEAN L DR
OBSERVATOIRE
DE MEUDON
F-92190 MEUDON
FRANCE

STEINITZ RAPHAEL PROF
PHYSICS DEPT
BG UNIVERSITY OF THE
NEGEV
BEER SHEVA
ISRAEL

STEINLIN ULI PROF
ASTRONOMISCHES INST
UNIVERSITAET BASEL
VENUSSTR 7
CH-4102 BINNINGEN
SWITZERLAND

STELLINGWERF ROBERT F DR
DEPT PHYSICS ASTRON
RUTGERS UNIV
PISCATAWAY NJ 08904
U.S.A.

STELLMACHER GOETZ
INST D ASTROPHYSIQUE
98 BIS BD ARAGO
F-75014 PARIS
FRANCE

STELLMACHER IRENE DR
BUREAU D LONGITUDES
77 AVENUE DENFERT-
ROCHEREAU
F-75014 PARIS
FRANCE

STENFLO JAN O DR
LUND OBSERVATORY
S-222 24 LUND
SWEDEN

STEPANIAN A A DR
CRIMEAN ASTROPHYSI-
CAL OBSERVATORY USSR
ACAD OF SCI NAUCHNIY
334413 CRIMEA
U.S.S.R.

STEPANIAN N N DR
CRIMEAN ASTROPHYSI-
CAL OBSERVATORY USSR
ACAD OF SCI NAUCHNIY
334413 CRIMEA
U.S.S.R.

STEPANOV V E PROF
SIBIZMIR
P B 4
664697 IRKUTSK 33
U.S.S.R.

STEPHENSON C BRUCE PROF
WARNER AND SWASEY
OBSERVATORY
1975 TAYLOR RD
E CLEVELAND OH 44112
U.S.A.

STEPHENSON F RICHARD DR
SCHOOL OF PHYSICS
UNIVERSITY
NEWCASTLE UPON TYNE
NE1 7RU
U.K.

STEPIEN KAZIMIERZ DR
ASTRONOMICAL
OBSERVATORY
AL UJAZDOWSKIE 4
00478 WARSZAWA
POLAND

STEPPE HANS DR
ASTRON INST BASEL
VENUSSTRASSE 7
CH-4102 BINNINGEN
SWITZERLAND

STERKEN CHRISTIAAN LEO DR
ASTROPHYSICAL INST
V U B
PLEINLAAN 2
B-1050 BRUSSELS
BELGIUM

STERNBERK BOHUMIL
ASTRON INST CSAV
BUDECSKA 6
120 23 PRAHA 2
CZECHOSLOVAKIA

STESHENKO N V DR
CRIMEAN ASTROPHYSI-
CAL OBSERVATORY USSR
ACAD OF SCI NAUCHNIY
334413 CRIMEA
U.S.S.R.

STEVENS GERARD A DR
LABORATORIUM VOOR
RUIMTEONDERZOEK
BENELUXLAAN 21
UTRECHT
THE NETHERLANDS

STEWART J M DR
INSTITUTE OF
ASTRONOMY
MADINGLEY ROAD
CAMBRIDGE CB3 0HA
U.K.

STEWART PAUL DR
MATHEMATICS DEPT
THE UNIVERSITY
MANCHESTER M13 9PL
U.K.

STEWART RONALD T MR
CSIRO DIVISION OF
RADIOPHYSICS
P O BOX 76
EPPING NSW 2121
AUSTRALIA

STEYAERT HERMAN PROF DR
STERRENKUNDIG INST
KRYGSLAAN 271 S9
B-9000 GENT
BELGIUM

STIBBS DOUGLAS W N PROF
UNIV OBSERVATORY
BUCHANAN GARDENS
ST ANDREWS
FIFE KY16 9L2
U.K.

STICKLAND DAVID J DR
ESA SAT TRACK STAT
VILLAFRANCA DEL CAST
P O BOX 54065
MADRID
SPAIN

STIFT M J DR
INST FUER ASTRONOMIE
TUERKENSCHANZSTR 17
A-1180 WIEN
AUSTRIA

STIX MICHAEL DR
KIEPENHEUER INST
FUER SONNENPHYSIK
SCHOENECKSTR 6
D-7800 FREIBURG
GERMANY, F.R.

STOBIE ROBERT S DR
INST OF ASTRONOMY
MADINGLEY ROAD
CAMBRIDGE CB3 0HA
U.K.

STOCK JURGEN
CENTRO DE INVESTIGA-
CION DE ASTRONOMIA
APARTADO 264
MERIDA
VENEZUELA

STOCKMAN HERVEY S JR DR
STEWARD OBS
UNIV OF ARIZONA
TUCSON AZ 85721
U.S.A.

STOCKTON ALAN N DR
INSTITUTE FOR
ASTRONOMY
2680 WOODLAWN DR
HONOLULU HI 96822
U.S.A.

STODOLKIEWICZ JERZY S DR
COPERNICUS ASTRONOMI
CAL CENTER
UL BARTYCKA 18
00 716 WARSAW
POLAND

STOHL JAN DR
ASTRONOMICAL INST
SAV
DUBRAVSKA CESTA A 11
89930 BRATISLAVA
CZECHOSLOVAKIA

STONE R G DR
CODE 693
LAB FOR EXTRATERRES
PHYSICS NASA-GSFC
GREENBELT MD 20771
U.S.A.

STONE REMINGTON P S DR
LICK OBSERVATORY
MT HAMILTON CA 95140
U.S.A.

STOYKO ANNA
OBSERVATOIRE
DE PARIS
61 AVE OBSERVATOIRE
F-75014 PARIS
FRANCE

STRAIZYS V PROF DR
ASTRONOMIJOS OBS
VILNIUS 31
CIURLIONIO 29
LITHUANIA
U.S.S.R.

STRAND KAJ AA DR
3202 ROWLAND PL NW
WASHINGTON DC 20390
U.S.A.

STRASSL HANS L PROF
OCHTRUPWEG 39
D-4400 MUENSTER
GERMANY, F.R.

STRAUSS FEDERICO M DR
CRAAM/ON/CNPQ
RUA PARA 277
01243 SAO PAULO SP
BRAZIL

STREL'NITSKIJ VLADIMIR DR
ASTRONOMICAL COUNCIL
USSR ACAD OF SCS
48 PYATNITSKAYA ST
109017 MOSCOW
U.S.S.R.

STRITTMATTER PETER A PROF
STEWARD OBSERVATORY
TUCSON AZ 85721
U.S.A.

STROBEL ANDRZEJ DR
INST OF ASTRONOMY
N COPERNICUS UNIV
CHOPINA 12/18
87-100 TORUN
POLAND

STROBEL WILLI DR
ASTRONOMISCHES
RECHEN INSTITUUT
MOENCHHOFSTR 12 14
D-6900 HEIDELBERG 1
GERMANY, F.R.

STROHMEIER WOLFGANG PROF
VOLKFELDSTRASSE 5
D-8600 BAMBERG
GERMANY, F.R.

STROM KAREN M
KITT PEAK NATIONAL
OBSERVATORY
P O BOX 26732
TUCSON AZ 85726
U.S.A.

STROM RICHARD G DR
RADIOSTERRENWACHT
OUDE HOOGEVEENSEDIJK
4
DWINGELOO-NL 7514
THE NETHERLANDS

STROM ROBERT G PROF
DEPT OF PLANETARY SC
UNIVERSITY
OF ARIZONA
TUCSON AZ 85721
U.S.A.

STROM STEPHEN E
GALACTIC EXTRAG PROG
KITT PEAK NAT OBS
P O BOX 26732
TUCSON AZ 85726
U.S.A.

STROMGREN BENGT PROF
GAMLE CARLSBERGV 15
DK-2500 COPENHAGEN
VALBY
DENMARK

STRONG IAN B DR
LOS ALAMOS SCI LAB
MAIL STOP 436
P O BOX 1663
LOS ALAMOS NM 87545
U.S.A.

STRONG JOHN D PROF
ASTRONOMY RESEARCH
FACILITY
UNIVERSITY OF MASS
AMHERST MA 01003
U.S.A.

STUMPFF PETER DR
MPI FUER RADIO
ASTRONOMIE
AUF DEM HUEGEL 69
D-5300 BONN
GERMANY, F.R.

STURCH CONRAD R DR
1942 LEWIS AVE
ROCKVILLE MD 20851
U.S.A.

STURROCK PETER A PROF
INST PLASMA RESEARCH
VIA CRESPI
STANFORD UNIVERSITY
STANFORD CA 94305
U.S.A.

SUDA KAZUO ASST PROF
DEPT OF ASTRONOMY
TOHOKU UNIVERSITY
ARAMAKI SENDAI 980
JAPAN

SUEMOTO ZENZABURO PROF DR
TOKYO ASTRONOMICAL
OBSERVATORY
OSAWA MITAKA
TOKYO 181
JAPAN

SUESS STEVEN T DR
SPACE ENVIRONMENT
LABORATORY
NOAA ERL
BOULDER CO 80302
U.S.A.

SUGAWA CHIKARA DR
INTERNATIONAL LATITU
DE OBSERVATORY OF
MIZUSAWA
MIZUSAWA-IWATE 023
JAPAN

SUGIMOTO DAIICHIRO PROF
DEPT EARTH SC ASTRO
COLL GENERAL EDUC
UNIVERSITY OF TOKYO
MEGURO-KU TOKYO 153
JAPAN

SUKHAREV LEONID A DR
PULKOVO OBSERVATORY
196140 LENINGRAD
U.S.S.R.

SULENTIC JACK W DR
DEPT PHYS ASTROPHYS
MICHIGAN ST UNIV
E LANSING MI 48824
U.S.A.

SULLIVAN WOODRUFF T IIIDR
DEPT ASTRONOMY FM 20
UNIVERSITY OF
WASHINGTON
SEATTLE WA 98195
U.S.A.

SULTANOV G F ACAD
SHEMAKHA ASTROPHYSI-
CAL OBSERVATORY
373243 AZERBAIDZAN
U.S.S.R.

SUMMERS HUGH P DR
DEPT OF APPLIED MATH
AND THEORETICAL PHYS
SILVER ST
CAMBRIDGE CB3 9EW
U.K.

SUZUKI YOSHIMASA PROF
23-1 NAKAJIMA
HIRONOMACHI
UJI SHI 611
JAPAN

SVALGAARD LEIF DR
INST FOR PLASMA RES
STANFORD UNIV
VIA CRESPI
STANFORD CA 94305
U.S.A.

SVATOS JAN DR
DEPT OF ASTRONOMY
CHARLES UNIVERSITY
SVEDSKA 8
150 00 PRAHA 5
CZECHOSLOVAKIA

SVECHNIKOV M A DR
ASTRONOMICAL DEPT OF
URALSKIJ
STATE UNIVERSITY
620083 SVERDLOVSK
U.S.S.R.

SVESTKA ZDENEK DR
SPACE RESEARCH LAB
BENELUXLAAN 21
UTRECHT
THE NETHERLANDS

SVOLOPOULOS SOTIRIOS PROF
DEPT OF ASTROPHYSICS
UNIVERSITY OF ATHENS
PANEPISTIMIOPOLIS
ATHENS 621
GREECE

SWANENBURG B N DR
COSMIC RAY WORKING
GROUP HUYGENS LAB
WASSENAARSEWEG 78
2300 RA LEIDEN
THE NETHERLANDS

SWARUP GOVIND PROF
RADIO ASTRONOMY
CENTRE OF TIFR
POST BOX 8
OOTACAMUND 643001
INDIA

SWEET PETER A PROF
DEPT OF ASTRONOMY
THE UNIVERSITY
GLASGOW G12 8QW
U.K.

SWEIGART ALLEN V DR
CODE 681
NASA GODDARD SPACE
FLIGHT CENTER
GREENBELT MD 20771
U.S.A.

SWENSON GEORGE W JR PROF
OBSERVATORY
UNIVERSITY
OF ILLINOIS
URBANA IL 61801
U.S.A.

SWENSSON JOHN W DR
DEPT OF PHYSICS
SOELVEGATAN 14
S-223 62 LUND
SWEDEN

SWERDLOV NOEL PROF
UNIV OF CHICAGO
CHICAGO IL 60637
U.S.A.

SWIHART THOMAS L DR
STEWARD OBSERVATORY
UNIVERSITY OF
ARIZONA
TUCSON AZ 85721
U.S.A.

SWINGS JEAN-PIERRE DR
9 ROUTE DE STRIVAY
B 4051 NEUPRE
BELGIUM

SWINGS P DR
AVE L SOUGUENET 23
B-4050 ESNEUX
BELGIUM

SWOPE HENRIETTA H DR
HALE OBSERVATORIES
813 SANTA BARBARA ST
PASADENA CA 91106
U.S.A.

SYKES J B DR
20 MILTON LANE
STEVENTON
ABINGDON
BERKSHIRE OX13 6SA
U.K.

SYKES-HART AVRIL B DR
DEPT OF ASTROPHYSICS
SOUTH PARKS ROAD
OXFORD OX1 3RQ
U.K.

SYKORA JULIUS DR
SKALNATE PLESO
OBSERVATORY
059 60 TATRANSKA
LOMNICA
CZECHOSLOVAKIA

SYUNYAEV R A DR
SPACE RESEARCH
INSTITUTE
USSR ACADEMY OF SCI
117810 MOSCOW
U.S.S.R.

SZABADOS LASZLO PH D
KONKOLY OBSERVATORY
P O BOX 67
H-1525 BUDAPEST
HUNGARY

SZAFRANIEC ROZALIA DR
ASTRONOMICAL OBS
UL ORLA 171
30 244 KRAKOW
POLAND

SZEBEHELY VICTOR G PROF
DEPT AEROSPACE ENG
UNIVERSITY OF TEXAS
AUSTIN TX 78712
U.S.A.

SZEIDL BELA DR
KONKOLY OBSERVATORY
BOX 67
1525 BUDAPEST
HUNGARY

SZKODY PAULA DR
DEPT OF ASTRONOMY
UNIV OF WASHINGTON
SEATTLE WA 98195
U.S.A.

TABARA HIROTO DR
FACULTY OF EDUCATION
UTSUNOMIYA UNIV
MINEMACHI
UTSUNOMIYA 320
JAPAN

TADEMARU E DR
UNIV OF MASS
AMHERST MA 01002
U.S.A.

TAFFARA SALVATORE PROF
VIA CALZA 5 BIS
I-35100 PADOVA
ITALY

TAGLIAFERRI GIUSEPPE PROF
OSSERVATORIO ARCETRI
LARGO FERMI 5
I-50125 FIRENZE
ITALY

TAI YUIN-KWEI PROF
NO 7 CHAO-CHOU ST
TAIPEI
TAIWAN

TAKAGI KOSHIRO PROF
DEPT OF PHYSICS
TOYAMA UNIVERSITY
3190 GOFUKU
TOYAMA 920
JAPAN

TAKAGI SHIGETSUGU DR
INTERNATIONAL LATITU
DE OBSERVATORY
MIZUSAWA IWATE 023
JAPAN

TAKAKUBO KEIYA PROF
DEPT OF ASTRONOMY
TOHOKU UNIVERSITY
ARAMAKI SENDAI 980
JAPAN

TAKAKURA TATSUO PROF
DEPT OF ASTRONOMY
UNIVERSITY OF TOKYO
BUNKYO-KU
TOKYO 113
JAPAN

TAKARADA KATSUO DR
KYOTO UNIVERSITY IND
ARTS TEXTILE FIBERS
MATSUGASAKI SAKYOKU
KYOTO 606
JAPAN

TAKASE BUNSHIRO PROF
TOKYO ASTRONOMICAL
OBSERVATORY
OSAWA MITAKA
TOKYO 181
JAPAN

TAKAYANAGI KAZUO PROF
INST SPACE AERON SC
UNIVERSITY OF TOKYO
KOMABA MEGURO KU
TOKYO 153
JAPAN

TAKEDA HIDENORI DR
DEPT OF AERONAUTICAL
ENGINEERING
KYOTO UNIVERSITY
SAKYOKU KYOTO 606
JAPAN

TAKENOUCHI TADAO DR
1-28-30 KICHIJYOJI
KITA MACHI
MUSASHINO TOKYO 180
JAPAN

TAKEUTI MINE DR
DEPT OF ASTRONOMY
TOHOKU UNIVERSITY
ARAMAKI SENDAI 980
JAPAN

TALBOT RAYMOND J JR PROF
SPACE PHYSICS AND
ASTRONOMY
RICE UNIVERSITY
HOUSTON TX 77001
U.S.A.

TALON R DR
CESR
9 AV DU COL ROCHE
B P 4336
F-31029 TOULOUSE CED
FRANCE

TALWAR SATYA P DR
PHYSICS DEPT
DELHI UNIVERSITY
DELHI 7
INDIA

TAMENAGA TATSUO DR
FACULTY OF EDUCATION
MIE UNIVERSITY
TSU SHI MIE 514
JAPAN

TAMMANN G ANDREAS PROF DR
ASTRONOMISCHES INST
UNIVERSITAET BASEL
VENUSST 7
CH-4102 BINNINGEN
SWITZERLAND

TAMURA SHINICHI DR
DEPT OF ASTRONOMY
TOHOKU UNIVERSITY
ARAMAKI SENDAI 980
JAPAN

TANABE HIROYOSHI DR
TOKYO ASTRONOMICAL
OBSERVATORY
OSAWA MITAKA
TOKYO 181
JAPAN

TANAKA HARUO PROF
TOKYO ASTRONOMICAL
OBSERVATORY
OSAWA MITAKA
TOKYO 181
JAPAN

TANAKA KATSUO DR
TOKYO ASTRONOMICAL
OBSERVATORY
OSAWA MITAKA
TOKYO 181
JAPAN

TANAKA RIICHIRO PROF
FACULTY GENERAL EDUC
NIIGATA UNIVERSITY
ASAHIMACHIDORI
NIIGATA 951
JAPAN

TANAKA YASUO DR
FACULTY OF EDUCATION
IBARAKI UNIVERSITY
BUNKYO MITO 310
JAPAN

TANAKA YASURO PROF
INST SPACE AERON SC
UNIVERSITY OF TOKYO
KOMABA MEGURO-KU
TOKYO 153
JAPAN

TANDBERG HANSSEN EINAR A
DR
NASA MARSHALL SPACE
FLIGHT CENTER ES51
HUNTSVILLE AL 35812
U.S.A.

TANDON JAGDISH NARAIN DR
DEPT OF PHYSICS AND
ASTROPHYSICS
UNIVERSITY OF DELHI
DELHI 110007
INDIA

TANNER RICHARD W
1046 FISHER AVENUE
OTTAWA ONT
CANADA

TAPLEY BYRON D DR
DEPT AEROSPACE ENGR
AND ENGR MECHANICS
UNIV TX-AUST WRW 402
AUSTIN TX 78712
U.S.A.

TARAFDAR S P DR
TATA INST FUND RES
HOMI BHABHA ROAD
BOMBAY 400 005
INDIA

TARENGHI MASSIMO DR
LABORATORIO DI FISI-
CA COSMICA
VIA CELORIA 16
I-20133 MILANO
ITALY

TARNSTROM GUY DR
MIT LINCOLN LAB
P O BOX 73
LEXINGTON MA 02173
U.S.A.

TARTER C BRUCE DR
LAWRENCE LIVERMORE
LABORATORY L-71
UNIVERSITY OF CALIF
LIVERMORE CA 94550
U.S.A.

TARTER JILL C DR
NASA-AMES RES CTR
MS 245-3
MOFFETT FLD CA 94035
U.S.A.

TASSOUL JEAN-LOUIS PROF
DEPT DE PHYSIQUE
UNIVERSITE MONTREAL
CP 6128
MONTREAL PQ H3C 3J7
CANADA

TASSOUL MONIQUE DR
C/O DEPT DE PHYSIQUE
UNIV DE MONTREAL
C P 6128
MONTREAL P Q H3C 3J7
CANADA

TATEVYAN S K DR
ASTRONOMICAL COUNCIL
USSR ACADEMY OF SCI
PYATNITSKAYA UL 48
109017 MOSCOW
U.S.S.R.

TATON RENE PROF
CENTRE ALEXANDRE
KOYRE
12 RUE COLBERT
F-75002 PARIS
FRANCE

TATUM JEREMY B DR
UNIVERSITY OF VICTO-
RIA OBSERVATORY
VICTORIA BC V8W 2Y2
CANADA

TAUBER GERALD E PROF
DEPT OF PHYSICS
TEL AVIV UNIVERSITY
TEL AVIV
ISRAEL

TAVARES J T L DR
AV DIAS DA SILVA
17 R/C ESQ
COIMBRA
PORTUGAL

TAVASTSHERNA KIRILL N DR
PULKOVO OBSERVATORY
196140 LENINGRAD
U.S.S.R.

TAWADROS MAHET JACOUB DR
HELWAN OBSERVATORY
HELWAN-CAIRO
ARAB REP. OF EGYPT

TAYLER ROGER J PROF
ASTRONOMY CENTRE
UNIVERSITY OF SUSSEX
BRIGHTON BN1 9QH
U.K.

TAYLOR DONALD J DR
DEPT OF PHYSICS AND
ASTRONOMY
UN OF NEBRASKA
LINCOLN NB 68588
U.S.A.

TAYLOR GORDON E
ROYAL GREENWICH
OBSERVATORY
HAILSHAM
EAST SUSSEX BN27 1RP
U.K.

TAYLOR JOSEPH H PROF
UNIV MASSACHUSETTS
GRADUATE RES CTR
TOWER B
AMHERST MA 01003
U.S.A.

TAYLOR KEITH DR
ROYAL GREENWICH
OBSERVATORY
HAILSHAM
EAST SUSSEX BN27 1RP
U.K.

TECH JACK L DR
B 306 METROLOGY BLDG
NATIONAL BUREAU OF
STANDARDS
WASHINGTON DC 20234
U.S.A.

TEERIKORPI VELI PEKKA DR
OBS AND ASTROPH LAB
UNIV OF HELSINKI
TAHTITORNINMAKI
SF-00130 HELSINKI 13
FINLAND

TEHERANY D
83 AVENUE REY
TEHERAN
IRAN

TEJFEL VIKTOR G DR
LAB OF LUNAR AND
PLANETARY PHYSICS
ASTROPHYSICAL INST
480068 ALMA-ATA
U.S.S.R.

TELEKI GEORGE DR
ASTRONOMSKA
OPSERVATORIJA
VOLGINA 7
11050 BEOGRAD
YUGOSLAVIA

TEMESVARY S DR
R STRAUSSSTRASSE 127
D-8000 MUENCHEN 80
GERMANY, F.R.

TEMPESTI PIERO PROF
OSSERVATORIO ASTRONO
MICO DI COLLURANIA
I-64100 TERAMO
ITALY

TENORIO-TAGLE G DR
ESO
C/O CERN
CH-1211 GENEVA 23
SWITZERLAND

TEPLITSKAYA R B DR
SIBIZMIR
P B 4
664697 IRKUTSK 33
U.S.S.R.

TER HAAR DIRK
MAGDALEN COLLEGE
OXFORD OX1 4AU
U.K.

TERASHITA YOICHI PROF
KANAZAWA INST TECHN
NONOICH CHO
MINAMIKYOKU
KANAZAWA 921
JAPAN

TEREBIZH VALERY YU DR
CRIMEAN STATION OF
STERNBERG
ASTRONOMICAL INST
NAUCHNY CRIMEA334413
U.S.S.R.

TERRAZAS MR L.R.
INST NAC DE ASTROFIS
OPT Y ELECTRONICA
APT POSTALES 216 51
PUEBLA
MEXICO

TERRIEN JEAN DIRECTEUR
B I P M
PAVILLON DE BRETEUIL
F-92310 SEVRES
FRANCE

TERZAN AGOP DR
OBSERVATOIRE DE LYON
F-69230 SAINT GENIS-
LAVAL
FRANCE

TERZIAN YERVANT PROF
SPACE SCIENCES BLDG
CORNELL UNIVERSITY
ITHACA NY 14853
U.S.A.

TESKE RICHARD G PROF
DEPT OF ASTRONOMY
UNIVERSITY OF
MICHIGAN
ANN ARBOR MI 48109
U.S.A.

TEXEREAU JEAN M
CERGA
AVENUE COPERNIC
F-06130 GRASSE
FRANCE

THADDEUS PATRICK PROF
INSTITUTE FOR
SPACE STUDIES
2880 BROADWAY
NEW YORK NY 10025
U.S.A.

THE PIK SIN
ASTRONOMICAL INST
ROETERSSTRAAT 15
1018 WB AMSTERDAM
THE NETHERLANDS

THERNOE KARL-AUGUST
LEKTOR
VINKELVEJ 36
DK-2800 LYNGBY
DENMARK

THIRY YVES R PROF
UNIVERSITE DE PARIS
VI TOUR 66
4 PLACE JUSSIEU
F-75230 PARIS CDX 05
FRANCE

THOMAS DAVID V DR
SCIENCE RES COUNCIL
CENTRAL OFFICE
P O BOX 18
SWINDON SN2 1ET WILT
U.K.

THOMAS HANS-CHRISTOPH DR
MPI FUER PHYSIK UND
ASTROPHYSIK
FOEHRINGER RING 6
D-8000 MUENCHEN 40
GERMANY, F.R.

THOMAS JOHN A DR
PHYSICS RAAF DEPT
UNIV OF MELBOURNE
PARKVILLE VIC 3052
AUSTRALIA

THOMAS JOHN H PROF
DEPT MECH AERO SC
UNIVERSITY ROCHESTER
ROCHESTER NY 14627
U.S.A.

THOMAS RICHARD N DR
LAT
98 BIS BLVD ARAGO
F-75014 PARIS
FRANCE

THOMAS ROGER J DR
CODE 682
NASA GODDARD SPACE
FLIGHT CENTER
GREENBELT MD 20771
U.S.A.

THOMASSON PETER DR
N R A L
JODRELL BANK
MACCLESFIELD
CHESHIRE SK11 9DL
U.K.

THOMPSON A RICHARD DR
NAT RADIO ASTRO OBS
VLA PROJECT
P O BOX 0
SOCORRO NM 87801
U.S.A.

THOMPSON G I DR
ROYAL OBSERVATORY
EDINBURGH EH9 3HJ
U.K.

THOMPSON LAIRD A DR
DEPT PHYSICS ASTRON
UNIV OF NEBRASKA
LINCOLN NB 68506
U.S.A.

THOMPSON ROGER I PROF
STEWARD OBSERVATORY
UNIVERSITY
OF ARIZONA
TUCSON AZ 85721
U.S.A.

THOMSEN BJARNE B PROF
INST OF ASTRONOMY
UNIVERSITY OF AARHUS
DK-8000 AARHUS C
DENMARK

THONNARD NORBERT DR
CARNEGIE INST WASH
DTM
5241 BROAD BR RD NW
WASHINGTON DC 20015
U.S.A.

THOREN VICTOR E PROF
130 GOODBODY HALL
INDIANA UNIVERSITY
BLOOMINGTON IN 47401
U.S.A.

THORNE KIP S PROF
130-33
CALTECH
PASADENA CA 91125
U.S.A.

THUAN TRINH X DR
DEPT OF ASTRONOMY
UNIV OF VIRGINIA
BOX 3818 UNIV STA
CHARLOTTESVL VA22903
U.S.A.

TIFFT WILLIAM G PROF
STEWARD OBSERVATORY
UNIVERSITY
OF ARIZONA
TUCSON AZ 85721
U.S.A.

TIFREA EMILIA DR
OBS DE BUCAREST
STR CUTITUL DE
ARGINT 5
75212 BUCAREST
ROUMANIA

TIMOTHY J GETHYN DR
LAB FOR ATMOSPHERIC
AND SPACE PHYSICS
UNIV OF COLORADO
BOULDER CO 80309
U.S.A.

TINBERGEN JAAP DR
STERREWACHT
POSTBUS 9513
2300 RA LEIDEN
THE NETHERLANDS

TING YEOU-TSWEN
ASTRONOMY SECTION
CENTRAL WEATHER BUR
64 KUNG YUEN ROAD
TAIPEI 100
TAIWAN

TINSLEY BEATRICE M PROF
YALE UNIVERSITY OBS
BOX 2023
YALE STATION
NEW HAVEN CT 06520
U.S.A.

TIURI MARTTI PROF
HELSINKI UNIV TECHN
RADIO LABORATORY
OTAKAARI 5 A
SF-02150 ESPOO 15
FINLAND

TJIN-A-DJIE HERMAN R E DR
ASTRONOMICAL INST
UNIVERSITY AMSTERDAM
ROETERSSTRAAT 15
AMSTERDAM 1004
THE NETHERLANDS

TLAMICHA ANTONIN DR
ASTRONOMICAL
OBSERVATORY
25165 ONDREJOV
CZECHOSLOVAKIA

TODORAN IOAN DR
ASTRONOMICAL
OBSERVATORY
STR REPUBLICII 109
3400 CLUJ-NAPOCA
ROUMANIA

TOFANI GIANNI PROF
OSSERVATORIO ARCETRI
LARGO FERMI 5
I-50125 FIRENZE
ITALY

TOLBERT CHARLES R DR
LEANDER MCCORMICK OB
BOX 3818
UNIV STATION
CHARLOTTESVL VA22901
U.S.A.

TOMASI PAOLO DR
LAB RADIOASTRONOMIA
VIA IRNERIO 46
I-40126 BOLOGNA
ITALY

TOMASKO MARTIN G DR
LUNAR PLANETARY LAB
UNIV OF ARIZONA
SPACE SCIENCES BLDG
TUCSON AZ 85721
U.S.A.

TOMBAUGH CLYDE W PROF
DEPT OF ASTRONOMY
BOX 4500
NM STATE UNIVERSITY
LAS CRUCES NM 88003
U.S.A.

TOMITA KENJI PROF
RESEARCH INSTITUTE F
THEORETICAL PHYSICS
HIROSHIMA UNIVERSITY
TAKEHARA 725
JAPAN

TOMITA KOICHIRO MR
TOKYO ASTRONOMICAL
OBSERVATORY
OSAWA MITAKA
TOKYO 181
JAPAN

TOOMRE ALAN DR
ROOM 2-371
MIT
77 MASSACHUSETTS AVE
CAMBRIDGE MA 02139
U.S.A.

TOOMRE JURI
DEPT ASTRO-GEOPHY-
SICS AND JILA
UNIV OF COLORADO
BOULDER CO 80309
U.S.A.

TOPAKTAS LATIF A DR
UNIVERSITY
OBSERVATORY
UNIVERSITY ISTANBUL
TURKEY

TORAO MASAHISA PROF
TOKYO SCIENCE UNIV
1-3 KAGURAZAKA
SHINJYUKU
TOKYO 162
JAPAN

TORELLI M DR
OSS ASTRONOMICO
VIA DEL P MELLINI 84
ROMA
ITALY

TOROSHLIDZE TEIMURAZ I DR
ABASTUMANI ASTRO-
PHYSICAL OBSERVATORY
383762 GEORGIA
U.S.S.R.

TORRES CARLOS ALBERTO DR
OBS NACIONAL
R GENERAL BRUCE 586
20000 RIO DE JANEIRO
BRAZIL

TORRES CARLOS DR
OBS ASTRON NACIONAL
UNIVERSIDAD DE CHILE
CASILLA 36-D
SANTIAGO
CHILE

TORRES PEIMBERT SILVIA DR
INST DE ASTRONOMIA
APDO POSTAL 70264
MEXICO 20 D F
MEXICO

TORROJA J PROF
CATEDRA DE ASTRON
FACULTAD DE CIENCIAS
UNIV COMPLUTENSE
MADRID 3
SPAIN

TOSA MAKOTO DR
DEPT OF PHYSICS
NAGOYA UNIVERSITY
FURUCHO CHIGUSA
NAGOYA 464
JAPAN

TOSHIAKI MASUURA OSCAR DR
RUA ITACEMA
APTO 43
SAO PAULO CEP 04530
BRAZIL

TOUSEY RICHARD DR
CODE 7140
NAVAL RESEARCH LAB
WASHINGTON DC 20375
U.S.A.

TOVMASSIAN H M DR
BYURAKAN ASTROPHYSI-
CAL OBSERVATORY
378433 ARMENIA
U.S.S.R.

TOWNES CHARLES HARD DR
DEPT OF PHYSICS
UNIVERSITY OF CALIF
RM 557 BIRGE HALL
BERKELEY CA 94720
U.S.A.

TOZER DAVID C DR
SCHOOL OF PHYSICS
UNIVERSITY NEWCASTLE
NEWCASTLE UPON TYNE
NE1 7RU
U.K.

TRAFTON LAURENCE M DR
ASTRONOMY DEPT
UNIVERSITY OF TEXAS
AT AUSTIN
AUSTIN TX 78712
U.S.A.

TRAN MINH N DR
OBS DE MEUDON
F-92190 MEUDON
FRANCE

TRAVING GERHARD PROF
INSTITUT FUER THEORE
TISCHE ASTROPHYSIK
NEUENHEIMER FELD 294
D-6900 HEIDELBERG
GERMANY, F.R.

TREDER H J PROF DR
ZNTRINST ASTROPHYSIK
STERNW BABELSBERG
ROSA LUXEMBURGST 17A
DDR-1502 POTSDAM
G.D.R.

TREFFTZ ELEONORE E DR
MPI F PHYSIK UND
ASTROPHYSIK
FOEHRINGER RING 6
D-8000 MUENCHEN 40
GERMANY, F.R.

TREFZGER CHARLES F DR
ASTRON INST BASEL
VENUSSTRASSE 7
CH-4102 BINNINGEN
SWITZERLAND

TREHAN SURINDAR K PROF
DEPT OF MATHEMATICS
PANJAB UNIVERSITY
CHANDIGARH 160014
INDIA

TRELLIS MICHEL DR
OBSERVATOIRE DE NICE
B P 252
F-06007 NICE CEDEX
FRANCE

TREMKO JOZEF DR
SKALNATE PLESO
OBSERVATORY
059 60 TATRANSKA
LOMNICA
CZECHOSLOVAKIA

TREUMANN RUDOLF DR
HOCHSTRASSE 58
CH-8044 ZUERICH
SWITZERLAND

TREXLER JAMES H MR
5609 OTTAWA STREET
LABORATORY
OXON HILL MD 20021
U.S.A.

TRIMBLE VIRGINIA L DR
DEPT OF PHYSICS
UNIVERSITY OF CALIF
IRVINE CA 92717
U.S.A.

TRITAKIS BASIL P DR
ASTRONOMY APPL MATHS
ACADEMY OF ATHENS
14 ANAGNOSTOPOULOU
ATHENS 136
GREECE

TRITTON KEITH P DR
ROYAL OBS EDINBURGH
BLACKFORD HILL
EDINBURGH EH9 3HJ
U.K.

TROITSKY V S PROF DR
RADIOPHYSICAL
RESEARCH INSTITUTE
603600 GORKIJ
U.S.S.R.

TRUEMPER JOACHIM PROF
MPI FUER EXTRATERRE-
STRISCHE PHYSIK
D-8046 GARCHING BEI
MUENCHEN
GERMANY, F.R.

TRULSEN JAN K PROF
P O BOX 953
UNIVERSITY OF TROMSO
N-9001 TROMSO
NORWAY

TRURAN JAMES W JR
DEPT OF ASTRONOMY
UNIV OF ILLINOIS
OBSERVATORY
URBANA IL 61801
U.S.A.

TRUTSE YU L DR
INSTITUTE OF PHYSICS
OF ATMOSPHERE
USSR ACADEMY OF SCI
109017 MOSCOW
U.S.S.R.

TSAI CHANG-HSIEN DIRECTOR
GENERAL SECRETARY
ASTRO SOCIETY CHINA
TAIPEI OBSERVATORY
TAIPEI 104
TAIWAN

TSAO M PROF
NO 47 SEC 3
HSIN-I ROAD
TAIPEI 106
TAIWAN

TSAP T T DR
CRIMEAN ASTROPHYSI-
CAL OBSERVATORY USSR
ACAD OF SCI NAUCHNIY
334413 CRIMEA
U.S.S.R.

TSCHARNUTER WERNER M DR
MPI FUER ASTROPHYSIK
FOEHRINGER RING 6
D-8000 MUENCHEN 40
GERMANY, F.R.

TSESEVICH V P PROF DR
ODESSA STATE UNIVER-
SITY ASTRONOMICAL
OBSERVATORY
270014 ODESSA
U.S.S.R.

TSIKOUDI VASSILIKI PH D
DEPT OF ASTRONOMY
UNIV OF IOANNINA
IOANNINA
GREECE

TSIOUMIS ALEXANDROS DR
DEPT OF GEODETIC
ASTRONOMY
UNIV OF THESSALONIKI
THESSALONIKI
GREECE

TSUBAKI TOKIO PROF
FACULTY OF EDUCATION
SHIGA UNIVERSITY
ISHIYAMAHIRATSUCHO
OHTSU 520
JAPAN

TSUBOKAWA IETSUNE DR
INTERNATIONAL LATITU
DE OBSERVATORY
MIZUSAWA IWATE 023
JAPAN

TSUCHIYA ATSUSHI DR PROF
TOKYO ASTRONOMICAL
OBSERVATORY
OSAWA
OSAWA MITAKA TOK 181
JAPAN

TSUJI TAKASHI
TOKYO ASTRONOMICAL
OBSERVATORY
MITAKA
TOKYO 181
JAPAN

TSURUTA SACHIKO DR
MPI PHYS UND ASTROPH
FOEHRINGER RING 6
POSTFACH 401212
D-8000 MUENCHEN 40
GERMANY, F.R.

TSVETKOV TSVETAN DR
DEPT OF ASTRONOMY
FACULTY OF PHYSICS
ANTON IVANOV STR 5
1126 SOFIA
BULGARIA

TUCKER ROY H
ROYAL GREENWICH
OBSERVATORY
HAILSHAM
EAST SUSSEX BN27 1RP
U.K.

TUCKER WALLACE H DR
CTR FOR ASTROPHYSICS
60 GARDEN STREET
CAMBRIDGE MA 02138
U.S.A.

TUFEKCIOGLU ZEKI DR
ASTRONOMY DEPT
UNIV OF ANKARA
FEN FAKULTESI
ANKARA
TURKEY

TULL ROBERT G
DEPT OF ASTRONOMY
UNIVERSITY OF TEXAS
AT AUSTIN RLM 15 212
AUSTIN TX 78712
U.S.A.

TULLY JOHN A DR
OBSERVATOIRE DE NICE
BP 252
F-06007 NICE CEDEX
FRANCE

TULLY RICHARD BRENT DR
UNIVERSITY OF HAWAII
INST OF ASTRONOMY
2680 WOODLAWN DR
HONOLULU HI 96822
U.S.A.

TUOMINEN ILKKA V DR
OBSERVATORY
TAHTITORNINMAKI
SF-00130 HELSINKI 13
FINLAND

TUOMINEN JAAKKO V PROF
PIHLAJATIE 49 B 20
SF-00270 HELSINKI 27
FINLAND

TURLO ZYGMUNT DR
ASTRONOMICAL CENTER
UL CHOPINA 12-18
87-100 TORUN
POLAND

TURNER BARRY E DR
NRAO
EDGEMONT RD
CHARLOTTESVL VA22901
U.S.A.

TURNER DAVID G DR
DAVID DUNLAP OBS
UNIV OF TORONTO
RICHMOND HL L4C 4Y6
CANADA

TURNER EDWIN L DR
PRINCETON UNIV OBS
PEYTON HALL
PRINCETON NJ 08540
U.S.A.

TURNER KENNETH C DR
DTM CARNEGIE INSTITU
TION OF WASHINGTON
5241 BROAD BR RD NW
WASHINGTON DC 20015
U.S.A.

TURNER MARTIN J L DR
X-RAY ASTRONOMY GR
PHYSICS DEPT
UNIV OF LEICESTER
LEICESTER LE1 7RH
U.K.

TURON PIERRE
INST D'ASTROPHYSIQUE
98 BIS BD ARAGO
F-75014 PARIS
FRANCE

TURTLE A J DR
DEPT OF PHYSICS
UNIVERSITY OF SYDNEY
SYDNEY NSW 2006
AUSTRALIA

TUTUKOV A V DR
ASTRONOMICAL COUNCIL
USSR ACADEMY OF SCI
PYATNITSKAYA UL 48
109017 MOSCOW
U.S.S.R.

TUVE MERLE A DR
CARNEGIE INSTITUTION
5241 BROAD BRANCH RD
WASHINGTON DC 20015
U.S.A.

TUZI KENOSUKE DR
2-2-20 OSAWA
MITAKA
TOKYO 181
JAPAN

TWISS R Q DR
NATIONAL PHYSICAL
LABORATORY
TEDDINGTON MIDDLESEX
U.K.

TYLER JR G L DR
CTR FOR RADAR ASTRON
STANFORD UNIV
STANFORD CA 94305
U.S.A.

TYSON JOHN A DR
BELL LABS RM 1D-316
600 MOUNTAIN AVE
MURRAY HILL NJ 07974
U.S.A.

TZVETKOV MILCHO K DR
BULGARIAN ACADEMY
OF SCIENCES
DEPT OF ASTRONOMY
1000 SOFIA
BULGARIA

UCHIDA JUICHI DR
TOHOKU GAKUEN
UNIVERSITY
TAGAJYOMACHI
MIYAGI 985
JAPAN

UCHIDA YUTAKA DR
TOKYO ASTRONOMICAL
OBSERVATORY
OSAWA
MITAKA TOKYO 181
JAPAN

UDAL'TSOV V A DR
PHYSICAL INSTITUTE
USSR ACADEMY OF
SCIENCES
117924 MOSCOW
U.S.S.R.

UENO SUEO PROF
KANAZAWA INSTITUTE
OF TECHNOLOGY
NONOICHIMACHI
ISHIKAWA 921
JAPAN

UESUGI AKIRA DR
DEPT OF ASTRONOMY
UNIVERSITY OF KYOTO
SAKYOKU KYOTO 606
JAPAN

ULFBECK OLE DR
NIELS BOHR INST
BLEGDAMSVEJ 17
DK-2100 COPENHAGEN-O
DENMARK

ULMER MELVILLE P PROF
DEARBORN OBSERVATORY
NORTHWESTERN
UNIVERSITY
EVANSTON IL 60201
U.S.A.

ULMSCHNEIDER PETER DR
INST FUER ASTRONOMIE
UND ASTROPHYSIK
AM HUBLAND
D-8700 WUERZBURG
GERMANY, F.R.

ULRICH BRUCE T PROF
ESO
C/O CERN
CH-1211 GENEVA 23
SWITZERLAND

ULRICH MARIE HELENE D DR
ESO C O CERN
CH-1211 GENEVA 23
SWITZERLAND

ULRICH ROGER K PROF
DEPT OF ASTRONOMY
UNIV OF CALIFORNIA
8931 MSB
LOS ANGELES CA 90024
U.S.A.

UNDERHILL ANNE B DR
CODE 680
GODDARD SPACE FL CTR
GREENBELT MD 20771
U.S.A.

UNDERWOOD JAMES H DR
INSTITUTE F PLASMA
RESEARCH
STANFORD UNIVERSITY
STANFORD CA 94305
U.S.A.

UNNO WASABURO PROF
DEPT OF ASTRONOMY
UNIVERSITY OF TOKYO
BUNKYO-KU
TOKYO 113
JAPAN

UNSOELD ALBRECHT PROF
INST THEO PHY STERNW
NEUE UNIV PHYSIK ZNT
OLSHAUSENST GB N 61C
D-2300 KIEL 1
GERMANY, F.R.

UPGREN ARTHUR R DR
VAN VLECK
OBSERVATORY
WESLEYAN UNIVERSITY
MIDDLETOWN CT 06457
U.S.A.

UPSON WALTER L II DR
PRINCETON UNIV OBS
PEYTON HALL
PRINCETON NJ 08540
U.S.A.

UPTON E K L DR
ASTRONOMY DEPT
UCLA
LOS ANGELES CA 90024
U.S.A.

URAS S DR
ISTITUTO DI
ASTRONOMIA
VIA OSPEDALE 72
I-09100 CAGLIARI
ITALY

URASIN L A DR
ENGELGARDT ASTRONOMI
CAL OBSERVATORY
422526 KAZAN
U.S.S.R.

URBARZ H DR
ASTRONOMISCHES INST
DER UNIVERSITAET
WALDHAUSER STR 64
D-7400 TUEBINGEN
GERMANY, F.R.

URECHE VASILE DR
UNIVERSITY OF CLUJ
NAPOCA
3400 CLUJ NAPOCA
ROUMANIA

UREY HAROLD C DR
7890 TORREY LANE
LA JOLLA CA 92037
U.S.A.

USHER PETER D DR
DEPT ASTRONOMY
PENNSYLVANIA ST UNIV
507 DAVEY LAB
UNIV PARK PA 16802
U.S.A.

UTSUMI KAZUHIKO DR
FAC OF GEN EDUCATION
HIROSHIMA UNIVERSITY
HIGASHI-SENDACHO
HIROSHIMA 730
JAPAN

UUS UNDO DR
IAPHA ESTONIAN ACAD
OF SCIENCE
TORAVERE
202444 TARTU
U.S.S.R.

VAGER ZEEV DR
DEPT OF PHYSICS
WEIZMANN INSTITUTE
REHOVOT
ISRAEL

VAGHI SERGIO DR
EUROPEAN SPACE
OPERATION CENTRE
ROBERT-BOSCH STR 5
D-6100 DARMSTADT
GERMANY, F.R.

VAIANA GIUSEPPE S DR
HARVARD COLLEGE OBS
HARVARD UNIVERSITY
60 GARDEN ST
CAMBRIDGE MA 02138
U.S.A.

VAIDYA P C PROF
VICE CHANCELLOR
GUJARAT UNIVERSITY
AHMEDABAD 380009
INDIA

VAINSTEIN L A DR
PHYSICAL INSTITUTE
USSR ACADEMY OF
SCIENCES
117924 MOSCOW
U.S.S.R.

VALBOUSQUET ARMAND DR
OBSERVATOIRE
DE STRASBOURG
11 RUE L UNIVERSITE
F-67000 STRASBOURG
FRANCE

VALLEE JACQUES P DR
ASTRONOMY GROUP
DEPT OF PHYSICS
QUEEN'S UNIVERSITY
KINGSTON ONT K7L 3N6
CANADA

VALNICEK BORIS DR
ASTRONOMICAL INST
CZECHO ACADEMY SC
OBSERVATORY
251 65 ONDREJOV
CZECHOSLOVAKIA

VALTIER J C DR
OBS DE NICE
B P 252
F-06007 NICE CEDEX
FRANCE

VALTONEN MAURI J DR
DEPT PHYSICS/ASTRON
UNIV OF ALABAMA
PO BOX 1921
UNIVERSITY AL 35486
U.S.A.

VAN AGT S L TH J DR
STERRENKUNDIG INST
TOERNOOIVELD
6525 ED NIJMEGEN
THE NETHERLANDS

VAN ALBADA TJEERD S DR
KAPTEYN LABORATORIUM
POSTBUS 800
GRONINGEN
THE NETHERLANDS

VAN ALLEN JAMES A PROF
DEPT OF PHYSICS AND
ASTRONOMY
UNIVERSITY OF IOWA
IOWA CITY IA 52242
U.S.A.

VAN ALTENA WILLIAM F PROF
YALE UNIVERSITY
OBSERVATORY BOX 2023
YALE STATION
NEW HAVEN CT 06520
U.S.A.

VAN BEEK F DR
LAB VOOR RUIMTE-
ONDERZOEK
BENELUXLAAN 21
UTRECHT
THE NETHERLANDS

VAN BLERKOM DAVID J PROF
ASTRONOMY DEPT
UNIVERSITY OF MASS
AMHERST MA 01002
U.S.A.

VAN BREDA IAN G DR
ROYAL GREENWICH
OBSERVATORY
HAILSHAM
EAST SUSSEX BN27 1RP
U.K.

VAN BUEREN HENDRIK G PROF
STERREKUNDIG INST
TE UTRECHT
ZONNENBURG 2
UTRECHT
THE NETHERLANDS

VAN CITTERS GORDON W DR
DEPT OF ASTRONOMY
UNIV OF TEXAS
AUSTIN TX 78712
U.S.A.

VAN DE HULST H C PROF DR
STERREWACHT
POSTBUS 9513
2300 RA LEIDEN
THE NETHERLANDS

VAN DE KAMP PETER
15 WELLESLEY ROAD
SWARTHMORE PA 19081
U.S.A.

VAN DE STADT HERMAN DR
ASTRONOMICAL
INSTITUTE
ZONNENBURG 2
3512 NL UTRECHT
THE NETHERLANDS

VAN DEN BERGH SIDNEY PROF
DIRECTOR DOMINION
ASTROPHYSICAL OBS
5071 W SAANICH ROAD
VICTORIA BC V8X 3X3
CANADA

VAN DEN HEUVEL EDWARD P J
ASTRONOMICAL INST
ROETERSSTRAAT 15
AMSTERDAM 1018
THE NETHERLANDS

VAN DER BORGHT RENE PROF
DEPT OF MATHEMATICS
MONASH UNIVERSITY
CLAYTON VICT 3168
AUSTRALIA

VAN DER HUCHT KAREL A DR
SPACE RESEARCH LAB
BENELUXLAAN 21
3527 HS UTRECHT
THE NETHERLANDS

VAN DER KRUIT PIETER C DR
KAPTEYN LABORATORIUM
POSTBUS 800
GRONINGEN
THE NETHERLANDS

VAN DER LAAN H PROF DR
STERREWACHT
POSTBUS 9513
2300 RA LEIDEN
THE NETHERLANDS

VAN DESSEL EDWIN LUDO DR
KON STERRENWACHT
RINGLAAN 3
B-1180 BRUSSELS
BELGIUM

VAN DIGGELEN J DR
STERREWACHT
ZONNENBURG 2
3512 NL UTRECHT
THE NETHERLANDS

VAN DORN BRADT H DR
MIT RM 37-581
CTR FOR SPACE RES
CAMBRIDGE MA 02139
U.S.A.

VAN DUINEN R DR
SPACE RESEARCH DEPT
P O BOX 800
GRONINGEN
THE NETHERLANDS

VAN FLANDERN THOMAS DR
US NAVAL OBSERVATORY
WASHINGTON DC 20390
U.S.A.

VAN GENDEREN A.M. DR.
STERRENWACHT LEIDEN
POSTBUS 9513
2300 RA LEIDEN
THE NETHERLANDS

VAN HERK G
STERREWACHT VAN
LEIDEN
POSTBUS 9513
2300 RA LEIDEN
THE NETHERLANDS

VAN HERK-KLUYVER H A DR
KORTE WAARDER 58
2415 AV NIEUWERBRUG
THE NETHERLANDS

VAN HOOF A PROF EM
ASTRONOMISCH
INSTITUUT KU LEUVEN
NAAMSESTRAAT 61
B-3000 LEUVEN
BELGIUM

VAN HORN HUGH M PROF
DEPT PHYSICS AND
ASTRONOMY
UNIVERSITY ROCHESTER
ROCHESTER NY 14627
U.S.A.

VAN HOUTEN C J DR
STERREWACHT
POSTBUS 9513
2300 RA LEIDEN
THE NETHERLANDS

VAN HOUTEN-GROENEVELD
INGRID DR
STERREWACHT LEIDEN
POSTBUS 9513
2300 RA LEIDEN
THE NETHERLANDS

VAN HOVEN GERARD DR
DEPT OF PHYSICS
UNIV OF CALIFORNIA
IRVINE CA 92717
U.S.A.

VAN LEER B DR
STERREWACHT
POSTBUS 9513
2300 RA LEIDEN
THE NETHERLANDS

VAN NIEUWKOOP J DR IR
STERREWACHT
SONNENBORGH
ZONNENBURG 2
3512 WK UTRECHT
THE NETHERLANDS

VAN PARADIJS J DR
ASTRONOMICAL INST
ROETERSSTRAAT 15
AMSTERDAM 1018
THE NETHERLANDS

VAN REGEMORTER HENRI DR
OBSERVATOIRE
DE MEUDON
F-92190 MEUDON
FRANCE

VAN RENSBERGEN WALTER DR
ASTROPHYSISCH INST
VRIJE UNIVERSITEIT
PLEINLAAN 2
B-1050 BRUSSEL
BELGIUM

VAN RIPER KENNETH A DR
DEPT OF PHYSICS
UNIV OF ILLINOIS
URBANA IL 61801
U.S.A.

VAN SPEYBROECK LEON P DR
CTR FOR ASTROPHYSICS
60 GARDEN ST
CAMBRIDGE MA 02138
U.S.A.

VAN T VEER FRANK DR
INST D ASTROPHYSIQUE
98 BIS BLVD ARAGO
F-75014 PARIS
FRANCE

VAN T VEER-MENNERET
CLAUDE DR
INST D ASTROPHYSIQUE
98 BIS BLVD ARAGO
F-75014 PARIS
FRANCE

VAN WOERDEN HUGO DR
KAPTEYN LABORATORIUM
POSTBUS 800
GRONINGEN
THE NETHERLANDS

VANDEN BOUT PAUL A
ASTRONOMY DEPT
RLM 15 212
UNIVERSITY OF TEXAS
AUSTIN TX 78712
U.S.A.

VANDERLINDEN HENRI L PROF
UNIVERSITEIT GENT
HERTOGENDREEF 15
B-1170 BRUSSEL
BELGIUM

VANDERVOORT PETER O DR
DEPT ASTRON ASTROPH
1100-14 E 58TH ST
CHICAGO IL 60637
U.S.A.

VANYSEK VLADIMIR PROF
ASTRONOMY DEPT
CHARLES UNIVERSITY
SVEDSKA 8
150 00 PRAHA 5
CZECHOSLOVAKIA

VAPILLON LOIC J DR
OBSERVATOIRE DE
PARIS-MEUDON
F-92190 MEUDON
FRANCE

VARDANIAN R A DR
BYURAKAN ASTROPHYSI-
CAL OBSERVATORY
378433 ARMENIA
U.S.S.R.

VARDYA M S DR
TATA INSTITUTE OF
FUNDAMENTAL RESEARCH
HOMI BHABHA RD
BOMBAY 400 005
INDIA

VARSAVSKY C M DR
LAS HERAS 1975 6 A
BUENOS AIRES
ARGENTINA

VARSHALOVICH D A DR
PHYSIKO-TECHNICAL
INSTITUTE USSR ACADE
MY OF SCIENCES
194021 LENINGRAD
U.S.S.R.

VASHKOV'YAK S N DR
STERNBERG STATE AS-
TRONOMICAL INSTITUTE
117234 MOSCOW
U.S.S.R.

VASILEV VLADIMIR M DR
PULKOVO OBSERVATORY
196140 LENINGRAD
U.S.S.R.

VASILEVA GALINA J DR
PULKOVO OBSERVATORY
196140 LENINGRAD
U.S.S.R.

VASILEVSKIS STANISLAUS PROF
455 GRAND AVE NO 16
PALO ALTO CA 94306
U.S.A.

VAUCLAIR GERARD P DR
OBSERVATOIRE DE PARIS-MEUDON
F-92190 MEUDON
FRANCE

VAUCLAIR SYLVIE D DR
OBSERVATOIRE DE PARIS-MEUDON
F-92190 MEUDON
FRANCE

VAUGHAN ARTHUR H DR
HALE OBSERVATORIES
813 SANTA BARBARA ST
PASADENA CA 91101
U.S.A.

VAZQUEZ MANUEL DR
INST ASTROFISICA DE CANARIAS
CANARIAS
SPAIN

VEEDER GLENN J DR
JPL
4800 OAK GROVE DRIVE
PASADENA CA 91103
U.S.A.

VEIS GEORGE PH D
GEODESY LAB
NAT TECHNICAL UNIV
ATHENS
GREECE

VEISMANN UNO DR
TARTU OBSERVATORY
202444 ESTONIA
U.S.S.R.

VELGHE ALBERT G PROF DR
OBSERVATOIRE ROYAL DE BELGIQUE
AVENUE CIRCULAIRE 3
B-1180 BRUSSELS
BELGIUM

VELTMANN ULO ILMAR DR
TARTU OBSERVATORY
202444 ESTONIA
U.S.S.R.

VELUSAMY T DR
RADIO ASTRONOMY CTR
P O BOX NO 8
OOTACAMUND 643 001
INDIA

VENUGOPAL V R DR
RADIO ASTRONOMY CENTRE OF TIFR
POST BOX 8
OOTACAMUND 643001
INDIA

VERBEEK PAUL DR
GEORGE MINNELAAN 50
9830 S MARTENS-LATEM
BELGIUM

VERDET JEAN-PIERRE DR
OBSERVATOIRE DE PARIS
61 AVE OBSERVATOIRE
F-75014 PARIS
FRANCE

VERGNANO A PROF
CSS ASTRONOMICO
10025 PINO TORINESE
ITALY

VERNIANI FRANCO PROF
CNR METEOR PHYSICS
VIA DEI CASTAGNOLI 1
I-40126 BOLOGNA
ITALY

VERON MARIE-PAULE DR
OBSERVATOIRE DE MEUDON
F-92190 MEUDON
FRANCE

VERON PHILIPPE DR
OBSERVATOIRE DE MEUDON
F-92190 MEUDON
FRANCE

VERSCHUUR GERRIT L PROF
FISKE PLANETARIUM
UNIVERSITY OF COLORADO
BOULDER CO 80302
U.S.A.

VESECKY J F DR
DEPT OF ASTRONOMY
THE UNIVERSITY
LEICESTER LE1 7RH
U.K.

VETESNIK MIROSLAV DR
DEPT OF ASTRONOMY
PURKYNE UNIVERSITY
KOTLARSKA 2
61137 BRNO
CZECHOSLOVAKIA

VEVERKA JOSEPH DR
312 SPACE SC BLDG
CORNELL UNIVERSITY
ITHACA NY 14853
U.S.A.

VICENTE RAIMUNDO O PROF
R MESTRE AVIZ 30 R C
P 1495 LISBOA
PORTUGAL

VIDAL J L DR
OBS DU PIC-DU-MIDI
F-65200 BAGNERES DE BIGORRE
FRANCE

VIDAL N V DR
DEPT OF PHYSICS AND ASTRONOMY
TEL-AVIV UNIVERSITY
TEL-AVIV
ISRAEL

VIDAL-MADJAR ALFRED DR
L P S P
BP 10
F-91370 VERRIERES LE BUISSON
FRANCE

VIGIER JEAN-PIERRE DR.
INSTITUT H.POINCARE
11 RUE P.ET M.CURIE
F-75005 PARIS
FRANCE

VIGOTTI MARIO
LABORATORIO DI RADIOASTRONOMIA
VIA IRNERIO 46
I-40126 BOLOGNA
ITALY

VIIK TONU DR
TORAVERE OBS
202444 TARTU RAJ EST
U.S.S.R.

VILA SAMUEL C PROF
DEPT OF ASTRONOMY
UNIVERSITY OF PENNA
33RD AND WALNUT STR
PHILADELPHIA PA19104
U.S.A.

VILHENA DE MORAES R DR
DEPTO DE ASTRONOMIA
ITA-CTA
12 200 SAO JOSE DOS CAMPOS SAO PAULO
BRAZIL

VILHU OSMI DR
OBS AND ASTROPHY LAB
UNIV OF HELSINKI
TAHTITORNINMAKI
SF-00130 HELSINKI 13
FINLAND

VINER MELVYN R DR
ASTRONOMY GROUP
DEPT OF PHYSICS
QUEEN'S UNIVERSITY
KINGSTON ONT K7L 3N6
CANADA

VINTI JOHN P DR
RM W91-202
M I T
MEASUREMENT SYST LAB
CAMBRIDGE MA 02139
U.S.A.

VIOTTI ROBERTO DR
CNR ASTROFISICA SPAZIALE C P 67
I-00044 FRASCATI
ROMA
ITALY

VIRGOPIA N DR
OSS ASTRONOMICO
VIA TRIONFALE 204
I-00136 ROMA
ITALY

VISVANATHAN NATARAJAN DR
MT STROMLO
OBSERVATORY
WODEN PO ACT 2606
AUSTRALIA

VITINSKIJ YURIJ I DR
PULKOVO OBSERVATORY
196140 LENINGRAD
U.S.S.R.

VITON MAURICE DR
L A S
ALLEE PEIRESC
F-13012 MARSEILLE
FRANCE

VIVES THEODORO JOSE DR
OBS DE PIEDADE
UFMG
30000 BELO HORIZONTE
BRAZIL

VLACHOS DEMETRIUS G PROF
UNIV THESSALONIKI
DEPT OF GEODESY
THESSALONIKI
GREECE

VOELK HEINRICH J PROF
MPI FUER KERNPHYSIK
POSTFACH 103 980
D-6900 HEIDELBERG
GERMANY, F.R.

VOGT NIKOLAUS DR
EUROPEAN SOUTHERN
OBSERVATORY
SCHLEISSHEIMERSTR 17
D-8046 GARCHING B M
GERMANY, F.R.

VOIGT HANS H PROF
UNIVERSITAETS-
STERNWARTE
GEISMARLANDSTR 11
D-3400 GOETTINGEN
GERMANY, F.R.

VOLLAND H DR
ASTRONOMISCHES INST
DER UNIVERSITAET
AUF DEM HUEGEL 71
D-5300 BONN
GERMANY, F.R.

VOLONTE SERGE DR
19 RUE CAPORAL
TRESIGNIES
B-7020 HYON
BELGIUM

VON BORZESZKOWSKI H H DR
ZENTR INST ASTROPHYS
DER ADW DER DDR
ROSA-LUXEMBURGST 17A
DDR-1502 POTSDAM
G.D.R.

VON DER HEIDE JOHANN DR
ALARDUSSTR 12
D-2000 HAMBURG 19
GERMANY, F.R.

VON HOERNER SEBASTIAN DR
NATIONAL RADIO ASTRO
NOMY OBSERVATORY
P O BOX 2
GREEN BANK WV 24944
U.S.A.

VON SOCHER HERMANN DR
INST FUER ASTRONOMIE
DER UNIV WIEN
TUERKENSCHANZSTR 17
A-1180 WIEN
AUSTRIA

VON WEIZSAECKER C F
PROF DR
MAX-PLANCK-INSTITUT
RIEMERSCHMID STR 7
D-8130 STARNBERG
GERMANY, F.R.

VONDRAK JAN DR
ASTRONOMICAL INST
CSAV
BUDECSKA 6
120 23 PRAHA 2
CZECHOSLOVAKIA

VORONTSOV-VEL'YAMINOV B A
PROF DR
STERNBERG STATE AS-
TRONOMICAL INSTITUTE
117234 MOSCOW
U.S.S.R.

VOROSHILOV V I DR
MAIN ASTRONOMICAL
OBSERVATORY UKRAINI-
AN ACAD OF SCIENCES
252127 KIEV
U.S.S.R.

VORPAHL JOAN A DR
AEROSPACE
CORPORATION
PO BOX 92957
LOS ANGELES CA 90009
U.S.A.

VRBA FREDERICK J DR
U S NAVAL OBS
P O BOX 1149
FLAGSTAFF AZ 86002
U.S.A.

VREUX JEAN MARIE DR
INST D ASTROPHYSIQUE
UNIVERSITE DE LIEGE
B-4200 COINTE OUGREE
BELGIUM

VSEKHSVYATSKIJ SERGIUS K
PROF
KIEV ASTRONOMICAL
OBSERVATORY
252053 KIEV PER 3
U.S.S.R.

VU D DR
BUREAU D LONGITUDES
77 AVENUE DENFERT
ROCHEREAU
F-75014 PARIS
FRANCE

VUJNOVIC VLADIS DR
INSTITUTE OF PHYSICS
OF THE UNIVERSITY
P O B 304
41001 ZAGREB
YUGOSLAVIA

VUKICEVIC K M PROF DR
DEPT OF ASTRONOMY
FACULTY OF SCIENCES
STUDENTSKI TRG 16
11000 BEOGRAD
YUGOSLAVIA

VYALSHIN GENNADIJ F DR
MAIN ASTRONOMICAL OB
PULKOVO M-140
196140 LENINGRAD
U.S.S.R.

WACHMANN A A PROF DR
SCHMIEDESBERG 2 B
D-2057 REINBEK
GERMANY, F.R.

WACKERNAGEL H BEAT DR
2939 COUNTRY CLUB DR
COLORADO SPRINGS
CO 80909
U.S.A.

WADDINGTON C JAKE PROF
SCHOOL PHYS ASTRON
UNIV MINNESOTA
116 CHURCH ST S.E.
MINNEAPOLIS MN 55455
U.S.A.

WADE CAMPBELL M DR
NAT RADIO ASTRON OBS
P O BOX 0
SOCORRO NM 87801
U.S.A.

WAGMAN NICHOLAS E PROF
3 ALLEGHENY CTR
APT 120
PITTSBURGH PA 15212
U.S.A.

WAGNER RAYMOND L DR
FORD AEROSPACE
2880 E.FOUNTAIN BLVD
F-10
COLO SPGS CO 80910
U.S.A.

WAGNER WILLIAM J DR
HIGH ALTITUDE
OBSERVATORY
P O BOX 3000
BOULDER CO 80307
U.S.A.

WAGONER ROBERT V PROF
DEPT OF PHYSICS
STANFORD UNIVERSITY
STANFORD CA 94305
U.S.A.

WAKAMATSU KEN-ICHI DR
JUNIOR TECHN COLLEGE
GIFU UNIVERSITY
KAGAMIGAHARA
GIFU 504
JAPAN

WAKO KOJIRO DR
INTERNATIONAL LATITU
DE OBSERVATORY
MIZUSAWA IWATE 023
JAPAN

WALBORN NOLAN R DR
CERRO TOLOLO INTER-
AMERICAN OBSERVATORY
CASILLA 63 D
LA SERENA
CHILE

WALDMEIER MAX PROF
SWISS FEDERAL OBS
SCHMELZBERGSTR 25
CH-8006 ZUERICH
SWITZERLAND

WALKER ARTHUR B C JR PROF
INST FOR PLASMA
RESEARCH
STANFORD UNIVERSITY
STANFORD CA 94305
U.S.A.

WALKER EDWARD N MR
ROYAL GREENWICH OBS
HERSTMONCEUX CASTLE
HAILSHAM
E SUSSEX BN27 1RP
U.K.

WALKER GORDON A H PROF
INST ASTRON SPACE SC
UNIVERSITY OF B C
2075 WESBROOK PLACE
VANCOUVER BC V6T 1W5
CANADA

WALKER MERLE F PROF
LICK OBSERVATORY
UNIVERSITY OF CALIF
SANTA CRUZ CA 95064
U.S.A.

WALKER R L
NAVAL OBSERVATORY
FLAGSTAFF STATION
BOX 1149
FLAGSTAFF AZ 86002
U.S.A.

WALKER ROBERT M A PROF
PHYSICS DEPT
BOX 1105
WASHINGTON UNIV
ST LOUIS MO 63130
U.S.A.

WALKER WILLIAM S G
14 APPLEYARD CRES
AUCKLAND 5
NEW ZEALAND

WALL JASPER V DR
MULLARD RADIO ASTRO-
OBS CAVENDISH LAB
MADINGLEY RD
CAMBRIDGE CB3 OHE
U.K.

WALLACE LLOYD V DR
KITT PEAK NATIONAL
OBSERVATORY
P O BOX 26732
TUCSON AZ 85726
U.S.A.

WALLACE PATRICK T DR
ANGLO-AUSTRALIAN OBS
P O BOX 296
EPPING N S W 2121
AUSTRALIA

WALLENQUIST AAKE A E PROF
NORRLANDSGATAN 34 D
S-752 29 UPPSALA
SWEDEN

WALLERSTEIN GEORGE PROF
ASTRONOMY DEPT FM 20
UNIVERSITY OF
WASHINGTON
SEATTLE WA 98195
U.S.A.

WALLIS MAX K. DR
DEPT APP MATH ASTRON
UNIVERSITY COLLEGE
P O BOX 78
CARDIFF CF1 1XL
U.K.

WALMSLEY C MALCOLM DR
MPI FUER RADIO-
ASTRONOMIE
AUF DEM HUEGEL 69
D-5300 BONN 1
GERMANY, F.R.

WALRAVEN TH DR
LEIDEN SOUTHERN
STATION P O BOX 13
BROEDERSTROOM
TRANSVAAL 0240
SOUTH AFRICA

WALSH DENNIS DR
NRAL
JODRELL BANK
MACCLESFIELD
CHESHIRE SK11 9DL
U.K.

WALTER HANS G DR
ASTRONOMISCHES
RECHEN-INSTITUT
MONCHHOFSTER 12-14
D-6900 HEIDELBERG 1
GERMANY, F.R.

WALTER KURT PROF DR
ASTRONOMISCHES INSTI
TUT DER UNIVERSITAET
WALDHAUSERSTR 64
D-7400 TUEBINGEN
GERMANY, F.R.

WAMPLER E JOSEPH PROF
LICK OBSERVATORY
UNIVERSITY OF CALIF
SANTA CRUZ CA 95064
U.S.A.

WAMSTEKER W DR
EUROPEAN SOUTHERN
OBSERVATORY
SCHLEISSHEIMERSTR 17
D-8046 GARCHING
GERMANY, F.R.

WANAS M I DR
FACULTY OF SCIENCE
CAIRO UNIVERSITY
GEZA ORMAN
ARAB REP. OF EGYPT

WANNIER PETER GREGORY DR
CALTECH
OWENS VLY RADIO OBS
PASADENA CA 91125
U.S.A.

WAPSTRA AALDERT H PROF
VAN WOENSEL KOOIJ-
LAAN 48
NAARDEN
THE NETHERLANDS

WARD RICHARD A DR
W K KELLOGG RADIAT
LAB 106-38
CALTECH
PASADENA CA 91125
U.S.A.

WARD WILLIAM R DR
JPL
4800 OAK GROVE DRIVE
PASADENA CA 91103
U.S.A.

WARDLE JOHN F C PROF
PHYSICS DEPT
BRANDEIS UNIVERSITY
WALTHAM MA 02154
U.S.A.

WARES GORDON W DR
73 PERKINS ST
WEST NEWTON MA 02165
U.S.A.

WARMAN J DR
INST DI ASTRONOMIA
UNAM
APT POSTAL 70- 64
MEXICO DF
MEXICO

WARNER BRIAN PROF
ASTRONOMY DEPT
UNIVERSITY OF CAPE
TOWN RONDEBOSCH
CAPE TOWN 7700
SOUTH AFRICA

WARNER JOHN W DR
NASA CODE TA-01
MARSHALL SFC AL35812
U.S.A.

WARNER PETER J DR
CAVENDISH LABS
MADINGLEY ROAD
CAMBRIDGE CB3 OHE
U.K.

WARREN WAYNE H JR DR
NASA-GSFC
CODE 601
GREENBELT MD 20771
U.S.A.

WARWICK JAMES W DR
ASTRO-GEOPHYSICS
DEPT
UNIV OF COLORADO
BOULDER CO 80309
U.S.A.

WARWICK ROBERT S DR
PHYSICS DEPT
UNIV OF LEICESTER
LEICESTER LE1 7RH
U.K.

WARZEE J DR
115 RUE MATTOT
B-1410 WATERLOO
BELGIUM

WASHIMI HARUICHI DR
RESEARCH INSTITUTE
OF ATMOSPHERICS
NAGOYA UNIVERSITY
TOYOKAWA AICHI 442
JAPAN

WASSERMAN LAWRENCE H DR
LOWELL OBS
P O BOX 1269
FLAGSTAFF AZ 86002
U.S.A.

WASSON JOHN T
DEPT CHEMISTRY
UNIVERSITY OF CALIF
LOS ANGELES CA 90024
U.S.A.

WATANABE TAKASHI DR
RESEARCH INSTITUTE
OF ATMOPHERICS
NAGOYA UNIVERSITY
TOYOKAWA 442
JAPAN

WATERFIELD REGINALD L DR
WOOLSTON OBSERVATORY
NORTH CADBURY NR
YEOVIL SOMERSET
U.K.

WATERWORTH MICHAEL DR
UNIVERSITY TASMANIA
GPO BOX 252C
HOBART 7001
AUSTRALIA

WATSON WILLIAM D PROF
PHYSICS DEPT
UNIVERSITY
OF ILLINOIS
URBANA IL 61801
U.S.A.

WATTENBERG D PROF
ARCHENHOLD STERNW
ALT TREPTOW 1
DDR 1193 BERLIN-
TREPTOW
G.D.R.

WAYMAN PATRICK A PROF
DUNSINK OBSERVATORY
CASTLEKNOCK
CO DUBLIN
IRELAND

WEAVER HAROLD F PROF
DEPT OF ASTRONOMY
UNIVERSITY OF
CALIFORNIA
BERKELEY CA 94720
U.S.A.

WEAVER THOMAS A DR
PHYSICS DEPT L-372
LAWRENCE LIVERM LAB
P O BOX 808
LIVERMORE CA 94550
U.S.A.

WEBBER JOHN C DR
UNIVERSITY OF ILLI-
NOIS OBSERVATORY
URBANA IL 61801
U.S.A.

WEBBINK RONALD F DR
DEPT OF ASTRONOMY
UNIV OF ILLINOIS
URBANA IL 61801
U.S.A.

WEBROVA LUDMILA DR
ASTRONOMICAL INST
CZECHO ACADEMY SC
BUDECSKA 6
120 23 PRAHA
CZECHOSLOVAKIA

WEBSTER ADRIAN S DR
14 CLARE HALL
HERSCHEL ROAD
CAMBRIDGE
U.K.

WEBSTER B LOUISE DR
ANGLO AUSTRALIAN
OBSERVATORY
P O BOX 296
EPPING NSW 2121
AUSTRALIA

WEEDMAN DANIEL W ASS PROF
DYER OBSERVATORY
VANDERBILT UNIV
BOX 1803
NASHVILLE TN 37235
U.S.A.

WEEKES TREVOR C DR
MOUNT HOPKINS
OBSERVATORY
P O BOX 97
AMADO AZ 85640
U.S.A.

WEHINGER PETER A DR
ROYAL GREENWICH OBS
HERSTMONCEUX CASTLE
HAILSHAM
SUSSEX BN27 1RP
U.K.

WEHLAU AMELIA DR
ASTRONOMY DEPT
UNIVERSITY OF
WESTERN ONTARIO
LONDON ONT N6A 5B9
CANADA

WEHLAU WILLIAM H PROF
ASTRONOMY DEPT
UNIVERSITY OF
WESTERN ONTARIO
LONDON ONT N6A 5B9
CANADA

WEHRSE RAINER DR
INSTITUT F THEORETI-
SCHE ASTROPHYSIK
IM NEUENHEIMER F 294
D-6900 HEIDELBERG
GERMANY, F.R.

WEIDEMANN VOLKER PROF
INST THEO PHY STERNW
NEUE UNIV PHYSIK ZNT
OLSHAUSENST GB N 61C
D-2300 KIEL 1
GERMANY, F.R.

WEIGERT ALFRED PROF
HAMBURGER STERNWARTE
GOJENBERGSWEG 112
D-2050 HAMBURG 80
GERMANY, F.R.

WEILER EDWARD J DR
PRINCETON UNIV OBS
PEYTON HALL
PRINCETON NJ 08540
U.S.A.

WEILER KURT W DR
MPI FUER RADIO-
ASTRONOMIE
AUF DEM HUEGEL 69
D-5300 BONN 1
GERMANY, F.R.

WEILL GILBERT M DR
CENTRE NATIONAL
D ETUDES SPATIALES
129 RUE L UNIVERSITE
F-75007 PARIS
FRANCE

WEIMER THEOPHILE P F DR
OBSERVATOIRE
DE PARIS
61 AV L'OBSERVATOIRE
F-75014 PARIS
FRANCE

WEINBERG J L DR
SPACE ASTRONOMY LAB
STATE UNIVERSITY OF
NY AT ALBANY
ALBANY NY 12203
U.S.A.

WEISHEIT JON C DR
PRINCETON UNIV OBS
PEYTON HALL
PRINCETON NJ 08540
U.S.A.

WEISS EDWARD W DR
VAN VLECK OBS
WESLEYAN UNIV
MIDDLETOWN CT 06457
U.S.A.

WEISS NIGEL O DR
DAMPT SILVER STREET
CAMBRIDGE CB3 9EW
U.K.

WEISS WERNER W DR
INST FUR ASTRON DER
UNIV WIEN
TUERKENSCHANZSTR 17
A-1180 WIEN
AUSTRIA

WEISSKOPF MARTIN CH DR
NASA/MSFC CODE ES-62
MARSHALL SFC AL35812
U.S.A.

WEISTROP DONNA W DR
LAB FOR ASTRONOMY
AND SOLAR PHYSICS
CODE 681 NASA GSFC
GREENBELT MD 20771
U.S.A.

WELCH GARY A DR
DEPT OF ASTRONOMY
ST MARY'S UNIVERSITY
HALIFAX NS B3H 3C3
CANADA

WELCH WILLIAM J PROF
RADIO ASTRONOMY LAB
UNIV OF CALIFORNIA
617 CAMPBELL HALL
BERKELEY CA 94720
U.S.A.

WELIACHEW LEONID DR
IRAM
53 RUE DES MARTYRS
F-38026 GRENOBLE
FRANCE

WELLER CHARLES S DR
NAVAL RESEARCH
LAB CODE 7124
WASHINGTON DC 20375
U.S.A.

WELLGATE G BERNARD MR
CANEHEATH HOUSE
ARLINGTON
POLEGATE
EAST SUSSEX BN26 6SJ
U.K.

WELLMANN PETER PROF DR
INST F ASTRON UND
ASTROPHYSIK
SCHEINER-STR 1
D-8000 MUENCHEN 80
GERMANY, F.R.

WELLS DONALD C III DR
KITT PEAK NAT OBS
P O BOX 26732
TUCSON AZ 85726
U.S.A.

WEMPE J PROF DR
ZNTRINST ASTROPHYSIK
ASTROPHYS OBS
TELEGRAFENBERG
DDR 15 POTSDAM
G.D.R.

WENDKER HEINRICH J PROF
HAMBURGER STERNWARTE
GOJENBERGSWEG 112
D-2050 HAMBURG 80
GERMANY, F.R.

WENIGER SCHAME DR
OBSERVATOIRE
DE MEUDON
F-92190 MEUDON
FRANCE

WENTZEL DONAT G DR
ASTRONOMY PROGRAM
UNIVERSITY MARYLAND
COLLEGE PARK MD20742
U.S.A.

WENZEL W DR
ZENTRALINSTITUT FUER
ASTROPHYSIK
STERNWARTE SONNEBERG
DDR-64 SONNEBERG
G.D.R.

WESSELINK ADRIAAN J DR
143 FALLS ROAD
BETHANY CT 06525
U.S.A.

WESSELIUS PAUL R DR
DEPT SPACE RESEARCH
P O BOX 800
GRONINGEN
THE NETHERLANDS

WEST RICHARD M DR
ESO
C-O CERN
CH-1211 GENEVA 23
SWITZERLAND

WESTERHOUT GART DR
SCIENTIFIC DIRECTOR
US NAVAL OBSERVATORY
WASHINGTON DC 20390
U.S.A.

WESTERLUND BENGT E PROF
ASTRONOMICAL
OBSERVATORY
BOX 515
S-751 20 UPPSALA
SWEDEN

WESTFOLD KEVIN C PROF
DEPT OF MATHEMATICS
MONASH UNIVERSITY
CLAYTON VICT 3168
AUSTRALIA

WESTPHAL JAMES A DR
CALTECH
170-25
169 SOUTH MUDD
PASADENA CA 91125
U.S.A.

WETHERILL GEORGE W
DEPT TERRESTR MAGNET
CARNEGIE INST WASH
5241 BROAD BR RD N W
WASHINGTON DC 20015
U.S.A.

WEYMANN RAY J PROF
STEWARD OBSERVATORY
UNIVERSITY OF
ARIZONA
TUCSON AZ 85721
U.S.A.

WHEELER J CRAIG
ASSOC PROF
DEPT OF ASTRONOMY
UNIVERSITY OF TEXAS
AUSTIN TX 78712
U.S.A.

WHEELER JOHN A DR
DEPT OF PHYSICS
UNIVERSITY OF TEXAS
AUSTIN TX 78712
U.S.A.

WHELAN JOHN A J DR
INST OF ASTRONOMY
CAMBRIDGE CB3 OHA
U.K.

WHIPPLE FRED L DR
CTR FOR ASTROPHYSICS
60 GARDEN ST
CAMBRIDGE MA 02138
U.S.A.

WHITAKER EWEN A
4332 E SIXTH STREET
TUCSON AZ 85711
U.S.A.

WHITE NATHANIEL M DR
LOWELL OBS
P O BOX 1269
FLAGSTAFF AZ 86002
U.S.A.

WHITE ORAN R DR
HIGH ALTITUDE
OBSERVATORY
BOX 3000
BOULDER CO 80307
U.S.A.

WHITE R STEPHEN PROF
IGPP
UNIVERSITY OF
CALIFORNIA
RIVERSIDE CA 92521
U.S.A.

WHITE RAYMOND E DR
STEWARD OBSERVATORY
UNIVERSITY OF
ARIZONA
TUCSON AZ 85721
U.S.A.

WHITEOAK J B DR
CSIRO
DIV OF RADIOPHYSICS
P O BOX 76
EPPING NSW 2121
AUSTRALIA

WHITFORD ALBERT E PROF
LICK OBSERVATORY
UNIVERSITY OF CALIF
SANTA CRUZ CA 95064
U.S.A.

WHITNEY BALFOUR S
1102 E MISSOURI
NORMAN OK 73069
U.S.A.

WHITNEY CHARLES A PROF
CENTER FOR ASTRO-
PHYSICS
60 GARDEN STREET
CAMBRIDGE MA 02138
U.S.A.

WHITROW GERALD JAMES PROF
41 HOME PARK RD
WIMBLEDON
LONDON SW19 7HS
U.K.

WHITTET DOUGLAS C B DR
UNIV OF LONDON
OBSERVATORY
MILL HILL PARK
LONDON NW7 2QS
U.K.

WICKRAMASINGHE D T DR
INSTITUTE OF
ASTRONOMY
MADINGLEY ROAD
CAMBRIDGE CB3 OHA
U.K.

WICKRAMASINGHE N C PROF
UNIVERSITY COLLEGE
DEPT OF APPLIED MATH
AND THEO PHYS
CARDIFF CF1 1XL
U.K.

WIDING KENNETH G DR
CODE 7144
US NAVAL RESEARCH
LABORATORY
WASHINGTON DC 20375
U.S.A.

WIDORN THOMAS DR
GRINZINGERSTR 93 6
A-1190 WIEN
AUSTRIA

WIEDLING TORD DR
VILLAVAEGEN 15
S-611 00 NYKOEPING
SWEDEN

WIEHR EBERHARD DR
UNIV STERNWARTE
GEISMARLAND STR 11
D-3400 GOETTINGEN
GERMANY, F.R.

WIELEBINSKI RICHARD PROF
MPI FUER RADIO-
ASTRONOMIE
AUF DEM HUEGEL 69
D-5300 BONN
GERMANY, F.R.

WIELEN ROLAND PROF DR
TECH UNIV BERLIN
INST FUR ASTROPHYS
ERNST REUTER PLATZ 7
D 1000 BERLIN 10
GERMANY, F.R.

WIESE WOLFGANG L DR
NATIONAL BUREAU OF
STANDARDS
WASHINGTON DC 20234
U.S.A.

WIETH KNUDSEN NIELS P DR
SVEND TROSTSVEJ 12
DK-1912 COPENHAGEN V
DENMARK

WILCOCK WILLIAM L PROF
SCHOOL PHY MOLEC SC
UNIV COLLEGE N WALES
BANGOR GWYNEDD
LL57 2UW WALES
U.K.

WILCOX JOHN M PROF
INSTITUTE FOR PLASMA
RESEARCH
STANFORD UNIVERSITY
STANFORD CA 94305
U.S.A.

WILD JOHN PAUL DR
CSIRO RADIOPHYSICS
PO BOX 76
EPPING NSW 2021
AUSTRALIA

WILD PAUL DR
ASTRONOMISCHES INST
UNIVERSITAET BERN
SIDLERSTR 5
CH-3012 BERN
SWITZERLAND

WILD PETER A T ASSOC PROF
ASTRONOMY DEPT
UNIVERSITY OF
CAPE TOWN
RONDEBOSCH 7700
SOUTH AFRICA

WILDEY ROBERT L PROF DR
NORTHERN ARIZONA
UNIVERSITY
ASTROPHYSICAL OBS
FLAGSTAFF AZ 86011
U.S.A.

WILKENING LAUREL L DR
DEPT PLANETARY SCS
UNIV OF ARIZONA
TUCSON AZ 85721
U.S.A.

WILKINS GEORGE A DR
ROYAL GREENWICH
OBSERVATORY
HAILSHAM
EAST SUSSEX BN27 1RP
U.K.

WILKINSON PETER N DR
N R A L
JODRELL BANK
MACCLESFIELD
CHESHIRE SK11 9DL
U.K.

WILL CLIFFORD M DR
INST OF THEOR PHYS
STANFORD UNIV
STANFORD CA 94305
U.S.A.

WILLIAMON RICHARD M
FERNBANK SCIENCE CTR
156 HEATON PARK DR
ATLANTA GA 30307
U.S.A.

WILLIAMS CAROL A
ASSOC PROF
DEPT OF ASTRONOMY
UNIVERSITY S FLORIDA
TAMPA FL 33620
U.S.A.

WILLIAMS D DR
RADIO-ASTRONOMY LAB
UNIV OF CALIFORNIA
BERKELEY CA 94720
U.S.A.

WILLIAMS DAVID A DR
MATHEMATICS DEPT
UMIST
MANCHESTER M60 1QD
U.K.

WILLIAMS IWAN P DR
APPLIED MATH DEPT
QUEEN MARY COLLEGE
MILE END RD
LONDON E1 4NS
U.K.

WILLIAMS JAMES G DR
JPL 264-720
4800 OAK GROVE DRIVE
PASADENA CA 91103
U.S.A.

WILLIAMS JOHN A DR
PHYSICS DEPT
ALBION COLLEGE
ALBION MI 49224
U.S.A.

WILLIAMS PEREDUR M DR
UK INFRARED
TELESCOPE UNIT
900 LEILANI ST
HILO HI 96720
U.S.A.

WILLIAMS ROBERT E PROF
STEWARD OBSERVATORY
UNIVERSITY OF
ARIZONA
TUCSON AZ 85721
U.S.A.

WILLIAMS THEODORE B DR
PRINCETON UNIV OBS
129 PEYTON HALL
PRINCETON NJ 08540
U.S.A.

WILLIS A G DR
DEPT OF PHYSICS
BRANDEIS UNIVERSITY
WALTHAM MA 02154
U.S.A.

WILLIS ALAN J DR
DEPT OF PHYSICS
AND ASTRONOMY
UNIV COLLEGE LONDON
LONDON
U.K.

WILLMORE A PETER PROF
SPACE RESEARCH DEPT
UNIV OF BIRMINGHAM
PO BOX 363
BIRMINGHAM B15 2TT
U.K.

WILLNER STEVEN PAUL DR
DEPT OF PHYSICS C011
UNIV OF CALIFORNIA
SAN DIEGO
LA JOLLA CA 92093
U.S.A.

WILLS BEVERLEY J DR
DEPT OF ASTRONOMY
RLM 15 212
UNIVERSITY OF TEXAS
AUSTIN TX 78712
U.S.A.

WILLS DEREK DR
ASTRONOMY DEPT
UNIVERSITY OF TEXAS
AUSTIN TX 78712
U.S.A.

WILLSON LEE ANNE DR
E W FICK OBSERVATORY
PHYSICS
IOWA ST UNIVERSITY
AMES IA 50011
U.S.A.

WILLSTROP RODERICK V DR
INST OF ASTRONOMY
MADINGLEY ROAD
CAMBRIDGE CB3 OHA
U.K.

WILSON ALBERT G DR
RESEARCH PROGRAM
STUDIES
P O BOX 113
TOPANGA CA 90290
U.S.A.

WILSON ANDREW S DR
ASTRONOMY CENTRE
UNIVERSITY OF SUSSEX
FALMER
BRIGHTON BN1 9QH
U.K.

WILSON BRIAN G PROF
SIMON FRASER
UNIVERSITY
BURNABY BC V5A 1S6
CANADA

WILSON JAMES R DR
LAWR LIVERMORE LAB
L-35
LIVERMORE CA 94550
U.S.A.

WILSON LIONEL DR
LUNAR AND PLANETARY
UNIT E S DEPT
LANCASTER UNIVERSITY
LANCASTER LA1 4YR
U.K.

WILSON OLIN C
HALE OBSERVATORIES
813 SANTA BARBARA ST
PASADENA CA 91106
U.S.A.

WILSON PETER R PROF
DEPT OF APPLIED
MATHEMATICS
UNIVERSITY OF SYDNEY
SYDNEY 2007
AUSTRALIA

WILSON RAYMOND H DR
C/O ANNIE WILSON
5829 NORTH 25TH ROAD
ARLINGTON VA 22207
U.S.A.

WILSON ROBERT E PROF
DEPT OF ASTRONOMY
UNIVERSITY OF
SOUTH FLORIDA
TAMPA FL 33620
U.S.A.

WILSON ROBERT PROF
DEPT PHYS ASTRONOMY
UNIV COLLEGE LONDON
GOWER ST
LONDON WC1E 6BT
U.K.

WILSON ROBERT W DR
BELL LABORATORIES
HOH L229
BOX 400
HOLMDEL NJ 07733
U.S.A.

WILSON THOMAS L DR
MPI FUER RADIO-ASTRO
NOMIE
AUF DEM HUEGEL 69
D-5300 BONN
GERMANY, F.R.

WILSON W J DR
UNIVERSITY OF TEXAS
DEPT OF ELECTRICAL
ENGINEERING
AUSTIN TX 78712
U.S.A.

WINCKLER JOHN R PROF
SCHOOL OF PHYSICS
AND ASTRONOMY
UNIV OF MINNESOTA
MINNEAPOLIS MN 55455
U.S.A.

WING ROBERT F PROF
ASTRONOMY DEPT
OHIO ST UNIVERSITY
174 WEST 18TH AVE
COLUMBUS OH 43210
U.S.A.

WINK JOERN ERHARD DR
MPI FUER
RADIOASTRONOMIE
AUF DEM HUEGEL 69
D-5300 BONN
GERMANY, F.R.

WINKLER FRANK P DR
DEPT OF PHYSICS
MIDDLEBURY COLLEGE
MIDDLEBURY VT 05753
U.S.A.

WINKLER GERNOT M R DR
US NAVAL OBSERVATORY
WASHINGTON DC 20390
U.S.A.

WINNBERG ANDERS DR
MPI FUER RADIO-
ASTRONOMIE
AUF DEM HUEGEL 69
D-5300 BONN
GERMANY, F.R.

WINNEWISSER GISBERT DR
MPI FUER RADIO-
ASTRONOMIE
AUF DEM HUEGEL 69
D-5300 BONN
GERMANY, F.R.

WISNIEWSKI WIESLAW Z
UNIVERSITY ARIZONA
LUNAR AND PLANETARY
LABORATORY
TUCSON AZ 85721
U.S.A.

WITHBROE GEORGE L DR
CTR F ASTROPHYSICS
HARVARD COLLEGE OBS
60 GARDEN ST
CAMBRIDGE MA 02138
U.S.A.

WITT ADOLF N DR
RITTER ASTROPHYSICAL
RESEARCH CENTER
UNIVERSITY OF TOLEDO
TOLEDO OH 43606
U.S.A.

WITTEN LOUIS PROF
DEPT OF PHYSICS
UNIVERSITY OF
CINCINNATI
CINCINNATI OH 45221
U.S.A.

WITTMANN AXEL DIETER PH D
UNIV STERNWARTE
GEISMARLANDSTR 11
D-3400 GOETTINGEN
GERMANY, F.R.

WITZEL ARNO DR
MPI FUER RADIO-
ASTRONOMIE
AUF DEM HUEGEL 69
D-5300 BONN
GERMANY, F.R.

WLERICK GERARD DR
OBSERVATOIRE PARIS
F-92190 MEUDON
FRANCE

WOEHL HUBERTUS DR
UNIVERSITAETS-
STERNWARTE
GEISMARLANDSTR 11
D-3400 GOETTINGEN
GERMANY, F.R.

WOLF BERNHARD PH D
MOMBERTSTRASSE 5
D-6900 HEIDELBERG
GERMANY, F.R.

WOLF RAINER E A DR
MPI FUER ASTRONOMIE
KOENIGSTUHL
D-6900 HEIDELBERG 1
GERMANY, F.R.

WOLFE ARTHUR M PROF
DEPT OF PHYSICS AND
ASTRONOMY
UNIV OF PITTSBURGH
PITTSBURGH PA 15260
U.S.A.

WOLFENDALE ARNOLD W PROF
PHYSICS DEPT
UNIVERSITY
SOUTH ROAD
DURHAM DH1 3LE
U.K.

WOLFF SIDNEY C DR
INSTITUTE FOR
ASTRONOMY
2680 WOODLAWN DRIVE
HONOLULU HI 96822
U.S.A.

WOLSTENCROFT RAMON D DR
ROYAL OBSERVATORY
BLACKFORD HILL
EDINBURGH EH9 3HJ
U.K.

WOLTJER LODEWIJK PROF.
EUROPEAN SOUTHERN
OBSERVATORY
C O CERN
CH-1211 GENEVA 23
SWITZERLAND

WOOD DAVID B DR
WOOD ASSOCIATES
5108 VIKING RD
BETHESDA MD 20014
U.S.A.

WOOD F BRADSHAW PROF
DEPT OF PHYSICS
AND ASTRONOMY
UNIVERSITY FLORIDA
GAINESVILLE FL 32611
U.S.A.

WOOD HARLEY W DR
178 KENTHURST RD
KENTHURST 2154
AUSTRALIA

WOOD III H J DR
DEPT OF ASTRONOMY
UNIV OF VIRGINIA
CHARLOTTESVL VA22903
U.S.A.

WOOD JOHN A DR
CTR FOR ASTROPHYSICS
60 GARDEN ST
CAMBRIDGE MA 02138
U.S.A.

WOOD PETER R DR
MT STROMLO AND
SIDING SPRING OBS
PRIVATE BAG
WODEN A C T 2606
AUSTRALIA

WOOD ROGER DR
ROYAL GREENWICH OBS
HERSTMONCEUX CASTLE
HAILSHAM
E SUSSEX BN27 1RP
U.K.

WOODWARD P R DR
STERREWACHT
POSTBUS 9513
2300 RA LEIDEN
THE NETHERLANDS

WOOLF NEVILLE J
STEWARD OBSERVATORY
UNIV OF ARIZONA
TUCSON AZ 85721
U.S.A.

WOOLFSON MICHAEL M PROF
UNIVERSITY OF YORK
YORK YO1 5DD
U.K.

WOOLLEY RICHARD SIR
MAGNOLIA HOUSE
HANKHAM NR PEVENSEY
EAST SUSSEX
U.K.

WOOLSEY E G
695 BRIERWOOD
OTTAWA ONT
CANADA

WOOSLEY S E PROF
LICK OBSERVATORY
UNIVERSITY OF CALIF
SANTA CRUZ CA 95064
U.S.A.

WORDEN SIMON P DR
A F GEOPHYSICS LAB
SACRAMENTO PEAK OBS
SOLAR RES BRANCH
SUNSPOT NM 88349
U.S.A.

WORLEY CHARLES E
ASTROMETRY AND ASTRO
PHYSICS DIVISION
US NAVAL OBSERVATORY
WASHINGTON DC 20390
U.S.A.

WORRALL GORDON DR
BIRDSWOOD
EARDISLEY HEREFDS
U.K.

WOSZCZYK ANDRZEJ DR
INST OF ASTRONOMY
CHOPINA 12-18
87100 TORUN
POLAND

WRAY JAMES D DR
DEPT OF ASTRONOMY
UNIVERSITY OF TEXAS
RL MOORE HALL 15-212
AUSTIN TX 78712
U.S.A.

WRIGHT ALAN E DR
ANGLO-AUSTRALIAN OBS
P O BOX 296
EPPING N S W 2121
AUSTRALIA

WRIGHT EDWARD L DR
DEPT OF PHYSICS
RM 26-345
MIT
CAMBRIDGE MA 02139
U.S.A.

WRIGHT FRANCES W DR
ASTRONOMY DEPT
HARVARD COLLEGE
60 GARDEN ST
CAMBRIDGE MA 02138
U.S.A.

WRIGHT GEOFFREY A E DR
INST F ASTROPHYS UND
EXTRATERR FORSCHUNG
AUF DEM HUEGEL 71
D-5300 BONN
GERMANY, F.R.

WRIGHT HELEN
579 FOREST LAKE DR
ANDOVER NJ 07821
U.S.A.

WRIGHT JAMES P DR
DIV ASTRONOMICAL SC
NAT SC FOUND RM 615
1800 G STREET N W
WASHINGTON DC 20550
U.S.A.

WRIGHT KENNETH O DR
DOMINION ASTROPHYSI-
CAL OBSERVATORY
5071 W SAANICH RD
VICTORIA BC V8X 3X3
CANADA

WRIGHT MELVYN C H DR
RADIO ASTRONOMY
LABORATORY
UNIVERSITY OF CALIF
BERKELEY CA 94720
U.S.A.

WRIXON GEORGE T DR
DEPT OF ELECTRICAL
ENGINEERING
UNIVERSITY COLLEGE
CORK
IRELAND

WU CHI CHAO DR
COMPUTER SCIENCES CO
NASA GODDARD SPACE
FLIGT CNTR CODE 685
GREENBELT MD 20771
U.S.A.

WU HSIN-HENG DR
DEPT OF PHYSICS
NAT CENTAL UNIV
CHUNG-LI
TAIWAN

WU SHI TSAN DR
SCHOOL SCIENCE ENGIN
UNIV OF ALABAMA
P O BOX 1247
HUNTSVILLE AL 35801
U.S.A.

WYATT STANLEY P JR PROF
DEPT OF ASTRONOMY
UNIV OF ILLINOIS
OBSERVATORY
URBANA IL 61801
U.S.A.

WYCKOFF SUSAN DR
PHYSICS DEPT
ARIZONA STATE UNIV
TEMPE AZ 85281
U.S.A.

WYLLER ARNE A PROF
STOCKHOLM
OBSERVATORY
S-133 00
SALTSJOEBADEN
SWEDEN

WYNN-WILLIAMS C G DR
INST FOR ASTRONOMY
UNIVERSITY OF HAWAII
2680 WOODLAWN DRIVE
HONOLULU HI 96822
U.S.A.

WYNNE CHARLES G PROF
ROYAL GREENWICH
OBSERVATORY
HAILSHAM
EAST SUSSEX BN27 1RP
U.K.

XANTHAKIS JOHN N PROF
R C A A M
ACADEMY OF ATHENS
14 ANAGNOSTOPOULOU
ATHENS 136
GREECE

YABUSHITA SHIN A PROF
DEPT OF APPLIED MATH
AND PHYSIC
KYOTO UNIVERSITY
SAKYOKU KYOTO 606
JAPAN

YABUUTI KIYOSHI PROF
20 TANAKA HIGASKI
HINOKUCH MACHI
SAKYOKU 606 KYOTO
JAPAN

YAHIL AMOS DR.
ASTRONOMY PROGRAM
ESS BUILDING
SUNY AT STONY BROOK
STONY BROOK NY 11794
U.S.A.

YAKHONTOVA N S DR
INSTITUTE OF THEORE-
TICAL ASTRONOMY
192187 LENINGRAD
U.S.S.R.

YAKOVKIN N A DR
KIEV STATE UNIV
ASTRONOMICAL OBS
252053 KIEV
U.S.S.R.

YALLOP BERNARD D DR
ROYAL GREENWICH
OBSERVATORY
HAILSHAM
EAST SUSSEX BN27 1RP
U.K.

YAMASAKI ATSUMA DR
DEPT EARTH SC ASTRON
UNIVERSITY OF TOKYO
KOMABA MEGURO-KU
TOKYO 153
JAPAN

YAMASHITA KOJUN DR
DEPT OF PHYSICS
NAGOYA UNIVERSITY
FUROCHO CHIKUSAKU
NAGOYA 464
JAPAN

YAMASHITA YASUMASA PROF
TOKYO ASTRONOMICAL
OBSERVATORY
OSAWA
MITAKA TOKYO 181
JAPAN

YAMAZAKI AKIRA DR
HYDROGRAPHIC DEPT
TSUKIJI 5 CHUO KU
TOKYO 104
JAPAN

YANKULOVA IVANKA DR
DEPT OF ASTRONOMY
FACULTY OF PHYSICS
ANTON IVANOV STR 5
1126 SOFIA
BULGARIA

YANOVITSKIJ EDGARDT G DR
MAIN ASTRONOMICAL OB
UKRAINIAN
252127 KIEV
U.S.S.R.

YAPLEE B S
6105 WESTLAND DR
HYATTSVILLE MD 20782
U.S.A.

YAROV-YAROVOJ M S DR
MATHEMATICS DEPT
OF MVTU
107005 MOSCOW
U.S.S.R.

YASUDA HARUO PROF DR
TOKYO ASTRONOMICAL
OBSERVATORY
OSAWA
MITAKA TOKYO 181
JAPAN

YATSKIV YA S DR
MAIN ASTRONOMICAL
OBS UKRAINIAN ACADE-
MY OF SCIENCES
252127 KIEV
U.S.S.R.

YAVNEL ALEXANDER A DR
METEORITE COMMITTEE
ACADEMY OF SCIENCES
UL M ULIANOVOJ 3 K 1
117313 MOSCOW
U.S.S.R.

YEIVIN Y PROF
TEL-AVIV UNIVERSITY
TEL-AVIV
ISRAEL

YEN YUI-LIN PROF
DEPT OF ELECTRICAL
ENGINEERING
UNIVERSITY TORONTO
ONT M5S 1A4
CANADA

YEOMANS DONALD K DR
JET PROPULSION LAB
4800 OAK GROVE DR
PASADENA CA 91103
U.S.A.

YILMAZ FATMA DR
OBS DE L'UNIVERSITE
ISTANBUL
TURKEY

YILMAZ NIHAL DR
FACULTY OF SCIENCE
DEPT OF ASTRONOMY
ANKARA
TURKEY

YOKOYAMA KOICHI DR
INTERNATIONAL LATITU
DE OBSERVATORY
MIZUSAWA IWATE 023
JAPAN

YONEYAMA TADAOKI DR
3-14-3-802
OHKUBO
SHINJUKU-KU
TOKYO 112
JAPAN

YORK DONALD G DR
PEYTON HALL
PRINCETON UNIVERSITY
PRINCETON NJ 08540
U.S.A.

YORKE HAROLD W DR
UNIV STERNWARTE
GEISMARLANDSTR 11
D-3400 GOETTINGEN
GERMANY, F.R.

YOSHIDA JUNZO PROF
DEPT OF PHYSICS
KYOTO SANGYO UNIV
KAMIGAMO KITA-KU
KYOTO 603
JAPAN

YOSHIMURA HIROKAZU DR
DEPT OF ASTRONOMY
UNIVERSITY OF TOKYO
YAYOI BUNKYO
TOKYO 113
JAPAN

YOSS KENNETH M DR
ASTRONOMY DEPT
UNIV OF ILLINOIS
OBSERVATORY
URBANA IL 61801
U.S.A.

YOUNG ANDREW T DR
1200 HOLLEMAN DR.
COLLEGE STATION
TX 77840
U.S.A.

YOUNG ARTHUR DR
ASTRONOMY DEPT
SAN DIEGO STATE UNIV
SAN DIEGO CA 92180
U.S.A.

YOUNG LOUISE GRAY DR
1200 HOLLEMAN DRIVE
COLLEGE STATION
TX 77840
U.S.A.

YOUSEF SHAHINAZ M DR
ASTRONOMY DEPT
FACULTY OF SCIENCE
CAIRO UNIVERSITY
CAIRO
ARAB REP. OF EGYPT

YOUSSEF NAHED H DR
ASTRONOMY DEPT
FACULTY OF SCIENCE
CAIRO UNIVERSITY
CAIRO
ARAB REP. OF EGYPT

YU KYUNG-LOH PROF
SEOUL NAT UNIVERSITY
SINLIM-DONG
KWANAK-KU
SEOUL 151
REPUBLIC OF KOREA

YUAN CHI PROF
DEPT OF PHYSICS
CITY COLLEGE OF N Y
138 ST CONVENE AVE
NEW YORK NY 10031
U.S.A.

YUASA MANABU DR
DEPT OF ASTRONOMY
UNIVERSITY OF TOKYO
BUNKYO-KU
TOKYO 113
JAPAN

YUMI S DR
TOYO UNIVERSITY
HAKUSAN 5-28-20
BUNKYO-KU
TOKYO 112
JAPAN

YUN HONG-SIK ASSOC PROF
SEOUL NAT UNIVERSITY
SINLIM-DONG
KWANAK-KU
SEOUL 151
REPUBLIC OF KOREA

ZABRISKIE F R PROF
R D 1
ALEXANDRIA PA 16611
U.S.A.

ZACHAROV IGOR DR
ASTRONOMICAL INST
CZECHO ACADEMY SC
OBSERVATORY
251 65 ONDREJOV
CZECHOSLOVAKIA

ZADUNAISKY PEDRO E PROF
OBS NAC FISICA COSM
AV MITRE 3100
1663 SAN MIGUEL
BUENOS AIRES
ARGENTINA

ZAHN JEAN-PAUL DR
OBSERVATOIRE DE NICE
BP 252
F-06007 NICE CEDEX
FRANCE

ZAITSEV VALERII V DR
INST OF APPIED PHYS
ULYANOVA ST 46
603600 GORKY
U.S.S.R.

ZANDER RODOLPHE DR
INST D ASTROPHYSIQUE
UNIVERSITE DE LIEGE
B-4200 COINTE OUGREE
BELGIUM

ZAPPALA ROSARIO ALDO DR
OSS ASTROFISICO
CITTA UNIVERSITARIA
I-95125 CATANIA
ITALY

ZAPPALA VINCENZO DR
OSSERVATORIO ASTRO-
NOMICO
PINO TORINESE
TORINO
ITALY

ZARE K DR
DEPT OF MATHEMATICS
TECHNICAL UNIVERSITY
P O BOX 6934
ISFAHAN
IRAN

ZASOV ANATOLE V DR
STERNBERG ASTRONOMI-
CAL INSTITUTE
117234 MOSCOW
U.S.S.R.

ZDANAVICIUS KAZIMERAS DR
ASTRONOMIJOS OBS
VILNIUS 31
CIURLIONIO 29
LITHUANIA
U.S.S.R.

ZEALY WILLIAM J DR
UK SCHMIDT TELESCOPE
PRIVATE BAG
COONBARABRAN
NSW 2857
AUSTRALIA

ZEILIK MICHAEL II DR
DEPT PHYSICS ASTRON
UNIV OF NEW MEXICO
800 YALE BLVD BE
ALBUQUERQUE NM 87131
U.S.A.

ZEIPPEN C DR
OBS DE MEUDON
F-92190 MEUDON
FRANCE

ZEL'DOVICH YA B ACAD
SPACE RESEARCH
INSTITUTE
USSR ACADEMY OF SCI
117810 MOSCOW
U.S.S.R.

ZEL'MANOV A L DR
STERNBERG STATE AS-
TRONOMICAL INSTITUTE
117234 MOSCOW
U.S.S.R.

ZELENKA ANTOINE DR
SWISS FEDERAL
OBSERVATORY
SCHMELZBERGSTR 25
CH-8006 ZUERICH
SWITZERLAND

ZELLNER BENJAMIN H DR
LUNAR AND PLANETARY
LABORATORY
UNIVERSITY ARIZONA
TUCSON AZ 85721
U.S.A.

ZHELEZNIAKOV VLADIMIR V
PROF APPLIED PHYSICS
INST ACADEMY SCIENCE
ULYANOV STREET 46
603600 GORKIJ
U.S.S.R.

ZHEVAKIN S A PROF DR
RADIOPHYSICAL
RESEARCH INSTITUTE
603600 GORKIJ
U.S.S.R.

ZHONGOLOVICH I D PROF DR
INSTITUTE OF THEORE-
TICAL ASTRONOMY
192187 LENINGRAD
U.S.S.R.

ZIEBA ANDRZEJ PROF
ASTRONOMICAL
OBSERVATORY
UL ORLA 171
30-244 KRAKOW
POLAND

ZIEBA STANISLAW DR
OBSERVATORIUM
ASTRONOMICZNE UJ
UL ORLA 171
30-244 KRAKOW
POLAND

ZIKIDES MICHAEL C DR
DEPT OF ASTRONOMY
PANEPISTIMIOPOLIS
ATHENS 621
GREECE

ZIMMERMAN HELMUT DR
UNIVERSITAETS
STERNWARTE JENA
SCHILLERGAESSCHEN 2
DDR-69 JENA
G.D.R.

ZINN ROBERT J DR
DEPT OF ASTRONOMY
YALE UNIVERSITY
BOX 2023 YALE STAT
NEW HAVEN CT 06520
U.S.A.

ZIOLKOWSKI J DR
COPERNICUS ASTRONOMI
CAL CENTER
UL BARTYCKA 18
00 716 WARSAW
POLAND

ZIOLKOWSKI KRZYSZTOF DR
MICKIEWICZA 4 M 17
01-517 WARSZAWA
POLAND

ZIRIN HAROLD DR
CAL TECH 264-33
PASADENA CA 91125
U.S.A.

ZIRKER JACK B DR
SACRAMENTO PEAK
OBSERVATORY
SUNSPOT NM 88349
U.S.A.

ZLOBEC PAOLO DR
S FLORIANA 1/A
I-34015 MUGGIA
ITALY

ZOMBECK MARTIN V DR
CTR FOR ASTROPHYSICS
60 GARDEN ST
CAMBRIDGE MA 02138
U.S.A.

ZWAAN CORNELIS DR
OBSERVATORY
ZONNENBURG 2
UTRECHT
THE NETHERLANDS

ZUCKERMAN BEN M DR
ASTRONOMY PROGRAM
UNIVERSITY MARYLAND
COLLEGE PARK MD20742
U.S.A.

ZVEREV MITROFAN S PROF DR
PULKOVO OBSERVATORY
196140 LENINGRAD
U.S.S.R.

REPORT OF THE EXECUTIVE COMMITTEE

Executive Committee 1976-1978

PRESIDENT

Professor Dr A. Blaauw, Sterrewacht, Postbus 9513,
2300 RA Leiden, The Netherlands.

VICE-PRESIDENTS

Mr J.G. Bolton, C.S.I.R.O., Division of Radiophysics, P.O. Box 276
Parkes, Epping, N.S.W. 2870, Australia.
Professor Ch. Fehrenbach, Directeur de l'Observatoire de Haute Provence,
Saint Michel l'Observatoire, 04300 Forcalquier, France.
Dr D.S. Heeschen, National Radio Astonomy Observatory,
Edgemont Road, Charlottesville, Virginia 22901, U.S.A.
Professor W. Iwanowska, Institute of Astronomy, Nicolaus Copernicus
University, ul. Chopina 12/18, Pl. 87-100 Toruń, Poland.
Academicien E.K. Kharadze, Director, Abastumani Astrophysical
Observatory, M.T. Canobili, Abastumani 383762, U.S.S.R.
Professor S. van den Bergh, Director, Dominion Astrophysical
Observatory, 5071 West Saanich Road, Victoria B.C., V8X 3X3, Canada.

GENERAL SECRETARY

Professor Edith A. Müller, Observatoire de Genève, CH-1290 Sauverny,
Switzerland.

ASSISSTANT GENERAL SECRETARY

Professor P.A. Wayman, Director, Dunsink Observatory,
Castleknock, Co. Dublin, Ireland.

ADVISERS TO THE EXECUTIVE COMMITTEE

Professor L. Goldberg, Kitt Peak National Observatory, P.O. Box 26732,
Tucson, AZ 85726, U.S.A. - former President
Professor Dr G. Contopoulos, Astronomy Department,
Panepistimiopolis, Athens 621, Greece.

INTRODUCTION

The present report covers the period from 1 January 1976 to 31 December 1978. The period under report includes the first eight months of the year 1976 (until 2 September) which fell under the responsibility of the former Executive Committee, but it does not include the first eight months of 1979, still within the responsibility of the present Executive Committee. An oral report for the period from 1 January until 13 August 1979 will be presented at the XVIIth General Assembly.

ADMINISTRATION

Executive Committee

The 41st meeting of the Executive Committee, in its former composition, was held in the course of the XVIth General Assembly with sessions on 23 and 31 August 1976. It was presided over by the former President of the IAU, Professor L. Goldberg, and attended by all members and the advisers. The former Vice-President, Prof. B.J. Bok, the President elect, Prof. A. Blaauw, and the Assistant General Secretary elect, Prof. P.A. Wayman, also participated in the meeting.

The Executive Committee, in its present composition, held its 42nd meeting in Grenoble, France, on 2 September 1976. It was chaired by the President of the Union, Prof. A. Blaauw. All members and the advisers attended the meeting. Further, the Executive Committee held the following meetings:

The 43rd meeting at the Observatoire de Genève, Sauverny, Switzerland, from 29 August to 2 September 1977. All members and the advisers were present.

The 44th meeting in Dublin, Ireland, from 15 to 19 August 1978. All members and the advisers participated in the meeting.

Officers' Meetings

The President of the Union, the General Secretary and the Assistant General Secretary (Officers) met as follows: in Leiden, The Netherlands, on 5 and 6 November 1976; in Norg, The Netherlands, on 1 and 2 May 1977; in Geneva, Switzerland, on 28 August 1977; in Geneva, Switzerland, on 10 and 11 May 1978; and in Dublin, Ireland, on 14 August 1978.

Adhering Countries

Iraq has been admitted to the IAU by the XVIth General Assembly as the Union's 48th Adhering Country.

Members of the IAU

The number of IAU Members was 3127 as on 1 April 1976. At the XVIth General Assembly 722 new Members were admitted and 16 names were deleted from the membership list. In the meantime the decease of 50 Members were communicated to the General Secretary. Thus the IAU had 3783 Members on 31 December 1978.

Consultants to IAU Commissions

The 40 IAU Commissions had a total of 192 consultants as at 31 December 1978.

Commissions of the IAU

The Transactions of the IAU, volume XVIIA, subtitled "Reports on Astronomy 1979", published in three parts, include the reports as prepared by each of the 40 Presidents of IAU Commissions, assisted by their Vice-Presidents, Organizing Committees, Chairmen of Working Groups formed within the Commissions, and by the Commission members themselves. They also contain the report of the Working Group for Planetary System Nomenclature, as prepared by the Chairman of that Working Group.

A number of Working Groups were formed in various Commissions, such as the Working Group on Nutation in Commission No. 4, the Working Group on Cartographic

Coordinates and Rotational Elements of Planets and Satellites jointly by Commissions 4 and 16, the Preliminary Working Group on Solar Activity Index in Commission 10, and the Working Group on the Solar Maximum Year also in Commission 10.

The Minor Planet Center, directed for over 30 years, in a most efficient and excellent way, by Dr P. Herget at the Cincinnati Observatory, was moved to the Smithsonian Observatory, Cambridge, Massachusetts and is now directed by Dr B.G. Marsden. The Executive Committee takes this opportunity to express to Dr P. Herget the thanks of the Union for his outstanding work and his great service to Astronomy.

Following the resignation of Professor M. Waldmeier from the editorship of the Quarterly Bulletin on Solar Activity (QBSA) this function has been entrusted to Dr F. Moriyama of the Tokyo Astronomical Observatory. The Executive Committee wishes to express to Professor M. Waldmeier the thanks of the Union for the service he rendered over many years to the scientific community studying the Sun and Solar-Terrestrial Relations.

On account of illness, Prof. B.E. Markarian resigned as President of IAU Commission No. 28. The Executive Committee appointed Prof. B.E. Westerlund, Vice-President of the Commission, to Acting President.

Dr G.A. Wilkins has succeeded Miss C.J.A. Penny as member of the Organizing Committee of Commission No. 31.

The President of IAU Commission No. 38, Prof. D.A. MacRae prepared rules for operating the project Exchange of Astronomers. These rules, approved by the Executive Committee, have been published in IAU Information Bulletin No. 39.

The Vice-President of IAU Commission No. 50, Prof. F.G. Smith has prepared a Report and Recommendations of Commission 50 regarding Identification and Protection of Existing and Potential Observation Sites. This printed report, produced with financial aid of UNESCO, has been distributed to National Committees of Astronomy and to a number of Observatories for whom the report might be useful. The report deals with the acceptable levels of artificial lighting, the acceptable levels of radio frequency interference, absorption due to atmospheric pollution, interference from aircraft, and contains practical recommendations.

Commissions have co-opted a number of new members from among IAU Members, and new consultants from among non-members. These co-options have been approved by the Executive Committee. Between the General Assemblies, many Commissions organized Symposia and Colloquia, and other projects financially supported by the IAU, as follows:

SYMPOSIA

IAU Symposia (held between 1 January 1976 and 31 December 1978)

IAU Symposium No. 74 "Radio Astronomy and Cosmology", Cavendish Laboratory, Cambridge, U.K., 16-20 August 1976

IAU Symposium No. 75 "Star Formation", University of Geneva, Geneva, Switzerland, 6-10 September 1976

IAU Symposium No. 76 "Planetary Nebulae", Ithaca, N.Y., U.S.A., 6-10 June 1977

IAU Symposium No. 77 "Structure and Properties of Nearby Galaxies", Bad Münstereifel, Germany F.R., 22-26 August 1977

IAU Symposium No. 78 "Nutation and the Earth's Rotation", Kiev, U.S.S.R., 23-28 May 1977

IAU Symposium No. 79 "The Large-Scale Structure of the Universe", Tallinn, U.S.S.R., 12-16 September 1977

IAU Symposium No. 80 "The H.R. Diagram", Washington, D.C., U.S.A., 2-5 November 1977

IAU Symposium No. 81 "Dynamics of the Solar System", Tokyo, Japan,

23-26 May 1978
IAU Symposium No. 82 "Time and the Earth's Rotation", Cadiz, Spain, 8-12 May 1978
IAU Symposium No. 83 "Mass-Loss and Evolution of O-Type Stars", Qualicum Beach, Vancouver Is., Canada, 5-9 June 1978
IAU Symposium No. 84 "The Large-Scale Characteristics of the Galaxy", College Park, Maryland, U.S.A., 12-17 June 1978
IAU Symposium No. 89 "Refractional Influences in Astronomy and Geodesy", Uppsala, Sweden, 1-5 August 1978

Symposia with IAU Participation

COSPAR/IAU/URSI Symposium "Infrared and Sub-millimeter Astronomy", Philadelphia, Pennsylvania, U.S.A., 8-12 June 1976
COSPAR/IAU/IUGG Symposium " Scientific Applications of Lunar Laser Ranging (SALUR)", Austin, Texas, U.S.A., 8-10 June 1976
SCOSTEP/COSPAR/IAU/IUGG Symposium "The Study of Travelling Interplanetary Phenomena", Tel-Aviv, Israel, 7-10 June 1977
COSPAR/IAU Symposium "New Instrumentation for Space Astronomy", Tel-Aviv, Israel, 8-10 June 1977
COSPAR/IAU/IUB/IUGG Symposium "Viking Experiments", Tel-Aviv, Israel, 9 and 10 June 1977
COSPAR/IAU/IUGG/IUGS/IUTAM/URSI Symposium "Planetary Surfaces and Atmospheres", Innsbruck, Austria, 29 May-1 June 1978
COSPAR/IAU Symposium "100 Years of Lunar Mapping", Lagonissi, Greece, 25-27 May 1978
SCOSTEP/IAU Symposium "Solar-Terrestrial Physics", Innsbruck, Austria, 29 May-3 June 1978
SBARMO/COSPAR/IAU/IUPAP Symposium "Scientific Use of Balloons and Related Technical Problems", Innsbruck, Austria, 8 and 9 June 1978

COLLOQUIA

IAU Colloquia (held between 1 January 1976 and 31 December 1978)

IAU Colloquium No. 35 "The Compilation, Critical Evaluation and Distribution of Stellar Data", Strasbourg, France, 19-21 August 1976
IAU Colloquium No. 36 "Energy Balance and Hydrodynamics of the Solar Chromosphere and Corona", Nice, France, 6-10 September 1976
IAU Colloquium No. 37 "Redshift and Expansion of the Universe", Paris, France, 6-9 September 1976
IAU Colloquium No. 38 "Problems of Stellar Convection", Nice, France, 16-20 August 1976
IAU Colloquium No. 39 "Relationship between Comets, Minor Planets and Meteorites", Lyon France, 17-20 August 1976
IAU Colloquium No. 40 "Astronomical Applications of Image Detector with Linear Response", Meudon, France, 6-8 September 1976
IAU Colloquium No. 41 "Dynamics of Planets and Satellites and Theories of their Motion", Cambridge, U.K., 17-19 August 1976
IAU Colloquium No. 42 "The Interaction of Variable Stars with their Environment", Bamberg, Germany F.R., 6-9 September 1977
IAU Colloquium No. 43 "Ultraviolet and X-Ray Spectroscopy of Astrophysical and Laboratory Plasmas" (Fifth Conference), London, U.K., 4-7 July 1977
IAU Colloquium No. 44 "Solar Prominences", Oslo, Norway, 14-17 August 1978
IAU Colloquium No. 45 "Chemical and Dynamical Evolution of the Galaxy", Toruń, Poland, 7-9 September 1977

IAU Colloquium No. 46 "Changing Trends in Variable-Star Research", Hamilton, New Zealand, 27 November-1 December 1978
IAU Colloquium No. 47 "Spectral Classification of the Future", Vatican City State, 11-15 July 1978
IAU Colloquium No. 48 "Modern Astrometry", Vienna, Austria, 12-14 September 1978
IAU Colloquium No. 49 "Formation of Images from Spatial Coherence Functions in Astronomy", Groningen, The Netherlands, 10-12 August 1978
IAU Colloquium No. 50 "High Angular Resolution Stellar Interferometry", College Park, Maryland, U.S.A., 30 August-1 September 1978
IAU Colloquium No. 52 "Protostars and Planets", Tuscon, Arizona, U.S.A., 3-7 January 1978

Specific Projects of Commissions

The list of Specific Projects of Commissions as approved by the XVIth General Assembly is given together with the corresponding expenditure in the report on the financial situation of the Union (page). These projects are continuations of previous projects and do not therefore need any further comments.

Regional Astronomy Meetings held under the auspices of the IAU

First Latin-American Regional Astronomy Meeting, Santiago, Chile, 16-21 January 1978
Fourth European Regional Astronomy Meeting, Uppsala, Sweden, 7-12 August 1978
First Asian-South Pacific Regional Astronomy Meeting, Wellington, New Zealand, 5-8 December 1978

Young Astronomers' Schools

IAU Commission No. 46 "Teaching of Astronomy" organized two schools for young astronomers in the course of 1977 and 1978.

The school at the National Observatory, Rio de Janeiro, Brazil, from 21 November to 22 December 1977 was attended by 29 astronomers from Argentia, Bolivia, Brazil, Colombia, Paraguay, Peru and Portugal. The curriculum included Astronomical Instruments and Technology, Stellar Astrophysics, Interstellar Matter, Variable Stars, Extra-Galactic Astronomy, Structure and Evolution of the Universe, and Data Processing.

The school at the University of Nigeria, Nsukka, Nigeria, from 31 July to 19 August 1978, was attended by 28 physicists (lecturers and professors) from Ghana, Kenya, Nigeria and Tanzania. The curriculum included Spherical Astronomy, Planetary System, Solar Physics, Solar Energy Utilization, Stars and Galaxies, Structure and Evolution of the Universe, Practical Astronomy, Seminars and General Discussions on astronomical problems.

IAU Publications

From the fall of 1976 on, all IAU publications, the Informations Bulletin included, have been produced by the offset method from camera-ready copies, which markedly reduced the retail price of the individual volumes. The Executive Committee wishes to express its thanks to all those who cooperated in the fine preparation of the camera-ready manuscripts.

The Secretariat continued to sell IAU Transactions, Highlights and Symposium volumes to 14 institutions in developing countries at 20% of the retail price of the volumes.

The Publications of the IAU continue to be produced by the D. Reidel Publishing Company, Dordrecht, Holland, which pays the Union a 5% royalty. The publisher also finances the production of the publications, with the exception of that of the Information Bulletin which is financed by the IAU.

Sales of IAU Publications

The Transactions of the IAU sold as follows in the course of the period under report:

	Number of Copies
Vol. XIIIA	20
Vol. XIIIB	13
Vol. XIVA	17
Vol. XIVB	17
Vol. XVA	24
Vol. XVB	54
Vol. XVIA	134
Vol. XVIB	538

The volumes of Highlights of Astronomy were sold as follows over the same period:

	Number of Copies
Highlights I	23
Highlights II	58
Highlights III	95
Highlights IV	877

The sales of Symposium volumes in 1976, 1977 and 1978 were:

Symposium volume No. (in number of copies

32-34	44-73	56-72	68-154
33-1	45-47	57-46	69-239
34-31	46-26	58-166	70-561
35-24	47-27	59-92	71-608
36-14	48-79	60-62	72-587
37-24	49-65	61-77	73-428
38-37	50-80	62-76	74-678
39-29	51-131	63-241	75-799
40-32	52-98	64-157	76-486
41-33	53-83	65-48	77-513
42-49	54-73	66-126	
43-30	55-69	67-238	79-555
			80-432

The Proceedings of Symposium No. 78 have not yet been published. For illustration, it should be noted that the total of volumes sold in the course of the period under report was 8.963 copies as against 15.459 copies sold in the preceding period of 3 years.

Information Bulletin

The Information Bulletin has been prepared, edited and typed for offset reproduction by the General Secretary and the staff of the IAU secretariat, and printed and distributed by the publisher of the Union. It has appeared twice a year and has been sent, free of charge, to IAU Members, consultants, members of the International Science Writers Association, and to selected international organizations and institutions.

RELATIONS TO OTHER ORGANIZATIONS

IAU Representation

1. International Council of Scientific Unions (ICSU). The General Secretary attended the 16th General Assembly of ICSU which together with the meeting of the General Committee was held in Washington, D.C., U.S.A. from 9 to 15 October 1976. The General Secretary also represented the Union at the meeting of the ICSU General Committee in Budapest, Hungary, from 7 to 9 September 1977 and at the 17th General Assembly of ICSU which was held in Athens, Greece, from 24 to 29 September 1978. At this General Assembly, Prof. C. de Jager, former General Secretary of the IAU, has been elected President of ICSU.

2. ICSU Organizations. The Union participates in the activities of a number of Special and Scientific Committees, and Inter-Union Commissions, formed and sponsored by ICSU. For more details please consult pages 13 and onwards of IAU Transactions, volume XIIC.

The following list only gives the present state of IAU representation in these organizations:

(a) Committee on Space Research (COSPAR): Edith Müller
(b) ICSU-Abstracting Board (IAB): J.C. Pecker
(c) Federation of Astronomical and Geophysical Services (FAGS): H. Enslin and Edith Müller which includes the following four services of interest to astronomy:
  (i) Bureau International de l'Heure (BIH): H. Enslin and Edith Müller
  (ii) International Polar Motion Service (IPMS): S. Yumi
  (iii) International Ursigram and World Day Service (IUWDS): V. Bumba
  (iv) Quarterly Bulletin on Solar Activity (QBSA): IAU Commission No. 10
(d) Inter-Union Committee on Frequency Allocation for Radio Astronomy and Space Science (IUCAF): G. Westerhout and R. Wielebinski
(e) Scientific Committee on Solar-Terrestrial Physics (SCOSTEP): Z. Svestka
(f) Scientific Committee on Problems of the Environment (SCOPE): R. Cayrel and J.C. Pecker
(g) Committee on the Teaching of Science (CTS): L.N. Houziaux
(h) Committee on Data on Science and Technology (CODATA): G.A. Wilkins
(i) Inter-Union Commission on Spectroscopy (IUCS): B. Edlén, J.G. Philips and M.J. Seaton

Other Organizations

The Union is represented in the following organizations:

(a) Le Comité Consultatif pour la Définition de la Seconde (CCDS) du Bureau International des Poids et Mesures: Wm. Markowitz
(b) Le Comité Consultatif pour la Définition du Mètre (CCDM) du Bureau International des Poids et Mesures: A.H. Cook
(c) Le Comité Consultatif International des Radiocommunications (CCIR): F.G. Smith and H.M. Smith

(d) Committee on Science and Technology in Developing Countries (COSTED): L.N. Houziaux
(e) European Physical Society - Conference Committee (EPS): P.A. Wayman
(f) Scientific Ballooning and Radiations Monitoring Organization (SBARMO): P. Simon

FINANCIAL REPORT

In accord with the policy adopted in 1961, the Executive Committee presents to the General Assembly a "Summarized Account of Receipts and Payments" for the years 1976, 1977 and 1978. The account is in Swiss francs, the figures in the U.S. dollar accounts and Dutch guilder accounts having been translated at the following ICSU conversion rates:

For the year 1976: $ U.S. = 2.51 Dfl = 2.44 Sfrs
For the year 1977: $ U.S. = 2.40 Dfl = 2.17 Sfrs
For the year 1978: $ U.S. = 2.08 Dfl = 1.73 Sfrs

The conversion rates varying over the years under report account for the disparities between the closing balances in 1975, 1976 and 1977 and the opening balances in 1976, 1977 and 1978 respectively (in Swiss francs):

Closing Balance	Opening balance	Difference
1975 - 309 564.84	1976 - 295 814.03	13 750.81
1976 - 242 055.98	1977 - 229 796.41	12 259.57
1977 - 251 682.97	1978 - 232 409.36	19 273.61

Certain bank operations, such as transfers from one IAU account to the other and reimbursements, were made at current rates of exchange and thus resulted in exchange differences.

The summarized account has been derived from 14 (5 each in 1976 and 1977, and 4 in 1978) accounts of receipts and payments audited and certified as correct by the Union's chartered accountant. These accounts will be available for examination by the Finance Committee at the XVIIth General Assembly.

SUMMARIZED ACCOUNT OF RECEIPTS AND PAYMENTS (in Swiss francs)
FOR THE TERM 1976 - 1978

RECEIPTS

		Swiss francs	Swiss francs
1.	Contribution from Adhering Countries		
	1976	292 693.96	
	1977	305 528.46	
	1978	320 119.94	918 342.36
2.	Revenue from IAU Publications		
	1976	29 331.75	
	1977	27 205.28	
	1978	21 163.69	77 700.72

3.	Interest on Current Accounts		
	1976	3 963.34	
	1977	3 518.75	
	1978	1 834.37	9 316.46
4.	Grant from UNESCO Subvention to ICSU		
	1976	29 280.00	
	1977	36 866.28	
	1978	20 750.81	86 897.09
5.	Other Receipts		
	1976	186 697.39	
	1977	43 953.85	
	1978	132 511.55	363 162.79
	Total Receipts		1 455 419.42
	TOTAL		1 455 419.42

PAYMENTS

			Swiss francs	Swiss francs
1.	Administrative Office			
	1976		133 002.81	
	1977		132 690.47	
	1978		137 879.86	403 573.14
2.	Contribution to ICSU			
	1976		7 221.35	
	1977		6 507.61	
	1978		7 638.21	21 367.17
3.	Commission Expenses			
	1976		-	
	1977		458.71	
	1978		2 895.01	3 353.72
4.	Specific Projects of Commissions Approved by the General Assembly			
	4.1 Exchange of Astronomers			
	1976	18 422.00		
	1977	12 694.50		
	1978	4 671.00	35 787.50	
	4.2 Other Projects			
	1976	7 157.37		
	1977	8 191.45		
	1978	5 839.82	21 188.64	56 976.14
5.	General Assembly			
	1976		84 843.80	
	1977		-	
	1978		-	84 843.80
6.	IAU Publications			
	1976		13 941.60	
	1977		16 850.41(5)	
	1978		25 848.50(4)	56 640.52

7. Distribution of IAU Publications (free and at reduced prices)		
1976	27.47	
1977	15 206.23	
1978	5 666.38	20 900.08
8. Executive Committee Meetings		
1976	-	
1977	22 470.89	
1978	22 064.48	44 535.37
9. Officers' Meetings		
1976	11 180.62	
1977	3 059.97	
1978	1 959.23	16 199.82
10. IAU Symposia and Colloquia		
1976	18 355.90	
1977	38 328.40	
1978	46 205.28	102 889.58
11. Inter-Union Commissions		
1976	6 954.00	
1977	5 644.98	
1978	5 034.30	17 633.28
12. Projects Authorized by the Executive Committee		
1976	2 440.00	
1977	3 982.00	
1978	-	6 422.00
13. Representation		
1976	1 284.64	
1977	5 795.35	
1978	3 802.83	10 882.82
14. Bank charges		
1976	388.64	
1977	179.81	
1978	271.13	839.58
15. Young Astronomers Schools		
1976	-	
1977	17 631.25	
1978	16 823.55	34 454.80
16. Regional Meetings		
1976	-	
1977	-	
1978	17 700.13	17 700.13
17. Other Payments		
1976	290 504.28	
1977	105 494.03	
1978	137 603.10	533 601.41
Total Payments		1 432 813.37
Excess of Receipts over Payments		22 606.05
TOTAL		1 455 419.42

BANK ACCOUNTS as on 1 January 1976

		Swiss francs
Savings accounts		194 513.60
Current accounts		295 814.03
	Total	490 327.63

BANK ACCOUNTS as on 31 December 1978

Savings accounts		361 740.19
Current accounts		286 886.90
	Total	648 627.09
TOTAL Balance on IAU accounts as on 31 December 1978		648 627.09

Note:

The difference between the closing and opening balances 1976/1977 and 1977/1978 of the current accounts, in total Sfrs 31 533.18, corresponds to the difference between the balances of the current accounts 1976-1978,

	8 927.13	(295 814.03 - 286 886.90)
and the excess of receipts over payments 1976-1978	22 606.05	
Total	31 533.18	

The statement on pages 504-507 is a simple account of receipts and payments collected for convenience under a few headings. The following explanatory notes provide more detailed information on each item of the accounts.

Receipts

1. Contributions from Adhering Countries. The XVIth General Assembly, Grenoble, France, 1976, increased the unit of contribution payable by Adhering Countries to the IAU to 1 465.00 Swiss francs a year.

The 48 countries adhering to the Union as at 31 December 1978 were paying 216 units of contribution in 1976 and 220 units of contribution both in 1977 and 1978. Eighteen countries adhered in category 1 with 1 unit of contribution, 14 countries in category 2 with 2 units of contribution, 6 countries in category 3 with 4 units of contribution, 2 countries in category 4 with 6 units of contribution, 3 countries (from 1977 on 2 countries) in category 5 with 10 units of contribution, 1 country (from 1977 2 countries) in category 6 with 14 units of contribution, 3 countries in category 7 with 20 units of contribution, and 1 country in category 8 with 30 units of contribution.

Forty-six countries had paid their dues for up to 31 December 1978; 1 country in category 1 is for 4 years in arear with its payments, and 1 country in category 1 owes the Union part of its dues for 1978.

2. Revenue from publications. About 93% of the receipts under this item are royalties paid to the IAU by the D. Reidel Publishing Company, Dordrecht, The Netherlands, for IAU publications sold; 7% of the income represents payments by institutions of developing countries for volumes sold to them at reduced prices. The royalties received from Academic Press, London, U.K., were negligible (SFrs. 12.45).

3. Interest on accounts. Sfrs 9 316.46 were earned on the Union's current accounts over 1976-1978. The interest earned on the Union's savings account is not shown in the summarized account. The Union has at present 7 savings accounts, that is 4 in the U.S.A., 2 in Switzerland and 1 in The Netherlands. The total

4. SPECIFIC PROJECTS OF COMMISSIONS (all figures in Swiss francs)

Item.	Comm.no.	Project	Paid in 1976	Allocation for 1977/79	Paid in 1977	Paid in 1978	Total paid 1977/1978	Total paid 1976/1978	Balance for 1979
4.1	38	Exchange of Astronomers	18 422.00	67 405.00	12 694.50	4 671.00	17 365.50	35 787.50	50 039.50
4.2		Other Projects							
	4	Ephemerides	715.74	2 485.00	828.33	828.33	1 656.66	2 372.40	828.34
	6	Telegram Bureau	1 610.40	5 595.00	1 658.61	1 486.83	3 145.44	4 755.84	2 449.56
	16	Planetary Documentation, Meudon	1 520.94	5 290.00	1 763.33	1 763.33	3 526.66	5 047.60	1 763.34
	17	Lunar Documentation, Meudon	1 520.94	5 285.00	1 761.33	1 761.33	3 522.66	5 043.60	1 762.34
	20	Minor Planet Center	1 789.32	6 220.00	1 843.92	-	1 843.92	3 633.24	4 376.08
	27	Variable Star Catalogue	-	9 330.00	335.93	-	335.93	335.93	8 994.07
TOTAL 4.2			7 157.34	34 205.00	8 191.45	5 839.82	14 031.27	21 188.61	20 173.73
TOTAL Specific Projects			25 579.34	101 610.00	20 885.95	10 510.82	31 396.77	56 976.11	70 213.23

interest earned on these accounts was Sfrs 40 442.46 over 1976-1978, while Sfrs 31 872.87 were earned in 1977 and 1978. Two of the four savings accounts in the U.S.A. and the two savings accounts in Switzerland were opened by the Executive Committee in office.

4. Grant from UNESCO subvention to ICSU. Detailed accounts of the expenditure of this grant are submitted to ICSU, for UNESCO, every year. The distribution of the grant is left to the IAU's discretion within specific UNESCO categories.

The allocations were: $ 12 000.-U.S. for 1976, $16 989.07 for 1977 (out of which $ 5 000.-for IAU Commission No. 50 "Protection of Existing and Potential Observatory Sites"), and $ 11 994.69 U.S. for 1978. $ 16.24 were lost on bank charges and exchange differences.

5. Other receipts. This heading shows internal bank transfers, transient items, reimbursements, etc. The following is an analysis of the amounts Sfrs 186 697.39, Sfrs 43 953.85 and Sfrs 132 511.55 for the years 1976, 1977 and 1978 respectively:

Other receipts 1976

		Swiss francs
Internal transfers from current accounts		175 400.01
Erroneous credits		3 909.86
Rectification of erroneous debits		2 245.58
Refunds of unused IAU grants		4 987.68
Refund of balance from "Working Account XVI IAU General Assembly Grenoble"		154.26
	Total	186 697.39

Other receipts 1977

Internal transfers		23 186.19
Receipt of undue payments		1 064.93
Refund of unused IAU travel grants		204.20
Erroneous credits		18 445.00
Rectification of erroneous debits		1 053.53
	Total	43 953.85

Other receipts 1978

Erroneous credits		87 236.40
Franklin Mint for Silver Map project (Sfrs 9 556.50 + 20 000 U.S. x 1.73)		44 156.50
Reimbursement of tax on interest		488.65
Transfer to Petty-Cash, Lausanne		400.00
Refund of unused travel grant		230.00
	Total	132 511.55

It is seen that the only effective major receipt under this heading is the donation of Sfrs 44 156.50 by the Franklin Mint Corporation. The other items are mostly internal transfers, erroneous credits and refunds which are balanced by corresponding payments and rectifications and cannot therefore be regarded as effective receipts.

The Franklin Mint donation will be used entirely for travel or subsistence grants to the XVIIth General Assembly in Montreal.

Payments

1. Administrative office. The expenses of the administrative office include salaries, social charges, health insurance and pension payments for the staff of the IAU Secretariat, audit fees, removal costs and small payments for stationery.

It should be noted that thanks to the Sterrewacht Leiden, the Dunsink Observatory and, in particular, to the Observatoire de Genève and the Institut d'Astronomie de l'Université de Lausanne, the Union paid no rent for office rooms, had free use of telephone, office material and copying facilities, paid no telegram charges and no postage. Thanks to the generous financial support of the Swiss Academy of Sciences (Société Hêlvétique des Sciences Naturelles), secretarial and typing help was provided to the Union. In the name of the IAU the Executive Committee wishes to express its gratitude to these institutions.

2. Subscription to ICSU. The annual dues of the IAU to ICSU were 2.5% of the contributions from Adhering Countries for the year preceding payment.

3. Commission expenses. The amount of Sfrs 3 353.72 was spent on travel and living expenses of IAU Members representing IAU Commissions at various scientific meetings.

4. Projects of Commissions
4.1 Exchange of Astronomers, and
4.2 Other Projects. Details of the allocations made by the XVIth General Assembly are on page 28 of IAU Transactions, volume XVIB for 1977-1979, and on page 50 of IAU Transactions, volume XVB for 1976. Actual payments are shown in the Table on page of the present Bulletin. The balances in favour of the Commissions in question may but need not be spent in 1979.

5. General Assembly. The expenses incurred with the XVIth General Assembly of the IAU are given in great detail in the 1976 accounts. The $33 000.--U.S. foreseen in the budget for the XVIth General Assembly as approved in August 1973 were worth Sfrs. 92 400.-- at that time ($1 U.S.= Sfrs. 2.80). They were worth Sfrs. 80 520.-- in August/September 1976 when the XVIth General Assembly was held ($1 U.S.=Sfrs. 2.44) and are, for comparison, worth Sfrs. 55 440.-- as of this day ($1 U.S.=Sfrs. 1.68). The amount actually spent was Sfrs. 84 843.80. It includes travel and living expenses of the members of the Executive Committee, contributions partially to cover travel expenses of Commission Presidents, young astronomers' grants, local expenses in Grenoble, etc.

6. IAU Publications. The IAU only financed the production and distribution of the Information Bulletin. The Transactions, Highlights and Symposium volumes were financed by the IAU publisher. The payments under this heading were:

	Swiss francs
in 1976 for production and distribution of IAU Information Bulletin, Nos. 34 and 35	13 941.60
in 1977 for production and distribution of IAU Information Bulletin, Nos. 36 and 37	16 850.42
in 1978 for production and distribution of IAU Information Bulletin, Nos. 38, 39 and 40	25 848.50
Total 1976-1978	56 640.52

7. Publications for developing countries. This item covers puchases by the Union of its publications at 60% of their retail prices for distribution as courtesy copies to members of the Executive Committee and Chairmen of the Scientific Organizing Committees of Symposia, and for sale at reduced prices to 14 institutions of some developing countries. Out of the amount of Sfrs. 20 900.08, roughly Sfrs. 9 753.-- were spent on volumes sent to developing countries from 1976 to 1978.

8. Executive Committee meetings. The costs of the 41st and 42nd meetings of the Executive Committee are included in those of the XVIth General Assembly under heading 4.

The amount of Sfrs. 44 535.37 spent for the 43rd and 44th meetings of the Executive Committee in 1977 and 1978 are well within the limit of Sfrs. 51 000.-- as in budget.

9. Officers' meetings. The sums spent for these meetings were Sfrs. 11 180.62 in 1976, Sfrs. 3 059.97 in 1977 and Sfrs. 1 959.23 in 1978. It is seen that the Officers made a special effort to keep expenses as low as possible. The total of Sfrs. 5 019.20 spent in 1977 and 1978 leaves a balance of Sfrs. 15 540.80 for 1979.

10. Symposia and Colloquia, cosponsored meetings included. Sfrs. 18 355.90 were spent in 1976 on Symposia 74 and 75, Colloquia 35 through 41, IUGG/COSPAR/IAU Symposium and as contribution toward the typing of a proceedings volume (Sfrs. 743.90). The Sfrs. 38 328.40 spent in 1977 cover Symposia 76 through 80, Colloquia 33, 42, 43, 45 and 52, and three cosponsored meetings. In 1978, Sfrs. 46 205.28 were spent on Symposia Nos. 81 through 84 and 89, Colloquia Nos. 44 and 46 through 50, and on 3 cosponsored meetings. Out of this amount Sfrs. 1 291.70 were contributions towards the production of 2 Symposium volumes.

11. Inter-Union Commissions. Annual contributions to IUCAF, ICSU-AB, ICSU-CTS and SCOSTEP amounted to Sfrs. 17 633.28 over the years under report, with an average of about Sfrs. 5 878.-- per year. In 1976, when the budget had been approved, the General Assembly expected additional (FAGS) and increased contributions to Inter-Union organizations. No such contributions were, however, paid in the period under report.

12. Executive Committee projects. Sfrs. 2 440.-- represent a grant towards the publication of the Atlas of Spectra of Stars of Types Later than GO, by P.C. Keenan and R.C. McNeil; Sfrs. 2 500.-- were a contribution towards the production of the Photographic Atlas of Be Stars by Commission No. 29; Sfrs. 1 085.-- were a subsidy towards the publication of "General History of Astronomy", and Sfrs. 397.-- were a travel grant to Prof. J. Kovalevsky for his participation in the IAU meeting on admission of new Members, Geneva, March 1977.

13. Representation. The Sfrs. 10 882.82 are travel and living expenses of IAU representatives at meetings of such organizations as EPS, FAGS, ICSU, COSPAR, Pic-du- Midi, etc.

14. Bank charges. They were less than foreseen in the approved budget.

15. Young Astronomers' Schools. Two such schools were organized, one in Brazil, November 1977, and one in Nigeria, Africa, August 1978. The amount spent on these schools leaves a balance of Sfrs. 4 395.20 for the year 1979.

16. Regional meetings. The expenses with the regional meetings in Santiago, Chile, and in Wellington, New Zealand, in 1978 were Sfrs. 17 700.13, by Sfrs. 12 299.87 less than foreseen in the budget.

17. Other payments. Reference is made to the comment in heading 6 under Receipts. The following is an analysis of the amounts of Sfrs. 290 504.28, Sfrs. 105 494.03 and Sfrs. 137 603.10 for the years 1976, 1977 and 1978 respectively:

Other payments	
1976	Sfrs.
Internal transfers to:	
current accounts	186 286.90
savings accounts	97 960.00

Rectification of erroneous credits		3 909.86
Erroneous debits		2 245.58
Tax on interest at UBS		101.94
	Total	290 504.28

1977		
Internal transfers to:		Sfrs.
current accounts		24 215.01
savings accounts		50 000.00
Transfers of ICSU grant to Commission No. 50		10 504.27
Transfer of undue amounts received		1 064.93
Erroneous debits		1 053.53
Transfer of interest from closed $US account at AMRO		1.54
Rectification of erroneous credits		18 445.00
Tax on interest in UBS account		209.75
	Total	105 494.03

1978		
Rectification of erroneous credits		87 236.40
Transfer to savings account at UBS		50 000.00
Tax on bank interest		331.50
Refund		35.20
	Total	137 603.10

Recapitulation:		
Other expenses in 1976		290 504.28
1977		105 494.03
1978		137 603.10
	Total 1976-1978	533 601.41

Audit for the Accounts 1976-1978

Messieurs,

Conformément à la mission que vous avez bien voulu nous confier, nous vous présentons le résultat de notre examen des comptes de l'UNION ASTRONOMIQUE INTERNATIONALE pour la période du 1er janvier 1976 au 31 décembre 1978.

L'examen, notamment, a porté sur les documents suivants:
-compte des recettes et dépenses pour l'année 1976, vérifiés par nous-mêmes
-compte des recettes et dépenses pour l'année 1977, vérifiés par nous-mêmes
-compte des recettes et dépenses pour l'année 1978, vérifiés par nous-mêmes
-compte des recettes et dépenses et situation des comptes vancaires pour la période de trois ans: 1976, 1977 et 1978.

Les comptes ont été comparés avec les cahiers de comptes, reçus, et pièces comptables diverses.
Nous certifions que ces comptes sont en accord complet avec les informations et explications que nous avons demandées pour nous permettre d'effectuer notre examen. Le compte ci-inclus des recettes et dépenses résumant les comptes des années 1976, 1977, 1978, est la représentation exacte des recettes et dépenses de l'UNION ASTRONOMIQUE INTERNATIONALE pour la période du premier janvier 1976 au trente et un décembre 1978.

A Paris, le 20 mars 1979
(signé)
ROGER BACLE H.E.C.

Diplômé Expert-Comptable

Conclusion

The balance of Sfrs. 286 886.90 as in the current accounts at 31 December 1978 is by Sfrs. 8 927.13 less than that on 1 January 1976. This is due, in the main, to the transfer of Sfrs. 197 960.-- from current to savings accounts and, in a lesser degree, to exchange losses. Clearly, however, the additional Sfrs. 197 960.-- on the IAU savings accounts show that the finances of the Union are in an excellent state.

Residual budget for 1979. The figures in the accounts for 1977 and 1978 subtracted from those of the budget approved for 1977-1979 give the residual budget for the year 1979 as follows:

Residual Budget for the Year 1979
(Sfrs. 1.-- = 0.8484 DFL = $0.5952 U.S. as in March 1979)

Receipts in Swiss francs

Balance as on 1 January 1979:	284 773.06
1. Contribution from Adhering Countries	319 276.60
2. Revenue from publications	11 631.03
3. Interest on accounts	14 926.86
4. UNESCO subvention through ICSU	34 902.91
Total receipts	665 510.46
Total payments	549 533.17
Excess of receipts over payments	115 977.29

Payments in Swiss francs

1. Administrative Office	179 329.67
2. Subscription to ICSU	9 479.18
3. Commission expenses	6 926.28
4. Projects of Commissions	
4.1 Exchange of Astronomers	50 039.50
4.2 Other projects	20 173.73
5. General Assembly	136 210.00
6. IAU Publications	4 148.91
7. Publications for developing countries	29 127.30
8. Executive Committe meetings	6 864.63
9. Officers' meetings	15 540.80
10. Symposia and Colloquia	33 006.32
11. Inter-Union Commissions	30 720.72
12. Projects of the Executive Committee	6 018.00
13. Representation	10 401.82
14. Bank charges	3 149.06
15. Young Astronomers' Schools	4 395.20
16. Regional meetings	12 299.87
Total	549 533.17

Budget estimate for 1979, as proposed by the Executive Committee in August 1978:

Receipts in Swiss francs

1. Contribution from Adhering Countries	314 975.00
2. Revenue from publications	20 000.00
3. Interest on accounts	13 308.00
4. UNESCO subvention through ICSU	22 320.00
Total receipts	370 603.00
Payments	-448 753.00
Excess of payments over receipts	78 150.00

Payments in Swiss francs

1. Administrative office	150 000.00
2. Subscription to ICSU	7 875.00
3. Commission expenses	3 430.00
4. Projects of Commissions	
4.1 Exchange of Astronomers	45 000.00
4.2 Other projects	11 450.00
5. General Assembly	136 210.00
6. Publications	12 850.00
7. Publications for developing countries	16 667.00

8. Executive Committee	0
9. Officers' meetings	6 550.00
10. Symposia & Colloquia	34 000.00
11. Inter-Union Commission	13 500.00
12. EC projects	3 334.00
13. Representation	6 667.00
14. Bank charges	1 220.00
15. Young Astronomers' Schools	0
16. Regional meetings	0
Total	448 753.00

The Executive Committee resolved at the same time that the imbalance be covered from the balance in hand on the IAU current accounts, and that the Young Astronomers' Grants to the XVIIth General Assembly be paid from the following funds:

-Franklin Mint donation ($1.-- U.S. = 1.9262 Sfrs. on the day of meeting of Executive Committee)	Sfrs. 48 156.50
-Budget item 5 (payments): General Assembly	Sfrs. 19 300.00
-Budget item 4.1 (payments): Exchange of Astronomers	Sfrs. 19 300.00
-Reserves	Sfrs. 19 300.00
Total Sfrs.	106 056.50

In view of this and of other factors the Executive Committee now proposes the following budgetary estimate for the year 1979:
(1 unit of contribution equal 1 465.--Swiss francs,
1 Sfrs. = 0.8484 Dfl. = $ 0.5952 U.S.)

Receipts in Swiss francs

Balance as on 1 January 1979:	284 773.06
1. Contribution from Adhering Countries	322 300.00
2. Revenue from publications	20 000.00
3. Interest on accounts	18 000.00
4. UNESCO subvention through ICSU	26 880.00
Total receipts	671 953.06
Total payments	524 121.31
Excess of receipts over payments	147 831.75*

*See item 4.2 Payments

Payments in Swiss francs

1.	Administrative office	170 000.00
2.	Subscription to ICSU	8 003.00
3.	Commission expenses	3 000.00
4.	Projects of Commissions	
4.1	Exchange of Astronomers	25 000.00
4.2	Other projects	6 629.51
5.	General Assembly	200 000.00
6.	IAU Publications	18 000.00
7.	Publications for developing countries	17 000.00
8.	Executive Committee meeting	0
9.	Officers' meeting	3 000.00
10.	Symposia and Colloquia	50 300.00
11.	Inter-Union Commissions	4 888.80
12.	Projects of the Executive Committee	2 000.00
13.	Representation	6 000.00
14.	Bank charges	300.00
15.	Young Astronomers' Schools	10 000.00
16.	Regional Meetings	0
	Total	524 121.31

Explanatory Notes:

Receipts:

The balance of Sfrs. 284 773.06 as on 1 January 1979 includes, among others, the total amount of Sfrs. 136 210.00 as foreseen for the General Assembly in the budget for 1977-1979 and for the Franklin Mint donation of Sfrs. 44 156.50.

1. The contribution of adhering countries for 1979 has been obtained by multiplying the number of 220 units of contribution payable in the course of the year by Sfrs. 1 465.--, which equals Sfrs. 322 300.--.

2. Revenue from publications. A preliminary report shows that the D. Reidel Publishing Company will pay the IAU Sfrs. 16 353.73 on royalties in 1979. It has already paid Sfrs. 991.46 on translation royalties. If it is assumed that the volume of sales by the IAU Secretariat will be Sfrs. 3 484.28 as it was in 1978, the resulting figure is Sfrs. 20 829.47.

3. Interest on accounts earned in 1978 was Sfrs. 17 400.--. In 1979, interest to accrue on the savings accounts has been estimated at Sfrs. 17 970.--. With interest to be earned on the current accounts, the amount of Sfrs. 18 000.--, as in the budget, is justified.

4. The UNESCO subvention through ICSU is $16 000.-- for 1979, out of which $4 000.-- have been earmarked for the Tenerife Young Astronomers' School (special contract). The total allocation multiplied by 1.68, the present rate of exchange of the U.S. dollar vs. the Swiss franc, gives Sfrs. 26 880.--.

Payments

1. The figure of Sfrs. 170 000.-- is by 9 329.67 Sfrs. less than the difference between the amount of Sfrs. 449 900.--, as in the approved budget for the present term of office, and the payments made in 1977 and 1978. It has to cover the salaries, social charges and health insurance of the present staff members until their termination of service and the staff members of the permanent seat in Paris from 15 September on, removal expenses of the personal belongings of the staff and the files of the office, the transportation of the IAU archives from Bruxelles, and Prague to Paris, the audit fee for the accounts 1978 and 1976-1978, expenses for stationery, office material, postage, telephone etc. at the new seat in Paris from September 15, 1979 on, and the furnishing, typewriters etc. of the new secretariat in Paris.

2. The subscription to ICSU of Sfrs. 8 003.-- for 1979 has already been paid.

3. A total of Sfrs. 3 353.72 were paid under this heading in 1977 and 1978. Thus, Sfrs. 3 000.-- is expected to be sufficient in 1979.

4.1 Commission No. 38 spent Sfrs. 18 422.--, 12 694.50 and 4 671.-- on the project Exchange of Astronomers in 1976, 1977 and 1978 respectively, with an average of Sfrs. 11 929.16 per year. The figures show that the allocations vary considerably from year to year, and that an upward trend may be possible in 1979. This project has always been considered the most important among IAU activities and it was therefore felt that an estimate of Sfrs. 25 000.-- for 1979 is justified.

4.2 Other Projects. The Sfrs. 6 629.51 already paid represent the total payments due for 1979. See also table on page .

Here it should be noted that the grants within Commission No. 27 towards the production of the Variable Stars Catalogue have systematically been saved up with a view of purchasing for tham an adequate printing equipment. A Rank Xerox 3107 copier has been bought in 1979 for shipment to Moscow. The price of the copier, transport included, in the amount of about 24 600.-- will be paid in 1979.

The grants saved have never been shown in the budget, as they were kept within the reserves of the IAU. Hence they are included in the balance as on 1 January 1979. The balance of Sfrs. 147 831.75, as expected on 31 December 1979, will therefore drop to about Sfrs. 123 231.75 (147 831.75 - 24 600.--)

5. General Assembly. The sum of Sfrs. 200 000.-- will be spent on travel and living expenses of the members of the Executive Committee and of 5 Officers elect, travel grants to Presidents of Commissions, travel and living expenses of the IAU Staff, local secretarial help, and on young astronomers grants for which the Executive Committee has set aside Sfrs. 106 056.--. Part of these Sfrs. 106 056.-- are covered by the Franklin Mint donation of Sfrs. 44 156.50 (in August 1978 - Sfrs. 48 156.50) and the savings under item 4.1.

6. IAU publications. The amount of Sfrs. 18 000.-- is to cover the production and distribution for IAU Information Bulletin Nos. 42 and 43.

7. Publications for developing countries. Transactions, Vol. XVIIA, and an increased number of Symposium volumes are expected to appear in 1979, which will require more funds for buying and distributing them to institutions of developing countries, members of the Executive Committee, Chairmen of the Scientific Organizing Committees of Symposia, UNESCO and the IAU Secretariat. The sum of Sfrs. 17 000.-- is therefore justified.

8. The expenses to incur with the 2 meetings of the Executive Committee during the XVIIth General Assembly are included in the costs of this Assembly.

9. It is expected that only 1 Officers' meeting will be held in 1979. The sum of Sfrs. 3 000.-- is adequate to cover the expenses of this meeting.

10. Sfrs. 37 800.-- is the IAU contribution to 7 Symposia and Sfrs. 7 500.-- are to cover 3 Colloquia in 1979. Sfrs. 5 000.-- will be spent on a cosponsored Symposium.

11. Inter-Union Commissions have been paid contributions in the amount of Sfrs. 4 888.80 in 1979. No other such contributions are due this year.

12. Sfrs. 2 000.-- are a reserve for unforeseen projects to be financed by the Executive Committee.

13. The trip of the General Secretary to the COSPAR Plenary meeting in Bangalore, her visit to Bulgaria and Czechoslovakia, and Dr Enslin's participation in the FAGS Council meeting in 1979 will cost at least Sfrs. 6 000.--.

14. Bank charges will not exceed Sfrs. 300.-- in 1979.

15. A UNESCO contract through ICSU provides $4 000.--, that is Sfrs. 6 720.--, for the Young Astronomers' School to be held in Tenerife in September 1979. A subsidy of Sfrs. 3 280.--, available from the sum foreseen in the approved budget, brings the total grant for this school to Sfrs. 10 000.--.

16. No regional meeting will be held in 1979.

The Finance Committee, to convene during the XVIIth General Assembly, will be asked to endorse the budgetary estimate for 1979, as proposed above, and to submit to the approval of the General Assembly the comprehensive budget prepared by the Executive Committee for the years 1980-1982.

APPENDIX

A. List of Countries Adhering to the IAU

The following is a list of the 48 countries that adhered to the Union in December 1978, giving also the year of admission and the Adhering Organization.

Country	Year	Adhering Organization
Arab Republic of Egypt	1925	Academy of Scientific Research and Technology Dept. Sci.Soc. and Internat. Unions 101, Kasr El Eini Street, Cairo
Argentina	1927	Comité Nacional de Astronomia Observatorio Astronomico "Felix Aguilar" Av. Benavidez 8175 Oeste 5407 Marquesado, San Juan

Australia	1939	Australian Academy of Science, P.O. Box 216 Civic Square, Canberra, A.C.T. 2608
Austria	1955	Oesterreichische Akademie der Wissenschaften Dr Ignaz Seipel-Platz 2, 1010 Vienna
Belgium	1920	Académie Royale de Belgique, Palais des Académies, 1, rue Ducale, Bruxelles 1
Brazil	1961	Comissao Brasileira de Astronomia (CBA) Conselho Nacional de Pesquisas Avenida Marechal Câmara 350, Rio de Janeiro GB
Bulgaria	1957	Bulgarian Academy of Sciences 7th November Street 1, Sofia
Canada	1920	National Research Council of Canada International Relations, Ottawa K1A OR6
Chile	1947	Universidad de Chile, Observatorio Astronomico Nacional, Casilla 36-D, Santiago de Chile
Colombia	1967	Academia Colombiana de Ciencias Exactas Fisicas y Naturales, Carrera 8a Apartado Nal. 2584, Bogota 1, D.E.
Cuba	1970	Instituto de Astronomia de la Academia de Ciencas de Cuba, Calle 174/1722, Cubanacan, Habana
Czechoslovakia	1922	Czechoslovak Academy of Sciences, Narodni 3, Praha 1, Czechoslovakia
Denmark	1922	Videnskabernes Selskab, Dantes Plads 5, DK-1556 Copenhagen V
Finland	1948	Académie des Sciences et Lettres Snellmaninkatu 9-11, Helsinki 17
France	1920	Académie des Sciences, Institut de France 23, Quai de Conti, 75006 Paris
Germany, D.R.	1951	Akademie der Wissenschaften der DDR Otto Nuschkestrasse 22/23, DDR 108 Berlin
Germany, F.R.	1951	Rat Westdeutscher Sternwarten, Institut für Theoretische Physik und Sternwarte der Universität Neue Universität, Olshausenstrasse, 2300 Kiel
Greece	1920	Academy of Athens, Panepistimiou Str. Athens
Hungary	1947	Hungarian Academy of Sciences, Roosevelt tér 9, Budapest V
India	1964	Indian National Science Academy Bahadur Shah Zafar Marg, New Delhi - 110001
Iran	1969	Université de Téhéran, Institut Géophysique Amirabad Shomali, Teheran

Iraq	1976	Fondation of Scientific Research P.O. Box 255, Baghdad
Ireland	1947	National Committee of Astronomy, The Royal Irish Academy, 19 Dawson Street, Dublin 2
Israel	1954	The Israel Academy of Sciences and Humanities, 43 Jabotinsky Road, P.O.B. 4040, Jerusalem 91 040
Italy	1921	Consiglio Nazionale delle Ricerche Piazale delle Scienze 7, Rome
Japan	1920	Science Council of Japan, 22-34 Roppongi, 7 chome, Minato-ku, Tokyo 106
Korea (DPR)	1961	Academy of Sciences, Pyongyang
Korea (Republic)	1973	Korean National Astronomical Observatory Yorksam-Dong, Kangnam-ku, Seoul, 134-03
Mexico	1921	Instituto de Astronomia, UNAM Apartado Postal 70-264, México, D.F.
Netherlands	1922	Koninklijke Nederlandse Akademie van Wetenschappen Kloveniersburgwal 29, Amsterdam-C
New Zealand	1964	The Royal Society of New Zealand, 6 Kelburn Parade P.O. Box 12249, Wellington
Norway	1922	Det Norske Videnskaps-Akademi i Oslo Drammensveien 78, Oslo 2
Poland	1922	Polaska Akademia Nauk, Palac Kultury i Nauki, Warsaw
Portugal	1924	Seccao Portuguesa das Unioes Internacionais Astronomica e Geodésia e Geofisica Praça de Estrela, Lisboa 2
Roumania	1928	Roumanian National Committee of Astronomy Astronomical Observatory, Str. Cutitul de Argint 5 Bucarest 28
South Africa	1938	Council for Scientific and Industrial Research Information and Research Services, Pretoria 0001
Spain	1922	Comision Nacional de Astronomia, Instituto Geografico y Catastral, General Ibanez 3, Madrid (3)
Sweden	1925	Kungl. Vetenskapsakademien, Fack, S-104 05 Stockholm 50
Switzerland	1923	Schweizerische Naturforschende Gesellschaft, Zentralsekretariat, Laupenstr. 10, 3000 Bern
Taiwan	1959	Academy of Sciences pf Taipei, No. 130 Yen Chiu Yuan Road, Section 11, Nankang-Taipei
Turkey	1961	Türk Astronomi Dernegi, University Rasathanesi Beyazit, Istanbul

United Kingdom	1920	The Royal Society, 6, Carlton House Terrace London, SW1Y 5AG
Uruguay	1970	Ministerio de Educacion y Cultura, Comité Nacional de Astronomia, Casilla de Correo 867, Montevideo
U.S.A.	1920	National Academy of Sciences, Office of the Foreign Secretary, 2101 Constitution Ave N.W., Washington, D.C. 20148
U.S.S.R	1935	Academy of Sciences of the U.S.S.R., Leninskij Prospekt 14, Moscow 71
Vatican City State	1932	Pontificia Academia delle Scienze Citta del Vaticano
Venezuela	1953	Asociacion Venezolana de Astronomia, Facultad de Ingenieria, Departamento de Geodesia, Universidad del Zulia, Maracaibo
Yugoslavia	1935	Savez društva matematičara, fizičara i astronoma Jugoslavije, Attn. Secretary General ul. Dr Ilije Duričica 4, Novi Sad.

B. List of Deceased Members

The Executive Committee regrets to record the loss by death of the following 51 Members:

ADAMOPOULOS Dr G. - 10 September 1977
BHATNAGAR Prof. P.L. - 5 October 1976
BLACKETT Prof. P.M.S. - F.R.S. - 13 November 1974
BUTLER Dr H.E. - 10 May 1978
CARPENTER G.J. - 1 June 1977
CHALONGE Dr D. - 28 November 1977
CHENAKAL Dr V.L. - 24 June 1977
CRUZ GONZALES C. - 8 April 1977
CUREA Prof. I. - 1 March 1977
DIMITROFF Prof. G.Z. - unknown date
DINGLE Prof. H. - 4 September 1978
ELVEY Dr C.T. - unknown date
EZERSKIJ Dr V.I. - 11 July 1978
FEDYNSKIJ Prof. V.V. - 17 June 1978
GENTILI DE GIUSEPPE M. - 2 July 1977
GORDON Dr J.E. - 5 May 1978
HAFFNER Prof. H. - 23 February 1977
HEARD Dr J.P. - 5 October 1976
HUANG Dr Su-Shu - 15 September 1977
ITZIGSOHN M. - 30 January 1978
KAPLAN Prof. S.A. - 11 June 1978
KOELBLOED Dr D. - 21 June 1977
KOROL' Dr A.K. - 10 March 1977
KUKARKIN Prof. B.V. - 15 September 1977
KUPO Dr I. - 1978 - unknown date
LACCHINI Prof. G.B. - unknown date

LACOMBE Dr C.G. - 16 September 1977
LALLEMAND Dr A. - 24 March 1978
LIMBER Dr D.N. - 16 April 1977
LIPSKIJ Dr Y.N. - 24 January 1978
MADEIRA Dr J.A. - 22 August 1976
MENZEL Dr D.H. - 14 December 1976
MICHKOVITCH Prof. V.V. - 25 November 1976
NEFED'EV Dr A.A. - 14 September 1976
NISSEN Prof. J.J. - 9 June 1978
OOSTERHOFF Prof. P. Th. - 14 March 1978
PALMER D.R. - August 1975
POPOVICI Prof. C. - 16 January 1977
POSTOIEV Dr A. - 21 July 1976
RIGHINI Prof. G. - 30 May 1978
RIIVES Dr V. - 28 May 1978
SEN Dr H.K. - 2 September 1976
SMERD Dr S.F. - 20 December 1978
STICKER Prof. B. - 30 August 1977
STIEFEL, Prof. E. - 1978 - unknown date
STOYKO Dr N. - 14 September 1976
THACKERAY Prof. A.D. - 21 February 1978
TREANOR Dr P.J. - 18 February 1978
UETA Dr J. - 13 November 1976
VANDEKERKHOVE Dr E. - September 1978

C. List of IAU Publications

Transactions of the IAU, Volume XVIA, "Reports on Astronomy, 1976", in set of 3 volumes (Part 1: pp. VI + 228; Part 2: pp. VI + 240; Part 3: pp. VI + 222), price Dfl 182 (Dfl 61 per volume - Members at 33.33% discount). Editor G. Contopoulos, Published by D. Reidel Publishing Company, Dordrecht, Holland, 1976.

Transactions of the IAU, Volume XVIB: "Proceedings of the Sixteenth General Assembly of the IAU", Grenoble, France, 1976, pp. X + 564, price cloth Dfl. 125 (Members at 33.33% discount). Edited by Edith A. Müller and A. Jappel. Published by D. Reidel Publishing Company, 1977.

Highlights of Astronomy, Volume No. 4, as presented at the XVIth General Assembly of the IAU, 1976, in 2 parts (Part 1: pp. VI + 370; Part 2: VI + 407), price Dfl 180. (Part 1 Dfl 85, Part 2 Dfl 95 - Members at 33.33% discount). Edited by Edith A. Müller. Published by D. Reidel Publishing Company, 1977.

IAU Symposium Volumes:

No. 70 "Be and Shell Stars", pp. XVI + 465, price: cloth Dfl 150, paper Dfl 115 (Members at 33.33% discount). Edited by A. Slettebak, published by D. Reidel Publishing Company, 1976.
No. 71 "Basic Mechanisms of Solar Activity", pp. XVI + 481, price Dfl 140, paper Dfl 105 (Members at 33.33% discount). Edited by V. Bumba and J. Kleczek, published by D. Reidel Publishing Company, 1976.
No. 72 "Abundance Effects in Classification", pp. XXVIII + 264, price Dfl 85, paper Dfl 60 (Members at 33.33% discount). Edited by B. Hauck and P.C. Keenan, published by D. Reidel Publishing Company, 1976.
No. 73 "Structure and Evolution of Close Binary Systems", pp. XII + 414, price Dfl 105, paper Dfl 85 (Members at 33.33% discount). Edited by P. Eggleton, S. Mitton and J. Whelan, published by D. Reidel Publishing Company, 1976.
No. 74 "Radio Astronomy and Cosmology", pp. XX + 398, price Dfl 95, paper Dfl 50

(Members at 33.33% discount). Edited by D.L. Jauncey, published by D. Reidel Publishing Company, 1977.

No. 75 "Star Formation", pp. XIV + 296, price Dfl 75, paper Dfl 45 (Members at 33.33% discount). Edited by T. de Jong and A. Maeder, published by D. Reidel Publishing Company, 1977.

No. 76 "Planetary Nebulae", pp. XVIII + 376, price Dfl 90, paper Dfl 60 (Members at 33.33% discount). Edited by Y. Terzian, published by D. Reidel Publishing Company, 1978.

No. 77 "Structure and Properties of Nearby Galaxies", pp. XVIII + 308, price Dfl 75, paper Dfl 50 (Members at 33.33% discount). Edited by E.M. Berhijsen and R. Wielebinski, published by the D. Reidel Publishing Company, 1978.

No. 79 "The Large-Scale Structure of the Universe", pp. XX +451, price Dfl 110, paper Dfl 70 (Members at 33.33% discount). Edited by M.S. Longair and J. Einasto, published by the D. Reidel Publishing Company, 1978.

No. 80 "The HR Diagram - The 100th Anniversary of Henry Norris Russell" pp. XXVI + 474, price Dfl 110, paper Dfl 70 (Members at 33.33% discount). Edited by A.G. Davis Philip and D.S. Hayes, published by D. Reidel Publishing Company, 1978.

IAU Colloquium Volumes:

No. 25 "Study of Comets", edited by B.D. Donn. Published as a NASA Special Publication, 1975.

No. 26 "Reference Coordinate System for Earth Dynamics", edited by B. Kolaczek and G. Weiffenbach. Published by Warsaw Technical University, June 1976.

No. 27 "X- and UV- Spectroscopy of Astrophysical and Laboratory Plasmas", edited by W.R.S. Garton. Published in Astrophysics and Space Science 38, 313-380, 1975.

No. 28 "Planetary Satellites", invited papers and other material edited by J.A. Burns and published by University of Arizona Press as "Planetary Satellites", 1975. Contributed papers published in ICARUS, April and July 1975, and in Celestial Mechanics, August 1975.

No. 29 "Multiple Periodic Variable Stars", volume 1 pp.348, volume 2 pp. 356. Edited by W.S. Fitch. Published by the Hungarian Academy of Sciences, 1976.

No. 30 "Jupiter", 36 papers in ICARUS, March 1976; Journ. Geophys. Research, July 1976; and ICARUS, October 1976. 44 review chapters in "Jupiter" edited by T. Gehrels, University of Arizona Press, 1978.

No. 31 "Interplanetary Dust and Zodiacal Light", edited by H. Elsässer and H. Fechtig. Published by Springer Verlag in the series "Lecture Notes in Physics", 1976.

No. 32 "Physics of Ap Stars", edited by W.W. Weiss. Published by Universitätssternwarte Wien, 1976.

No. 33 "Observational Parameters and Dynamical Evolution of Multiple Stars", edited by O.G. Franz and P. Pismis. Published in the May 1977 issue, Vol. 3 of Revista Mexicana de Astronomia y Astrofisica.

No. 35 "The Compilation, Critical Evaluation, and Distribution of Stellar Data", edited by C. Jaschek and G.A. Wilkins. Published by D. Reidel Publishing Company (Astrophysics and Space Science Library 64), 1977.

No. 36 "Energy Balance and Hydrodynamics of the Solar Chromosphere and Corona", edited by R.M. Bonnet and Ph. Delache. Published by Observatoire de Nice, France, 1977.

No. 37 "Redshifts and the Expansion of the Universe", followed by "The Evolution of Galaxies and its Cosmological Implications", C.N.R.S. Colloquium No. 263. Edited by B.E. Westerlund and C. Balkowski. Published in the editions du C.N.R.S., 15 Quai Anatole France, 75007 Paris, France, 1978.

No. 38 "Problems of Stellar Convection", edited by E.A. Spiegel and J.P. Zahn. Published by Springer Verlag (Lecture Notes in Physics, Vol. 71, Berlin), 1977.

No. 39 "Relationship between Comets, Minor Planets and Meteorites". Proceedings included in volume "Comets, Asteroids, Meteorites" edited by A.H. Delsemme. Published by University of Toledo, 1977 (Toledo University Bookstore, Toledo, Ohio 43606).

No. 40 "Astronomical Applications of Image Detectors with Linear Response", edited by M. Duchesne and G. Lelièvre. Published by Observatoire de Meudon, France, 1978.

No. 41 "Dynamics of Planets and Satellites, and Theories of their Motion", edited by V.G. Szebehely. Published by D. Reidel Publishing Company, 1978 (Astrophysics and Space Science Library 72).

No. 42 "The Interaction of Variable Stars with their Environments", edited by R. Kippenhahn, J. Rabe and W. Strohmeier. Published by Verlag der Remeis-Sternwarte, Bamberg, Bd. XI, No. 121, 1977.

No. 43 "Ultraviolet and X-Ray Spectroscopy of Astrophysical and Laboratory Plasmas". Abstracts only. Prepared by Physics Department, Imperial College, London SW7-2BZ, U.K., for limited distribution.

No. 45 "Chemical and Dynamical Evolution of our Galaxy", edited by E. Basinski, Grzezik and M. Mayor. Published by Geneva Observatory, 1290 Sauverny, Switzerland, 1977.

Other Publications

Proceedings of the Third European Astronomy Meeting, held under the auspices of the IAU, "Stars and Galaxies from Observational Points of View". Edited by E.K. Kharadze. Published by Academy of Sciences of the Georgian SSR, Abastumani Astrophysical Observatory, U.S.S.R. For sale outside the Socialist countries by D. Reidel Publishing Company, 1976.

Proceedings of a Special IAU Session on "CNO Isotopes in Astrophysics", edited by J. Adouze. Published by D. Reidel Publishing Company, 1977.

Proceedings of a Special IAU Session on "Supernovae", edited by D.N. Schramm. Published by D. Reidel publishing Company, 1977.

Proceedings of Joint Session of IAU Commission No. 34 with several other Commissions "Topics in Interstellar Matter", edited by H. van Woerden. Published by D. Reidel Publishing Company, 1977.

Publications by, or on behalf of, IAU Commissions and by Astronomical Services

Commission No. 6 IAU (Telegram Bureau) circulars Nos 2876 - 3317. Issued by the IAU Central Bureau for Astronomical Telegrams.

Commission No. 10 Cartes Synoptiques de la Chromosphère Solaire. Prepared by Observatoire de Paris, Meudon.

Quarterly Bulletin on Solar Activity(QBSA). Published by the Swiss Federal Observatory, Zürich, Switzerland, up to No. 196 and by the Tokyo Astronomical Observatory, Mitaka, Tokyo, from No. 197 on.

Commission No. 19 Monthly Notes of the IPMS Nos. 11 (1975) - 11 (1978). Published by the International Polar Motion Service, Mizusawa-shi, Japan.

Annual Report of the International Polar Motion Service for the Year 1975, by S. Yumi, published by the Central Bureau of the International Polar Motion Service, 1977.

Results of the International Latitude Service 1962. 0 - 1967. 0, Vol. XII with one Appendix, by S. Yumi and K. Yokoyama. Published by Central Bureau of the International Polar Motion Service, International Latitude Observatory of Mizusawa.

Commission No. 20 Minor Planet Circulars Nos. 3909 - 4582. Mimeographed circulars issued by the Minor Planet Center, Cincinnati, now Cambridge, Mass., U.S.A.

Commission No. 24 Astrographic Catalogues. Information regarding the different sections of the Catalogue can be obtained from the Observatories listed in IAU Transactions, Vol. XIVB, pp. 176, 177, 1971.

Commission No. 26 Circulaire d'Information Nos. 68-76. President: P. Muller, CERGA, Grasse, France.
Commission No. 27 Information Bulletin on Variable Stars. Prepared by Konkoly Observatory, Budapest, Hungary.
Commission No. 31 Annual Report of the BIH. Published by the Bureau International de l'Heure, Paris, France.
Commission No. 41 Information Circular Nos. 26-28. Published by the President of the Commission.
IAU Information Bulletin Nos. 35 - 40. Prepared by the General Secretary.

APPENDIX IV

The following letters were exchanged between the President of the IAU, Professor A. Blaauw, and the President of the Chinese Astronomical Society at Nanking, Professor Y.C. Chang.

Montreal, August 22, 1979

To Professor A. BLAAUW, President
of the International Astronomical Union
c/o the General Assembly of the IAU 1979
Montreal, Quebec, Canada

Dear Professor Blaauw,

On behalf of the Chinese Astronomical Society, whose delegates have convened with IAU Officers during the 17th General Assembly of IAU at Montreal, I herewith confirm that,

considering that both the Chinese Astronomical Society and the IAU believe that the question of China's representation in the IAU needs urgent attention;

noting with satisfaction that the IAU wishes to find a solution to the question on the principle of one China, embracing the Mainland and Taiwan;

understanding the nature of complexity of such an endeavor in the present circumstances;

considering also the desire expressed by the IAU that the Chinese Astronomical Society is welcome to return without preventing the astronomical organization as well as astronomers in Taiwan from participating in the IAU;

reiterating that the Chinese Astronomical Society opposes any solutions which can suggest the existence of "two Chinas" or "one China, one Taiwan";

the Chinese Astronomical Society proposes to the IAU Executive Committee

a) that for the time being dual adherence should be introduced, the adherence and the adhering bodies to be specified as:

China:
- Chinese Astronomical Society (Purple Mountain Observatory, Nanking)
- (a name for the adhering organization representing the astronomical community of Taiwan; for this name we propose "Chinese Taiwan Astronomical Society" or, provisionally, as reference "Astronomy Taiwan, China"; the name to be established by the IAU Executive Committee in consultation with the two adhering organizations);

b) that for the time being the financial contributions of China be composed of one component due from the Chinese Astronomical Society (Nanking) and one component due from (the name to be defined as under (a));

c) that for the time being each of these adhering bodies nominates a representative authorized to vote in the General Assembly. In accordance with article VII, 11 of the IAU Statutes, on questions involving the budget of the Union, the number of votes of these representatives will be one greater than the number of its category as defined in article 8, and on questions of administration, not involving the budget, each representative will have one vote.

Yours sincerely,

Y.C. Chang, President
Chinese Astronomical Society

August 22, 1979

To Prof. Y.C. CHANG, President
of the Chinese Astronomical Society
c/o IAU General Assembly 1979
Montreal, Canada

Dear Professor Chang,

In reply to your letter of August 22, 1979 I have the pleasure to inform you as follows:

The Executive Committee has taken note with great satisfaction of the proposal contained in your letter, recognizing the Chinese Astronomical Society's view leading to this proposal. It wishes to express its high appreciation for the constructive spirit in which the delegation of your Society has contributed towards restoring full participation of China in the IAU.

The Executive Committee therefore will be glad to take up the problem of deciding on the name to be used for the adhering body on behalf of the astronomical community of Taiwan. It will do so in the near future in consultation with the astronomical community of Taiwan and with your Society, and possibly benefiting also from the experience gained meanwhile by other scientific unions working similarly towards the full participation of China.

The Executive Committee, in view of the discussions held during this Assembly, hopes and expects that this course will soon lead to the formal restoration of full Chinese participation and would be extremely pleased to see the valuable participation of members of your Society in the scientific discussions at this Assembly, continued at the forthcoming scientific meetings of the IAU. To this purpose, the General Secretariat will supply your Society with all relevant documentation.

Yours respectfully,

A. Blaauw, President
International Astronomical Union